Macrocyclic and Supramolecular Chemistry

Macrocyclic and Supramolecular Chemistry

How Izatt–Christensen Award Winners Shaped the Field

Edited by

REED M. IZATT

IBC Advanced Technologies, Inc., American Fork;
Department of Chemistry and Biochemistry, Brigham Young University, Provo, UT, USA

WILEY

This edition first published 2016

Registered Office
John Wiley & Sons, Ltd, The Atrium, Southern Gate, Chichester, West Sussex, PO19 8SQ, United Kingdom

For details of our global editorial offices, for customer services and for information about how to apply for permission to reuse the copyright material in this book please see our website at www.wiley.com.

Library of Congress Cataloging-in-Publication Data

Names: Izatt, Reed M., 1926– editor.
Title: Macrocyclic and supramolecular chemistry : how Izatt–Christensen Award winners shaped the field / edited by Reed M. Izatt.
Description: Chichester, West Sussex : John Wiley & Sons, Inc., 2016. | Includes bibliographical references and index.
Identifiers: LCCN 2016016106| ISBN 9781119053842 (cloth) | ISBN 9781119053866 (epub) | ISBN 9781119053873 (epdf)
Subjects: LCSH: Izatt–Christensen Award in Macrocyclic and Supramolecular Chemistry. | Supramolecular chemistry–Awards. | Chemistry–Awards. | Macrocyclic compounds. | Izatt, Reed M., 1926– | Christensen, James J., 1931–
Classification: LCC QD878 .M327 2016 | DDC 541/.226–dc23
LC record available at https://lccn.loc.gov/2016016106

A catalogue record for this book is available from the British Library.

Set in 10/12pt Times by SPi Global, Pondicherry, India

1 2016

Contents

List of Contributors

Eric V. Anslyn
Department of Chemistry and Biochemistry, The University of Texas at Austin, Austin, TX, USA

Shin Aoki
Faculty of Pharmaceutical Sciences, Tokyo University of Science, Noda, Japan

Jerry L. Atwood
Department of Chemistry, University of Missouri–Columbia, Columbia, MO, USA

Paul D. Beer
Chemistry Research Laboratory, Department of Chemistry, University of Oxford, Oxford, UK

Jerald S. Bradshaw
IBC Advanced Technologies, Inc., American Fork; Department of Chemistry and Biochemistry, Brigham Young University, Provo, UT, USA

Daryle H. Busch
Department of Chemistry, University of Kansas, Lawrence, KS, USA

Ryan Cantwell
Center for Nanoscience, Departments of Chemistry & Biochemistry and Biology, University of Missouri–St. Louis, St. Louis, MO, USA

Pongphak Chidchob
Department of Chemistry, McGill University, Montreal, Québec, Canada

Carol A. Deakyne
Department of Chemistry, University of Missouri–Columbia, Columbia, MO, USA

Vincent Duplan
Institut de Science et d'Ingénierie Supramoléculaires (ISIS), Université de Strasbourg et CNRS, Strasbourg, France

Johannes A.A.W. Elemans
Radboud University, Institute for Molecules and Materials, Nijmegen, The Netherlands

Luigi Fabbrizzi
Dipartimento di Chimica, Università di Pavia, Pavia, Italy

George W. Gokel
Center for Nanoscience, Departments of Chemistry & Biochemistry and Biology, University of Missouri–St. Louis, St. Louis, MO, USA

Michael R. Gokel
Center for Nanoscience, Departments of Chemistry & Biochemistry and Biology, University of Missouri–St. Louis, St. Louis, MO, USA

Andrew D. Hamilton
Chemistry Research Laboratory, Department of Chemistry, University of Oxford, Oxford, UK

Akira Harada
Graduate School of Science, Osaka University, Osaka, Japan

Roger G. Harrison
Department of Chemistry and Biochemistry, Brigham Young University, Provo, UT, USA

Mir Wais Hosseini
Institut Le Bel, Université de Strasbourg & CNRS, Strasbourg, France

Reed M. Izatt
IBC Advanced Technologies, Inc., American Fork; Department of Chemistry and Biochemistry, Brigham Young University, Provo, UT, USA

Steven R. Izatt
IBC Advanced Technologies, Inc., American Fork, UT, USA

Kimoon Kim
Center for Self-assembly and Complexity, Institute for Basic Science (IBS); Department of Chemistry and Division of Advanced Materials Science, Pohang University of Science and Technology (POSTECH), Pohang, Republic of Korea

Eiichi Kimura
Hiroshima University, Hiroshima, Japan

Peter C. Knipe
Chemistry Research Laboratory, Department of Chemistry, University of Oxford, Oxford, UK

Tohru Koike
Institute of Biomedical and Health Sciences, Hiroshima University, Hiroshima, Japan

Harshita Kumari
James L. Winkle College of Pharmacy, University of Cincinnati, Cincinnati, OH, USA

Matthew J. Langton
Chemistry Research Laboratory, Department of Chemistry, University of Oxford, Oxford, UK

David A. Leigh
School of Chemistry, The University of Manchester, Manchester, UK

Paul R. McGonigal
Department of Chemistry, Durham University, Durham, UK

Margaret K. Meadows
Department of Chemistry and Biochemistry, The University of Texas at Austin, Austin, TX, USA

Joseph W. Meisel
Center for Nanoscience, Departments of Chemistry & Biochemistry and Biology, University of Missouri–St. Louis, St. Louis, MO, USA

Saeedeh Negin
Center for Nanoscience, Departments of Chemistry & Biochemistry and Biology, University of Missouri–St. Louis, St. Louis, MO, USA

Frédéric Niess
Institut de Science et d'Ingénierie Supramoléculaires (ISIS), Université de Strasbourg et CNRS, Strasbourg, France

Roeland J.M. Nolte
Radboud University, Institute for Molecules and Materials, Nijmegen, The Netherlands

Kyeng Min Park
Center for Self-assembly and Complexity, Institute for Basic Science (IBS); Department of Nanomaterials Science and Engineering, University of Science and Technology (UST), Pohang, Republic of Korea

Mohit B. Patel
Department of Biology, University of Missouri–St. Louis, St. Louis, MO, USA

David N. Reinhoudt
MESA+ Institute for Nanotechnology, University of Twente, Enschede, The Netherlands

Alan E. Rowan
Australian Institute for Bioengineering and Nanotechnology (AIBN), The University of Queensland, Brisbane, Australia

Jeremy K.M. Sanders
University Chemical Laboratory, University of Cambridge, Cambridge, UK

Jean-Pierre Sauvage
Institut de Science et d'Ingénierie Supramoléculaires (ISIS), Université de Strasbourg et CNRS, Strasbourg, France

Alexander Schoedel
Department of Chemistry, University of California; Materials Sciences Division, Lawrence Berkeley National Laboratory; Kavli Energy Nanoscience Institute, Berkeley, CA, USA

Jonathan L. Sessler
Department of Chemistry, The University of Texas at Austin, Austin, TX, USA

Dinesh Shetty
Center for Self–assembly and Complexity, Institute for Basic Science (IBS), Pohang, Republic of Korea

Seiji Shinkai
Institute for Advanced Study, Kyushu University, Fukuoka; Nanotechnology Laboratory, Institute of Systems, Information Technologies and Nanotechnologies (ISIT), Fukuoka; Department of Nanoscience, Faculty of Engineering, Sojo University, Kumamoto, Japan

Hanadi Sleiman
Department of Chemistry, McGill University, Montreal, Québec, Canada

J. Fraser Stoddart
Department of Chemistry, Northwestern University, Evanston, IL,USA

Sam Thompson
Chemistry Research Laboratory, Department of Chemistry, University of Oxford, Oxford, UK

Omar M. Yaghi
Department of Chemistry, University of California, Berkeley; Materials Sciences Division, Lawrence Berkeley National Laboratory, Berkeley; Kavli Energy Nanoscience Institute, Berkeley, CA, USA; King Fahd University of Petroleum and Minerals, Dhahran, Saudi Arabia

Preface

The International Izatt–Christensen Award in Macrocyclic and Supramolecular Chemistry was initiated in 1991. It recognizes excellence in these fields of chemistry and is presented annually at the International Symposium on Macrocyclic and Supramolecular Chemistry (ISMSC). Recipients are representative of those who are doing fine work in this vibrant and expanding field of chemistry, which has developed into one of the most active and promising research areas in science.

In 2016, the 25th International Izatt–Christensen Award will be presented at the 11th ISMSC in Seoul, Korea. I conceived the idea of a book featuring chapters from Award recipients in August 2015. Wiley editor Sarah Higginbotham agreed that the idea was sound and 21 recipients of the Award agreed to prepare the chapters found in this volume. In Chapter 1, recipients of the Award from 1991 to 2016 are given, together with titles of their ISMSC presentations.

The work presented in this volume covers a wide range of subject matter. This range reflects the growth of the field during the past half century. Several chapters present historical aspects providing insight into the beginnings of work on macrocyclic and supramolecular chemistry in the 1960s. A recurring theme in the chapters is molecular recognition, which has played an important role from the beginning and has led investigators along many different paths. This diversity is evident in the work presented in the various chapters.

Backgrounds of the award recipients are varied and their contributions to the field are different. Not all authors present up-to-date research results although the book contains many of these. Several of the authors are retired and not as active in research as they once were. However, their contributions were cutting-edge and a review of their work and, perhaps, how their studies added to our present knowledge base is of great interest. I feel that it is important for the younger generation to have an appreciation for origins of the science they now enjoy and of the people who created it. How better can this be obtained than through reading first-hand accounts of the work of those who went before. The material in this book provides an opportunity to convey much useful information to the next generation. It has thrilled me to see the continuing enthusiasm of students and young faculty members in this field. This interest is expressed each year in locations worldwide by the large number of enthusiastic young people who attend the ISMSC meetings where the Izatt–Christensen Award is made. I suspect that most of these individuals have little knowledge of the history behind the chemistry presented at these symposia. This book will fill that void to some degree.

The authors are prominent among those whose contributions have shaped, and continue to shape, macrocyclic and supramolecular chemistry. Prospects for the future in this rapidly developing field are part of many chapters. Coincident with the development in the 1960s and 1970s of the concept of molecular recognition into the current rapidly expanding field of supramolecular chemistry was the following statement by Richard Feynman (*Eng. Sci.* 1960, 23, 22–36; 1965 Physics Nobel Laureate) about rearranging atoms at the molecular level.

> *What would the properties of materials be if we could really arrange the atoms the way we want them? They would be very interesting to investigate theoretically. I can't see exactly what would happen, but I can hardly doubt that when we have some control of the arrangement of things on a small scale we will get an enormously greater range of possible properties that substances can have, and of different things that we can do.*

I believe that this prediction of Feynman's has seen its fulfillment in the demonstrated ability of scientists, including those in supramolecular chemistry, to do exactly what he describes. Design at the molecular level has enabled the production of hosts with impressive selectivity for guest molecules. In the pages of this volume, many examples are given and I predict that the reader will have fun discovering them. One of the chapters is titled "Serendipity." Several authors use the word serendipity to describe some of their findings. Charles J. Pedersen's discovery of crown ethers (Chapter 1) was an example of serendipity. The future of macrocyclic and supramolecular chemistry is bright with opportunity even though we cannot *see exactly what will happen.*

Reed M. Izatt,
Provo, Utah
March 2016

Acknowledgements

I appreciate and thank the authors of each chapter who have worked diligently to deliver high-quality content for this volume. I have enjoyed the constant guidance of a terrific set of Wiley editors who have provided help whenever it was needed for me and for the authors. My computer-literate daughter, Anne Marie Izatt, has been of immeasurable assistance throughout this editing experience. Knowing she was there when needed, which was often, brought me great comfort. Finally, I thank my wife, Janet, for her patience, understanding and support throughout the preparation of this book. Janet has been a valuable sounding-board and has made many helpful suggestions.

Reed M. Izatt
March 2016

1

The Izatt–Christensen Award in Macrocyclic and Supramolecular Chemistry: A 25-Year History (1991–2016)

Reed M. Izatt,[1,2] Jerald S. Bradshaw,[1,2] Steven R. Izatt,[1] and Roger G. Harrison[2]

[1] *IBC Advanced Technologies, Inc., American Fork, UT, USA*
[2] *Department of Chemistry and Biochemistry, Brigham Young University, Provo, UT, USA*

1.1 Introduction

The Izatt–Christensen Award (I–C Award) recognizes excellence in macrocyclic and supramolecular chemistry. It has been presented annually since 1991 by the International Symposium on Macrocyclic and Supramolecular Chemistry (ISMSC). A common theme to both of these fields is molecular recognition. The search for underlying principles governing molecular recognition or how molecules recognize each other began in earnest in the early 1960s. Working independently, several individuals who later became prominent in the emerging fields of macrocyclic chemistry and supramolecular chemistry, made important early contributions to molecular recognition. Four of these were Charles J. Pedersen (1904–1989), Daryle H. Busch, Jean-Marie Lehn, and Donald J. Cram (1919–2001). Prior to the 1960s, no concentrated effort had been made to investigate chemical selectivity involving macrocyclic compound interactions with metal ions or other guest molecules [1].

Charles Pedersen while employed at du Pont serendipitously discovered the compound that later came to be known as dibenzo-18-crown-6 (DB18C6). Pedersen isolated DB18C6 in a 0.4% yield from a "brownish goo" while attempting to prepare a completely different compound [2]. The decision to expend the effort needed to isolate, purify, and characterize the compound that became known as DB18C6 represents a true example of scientific creativity and luck. The story of Pedersen's discovery, reported in 1967, his identification of the many new cyclic polyether macrocyclic compounds he synthesized, his characterization of their selective complexation with alkali metal ions, and his own account of the events surrounding the discovery make fascinating reading [2, 3].

Macrocyclic and Supramolecular Chemistry: How Izatt–Christensen Award Winners Shaped the Field, First Edition.
Edited by Reed M. Izatt.

Daryle Busch remembers that his first ideas of synthesizing macrocycles occurred while a graduate student with John Bailar at the University of Illinois in the early1950s. His account of these first ideas of forming macrocycles from bidentate amines involved in copper(II) chelation illustrates the workings of a creative mind. It was several years later in 1962, as a Professor of Inorganic Chemistry at Ohio State University, that he reported the first synthesis of a macrocycle using a metal template [1, 5] He received the I–C Award in 1994 and is the author of a chapter in this book [4], in which he gives a first-hand account of his work.

Jean-Marie Lehn reported the synthesis of macrobicyclic polyethers containing three polyether strands joined by two bridgehead nitrogen atoms [6] in 1969, shortly after Pedersen's initial paper. Lehn later coined the term, "supramolecular chemistry," to describe the broadening of the scope of host–guest chemistry which he and his research group had spearheaded [7]. To quote Professor Lehn, "Beyond molecular chemistry, supramolecular chemistry aims at constructing highly complex, functional chemical systems from components held together by intermolecular forces." These components can be visualized as host–guest systems bonded by intermolecular forces, which are much weaker than covalent chemical bonds. The guest systems may include organic guests as well as metal ions. The number and variety of hosts synthesized has expanded far beyond macrocyclic compounds. Lehn has provided an account of his early work [7].

Donald Cram was a prominent organic chemist in the 1960s. John Sherman [8], one of his Ph.D. students, describes him as "definitely old school. Eccentric. Hard driven. Strong-willed. Spirited. Fearless." Cram's accomplishments included a major research program in organic chemistry, co-author of three major organic chemistry textbooks, and instructor at UCLA of several generations of organic chemistry students. His first acquaintance with macrocyclic chemistry was recorded by Roeland Nolte who remembers [9] that during a stay as a visiting scientist in Cram's laboratory at UCLA in 1981, Cram told him "after reading Pedersen's paper he had become so excited that he had made the decision to completely change his research program." Nolte goes on to say, "After having seen the potential of host–guest chemistry and the way it was approached by Cram, i.e., by designing compounds with the help of space-filling (CPK) models, we became fascinated and concluded that we should start a line of research in the Netherlands in which this new type of chemistry was incorporated." This attitude was contagious, and transfer of the excitement to others was responsible for the explosion of interest in macrocyclic chemistry, which characterized the field in the 1970s and 1980s.

As was the case with Nolte, many of the I–C Award winners spent time in the laboratories of Donald Cram, Daryle Busch, or Jean-Marie Lehn. A number of researchers, including one of us (RMI) and James J. Christensen, had close contact with Charles Pedersen, who influenced our early work in macrocyclic chemistry [10]. The influence of these early pioneers on the field through their own work and the work they inspired in others has been remarkable. The list of I–C Award recipients includes many of the early workers in the field who had close association with these individuals.

1.2 International Izatt–Christensen Award in Macrocyclic and Supramolecular Chemistry

In 1991, Jerald S. Bradshaw and Steven R. Izatt, President of IBC Advanced Technologies, Inc. (IBC), conceived the idea that it would be appropriate to initiate an annual award, titled the Izatt–Christensen Award, recognizing the vision of Reed M. Izatt and James J. Christensen in organizing the First Symposium on Macrocyclic Chemistry in 1977. From 1977 to 1991, the field had broadened, resulting in the design, synthesis and characterization of increasingly more complex organic ligands and their application to new fields of chemistry that were scarcely envisioned decades earlier. This trend is illustrated by the titles of the lectures presented by the I–C awardees.

The I–C Award was instituted in 1991 by IBC. This competitive annual award recognizes excellence in macrocyclic chemistry and is given to individuals who have not received a major award in chemistry. The awardee receives a small honorarium and a travel grant, provided by IBC, and is expected to present an invited lecture at the Symposium in the year of the award. The recipients of the I–C Award from 1991 through 2016 are listed in Table 1.1, together with the locations and titles of their Award lectures.

Table 1.1 *Izatt–Christensen Awardees in Macrocyclic and Supramolecular Chemistry 1991–2016*

Year	Awardee	Location	Title of Award Lecture
1991	Jean-Pierre Sauvage	Sheffield, UK	Synthetic Molecular Knots
1992	Eiichi Kimura	Provo, Utah	Role of Zinc(II) in Zinc Enzymes
1993	J. Fraser Stoddart	Enschede, The Netherlands	Self-Assembly in Unnatural Product Synthesis
1994	Daryle H. Busch	Lawrence, Kansas	A Sampling of Multi-receptor Supramolecular Systems
1995	David N. Reinhoudt	Jerusalem, Israel	Synthesis and Self-assembly of Supramolecular Structures for Switches and Sensors
1996	George W. Gokel	Montecatini, Terme, Italy	Synthetic Models for Cation Channel Function
1997	Alan M. Sargeson	Seoul, Korea	Outer-sphere Electron Transfer Reactions of Macro-bicyclic Complexes
1998	Seiji Shinkai	Turtle Bay, Oahu, Hawaii	Dynamic Control of Ion and Molecule Recognition Processes in Macrocyclic Host–guest Systems
1999	Fritz Vögtle	Barcelona, Spain	Rotaxanes, Catenanes, Pretzelanes–Template Synthesis and Chirality
2000	Jerry L. Atwood	St. Andrews, UK	Macrocycles as Building Blocks for Large Supramolecular Assemblies
2001	Jonathan L. Sessler	Fukuoka, Japan	Novel Polypyrrole Macrocycles
2002	C. David Gutsche	Park City, Utah	The Cornucopia of Calixarene Chemistry
2003	Jeremy K. M. Sanders	Gdansk, Poland	The Ins and Outs of Templating: A Dynamic Future for Macrocyclic Chemistry
2004	Makoto Fujita	Cairns, Australia	Self-assembly and Function of Metal-linked Macrocyclic and Cage-like Molecular Frameworks
2005	Kenneth N. Raymond	Dresden, Germany	Chemistry in Chiral, Nanoscale Flasks
2006	Roeland J. M. Nolte	Victoria, Canada	Supramolecular Catalysts and Materials from Macrocyclic Building Blocks
2007	David A. Leigh	Salice Terme, Italy	Exercising Demons: Synthetic Molecular Motors and Machines
2008	Akira Harada	Las Vegas, Nevada	Cyclodextrin-based Supramolecular Architectures, Dynamics, and Functions
2009	Omar M. Yaghi	Maastricht, The Netherlands	Reticular Chemistry and Frameworks Replete with Large Macrocycles
2010	Luigi Fabbrizzi	Nara, Japan	Put the Anion into the Cage–If You Can
2011	Andrew D. Hamilton	Brighton, UK	Protein Surface Recognition: A Supramolecular Approach to Controlling Biological Function
2012	Kimoon Kim	Otago, New Zealand	Cucurbituril-based Functional Materials
2013	Eric V. Anslyn	Arlington, Virginia	Three Short Stories of Analytical Supramolecular Chemistry
2014	Mir Wais Hosseini	Shanghai, China	Perspectives in Molecular Tectonics
2015	Paul D. Beer	Strasbourg, France	Interlocked Host Molecules for Anion Recognition and Sensing
2016	Hanadi Sleiman	Seoul, Korea	Supramolecular Chemistry with DNA: Towards Biological and Materials Applications

1.3 International Symposium on Macrocyclic and Supramolecular Chemistry

The First Symposium on Macrocyclic Compounds was organized by Reed M. Izatt and James J. Christensen and was held August 15–17, 1977 at Brigham Young University (BYU) in Provo, Utah [11]. Seventy-nine persons attended, 13 of them from ten countries outside of the USA. Of those attending from the USA, 23 were from BYU. Sixteen of the attendees came from 13 industrial companies. Twenty-eight universities were represented. The expenses for the symposium totaled $9500. The Provo symposia were held annually through 1981.

Izatt and Christensen envisioned the value of an annual symposium to provide a forum for the presentation and discussion of research activities in the field of macrocyclic chemistry. They saw a need in this new and rapidly evolving field to bring together persons from a variety of chemical and non-chemical fields who had an interest in macrocyclic chemistry, but who were not personally acquainted with each other. It was already apparent that the number of workers in the field was increasing rapidly and that interest spanned chemistry, physics, biology, and pharmacy. It was felt that an annual symposium could be the means to catalyze growth in the field and lead to the exploration of new areas of chemistry. It was visualized that both theoretical and experimental aspects of the properties and behavior of synthetic and naturally occurring macrocyclic compounds would be covered in a series of invited lectures as well as accepted contributed papers.

In 1980, the First European Symposium on Macrocyclic Compounds was held in Basel, Switzerland with Thomas Kaden as Chair. In 1982, the Second European Symposium on Macrocyclic Compounds was held in Strasbourg, France. At this meeting, informal discussions were held on the possibility of combining these two meetings into an annual symposium, which would be international in nature. It was agreed that the 1983 Symposium on Macrocyclic Compounds in Provo and the 1984 European Symposium in Stirling, Scotland would be held as scheduled. The 1985 meeting in Provo would be the first to be held under the new title of International Symposium on Macrocyclic Chemistry (ISMC). The ISMC meetings were held on an annual basis from 1985 until 2005.

In the early 2000s, it was recognized by several individuals that the fields of macrocyclic and supramolecular chemistry were growing together and becoming intertwined. As a result, it was proposed that the conferences in the two areas, the ISMC and the International Symposium on Supramolecular Chemistry (ISSC) be combined. Scientists found themselves attending both conferences to learn of new findings and meet colleagues. Also, it was evident that macrocycles were being used in many supramolecular structures, as can be seen in many of the chapters in this book. The committees for the two conferences decided to join the two conferences into one, to be called the International Symposium on Macrocyclic and Supramolecular Chemistry (ISMSC). Thomas Fyles from the University of Victoria organized the first joint meeting, which was held in June 2006 in Victoria, British Columbia, Canada. The ISMSC meetings continue the tradition of previous meetings of having excellent presenters who present groundbreaking discoveries in the fields of macrocyclic and supramolecular chemistry.

After being held in Canada, the meeting moved to Italy (2007) and was hosted by Luigi Fabbrizzi. Next, the meeting went back to North America and to Las Vegas, Nevada (2008) and was hosted by Jonathan Sessler and Eric Anslyn. Again it returned to Europe to Maastricht, The Netherlands (2009) and was hosted by Roeland Nolte and Alan Rowan. Asia came next with the meeting in Nara, Japan (2010) hosted by Makoto Fujita and Yoshihisa Inoue. Back to Europe the meeting went to Brighton, United Kingdom (2011) under Philip Gale. After the United Kingdom, it went to Otago, New Zealand (2012) where Sally Brooker hosted it. Next it went to Arlington, Virginia (2013), where Lyle Isaacs, Jeffery Davis, and Amar Flood were hosts. After Virginia, it went to Shanghai, China (2014) hosted by Zhanting Li. The 10th meeting was in Strasbourg, France (2015), where it was hosted by Luisa De Cola. The 2016 meeting will be held in Seoul, Korea under the direction of Kimoon Kim, Jong Seung Kim, and Juyoung Yoon. Future meetings are scheduled to be held in the United Kingdom, Canada, and Italy.

Table 1.2 *Symposia involving Macrocyclic and Supramolecular Chemistry (1977–2016). Abbreviations used: ISMC (International Symposium on Macrocyclic Chemistry); ISMSC (International Symposium on Macrocyclic and Supramolecular Chemistry)*

Year	Title	Location	Dates	Chair(s)
1977	First Symposium on Macrocyclic Compounds	Provo, Utah	15–17 August	R.M. Izatt, J.J. Christensen
1978	Second Symposium on Macrocyclic Compounds	Provo, Utah	14–16 August	R.M. Izatt, J.J. Christensen
1979	Third Symposium on Macrocyclic Compounds	Provo, Utah	6–8 August	R.M. Izatt, J.J. Christensen
1980	First European Symposium on Macrocyclic Compounds	Basel, Switzerland	2–4 July	T.A. Kaden
1980	Fourth Symposium on Macrocyclic Compounds	Provo, Utah	11–13 August	R.M. Izatt, J.J. Christensen
1981	Fifth Symposium on Macrocyclic Compounds	Provo, Utah	10–12 August	R.M. Izatt, J.J. Christensen
1982	Second European Symposium on Macrocyclic Compounds	Strasbourg, France	30 August–1 September	M-J. Schwing
1983	Seventh Symposium on Macrocyclic Compounds	Provo, Utah	8–10 August	R.M. Izatt, J.J. Christensen
1984	Third European Symposium on Macrocyclic Compounds	Stirling, UK	29–31August	R.W. Hay, R.M. Clay
1985	X ISMC	Provo, Utah	5–7 August	R.M. Izatt, J.J. Christensen
1986	XI ISMC	Florence, Italy	1–4 September	P. Paoletti L. Fabbrizzi
1987	XII ISMC	Hiroshima, Japan	20–23 July	E. Kimura
1988	XIII ISMC	Hamburg, Germany	4–8 September	A. Knöchel
1989	XIV ISMC	Townsville, Australia	25–28 June	L.F. Lindoy
1990	XV ISMC	Odessa, Ukraine	3–8 September	S.A. Andronati, V.P. Kukhar, N.G. Lukyanenko
1991	XVI ISMC	Sheffield, UK	1–6 September	J.F. Stoddart, D. Fenton
1992	XVII ISMC	Provo, Utah	9–14 August	R.M. Izatt, J.S. Bradshaw
1993	XVIII ISMC	Enschede, The Netherlands	27 June–2 July	D.N. Reinhoudt
1994	XIX ISMC	Lawrence, Kansas	12–17 June	D.H. Busch, K. Bowman-James
1995	XX ISMC	Jerusalem, Israel	2–7 July	A.Shanzer, D. Meyerstein
1996	XXI ISMC	Montecatini Terme, Italy	23–28 June	P. Paoletti, A. Bianchi
1997	XXII ISMC	Seoul, Korea	3–8 August	S.J. Kim
1998	XXIII ISMC	Turtle Bay, Oahu, Hawaii	7–12 June	J.L. Sessler, E.V. Anslyn
1999	XXIV ISMC	Barcelona, Spain	18–23 July	J. Casabo, E. Garcia-España
2000	XXV ISMC	St. Andrews, UK	2–7 July	R.W. Hay (posthumous) M. Schröder

(Continued)

Table 1.2 *(Continued)*

Year	Title	Location	Dates	Chair(s)
2001	XXVI ISMC	Fukuoka, Japan	15–20 July	S. Shinkai
2002	XXVII ISMC	Park City, Utah	23–27 June	J.D. Lamb
2003	XXVIII ISMC	Gdansk, Poland	13–18 July	J.F. Biernat
2004	XXIX ISMC	Cairns, Australia	4–8 July	L.F. Lindoy, F.R. Keene
2005	XXX ISMC	Dresden, Germany	17–21 July	K. Gloe
2006	1st ISMSC	Victoria, Canada	25–30 June	T.M. Fyles
2007	2nd ISMSC	Salice Terme, Italy	24–28 June	L. Fabbrizzi
2008	3rd ISMSC	Las Vegas, Nevada	13–18 July	E.V. Anslyn, J.L. Sessler
2009	4th ISMSC	Maastricht, The Netherlands	21–25 June	R.J.M. Nolte, A.E. Rowan
2010	5th ISMSC	Nara, Japan	6–10 June	M. Fujita, Y. Inoue
2011	6th ISMSC	Brighton, UK	2–7 July	P.A. Gale
2012	7th ISMSC	Otago, New Zealand	29 January–2 February	S.A. Brooker
2013	8th ISMSC	Arlington, Virginia	7–11 July	A.H. Flood, J.T. Davis L.D. Isaacs
2014	9th ISMSC	Shanghai, China	7–11 June	Z. Li
2015	10th ISMSC	Strasbourg, France	28 June–2 July	L. De Cola
2016	11th ISMSC	Seoul, Korea	10–14 July	K. Kim, J.S. Kim, J. Yoon

From the first joint meeting in 2006, the ISMSC has attracted top scientists and many students interested in the fields of macrocyclic and supramolecular chemistry, as was the case in its predecessor meetings. Presentations on molecular machines, metal organic frameworks, and supramolecular polymers as well as traditional topics such as host–guest binding and new macrocycles have been given. The 2015 ISMSC meeting in Strasbourg, France attracted 550 participants, two thirds of whom were students. The number of students attending the Symposium has always been high. For example, at the 8th ISMSC in Virginia, over 50% of the 350 attendees were students. Attendance at the meeting is normally around 350 as it was at the 3rd ISMSC in Las Vegas, Nevada, and at the 6th ISMSC in Brighton, U.K. The attendance at the 5th ISMSC in Nara, Japan was over 420. At the 2015 ISMSC in Strasbourg, the number of speakers was 50 and the number of poster presentations was 200. The large number of students in attendance at the symposia augurs well for the future of the field. The interest that characterized the early development of the field in the 1960s and 1970s continues to stimulate young people today. A summary of the symposia held from 1977 through 2016 is given in Table 1.2. Symposium titles, chairs, locations, and dates are presented.

1.4 Izatt–Christensen award sponsor: IBC Advanced Technologies, Inc.

IBC shares a common interest with the ISMSC in promoting molecular recognition as a guiding principle in developing new chemistry. This interest stems from a strong belief that scientific and engineering excellence in this field should be encouraged and will result in untold benefits for future generations. IBC has made an important contribution to the ISMSC for 25 years by providing financial support for the I–C Award. The chapters in this book provide insight into the way I–C Award winners have influenced progress in the fields of macrocyclic and supramolecular chemistry over this period.

The study of molecular recognition over the past half century has led to important discoveries and many applications, particularly in the medical, pharmacological, metallurgical and radiochemical sciences. The

1987 Nobel Prize in Chemistry was awarded to Charles Pedersen, Donald Cram, and Jean-Marie Lehn for "development and use of molecules with structure-specific interactions of high selectivity [12]." Two of the founders of IBC (Izatt and Bradshaw) received the American Chemical Society National Award in Separations Science and Technology in 1996 [13]. Inscribed on the award plaque were the words: "For advancing the separations science of metals and for new technology to forward industrial-scale recovery of metals from aqueous solutions." Many of the I–C Award recipients have received significant prizes and/or awards for their contributions to understanding and advancing the concepts of molecular recognition.

IBC was founded in 1988 by, and named after, Reed M. Izatt, Jerald S. Bradshaw, and James J. Christensen, three early workers in the field. These professors received much stimulus from contacts in the macrocyclic and supramolecular chemistry community. They made use of the ideas evolving in this scientific community during the 1970s and 1980s to design and prepare ligands with high affinity and high selectivity for specific metal ions in the presence of other competing metal ions that often have chemical properties very similar to those of the target metal ion. The resulting high selectivity of the ligands enabled IBC to develop a series of products (trade named SuperLig® and AnaLig®) using solid-supported ligands that could perform difficult metal ion separations at both the process and analytical scales [14]. The term Molecular Recognition Technology (MRT) was formulated in 1989 by Steven R. Izatt, President and CEO of IBC, to describe the practical application of molecular recognition in engineered systems for which customers receive value (i.e., for which they will pay money). MRT is well known today in extractive metallurgy [15–17] radionuclide separations [18], and chemical analysis [19]. Some early successes of MRT were its adoption by Impala Platinum Limited to process the palladium produced at its Springs Refinery in South Africa; its adoption by Tanaka Kikinzoku, Kogyo K.K. in Japan to recover rhodium from spent precious metal wastes [16, 17]; and the development, by IBC, 3M and Argonne National Laboratory, of Empore™ Rad Disks [18] marketed worldwide by 3M for analysis of radionuclides such as Sr and Ra. This development of Rad Disks was recognized by R&D 100 awards in 1996 and 1999 as well as the Federal Laboratory Consortium Award for Excellence in Technology Transfer in 1997 [18]. A hallmark of these processes is that they are based on green chemistry principles [15–17] resulting in significant economic and environmental advantages to the customer. The latest achievement made using MRT has been the demonstrated green chemistry separation in early 2015 of individual rare earth metals at the laboratory scale [20, 21]. Scale-up of the REE separations is underway and a pilot plant is expected to be operational in early 2016 [21, 22] that will be capable of producing individual rare earth metals at >99% recovery and >99% purity, with minimal waste generation.

IBC is committed to the principles of supramolecular chemistry which are central to the development of its highly selective separation systems that operate at the molecular level. Continuing support by IBC of the I–C Award is predicated on the belief that there are individuals in the macrocyclic and supramolecular chemistry community that will visualize and carry to fruition applications that will benefit society. The experience of IBC and the ideas expressed in this volume may be of value in stimulating others to commercialize their findings to the benefit of the larger community.

1.5 Summary

Twenty-five scientists have received the I–C Award since its inception in 1991. These individuals have made remarkable contributions to the fields of macrocyclic and supramolecular chemistry. The generosity of IBC in funding the Award and the willingness of ISMSC to host the Awardees at their annual symposia have made the Award possible. The ISMSC and its predecessor symposia have made significant contributions to the development of the fields of macrocyclic and supramolecular chemistry over the past nearly four decades. These symposia have provided a venue for the presentation of new results, for the discussion of new ideas, and for the development of new collaborations among researchers young and old. As the meetings move from country to country, opportunity is afforded for younger scientists to meet and interact with senior scientists in the field. The scientific programs of the Symposium have changed over the years and reflect the changing nature of the field. A strength of macrocyclic and supramolecular chemistry that has emerged over the decades has been its use in the hands of creative and skilled scientists to explore new areas of chemistry, as exemplified by the I–C Awardees. For example, the creation of host molecules of predesigned shapes enables one to mimic, on the molecular scale, components of simple machines such as molecular on–off switches, molecular axles, and molecular wires. Macrocycles pre-designed to interact selectively with target inorganic or organic guests have also had an important impact in separations chemistry. The future of the macrocyclic chemistry field is limited only by the imagination and creativity of its practitioners. It is expected that the ISMSC will continue to play an important role in facilitating personal interactions, exchanges of ideas, and the discovery of new chemistry.

References

1. Melson, G.A. (1979) General Introduction, In *Coordination Chemistry of Macrocyclic Compounds*, Melson, G.A. (Ed), Plenum Press, New York.
2. (a) Pedersen, C.J. (1967) Cyclic polyethers and their complexes with metal salts, *Journal of the American Chemical Society*, **89**, 7017–7036; (b) Pedersen, C.J. (1970) New macrocyclic polyethers, *Journal of the American Chemical Society*, **92**, 391–394.
3. (a) Izatt, R.M. (2007) Charles J. Pedersen: Innovator in macrocyclic chemistry and co-recipient of the 1987 Nobel Prize in Chemistry, *Chemical Society Reviews*, **36**, 143–147; (b) Izatt, R.M., Bradshaw, J.S. (1992) Charles J. Pedersen (1904–1989), Nobel Laureate in Chemistry (1987), *Journal of Inclusion Phenomena and Molecular Recognition Chemistry*, **12**, 1–6; (c) Izatt, R.M., Bradshaw, J.S. (Eds) (1992) *The Pedersen Memorial Issue. Advances in Inclusion Science*, Vol. **7**.
4. Busch, D.H. (2016) Synthesis of macrocyclic complexes using metal ion templates, This volume, Chapter 19.
5. (a) Thompson, M.C., Busch, D.H. (1962) Chemical & Engineering News, *September* **17**, 57; (b) Thompson, M.C., Busch, D.H. (1964) Reactions of coordinated ligands. IX. Utilization of the template hypothesis to synthesize macrocyclic ligands *in situ*, *Journal of the American Chemical Society*, **86**, 3651–3656.
6. (a) Dietrich, B., Lehn, J.-M., Sauvage, J.-P. (1969) Diaza-polyoxa-macrocycles et Macrobicycles, *Tetrahedron Letters*, 2885–2888; (b) Dietrich, B., Lehn, J.-M., Sauvage, J.-P. (1969) Les Cryptates, *Tetrahedron Letters*, 2889–2892.
7. Lehn, J.-M. (2007) From supramolecular chemistry towards constitutional dynamic chemistry and adaptive chemistry, *Chemical Society Reviews*, **36**, 151–160.
8. Sherman, J.C. (2007) Donald J. *Cram, Chemical Society Reviews*, **36**, 148–150.
9. Nolte, R.J.M., Rowan, A.E., Elemans, J.A.A.W. (in press) Clipping angel's wings, This volume, Chapter 12.
10. (a) Izatt, R.M., Rytting, J.H., Nelson, D.P., Haymore, B.L., Christensen, J.J. (1969) Binding of alkali metal ions by cyclic polyethers: Significance in ion transport processes, *Science*, **164**, 443–444; (b) Izatt, R.M., Nelson, D.P., Rytting, J.H., Haymore, B.L., Christensen, J.J. (1971) A calorimetric study of the interaction in aqueous solution of several uni- and bi-valent metal ions with the cyclic polyether dicyclohexyl-18-crown-6 at 10, 25, and 40°, *Journal of the American Chemical Society*, **93**, 1619–1623.
11. Izatt, R.M., Pawlak, K., Bradshaw, J.S. (2005) Contributions of the International Symposium on Macrocyclic Chemistry to the development of macrocyclic chemistry, In *Macrocyclic Chemistry: Current Trends and Future Perspectives*, Gloe, K. (Ed), Springer, Dordrecht.
12. Nobel Foundation (1987) The Nobel Prize in Chemistry 1987, <http://www.nobelprize.org/nobel_prizes/chemistry/laureates/1987/>, accessed February 14, 2016.
13. Izatt, R.M., Bradshaw, J.S. (1996) Joint recipients of the 1996 American Chemical Society Award in Separations Science and Technology, *Chemical & Engineering* News, January **22**, 56.
14. Izatt, R.M., Bruening, R.L., Bruening, M.L., Tarbet, B.J., Krakowiak, K.E., Bradshaw, J.S., Christensen, J.J. (1988) Removal and separation of metal ions from aqueous solutions using a silica-gel-bonded macrocycle system, *Analytical Chemistry*, **60**, 1825–1826.

15. Izatt, R.M., Izatt, S.R., Izatt, N.E., Krakowiak, K.E., Bruening, R.L., Navarro, L. (2015) Industrial applications of molecular recognition technology to separations of platinum group metals and selective removal of metal impurities from process streams, *Green Chemistry*, **17**, 2236–2245.
16. Izatt, N.E., Bruening, R.L., Krakowiak, K.E., Izatt, S.R. (2000) Contributions of Professor Reed M. Izatt to molecular recognition technology: from laboratory to commercial application, Festschrift issue of *Industrial & Engineering Chemistry Research*, **39**, 3405–3411.
17. Izatt, S.R., Bruening, R.L., Izatt, N.E. (2012) Status of metal separation and recovery in the mining industry, *Journal of Metals*, **64**, 1279–1284.
18. (a) Goken, G.L., Bruening, R.L., Krakowiak, K.E., Izatt, R.M. (1999) Metal-ion separations using SuperLig® or AnaLig materials encased in Empore cartridges and disks, In *Metal-Ion Separation and Preconcentration: Progress and Opportunities*, Bond, A.H., Dietz, M.L., Rogers, R.D. (Eds), American Chemical Society, Washington, DC; (b) Izatt, S.R., Bruening, R.L., Krakowiak, K.E., Izatt, R.M. (2003) The selective separation of anions and cations in nuclear waste using commercially available molecular recognition technology (MRT) products, paper presented at the Waste Management '03 Conference, Tuscon, AZ; (c) (1996) Rad disks simplify sample prep, 1996 R&D 100 Awards, *R&D Magazine*, **38**, 21;(d)(1999) Empore Rad disks, Top 40 R&D 100 Winners, Top 1% of all R&D 100 Awards Ever Given, *R&D Magazine*, **41**, 159;(e)Federal Laboratory Consortium (1997) FLC Awards Archive, <http://www.federallabs.org/awards/1997/>, accessed February 15, 2016.
19. (a) Rahman, I.M.M., Furusho, Y., Begum, Z.A., Sato, R., Okumura, H., Honda, H., Hasegawa, H. (2013) Determination of lead in solution by solid phase extraction, elution, and spectrophotometric detection using 4-(2-pyridylazo)-resorcinol, *Central European Journal of Chemistry*, **11**, 672–678; (b) Izatt, R.M., Bradshaw, J.S., Bruening, R.L., Bruening, M.L. (1994) Solid phase extraction of ions of analytical interest using molecular recognition technology, *American Laboratory*, **26**, 28C–28M.
20. (a) Molecular recognition technology: environmentally friendly, cost efficient separation of each individual rare earth element, http://mrt.ucore.com, accessed February 15, 2016; (b) Izatt, R.M. (2015) Molecular recognition technology: clean chemistry applied to 21st century rare earth separation, *InvestorIntel*, May 15, <http://investorintel.com/technology-metals-intel/molecular-recognition-technology-clean-chemistry-applied-to-21st-century-rare-earth-separation/>, accessed February 15, 2016; (c) Izatt, R.M. (2015) Green chemistry in modern mining and rare earth beneficiation, *InvestorIntel*, July 9, <http://investorintel.com/technology-metals-intel/green-chemistry-in-modern-mining-rare-earth-beneficiation/>, accessed February 15, 2016.
21. Izatt, S.R., Izatt, R.M., Izatt, N.E., *et al.* (in press) Selective recovery of platinum group metals and rare earth metals from complex matrices using a green chemistry–molecular recognition technology approach, In *Metal Sustainability: Global Challenges, Consequences, and Prospects*, Izatt, R.M. (Ed), Wiley-Blackwell, Oxford.
22. Ucore commissions design and construction of SuperLig® One Pilot Plant, <http://ucore.com/ucore-commissions-design-and-construction-of-superlig-one-pilot-plant>, accessed February 15, 2016.

2

Supramolecular Chemistry with DNA

Pongphak Chidchob and Hanadi Sleiman

Department of Chemistry, McGill University, Montreal, Québec, Canada

2.1 Introduction

DNA has emerged as a powerful guiding molecule to achieve supramolecular organization [1]. The fundamental features of this fascinating molecule include highly selective assembly, programmable sequence design, and a well-defined, rigid structure. These properties make DNA one of the most programmable building blocks for supramolecular chemistry. The ability of DNA assembly to synthetically generate any arbitrary object with nanoscale precision is unparalleled by other types of materials. In addition, the dynamic character of DNA can result in functional systems that perform precise tasks in response to specific stimuli. The aim of this chapter is to provide an overview on the development of DNA assembly, generally known as the field of DNA nanotechnology, with an emphasis on the contributions from our research group in using synthetic molecules to guide DNA self-assembly. This latter area is termed supramolecular DNA assembly [2], and opens up a new direction in the assembly of DNA-based materials.

2.2 Motifs in structural DNA nanotechnology

2.2.1 DNA structural properties

DNA is a biopolymer of four nucleotide monomers. It consists of nitrogen-containing bases (nucleobases) attached to five-membered deoxyribose units and connected by phosphate groups. Adenine (A) and guanine (G) belong to one type of nucleobases called purines and can form specific hydrogen bonds to thymine (T) and cytosine (C), which are pyrimidine bases (Figure 2.1a). The A:T and G:C hydrogen-bond motifs are called Watson–Crick base-pairs. These interactions, as well as π–π stacking of the nucleobases and hydrophobic

Macrocyclic and Supramolecular Chemistry: How Izatt–Christensen Award Winners Shaped the Field, First Edition.
Edited by Reed M. Izatt.

effects, result in the cooperative association of two DNA strands, termed DNA hybridization into a double helix (or duplex). The two strands in this duplex are antiparallel, meaning that the 3′-sugar end of one strand aligns with the 5′-end of the other strand. In its most common B-form, the duplex has a diameter of 2 nm and ~10.5 bases per helical turn, with a pitch length of 3.4 nm (Figure 2.1a). B-DNA behaves like a rigid rod over about three helix turns or ~30 base-pairs. Its persistence length, another indication of stiffness, is about 50 nm. All these properties make DNA an excellent building block for supramolecular organization. DNA strands can now be generated very easily by automated DNA synthesis, in a near-infinite number of sequences, each capable of selective assembly with its complementary strand. Thus, unlike many supramolecular building blocks, DNA is a highly programmable molecule that has the potential to assemble into symmetric, as well as asymmetric and anisotropic structures.

2.2.2 The beginning of DNA nanotechnology: DNA tile assembly

The idea of building nanoscale structures from DNA was first proposed by Nadrian Seeman in the 1980s. It relied on two concepts borrowed from molecular biology. The first is the DNA crossover junction, similar to the Holliday junction found in DNA recombination. In this motif, a strand starts from one DNA helix and switches over to the next, connecting two DNA double helices together (Figure 2.1b) [3]. This gives the branched DNA units necessary for 2D- and 3D-construction. The second concept is sticky-end cohesion. If a DNA duplex has a short single stranded component at its end (sticky-end or overhang), it can come together with another duplex having the complementary overhang via base-pairing (Figure 2.1b). This interaction allows the DNA units to be selectively connected together into nanostructures.

Initially, branched molecules consisting of four DNA strands were used. To increase rigidity of the motifs and enable the construction of more robust materials, DNA double-crossover (DX) junctions, containing two double helices connected to each other twice through crossover points, were developed [4]. Together with sticky-end cohesion, DX molecules were successfully used to generate periodic two-dimensional crystals (Figure 2.1c) [5]. Since then, other motifs with increased complexity such as triple crossover [6], paranemic crossover [7], cross-shaped tiles [8], tensegrity triangle [9], three-point star [10], and six-helix bundles [11] have been introduced and applied to create 2D- and 3D-DNA structures. Seeman and Mao reported the first construction of crystalline 3D-DNA assemblies in 2009, from rigid triangular units and very short sticky ends [12].

2.2.3 DNA origami and single-stranded tiles

A scaffolding strategy referred to as "DNA origami" was reported by Rothemund in 2006 [13]. This approach significantly increased the complexity and range of DNA nanostructures. In this method, a 7-kilobase single-stranded plasmid DNA is folded with the help of hundreds of DNA single strands. These act as "staples" to guide and connect the scaffold into a variety of two-dimensional objects, such as rectangles, stars, and smiley faces (Figure 2.1d) [13]. The objects are designed with the help of a computer interface. Patterning on this origami object, such as drawing a map of the Americas, can be easily done because the sequences of staple strands are all unique. Yan, Shih, Gothelf, Sugiyama and others extended this approach into three-dimensional construction [15–17]. Twist and curvature in origami objects can also be controlled, by insertion/deletion of base pairs to alter the distance between crossover points [18] or by modification of positions and patterns of crossover points of the whole structure [19].

As a complementary approach to DNA origami, Yin introduced the concept of "single-stranded tile assembly," which creates objects with similar complexity to origami, without the need of a long scaffold [14]. The motifs are single-stranded DNA containing four modular domains which are designed to form interconnected staggered duplexes with one another, resulting in DNA lattices. As sequences are all unique, these motifs can

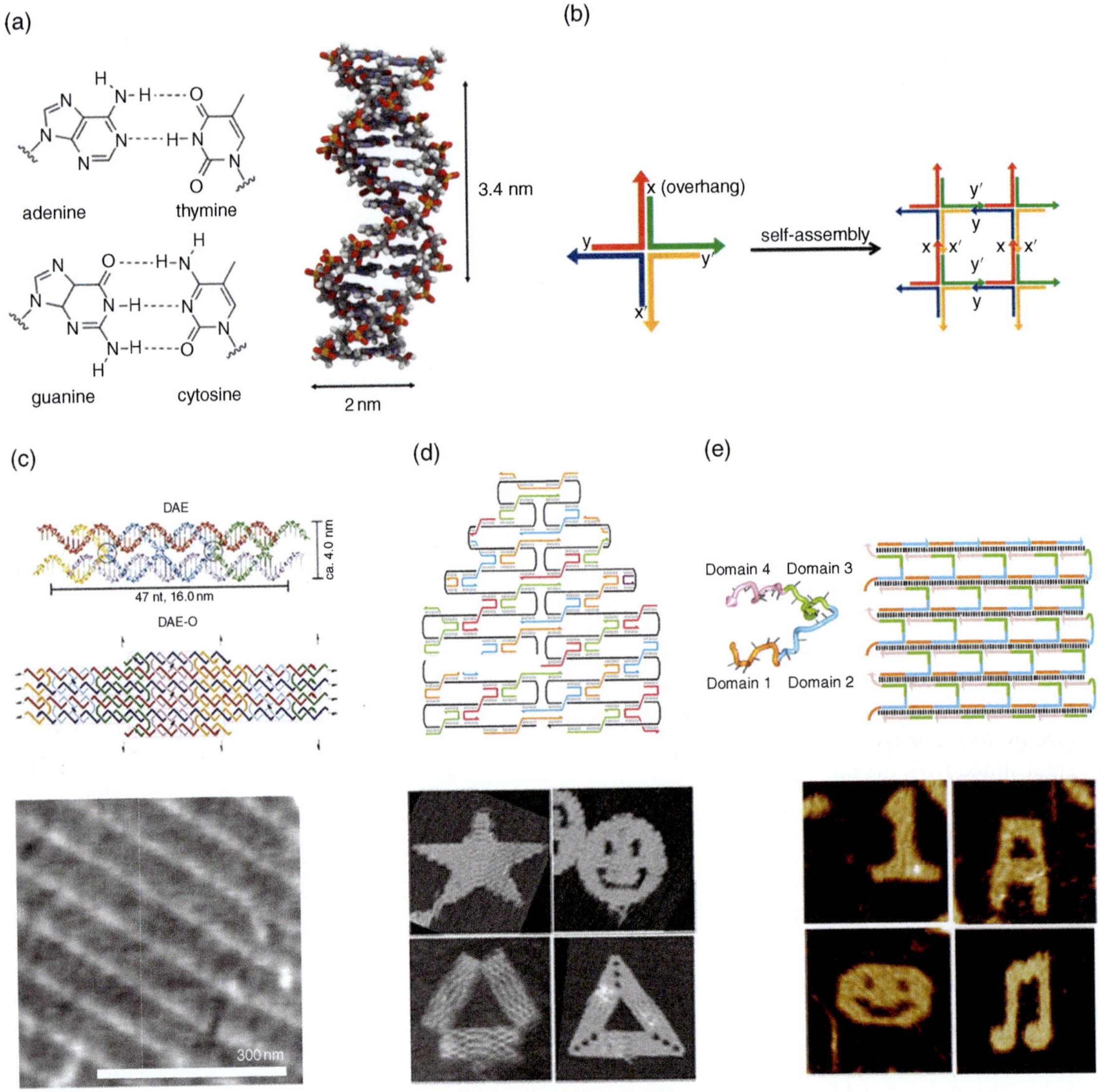

Figure 2.1 *Motifs in DNA nanotechnology. (a) Left. Hydrogen-bond motifs of A:T and C:G; right. B-form DNA duplex. (b) DNA crossover junction and its self-assembly into periodic arrays. (c) Double-crossover molecule for tile-based assembly into two-dimensional crystals [5]; adapted by permission from Macmillan Publishers Ltd:* Nature, *copyright © 1998. (d) DNA origami assembly which uses long viral DNA strands as scaffolds to achieve complex objects [13]; adapted by permission from Macmillan Publishers Ltd:* Nature, *copyright © 2006. (e) Single-stranded tile approach for two-dimensional construction without a need of scaffold [14]; adapted by permission from Macmillan Publishers Ltd:* Nature, *copyright © 2012*

be used as a molecular canvas where one can make any arbitrary shape by selecting a set of strands which defines the structure (Figure 2.1e). This approach has been extended to three-dimensional DNA discrete and periodic structures [20].

2.2.4 Perspective

DNA origami and single-stranded tile assembly have been used to make a range of 2D and 3D structures, with impressive complexity. A distinctive feature of these methods is their need for hundreds of component DNA strands, each with a unique sequence. The techniques are currently expensive, and difficult to apply *in vivo* for medicinal applications, as each component DNA strand would need to be tested for its toxicity and immune response. On the other hand, DNA tile structures can be constructed from very few starting strands (sometimes only one) [21]. However, they are mostly symmetrical and periodic. Thus, one important challenge in DNA nanotechnology is to balance design simplicity with complexity, such that the structures created have a small number of DNA strands and are easy to construct, but are highly controlled in their geometry and size, and allow anisotropic and addressable organization. In addition, nearly all of the nanostructures constructed by tile or origami assembly are densely packed with double-stranded and rigid DNA helices. Many of them are unstable in biologically relevant environments, because their highly negatively charged nature requires a high cation concentration.

2.3 Dynamic assembly and molecular recognition with DNA

Strand displacement is an efficient concept which has been widely applied to program dynamic motion in DNA structures, and led to development of many stimuli-responsive DNA systems, DNA machines and DNA computing tools [22]. DNA strand displacement is shown in Figure 2.2a. If a DNA strand **a–b** is hybridized to a shorter complementary strand **a′**, then a duplex will form with single-stranded "overhang" or "toehold" **b**. When a fully complementary "input" strand **b′–a′** is added, "output" strand **a′** is displaced to yield a fully complementary, longer duplex. This process begins with the hybridization of input strand to the single-stranded overhang on the duplex. Branch migration, a random walk which involves single nucleotide unzipping of output strand from the duplex and hybridization of input strand to and from the new duplex, will generate the more thermodynamically stable duplex. This process occurs rapidly (seconds) and in quantitative yields with overhangs above six bases [23].

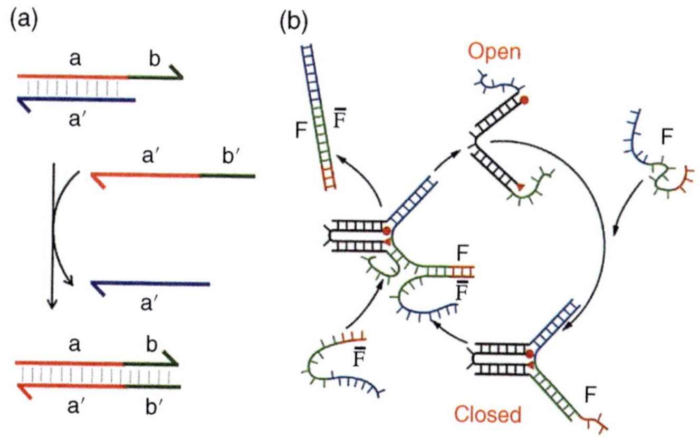

Figure 2.2 *Dynamic assembly of DNA. (a) Strand displacement strategy where b′–a′ input strand binds to duplex a′: a–b and removes a′, to form new duplex b′–a′: a–b. (b) Stimuli-responsive molecular DNA tweezer that operates via strand displacement mechanism; adapted by permission from Macmillan Publishers Ltd:* Nature *[24], copyright © 2000*

Yurke and Turberfield demonstrated the first use of strand displacement to operate conformational change of molecular DNA tweezer [24]. The system is composed of central strand hybridizing to two strands with overhangs as the two arms. This tweezer is closed by addition of input strand F which binds to the two arms. Addition of another input strand, $\bar{F}$, complementary to F will remove the first input strand, resulting in opening of the tweezer (Figure 2.2b). Seeman extended this concept to perform larger-scale mechanical motion on rotatable DNA device where four DNA structures are joined on one-dimensional arrays, spacing by DNA junctions which can change between two conformations [25]. Addition of appropriate sets of DNA inputs will trigger the reversible interconversion between the two states which in turns rotates the DNA structures up and down. Since then, even more complicated motions of DNA machines have been reported, such as a DNA walker that can move and collect cargos [26] or perform a series of organic syntheses [27] along the track.

DNA stimuli can be manipulated in an analogous way to electronic signal processing by dynamic DNA assembly. Pierce demonstrated signal amplification by introducing the concept of hybridization chain reaction [28]. The transducer contains two species of DNA hairpins which can assemble together repeatedly with the trigger from input strand. DNA devices that can respond to specific sets of stimuli conditionally similar to electronic logic circuits were reported by Winfree [29].

2.4 Supramolecular assembly with hybrid DNA materials: increasing the letters of the alphabet

The power of structural DNA nanotechnology lies in the predictability and programmability of DNA structures. In a sense, this research field has reduced the self-assembly space into four letters: A, T, G, and C. By using DNA base-pairing as the major interaction, and with an intimate knowledge of DNA structure, one can set out to create a specific assembly, with a good chance of successfully obtaining this design in the laboratory. To generate complexity in DNA nanotechnology, we increase the number of unique DNA strands that make up a structure.

Nature relies on many more interactions to achieve order, and build its functional structures. For example, if one looks at protein assembly, hydrogen bonding between the peptide units is not the only interaction used. Instead, a hierarchical interplay of hydrogen bonding, hydrophobic, electrostatic, and metal-coordinating interactions work together to fold the protein into its active form. Over the last 50 years, supramolecular chemistry has taken advantage of these multiple non-covalent interactions to assemble materials with exquisite control over geometry and function.

Can we bring these orthogonal interactions into DNA assembly? In this way, we would blend the approaches of supramolecular chemistry with the programmability of DNA. This can then give access to new motifs in DNA nanotechnology and can introduce functions such as photoactivity, redox, or catalytic behavior into the typically passive DNA structures. Now, complexity can be generated without the need to use a large number of DNA sequences.

From the point of view of supramolecular chemistry, bringing the DNA toolbox into the fold allows us to create asymmetric, anisotropic structures with arbitrary placement of functional groups. DNA scaffolds bring a wealth of biological activity into supramolecular chemistry, with such potential applications as gene-silencing therapeutics, aptamers that selectively bind proteins, small biomolecules and even cells, sensing and diagnostic units, among many others.

2.4.1 Three-dimensional structures: DNA cages

One of the starting points of our research program was the introduction of synthetic molecules within DNA strands, and their use as vertices for nanostructures. By equipping a molecule with a phosphoramidite unit and a protected alcohol, it can be directly incorporated in any position of the DNA strand through automated

DNA synthesis (Figure 2.3a). Coupling conditions need to be adjusted each time, and groups that are incompatible with DNA solid-phase synthesis need to be protected.

2.4.1.1 Synthesis

To create prismatic cages, we set out to build the component DNA polygons. We used a triphenylene molecule as the corner unit [30]. To build a DNA triangle, we synthesized a DNA strand with three triphenylene insertions via automated phosphoramidite chemistry. We then used a short DNA template that is complementary to the two strand ends, thus bringing these ends together upon hybridization. This was followed by the covalent joining of the two ends via a phosphodiester linkage, which can be performed either enzymatically (via a DNA ligase) or chemically. This gave a DNA triangle with three single-stranded arms of different sequences and three rigid organic corner units (Figure 2.3b). Similar strategies gave DNA squares, pentagons and hexagons. We then aligned the polygons on top of one another and connected them via hybridization with three linking strands. The linking strands have three segments: a top segment complementary to the first triangle edge, a bottom segment complementary to the second triangle edge, and a middle segment. When the linking strands are hybridized to these polygons, they bring them together with three single-stranded vertical arms (the middle segments) connecting them. In the next step, we added the strands complementary to these vertical arms, thus closing the DNA prism (Figure 2.3c).

There were three previous strategies for making DNA cages at the time. Seeman demonstrated the first example of DNA cube from a three-arm DNA junction by repetitive ligation and hybridization of ten DNA strands [31]. The same group then showed the assembly of truncated octahedron, composed of six squares and eight hexagons [32]. However, these assemblies suffered from multistep preparation and low yields. Without the need of ligation, Turberfield designed four DNA strands that could efficiently fold in a single step into a tetrahedral cage with three-arm junction at the vertex [33, 34]. Unhybridized thymine-spacer hinges were incorporated at the vertex to increase flexibility and relieve any structural strain resulting from bending of DNA duplexes at the vertices. To date, this strategy has been used to build two triangulated cages, a tetrahedron and a bipyramid. Finally, Joyce reported the folding of a long DNA strand with the help of a number of other strands into an octahedron, in a method that undoubtedly inspired the DNA origami approach [35]. Beyond the Turberfield tetrahedron, 3D construction with DNA was difficult and low-yielding; there were no reported methods to generate cages with deliberate variation of geometry.

Within this context, we were surprised to find that our DNA prisms assembled in quantitative, or near quantitative yields for all geometries: triangular prism, cube, pentameric and hexameric prisms, heteroprisms and biprisms. We later discovered that the key to the success of our strategy was the use of linking strands with single-stranded middle segments. These flexible portions of the assembly likely reduce the strain upon full hybridization and closure of the prism; they can then be converted to the double-stranded form after prism formation.

In addition, and importantly, this assembly does not need to be carried out in a stepwise manner. We can engineer the thermal denaturation temperature (T_m) of the DNA double strands by adjusting their sequences; for example, increasing the G–C content can raise the T_m. By making the hybridization of the linking strands (Figure 2.3c) to the polygon more stable (higher T_m) than the hybridization of the vertical strands to their complements, we can carry out a single-pot assembly. The system is heated to ~95 °C, upon which all double strands dissociate into their single-stranded form. Upon slow cooling, we first reach the T_m of the linking strand to polygons, leading to cohesion of the prism; upon further cooling, we reach the T_m of the vertical strands, thus closing the prism in a second step. This hierarchical assembly is another attractive feature of DNA construction.

The DNA cages obtained here were shown to be dynamic and stimuli responsive. Their controlled and reversible switching was demonstrated using strand displacement [30]. By changing the length of linking regions with added DNA strands, a dynamic triangular prism was able to oscillate between three predetermined sizes (Figure 2.3d). This was the first demonstration of dynamic and reconfigurable character in 3D DNA structures.

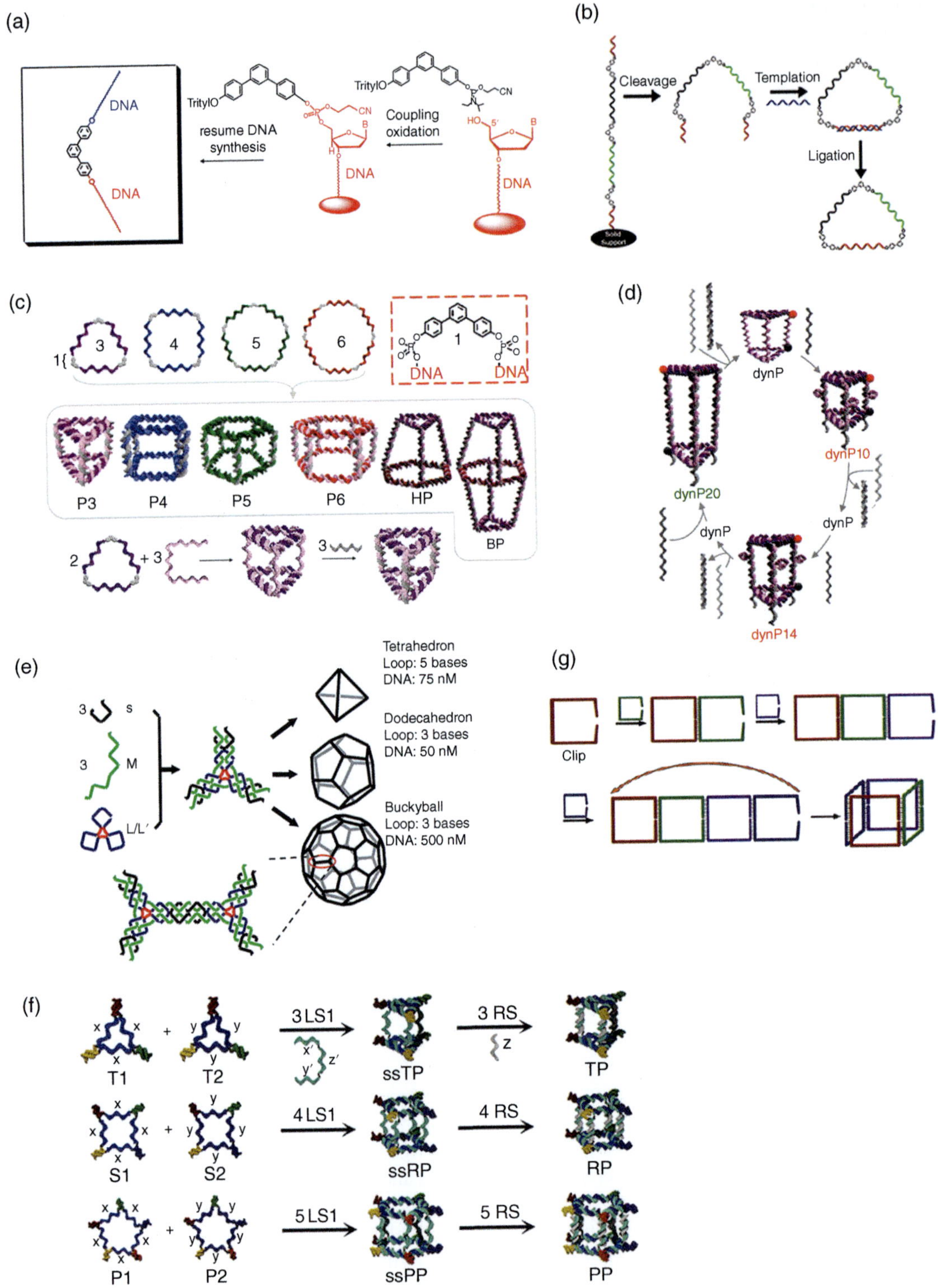
(a)
DNA
DNA
TritylO
resume DNA synthesis
CN
B
H
DNA
TritylO
CN
Coupling oxidation
HO
5′
B
DNA
(b)
Cleavage
Templation
Ligation
(c)
1{
3
4
5
6
1
DNA
DNA
P3
P4
P5
P6
HP
BP
2
+ 3
3
(d)
dynP
dynP10
dynP
dynP14
dynP
dynP20
(e)
3
S
3
M
L/L′
Tetrahedron
Loop: 5 bases
DNA: 75 nM
Dodecahedron
Loop: 3 bases
DNA: 50 nM
Buckyball
Loop: 3 bases
DNA: 500 nM
(g)
Clip
(f)
T1
T2
3 LS1
x′
y′
z′
ssTP
3 RS
z
TP
S1
S2
4 LS1
ssRP
4 RS
RP
P1
P2
5 LS1
ssPP
5 RS
PP

Mao then reported the use of a three-point star tile, composed of seven DNA strands connected together by three four-arm junctions, to selectively create polyhedra including a tetrahedron, dodecahedron, and buckyball, simply by altering the flexibility and concentration of the building blocks (Figure 2.3e) [36]. The higher the concentration and stiffness of the building blocks, the larger the structures that one can obtain. The polyhedra were fully symmetrical, having been generated from identical tiles.

2.4.1.2 Simplified and DNA-minimal design of DNA cages

Once we developed the face-centered strategy, we set out to simplify it so that it could be used by other laboratories that do not have access to a DNA synthesizer. DNA polygons were prepared from single strands with three segments (Figure 2.3f) [37]. The outer segments of the strands are complementary to one another, such that they guide the formation of the DNA polygons. The inner segments (*x* or *y*) will then constitute the sides of these polygons. A linking-strand strategy as described above (Figure 2.3c) assembles the prismatic cages (Figure 2.3f). Thus, the strategy does not require ligation or chemical modification. To further simplify the strategy, we found that the sides of each polygon can be made identical to each other in sequence, without compromising the quantitative yield of the DNA cage assembly. This reduced the number of unique DNA sequences that make up the structure, thus reducing cost and assembly errors.

A particularly versatile strategy developed in our laboratory starts from a minimum number of DNA strands, but maximizes the single-stranded regions on the final assemblies [38]. For example, to assemble a rectangular prism, we start with four unmodified DNA strands (clips, Figure 2.3g). Each clip is designed so that its two ends can be brought together by hybridization with the back edge of the next clip. To form a closed prism, the two ends of the fourth clip are designed to be complementary to the back edge of the first strand. For proper folding of each DNA strand and efficient assembly, we inserted a thymine spacer of four nucleotides in length (T4) to help mitigate helical distortion when placed at the vertex positions within 3D structures. To make a pentagonal prism, three of the original clip strands can be retained, and two additional clips of the proper sequence are added. Thus a cube can be made from four unmodified strands (Figure 2.3g). We used this strategy to make triangular, square, pentagonal, hexagonal, heptagonal and octagonal prisms, all in quantitative or near-quantitative yields. Another feature of this method is that the top and bottom single-stranded faces of prism can be designed to have unique sequences, which are useful as a three-dimensional scaffold for site-specific organization of other materials.

2.4.1.3 Perspective

We have presented modular and DNA economic syntheses of DNA cages, with deliberate variation of size, shape, and site-specific addressability. Unlike other 3D-structures obtained with conventional approaches (such as tile or origami assembly), they are constructed from a minimum number of DNA strands, and they have intrinsic dynamic character: we showed their controlled size switching with added DNA strands. DNA cages have great potential for encapsulating and delivering drugs, proteins and nanomaterials. Their confined spaces allow the exploration of new reactivity and catalysis for guest molecules, and can serve as templates to

Figure 2.3 *(a) Insertion of synthetic molecule within DNA strands by phosphoramidite chemistry. (b) Construction of single-stranded DNA triangle with three rigid organic corner units. (c) Prismatic cages constructed by face-centered strategy; Adapted with permission from [30], copyright © 2007 American Chemical Society (d) Dynamic triangular prisms that can oscillate between three predetermined lengths via strand displacement; adapted with permission from [30], copyright © 2007 American Chemical Society. (e) Hierarchical assembly of polyhedra from three-point star motifs; adapted by permission from Macmillan Publishers Ltd:* Nature *[36], copyright © 2008. (f) Simplified strategy for an assembly of prismatic cages from single-stranded DNA without the need of ligation; adapted from [37] with permission from The Royal Society of Chemistry. (g) DNA-minimal design of cube from only four single-stranded DNA strands; Adapted with permission from [38], copyright © 2012 American Chemical Society*

grow supramolecular materials. In addition to biological applications, DNA cages can be used as 3D scaffolds to position numerous other supramolecular materials (light-harvesting units, magnetic materials, nano-electronic components) into functional nanostructures.

2.4.2 Three-dimensional structures: DNA nanotubes

DNA nanotubes are also particularly interesting, for applications in the growth of nanowires of controlled shape and length, to control loading and release of drugs, and as programmable tracks for molecular motors.

2.4.2.1 Synthesis

Rothemund and Winfree showed that DNA nanotubes can be constructed by employing the intrinsic curvature of DX molecules which close up the nanotube [39]. Liu and LaBean reported the construction of DNA nanotubes from crossover tiles in which three double helices are connected with four crossover points [40]. A reactive thiol group was incorporated in the motifs to roll the sheet of tiles into nanotube, which can be metallized to generate conductive nanowire. Turberfield created DNA nanotubes from two different DX molecules [41]. Depending on the connectivity of the sheet, either alternating or nested patterns were obtained. Seeman reported the construction of DNA helix bundles, containing a defined number of parallel DNA duplexes, and assembled them into nanotubes of defined diameters [11].

We developed a modular approach to construct DNA nanotubes which finely controls the geometry, size, and persistence length of these assemblies [42]. First, we cyclized single-stranded DNA with the appropriate number of organic vertices into triangles or squares, as described above (Figure 2.3b). Triangular and rectangular nanotubes can be generated by joining these polygons top to bottom with linking strands (Figure 2.4a). Depending on the linking strands, these nanotubes can be generated in single-stranded and open forms, as well as double-stranded and closed versions of dramatically different stiffness (Figure 2.4b). With the aid of single-molecule fluorescence imaging, stepwise growth of the surface-grafted nanotubes can be monitored *in situ* and controlled in real time [43]. Recently, we designed similar DNA polygons without using organic vertices and without ligation [44]. For example, a triangular unit that assembles into nanotubes can be quantitatively generated from six unmodified DNA strands, allowing facile nanotube synthesis.

2.4.2.2 Control of nanotube length and stability

Most nanotube assembly methods rely on step-growth polymerization of the building blocks. To precisely control the length of DNA nanotube and yield monodisperse nanotubes is more challenging. Our group employed a template approach to create a long guiding strand containing repeating regions which are available for binding with DNA triangles (Figure 2.4c) [45]. This method produced nanotubes with length up to 1 μm with narrow size distribution. Encapsulation of gold nanoparticles within the periodic cavity of the nanotubes yielded lines of the nanoparticles all of the same length. However, the method required a relatively large number of starting strands (see below). As an alternative to the use of multiple short DNA strands, we prepared a long template strand by rolling circle amplification (RCA). This method uses a polymerase enzyme to synthesize continuous single stranded DNA with repeating sequence from a circular template. Attaching DNA polygons created nanotubes with a relatively defined length, and increased stability (Figure 2.4d) [44, 46]. The Willner group used DNA single strands produced by RCA to template the joining of two-dimensional hexagonal arrays into nanotubes with nearly monodisperse diameters [47]. Shih and co-workers were able to "roll" DNA origami into well-defined nanotubes [48], and Yin and LaBean used single-stranded tiles to control DNA nanotube circumferences [49].

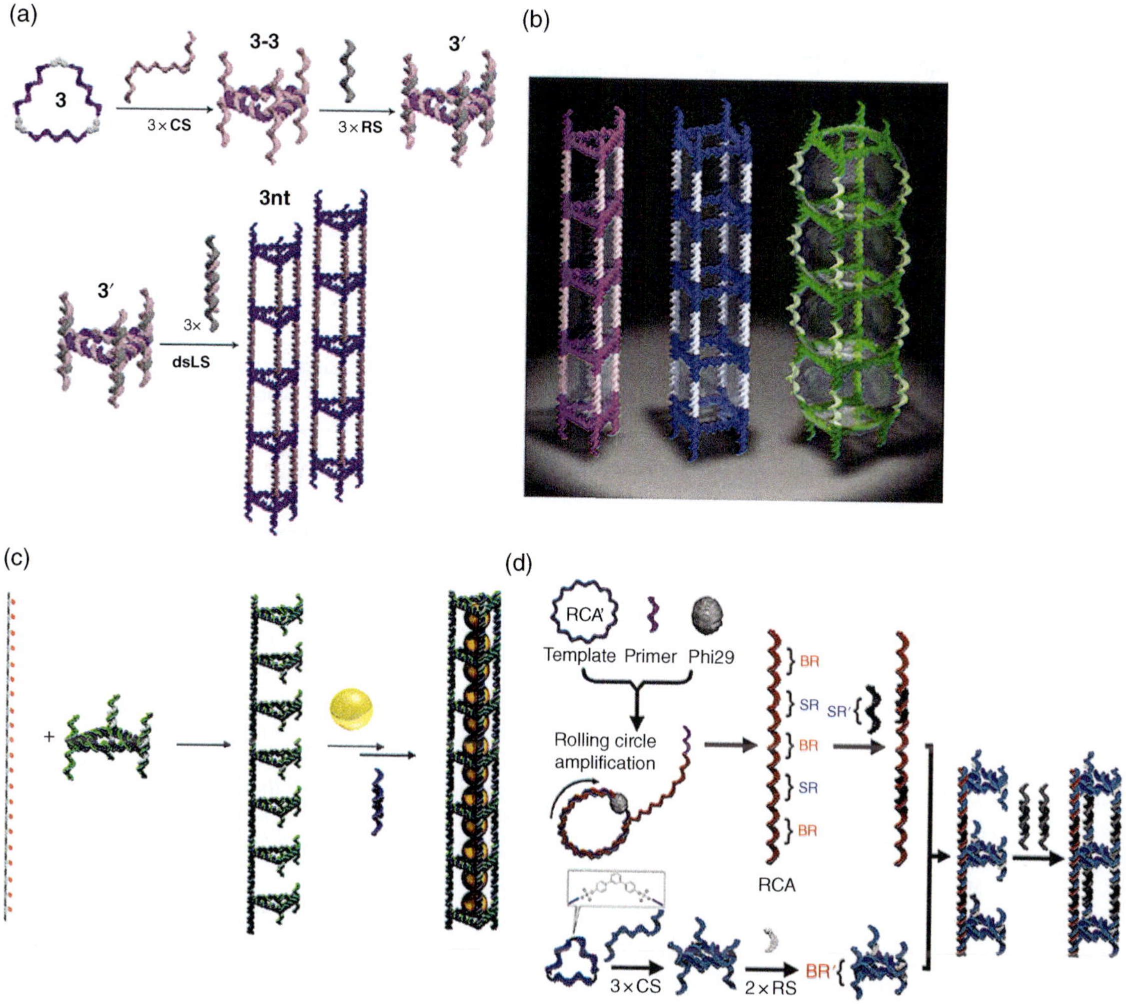

Figure 2.4 (a) Modular construction of DNA nanotubes from single-stranded polygons, containing organic vertices, and linking strands; adapted by permission from Macmillan Publishers Ltd: Nature Nanotechnology [42], copyright © 2009. (b) DNA nanotubes in their double-stranded closed form and single-stranded open form; adapted by permission from Macmillan Publishers Ltd: Nature Nanotechnology [42], copyright © 2009. (c) DNA nanotubes of predetermined length generated from template strand created from multiple short DNA strands; reprinted with permission from [45]; copyright © 2010 American Chemical Society. (d) Construction of DNA nanotubes of predetermined length from template strand prepared by enzymatic method (Phi29 DNA polymerase); Adapted with permission from [46], copyright © 2012 American Chemical Society

2.4.2.3 Temporal growth of DNA assembly

An important challenge in DNA nanotechnology is the question of scalability. On the one hand, methods such as DNA origami provide exquisite anisotropic control over matter; however, they come at a high cost – hundreds of different DNA strands to build a single object. On the other hand, one can use a small number of building blocks with identical DNA sequences, although structural complexity is significantly diminished.

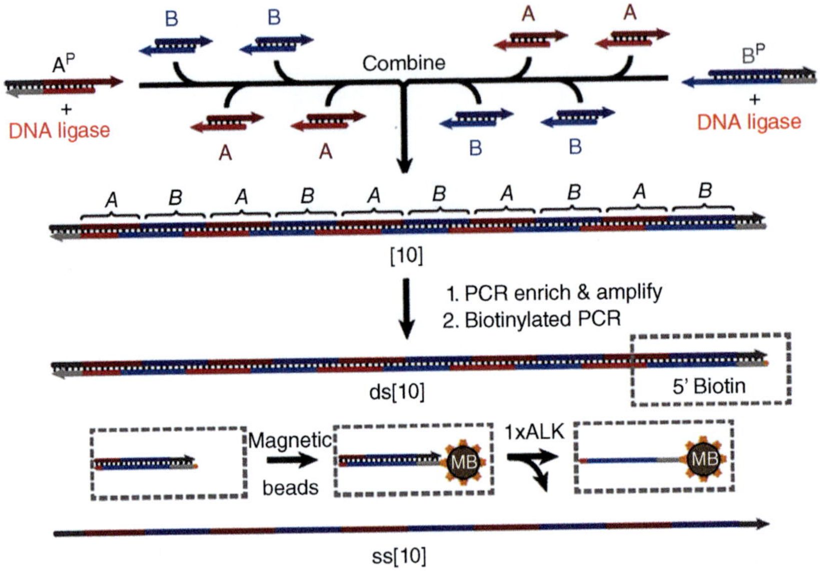

Figure 2.5 Temporal growth of single-stranded long DNA strands with user-defined length and sequence pattern. The synthesis can be initiated by joining a seed duplex AP, which contains a primer region and A, with B by using DNA ligase. Sequential addition of A, B and A to the AP-B mixture in the presence of the ligase can generate a DNA duplex with alternating A-B-A-B-A pattern, where A and B are two different sequences. Similarly, B-A-B-A-B pattern can be generated from BP seed duplex. Ligation of the two pentameric products results in a decameric product [10], which can be enriched by polymerase chain reaction (PCR) where the antisense strand is simultaneously biotinylated at its 5′-end. Single-stranded sense strand (ss[10]) can be then isolated by incubating the duplex ds[10] with streptavidin-coated magnetic beads (MB) under alkaline (ALK) condition; adapted by permission from Macmillan Publishers Ltd: Nature Communications [50], copyright © 2015

We recently presented a solution to this problem. We demonstrated that one can grow a long DNA strand with any desired length and sequence pattern by using a small number of building blocks with identical DNA sequences. For this, we took inspiration from the automated DNA synthesizer, which adds nucleotides one by one to sequence. We thus used *time*, or order of addition, as a parameter to define structural complexity [50]. DNA building blocks that have complementary regions are sequentially added with ligation *in situ*, so that the growth of a particular DNA sequence is controlled. Using DNA polymerases enables the enzymatic enrichment of the desired fully connected DNA strand. This yields a monodisperse, single-stranded long product (e.g. 1000 bases) with user-defined length and sequence pattern (Figure 2.5). This time-controlled addition allows the use of repeating units to define structure, but any number of unique strands can be inserted to create addressable regions. The resulting DNA scaffolds were used to organize DNA polygons into nanotubes with complete control of length and sequence, starting from a small number of strands. Thus, new DNA materials can be made with minimal cost, but anisotropy and site-specific functionalization are maintained. This method allows construction of "designer" peptides and proteins.

2.4.2.4 Perspective

We reported the modular synthesis of DNA nanotubes, with ready control of geometry, diameter and stiffness, and the ability to generate these in single-stranded "open," and double-stranded "closed" forms. These structures can be generated from a small number of unmodified strands that can be commercially and inexpensively ordered, allowing their use by any laboratory with no required expertise in DNA synthesis. As will be shown below, the nanotubes can be used for encapsulation and release of guests. Their shape is especially interesting for drug delivery applications, as elongated nanostructures have been shown to circulate much longer in living organisms than spherical objects. We also introduced the concept of *temporal growth* into DNA self-assembly. To create a complex structure, we can build it up by self-assembly of building blocks *one after the other* in a sequential manner: for this to be successful, we assemble each component with the next using non-covalent DNA hybridization for recognition, but this needs to be immediately followed by covalent ligation. This yields long DNA strands with user-defined sequences and patterns, from a minimum number of starting units.

2.4.3 Biological applications of DNA cages and nanotubes

2.4.3.1 Tools for conditional drug delivery

We set out to explore the applications of our 3D-structures as drug-delivery vehicles. For this to be a viable strategy, we needed to demonstrate their ability to encapsulate and selectively release cargo. A triangular DNA nanotube with alternating small and large pore size was used to encapsulate gold nanoparticles of specific sizes, thus forming nanoparticle "pea-pod" lines. When a specific DNA strand was added, it caused release of the strands that close the nanotube via strand displacement, and this nanoparticle cargo was released quantitatively in less than 15 minutes (Figure 2.6a) [51]. This example is the first one involving encapsulation and release of guests in DNA nanotechnology.

To demonstrate the potential in biomedical applications, we presented a DNA cube that recognizes a prostate-cancer-specific RNA strand, thereby unzipping and opening into a two-dimensional structure. This molecule is thus capable of revealing or releasing cargo conditionally, provided a cellular cancer marker is present (Figure 2.6b) [52].

2.4.3.2 Biological stability and gene-silencing activity

Some of the major barriers for the implementation of nucleic acid therapeutics in the clinic are their inability to penetrate cells, and their susceptibility to degradation by nucleases. Interestingly, we discovered that our DNA cages and nanotubes readily enter human cervical cancer cells, as well as other mammalian cells, to a significantly greater extent than regular DNA strands (Figure 2.6c) [46, 52]. Also, importantly, we showed that their 3D-morphology protects them from nuclease degradation and, with simple end-modifications, increases their stability in serum from less than 20 minutes for simple DNA, to multiple days [53]. We then integrated antisense strands within a DNA prismatic structure, and showed the ability of these constructs to silence gene expression to a significantly greater extent than the single-stranded antisense oligonucleotides themselves (Figure 2.6d) [54].

2.4.3.3 Interaction with lipid bilayers

We carried out fundamental studies to examine the interaction of these DNA cages with lipid bilayers, as models of the cellular membrane. Cages with cholesterol functional groups can efficiently land on the bilayer membranes and can be efficiently "lifted-off" strand displacement (Figure 2.6e) [55]. They showed significant 2D mobility on the membrane, and, using different cholesterol-substitution patterns, we can embed them deeper within the bilayer membrane, thus protecting them from DNA-cleaving enzymes.

2.4.3.4 Perspective

We developed a new class of 3D-DNA host structures that encapsulate and selectively release drugs and materials, and that accomplish anisotropic 3D organization. Unlike other DNA nanostructures, they are built from a minimum number of DNA strands, and are intrinsically reconfigurable. Their biological studies now

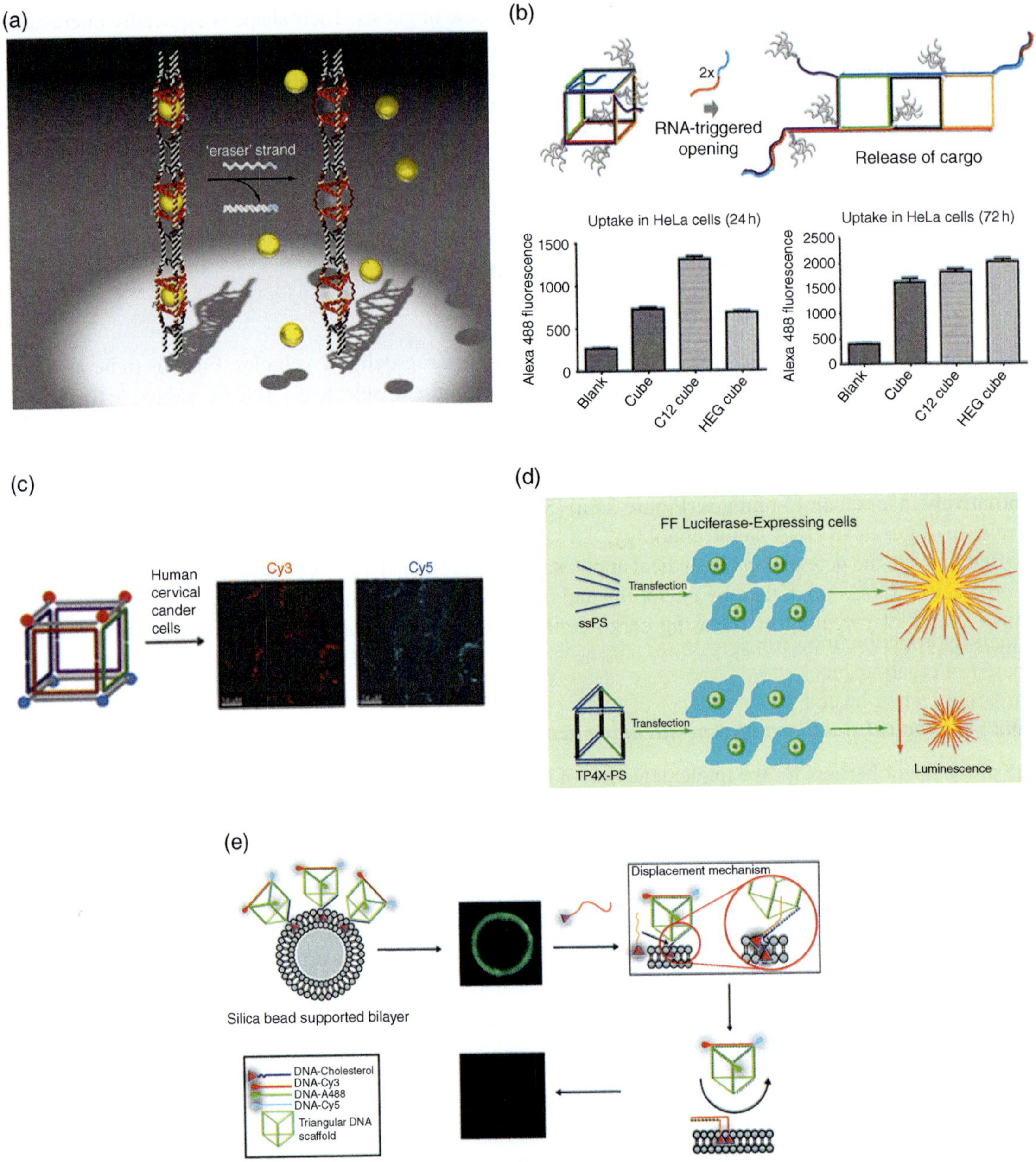

Figure 2.6 *(a) Release of gold nanoparticles encapsulated in DNA nanotube with DNA input; adapted by permission from Macmillan Publishers Ltd:* Nature Chemistry *[51], copyright © 2010. (b) Top: DNA cube that can recognize and open to release its cargo in the presence of cellular cancer marker; bottom: cellular uptake of the cube in its naked form, and decorated with hexa(ethylene) or oligoethylene glycol (HEG) chains; adapted from [52] with permission from The Royal Society of Chemistry. (c) Cellular uptake of DNA cages in cancer cells. (d) Gene-silencing DNA prisms that integrate antisense oligonucleotide therapeutics; adapted with permission from [54], copyright © 2014 American Chemical Society. (e) Interaction of cholesterol-anchored DNA cages with lipid bilayer adapted with permission from [55], copyright © 2014 American Chemical Society*

raise interesting questions on the mechanism of cell uptake of DNA structures, and allow their application as cellular reporters, imaging agents and drug delivery tools. In particular, we are systematically modifying the size, shape and substitution patterns of the 3D-objects to probe the effect of these structural parameters on cellular uptake and biological activity.

2.4.4 Effect of small-molecule insertions on DNA structure and stability

The insertion of small molecules into DNA strands (Figure 2.3a) can have a profound effect on DNA properties. We discovered that an organic small molecule placed at a corner unit between two DNA strands plays a major role in increasing the stability of DNA duplexes and directing the assembled product distribution [56]. Two DNA strands were connected via flexible molecules including a hexane–diol linker and a four-thymidine linker. Assembly of these structures with their complement favored dimer formation. On the other hand, rigid, small linkers such as a triphenylene unit resulted in oligomers (Figure 2.7a). We showed that assembly selectivity arises from a combination of: (a) linker geometry and size, (b) linker to DNA connectivity (i.e., whether the linker is connected to the 5′- or 3′-DNA end), and (c) the structural requirements of the DNA double helix. Knowing these parameters, we can predict the assembly outcome of these hybrid materials. It is remarkable that a small organic molecule (as small as a single nucleotide) can significantly stabilize DNA hybridization and direct product formation.

Small molecules also improve biological properties of DNA. We showed that simple end-modifications such as hexa(ethylene glycol) and 1,6-hexanediol insertions can increase the stability of DNA cages in serum from less than 20 min for simple DNA, to multiple days [53]. This finding will be important to overcome some of the major barriers for the implementation of nucleic acid therapeutics in the clinic.

Small hydrophobic molecules have been used to control the change in conformation of DNA objects. Li prepared alternating block copolymer of perylene tetracarboxylic diimide and single-stranded DNA [57]. Interestingly when heated, this polymer exhibited unusual folding by stacking π-conjugated rings together. Simmel showed that intramolecular folding of rectangular origami sheet can be achieved by decorating the sheets with cholesterol–DNA conjugates, and external stimuli such as surfactant and lipid membrane can unfold the sheets [58]. The formation of intramolecular or intermolecular interactions between origami sheets can be selective by changing the numbers and position of hydrophobic units on the surface.

Intense research from a number of laboratories has resulted in artificial DNA bases that show hydrogen-bonding motifs that differ from regular Watson–Crick base-pairing. In addition, molecules that do not hydrogen-bond but exhibit hydrophobic and π-stacking interactions have been used to replace the A, T, G and C bases. These have been replicated by polymerases, and recently an organism (*E. coli*) was shown to replicate and propagate this expanded genetic alphabet [59].

2.4.5 Non-covalent small molecule-mediated DNA assembly

An assembly instruction from the four DNA letters can lead not only to desired structures but also to the possibility of having by-products and misfolded structures. Assembly errors arise as DNA designs become more complex and require the use of a large number of strands. They are particularly challenging when one needs to design DNA nanostructures with symmetrical or repeating DNA sequences. Slow thermal annealing has been optimized such that the more thermodynamically stable structures are dominant over kinetic products.

Interestingly, we discovered that small molecule guests impart selectivity in DNA self-assembly. In one study, addition of $[Ru(bpy)_3]^{2+}$ to a mixture of DNA molecules with identical arms branching from rigid organic vertex can mediate the formation of only one product, a DNA square, from a dynamic library of multiple structures (Figure 2.7b) [60]. This is the first application of dynamic combinatorial chemistry to DNA nanotechnology, and can correct errors in DNA self-assembly.

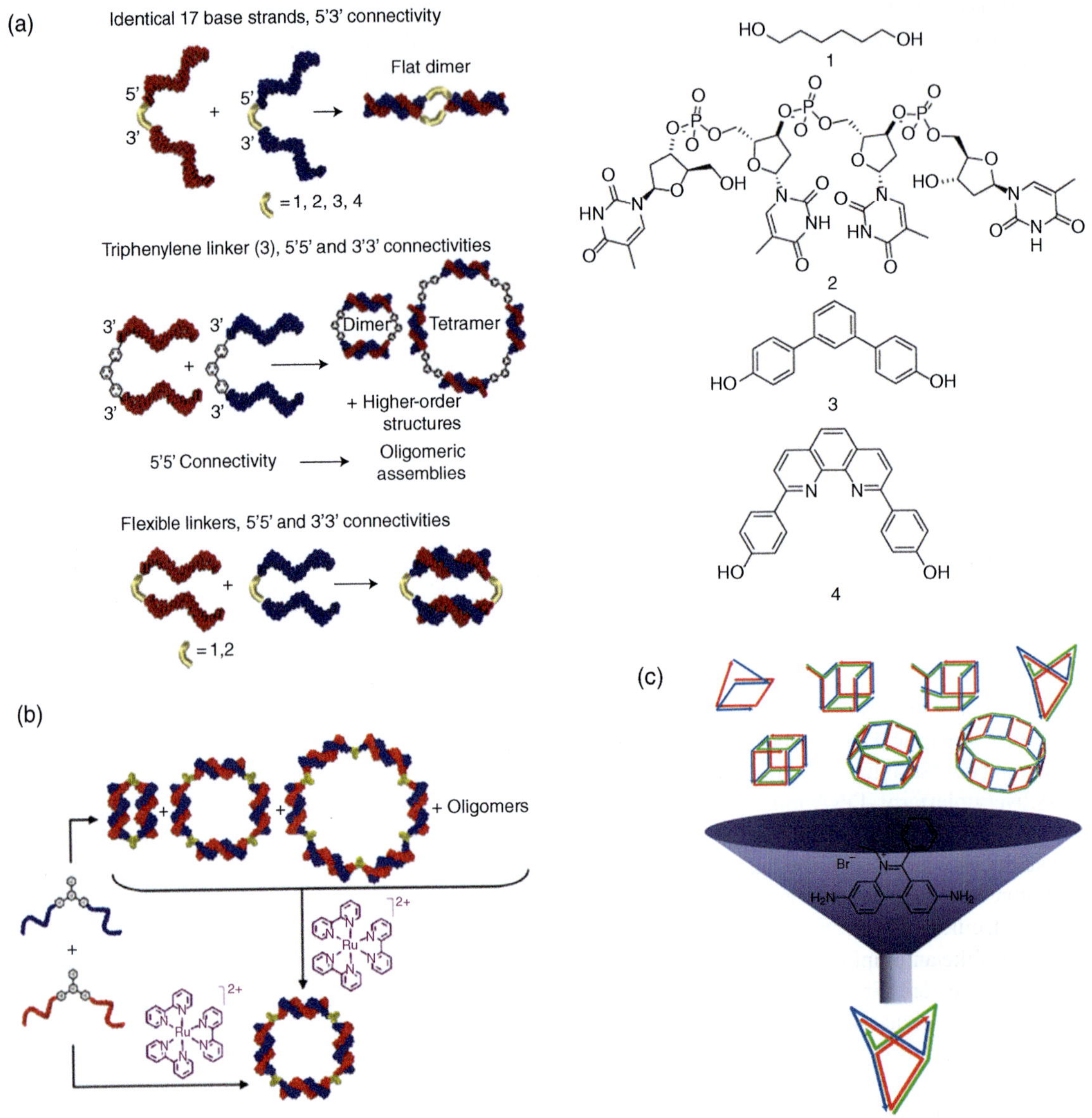

Figure 2.7 Effect of small molecules on DNA assembly. (a) Insertion of organic molecules between two DNA strands to increase duplex stability and direct product distribution; adapted with permission from [56], copyright © 2012 American Chemical Society. (b) Selectivity and error-correction in DNA assembly induced by addition of $[Ru(bpy)_3]^{2+}$ guest; adapted with permission from [60], copyright © 2007 American Chemical Society. (c) Use of ethidium bromide as a molecular chaperone in DNA assembly to achieve the smallest and most stable structure during the assembly process; adapted with permission from [61], copyright © 2013 American Chemical Society

We also showed that DNA intercalators can act as molecular chaperones in DNA assembly. DNA intercalators, such as ethidium bromide are small molecules which can insert their extended aromatic units between base-pairs in a DNA duplex. With the addition of ethidium bromide, a large library of 3D structures converged into the smallest, most stable and fully duplexed product (Figure 2.7c) [61]. Two factors involved in the sorting process are (i) intercalators change the helical twist in a DNA structure; they thus modify the alignment of the DNA strand ends; (ii) intercalators stabilize double-stranded DNA and favor the formation of closed structures where all single strands have hybridized into duplexes. These findings allow researchers to optimize an assembly of target structures from by-products, leading the way towards error-free DNA assembly.

2.4.5.1 Perspective

Small molecules can profoundly impact DNA stability and self-assembly, by introducing new structural requirements and orthogonal interactions. We discovered that DNA binding molecules can bias a dynamic library of DNA structure into a single product. There are hundreds of molecules whose interaction with DNA is well-defined, and many of them are already used for therapeutic applications. This means that we can use many of these molecules to change the outcome and selectivity of DNA self-assembly in a predictable manner.

2.4.6 DNA–polymer and DNA–lipid conjugates

Block copolymers are unique materials that can assemble into numerous structures, ranging from spherical and cylindrical micelles to bilayers, vesicles and many other morphologies [62]. Combining DNA with synthetic polymers is attractive, because it brings the ability of polymers to achieve long-range organization, with the programmability and anisotropy provided by DNA.

2.4.6.1 Synthesis

Park reported an early example of amphiphilic DNA-polymer assembly [63]. Hydrophobic poly(D,L-lactic-co-glycolic acid) was conjugated to DNA and demonstrated self-assembly in aqueous solution into spherical micelles. Due to the biodegradability of PLGA, these micelles exhibited sustained release of DNA over time, which is a useful delivery platform. Reversible control of aggregation of micelles has been demonstrated by Mirkin [64]. Adding a DNA strand complementary to the strands on a micelle of polystyrene–DNA conjugates can bring together individual micelles into an aggregated state. This transition can be controlled by changing temperature. Heating to a temperature higher than the melting temperature of the duplexes will disrupt this aggregation.

Lipid molecules have been incorporated with DNA and showed interesting morphologies. Tan prepared phospholipid-like amphiphile-containing DNA as a hydrophilic head and two C_{18} alkyl tails [65]. The size of homogenous micelles can be tuned by changing the length of DNA. These micelles were able to interact with cell membranes, thus allowing them to internalize into the cell by endocytosis.

Our group attached dendritic oligo(ethylene glycol) units on to short DNA duplexes, which dramatically changed their long-range self-assembly: these hybrid duplexes assemble end-to-end, forming long fibers and 2D-networks in selective solvents (Figure 2.8a) [66]. Liu reported the assembly of DNA with hydrophobic poly(benzyl ether) dendrons into a spherical micelle, which can then reversibly transform into long nanofibers (more thermodynamically stable) by changing assembly conditions [67].

2.4.6.2 Dynamic behavior

Dynamic morphological change in the assembly of DNA–polymer conjugates has also been reported. Herrmann showed that spherical micelles of DNA-b-poly-propyleneoxide displayed morphological changes in the presence of stimuli (Figure 2.8b) [68]. Hybridization with long DNA template encoding multiple copies of complementary sequence resulted in rod-like micelles the length of which can be controlled by length of template strand. Gianneschi demonstrated morphological control of micelle by using two stimuli [69]. DNA–brush copolymers assembled into spherical micelles in aqueous buffer. DNAzymes are DNA strands that are engineered to cleave specific DNA sequences. Upon addition of a DNAzyme, these micelles transformed into

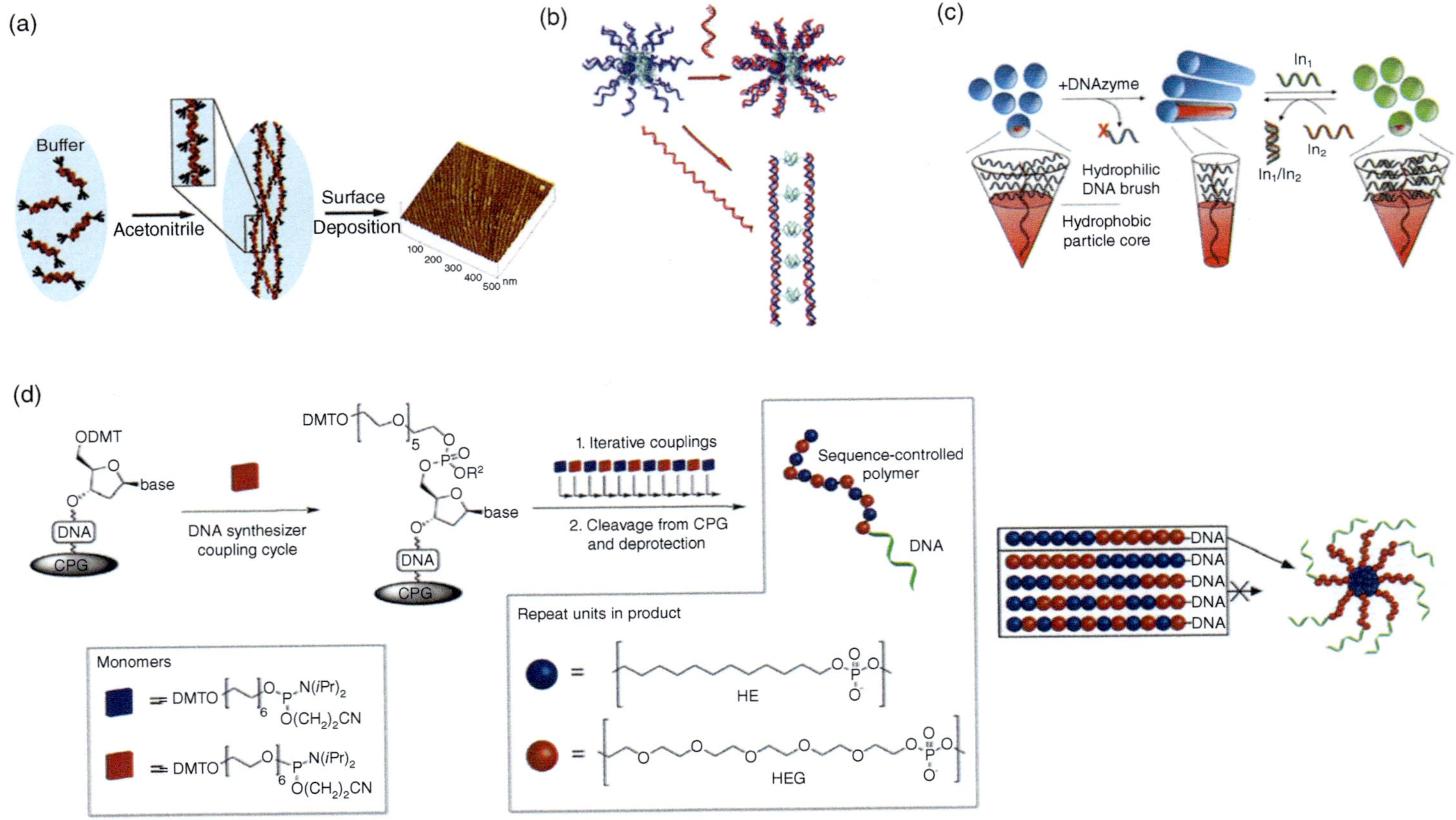

Figure 2.8 *DNA conjugates and their self-assembly behavior. (a) Long-range assembly into fiber and two-dimensional networks from DNA duplexes with dendritic oligo(ethylene glycol); adapted with permission from [66], copyright © 2010 American Chemical Society. (b) Morphological change of DNA–polymer micelles in response to DNA stimuli [68]; reproduced with permission of John Wiley & Sons, Ltd. (c) Programmable morphological transition of DNA–polymer micelles by using two stimuli [69]. Reproduced with permission of John Wiley & Sons, Ltd.] (d) Solid-phase synthesis of DNA–polymer conjugates with sequence control and monodisperse character via phosphoramidite chemistry CPG is controlled pore glass and DMT is 4,4′-dimethoxytrityl group; sequence-dependent self-assembly in solution [70]. Reproduced with permission of John Wiley & Sons, Ltd*

a cylinder. Hybridization of a complementary strand to shorten strands on the cylinder induced cylinder-to-micelle transition, and this was reversibly controlled by adding other strands which had full complementary sequence to the previously added strands (Figure 2.8c).

2.4.6.3 *Sequence-defined DNA–polymer conjugates*

Sequence-defined macromolecules, where monomer units are arranged one by one in an ordered fashion, are a particularly interesting class of polymer. This sequence regulation plays a fundamental role in structural control and molecular recognition for biopolymers such as DNA and polypeptides. Non-biological synthesis of such polymers is, however, very challenging, and has been touted as the "next holy grail in polymer science [71]." Given our success in incorporating non-natural molecules into DNA via phosphoramidite synthesis, we asked the following question: Can a large number of non-natural phosphoramidites (not just one), be attached together one after the other by sequential automated solid-phase synthesis?

We found that this approach is extremely successful, allowing the development of a simple, high-yield synthesis of monodisperse and fully sequence-controlled DNA–polymer conjugates [70]. Two synthetic monomers were chosen: hexa(ethylene) or HE and hexa(ethylene glycol) or HEG, and their sequence along the polymer chain was deliberately varied. Both the length of hydrophobic HE block and sequence order of monomer units are important parameters to determine their self-assembly behavior and their ability to encapsulate hydrophobic guests (Figure 2.8d). In principle, any functional monomer which is compatible with automated DNA synthesis can be incorporated into specific positions on polymer chains, allowing the possibility to introduce various functionalities into DNA as well as non-biological synthetic polymers.

2.4.6.4 *Perspective*

Block copolymers containing DNA are an attractive class of molecules, both for long-range self-assembly and biological applications. They are typically challenging to synthesize, particularly when the polymer block is hydrophobic. We developed a simple, automated method to synthesize monodisperse, sequence-controlled polymers attached to DNA. We are currently examining their applications as dual therapeutics: their micelle cores can encapsulate small molecule drugs, and their DNA exterior can silence the expression of disease-causing proteins.

2.4.7 3D organization of polymers with DNA nanostructures

DNA nanotechnology provides ready access to three-dimensional structures that can serve as well-defined scaffolds; we can also synthesize DNA–polymer or DNA–lipid conjugates as described above. We were interested in the use of DNA scaffolds to arrange polymers/lipid chains into arbitrarily chosen 3D-patterns, opening a new level of protein-inspired self-assembly.

We reported DNA-mediated block copolymer organization in three-dimensional space [38]. DNA cages, containing up to eight binding sites, were used as scaffolds, in order to site-specifically position single polymer chains prepared with ring-opening metathesis polymerization (ROMP). We also decorated DNA nanotubes with linear and graft PEG–DNA conjugates [72]. Spherical aggregates of PEG–DNA can be regularly aligned on DNA nanotubes, creating a striped structure, and can be displaced from the nanotube by adding DNA strands complementary to PEG–DNA conjugates (Figure 2.9a). This demonstrates that write–erase properties were maintained in DNA cage-polymer designs.

Remarkably, we discovered an entirely new mode of protein-inspired interaction. Dendritic hexa(ethylene)–DNA conjugates were hybridized to different positions of a DNA cube [73]. When four conjugates were organized on one face of the cube, they engaged in an intermolecular "handshake" across two cubes, resulting

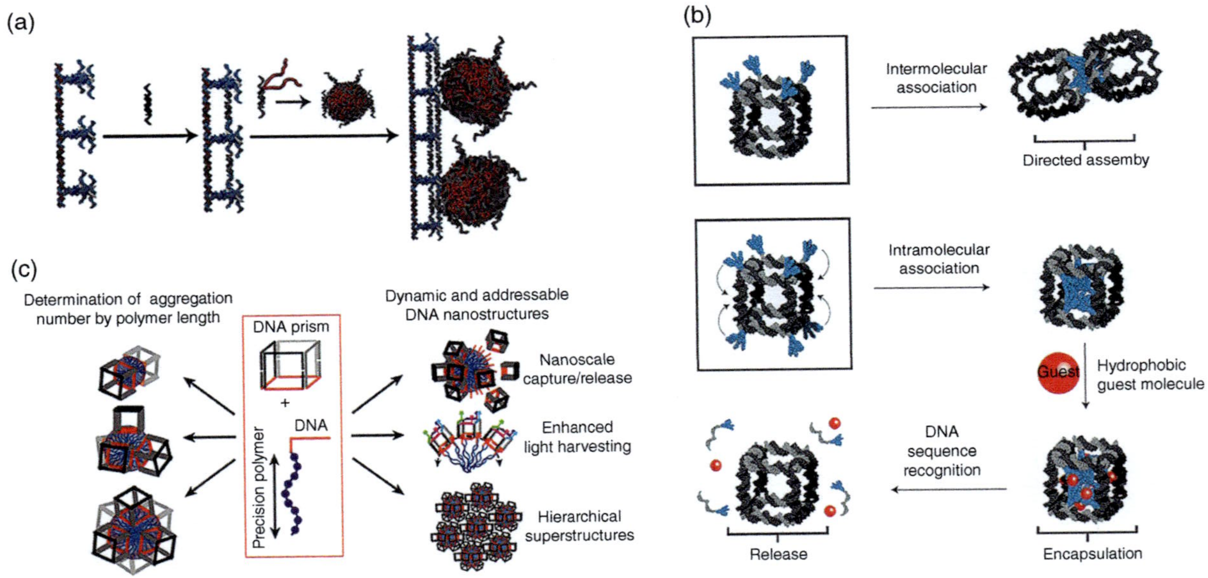

Figure 2.9 *Organization of polymer on three-dimensional DNA scaffolds. (a) Periodic decoration of PEG–DNA micelles on DNA nanotubes; adapted from [72] with permission from The Royal Society of Chemistry. (b) Inter- and intramolecular "handshake" of hydrophobic side chain of dendritic hexa(ethylene)–DNA conjugates decorated on DNA cubes; adapted by permission from Macmillan Publishers Ltd:* Nature Chemistry *[73], copyright © 2013. (c) quantized self-assembly of DNA cubes decorated with linear hexa(ethylene)–DNA conjugates and its applications; adapted with permission from [74], copyright © 2014 American Chemical Society*

exclusively in a cube dimer. This interaction is similar to a protein coiled-coil motif. Interestingly, when eight conjugates were organized on the top and bottom faces of the cube, they engaged in a "handshake" inside the cube. This forms the first example of a monodisperse micelle within a DNA cage, which can encapsulate small hydrophobic molecules, and releases them with added DNA strands (Figure 2.9b).

We then attached linear hexa(ethylene) polymers to three-dimensional DNA prisms [74]. By changing the number of hydrophobic repeat units on the polymer, unprecedented quantized self-assembly was observed (Figure 2.9c). While very short hydrophobic chains collapsed within monomeric DNA cubes, longer polymers yielded a dimer. With more repeat units, discrete assemblies with increasing but still finite aggregation number formed. With twelve repeat units, a monodisperse spherical "DNA prism micelle" was obtained. The DNA prisms can be released from this micelle at will by strand displacement, and enabled the formation of an enhanced light-harvesting cascade (Figure 2.9c).

2.4.7.1 *Perspective*

Protein folding is a fascinating process, which uses multiple interactions to create precise tertiary structure. In a protein, hydrogen-bonded peptide motifs such as α-helices or β-sheets display side chains with orthogonal interactions (hydrophobic, electrostatic, metal-binding) in a well-defined pattern; this provides a low-energy folding pathway towards a single protein structure. Our group has demonstrated that when supramolecular interactions are anisotropically organized on DNA scaffolds, new, cooperative, directed interactions emerge. We have only worked with hydrophobic interactions so far, but the potential for structural complexity and protein-inspired folding is tremendous when additional interactions are introduced (fluorophilic, metal-binding, etc.).

2.4.8 Metal–DNA assemblies

The organization of metals one by one into arbitrarily designed asymmetric patterns is a challenging goal. This is what one may need, in order to assemble, e.g., a heterometallic metal–organic framework (MOF), an artificial photosynthetic system, or a multimetallic catalytic factory. In principle, this can be achieved via the use of DNA as a scaffold. However, there are some challenges to combining metals and DNA: the non-specific binding of metals with the nitrogen-rich DNA bases or with the negatively charged phosphate backbone, the activity of some metal complexes towards DNA degradation, and the decomposition of metal complexes in aqueous buffer solutions, or under the automated DNA synthesis conditions.

2.4.8.1 *Strategy for incorporation of metal into DNA*

Two main approaches have been developed to site-specifically incorporate metals into DNA. The first approach involves modification of DNA bases with metal-binding ligands. Shionoya replaced natural nucleobases with a hydroxypyridone nucleobase [75]. In the presence of Cu^{2+}, DNA duplexes containing 1–5 modified nucleobases can form stable square-planar complexes with Cu^{2+} inside the duplexes. They also demonstrated that different metal ions can be stacked on top of each other inside the duplexes by various types of modification of nucleobases [76].

The second approach is to covalently attach metal-binding units to the DNA backbone. Our group reported the synthesis of transition metal-linked branched DNA molecules using a ruthenium complex [77]. We then showed that branched DNA molecules containing a ruthenium bipyridine complex can assemble into discrete cyclic metal–DNA nanostructures with metals positioned at the vertices (Figure 2.10a) [78]. Using a similar concept, McLaughlin constructed four tethered DNA strands on a nickel core and showed the formation of higher-order assemblies [79]. The same group also prepared six-arm branched DNA molecule from a ruthenium bipyridine core by attaching DNA strands at 4 and 4′ position of bpy [80].

2.4.8.2 Metal–DNA nanostructures

Our group used DNA organization to create optimized ligand environments for the site-selective incorporation of different metal ions into assemblies. This allowed the introduction of even reactive transition metals such as Cu^+, Cu^{2+} and Ag^+ into DNA junctions with bis(2,9-diphenyl)-1,10-phenanthroline (ddp) ligand (Figure 2.10b and c) [81]. The method created an architecture in which the ligand environment is in close contact with DNA, allowing the DNA double helix and the metal complex to stabilize each other synergistically. The metal acquired the chirality of the DNA helix, and significantly stabilized DNA hybridization: a melting temperature increase from 40 °C to 80 °C occured upon single metal substitution of a DNA 10-base duplex – the highest recorded T_m increase upon a single DNA substitution. We then constructed multimetallic 2D structures with three Cu^+ complexes at the corner, where the distance between two metal centers can be controlled by an added DNA strand (Figure 2.10b).

The real potential of DNA lies in its ability to organize different transition metals within a nanostructure. We showed the DNA-templated creation of three different ligand environments, each selective for a specific transition metal ion [82]. DNA strands were terminated with either terpyridine (tpy) or diphenylphenanthroline (dpp) derivatives. They were hybridized together to create three specific ligand environments: $(dpp)_2$, tpy:dpp, or $(tpy)_2$ (Figure 2.10c). These DNA–ligand environments had strong preference for tetrahedral Cu(I), five-coordinate Cu(II) and octahedral Fe(II), respectively. Furthermore, the specificity of the complexes allowed "error correction" if a metal was placed into the "incorrect" environment. For example, Cu^+ spontaneously oxidized to Cu^{2+} if added to the tpy:dpp duplex, and Cu^{2+} was spontaneously reduced to Cu^+ if added to the $(dpp)_2$ duplex. The preference of Fe^{2+} for the $(tpy)_2$ ligand sphere was so strong that if it was added to the tpy:dpp duplex a rearrangement of the DNA occurred to form dimers linked by $Fe(tpy)_2^{2+}$.

Metal complexes capable of heteroleptic and anisotropic assembly are highly desirable building blocks. We used a DNA template to bring together a $Cu(dpp)_2$ complex with four different DNA arms on it [83]. The template was able to introduce chirality into the metal junction, which was preserved after its removal (Figure 2.10d). These chiral metal junctions were then able to assemble into metalated nanotubular structures. We recently showed the ability of the metal complexes within our DNA strands to mediate charge transport. When attached to an electrode, copper(I) centers can be oxidized to Cu(II) even if they are separated from the electrode by a large number of base-pairs. On the other hand, copper oxidation does not occur if there is a mismatch in the DNA duplex, suggestive of a DNA-mediated charge transport phenomenon [85].

Finally, we reported the assembly of a metal–DNA three-dimensional cage, with site-specific incorporation of a range of transition metals in the vertices of this structure [84]. Assembly of three DNA strands having two internal dpp ligands created a triangle with preorganized ligand environments suitable for complexation with metal ions. Connecting two triangles generated a cage with well-defined metal–DNA junctions, allowing precise three-dimensional positioning of transition metals (Figure 2.10e).

2.4.8.3 Perspective

We reported the site-specific incorporation of transition metals into DNA junctions, allowing the programmable positioning of different metals within a structure. We constructed a metal-DNA cage and the ability of metal-DNA structures to mediate charge transport. The incorporation of transition metals into DNA nanoassemblies can contribute important functionalities to otherwise passive DNA templates. They can impart added stability, photo- or redox activity, magnetic or catalytic properties. They introduce new energy or charge transport pathways into the DNA double helix. Structurally, the diverse coordination geometries of transition metal complexes (tetrahedral, octahedral, square planar and trigonal-bipyramidal) can give rise to linker angles and geometries that are difficult, or simply inaccessible, to DNA itself. A particularly attractive goal will be the creation of "metal–nucleic acid frameworks" in analogy to metal–organic frameworks, in

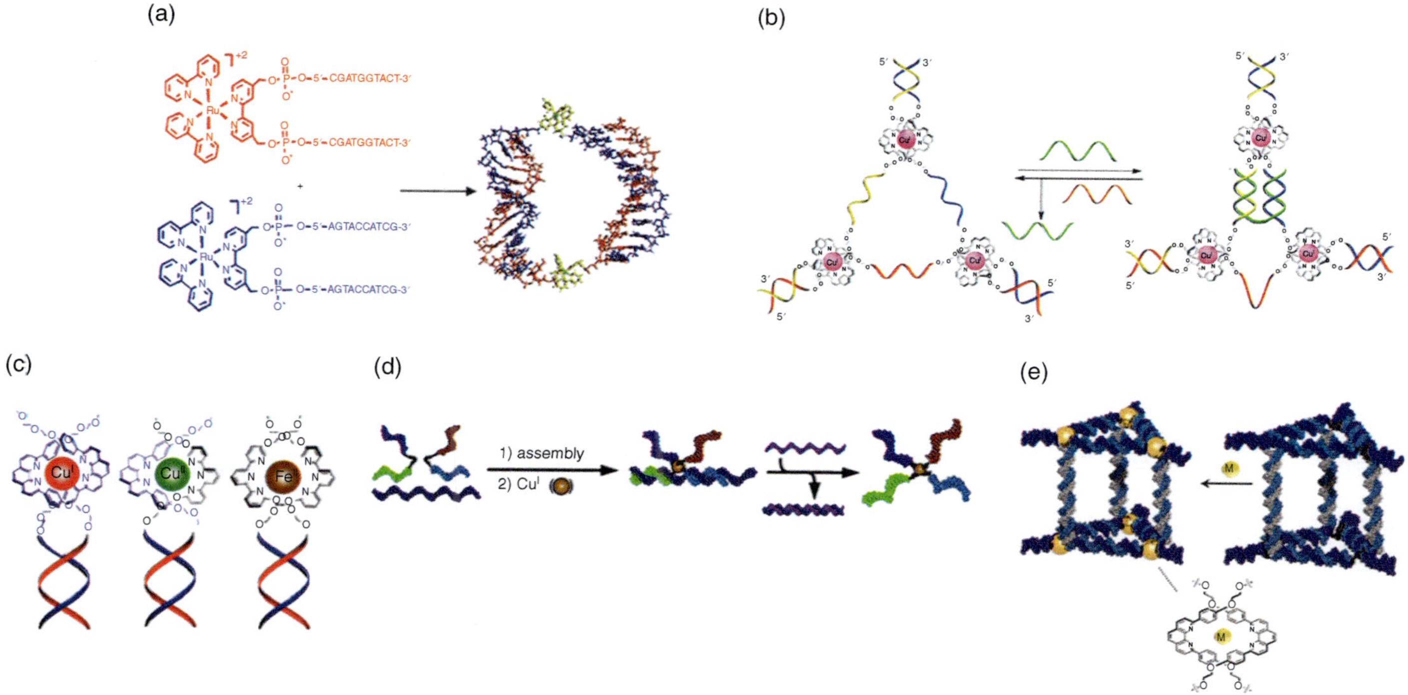

Figure 2.10 Metal–DNA assemblies. (a) Discrete cyclic nanostructures constructed from transition metal-linked branched DNA molecules [78]; reproduced with permission of John Wiley & Sons, Ltd. (b) Multimetallic DNA nanostructure containing three Cu^+ complexes at the corners [81]; reproduced with permission of John Wiley & Sons, Ltd. (c) DNA-templated creation of three ligand environments for incorporation of three different reactive transition metals; [82] reproduced with permission of John Wiley & Sons, Ltd. (d) Chiral metal–DNA four-arm junction [83]; (e) Metal–nucleic acid cages with site-specific incorporation of transitional metals in the vertices; adapted by permission from Macmillan Publishers Ltd: Nature Chemistry [84], copyright © 2009

which the features are programmably arranged with DNA, rather than symmetrical or periodic. These frameworks will combine metal-mediated function with the programmability and biological properties of DNA.

2.4.9 Gold nanoparticle–DNA structures

Gold nanoparticles are promising in optical, electronic, catalytic, and sensing applications, with unique properties that differentiate them from their bulk counterparts. In particular, their ability to exhibit single-electron transport and to act as optical "antennae" for focusing light into nanoscale volumes have generated great interest. Many of these properties arise directly from the geometric arrangement of the particles within an assembly.

Two main approaches to organize gold nanoparticles with DNA have been developed. Long-range assembly of gold nanoparticles mediated by DNA hybridization has been pioneered by Mirkin [86] who reported the reversible aggregation of polyvalent DNA-functionalized gold nanoparticles when a specific DNA strand was added. Gang and Mirkin showed that DNA can be used to control the crystallization of nanoparticles into different patterns [87, 88]. Systematic investigation of the effect of linker strands, nanoparticle size and interparticle distance and showed that nanoparticle lattices of a variety of crystal patterns can be deliberately prepared [89].

Another approach is to use DNA constructs to template gold nanoparticles. Alivisatos and Schultz reported one-dimensional discrete structures, where predetermined numbers of gold clusters conjugated to one single-stranded DNA were organized on singled-stranded DNA template [90]. Our group demonstrated the assembly of discrete gold nanoparticle two-dimensional structure [91]. The rigid vertex, 1,3-bis (4-hydroxyphenyl)benzene, was inserted in the middle of two DNA strands. Assembly of six such DNA molecules produced DNA hexagon, where gold nanoparticles can be precisely positioned on scaffold by labelling each molecule with one nanoparticle (Figure 2.11a). We then reported a simple method to append one or more gold nanoparticles at both internal and terminal positions of DNA strands by using cyclic disulfide groups, and showed that these DNA–gold conjugates can assemble into well-defined rectangles and tapered tetrameric linear structures [92]. Finally, we constructed DNA polygons with single stranded arms, as dynamic scaffolds on to which nanoparticles can be organized [93]. This method allows the modular assembly of any pattern of nanoparticles, with deliberate control of (i) geometry, (ii) number of particles, (iii) relative positions and distances, (iv) assembly of *different* particles at precise locations, and (v) writing, erasing and modification of structural features in real time using specific external molecules (Figure 2.11b).

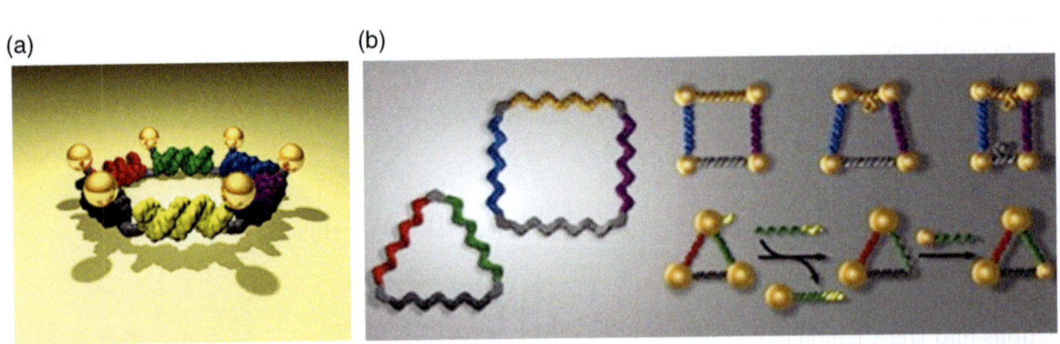

Figure 2.11 *(a) DNA hexagon with precisely positioned gold nanoparticles [91]; reproduced with permission of John Wiley & Sons, Ltd. (b) Cyclic DNA templates for the assembly of nanoparticles into controlled and dynamic geometries; reprinted with permission from [93], copyright © 2007 American Chemical Society*

Alivisatos [94] arranged gold nanoparticles into 3D chiral tetrahedral structures from four gold nanoparticles monoconjugated with single-stranded DNA. More recently, Mao introduced symmetry-breaking by first encapsulating gold nanoparticles inside three-dimensional DNA polyhedral cages, which were then used to provide connectivity with another set of gold nanoparticles [95]. Patterning of nanoparticles into materials with novel optical properties can also be achieved by using DNA origami objects [96, 97].

2.5 Conclusion

DNA nanotechnology has revolutionized many research areas by offering powerful tools to create objects with arbitrary control of size and shape. Unique sequences of DNA strands within a nanostructure provide a precise "address" to position molecules or materials into any geometry. The materials arranged by DNA include gold particles, polymers, lipids, enzymes, transition metal complexes, or small organic molecules. Moreover, dynamic DNA binding is key to the design of molecular-responsive switching, providing a platform for the construction of artificial mimics of biological machines. DNA nanotechnology is starting to examine applications in nanophotonics, biophysics, enzyme-mimetics, drug delivery systems, synthetic biology, and molecular computation. Some of the major challenges in the field include the cost of DNA synthesis, which makes scaling-up of the technology difficult, biocompatibility of DNA nanostructures which needs be investigated before translation into clinical applications, and finding ways to increase system complexity without introducing errors.

Our research group is examining the interface of supramolecular chemistry and DNA nanotechnology, as a growing area with unique potential. We have shown simple and modular approaches to 3D-cages and nanotubes from a minimum number of DNA strands, with fine control of size and shape. These can encapsulate gold nanoparticles, as well as small molecules and release them when a specific DNA strand is added. A cage that conditionally unzips and opens with an RNA sequence only found in prostate cancer cells was designed. These structures can readily penetrate into a range of mammalian cells, and they show increased nuclease resistance and enhanced gene-silencing ability. We thus can envisage the use of these DNA cages as dual therapeutics, that transport gene silencing oligonucleotides as well as small molecule drugs into diseased cells. The cages would then release their cargo conditionally in these cells upon encountering a molecule that is overexpressed in the disease environment. Their targeting ability can be tuned by attaching aptamers (DNA or RNA strands that recognize specific molecules) or other ligands on their periphery, and their biological compatibility can be adjusted with polymer modifications on the exterior. On a fundamental level, these molecules allow the systematic examination of how size, shape and presentation of targeting ligands affect the biodistribution, cellular uptake, and gene silencing abilities of nanomaterials.

We generated a new class of polymer-DNA conjugates, in which the polymer block is monodisperse and fully sequence controlled, by solid-phase automated synthesis. Organization of these individual polymer chains on DNA cages leads to emergent, protein-inspired behavior: depending on their spatial orientation, the chains can undergo a "handshake" to bring the DNA cages into quantized objects, or they can fold on the inside of the cage to create a well-defined hydrophobic core. We showed that small synthetic molecules can change the course of DNA assembly, for example converting a library of equilibrating DNA structures to a single product. We created metal–DNA nanostructures, in which DNA organizes different transition metals, one by one in 2D and 3D space. The metal complexes dramatically stabilize the DNA double helix and add photophysical and electron transport behavior.Introducing orthogonal interactions into DNA nanotechnology leads to completely new structures and functions, while reducing the number of required DNA sequences. The potential of this growing area to increase scalability in the field, to render DNA structures more robust and more biocompatible, and to increase functionality and applications is tremendous.

References

1. Aldaye, F.A., Palmer, A.L., Sleiman, H.F. (2008) Assembling materials with DNA as the guide, *Science*, **321** (5897), 1795–1799.
2. McLaughlin, C.K., Hamblin, G.D., Sleiman, H.F. (2011) Supramolecular DNA assembly, *Chemical Society Reviews*, **40** (12), 5647–5656.
3. Seeman, N.C. (1982) Nucleic acid junctions and lattices, *Journal of Theoretical Biology*, **99** (2), 237–247.
4. Fu, T.J., Seeman, N.C. (1993) DNA double-crossover molecules, *Biochemistry*, **32** (13), 3211–3220.
5. Winfree, E., Liu, F., Wenzler, L.A., Seeman, N.C. (1998) Design and self-assembly of two-dimensional DNA crystals, *Nature*, **394** (6693), 539–544.
6. LaBean, T.H., Yan, H., Kopatsch, J., *et al.* (2000) Construction, analysis, ligation, and self-assembly of DNA triple crossover complexes, *Journal of the American Chemical Society*, **122** (9), 1848–1860.
7. Seeman, N.C. (2001) DNA nicks and nodes and nanotechnology, *Nano Letters*, **1** (1), 22–26.
8. Yan, H., Park, S.H., Finkelstein, G., Reif, J.H., LaBean, T.H. (2003) DNA-templated self-assembly of protein arrays and highly conductive nanowires, *Science*, **301** (5641), 1882–1884.
9. Liu, D., Wang, M., Deng, Z., Walulu, R., Mao, C. (2004) Tensegrity: construction of rigid DNA triangles with flexible four-arm DNA junctions, *Journal of the American Chemical Society*, **126** (8), 2324–2325.
10. He, Y., Chen, Y., Liu, H., Ribbe, A.E., Mao, C. (2005) Self-assembly of hexagonal DNA two-dimensional (2D) arrays, *Journal of the American Chemical Society*, **127** (35), 12202–12203.
11. Mathieu, F., Liao, S., Kopatsch, J., *et al.* (2005) Six-helix bundles designed from DNA, *Nano Letters*, **5** (4), 661–665.
12. Zheng, J., Birktoft, J.J., Chen, Y., *et al.* (2009) From molecular to macroscopic via the rational design of a self-assembled 3D DNA crystal, *Nature*, **461** (7260), 74–77.
13. Rothemund, P.W. (2006) Folding DNA to create nanoscale shapes and patterns, *Nature*, **440** (7082), 297–302.
14. Wei, B., Dai, M., Yin, P. (2012) Complex shapes self-assembled from single-stranded DNA tiles, *Nature*, **485** (7400), 623–626.
15. Douglas, S.M., Dietz, H., Liedl, T., *et al.* (2009) Self-assembly of DNA into nanoscale three-dimensional shapes, *Nature*, **459** (7245), 414–418.
16. Ke, Y., Douglas, S.M., Liu, M., *et al.* (2009) Multilayer DNA origami packed on a square lattice, *Journal of the American Chemical Society*, **131** (43), 15903–15908.
17. Ke, Y., Voigt, N.V., Gothelf, K.V., Shih, W.M. (2012) Multilayer DNA origami packed on hexagonal and hybrid lattices, *Journal of the American Chemical Society*, **134** (3), 1770–1774.
18. Dietz, H., Douglas, S.M., Shih, W.M. (2009) Folding DNA into twisted and curved nanoscale shapes, *Science*, **325** (5941), 725–730.
19. Han, D., Pal, S., Nangreave, J., *et al.* (2011) DNA origami with complex curvatures in three-dimensional space, *Science*, **332** (6027), 342–346.
20. Ke, Y., Ong, L.L., Shih, W.M., Yin, P. (2012) Three-dimensional structures self-assembled from DNA bricks, *Science*, **338** (6111), 1177–1183.
21. Liu, H., Chen, Y., He, Y., Ribbe, A.E., Mao, C. (2006) Approaching the limit: can one DNA oligonucleotide assemble into large nanostructures?, *Angewandte Chemie International Edition*, **45** (12), 1942–1945.
22. Zhang, D.Y., Seelig, G. (2011) Dynamic DNA nanotechnology using strand-displacement reactions, *Nature Chemistry*, **3** (2), 103–113.
23. Zhang, D.Y., Winfree, E. (2009) Control of DNA strand displacement kinetics using toehold exchange, *Journal of the American Chemical Society*, **131** (47), 17303–17314.
24. Yurke, B., Turberfield, A.J., Mills, A.P., Simmel, F.C., Neumann, J.L. (2000) A DNA-fuelled molecular machine made of DNA, *Nature*, **406** (6796), 605–608.
25. Yan, H., Zhang, X., Shen, Z., Seeman, N.C. (2002) A robust DNA mechanical device controlled by hybridization topology, *Nature*, **415** (6867), 62–65.
26. Gu, H., Chao, J., Xiao, S.-J., Seeman, N.C. (2010) A proximity-based programmable DNA nanoscale assembly line, *Nature*, **465** (7295), 202–205.
27. He, Y., Liu, D.R. (2010) Autonomous multistep organic synthesis in a single isothermal solution mediated by a DNA walker, *Nature Nanotechnology*, **5** (11), 778–782.

28. Dirks, R.M., Pierce, N.A. (2004) Triggered amplification by hybridization chain reaction, *Proceedings of the National Academy of Sciences*, **101** (43), 15275–15278.
29. Seelig, G., Soloveichik, D., Zhang, D.Y., Winfree, E. (2006) Enzyme-free nucleic acid logic circuits, *Science*, **314** (5805), 1585–1588.
30. Aldaye, F.A., Sleiman, H.F. (2007) Modular access to structurally switchable 3D discrete DNA assemblies, *Journal of the American Chemical Society*, **129** (44), 13376–13377.
31. Chen, J.H., Seeman, N.C. (1991) Synthesis from DNA of a molecule with the connectivity of a cube, *Nature*, **350** (6319), 631–633.
32. Zhang, Y., Seeman, N.C. (1994) Construction of a DNA-truncated octahedron, *Journal of the American Chemical Society*, **116** (5), 1661–1669.
33. Goodman, R.P., Berry, R.M., Turberfield, A.J. (2004) The single-step synthesis of a DNA tetrahedron, *Chemical Communications*, **12**, 1372–1373.
34. Goodman, R.P., Schaap, I.A., Tardin, C.F., *et al.* (2005) Rapid chiral assembly of rigid DNA building blocks for molecular nanofabrication, *Science*, **310** (5754), 1661–1665.
35. Shih, W.M., Quispe, J.D., Joyce, G.F. (2004) A 1.7-kilobase single-stranded DNA that folds into a nanoscale octahedron, *Nature*, **427** (6975), 618–621.
36. He, Y., Ye, T., Su, M., *et al.* (2008) Hierarchical self-assembly of DNA into symmetric supramolecular polyhedra, *Nature*, **452** (7184), 198–201.
37. McLaughlin, C.K., Hamblin, G.D., Aldaye, F.A., Yang, H., Sleiman, H.F. (2011) A facile, modular and high yield method to assemble three-dimensional DNA structures, *Chemical Communications*, **47** (31), 8925–8927.
38. McLaughlin, C.K., Hamblin, G.D., Hanni, K.D., *et al.* (2012) Three-dimensional organization of block copolymers on "DNA-minimal" scaffolds, *Journal of the American Chemical Society*, **134** (9), 4280–4286.
39. Rothemund, P.W., Ekani-Nkodo, A., Papadakis, N., *et al.* (2004) Design and characterization of programmable DNA nanotubes, *Journal of the American Chemical Society*, **126** (50), 16344–16352.
40. Liu, D., Park, S.H., Reif, J.H., LaBean, T.H. (2004) DNA nanotubes self-assembled from triple-crossover tiles as templates for conductive nanowires, *Proceedings of the National Academy of Sciences*, **101** (3), 717–722.
41. Mitchell, J.C., Harris, J.R., Malo, J., Bath, J., Turberfield, A.J. (2004) Self-assembly of chiral DNA nanotubes, *Journal of the American Chemical Society*, **126** (50), 16342–16343.
42. Aldaye, F.A., Lo, P.K., Karam, P., *et al.* (2009) Modular construction of DNA nanotubes of tunable geometry and single- or double-stranded character, *Nature Nanotechnology*, **4** (6), 349–352.
43. Hariri, A.A., Hamblin, G.D., Gidi, Y., Sleiman, H.F., Cosa, G. (2015) Stepwise growth of surface-grafted DNA nanotubes visualized at the single-molecule level, *Nature Chemistry*, **7** (4), 295–300.
44. Hamblin, G.D., Hariri, A.A., Carneiro, K.M., *et al.* (2013) Simple design for DNA nanotubes from a minimal set of unmodified strands: rapid, room-temperature assembly and readily tunable structure, *ACS Nano*, **7** (4), 3022–3028.
45. Lo, P.K., Altvater, F., Sleiman, H.F. (2010) Templated synthesis of DNA nanotubes with controlled, predetermined lengths, *Journal of the American Chemical Society*, **132** (30), 10212–10214.
46. Hamblin, G.D., Carneiro, K.M., Fakhoury, J.F., Bujold, K.E., Sleiman, H.F. (2012) Rolling circle amplification-templated DNA nanotubes show increased stability and cell penetration ability, *Journal of the American Chemical Society*, **134** (6), 2888–2891.
47. Wilner, O.I., Orbach, R., Henning, A., *et al.* (2011) Self-assembly of DNA nanotubes with controllable diameters, *Nature Communications*, **2** (540).
48. Douglas, S.M., Chou, J.J., Shih, W.M. (2007) DNA-nanotube-induced alignment of membrane proteins for NMR structure determination, *Proceedings of the National Academy of Sciences*, **104** (16), 6644–6648.
49. Yin, P., Hariadi, R.F., Sahu, S., *et al.* (2008) Programming DNA tube circumferences, *Science*, **321** (5890), 824–826.
50. Hamblin, G.D., Rahbani, J.F., Sleiman, H.F. (2015) Sequential growth of long DNA strands with user-defined patterns for nanostructures and scaffolds, *Nature Communications*, **6** (7065).
51. Lo, P.K., Karam, P., Aldaye, F.A., *et al.* (2010) Loading and selective release of cargo in DNA nanotubes with longitudinal variation, *Nature Chemistry*, **2** (4), 319–328.
52. Bujold, K.E., Fakhoury, J., Edwardson, T.G.W., *et al.* (2014) Sequence-responsive unzipping DNA cubes with tunable cellular uptake profiles, *Chemical Science*, **5** (6), 2449–2455.

53. Conway, J.W., McLaughlin, C.K., Castor, K.J., Sleiman, H. (2013) DNA nanostructure serum stability: greater than the sum of its parts, *Chemical Communications*, **49** (12), 1172–1174.
54. Fakhoury, J.J., McLaughlin, C.K., Edwardson, T.W., Conway, J.W., Sleiman, H.F. (2014) Development and characterization of gene silencing DNA cages, *Biomacromolecules*, **15** (1), 276–282.
55. Conway, J.W., Madwar, C., Edwardson, T.G., *et al.* (2014) Dynamic behavior of DNA cages anchored on spherically supported lipid bilayers, *Journal of the American Chemical Society*, **136** (37), 12987–12997.
56. Greschner, A.A., Toader, V., Sleiman, H.F. (2012) The role of organic linkers in directing DNA self-assembly and significantly stabilizing DNA duplexes, *Journal of the American Chemical Society*, **134** (35), 14382–14389.
57. Wang, W., Wan, W., Zhou, H.H., Niu, S., Li, A.D. (2003) Alternating DNA and pi-conjugated sequences. Thermophilic foldable polymers, *Journal of the American Chemical Society*, **125** (18), 5248–5249.
58. List, J., Weber, M., Simmel, F.C. (2014) Hydrophobic actuation of a DNA origami bilayer structure, *Angewandte Chemie International Edition*, **53** (16), 4236–4239.
59. Malyshev, D.A., Dhami, K., Lavergne, T., *et al.* (2014) A semi-synthetic organism with an expanded genetic alphabet, *Nature*, **509** (7500), 385–388.
60. Aldaye, F.A., Sleiman, H.F. (2007) Guest-mediated access to a single DNA nanostructure from a library of multiple assemblies, *Journal of the American Chemical Society*, **129** (33), 10070–10071.
61. Greschner, A.A., Bujold, K.E., Sleiman, H.F. (2013) Intercalators as molecular chaperones in DNA self-assembly, *Journal of the American Chemical Society*, **135** (30), 11283–11288.
62. Mai, Y., Eisenberg, A. (2012) Self-assembly of block copolymers, *Chemical Society Reviews*, **41** (18), 5969–5985.
63. Jeong, J.H., Park, T.G. (2001) Novel polymer-DNA hybrid polymeric micelles composed of hydrophobic poly(D,L-lactic-co-glycolic acid) and hydrophilic oligonucleotides, *Bioconjugate Chemistry*, **12** (6), 917–923.
64. Li, Z., Zhang, Y., Fullhart, P., Mirkin, C.A. (2004) Reversible and chemically programmable micelle assembly with DNA block-copolymer amphiphiles, *Nano Letters*, **4** (6), 1055–1058.
65. Liu, H., Zhu, Z., Kang, H., *et al.* (2010) DNA-based micelles: synthesis, micellar properties and size-dependent cell permeability, *Chemistry*, **16** (12), 3791–3797.
66. Carneiro, K.M., Aldaye, F.A., Sleiman, H.F. (2010) Long-range assembly of DNA into nanofibers and highly ordered networks using a block copolymer approach, *Journal of the American Chemical Society*, **132** (2), 679–685.
67. Wang, L., Feng, Y., Yang, Z., *et al.* (2012) Reversibly controlled morphology transformation of an amphiphilic DNA-dendron hybrid, *Chemical Communications*, **48** (31), 3715–3717.
68. Ding, K., Alemdaroglu, F.E., Borsch, M., Berger, R., Herrmann, A. (2007) Engineering the structural properties of DNA block copolymer micelles by molecular recognition, *Angewandte Chemie International Edition*, **46** (7), 1172–1175.
69. Chien, M.P., Rush, A.M., Thompson, M.P., Gianneschi, N.C. (2010) Programmable shape-shifting micelles, *Angewandte Chemie International Edition*, **49** (30), 5076–5080.
70. Edwardson, T.G., Carneiro, K.M., Serpell, C.J., Sleiman, H.F. (2014) An efficient and modular route to sequence-defined polymers appended to DNA, *Angewandte Chemie International Edition*, **53** (18), 4567–4571.
71. Lutz, J.-F., Ouchi, M., Liu, D.R., Sawamoto, M. (2013) Sequence-controlled polymers, *Science*, **341** (6146).
72. Carneiro, K.M.M., Hamblin, G.D., Hanni, K.D., *et al.* (2012) Stimuli-responsive organization of block copolymers on DNA nanotubes, *Chemical Science*, **3** (6), 1980–1986.
73. Edwardson, T.G., Carneiro, K.M., McLaughlin, C.K., Serpell, C.J., Sleiman, H.F. (2013) Site-specific positioning of dendritic alkyl chains on DNA cages enables their geometry-dependent self-assembly, *Nature Chemistry*, **5** (10), 868–875.
74. Serpell, C.J., Edwardson, T.G., Chidchob, P., Carneiro, K.M., Sleiman, H.F. (2014) Precision polymers and 3D DNA nanostructures: emergent assemblies from new parameter space, *Journal of the American Chemical Society*, **136** (44), 15767–15774.
75. Tanaka, K., Tengeiji, A., Kato, T., Toyama, N., Shionoya, M. (2003) A discrete self-assembled metal array in artificial DNA, *Science*, **299** (5610), 1212–1213.
76. Tanaka, K., Clever, G.H., Takezawa, Y., *et al.* (2006) Programmable self-assembly of metal ions inside artificial DNA duplexes, *Nature Nanotechnology*, **1** (3), 190–194.
77. Vargas-Baca, I., Mitra, D., Zulyniak, H.J., Banerjee, J., Sleiman, H.F. (2001) Solid-phase synthesis of transition metal linked, branched oligonucleotides, *Angewandte Chemie International Edition*, **40** (24), 4629–4632.

78. Mitra, D., Di Cesare, N., Sleiman, H.F. (2004) Self-assembly of cyclic metal-DNA nanostructures using ruthenium tris(bipyridine)-branched oligonucleotides, *Angewandte Chemie International Edition*, **43** (43), 5804–5808.
79. Stewart, K.M., McLaughlin, L.W. (2004) Four-arm oligonucleotide ni(II) – cyclam-centered complexes as precursors for the generation of supramolecular periodic assemblies, *Journal of the American Chemical Society*, **126** (7), 2050–2057.
80. Stewart, K.M., Rojo, J., McLaughlin, L.W. (2004) Ru(II) tris(bipyridyl) complexes with six oligonucleotide arms as precursors for the generation of supramolecular assemblies, *Angewandte Chemie International Edition*, **43** (43), 5808–5811.
81. Yang, H., Sleiman, H.F. (2008) Templated synthesis of highly stable, electroactive, and dynamic metal-DNA branched junctions, *Angewandte Chemie International Edition*, **47** (13), 2443–2446.
82. Yang, H., Rys, A.Z., McLaughlin, C.K., Sleiman, H.F. (2009) Templated ligand environments for the selective incorporation of different metals into DNA, *Angewandte Chemie International Edition*, **48** (52), 9919–9923.
83. Yang, H., Altvater, F., de Bruijn, A.D., *et al.* (2011) Chiral metal-DNA four-arm junctions and metalated nanotubular structures, *Angewandte Chemie International Edition*, **50** (20), 4620–4623.
84. Yang, H., McLaughlin, C.K., Aldaye, F.A., *et al.* (2009) Metal-nucleic acid cages, *Nature Chemistry*, **1** (5), 390–396.
85. Dauphin-Ducharme, P., Rosati, F., Greschner, A., *et al.* (2015) Modulation of charge transport across double-stranded DNA by the site-specific incorporation of copper bis-phenanthroline complexes, *Langmuir*, **31** (5), 1850–1854.
86. Mirkin, C.A., Letsinger, R.L., Mucic, R.C., Storhoff, J.J. (1996) A DNA-based method for rationally assembling nanoparticles into macroscopic materials, *Nature*, **382** (6592), 607–609.
87. Nykypanchuk, D., Maye, M.M., van der Lelie, D., Gang, O. (2008) DNA-guided crystallization of colloidal nanoparticles, *Nature*, **451** (7178), 549–552.
88. Park, S.Y., Lytton-Jean, A.K., Lee, B., *et al.* (2008) DNA-programmable nanoparticle crystallization, *Nature*, **451** (7178), 553–556.
89. Macfarlane, R.J., Lee, B., Jones, M.R., *et al.* (2011) Nanoparticle superlattice engineering with DNA, *Science*, **334** (6053), 204–208.
90. Alivisatos, A.P., Johnsson, K.P., Peng, X., *et al.* (1996) Organization of "nanocrystal molecules" using DNA, *Nature*, **382** (6592), 609–611.
91. Aldaye, F.A., Sleiman, H.F. (2006) Sequential self-assembly of a DNA hexagon as a template for the organization of gold nanoparticles, *Angewandte Chemie International Edition*, **45** (14), 2204–2209.
92. Wen, Y., McLaughlin, C.K., Lo, P.K., Yang, H., Sleiman, H.F. (2010) Stable gold nanoparticle conjugation to internal DNA positions: facile generation of discrete gold nanoparticle-DNA assemblies, *Bioconjugate Chemistry*, **21** (8), 1413–1416.
93. Aldaye, F.A., Sleiman, H.F. (2007) Dynamic DNA templates for discrete gold nanoparticle assemblies: control of geometry, modularity, write/erase and structural switching, *Journal of the American Chemical Society*, **129** (14), 4130–4131.
94. Mastroianni, A.J., Claridge, S.A., Alivisatos, A.P. (2009) Pyramidal and chiral groupings of gold nanocrystals assembled using DNA scaffolds, *Journal of the American Chemical Society*, **131** (24), 8455–8459.
95. Li, Y., Liu, Z., Yu, G., Jiang, W., Mao, C. (2015) Self-assembly of molecule-like nanoparticle clusters directed by DNA nanocages, *Journal of the American Chemical Society*, **137** (13), 4320–4323.
96. Ding, B., Deng, Z., Yan, H., *et al.* (2010) Gold nanoparticle self-similar chain structure organized by DNA origami, *Journal of the American Chemical Society*, **132** (10), 3248–3249.
97. Pal, S., Deng, Z., Ding, B., Yan, H., Liu, Y. (2010) DNA-origami-directed self-assembly of discrete silver-nanoparticle architectures, *Angewandte Chemie International Edition*, **49** (15), 2700–2704.

3

Anion, Cation and Ion-Pair Recognition by Macrocyclic and Interlocked Host Systems

Paul D. Beer and Matthew J. Langton

Chemistry Research Laboratory, Department of Chemistry, University of Oxford, Oxford, UK

3.1 Introduction

Over the past three decades, my group has made major contributions to macrocyclic and supramolecular chemistry in a number of fields, some of which are reviewed throughout this chapter. Chronologically, in the late 1980s the research interests of the Beer group began with the topic of electrochemical molecular recognition (Section 3.2) where metallocene redox-active hosts were designed for cation, anion, and neutral guest species electrochemical sensing applications. Around this time the emerging field of anion coordination chemistry stimulated our interest in the development of transition metal-based photo-active macrocyclic receptors for anion sensing and led on to the anion-templated construction of mechanically bonded molecular frameworks (Section 3.3). This strategic anion templation methodology proved notably significant in advancing interlocked host anion recognition and sensing capabilities and has recently been exploited further with halogen-bonding anion recognition (Section 3.4). We have also made significant contributions to the fields of ion-pair recognition (Section 3.5) and metal-directed self-assembly (Section 3.6).

3.2 Electrochemical molecular recognition

With the objective of developing electrochemical molecular sensors, during the early part of my academic career my group designed and constructed a series of novel redox-active acyclic, macrocyclic, and macropolycyclic ligand systems that were capable of binding selectively and sensing via voltammetric electrochemical

Macrocyclic and Supramolecular Chemistry: How Izatt–Christensen Award Winners Shaped the Field, First Edition.
Edited by Reed M. Izatt.

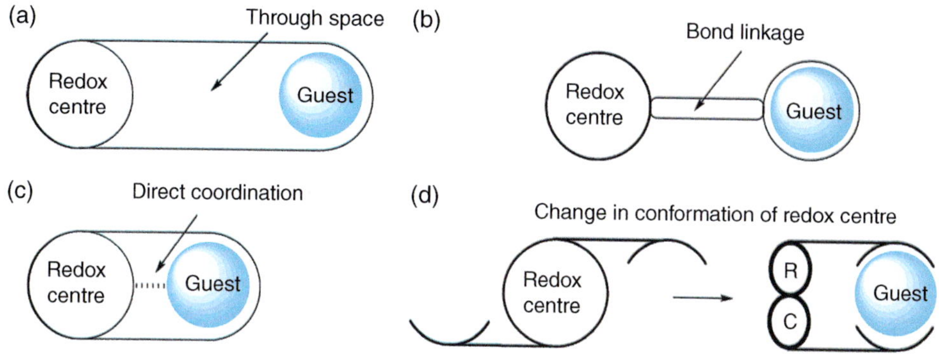

Figure 3.1 *Schematic illustrations of four different intramolecular interaction pathways for electrochemical molecular recognition of guest species using redox-responsive receptor molecules*

methodologies, target cationic and anionic guest species of biological and environmental importance [1, 2]. The redox-active host or receptor design structure consisted of a binding site in proximity to a redox-responsive center. Upon guest binding, electron transfer to or from the host's redox center was achieved at a significantly different potential from that of the host in the absence of the guest species.

Having demonstrated the principle of electrochemical molecular recognition, we elucidated the various electrostatic intramolecular signaling mechanisms (through space, through conjugated bonds, direct coordination, conformational changes of the redox center) by which organometallic ferrocene, cobaltocenium, and organic (quinone, bipyridinium) redox-active containing receptor molecules detected electrochemically Groups 1 and 2, ammonium, and transition metal cations and anions in polar organic and aqueous solutions (Figure 3.1) [3].

Examples of redox-active receptors which exhibit each of these mechanisms are now discussed in the following sections according to the nature of the target guest species.

3.2.1 Electrochemical cation recognition

In the early 1990s we reported the syntheses of a variety of ferrocene functionalized crown ethers that contained conjugated and saturated methylene linkages between the crown ether and ferrocene redox-center (Figure 3.2) [4–7]. Electrochemical cyclic voltammetric (CV) experiments, typically undertaken in acetonitrile solutions, demonstrated significant anodic perturbations of the respective ferrocene/ferrocenium (Fc/Fc+) oxidation wave resulted upon addition of Group 1 and 2 metal cations to the conjugated receptor systems (**1a, b, 2a, b**), for example of up to 110 mV with magnesium cations. With the saturated analogue (**3**) however, much smaller perturbations were observed (<20 mV with Mg^{2+}) which suggested a through-bond mode of communication (Figure 3.1b) as the primary mechanism of electrochemical recognition with this class of conjugated redox-active ionophore. The addition of ammonium ions to electrochemical acetonitrile solutions of the tris-ferrocene functionalized aza crown ether receptor (**4**) produced a remarkable anodic perturbation of 220 mV of the receptor's Fc/Fc + redox couple, due to a combination of through-space and ammonium cation $N–H^{+}\cdot\cdot$ amide carbonyl bonding interactions [5].

In view of the environmental concern for monitoring radioactive cesium in nuclear waste streams and the potential use of rubidium isotopes in radiopharmaceutical reagents, we constructed novel redox-active bis-calix[4]diquinone receptors [8] based on our previously reported calix[4]tube ligand design (Figure 3.3a) [9, 10].

UV-visible titration experiments in aqueous DMSO solutions with Group 1 metal cations established the propylene spaced receptor (**5**) was selective for rubidium, whereas the receptor containing the longer butylene

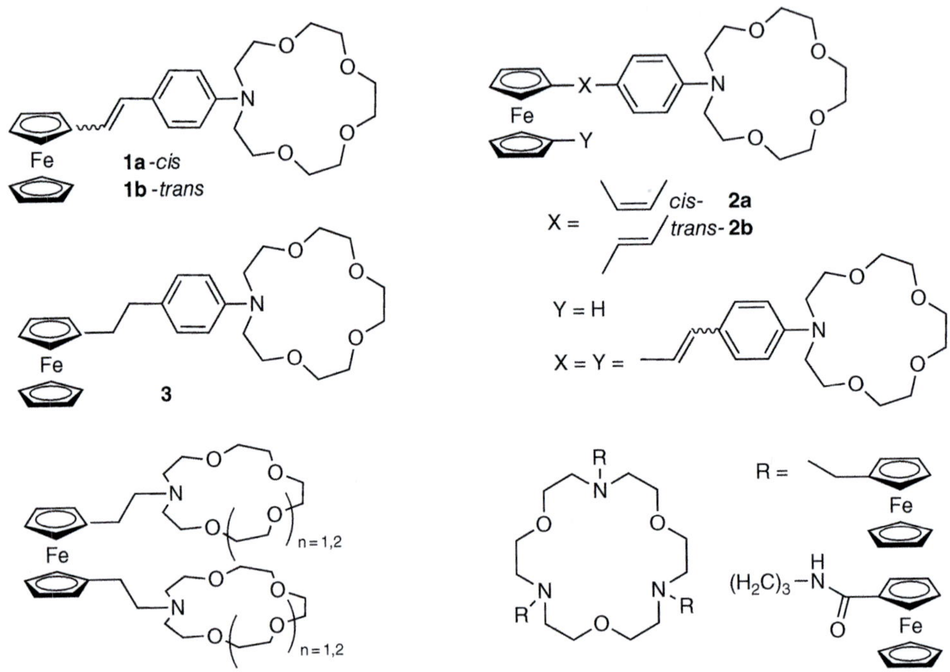

Figure 3.2 Ferrocene crown ether species

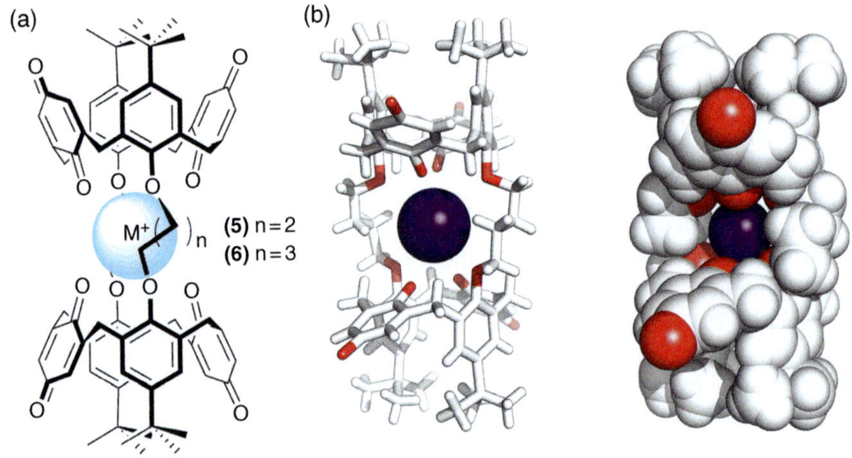

Figure 3.3 (a) Rubidium- and cesium-selective redox-active ionophores, and (b) Perspective (left) and space-filling representation (right) of the X-ray crystal structure of the cesium complex of bis-calix[4]diquinone receptor (**6**)

spacers (**6**) exhibited a remarkably high selectivity for cesium ions over sodium and potassium cations which were not bound. Solid state X-ray structural analysis of the cesium complex of butylene-bridged bis-calix[4] diquinone receptor revealed a complementary size match with the metal cation complexed to all eight oxygen

Table 3.1 *Anodic shifts of the respective quinone reduction couple of bis-calix[4] diquinone receptors (**5** and **6**) upon addition of Group 1 metal cations*

	(**5**)	(**6**)
ΔE (Na^+, mV)	350	330
ΔE (K^+, mV)	250	250
ΔE (Rb^+, mV)	250	210
ΔE (Cs^+, mV)	240	190

ΔE determined upon addition of 5-equivalents of Group 1 metal cations in 4:1 CH_2Cl_2/CH_3CN solution containing 0.1 M Bu_4NBF_4 as the supporting electrolyte.

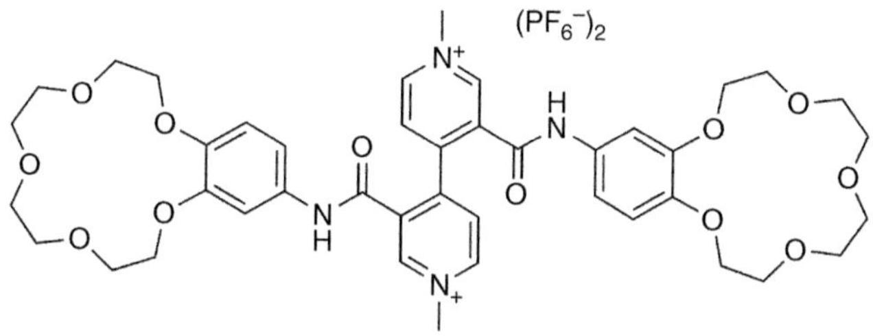

Figure 3.4 *Redox-active bipyridinium bis(benzo)-crown ether (7)*

donor atoms of the receptor's cavity (Figure 3.3b). Voltammetric investigations in 4:1 dichloromethane: acetonitrile solutions demonstrated both receptors to electrochemically sense Group 1 metal cations via the direct coordination electrochemical recognition pathway (Figure 3.1c); with substantial anodic perturbations of the respective quinone reduction couples (Table 3.1). Importantly, electrochemical competition experiments in the presence of a large excess of sodium cations, typically found in nuclear waste, revealed receptor (**6**) to be capable of selectively sensing cesium ions [8].

We discovered that guest-induced conformational changes in redox-active molecules also provides a mechanism to induce perturbations in electrochemical behavior, for example with the bipyridinium bis(benzo) crown ether (**7**) (Figure 3.4) [11].

Our previous work on the redox-responsive properties of ferrocene bis(benzo) crown ethers had shown that the amide bond linkage was insulating with respect to crown ether metal cation sensing [4]; therefore any observed perturbation of the redox behavior of (**7**) would not be caused by through-bond interactions. NMR, UV/Vis spectroscopy and X-ray crystallography established the receptor to bind Group 1 and 2 metals and ammonium cations, forming 1:1 intramolecular sandwich complexes with Ba^{2+}, K^+ and NH_4^+ and 2:1 complexes with Mg^{2+} and Na^+ with a cation bound in each crown ether motif. Upon 1:1 sandwich complex formation, a significant twist and conformational change of the 4,4′-bipyridinium redox-center occurs. However, formation of the Na^+ complex does not induce such a dramatic change in the conformation of the 4,4′-bipyridinium motif. Upon addition of Ba^{2+} cations, the 2+/1+ bipyridinium redox couple was found to shift anodically by 45 mV and the 1+/0 couple shift cathodically by 10 mV. Potassium ion and NH_4^+ produced similar effects. However addition of Na^+ cations caused contrasting redox responsive behavior with a small cathodic shift of the 2+/1+ couple and an anodic perturbation of the 1+/0 couple (Table 3.2). These results therefore supported the proposal that the conformational change signaling pathway for coupling complexation and redox reactions was operating in this case.

Table 3.2 Electrochemical data for (7) in acetonitrile containing Bu_4NBF_4 as the supporting electrolyte (4 equivalents of metal cation salt added)

	Redox couple	
	2+/1+	1+/0
$E_{1/2}$ (V)	−0.73	−0.87
ΔE_p (mV)	70	70
ΔE (Ba^{2+}, mV)	45	−10
ΔE (K^+, mV)	10	−40
ΔE (NH_4^+, mV)	10	−40
ΔE (Na^+, mV)	−10	30

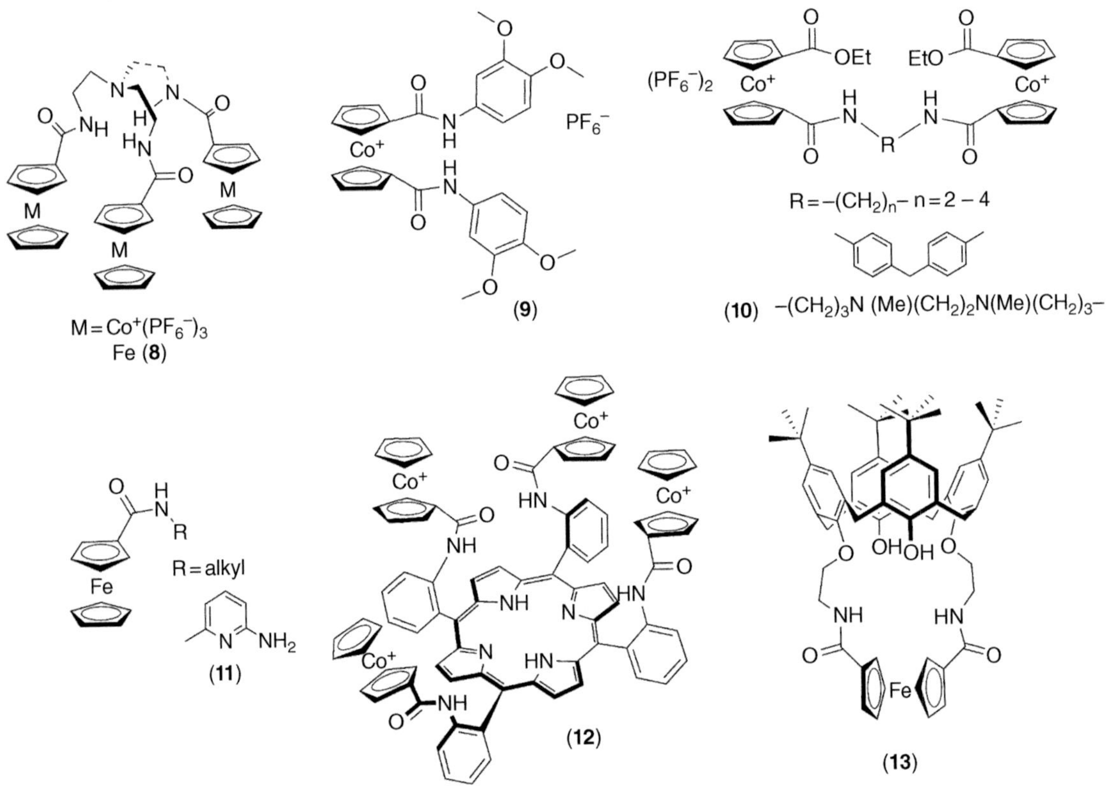

Figure 3.5 Redox-responsive cobaltocenium and ferrocene anion receptors

3.2.2 Electrochemical anion recognition

Through the syntheses of a variety of acyclic, macrocyclic and calixarene amide functionalized cobaltocenium and ferrocene based receptor systems (Figure 3.5), we reported the first electrochemical voltammetric responsive classes of redox-active ionophore to sense anionic guest species [12–17]. Typically, anion binding caused significant *cathodic* perturbations of the metallocene's redox-couple via, with amide receptors, a favorable anion-amide hydrogen bonding through bond communication mechanism.

Table 3.3 Cathodic shifts (mV) of the respective metallocene redox couple upon addition of anions

Receptor[a]	ΔE (Cl^-)	ΔE (Br^-)	ΔE (HSO_4^-)	ΔE ($H_2PO_4^-$)
(**8**)	20	–	50	180
(**9**)	90	40	–	240
(**10**)	50	–	115	250
(**11**)	5	–	40	100
(**12**)	40	35	75	225
(**13**)	40	–	<5	110

[a]Obtained in CH_3CN solution containing Bu_4NBF_4 as the supporting electrolyte (5 equiv. TBA salt added).

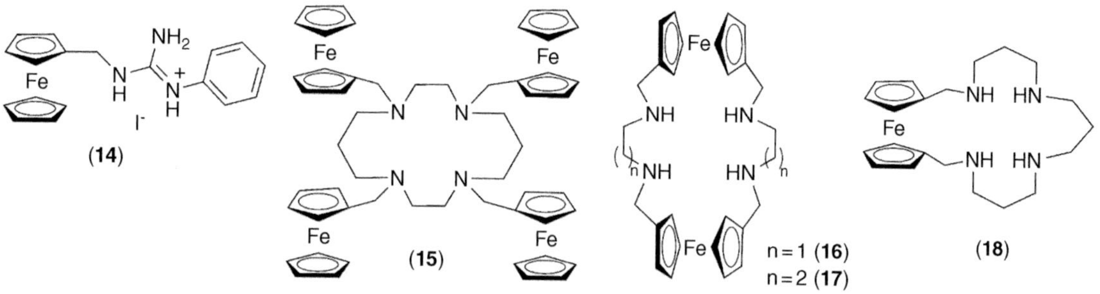

Figure 3.6 *Guanidinium and polyazamacrocycle ferrocene receptors*

For example the observed cathodic shifts of the respective metallocene redox couples of receptors (**8–13**) on addition of halides and oxoanions (Table 3.3) are due to the stabilization of the ferrocenium or cobaltocenium cation by the bound anion which causes the redox couple to shift to a more negative potential. It is noteworthy that chloride, by virtue of its higher charge density, causes relatively larger cathodic perturbations than bromide. Correlating with quantitative anion association constant data determined by ^{1}H NMR titration experiments, the tren-based amide tripodal ferrocene receptor (**8**) is selective for the tetrahedral shaped anion dihydrogen phosphate which produces the largest magnitude of cathodic shift.

The guanidinium ferrocene functionalized receptor (**14**) (Figure 3.6) was shown to be capable of sensing pyrophosphate anions in aqueous methanol solutions via a cathodic shift of 70 mV in the Fc/Fc+redox couple [18]. In collaboration with Ramon Martinez-Manez, we achieved the selective electrochemical recognition of sulfate over phosphate, and phosphate over sulfate, using polyaza-ferrocene macrocyclic receptors in aqueous solution [19]. Receptors (**15**)–(**18**) (Figure 3.6) can, through an electrochemical response, selectively detect at certain pH values sulfate and phosphate in the presence of competing anions in an aqueous environment. Maximum selective redox cathodic potential shifts of 54 and 50 mV were observed for sulfate and phosphate, using receptors (**16**) and (**18**) at pH 4 and 7, respectively.

With the objective of fabricating a prototype anion sensory device, we used the 1,1-bis(alkyl-*N*-amido) disulfide functionalized ferrocene derivative (**19**) to form stable self-assembled monolayers (SAMs) on gold electrodes (Figure 3.7) [20]. Importantly, comparing the diffusive electrochemical anion recognition behavior of (**19**) with the SAM of the receptor, revealed the consistent observation of "surface-sensing amplification" where the anion binding-induced cathodic shift magnitudes were significantly larger than that observed diffusively. This was attributed to the preorganization of a pseudo-macrocyclic host on the gold electrode surface. For example, redox responses were more cathodically shifted by *ca.* 90 mV and 60 mV for dihydrogen phosphate and chloride respectively. Anion competition experiments demonstrated

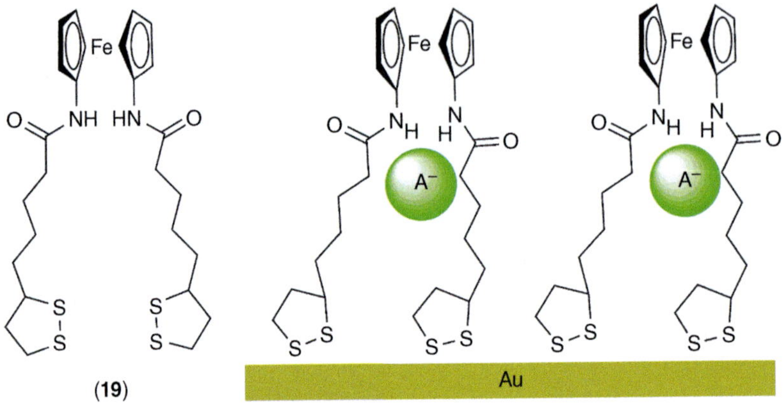

Figure 3.7 An illustration of a SAM of (**19**)

the capability of (**19**) and the surface confined material to selectively sense dihydrogen phosphate in the presence of a 100-fold excess of halide.

3.3 Anion recognition and sensing by macrocyclic and interlocked hosts

Anions play a fundamental role in many biological, environmental and medical processes [21]. Indeed, around 70% of biological substrates are negatively charged, whilst a plethora of inorganic anions play indispensable roles in many processes *in vivo*. For instance, the chloride anion is found in high concentrations in extracellular fluid and yet its misregulation has been linked to diseases such as cystic fibrosis; whilst iodide is crucial to hormone biosynthesis by the thyroid gland. From an environmental perspective, anions such as nitrate and phosphates in excess can lead to eutrophication and disruption of aquatic life cycles, whilst pertechnetate (a by-product of nuclear waste reprocessing) causes further pollution problems in the nuclear fuel cycle. With this in mind, there has been an increasing interest in the past two decades or so in designing synthetic hosts capable of anion recognition, with the aim of binding, sensing and sequestering anionic species in solution [22].

3.3.1 Transition metal bipyridyl-based macrocyclic receptors for anion recognition and sensing

In the mid 1990s we initiated a research program to develop optical and dual optical/electrochemical sensory probes for anionic species, based around both acyclic and macrocyclic anion receptors. Whilst we and others had incorporated transition metal and organic redox-active groups into cryptand and crown-ether frameworks for cation-sensing applications, the detection of anions by such systems was at the time only just being realized.

With the aim of elucidating new anion selectivity trends and imparting enhanced thermodynamic binding affinities, we prepared a series of ruthenium(II) and osmium(II) bipyridyl-, ferrocene-, and cobaltocenium-containing macrocyclic anion hosts to simultaneously exploit the well-understood luminescence and electrochemical properties of Ru(II) and Os(II) bipyridine complexes (Figure 3.8) [23]. An extensive investigation into the anion coordination and sensing capabilities of such macrocyclic systems, using a combination of NMR spectroscopy, electrochemistry, X-ray crystallography and photophysical investigations revealed that remarkable levels of selectivity for Cl^-, AcO^- or $H_2PO_4^-$ could be dictated by modifying the topology and relative dimensions of the macrocyclic binding domain.

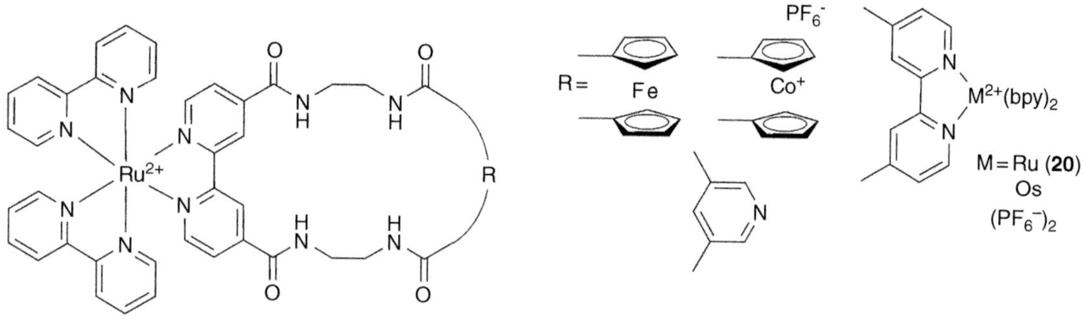

Figure 3.8 *Macrocyclic transition metal receptor systems for anion sensing by dual optical and electrochemical means*

^{1}H NMR anion binding titrations in d_6-DMSO solution revealed enhanced binding affinities for Cl^- by the macrocyclic hosts (~10^4 M^{-1}, *ca.* two orders of magnitude stronger than the analogous acyclic systems) and a rare selectivity for chloride over the more basic oxoanions AcO^- and $H_2PO_4^-$. Most notably, the bis-Ru(bipy)$_3$-based macrocycle **20** displayed outstanding Cl^- selectivity over $H_2PO_4^-$; an association constant of 4×10^4 M^{-1} compared to no binding of $H_2PO_4^-$ in DMSO solution. The solid state crystal structure of the chloride complex of the pyridine 3,5-bis-amide-derived macrocycle revealed that the halide anion is bound within the macrocyclic cavity, through convergent hydrogen bonding interactions. Interestingly, increasing the size of the macrocycle by lengthening the alkane spacer unit reverses the selectivity trend in favor of $H_2PO_4^-$ over Cl^-, highlighting the crucial importance of topology, degree of preorganization and host–guest complementarity in achieving anion guest selectivity. Cyclic voltammetry studies demonstrated the receptors to be capable of electrochemically sensing the anion binding event via cathodic perturbations of the respective bipyridyl-based and metallocene redox couples. Furthermore, photophysical investigations revealed a marked increase in the emission intensity and lifetimes of the Ru(bipy)$_3$-based macrocyclic receptors upon anion binding.

3.3.2 Interlocked host structures for anion recognition and sensing

The exquisite anion-binding affinity and selectivity achieved by natural sulfate and phosphate anion-binding proteins is achieved by binding the oxoanion guest through a convergent network of hydrogen bonds, in a cavity buried deep within the protein and shielded from the aqueous solvent [24, 25]. This suggested to us that in order to significantly "raise the bar" in the levels of anion selectivity exhibited by abiotic acyclic and macrocyclic receptors, the design and synthesis of more elaborate three-dimensional receptors were required, in which the anion is bound in a fully encapsulated microenvironment. To respond to this challenge, we pioneered the anion template-directed synthesis of interlocked structures, such as rotaxanes and catenanes, paving the way for utilizing such systems for sophisticated anion recognition and sensing applications [26]. This step-changing method uses anions to template the assembly of interpenetrating and interlocked host molecules based on the coupling of anion recognition with ion-pairing. This involves the threading of a one-dimensional, acyclic receptor through the anulus of a two-dimensional, macrocyclic receptor which creates a unique topological three-dimensional cavity for anion recognition.

Building on our discovery that an orthogonal complex could be formed between a hydrogen bond-donating 3,5-bis-amide pyridinium motif and a neutral isophthalamide motif via chloride anion coordination (Figure 3.9a), we demonstrated that a chloride anion templated pseudorotaxane could be assembled in acetone solution by incorporating the isophthalamide motif within a macrocycle (Figure 3.9b, c) [27]. Further stabilization of the

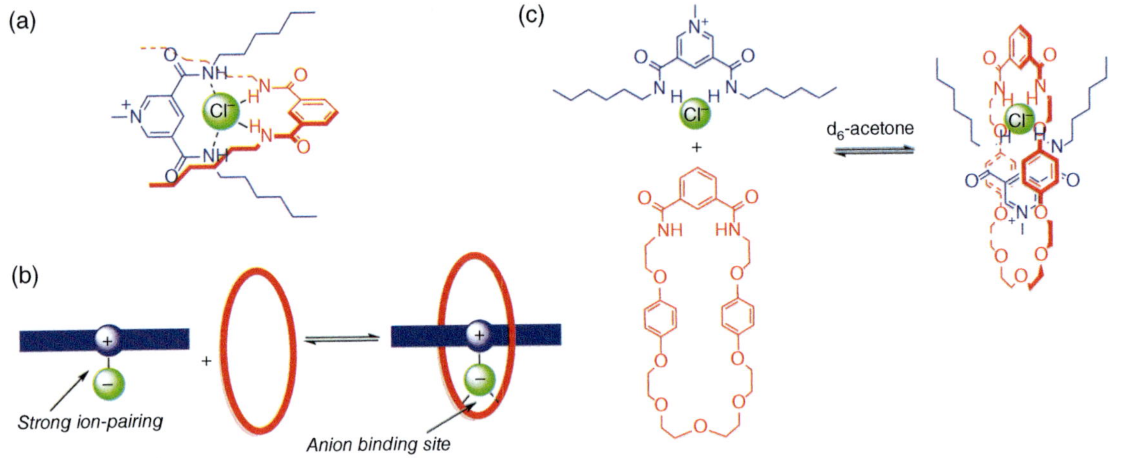

Figure 3.9 (a) Orthogonal association of hydrogen-bonding components about a chloride anion. (b and c) Anion templation strategy for the assembly of interpenetrative pseudorotaxanes

interpenetrative assembly is achieved through secondary supramolecular interactions: aromatic donor–acceptor interactions and pyridinium-N-methyl-polyether hydrogen bonding. This templation strategy is efficient in organic solvent media, in which the ion-pairing is strong and, importantly, the anion itself remains coordinatively unsaturated. The success of this general anion template procedure was illustrated further with the halide anion directed assembly of a series of pseudorotaxanes containing pyridinium-, pyridinium nicotinamide-, imidazolium-, benzimidazolium-, guanidinium-, and triazolium-threading components [28, 29].

Exploiting this work led to the chloride anion templated assembly of the first interlocked anion host molecule in the form of a [2]rotaxane via a ring-closing metathesis (RCM) approach by cyclizing a bis-vinyl-appended macrocycle precursor around a terphenyl-stoppered pyridinium chloride axle (Figure 3.10) [30]. Subsequently, the first example of an anion templated [2]catenane synthesis was demonstrated, through a related monocyclization protocol [31]. More recently, we have also highlighted an alternative high-yielding route for anion-templated interlocked molecular construction in which a chloride anion templates the formation of a macrocycle in the presence of a pyridinium chloride axle using an acid chloride–amine condensation strategy [32].

An impressive feature of this method of rotaxane and catenane assembly is that the resultant interlocked molecular structure retains a degree of functionality based on the anion template itself. After template removal, the rotaxane and catenane host systems, by virtue of the superstructures' interlocked and topologically restrained anion recognition site, exhibit a high degree of selectivity for chloride (the templating anion) in preference to competitive basic oxoanions which are too large to be able to penetrate the respective buried interlocked binding pocket. The power of this interweaving anion templation methodology has since been exploited with a range of other anions, including bromide [29], sulfate [33], nitrate [34, 35] and nitrite [36]. The use of nitrate as a templating species for both [2]rotaxane [34] (Figure 3.11) and [2]catenane [35] synthesis resulted in interlocked hosts that demonstrated remarkable and unprecedented binding affinities and selectivity for the templating nitrate anion, even in aqueous-organic solvent mixtures. Importantly, the interlocked hosts display superior levels of recognition, including enhanced selectivity and binding affinities compared to that of an analogous non-interlocked host. Of particular note is the unusual selectivity of the rotaxane for nitrate over acetate (an association constant of 430 M^{-1} compared to no binding of acetate in a chloroform/methanol solution containing 10% water), despite acetate being five orders-of-magnitude more basic. This example exemplifies the crucial role of a complementary, geometrically constrained binding domain for achieving high levels of anion binding selectivity.

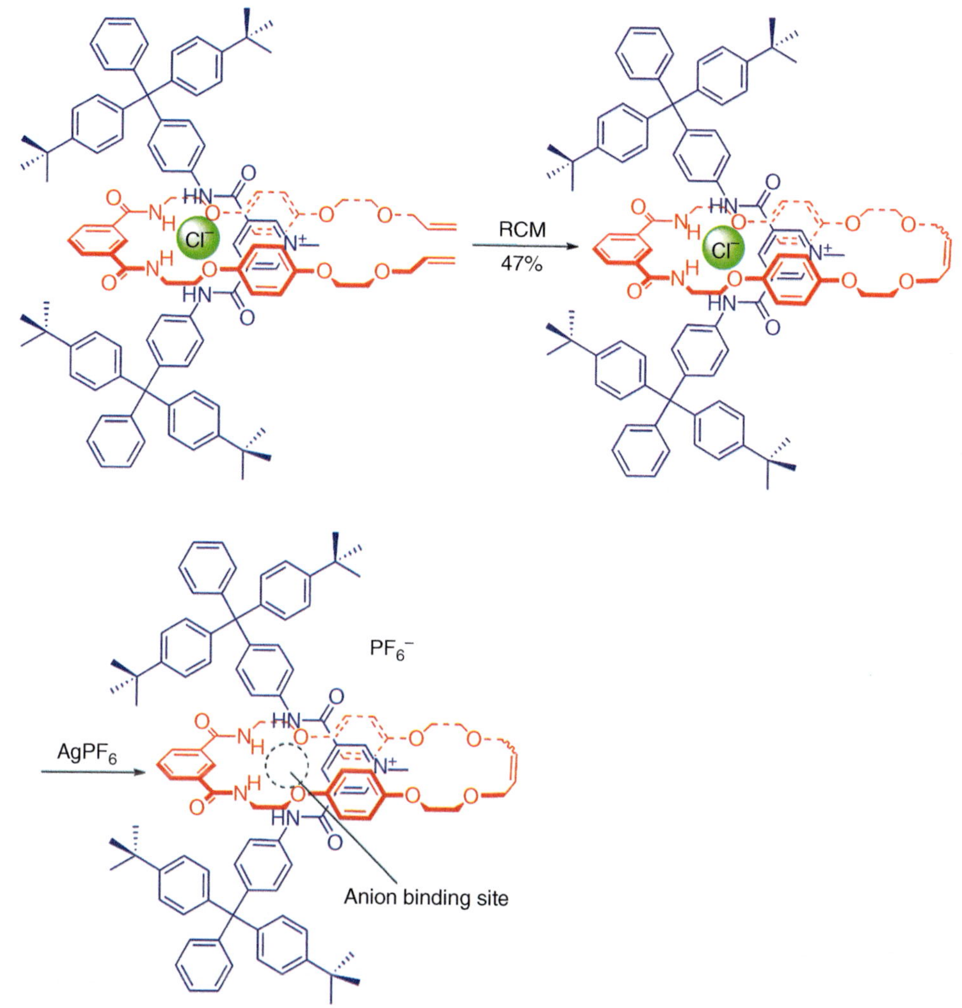

Figure 3.10 *Anion templated [2]rotaxane synthesis via a ring closing metathesis (RCM) strategy*

In addition, this anion templation methodology has been applied to the synthesis of exotic higher-order interlocked structures, such as [3]rotaxanes, [37, 38] a triply-interlocked cage [39], a "handcuff" catenane (Figure 3.12) [40] and a "Janus" rotaxane [41].

In 2005 we reported a new family of anion receptors, the indolo[2,3-a]carbazoles, which bind anions strongly via their two preorganized hydrogen bond donating pyrrole groups [42]. The real potential for manipulating the unique topological cavities of mechanically bonded molecules for anion recognition applications was further exemplified by the synthesis of the first indolocarbazole-containing interlocked structure [43]. The [2]rotaxane comprises an indolocarbazole axle component and a novel tetra-cationic macrocycle functionalized with an isophthalamide anion binding motif (Figure 3.13). The interlocked host exhibits enhanced selectivity for chloride and bromide anions in comparison with the individual axle and macrocyclic components. Computational studies suggested that, whilst chloride and bromide anions are of complementary size and shape to interact cooperatively with the rotaxane's orthogonally disposed indolocarbazole axle

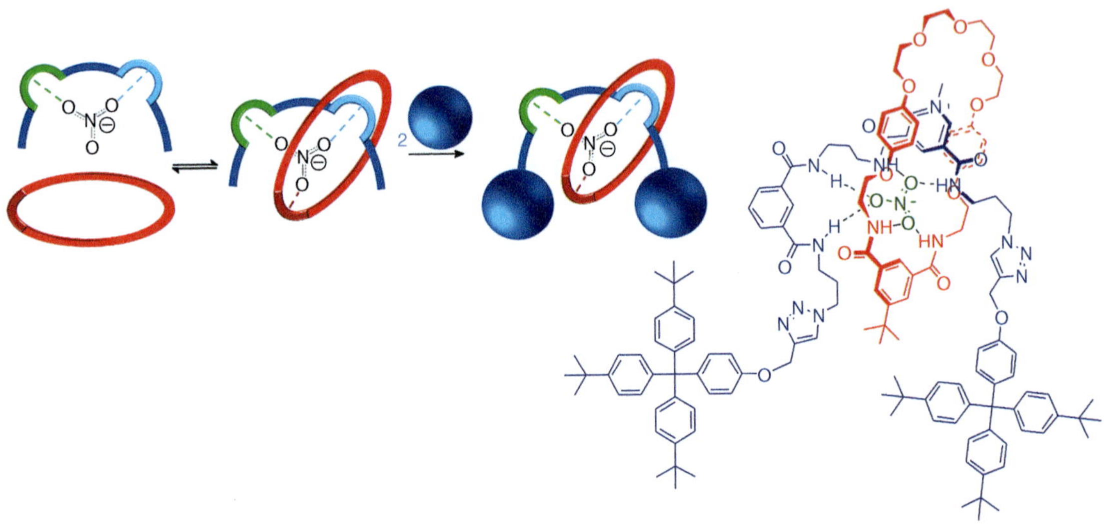

Figure 3.11 Nitrate-templated synthesis of a [2]rotaxane, in which two hydrogen bonding recognition sites are built into the axle (blue and green) and one into the macrocycle (red), forming a complementary binding site for the trigonal nitrate anion between the interlocked components

and isophthalamide macrocycle convergent hydrogen bond donor groups, the larger iodide and nitrate anions are unable to penetrate the interlocked binding cavity.

3.3.3 Rotaxane and catenane host structures for sensing anions

Rotaxane and catenane anion host structures – by virtue of their superior recognition and selectivity capabilities compared to acyclic and macrocyclic hosts – are ideal candidates for chemical sensor design. By incorporating electrochemical or optical reporter groups into such mechanically bonded hosts, we have developed a range of systems capable of highly selective sensing of anion guest species [44].

Our first work in this area explored the incorporation of luminescent transition metal complexes into the macrocycle components of rotaxane-based anion receptors. In 2005 we reported a Re(I)-bipyridyl-based rotaxane, in which the luminophore is incorporated into a macrocycle which itself is interlocked with a calix[4]arene-stoppered 3,5-bis-amide pyridinium axle component (Figure 3.14a) [45]. Anion binding in acetone resulted in enhancement of the MLCT emission band of the Re(I) center, which was postulated to arise from the increased rigidity of the rotaxane–anion complex, which disfavors non-radiative decay processes. The system was found to be selective for HSO_4^-, an analyte that has proved particularly difficult to bind and sense due to its poor hydrogen-bond-acceptor ability.

We later incorporated Re(I) and Ru(II) into bipyridyl macrocycle-based rotaxanes which were capable of sensing Cl^- anions in highly competitive aqueous–organic solvent mixtures, with selectivity over both $H_2PO_4^-$ and AcO^-, by means of enhancement in the MLCT emission intensity of both rotaxanes [46].

Luminescent lanthanide complexes are attractive candidates for anion sensing, imaging and assay, due to their long-lived luminescence [47]. In 2012 we reported the seminal example of a lanthanide-functionalized interlocked structure, by appending a DO3A-lanthanide complex to a preformed rotaxane species (Figure 3.14b) with the desire to exploit time-gated luminescence sensing techniques to detect anions with high sensitivity [48]. The luminescence of the europium-containing rotaxane was found to be chloride responsive in dichloromethane solution, with initial chloride coordination directly at the lanthanide metal

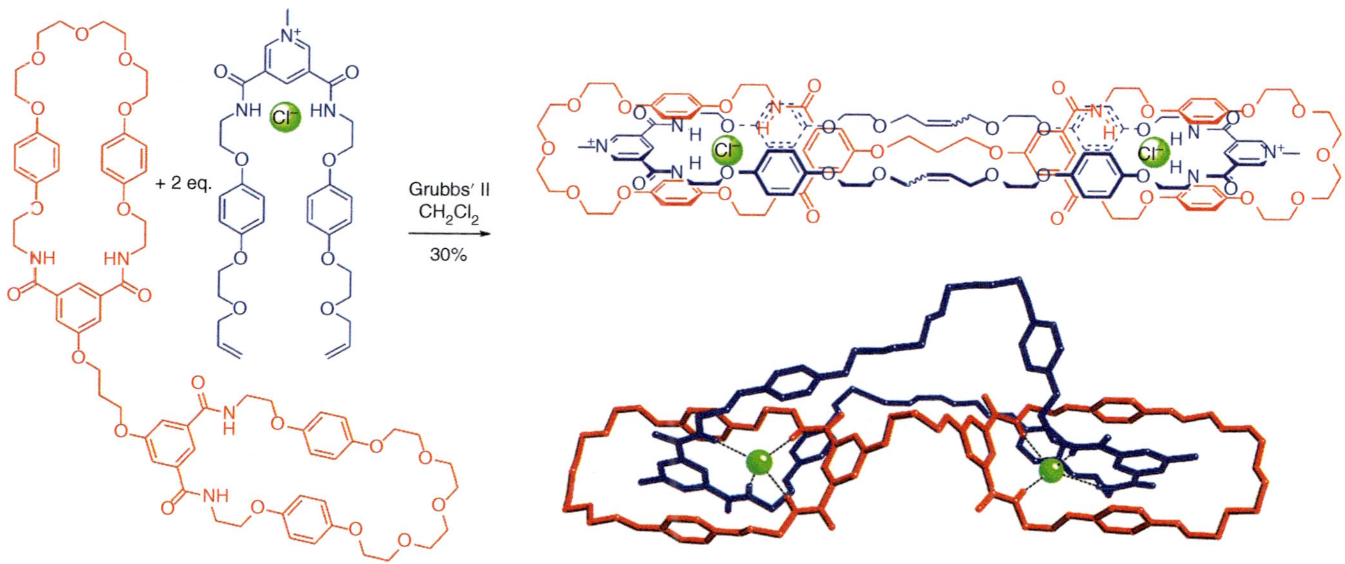

Figure 3.12 *Chloride-templated synthesis of a handcuff catenane and X-ray crystal structure*

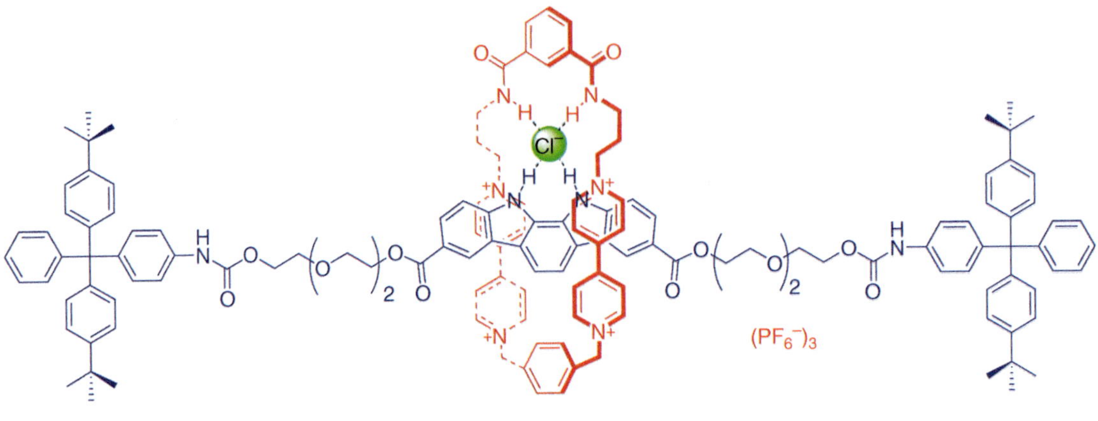

Figure 3.13 *Interlocked host anion recognition by an indolocarbazole-containing rotaxane*

center leading to a dramatic quenching of the hypersensitive $\Delta J = 2$ transition, and subsequent binding of the anion within the interlocked anion binding domain resulting in a gradual restoration of the emission intensity. In a preliminary step towards integrating the lanthanide center as part of the interlocked anion binding domain, we later demonstrated the first lanthanide–cation templated synthesis of a mechanically bonded architecture (see Section 3.6 on metal-directed self-assembly) [49]. To simultaneously exploit direct lanthanide–anion coordination *and* the unique selectivities afforded by such mechanically bonded host systems we exploited a nitrite anion template to assemble lanthanide-based [2]rotaxanes in remarkably high yields (~50% isolated), which were demonstrated to selectively sense fluoride anions in an acetone-water mixture (Figure 3.14c) [36].

The unique properties of mechanically interlocked molecular architectures, such as shuttling and molecular motion, conformational flexibility and stimuli-responsive behavior, have ensured that they remain the subject of intense research, driven by the desire to employ such species in prototypical molecular machines and devices [50]. In the context of molecular recognition using mechanically bonded structures, we have developed a novel paradigm for anion sensing in which the anion-binding event modulates the dynamic interactions between the mechanically bonded components, resulting in a measurable optical response. Our prototype system that demonstrated this method of signal transduction consisted of the chloride anion templated assembly of a pseudorotaxane formed between a Re(I)bipyridyl-functionalized macrocycle and a luminescent lanthanide complex appended to a chloride-benzimidazolium threading component (Figure 3.15a) [51]. Upon interpenetration of the two species, the rhenium(I) and neodymium(III) photoactive motifs are placed in close spatial proximity, resulting in the turn-on of the distance-dependent Forster energy-transfer process between the MLCT excited state of the Re(I)-bipyridyl species on the macrocycle and the lanthanide threaded complex, which results in the concomitant appearance of a near-IR lanthanide emission band. Crucially, this energy-transfer process only occurs with the formation of the pseudorotaxane assembly, which in turn is formed only in the presence of a chloride anion.

To extend this concept, we have prepared a [2]rotaxane anion-responsive molecular shuttle in which addition of Cl^- or Br^- initiates unidirectional shuttling behavior in the system (Figure 3.15b) [52]. The macrocycle initially resides over the naphthalimide group due to aromatic donor–acceptor interactions with the electron-rich hydroquinones on the macrocycle. Upon addition of a halide anion the macrocycle translocates to the axle's triazolium group to facilitate the formation of the anion-binding cavity. This shuttling response is signaled through changes in the rotaxane's UV–Vis absorption band, which occur due to changes in the naphthalimide–hydroquinone donor – acceptor interactions upon macrocycle translocation.

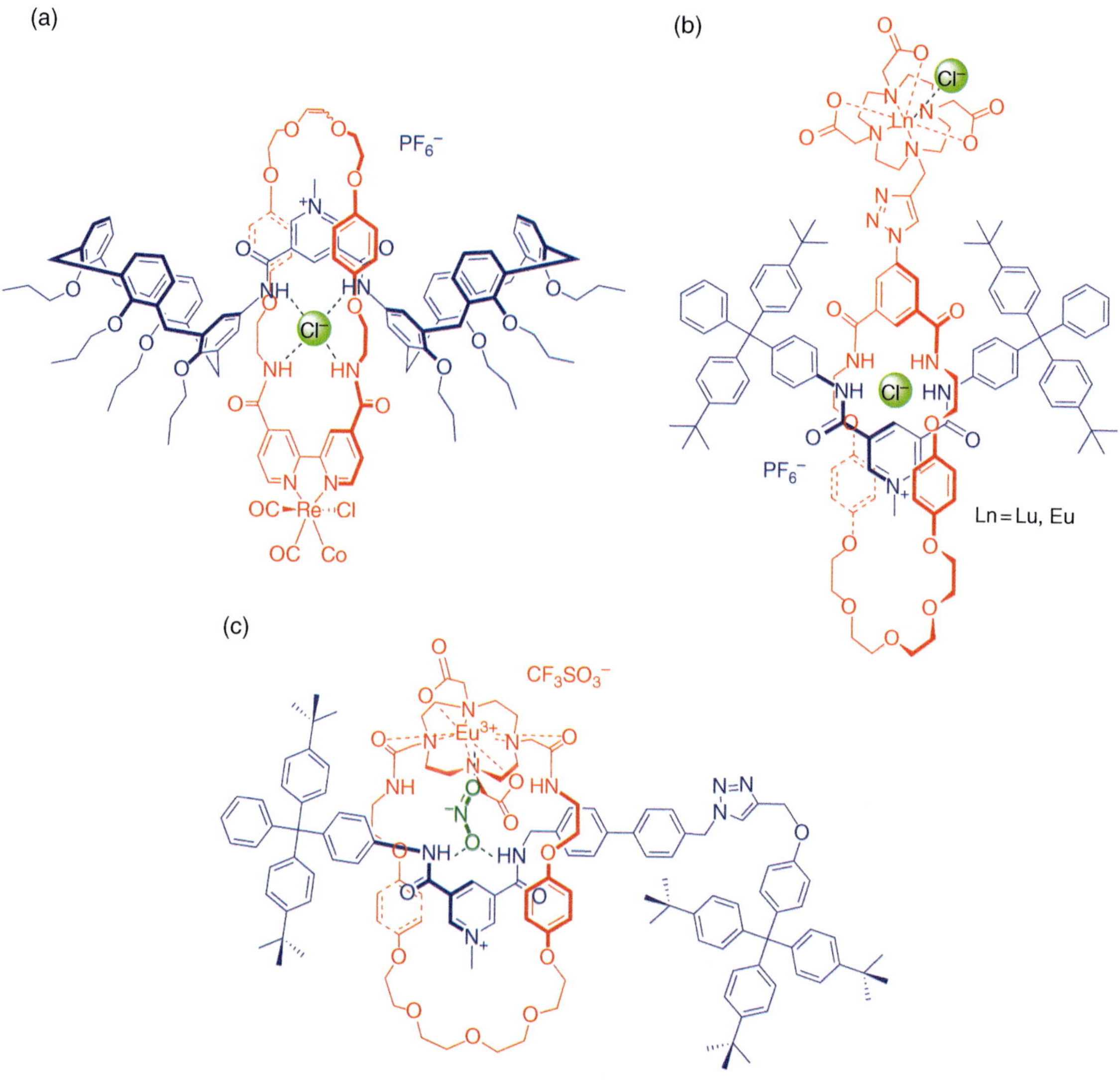

Figure 3.14 *Transition-metal- and lanthanide-based luminescent rotaxane hosts*

We have also exploited the conformational flexibility of higher-order interlocked structures as a novel means of signal transduction. We reported a fluorescent naphthalene-derived [3]rotaxane which selectively senses sulfate anions in solution [38]. Upon sulfate binding, the rotaxane undergoes a conformational change in which the sulfate is bridged between the two macrocycle components and placed in close proximity to the naphthalene fluorophore (Figure 3.16). This leads to a selective quenching response in the emission spectrum. In contrast, halide anions – which bind within the cavities formed between the respective macrocycles and axle – do not lead to folding of the structure, and their binding is signaled by an enhancement of the naphthalene emission.

We described in Section 1 the development of electrochemical anion sensors, incorporating ferrocene and cobaltocenium motifs within acyclic and macrocyclic anion receptor systems. More recently, we have

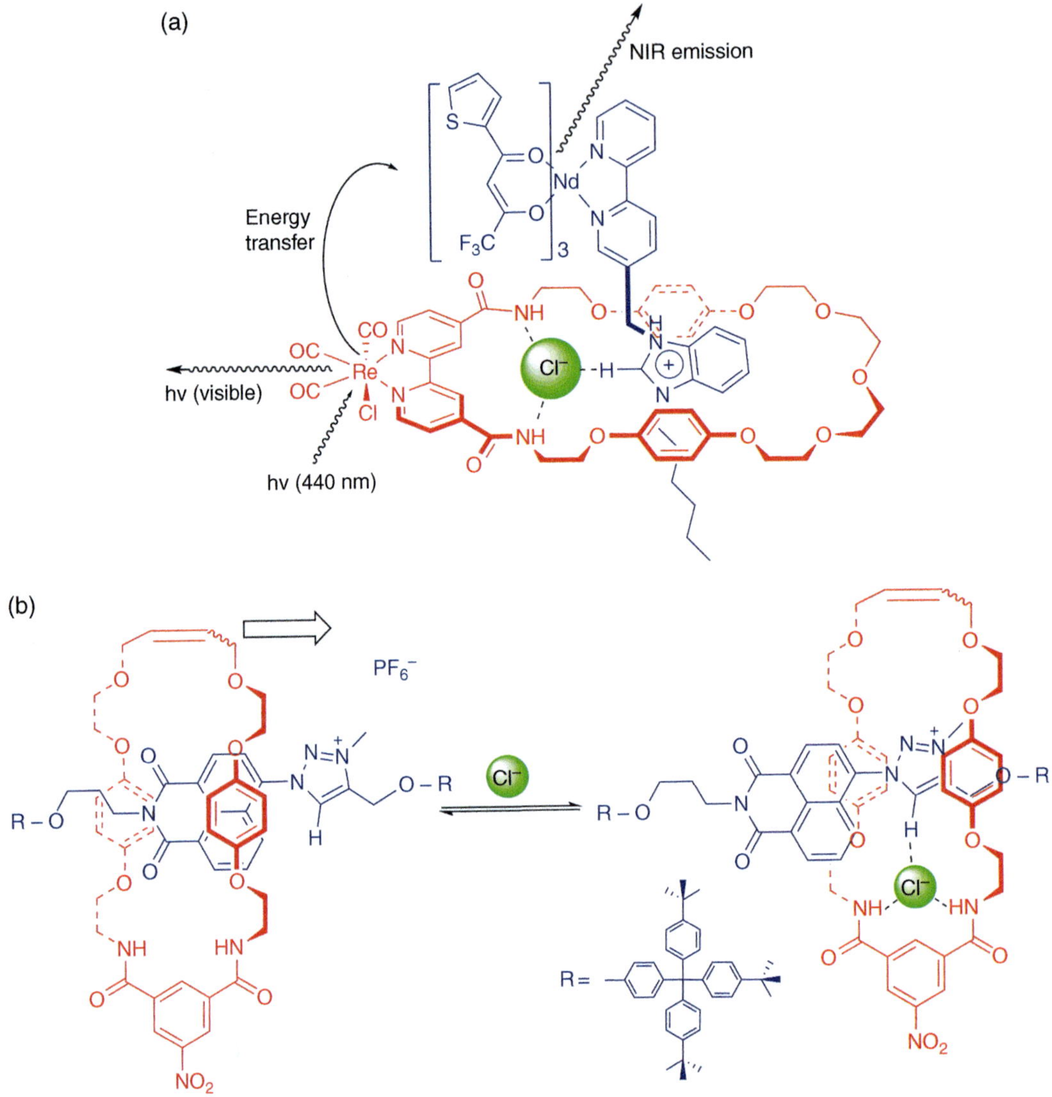

Figure 3.15 Dynamic optical anion sensors

extended this concept to include redox-active ferrocene-functionalized interlocked anion receptors to exploit the unique anion-recognition capabilities of such systems. We have incorporated ferrocene motifs on the periphery of the macrocycle component of rotaxanes [53] and catenanes [54], and, in the form of pentaphenylferrocene, as the axle's stopper components (Figure 3.17a) [55]. Such systems exhibit a common selectivity trend for binding chloride over a range of mono-anionic oxo-anions, and this selectivity is also manifested in the electrochemical anion-sensing behavior. In these examples, the maximum cathodic shift of the ferrocene/ferrocenium redox couple in acetonitrile solution occurred after the addition of one equivalent of chloride, whilst in contrast, a larger excess of oxo-anion is required to elicit the maximum redox response.

In order to develop functional anion sensory devices for proposed applications in sensing or information storage, the interfacing of solution phase anion sensors with optically transparent or electrochemically active

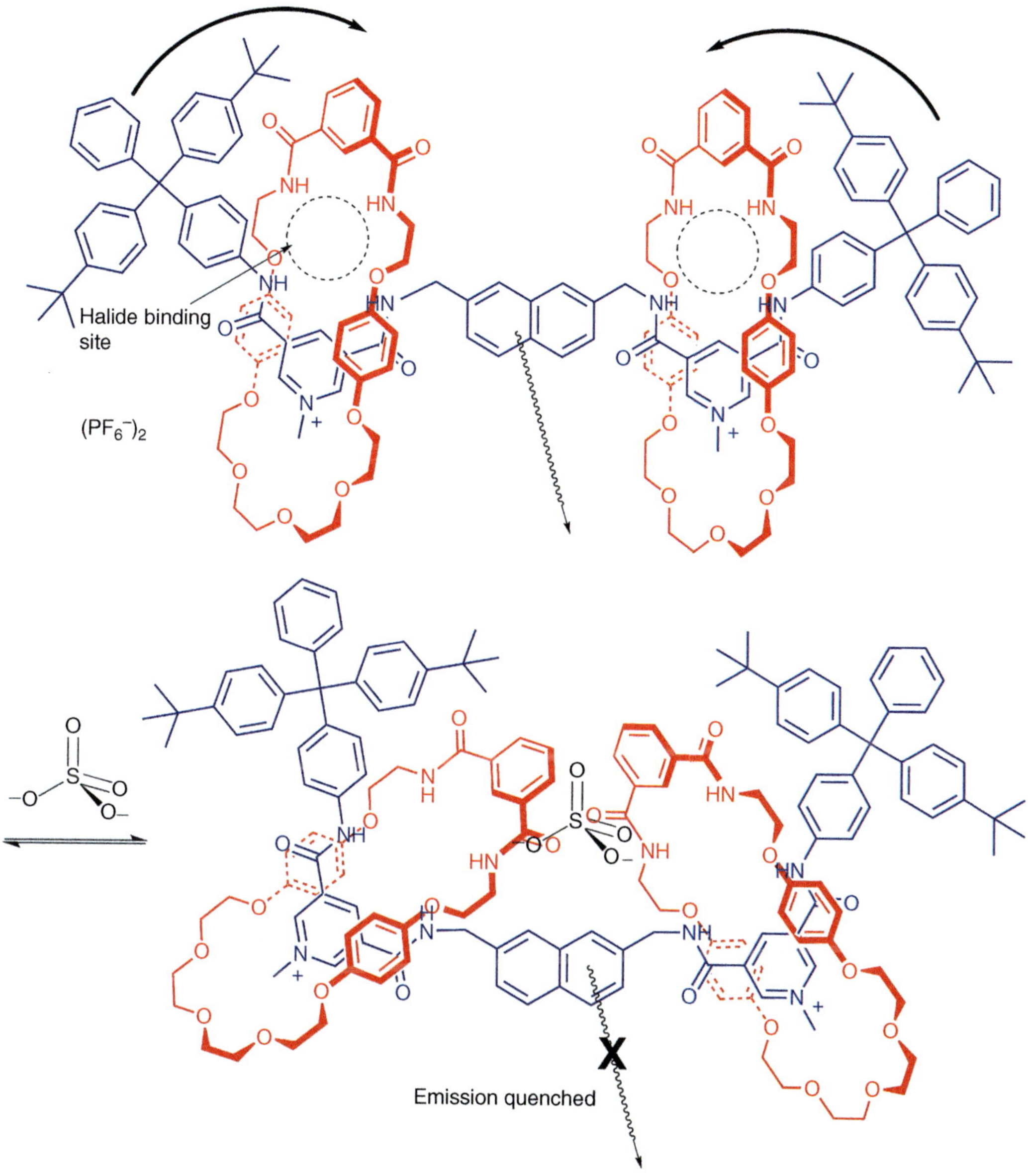

Figure 3.16 *Sulfate-induced conformational folding and fluorescence quenching in a [3]rotaxane host*

surfaces is required. The construction of two-dimensional self-assembled monolayers (SAMs) offers the exciting possibility of generating robust, reusable anion-responsive surface materials, with the capability of amplifying the molecular-level properties of the system to a macroscopic scale. In Section 1 we described a prototype anion-sensory device that consisted of SAMs of a 1,1-bis(alkyl-*N*-amido) disulfide functionalized ferrocene derivative capable of selectively sensing dihydrogen phosphate via substantial surface-amplified cathodic perturbation of the metallocene redox couple.

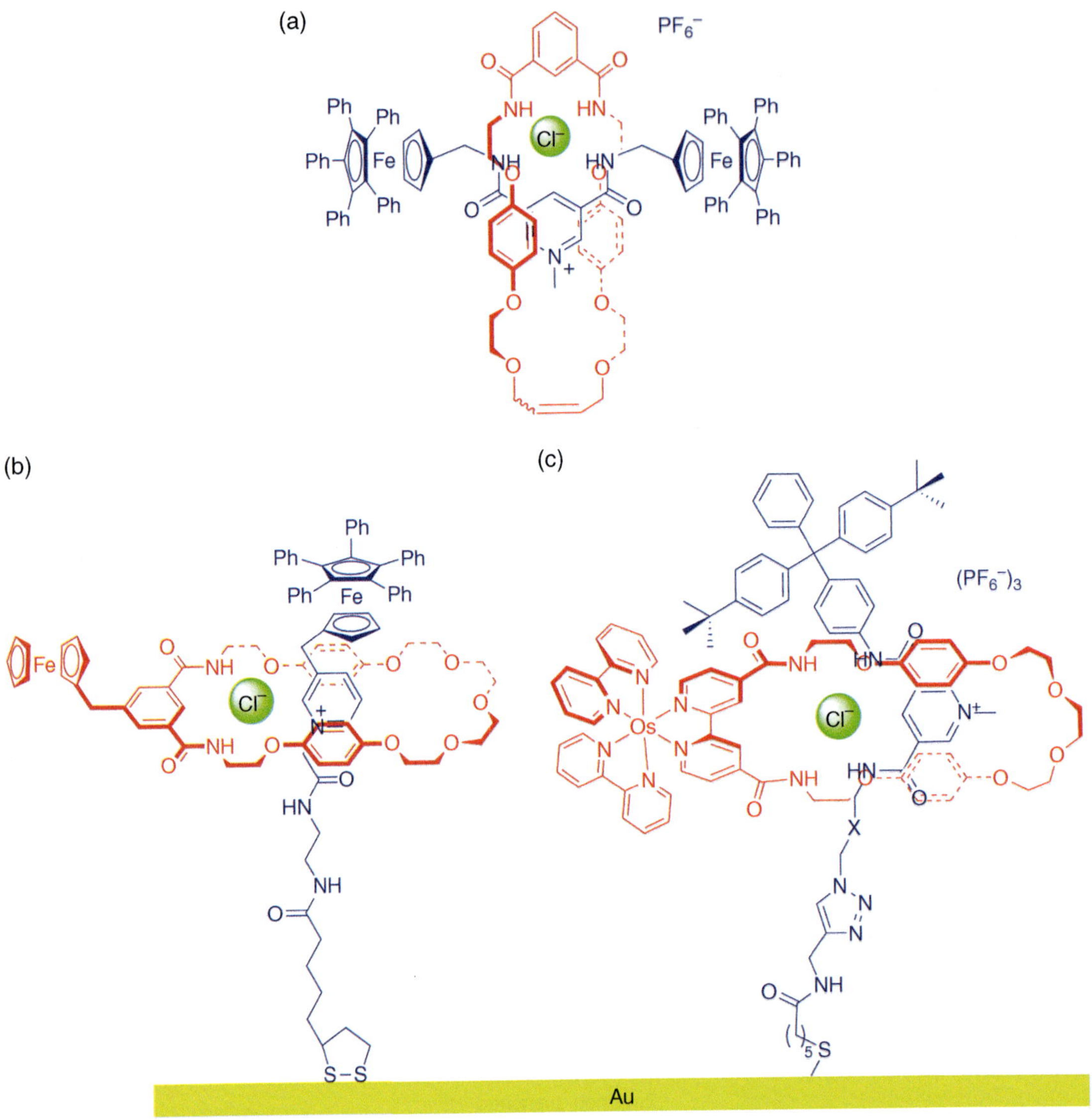

Figure 3.17 Solution-based and surface-appended interlocked electrochemical anion sensors

In 2007 we reported the first example of an anion-responsive interlocked rotaxane SAM, in which a chloride-templated pseudorotaxane assembly was tethered to a gold surface (Figure 3.17b) [56]. The so-formed surface-attached rotaxane consisted of a pentaphenylferrocene-stoppered axle, and a ferrocene-appended macrocycle component. Removal of the chloride anion template was achieved by washing the ad-layer with aqueous ammonium hexafluorophosphate solution, to reveal the chloride-responsive SAM. Pleasingly, a selective voltammetric response to the presence of Cl^- in a 0.1 M NH_4PF_6/CH_3CN electrolyte solution was observed for the macrocycle's ferrocene motif, even in the presence of a 100-fold excess of oxo-anions. Interestingly, the free macrocycle in solution displayed a selective response for oxo-anions, such as dihydrogenphosphate – the reverse of the rotaxane selectivity observed on the surface – highlighting the crucial

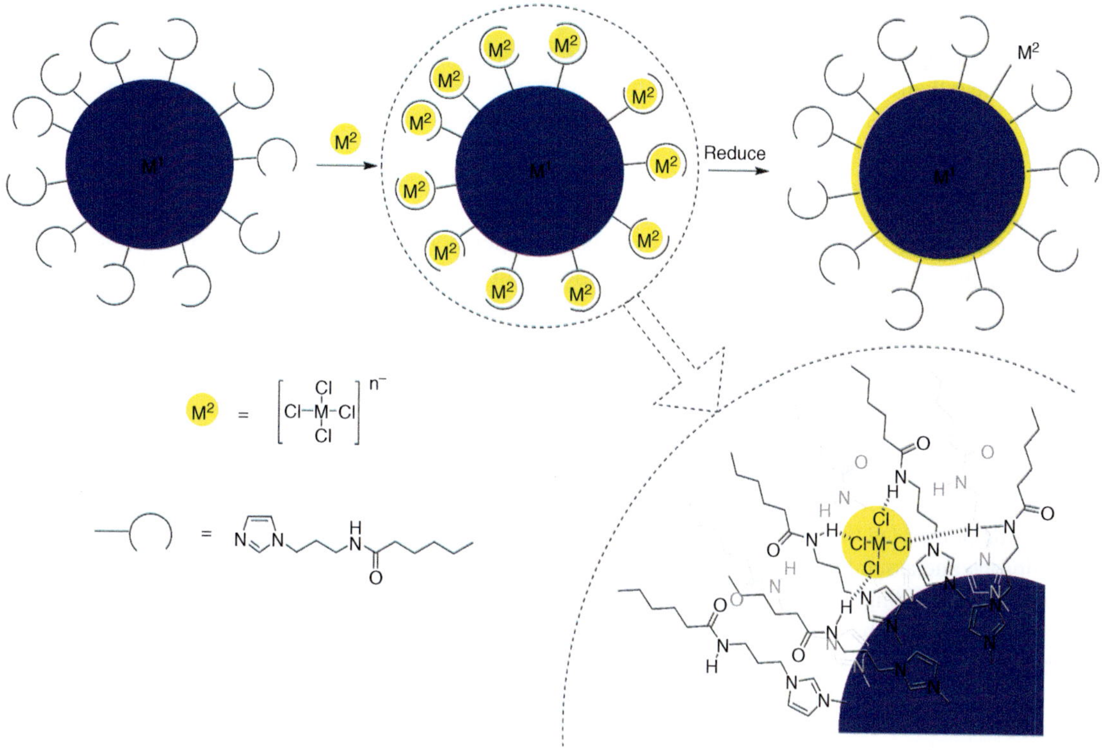

Figure 3.18 Core@shell bimetallic nanoparticle synthesis via anion coordination

nature of the interlocked SAM for achieving selective chloride sensing. We have also reported a redox-active surface-bound catenane [54], and very recently a series of rotaxane hosts that contain an osmium(II) bipyridyl-derived macrocycle, which sense anions via both luminescence and electrochemistry in solution and on a surface (Figure 3.17c) [57].

In a new research direction that further exploits anion templation at surfaces, we have introduced a novel strategy for the preparation of structured bimetallic core@shell nanoparticles for catalytic applications [58]. Using ligand-based amide–chlorometallate anion hydrogen bonding, the controlled deposition of gold, palladium, or platinum can be effected selectively on the surface of a different metal (Figure 3.18). In the case of the Au/Pd nanoparticles, the mutual electronic influence of these metals and their structural relationship serves to dramatically modify the catalytic properties of the resultant core@shell nanoparticles. For example, the selective reduction of 2-chloronitrobenzene to 2-chloroaniline can be vastly improved with respect to the undesirable side-product aniline using Au@Pd nanoparticles.

3.4 Halogen-bonding anion recognition

Halogen bonding (XB) is the non-covalent bonding interaction between halogen atoms which function as electrophilic centers (Lewis acids) and neutral or anionic Lewis bases. The Lewis-acidic electropositive region, which is known as the sigma-hole, is greatest in the heavier, more polarizable halogens and is enhanced by a covalently attached electron-withdrawing group. This region of electropositive electron density is located at the pole of the polarized halogen atom; the corollary of which is that the interaction of this region with a Lewis base is strongest when the two components are linearly deposed, and thus XB interactions are more stringently linear than the hydrogen bond.

Of the many non-covalent interactions commonly utilized in solution phase supramolecular assemblies, halogen bonding is arguably the least exploited, which is surprising given its potentially powerful analogy to ubiquitous hydrogen bonding (HB), and its more stringent linearity. In a significant step forward to highlighting the potential importance of halogen bonding in anion supramolecular chemistry, in recent years we have demonstrated the first examples of solution phase halogen bonding being exploited to recognize anions in aqueous media, and to control and facilitate the anion-templated assembly of interpenetrated and rotaxane and catenane interlocked structures.

Our first foray into this field was in 2010 when we reported the seminal use of halogen bonding as an intermolecular interaction for directing the formation of solution-phase interlocked supramolecular assemblies, through a halogen-bond-mediated anion-templated pseudorotaxane formation [59]. A halogen bonding bromo-imidazolium threading component was shown to interpenetrate within an isophthalamide macrocycle by chloride anion templation, and, notably, the stability of the interpenetrated assembly is enhanced for the halogen-bonded pseudorotaxane as compared to hydrogen-bonded pseudorotaxane analogues (Figure 3.19a).

We later demonstrated that iodo-triazolium threading components could also be employed in a similar way for pseudorotaxane assembly, which led directly to the synthesis of the first interlocked structure assembled through halogen bonding, in the form of an iodo-triazolium axle-containing [2]rotaxane (Figure 3.19b) [60]. The incorporation of the halogen atom into the rotaxane host cavity dramatically enhances the anion-recognition capabilities of the interlocked receptor, which exhibits iodide selectivity in a competitive organic–aqueous solvent medium. Furthermore, the recognition capability of the halogen-bonding rotaxane is significantly enhanced compared to the analogous proto-triazolium axle-containing hydrogen bonding analogue. We have further expanded the use of the halogen-bonding interaction in the template-directed synthesis of interlocked structures, through the synthesis of a fluorescent all-halogen-bonding catenane capable of halide sensing (Figure 3.20a) [61], and a catenane assembled by a single halogen bond formed between an 4-iodopyridinium (XB donor) in a threading component and a pyridine (XB acceptor) in the macrocycle (Figure 3.20b) [62].

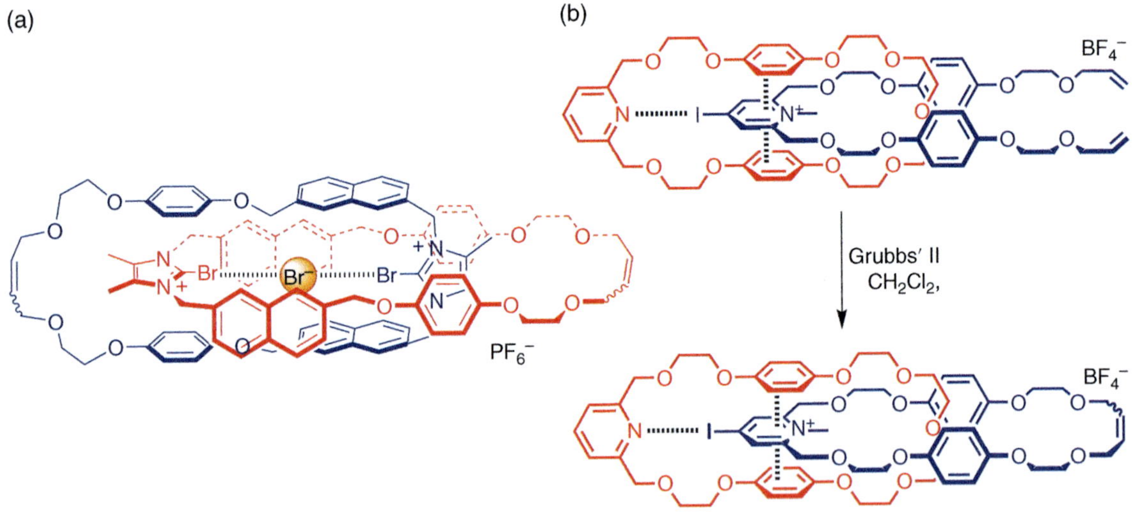

Figure 3.19 Halogen bonding catenanes: (a) A fluorescent XB catenane capable of halide anion sensing; and (b) XB-templated catenane synthesis

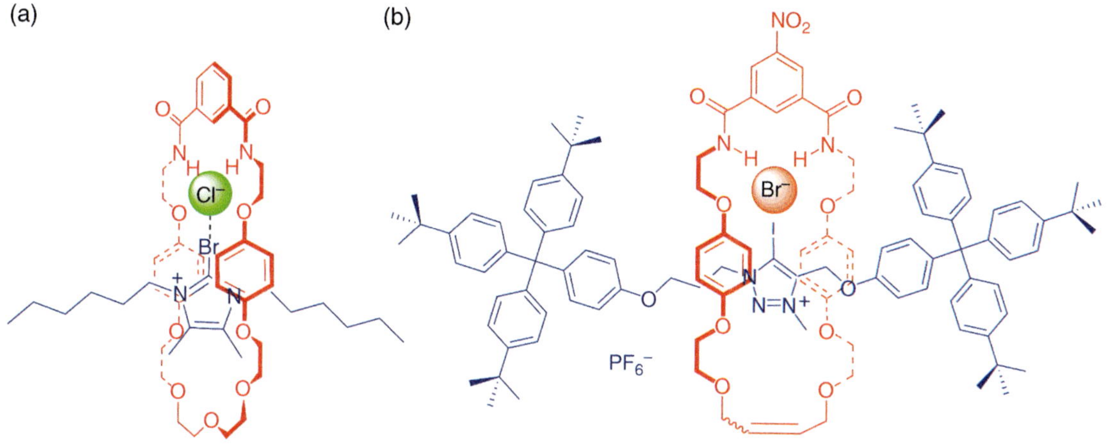

Figure 3.20 *Halogen bonding pseudorotaxane and rotaxane prepared through anion templation*

In a series of acyclic, macrocyclic and interlocked anion hosts we have demonstrated, for the first time, the superiority of halogen bonding over hydrogen bonding for anion recognition in highly competitive, protic aqueous solvent mixtures. Our first system capable of anion binding in organic-aqueous solvents using halogen bonding was a bidentate halogen bonding bromoimidazoliophane (Figure 3.21a) [63]. The integration of the bromine atoms into the macrocyclic anion receptor framework at the expense of hydrogen bonding dramatically influenced the receptor's anion-recognition behavior from both electronic and geometric considerations. The bromoimidazoliophane was revealed to bind bromide selectively by two near-linear halogen bonds above the plane of the macrocycle in CD_3OD–10% D_2O solution. Related macrocyclic host systems incorporating fluorescent naphthalene motifs [64], and the first all-halogen bonding rotaxane incorporating four convergent halogen bond donors (Figure 3.21b) [65], have also been prepared, which bind and sense bromide and iodide selectively in highly competitive organic-aqueous solvent mixtures containing up to 50% water.

Furthermore, we have exploited the contrasting anion-recognition properties of halogen and hydrogen bonding motifs to demonstrate the first example of a molecular shuttle that utilizes XB–anion interactions to facilitate molecular motion in a bistable [2] rotaxane [66]. Upon iodide recognition by the XB-donor iodotriazolium motif, the macrocycle component shuttles from the hydrogen-bonding station to the halogen-bonding station (Figure 3.21c). In contrast, chloride recognition reinforces the macrocycle to reside over the hydrogen-bonding station.

Our investigations into the anion-recognition capabilities of halogen-bonding receptors suggested that, overall, in polar, protic organic solvents the strength of the anion–receptor complex is strongest in the halogen-bonding cases compared to the hydrogen-bonding analogues. This led us to propose that XB may be an effective intermolecular interaction for anion recognition in pure water – a field, as discussed previously, that is particularly challenging.

In 2014, in a groundbreaking development, we reported the seminal exploitation and quantification of XB in water, in the context of anion recognition [67]. Water-soluble acyclic and rotaxane hosts were prepared using mono-functionalized permethylated β-cyclodextrin stopper derivatives, and the incorporation of XB-donors into such receptors resulted in a dramatic enhancement of halide anion binding in water compared to the equivalent HB analogues, with up to two orders of magnitude enhancement observed in the interlocked XB rotaxane host system (Figure 3.22). Furthermore, this XB-mediated augmentation of anion association was also observed in acyclic receptors, resulting in remarkable binding affinity despite the low charge and simplicity of the receptor. Strong iodide binding by the XB-rotaxane was demonstrated to be driven exclusively

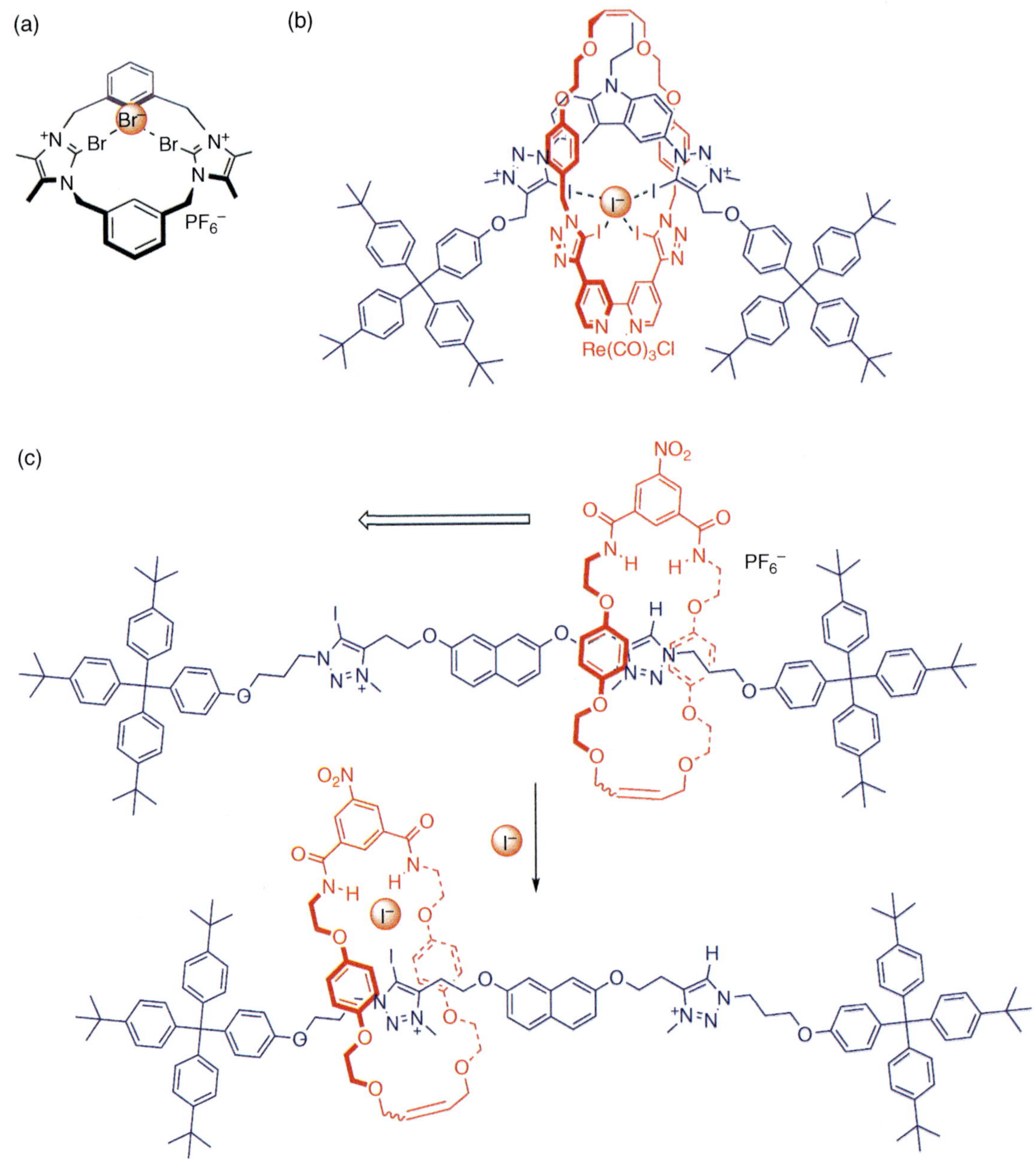

Figure 3.21 *All halogen-bonding macrocyclic (a) and rotaxane-based (b) anion receptors capable of halide binding in organic–aqueous solvent mixtures; and (c) a two-station halogen-bonding and hydrogen-bonding molecular shuttle that is responsive to iodide*

by favorable enthalpic contributions arising from the XB interactions, whereas weaker association with the HB rotaxanes is entropically driven. In collaboration with Pierre Kennepohl, we very recently investigated the origin of these remarkable halogen bond strengths using Cl and Br K-edge X-ray Absorption Near Edge Spectroscopy to probe the charge transfer contribution to the halogen bond in a series of iodo-triazole and

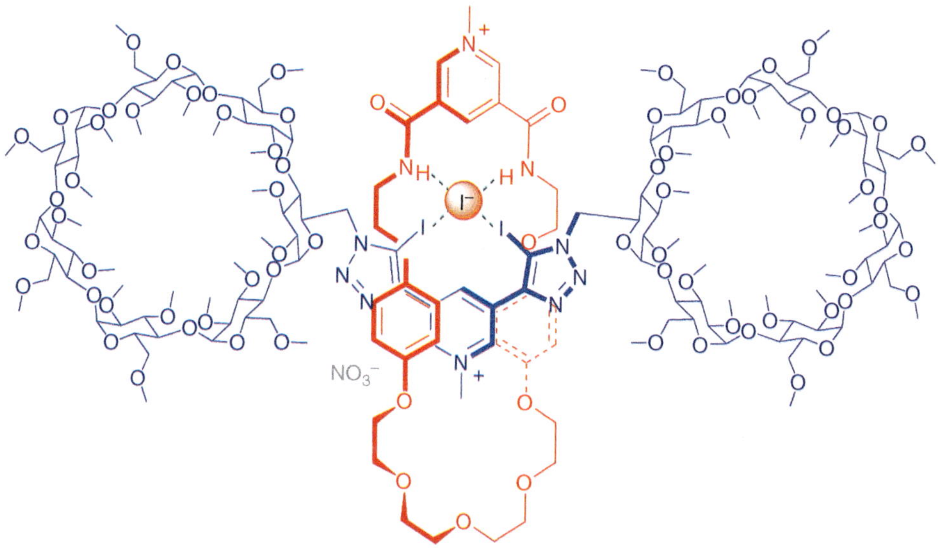

Figure 3.22 *A water-soluble halogen bonding rotaxane capable of halide anion binding in water*

iodo-triazolium derived acyclic and catenane-based receptors [68]. This technique can be used to quantify the charge transfer from the halide to its bonding partner ($\sigma_{A*X\leftarrow X^-} \leftarrow X_{1s}$), and thus provide a direct measure of the degree of covalency in the halogen bond. Notably, these investigations reveal that the degree of covalency is similar to that which is observed in transition metal coordinate bonds.

Taken together, these results highlight the superiority of XB as an intermolecular interaction for anion binding in water, and exemplify the exciting potential of XB in other applications requiring molecular assembly in aqueous environments that have been traditionally dominated by the ubiquitous hydrogen bond, including self-assembly, green-chemistry, catalysis, and drug discovery.

3.5 Ion-pair recognition

The design and synthesis of heteroditopic receptors for the recognition of ion-pairs has been an emerging area of supramolecular chemistry in recent years [69]. This has been stimulated by the observation that binding affinities of ions within monotopic receptors can be heavily dependent on the nature of the counter ion, and the possibility of exploiting cooperative interactions between the individual cation and anion to achieve superior levels of recognition. Ion-pair recognition also has important applications in the areas of salt extraction, membrane transport, sensors, and salt-solubilization agents.

We first entered this field in 1995, by designing calix[4]arene derivatives capable of simultaneously binding both cations and anions in solution [70]. For instance, we reported a calix[4]arene receptor functionalized with an upper-rim bis(benzo[15]crown-5 ether) amide which – depending on the nature of the ion-pair – displays both cooperative and anti-cooperative binding [71]. In the presence of one equivalent of potassium, binding of Cl^-, BzO^-, and $H_2PO_4^-$ is enhanced by an order of magnitude in acetonitrile/DMSO mixtures, due to the formation of the bis-crown ether:potassium 1:1 stoichiometric sandwich complex which preorganizes the amide groups of the receptor, and favorable electrostatic interactions between the metal cation and anion guests

(Figure 3.23a). In contrast, binding of two (smaller) sodium cations within the individual crown ether motifs led to anti-cooperative behavior, resulting from mutual metal cation repulsion which prevents the formation of a favorable conformation for anion binding (Figure 3.23b).

Having previously shown that ferrocene and transition metal bipyridyl-based amide systems can bind anions through electrostatics and hydrogen bonding interactions, we incorporated such motifs within calix[4]arene and crown ether heteroditopic ion-pair receptors, with the aim of both enhancing the recognition capability and sensing the ion-pair binding event through either an electrochemical or luminescence response. For instance, the ferrocene-incorporated calix[4]arene shown in Figure 3.24a exhibits enhanced iodide binding in acetonitrile solution in the presence of alkali metal cations, which bind in a 2:1 stoichiometry within the

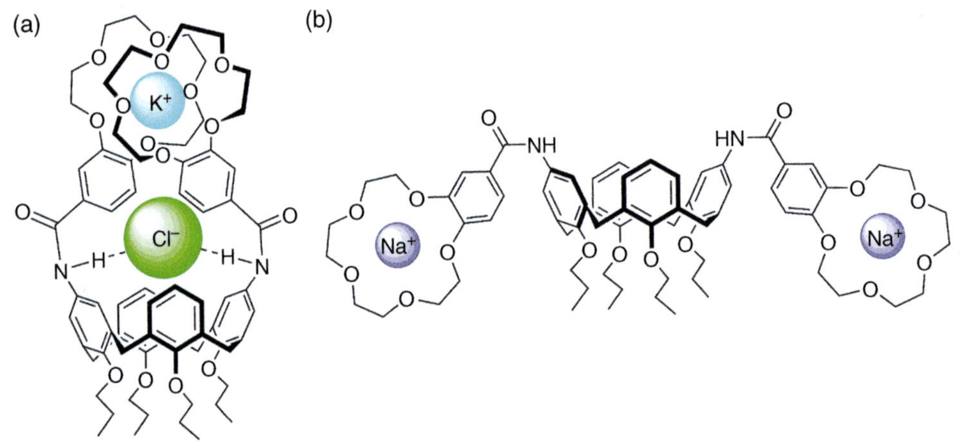

Figure 3.23 Calix[4]arene-based heteroditopic receptor. (a) Cooperative binding of KCl; (b) anti-cooperative binding in the presence of sodium cations

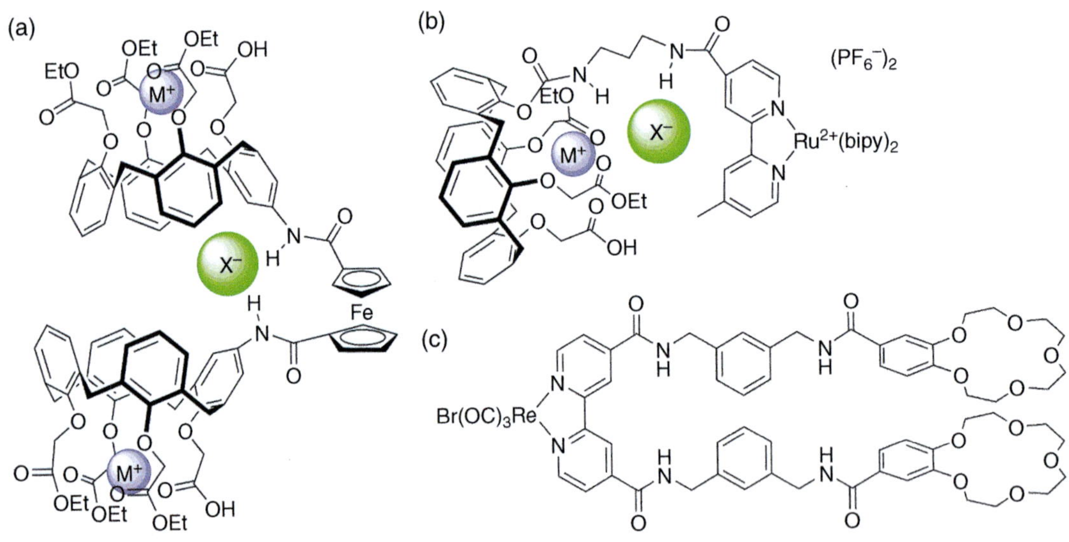

Figure 3.24 Transition metal-based heteroditopic ion-pair receptors

lower rim of the calix [72]. This enhancement is also manifested in the electrochemistry, such that in the presence of lithium cations the cathodic shift of the ferrocene/ferrocenium redox potential in response to bromide binding is significantly amplified (from 55 mV to 130 mV). Similarly, both the analogous Ru(bipy)$_2$-functionalized calix[4]arene (Figure 3.24b) [73], and the related crown ether functionalized systems (Figure 3.24c) [74, 75] display cooperative alkali metal cation–halide ion-pair binding. In the latter case, this Re(I)bipyridyl appended heteroditopic receptor cooperatively binds KCl and KOAc salts, with concomitant modulation of the Re(I)-centered MLCT emission.

In 2006 we reported the first contact ion-pair heteroditopic receptors which require both the anion and the cation guest to be present in order for ion-pair recognition to occur [76, 77]. The receptor shown in Figure 3.25a does not bind either ammonium cations or chloride anions alone, but binds the contact ion-pair extremely strongly ($K_a > 10^4\,\mathrm{M}^{-1}$ in 98:2 CD_3CN/D_2O). This was at the time an unprecedented example of Boolean AND logic in an ion-pair recognition process. The lack of "free" anion or cation recognition was ascribed to the presence of an intramolecular hydrogen bond between the quinone oxygen of the calix[4] diquinone and the amide (as observed in the solid state crystal structure), such that only the binding of an ion-pair can compensate for the energetic cost of breaking this interaction. We later incorporated triazole motifs within the calix[4]arene-based ion-pair receptors (Figure 3.25a, b), which are capable of binding simultaneously to the metal (through the nitrogen atom) and the anion (through a C–H hydrogen bond) [78]. In this case, NaCl is bound in a separated ion-pair fashion. Very recently, we reported a [2]rotaxane heteroditopic ion-pair receptor, in which alkali metal cation–halide ion-pairs are bound in a cooperative axle-separated fashion, as depicted in Figure 3.25c [79].

As mentioned previously, the possibility of binding ion-pairs offers practical applications in a range of areas. One such function of heteroditopic receptors is as salt-extraction agents, and we demonstrated in 1999 that a tren-based ion-pair receptor (Figure 3.26a) could be used to bind and extract sodium pertechnetate [80]. Binding studies in $CDCl_3$ solution revealed cooperative binding of chloride, iodide, and perrhenate (ReO_4^-) in the presence of one equivalent of sodium cations, with up to 20-fold enhancement of anion binding in the

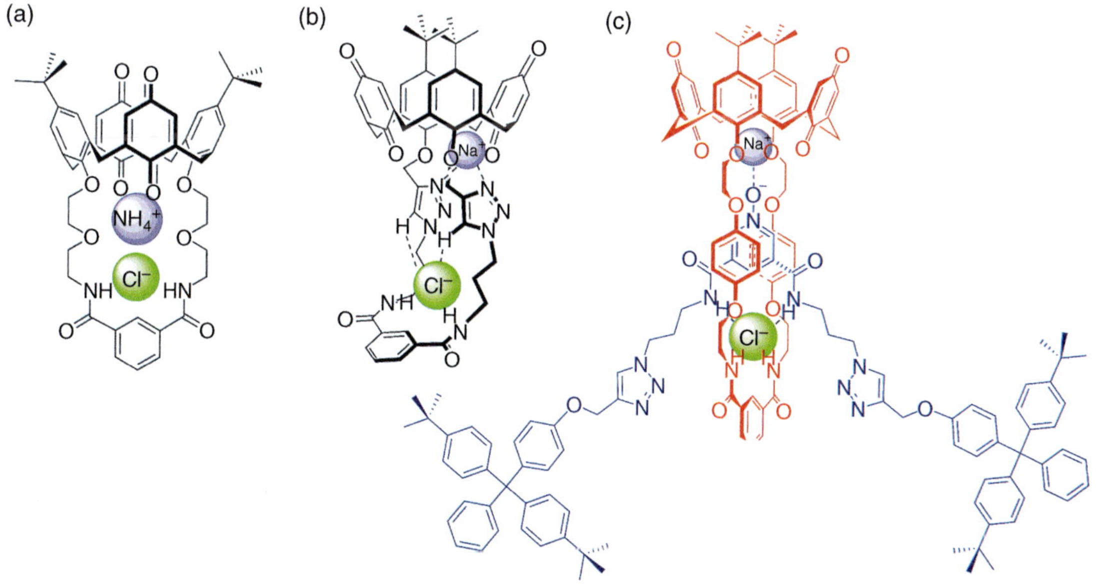

Figure 3.25 *(a) Cooperative AND contact ion-pair receptor. Separated ion-pair recognition in (b) a triazole-containing macrocycle, and (c) a neutral N-oxide [2]rotaxane*

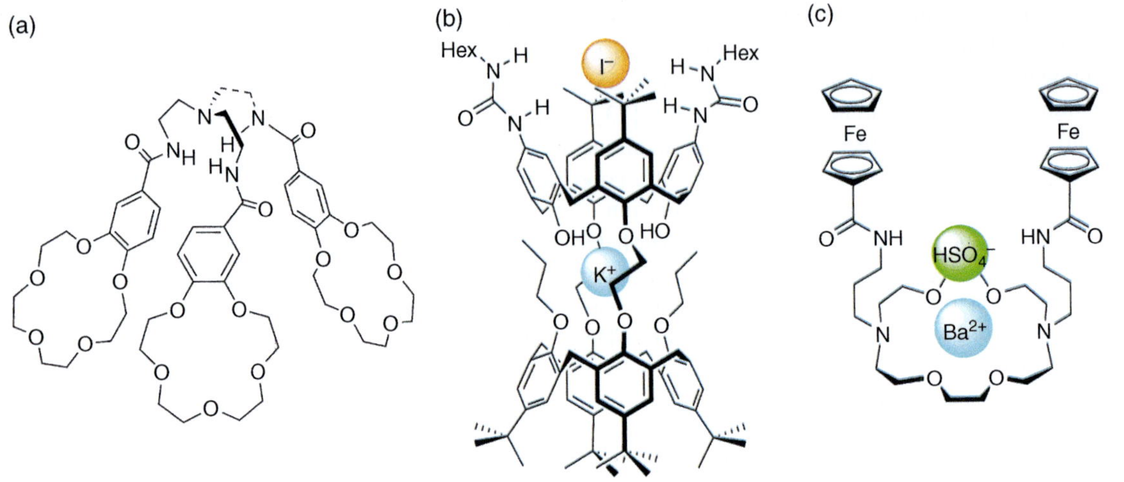

Figure 3.26 Applications of ion-pair receptors: (a) Pertechnetate extraction; (b) salt-solubilization; and (c) ion-pair sensing

case of perrhenate. The perrhenate anion is structurally similar to pertechnetate (TcO_4^-), which is a toxic radioactive by-product found in nuclear waste. Subsequent pertechnetate extraction experiments demonstrated that the receptor was able to selectively and efficiently extract $NaTcO_4$ from an aqueous phase that closely simulates the conditions found in nuclear waste streams.

Salt-solubilization is also an important application of ion-pair receptors, and we reported in 2003 a calix[4] semitube receptor (Figure 3.26b) which showed remarkable affinity for potassium cations, and cooperatively binds alkali-metal halide ion-pairs in chloroform solution [81]. Notably, the heteroditopic receptor was able to solubilize such salts (which were added as a solid to a solution of the host in $CDCl_3$), with up to 95% solubilization observed in the case of KI.

Finally, the development of ion-pair sensors is of importance for the detection of environmentally and biologically important ion-pair species, and can be achieved by incorporating optical or electrochemical reporter groups into the heteroditopic host architecture. For instance we have used both copper(II) dithiocarbamate (dtc) [82] (see Section 3.6) and ferrocene moieties as redox probes, such as in the ferrocene-based ion-pair receptor shown in Figure 3.26c, which exhibits anodic shifts in the ferrocene/ferrocenium redox couple (in 10:1 CH_3CN/CH_2Cl_2) in the presence of barium cations, and contrasting cathodic perturbations in the presence of anions such as chloride or hydrogen sulfate [6]. Importantly, the simultaneous binding of the ion-pair was also evident from electrochemical studies, corroborating the results obtained through ^1H NMR binding titration experiments.

3.6 Metal-directed self-assembly

With the potential applications in host–guest chemistry, catalysis, sensing, and drug delivery, the metal-directed self-assembly of macrocycles, three-dimensional polyhedral cage-like structures and spectacular supramolecular architectures of nanoscale dimensions continues to be an intense area of research [83]. Our contribution to the field has focused on exploiting for the first time the dithiocarbamate (dtc) ligand in the metal-directed assembly of a variety of macrocycles, cryptands, catenanes, and nano-sized polymetallic architectures. Taking into account its relative ease of synthesis, well-documented and varied coordination chemistry that includes useful electrochemical and optical properties with redox-active transition metals, it was somewhat surprising the dithiocarbamate anion had not been utilized for metal-directed self-assembly before.

A range of dinuclear macrocyclic and macrobicyclic transition metal complexes of varying dimensions were constructed using appropriate bis-substituted dtc ligands containing various spacer groups (Figure 3.27) [82, 84–90]. The stereochemical preference of the transition metal dictates the cyclic structure that is formed. Whereas square planar directing metal centers such as Cu(II), Ni(II), Au(III) produce a macrocyclic dinuclear complex, octahedral stereochemical directing metals Fe(III) and Co(III) afford macrobicyclic "cryptand-like" architectures. It is noteworthy that mild oxidation of the dinuclear macrocyclic Ni(II) complexes leads to the formation of the dinuclear cryptand Ni(IV) structure.

Dependent upon the nature of the spacer group, such assemblies were demonstrated to be capable of binding and electrochemically sensing cation and anion guest species. The redox-active polyether containing cryptands (Figure 3.28a) were shown initially to complex Group 1 metal cations by electrospray mass spectrometry [87]. Significant anodic shifts of the Co(IV/III) dtc redox couple of the bis-cobalt(III) cryptands were observed upon Group 1 metal cation binding. Notably, the magnitude of this shift correlated with metal cation:cryptand size complementarity between the host and the guest. For example, with the large tris(ethylene

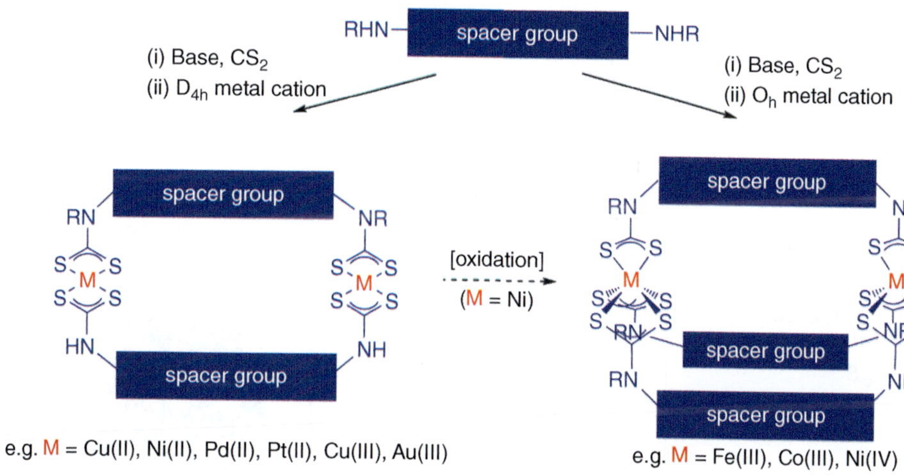

Figure 3.27 Formation of dinuclear macrocyclic and bimacrocyclic complexes

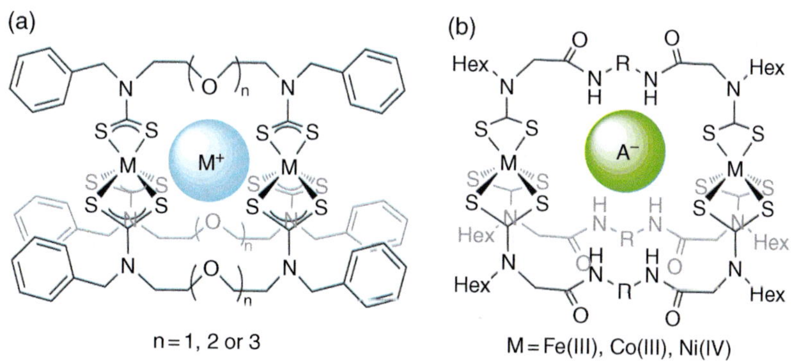

Figure 3.28 (a) Polyether metal dtc cryptands; and (b) amide metal dtc cryptands

glycol)-spaced cryptand, the addition of cesium cations in 1:1 CH_2Cl_2:CH_3CN solution caused the largest perturbation (45 mV), whereas potassium and sodium cations induced the largest anodic shifts with the smaller cryptands.

The integration of amides, Lewis acid centers, or positively charged groups into the structural framework of dtc-based macrocycles and cryptands produced host systems designed to recognize and sense anionic guest species (Figure 3.28b) [82, 91–94]. Chloride, nitrate and dihydrogen phosphate addition to electrochemical solutions of the bimetallic amide-containing macrocycles and cryptands caused cathodic perturbations of the respective Co(IV/III)/dtc and Ni(IV/III)/dtc redox couples [82].

In a serendipitous discovery, a novel mixed-valence tetranuclear Cu(II)/Cu(III) dithiocarbamate [2]catenane was prepared in near-quantitative yield by the partial oxidation of a naphthalene-spaced dinuclear Cu(II)/dtc macrocycle (Figure 3.29) [95]. Formation of the [2]catenane architecture was confirmed by ESIMS, magnetic susceptibility measurements, elemental analysis and X-ray crystallography. Examination of the bond lengths within the X-ray crystal structure provided evidence that a mixed-valence complex was produced, with alternate copper(II) and copper(III) metal centers within the [2]catenane.

The lability of the copper(II)–dtc coordinate bonds together with favorable charge-transfer stabilization effects were proposed to be responsible for the near-quantitative yield of the interlocked structure. This hypothesis was substantiated by the preparation of a heteropolymetallic catenane architecture. Combining a kinetically inert dinuclear gold(III) dithiocarbamate naphthalene spaced macrocycle with the equivalent labile copper(II) complex instantaneously produced the novel heteropolymetallic interlocked structure, which was isolated in high yield (Figure 3.30) [96].

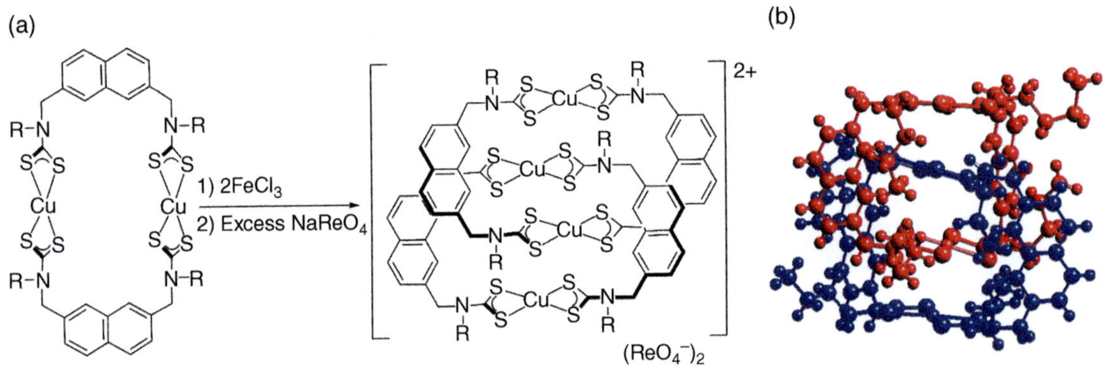

Figure 3.29 (a) Mixed valence Cu(II)/(III) catenane formation; and (b) X-ray crystal structure

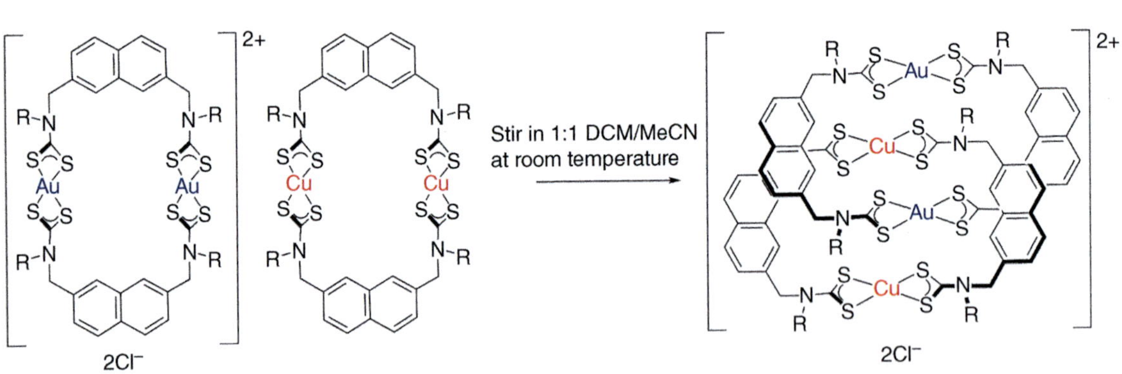

Figure 3.30 Heteropolymetallic Au(III)/Cu(II) catenane

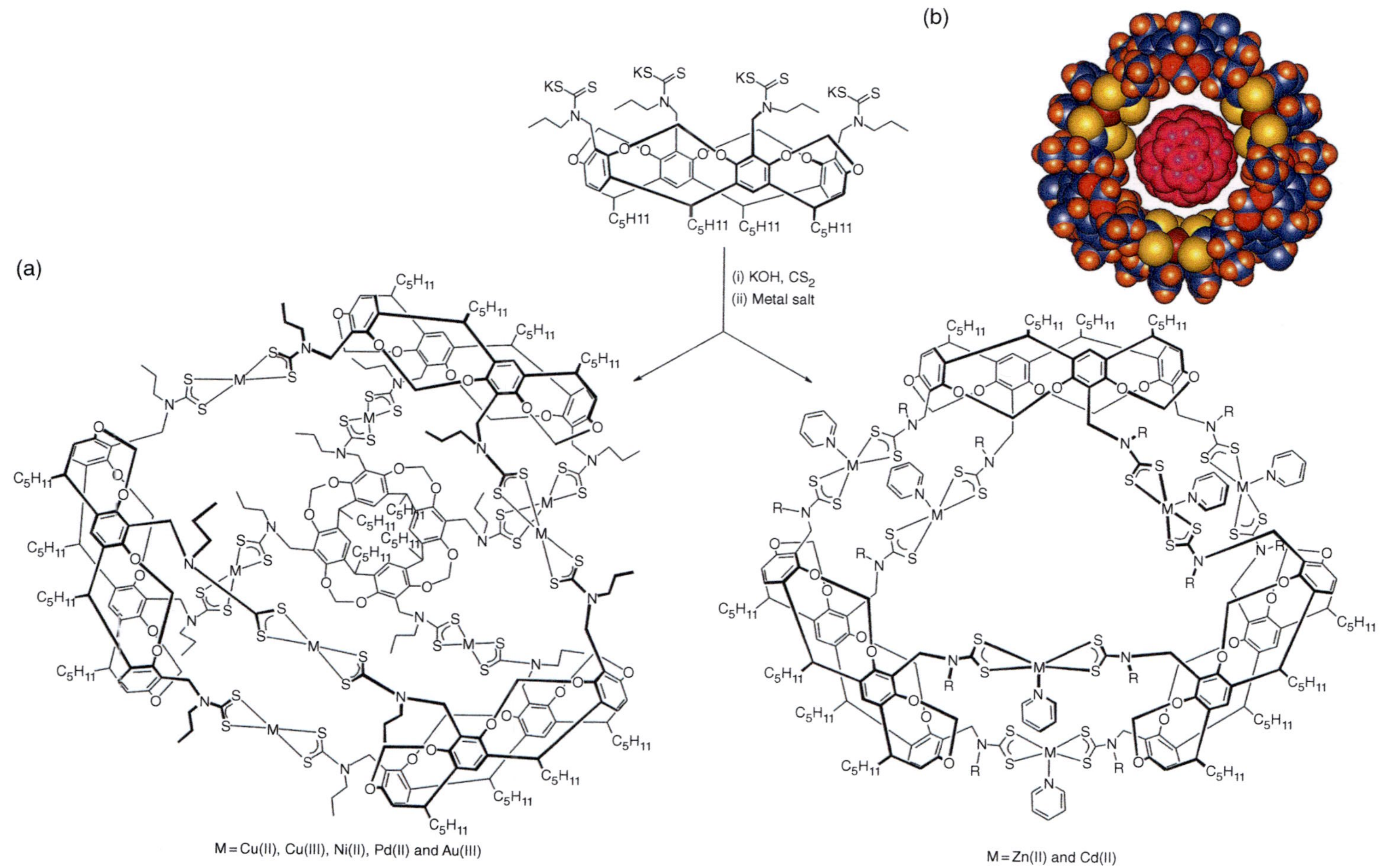

Figure 3.31 *Preparation of nano-sized resorcinarene-based assemblies. Inset – structure of C_{60} complex of the Zn(II) terimeric assembly*

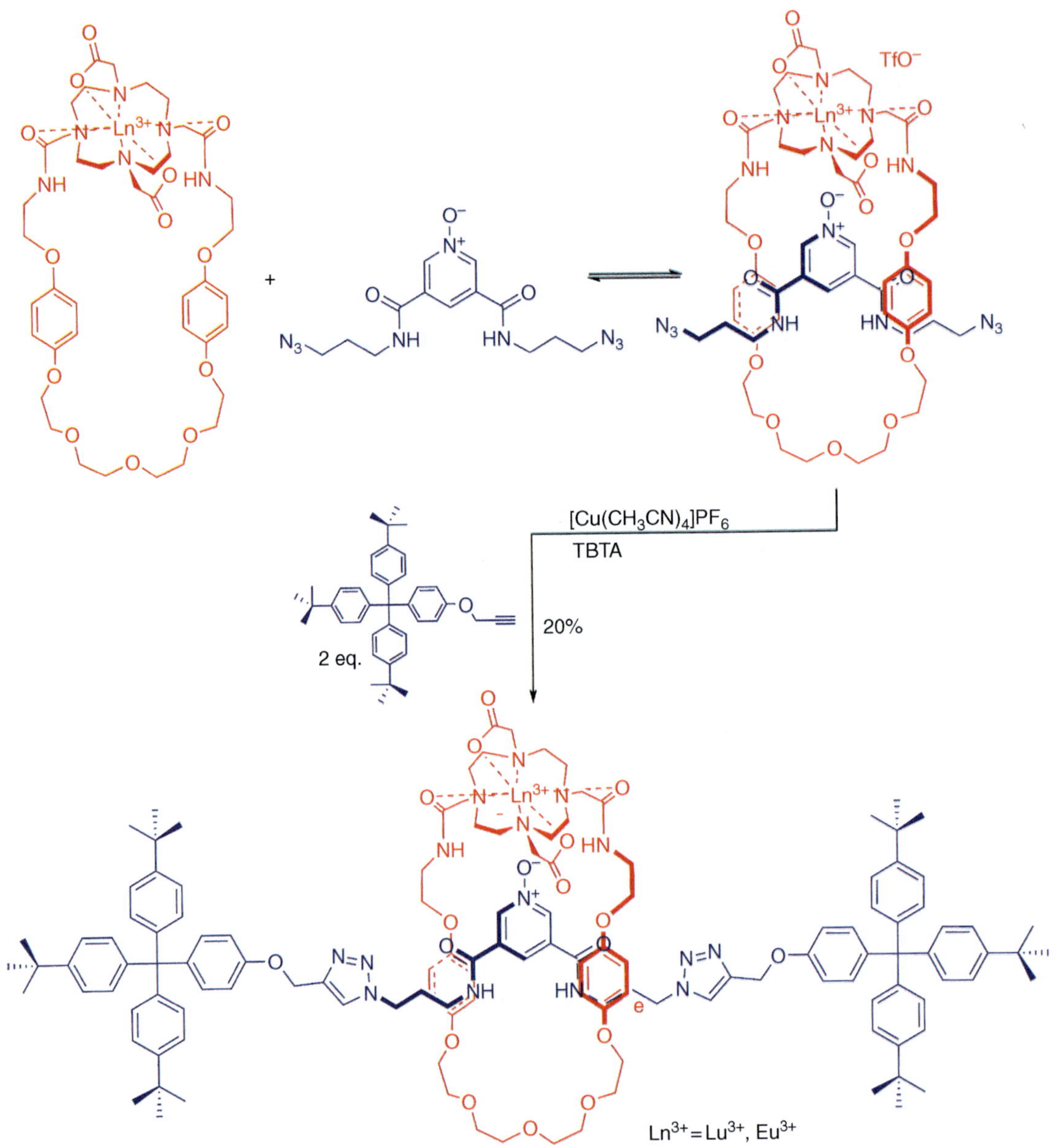

Figure 3.32 Lanthanide cation template rotaxane synthesis

Nano-sized trimeric and tetrameric polymetallic architectures were assembled using tetrakis dithiocarbamate-functionalized resorcinarenes [97–99]. The nuclearity of these assemblies was found to be dependent upon the geometry of the metal center. For example, with square planar metal centers such as Cu(II), Cu(III), Ni(II), Pd(II) and Au(III), an octanuclear tetrameric structure was the thermodynamic product (Figure 3.31a). In contrast, zinc(II) and cadmium(II) ions directed the assembly of a hexanuclear "molecular loop" (Figure 3.31b). Both classes of resorcinarene structures contain large internal cavities which were shown to be of suitable nanodimensions for binding C_{60} and C_{70} strongly through size and electronic complementarity between the electron-deficient fullerenes and the electron-rich cavity of the host complexes [98, 99].

Although transition metal cation, and to a lesser extent alkali and alkaline earth metal cation, templates had been imaginatively exploited in the synthesis of rotaxanes and catenanes, the utilization of lanthanides as potential templating reagents for interlocked molecular framework assembly was unprecedented. The first lanthanide cation template-directed assembly of an interlocked structure was achieved by us through the preparation of lutetium and europium templated [2]rotaxanes [49]. Adapting our recently reported sodium and barium cation templated synthesis of a rotaxane using the pyridine N-oxide motif [100], the synthetic strategy undertaken for the preparation of the target lanthanide-[2]rotaxane is shown in Figure 3.32. A kinetically stable lanthanide complexed DOTA cyclen derivative integrated into a macrocyclic structural framework, formed initially a pseudorotaxane assembly with an appropriately functionalized pyridine N-oxide threading component, where the pyridine N-oxide ligand serves to satisfy the lanthanide cation's coordination sphere. Subsequent stoppering of the interpenetrated assembly produced the novel lanthanide containing interlocked structure which was characterized by high-resolution electrospray mass spectrometry, ^{1}H NMR and luminescence spectroscopy.

3.7 Conclusions

As this chapter illustrates, Beer's research career to date has covered many areas of macrocyclic and supramolecular chemistry. The group's future research directions will continue to focus on the manipulation of the unique topological cavities of mechanically bonded molecules for anion and ion-pair recognition applications, together with the exploitation of the hugely undeveloped field of aqueous halogen-bonding molecular recognition, and the supramolecular surface fabrication of functional materials for sensing, imaging and catalysis.

3.8 Acknowledgements

P.D.B acknowledges the great honor in being the 2015 winner of the Izatt–Christensen award. P.D.B warmly thanks the many graduates, postdoctoral co-workers and collaborators who have over the years made invaluable contributions to the research undertaken in the Beer group. P.D.B gratefully acknowledges funding from the EPRSC, the European Research Council, and European Union Marie Curie, the Royal Commission for the Exhibition of 1851 and NSERC (Canada) postdoctoral fellowship funding.

References

1. Beer, P.D. (1989) Meldola Medal Lecture. Redox responsive macrocyclic receptor molecules containing transition metal redox centres, *Chemical Society Reviews*, **18**, 409–450.
2. Beer, P.D. (1992) Transition metal and organic redox-active macrocycles designed to electrochemically recognize charged and neutral guest species, In *Advances in Inorganic Chemistry*, Sykes, A.G. (Ed), Academic Press, New York.
3. Beer, P.D., Gale, P.A., Chen, G.Z. (1999) Electrochemical molecular recognition: pathways between complexation and signaling, *Journal of the Chemical Society, Dalton Transactions*, **12**, 1897–1910.
4. Beer, P.D., Blackburn, C., McAleer, J.F., Sikanyika, H. (1990) Redox-responsive crown ethers containing a conjugated link between the ferrocene moiety and a benzo crown ether, *Inorganic Chemistry*, **29**, 378–381.
5. Beer, P.D., Crowe, D.B., Ogden, M.I., Drew, M.G.B., Main, B. (1993) Ammonium redox-responsive receptors containing multiple ferrocene and quinone redox-active centres attached to di- and tri-aza crown ether macrocycles, *Journal of the Chemical Society, Dalton Transactions*, 2107–2116.

6. Beer, P.D., Chen, Z., Ogden, M.I. (1995) Voltammetric and NMR studies of a bis(ferrocenecarboxamide)-substituted diaza 18-crown-6 receptor that simultaneously complexes and electrochemically recognises both cations and anions, *Journal of the Chemical Society, Faraday Transactions*, **91**, 295–302.
7. Chen, Z., Pilgrim, A.J., Beer, P.D. (1995) Asymmetrical complexation and simultaneous electrochemical recognition of two different group 2 metal cations by a 1,1′-ferrocene-bis(methylene aza-18-crown-6) receptor, *Journal of the Chemical Society, Faraday Transactions*, **91**, 4331–4333.
8. Webber, P.R.A., Beer, P.D., Chen, G.Z., Felix, V., Drew, M.G.B. (2003) Bis(calix[4]diquinone) receptors: cesium- and rubidium-selective redox-active ionophores, *Journal of the American Chemical Society*, **125**, 5774–5785.
9. Schmitt, P., Beer, P.D., Drew, M.G.B., Sheen, P.D. (1997) Calix[4]tube: a tubular receptor with remarkable potassium ion selectivity, *Angewandte Chemie International Edition*, **36**, 1840–1842.
10. Matthews, S.E., Schmitt, P., Felix, V., Drew, M.G.B., Beer, P.D. (2002) Calix[4]tubes: a new class of potassium-selective ionophore, *Journal of the American Chemical Society*, **124**, 1341–1353.
11. Beer, P.D., Chen, Z., Grieve, A., Haggitt, J. (1994) A new bipyridinium bis benzo crown ether ligand whose redox properties are dependent upon complexed cation induced conformational switching effects, *Journal of the Chemical Society, Chemical Communications*, 2413–2414.
12. Beer, P.D., Keefe, A.D. (1989) A new approach to the coordination of anions, novel polycobalticinium macrocyclic receptor molecules, *Journal of Organometallic Chemistry*, **375**, 40–42.
13. Beer, P.D., Hesek, D., Hodacova, J., Stokes, S.E. (1992) Acyclic redox responsive anion receptors containing amide linked cobalticinium moieties. *Journal of the Chemical Society, Chemical Communications*, 270–272.
14. Beer, P.D., Chen, Z., Goulden, A.J., Graydon, A., Stokes, S.E., Wear, T. (1993) Selective electrochemical recognition of the dihydrogen phosphate anion in the presence of hydrogen sulfate and chloride ions by new neutral ferrocene anion receptors, *Journal of the Chemical Society, Chemical Communications*, 1834–1836.
15. Beer, P.D., Drew, M.G.B., Graydon, A.R., Smith, D.K., Stokes, S.E. (1995) Quantitative and structural investigations of hydrogen bonding interactions in anion binding of mono- and 1,1′-bis-substituted aryl cobaltocenium receptors, *Journal of the Chemical Society, Dalton Transactions*, 403–408.
16. Beer, P.D., Hesek, D., Kingston, J.E., Smith, D.K., Stokes, S.E., Drew, M.G.B. (1995) Anion recognition by redox-responsive ditopic bis-cobaltocenium receptor molecules including a novel calix[4]arene derivative that binds a dicarboxylate dianion, *Organometallics*, **14**, 3288–3295.
17. Beer, P.D., Drew, M.G.B., Hesek, D., Jagessar, R. (1995) Spectral and electrochemical anion sensing by a novel 5,10,15,20-tetrakis(R-substituted) porphyrin receptor ($R=C_6H_4NHC(O)C_5H_4CoC_5H_5^+PF_6^-$), *Journal of the Chemical Society, Chemical Communications*, 1187–1189.
18. Beer, P.D., Drew, M.G.B., Smith, D.K. (1997) Selective electrochemical recognition of bidentate anionic guests in competitive solvents using novel ferrocenyl thiourea and guanidinium receptors, *Journal of Organometallic Chemistry*, **543**, 259–261.
19. Beer, P.D., Cadman, J., Lloris, J.M., Martínez-Máñez, R., Padilla, M.E., Pardo, T., *et al.* (1999) Selective electrochemical recognition of sulfate over phosphate and phosphate over sulfate using polyaza ferrocene macrocyclic receptors in aqueous solution, *Journal of the Chemical Society, Dalton Transactions*, 127–134.
20. Beer, P.D., Davis, J.J., Drillsma-Milgrom, D.A., Szemes, F. (2002) Anion recognition and redox sensing amplification by self-assembled monolayers of 1,1′-bis(alkyl-N-amido)ferrocene, *Journal of the Chemical Society, Chemical Communications*, 1716–1717.
21. Sessler, J.L., Gale, P.A., Cho, W.-S. (2006) *Anion Receptor Chemistry*, RSC, Cambridge.
22. Evans, N.H., Beer, P.D. (2014) Advances in anion supramolecular chemistry: from recognition to chemical applications, *Angewandte Chemie International Edition*, **53**, 11716–11754.
23. Beer, P.D., Szemes, F., Balzani, V., Salà, C.M., Drew, M.G.B., Dent, S.W., *et al.* (1997) Anion selective recognition and sensing by novel macrocyclic transition metal receptor systems. ^1H NMR, electrochemical, and photophysical investigations, *Journal of the American Chemical Society*, **119**, 11864–11875.
24. Pflugrath, J.W., Quiocho, F.A. (1985) Sulphate sequestered in the sulphate-binding protein of Salmonella typhimurium is bound solely by hydrogen bonds, *Nature*, **314**, 257–260.
25. Luecke, H., Quiocho, F.A. (1990) High specificity of a phosphate transport protein determined by hydrogen bonds, *Nature*, **347**, 402–406.

26. Spence, G.T., Beer, P.D. (2013) Expanding the scope of the anion templated synthesis of interlocked structures, *Accounts Chemical Research*, **46**, 571–586.
27. Wisner, J.A., Beer, P.D., Drew, M.G.B. (2001) A demonstration of anion templation and selectivity in pseudorotaxane formation, *Angewandte Chemie International Edition*, **40**, 3606–3609.
28. Sambrook, M.R., Beer, P.D., Wisner, J.A., Paul, R.L., Cowley, A.R., Szemes, F., *et al.* Anion-templated assembly of pseudorotaxanes: importance of anion template, strength of ion-pair thread association, and macrocycle ring size, *Journal of the American Chemical Society*, **127**, 2292–2302.
29. Mullen, K.M., Mercurio, J., Serpell, C.J., Beer, P.D. (2009) Exploiting the 1,2,3-triazolium motif in anion-templated formation of a bromide-selective rotaxane host assembly, *Angewandte Chemie International Edition*, **48**, 4781–4784.
30. Wisner, J.A., Beer, P.D., Drew, M.G.B., Sambrook, M.R. (2002) Anion-templated rotaxane formation, *Journal of the American Chemical Society*, **124**, 12469–12476.
31. Sambrook, M.R., Beer, P.D., Wisner, J.A., Paul, R.L., Cowley, A.R. (2004) Anion-templated assembly of a [2]catenane, *Journal of the American Chemical Society*, **126**, 15364–15365.
32. Hancock, L.M., Beer, P.D. (2008) Chloride recognition in aqueous media by a rotaxane prepared via a new synthetic pathway, *Chemistry – A European Journal*, **15**, 42–44.
33. Huang, B., Santos, S.M., Felix, V., Beer, P.D. (2008) Sulfate anion-templated assembly of a [2]catenane, *Chemical Communications*, 4610–4612.
34. Langton, M.J., Duckworth, L.C., Beer, P.D. (2013) Nitrate anion templated assembly of a [2]rotaxane for selective nitrate recognition in aqueous solvent mixtures, *Chemical Communications*, **49**, 8608–8610.
35. Langton, M.J., Beer, P.D. (2014) Nitrate anion templated synthesis of a [2]catenane for nitrate recognition in organic–aqueous solvent media, *Chemical Communications*, **50**, 8124–8127.
36. Langton, M.J., Blackburn, O.A., Lang, T., Faulkner, S., Beer, P.D. (2014) Nitrite-templated synthesis of lanthanide-containing [2]rotaxanes for anion sensing, *Angewandte Chemie International Edition*, **53**, 11463–11466.
37. Evans, N.H., Serpell, C.J., Beer, P.D. (2011) A redox-active [3]rotaxane capable of binding and electrochemically sensing chloride and sulfate anions, *Chemical Communications*, **47**, 8775–8777.
38. Langton, M.J., Beer, P.D. (2012) Sulfate-selective binding and sensing of a fluorescent [3]rotaxane host system, *Chemistry – A European Journal*, **18**, 14406–14412.
39. Li, Y., Mullen, K.M., Claridge, T.D.W., Costa, P.J., Felix, V., Beer, P.D. (2009) Sulfate anion templated synthesis of a triply interlocked capsule, *Chemical Communications*, 7134.
40. Evans, N.H., Serpell, C.J., Beer, P.D. (2011) Chloride anion templated synthesis and crystal structure of a handcuff catenane, *Angewandte Chemie International Edition*, **50**, 2507–2510.
41. Evans, N.H., Beer, P.D. (2011) A janus [2]rotaxane synthesized by using an anion-templated clipping methodology, *Chemistry – A European Journal*, **17**, 10542–10546.
42. Curiel, D., Cowley, A., Beer, P.D. (2005) Indolocarbazoles: a new family of anion sensors, *Chemical Communications*, 236–238.
43. Brown, A., Mullen, K.M., Ryu, J., Chmielewski, M.J., Santos, S.M., Felix, V., *et al.* (2009) Interlocked host anion recognition by an indolocarbazole-containing [2]rotaxane, *Journal of the American Chemical Society*, **131**, 4937–4952.
44. Langton, M.J., Beer, P.D. (2014) Rotaxane and catenane host structures for sensing charged guest species, *Accounts of Chemical Research*, **47**, 1935–1949.
45. Curiel, D., Beer, P.D. (2005) Anion directed synthesis of a hydrogensulfate selective luminescent rotaxane, *Chemical Communications*, 1909–1911.
46. Hancock, L.M., Marchi, E., Ceroni, P., Beer, P.D. (2012) Anion sensing in aqueous media by photo-active transition-metal bipyridyl rotaxanes, *Chemistry – A European Journal*, **18**, 11277–11283.
47. Allain, C., Faulkner, S. (2010) Photophysical approaches to responsive optical probes, *Future Medicinal Chemistry*, **2**, 339–350.
48. Allain, C., Beer, P.D., Faulkner, S., Jones, M.W., Kenwright, A.M., Kilah, N.L., *et al.* (2013) Lanthanide appended rotaxanes respond to changing chloride concentration, *Chemical Science*, **4**, 489–493.
49. Zapata, F., Blackburn, O.A., Langton, M.J., Faulkner, S., Beer, P.D. (2013) Lanthanide cation-templated synthesis of rotaxanes, *Chemical Communications*, **49**, 8157–8159.

50. Kay, E.R., Leigh, D.A., Zerbetto, F. (2007) Synthetic molecular motors and mechanical machines, *Angewandte Chemie International Edition*, **46**, 72–191.
51. Sambrook, M.R., Curiel, D., Hayes, E.J., Beer, P.D., Pope, S.J.A., Faulkner, S. (2006) Sensitised near infrared emission from lanthanides via anion-templated assembly of d–f heteronuclear [2]pseudorotaxanes, *New Journal of Chemistry*, **30**, 1133–1136.
52. Spence, G.T., Pitak, M.B., Beer, P.D. (2012) Anion-induced shuttling of a naphthalimide triazolium rotaxane, *Chemistry – A European Journal*, **18**, 7100–7108.
53. Evans, N.H., Beer, P.D. (2010) A ferrocene functionalized rotaxane host system capable of the electrochemical recognition of chloride, *Organic and Biomolecular Chemistry*, **9**, 92–100.
54. Evans, N.H., Rahman, H., Leontiev, A.V., Greenham, N.D., Orlowski, G.A., Zeng, Q., *et al.* (2012) Solution and surface-confined chloride anion templated redox-active ferrocene catenanes, *Chemical Science*, **3**, 1080–1089.
55. Evans, N.H., Serpell, C.J., White, N.G., Beer, P.D. (2011) A 1,2,3,4,5-pentaphenylferrocene-stoppered rotaxane capable of electrochemical anion recognition, *Chemistry – A European Journal*, **17**, 12347–12354.
56. Bayly, S.R., Gray, T.M., Chmielewski, M.J., Davis, J.J., Beer, P.D. (2007) Anion templated surface assembly of a redox-active sensory rotaxane, *Chemical Communications*, 2234–2236.
57. Lehr, J., Lang, T., Blackburn, O.A., Barendt, T.A., Faulkner, S., Davis, J.J., *et al.* (2013) Anion sensing by solution- and surface-assembled osmium(II) bipyridyl rotaxanes, *Chemistry – A European Journal*, **19**, 15898–15906.
58. Serpell, C.J., Cookson, J., Ozkaya, D., Beer, P.D. (2011) Core@shell bimetallic nanoparticle synthesis via anion coordination, *Nature Chemistry*, **3**, 478–483.
59. Serpell, C.J., Kilah, N.L., Costa, P.J., Félix, V., Beer, P.D. (2010) Halogen bond anion templated assembly of an imidazolium pseudorotaxane, *Angewandte Chemie International Edition*, **49**, 5322–5326.
60. Kilah, N.L., Wise, M.D., Serpell, C.J., Thompson, A.L., White, N.G., Christensen, K.E., *et al.* (2010) Enhancement of anion recognition exhibited by a halogen-bonding rotaxane host system, *Journal of the American Chemical Society*, **132**, 11893–11895.
61. Caballero, A., Zapata, F., White, N.G., Costa, P.J., Félix, V., Beer, P.D. (2012) A halogen-bonding catenane for anion recognition and sensing, *Angewandte Chemie International Edition*, **51**, 1876–1880.
62. Gilday, L.C., Lang, T., Caballero, A., Costa, P.J., Félix, V., Beer, P.D. (2013) A catenane assembled through a single charge-assisted halogen bond, *Angewandte Chemie International Edition*, **52**, 4356–4360.
63. Caballero, A., White, N.G., Beer, P.D. (2011) A bidentate halogen-bonding bromoimidazoliophane receptor for bromide ion recognition in aqueous media, *Angewandte Chemie International Edition*, **50**, 1845–1848.
64. Zapata, F., Caballero, A., White, N.G., Claridge, T.D.W., Costa, P.J., Félix, V., *et al.* (2012) Fluorescent charge-assisted halogen-bonding macrocyclic halo-imidazolium receptors for anion recognition and sensing in aqueous media, *Journal of the American Chemical Society*, **134**, 11533–11541.
65. Mullaney, B.R., Thompson, A.L., Beer, P.D. (2014) An all-halogen bonding rotaxane for selective sensing of halides in aqueous media, *Angewandte Chemie International Edition*, **53**, 11458–11462.
66. Caballero, A., Swan, L., Zapata, F., Beer, P.D. (2014) Iodide-induced shuttling of a halogen- and hydrogen-bonding two-station rotaxane, *Angewandte Chemie International Edition*, **53**, 11854–11858.
67. Langton, M.J., Robinson, S.W., Marques, I., Félix, V., Beer, P.D. (2014) Halogen bonding in water results in enhanced anion recognition in acyclic and rotaxane hosts, *Nature Chemistry*, **6**, 1039–1043.
68. Robinson, S.W., Mustoe, C.L., White, N.G., Brown, A., Thompson, A.L., Kennepohl, P., *et al.* (2015) Evidence for halogen bond covalency in acyclic and interlocked halogen-bonding receptor anion recognition, *Journal of the American Chemical Society*, **137**, 499–507.
69. McConnell, A.J., Beer, P.D. (2012) Heteroditopic receptors for ion-pair recognition, *Angewandte Chemie International Edition*, **51**, 5052–5061.
70. Beer, P.D., Drew, M.G.B., Knubley, R.J., Ogden, M.I. (1995) Synthesis and co-ordination chemistry of a novel bis(benzo crown ether) substituted calix[4]arene that can simultaneously complex cations and anions, *Journal of the Chemical Society, Dalton Transactions*, 3117–3123.
71. Evans, A.J., Beer, P.D. (2003) Potassium cation cooperative anion recognition by heteroditopic calix[4] arene bis(benzo-15-crown-5) receptor molecules, *Journal of the Chemical Society, Dalton Transactions*, 4451–4456.

72. Cooper, J.B., Drew, M.G.B., Beer, P.D. (2000) Alkali metal cation cooperative anion recognition by heteroditopic bis(calix[4]arene) rhenium(I) bipyridyl and ferrocene receptor molecules, *Journal of the Chemical Society, Dalton Transactions*, 2721–2728.
73. Cooper, J.B., Drew, M.G.B., Beer, P.D. (2001) Heteroditopic rhenium(I) and ruthenium(II) bipyridyl calix[4]arene receptors for binding cation–anion ion-pairs, *Journal of the Chemical Society, Dalton Transactions*, 392–401.
74. Redman, J.E., Beer, P.D., Dent, S.W., Drew, M.G.B. (1998) Cooperative binding of potassium cation and chloride anion by novel rhenium(I) bipyridyl amide crown ether receptors, *Chemical Communications*, 231–232.
75. Uppadine, L.H., Redman, J.E., Dent, S.W., Drew, M.G.B., Beer, P.D. (2001) Ion-pair cooperative binding of potassium salts by new rhenium(i) bipyridine crown ether receptors, *Inorganic Chemistry*, **40**, 2860–2869.
76. Lankshear, M.D., Cowley, A.R., Beer, P.D. (2006) Cooperative AND receptor for ion-pairs, *Chemical Communications*, 612–614.
77. Lankshear, M.D., Dudley, I.M., Chan, K.-M., Cowley, A.R., Santos, S.M., Felix, V., *et al.* (2008) Cooperative AND ion-pair recognition by heteroditopic calix[4]diquinone receptors. *Chemistry – A European Journal*, **14**, 2248–2263.
78. Picot, S.C., Mullaney, B.R., Beer, P.D. (2012) Ion-pair recognition by a heteroditopic triazole-containing receptor, *Chemistry – A European Journal*, **18**, 6230–6237.
79. Knighton, R.C., Beer, P.D. (2014) Axle component separated ion-pair recognition by a neutral heteroditopic [2] rotaxane, *Chemical Communications*, **50**, 1540–1542.
80. Beer, P.D., Hopkins, P.K., McKinney, J.D. (1999) Cooperative halide, perrhenate anion–sodium cation binding and pertechnetate extraction and transport by a novel tripodal tris(amido benzo-15-crown-5) ligand, *Chemical Communications*, 1253–1254.
81. Webber, P.R.A., Beer, P.D. (2003) Ion-pair recognition by a ditopic calix[4]semitube receptor, *Journal of the Chemical Society, Dalton Transactions*, 2249–2252.
82. Cookson, J., Vickers, M.S., Paul, R.L., Cowley, A.R., Beer, P.D. (2008) Amide functionalised dithiocarbamate ruthenium(II) bis-bipyridyl receptors: a new class of redox-responsive anion sensor, *Inorganica Chimica Acta*, **361**, 1689–1698.
83. Fujita, M., Tominaga, M., Hori, A., Therrien, B. (2005) Coordination assemblies from a Pd(II)-cornered square complex, *Accounts of Chemical Research*, **38**, 369–378.
84. Beer, P.D., Berry, N., Drew, M.G.B., Fox, O.D., Padilla-Tosta, M.E., Patell, S. (2001) Self-assembled dithiocarbamate–copper(II) macrocycles for electrochemical anion recognition, *Chemical Communications*, 199–200.
85. Lai, S.-W., Drew, M.G.B., Beer, P.D. (2001) Metal-directed assembly of polyferrocenyl transition metal dithiocarbamate macrocyclic molecular boxes, *Journal of Organometallic Chemistry*, **637**, 89–93.
86. Uppadine, L.H., Weeks, J.M., Beer, P.D. (2001) Metal-directed self-assembly of terphenyl based dithiocarbamate ligands, *Journal of the Chemical Society, Dalton Transactions*, 3367–3372.
87. Beer, P.D., Berry, N.G., Cowley, A.R., Hayes, E.J., Oates, E.C., Wong, W.W.H. (2003) Metal-directed self-assembly of bimetallic dithiocarbamate transition metal cryptands and their binding capabilities, *Chemical Communications*, 2408–2409.
88. Webber, P.R.A., Drew, M.G.B., Hibbert, R., Beer, P.D. (2004) Transition metal-directed self-assembly of calix[4] arene based dithiocarbamate ligands, *Journal of the Chemical Society, Dalton Transactions*, 1127–1135.
89. Wong, W.W.H., Curiel, D., Cowley, A.R., Beer, P.D. (2005) Dinuclear zinc(II) dithiocarbamate macrocycles: ditopic receptors for a variety of guest molecules, *Journal of the Chemical Society, Dalton Transactions*, 359–364.
90. Wong, W.W.H., Curiel, D., Lai, S.-W., Drew, M.G.B., Beer, P.D. (2005) Ditopic redox-active polyferrocenyl zinc(II) dithiocarbamate macrocyclic receptors: synthesis, coordination and electrochemical recognition properties, *Journal of the Chemical Society, Dalton Transactions*, 774–781.
91. Berry, N.G., Pratt, M.D., Fox, O.D., Beer, P.D. (2001) Transition metal self-assembly of dithiocarbamate based anion receptors, *Supramolecular Chemistry*, **13**, 677–682.
92. Berry, N.G., Shimell, T.W., Beer, P.D. (2002) Heteroditopic transition metal dithiocarbamate receptors for binding cation-anion ion-pairs, *Journal of Supramolecular Chemistry*, **2**, 89–92.
93. Pratt, M.D., Beer, P.D. (2004) Heterodinuclear ruthenium(II) bipyridyl-transition metal dithiocarbamate macrocycles for anion recognition and sensing, *Tetrahedron*, **60**, 11227–11238.
94. Wong, W.W.H., Phipps, D.E., Beer, P.D. (2004) Acyclic and macrocyclic transition metal dithiocarbamate complexes containing imidazolium moieties for anion binding, *Polyhedron*, **23**, 2821–2829.

95. Padilla-Tosta, M.E., Fox, O.D., Drew, M.G.B., Beer, P.D. (2001) Self-assembly of a mixed-valence copper(II)/copper(III) dithiocarbamate catenane, *Angewandte Chemie International Edition*, **40**, 4235–4239.
96. Wong, W.W.H., Cookson, J., Evans, E.A.L., McInnes, E.J.L., Wolowska, J., Maher, J.P., *et al.* (2005) Heteropolymetallic copper(II)–gold(III) dithiocarbamate [2]catenanes via magic ring synthesis, *Chemical Communications*, 2214–2216.
97. Fox, O.D., Drew, M.G.B., Beer, P.D. (2000) Resorcarene-based nanoarchitectures: metal-directed assembly of a molecular loop and tetrahedron, *Angewandte Chemie International Edition*, **39**, 135–140.
98. Fox, O.D., Drew, M.G.B., Wilkinson, E.J.S., Beer, P.D. (2000) Cadmium- and zinc-directed assembly of nano-sized,resorcarene-based host architectures which strongly bindC60, *Chemical Communications*, 391–392.
99. Fox, O.D., Cookson, J., Wilkinson, E.J.S., Drew, M.G.B., MacLean, E.J., Teat, S.J., *et al.* (2006) Nanosized poly-metallic resorcinarene-based host assemblies that strongly bind fullerenes, *Journal of the American Chemical Society*, **128**, 6990–7002.
100. Hancock, L.M., Beer, P.D. (2011) Sodium and barium cation-templated synthesis and cation-induced molecular pirouetting of a pyridine N-oxide containing [2]rotaxane, *Chemical Communications*, **47**, 6012.

4

Perspectives in Molecular Tectonics

Mir Wais Hosseini

Institut Le Bel, Université de Strasbourg & CNRS, Strasbourg, France

This chapter is dedicated to the memory of Guy Ourisson and John Osborn, my professors and mentors.

4.1 Preamble: dreams and pathway

Although I was awarded the Izatt–Christensen Award in Macrocyclic Chemistry rather late in my career, I am extremely happy and honored. Being recognized by members of my intellectual community is a great feeling.

I was born in Kabul, Afghanistan, and moved to France at the age of 17. After two years spent at Lycée Kleber, the famous boarding school in Strasbourg, I obtained my baccalaureate. I then was educated at the next higher level at the University Louis Pasteur, which became University of Strasbourg, and obtained BS and MS degrees in physical chemistry. During my studies, I had the great pleasure of following outstanding courses by G. Ourisson, J.-M. Lehn, J. Osborn and R. Weiss. Their visions of chemistry and education played a major role in my passion for research in chemistry. I had the unique opportunity to start my research career under the supervision of Jean-Marie Lehn and worked with him for 10 years in the area of physical organic chemistry. During that period, I had the privilege of interacting with many talented scientists coming from all over the world. In particular, in addition to Bernard Dietrich with whom I worked on anion recognition, I had the pleasure of discovering Jean-Pierre Sauvage, a great and creative chemist with whom I still continue to discuss science and other matters. Research in chemistry, although ultimately an individual realization, is performed within a team. Crossing the path of Jean-Marie Lehn, with whom I learned about intuition and imagination, scientific rigor, precision, and hard work, was essential to my dedication to chemistry, which I consider as a domain of expression (Merci Jean-Marie). In 1985–86, I had the privilege of joining the group of K. N. Raymond at Berkeley as a postdoctoral fellow. Ken offered me the possibility of working on abiotic models of siderophores (natural iron

Macrocyclic and Supramolecular Chemistry: How Izatt–Christensen Award Winners Shaped the Field, First Edition.
Edited by Reed M. Izatt.

transporters) and thus completed my education in coordination chemistry. My period at Berkeley, one of the top academic institutions, home to many high-quality researchers with interesting characters, was a great opportunity for me to further complete my understanding of chemistry (Thanks Ken). On the professional side, very fortunately, at the age of 26, I was offered a permanent position by the French National Research Centre (CNRS) as a research associate. In 1990, at the age of 35, I was offered a full professorship by the University Louis Pasteur. Two years later, I was elected as a junior member to the Institut Universitaire de France (IUF), one of the most prestigious and selective institutions in France. In 2004, I became a senior member of the same institution and created the chair of molecular tectonics. Finally, I had the privilege of being extended for another period of five years.

My scientific career may be divided into two periods. The first part (1979–90) was in the "Labo Lehn". While at that time the vast majority of supramolecular chemists were exclusively working on cation recognition, I worked on anion recognition and defended the first PhD thesis on that topic [1]. Furthermore, during that period we moved to supramolecular catalysis, another rather unexplored topic at that time. In this challenging area, today called organocatalysis, we developed a series of artificial enzymes, in particular ATPase and kinase mimics. We first studied the catalysis of the hydrolysis of nucleotides (ATP and ADP) by macrocyclic polyamines, and demonstrated that the hydrolytic reaction could be accelerated by factor of 500 under physiological conditions in water (pH 7, 37 °C). A detailed mechanistic study revealed that the catalytic process takes place through an intracomplex nucleophilic attack of the terminal phosphoryl group of either ATP or ADP, leading to an intermediate phosphoramidate [2]. We further explored the reactivity of the intermediate and developed an artificial kinase capable of catalyzing the synthesis of ATP from ADP and acetylphosphate as phosphoryl source under physiological conditions [3]. We have also exploited this bond-making process for the synthesis of several other phosphorylated substrates [4]. This kinase-type activity still remains unequalled today. Finally, we reported the first example of ATP binding through simultaneous recognition of its nucleic base and the polyphosphate chain by an acridine-appended macrocyclic polyamine [5].

The second period of my research career started in 1990 when I became a professor at the University of Strasbourg. I started my own research group and dropped my activities in molecular recognition of anions and in supramolecular catalysis. Although my background was in organic and supramolecular chemistry oriented towards biology, I started a new area of research called molecular tectonics dealing with solid state chemistry and crystal engineering. It is worth noting that at that time only few groups were interested in the latter topic. In this chapter, I shall mainly present a small part of our investigations in this domain.

In addition to development of molecular tectonics, we explored several other areas. We reported the first example of silacrown ethers [6]. We developed the chemistry of borocryptands, mimics of naturally occurring antibiotics boromycine, and designed a series of specific borocryptands for Li^+, Na^+, K^+, and Cs^+ with substantially higher affinity when compared to cryptands [7]. The most relevant contribution in this topic concerned the peristatic chirality describing the formation of *R* and *S* Cs^+ (chirality centered on the cation) by quadrupolar splitting constants measurements in a chiral cholesteric phase [8]. In the domain of bioinorganic chemistry, we designed an artificial DNAse based on a manganese–porphyrin backbone bearing two acridine moieties as intercalating units [9]. Along with D. Gutsche, we contributed to the synthesis and chemistry of mercaptocalix[4]arene [10]. Concomitantly with Sone and Myano, we have reported the synthesis and structural investigations on thiacalix[4]arene [11]. We have also described the synthesis and X-ray structure of mercaptothiacalix[4]arene [12].

We used the thiacalix[4]arene backbone for the first synthesis of discrete tetranuclear clusters [13] and the formation of nanoclusters composed of 32 Ni or Co atoms [14]. We also used the mercaptocalix[4]arene platform for the design of a selective extracting agent for Hg^{2+} cation [15]. More recently, we exploited the formation of atroposiomers for the design of porphyrin based ligands bearing isoquinoline appended coordinating sites and demonstrated the first example of near infra red (NIR) emission of neodymium through excitation of the Soret band of the porphyrin in the visible domain [16]. Last but not least, in the last decade we have initiated another research project dealing with the control of molecular movements. In particular, we have designed Sn–porphyrin-based molecular turnstiles [17] and extended the concept to oscillating systems [18]. We have also reported the design of turnstiles based on strapped porphyrins [19]. Along the same lines, we have reported the design and dynamics of Pt(II)-based organometallic turnstiles [20]. Finally, we have designed an organic luminescent turnstile with optical reading of its closed and open states [21].

4.2 Introduction

Molecular crystals, periodic entities, are defined by the nature of their molecular components and interactions between them in the solid state. Although crystals are described by translation of the unit cell into all three directions of space, nevertheless, by considering them as supramolecular entities [22, 23], they may be described as molecular lattices (a particular case of networks) by analyzing intermolecular interactions (specific recognition patterns) and their geometrical features. The term molecular network, widely used instead of molecular lattice, will be employed here. In marked contrast with discrete molecules, molecular networks display translational symmetry. Thus, they may be regarded as infinite molecular assemblies for which specific interaction patterns are repeated through space. In other terms, the recognition motifs or supramolecular synthons [24] taking place between molecular components comprising the crystal become assembling nodes of the network. The dimensionality of molecular network depends on the number of translations operating on the assembling nodes [25, 26]. Thus, 1D or α-networks are formed when a single translation takes place. Similarly, β- (2D) and γ-networks (3D) are defined when two or three translations respectively of the same or different assembling nodes take place. Although, in principle, molecular networks may be generated in solution, gel, or solid state, the crystalline phase has been extensively used because crystal structures may be studied by X-ray diffraction methods.

Molecular tectonics [27–30] (Figure 4.1) is an approach describing molecular crystals as networks generated by mutual interconnections of tectons [31], which are structurally and energetically defined active molecular building blocks bearing within their backbone an assembling program based on molecular recognition processes. In addition to the ability to describe crystals as networks, more interestingly, molecular tectonics enables one to program the formation of molecular networks both in terms of dimensionality and connectivity through the nature and localization of recognition/interaction sites within the structure of tectons.

One may draw a parallel between molecular synthesis, i.e. connection of atoms by covalent bonds, and supramolecular synthesis [23] of molecular networks by reversible intermolecular interactions under self-assembly processes [32]. However, it is noteworthy that the main concern of molecular tectonics is the design of molecular networks in the crystalline phase and not crystal engineering [24] based on prediction of crystal structures [33]. Indeed, the formation of crystals results from the packing of networks in the condensed phase and, within the network, the arrangement and conformation of tectons. Owing to our limited knowledge of all possible intermolecular interactions governing the packing and conformational space of tectons, it is currently impossible to fully predict the formation of crystals.

4.3 From tectons to networks

Molecular networks in the crystalline phase are generated by self-assembly processes, either by a single self-complementary, or several complementary, tectons. Although a mono component system is attractive in terms of atoms and synthetic economy, in practice, often, such a system produces insoluble amorphous powders.

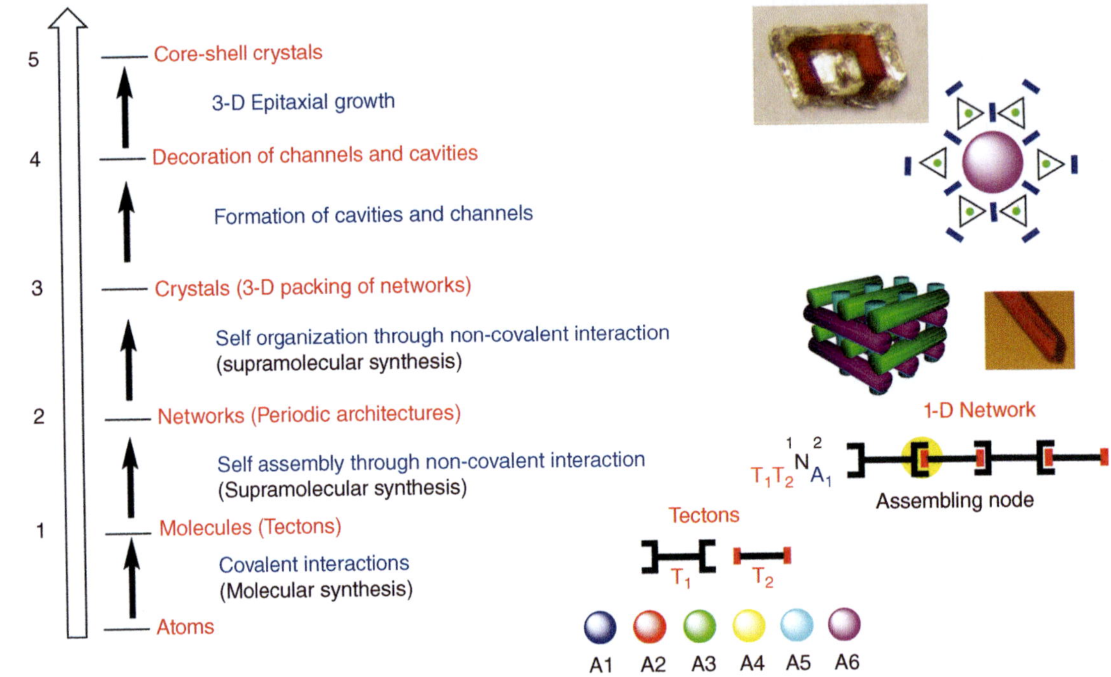

Figure 4.1 *Different levels of organization and complexity reached by the molecular tectonics approach*

The use of multicomponent systems is more viable. Indeed, for such combinations, through the control of thermodynamic and kinetic parameters, crystals may be obtained and structurally studied by X-ray diffraction methods. It is however noteworthy that increasing the number of components leads to a larger possibility space and thus requires precise design of tectons and recognition patterns.

For simplicity of illustration, we will focus on 1D and 2D networks. The generation of 1D networks requires tectons bearing at least two divergently oriented interaction sites. The divergent orientation is essential for avoiding the formation of discrete species. A 1D network may be formed using a self-complementary tecton (Figure 4.2a). Such a network may also be designed using a two-component system based on a combination of two complementary tectons (Figure 4.2b and c). For this design, one may use either two centric tectons leading to a single recognition pattern (Figure 4.2b) or two non-centric tectons leading to two different assembling nodes (Figure 4.2c). One may increase the number of tectons by combining three different tectons, leading to two different recognition patterns for example (Figure 4.2d). A tetra-component system based on four different tectons leading to two different assembling nodes may also be designed (Figure 4.2e, f). The analysis given here is not exhaustive. One may imagine many other cases. The only limitations are the imagination and the synthetic ability of chemists to design and prepare the tectons and to find the proper conditions for self-assembly processes.

As stated above, for 1D networks, the information coded within the structure of tectons concerns mainly the recognition event generating the assembling nodes and the repetition of the process leading to its translation into one direction of space. In principle, arrangements in the other two directions of space leading to the formation of a crystal are not, a priori, controlled. In other words, the system is deterministic in only one direction of space.

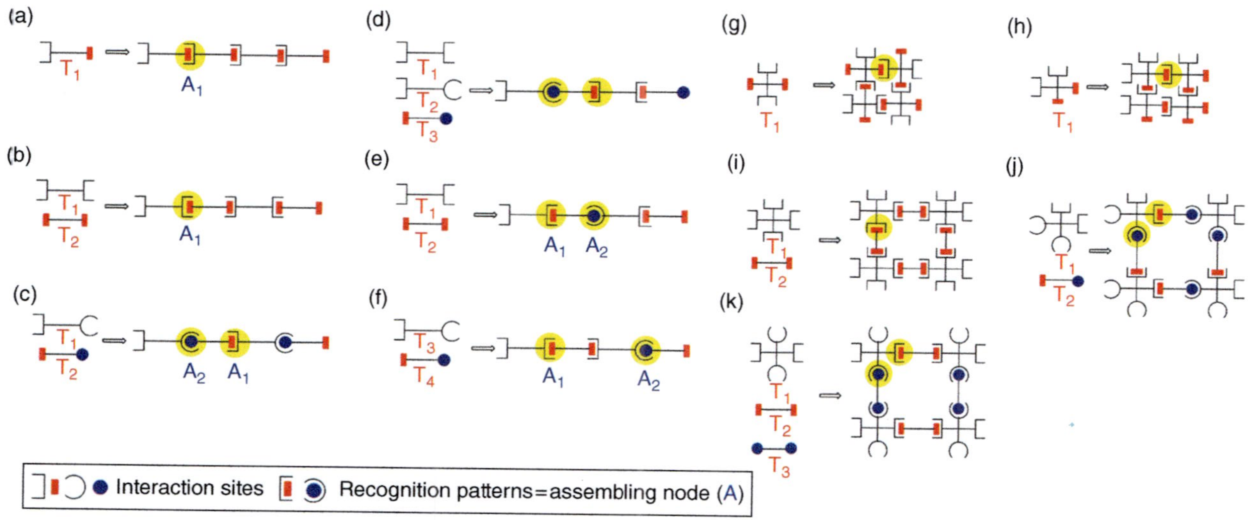

Figure 4.2 *A non-exhaustive schematic representation of a variety of 1D (a–f) and 2D (g–k) molecular networks based on self-assembly of either self-complementary tectons (a, g) or complementary tectons (b–f, h–k).* T_1, T_2, T_3 *represent tectons,* A_1, A_2 *represent assembling nodes*

For the design of 2D networks, tectons bearing at least three divergently oriented recognition sites are required. Again bidimensional networks may either be generated using a single-component system (Figure 4.2 g and h) or polycomponent combinations based on several complementary tectons (Figure 4.2i–k). For example, depending on the number of recognition sites, a two-component system may be based on two complementary tectons leading to a single assembling node (Figure 4.2i) or to two different recognition patterns (Figure 4.2j). As in the case of 1D networks, the number of components may be increased. For example, one may design a combination based on three tectons generating two different assembling nodes (Figure 4.2 k). Again, 2D networks possess lower dimensionality that the crystal. The information coded within the structure of tectons lead to bidimensional architectures through the control of two space dimensions (a deterministic system in two dimensions). Consequently, the arrangement in the third dimension, i.e. the packing of 2D sheets, is not programmed and thus cannot be, in principle, predicted. Only in the case of 3D networks is the connectivity between tectons in all three dimensions of space programmed.

The difference between the size of tectons (*ca.* nanometer) and crystals (*ca* millimeter) is a factor of 10^6. In other words, the formation of millimetric crystals requires some one million interconnections between tectons in one dimension and thus *ca.* 10^{18} operations. This clearly cannot be achieved through step-by-step strategies; it requires self-assembly processes. In order to avoid the generation of diversity under self-assembly conditions, i.e. a large collection of structures, reversible interactions allowing self-correcting processes must be used. In other words, the system composed of tectons should find its own way to the most stable situation under given conditions (temperature, pressure, concentration, solvent etc.). The recognition event between complementary sites may take place through a variety of reversible attractive intermolecular interactions (Figure 4.3).

The supramolecular toolbox contains van der Waals interactions, charge-assisted electrostatic interactions, π–π interactions, charge-transfer interactions, metal–metal interactions, H-bonding, halogen-bonding, and coordination bonding.

As stated above, although in principle any type of reversible intermolecular interaction may be used for the design of molecular networks, the majority of cases reported are either based on hydrogen bonding [24–26, 28, 31–34] or on coordination bonds [35]. A few examples of inclusion networks, mainly based on van der Waals interactions, in the solid state between concave and convex tectons have been also reported [36, 37].

In the following section, we present some selected examples of challenging strategies and networks following an increasing order of energy of interaction.

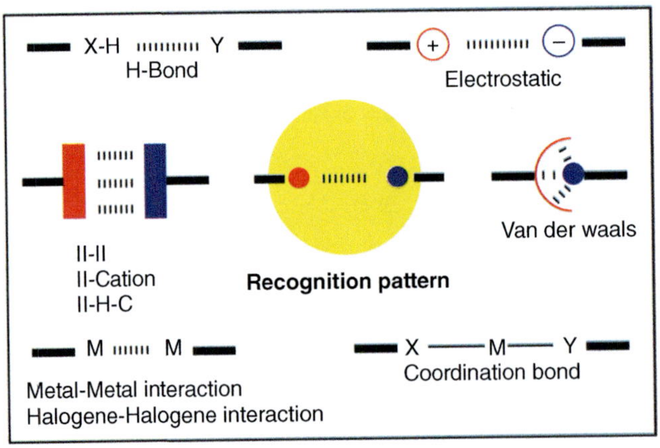

Figure 4.3 Schematic representation of the supramolecular chemistry toolbox of reversible intermolecular interactions

4.3.1 Inclusion networks: an interplay between concave and convex tectons

Inclusion processes between concave and convex units lead to molecular complexes in solution and/or the solid state. Many concave receptors have been designed to bind substrates by inclusion processes. We have extended the concept of inclusion in solution to the solid state by designing inclusion networks called koilates [36, 37]. The design of the latter is based on the use of koilands (from Greek *koilos*, hollow), [38] multicavity tectons possessing at least two cavities arranged in a divergent fashion, and convex connectors capable of being included within the cavities of koilands. Attractive van der Waals interactions are the driving forces for the formation of inclusion networks. For the design of koilates, several possibilities, both in terms of geometry and dimensionality, have been investigated [37]. For 1D koilates, as for any 1D network, two possibilities (directional or non-directional) may be envisaged (Figure 4.4). A combination of centric koilands and connectors should lead to non-directional koilates due to the translation of centers of symmetry (Figure 4.4a). In order to impose directionality, one may envisage combinations of non-symmetric koilands and/or connectors (Figure 4.4b–f). When combining centric koilands and non-centric connectors, either directional (Figure 4.4b) or non-directional (Figure 4.4c) koilates may be generated. Similarly, when using non-centric koilands and symmetric connectors, directional (Figure 4.4d) or non-directional (Figure 4.4e) koilates may be obtained. Finally, the most promising strategy, based on specific recognition between koilands and connectors, is a combination of non-centric koilands and connectors (Figure 4.4f).

Calix[4]arene derivatives [39] are candidates of choice for the design of koilands. Indeed, these compounds, in their cone conformation, offer a preorganized and tunable hydrophobic pocket as well as four hydroxy groups for further functionalization. Koilands based on fusion of two calix[4]arene derivatives by two silicon atoms have been prepared. Four examples of koilands are shown in Figure 4.5.

Using koilands **4–7** and connectors **8–10**, the formation of koilates was achieved (Figure 4.6). The formation of non-directional koilates in the crystalline phase was demonstrated by combining the centric koilands **4** (Figure 4.6a) [40] or **5** (Figure 4.6b) [41] with the rod-type molecule **8**. The latter was also used to generate directional koilates with **6** (Figure 4.6c). However, the parallel and centric packing of koilates within the crystal leads to a non-directional crystal. The formation of directional koilates was also achieved using combinations of the connectors **9** or **10** with koiland **7** (Figure 4.6d and e) [42]. Unfortunately, again, the directional linear koilates are packed in antiparallel fashion, thus there is a loss of directionality.

4.3.2 H-bonded networks

H-bonding is one of the most commonly used interactions for the design of molecular networks [24–26,28,31–34]. The classical hydrogen bond (DH…A), formed between two electronegative atoms D (donor) and A (acceptor), is defined by the nature of D and A, the distance between them and by the angle DHA. For the design of

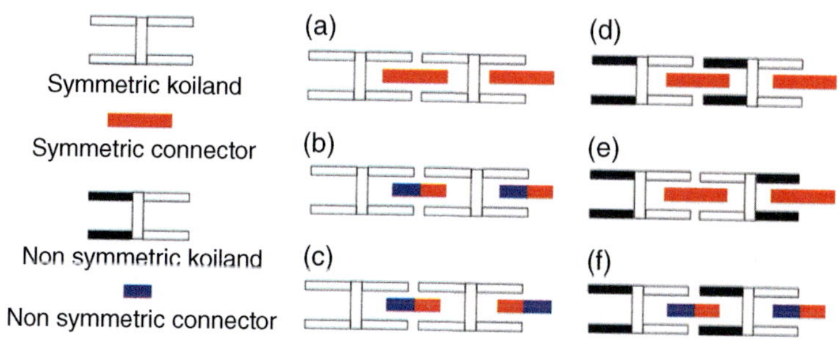

Figure 4.4 *Schematic representations of 1D inclusion networks based on interconnection of koilands and connector molecules*

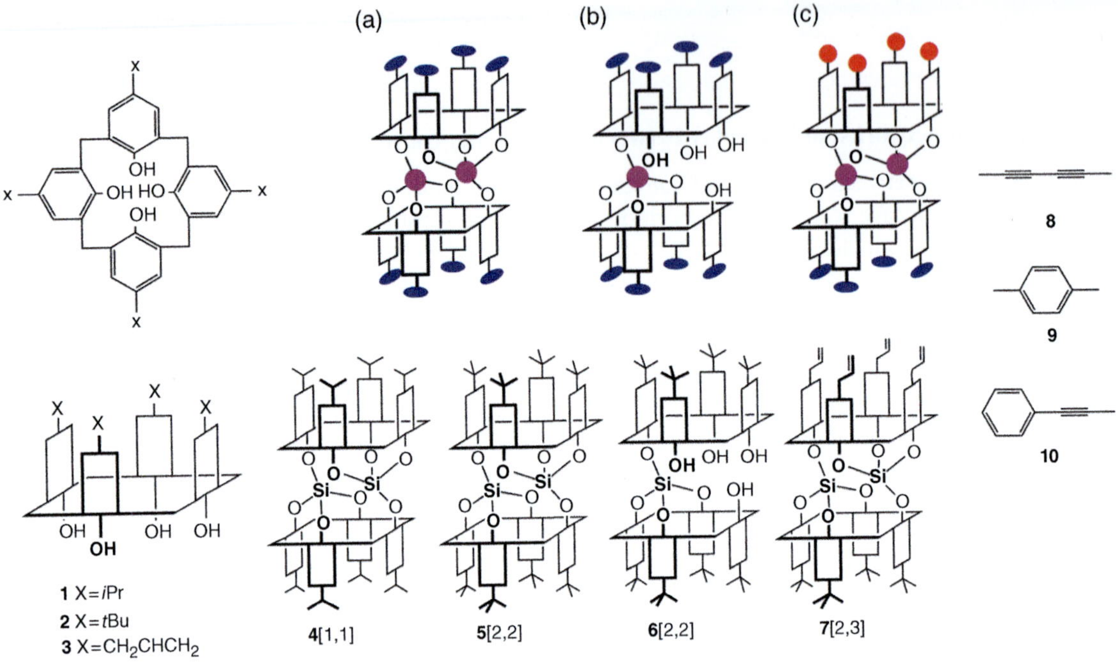

Figure 4.5 *Schematic representation of centric (a) and non-centric (b and c) koilands; 4–7 are based on fusion of calix[4]arene derivatives 1–3 with connectors 8–10 used for the formation of inclusion networks*

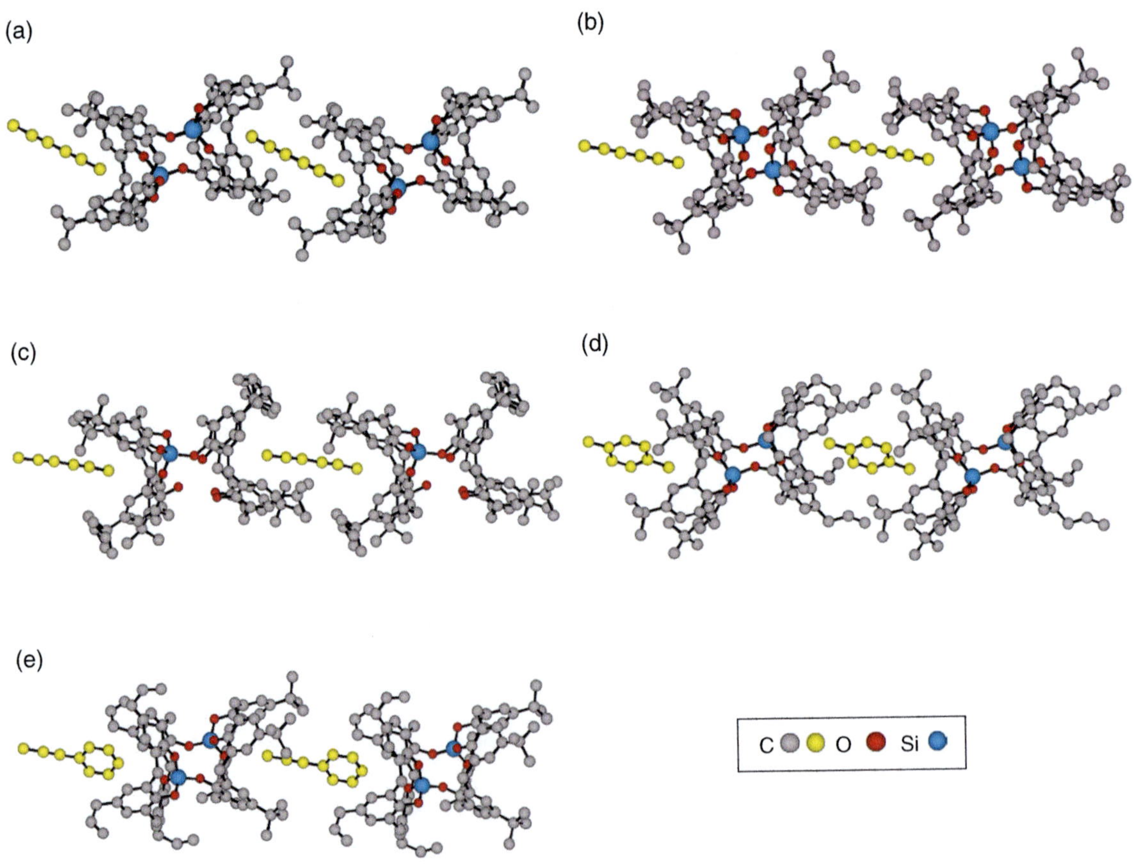

Figure 4.6 *Portions of the X-ray structures of koilates formed between koiland* **4** *(a),* **5** *(b) and* **6** *(c) and the connector* **8***, and between koiland* **7** *with connectors* **9** *(d) and* **10** *(e). For the sake of clarity, the carbon atoms of the connector are colored in yellow and hydrogen atoms are not represented*

H-bonded networks, one may use different modes of interaction. Owing to the large angular (DHA) distribution, [43] the use of the monohapto mode is not always a viable strategy because it often does not allow fine control of geometrical features of the recognition pattern and thus the prediction of the connectivity of the network. As exploited by nature, the number of possible geometrical arrangements may be reduced by the use of a di- or tri-hapto mode of H-bonding.

4.3.3 Neutral 1D helical H-bonded networks

Although few cases of enantiomerically pure H-bonded helical networks have been published, [44] the majority of examples reported deal with achiral, or racemic mixtures of, tectons leading to the formation of racemates. Neutral chiral tectons **11** and **12** (Figure 4.7) are based on the optically active isomannide backbone bearing both H-bond donor (OH) and acceptor (pyridyl) sites. Imposed by the stereochemistry of the isomannide unit, the pyridyl and OH groups are divergently oriented towards the concave face of the bicyclic unit in both cases. The difference between enantiomerically pure tectons **11** and **12** resides in the orientation of the nitrogen atom of the pyridyl ring [45].

Owing to their self-complementarily and enantiomerically pure nature, both tectons **11** (Figure 4.7a–b) and **12** (Figure 4.7c–d) lead to the formation of single-stranded H-bonded helical networks. The mutual interconnection of consecutive tectons indeed takes place through H-bonding between the OH moiety and the N atom of the pyridyl group. For both helical networks, the *P* helicity is imposed by the predefined stereochemistry of isomannide.

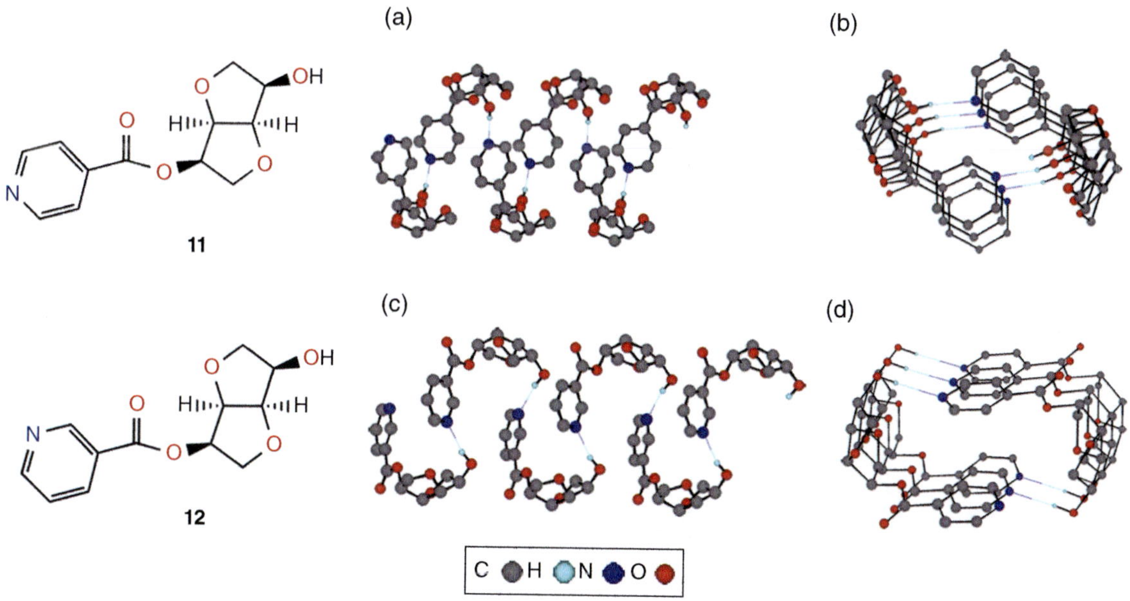

Figure 4.7 *Portions of the structure of enantiomerically pure 1D helical H-bonded networks formed upon self-bridging of tectons* **11** *(a, b) or* **12** *(c, d). H atoms, except those involved in H-bonding, are omitted for clarity*

4.3.4 Charge-assisted H-bonded networks

The majority of reported H-bonded molecular networks are based on non-ionic hydrogen bonds [34]. A possible way to increase the robustness of the recognition pattern between H-bond donors and acceptors is to combine directional hydrogen bonds with strong but less directional ionic interactions [29, 46–48].

Bis-amidinium dications, such as **13–16** (Figure 4.8), are interesting building blocks for the design of charge-assisted H-bonded networks. These dicationic tectons, owing to the presence of four acidic N–H protons oriented in a divergent fashion, act as tetra H-bond donors in the presence of H-bond acceptors such as carboxylates, phosphonates, sulfonates, and polycyanometallates. By tuning the distance between the acidic hydrogen atoms on the same side of the tecton, it is possible to recognize anions through a dihapto mode of interaction.

The formation of 1D H-bonded networks based on dicarboxylate anions was demonstrated using tectons such as **15** (Figure 4.9) [29, 46]. Owing to the proper spacing of the two cyclic amidinium moieties ($(CH_2)_2$), compound **15** recognizes on each of its faces the carboxylate moiety by a dihapto mode of H-bonding. In all cases, the common features are: the neutral 1D H-bonded networks are formed through the mutual interconnection of dicationic and dianionic tectons. The dicationic tecton **15** adopts a centric conformation. The recognition of the carboxylates by **15** takes place on each side of the tecton.

Using other combinations of cationic and anionic tectons, other 1D as well as 2D and 3D hydrogen-bonded networks have also been obtained [48e,f, 49].

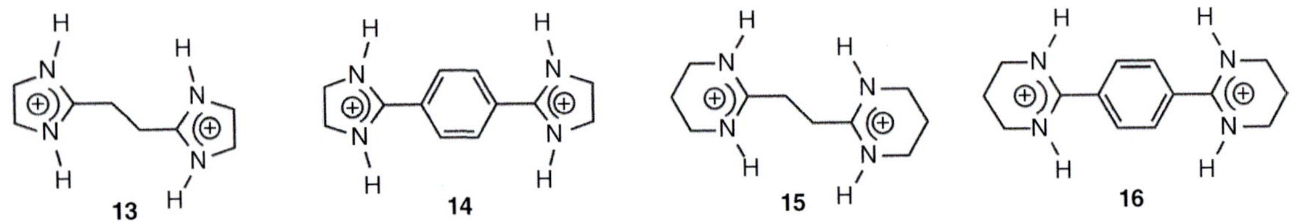

Figure 4.8 *Dicationic bisamidinium type tectons as tetra H-bond donors*

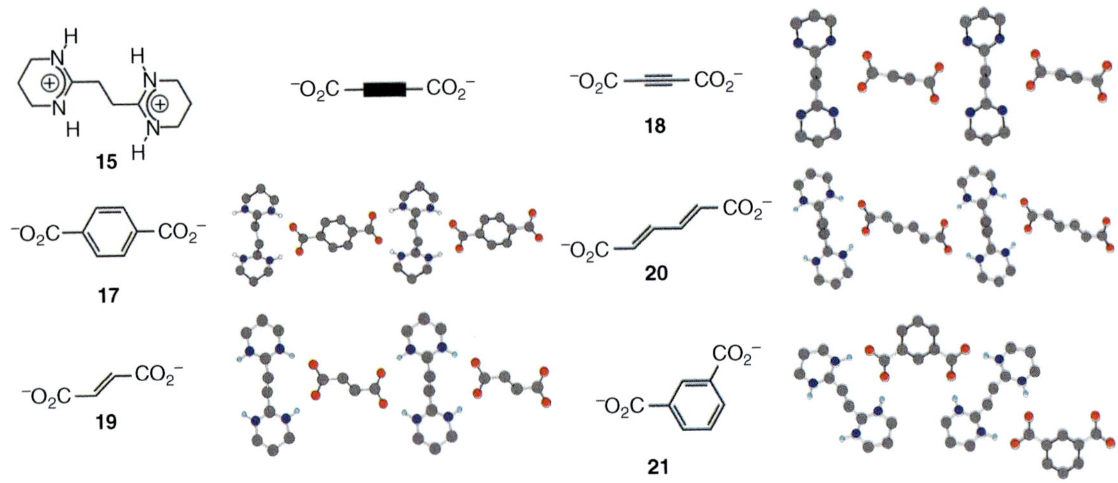

Figure 4.9 *Portions of structures of the 1D networks formed between* **15** *and dianions* **17–21**. *H atoms, except those involved in H-bonding, are omitted for clarity*

4.3.5 H-bonded networks based on recognition of hexacyanometallates

Generation of charge-assisted H-bonded networks was also achieved using polycyanometallate-type anions [50]. In particular, combinations of the dicationic tecton **16** with $M(CN)_6]^{3-}$ (M = Co, Fe, Cr) anions leads to the formation of 2D isostructural neutral networks with a **16**/$[M(CN)_6]^{3-}$ ratio of 3/2 (Figure 4.10).

A combination of the tecton **16** with $[Fe(CN)_6]^{3-}$ (Figure 4.11a) or $[Co(CN)_6]^{3-}$ (Figure 4.11b) leads indeed to the formation of the expected 2D networks [51]. The sheet-type architecture is formed by interconnection of dicationic and dianionic units through dihapto modes of H-bonding. In both cases, the packing of the neutral 2D networks leads to channels filled with water molecules. The latter, in both cases, form hydrogen-bonded 1D networks composed of hexagons interconnected by tetragons. Owing to the dihapto mode of hydrogen bonding between **16** and $[M(CN)_6]^{3-}$, supramolecular chirality, taking place within the second coordination sphere around the metal, is generated. By analogy with Δ and Λ chirality for octahedral metal

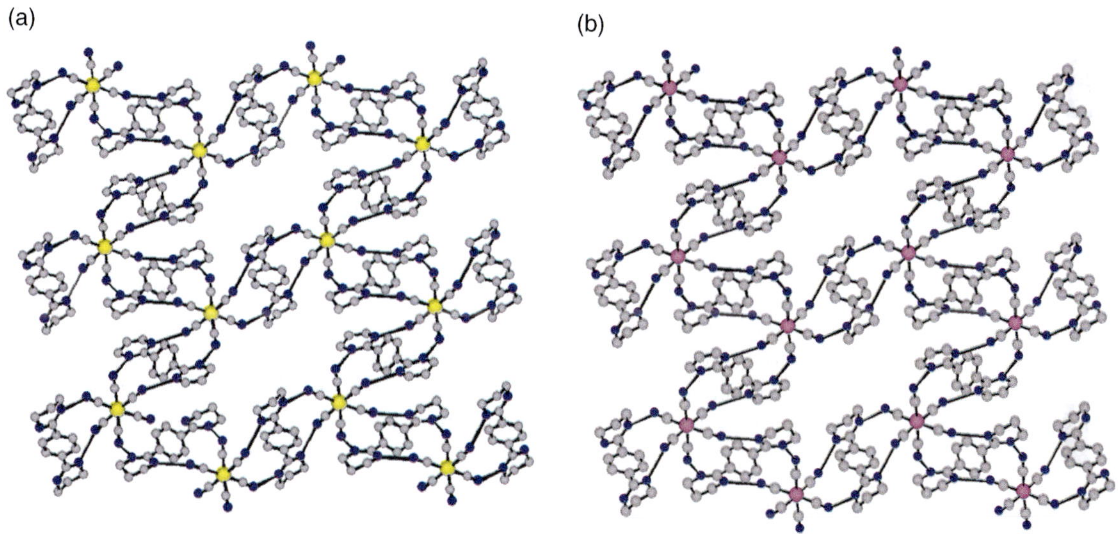

Figure 4.10 *Portions of X-ray structures of 2D H-bonded networks formed between* ***16*** *and (a)* $[Fe(CN)_6]^{3-}$ *and (b)* $[Co(CN)_6]^{3-}$ *anions. H atoms are omitted for clarity*

Figure 4.11 *Left to right: X-ray structure of the 2D networks formed between tecton* ***16*** *and* $M(CN)_6^{4-}$ *(M = Ru or Fe); pictures of isostructural crystals; (a) schematic representation of epitaxial growth of core–shell crystals; and (b) picture of the B@A core–shell crystal. See text for the definition of A and B and growth strategy*

complexes surrounded by two or three chelating ligands, the Δ′ and Λ′ supramolecular chirality occurs within the second coordination sphere. The geometrical features of both the organic tecton **16** and the anionic $[M(CN)_6]^{3-}$ complex are almost identical, thus leading to the formation of isostructural and almost isometric crystals.

4.3.6 From hydrogen-bonded networks to core–shell crystals

Upon combining **16** with $[Ru(CN)_6]^{4-}$ or $[Fe(CN)_6]^{4-}$, again isostructural 2D networks (Figure 4.11) are formed. The packing of sheets leads again to isostructural and almost isometric crystals **A** (Ru(II), colorless) and **B**, (Fe(II), orange). These two crystalline systems have been used for the formation of core–shell crystals **B@A** or **A@B** by 3D epitaxial growth (Figure 4.11a) [52]. Because of the difference in color between crystals **A** and **B**, the growth process could be monitored by changes in color at the interface between the two crystalline zones (Figure 4.11). Starting with **A**, the growth of **B** on **A** generates an orange crystal. Cutting the latter clearly shows the formation of a core–shell crystal (**B@A**) (Figure 4.11b). X-ray diffraction on the core–shell (**B@A**) crystal revealed the continuation of the crystalline order. The same holds for the reverse, i.e. growth of **A** on **B**, leading to the formation of (**A@B**) crystals. The study was further extended to the fabrication of higher generation of core–shell crystals. A similar strategy has been used earlier by J. C. MacDonald to fabricate other types of core–shell crystals [53].

4.3.7 Coordination networks: an interplay between organic and metallic tectons

Coordination networks are infinite organometallic architectures resulting from mutual bridging between organic and metallic tectons by coordination bonds. The design of this type of networks with predefined dimension (1D, 2D or 3D) is based on the interplay between both electronic and geometric requirements of the organic and metallic tectons. Research in this area is very active and varieties of networks have been reported [35]. Although we have reported many examples of 1D, 2D and 3D coordination networks, here we shall only describe the design principles dealing with the formation of challenging 1D networks.

4.3.8 Helical coordination networks

The design and formation of helical networks are challenging and of interest. Tectons **22–24** are bis-monodentate ligands based on two coordinating pyridyl units interconnected by oligoethylenglycol spacers (Figure 4.12).

The combination of **22–24** with Ag^+ leads to the formation of double-stranded interwoven infinite linear networks (Figure 4.12) [54]. The driving force for the formation of these assemblies is related to the loop-type disposition (pseudo-crown ether) of the oligoethyleneglycol units allowing interactions between silver cations belonging to one strand and ether oxygen atoms of the other strand. As it can be seen in Figure 4.12, the pitch of the helical arrangement may be controlled by the length of the oligoethyleneglycol spacer and by the orientation of the nitrogen atom of the pyridyl ring.

All three examples of helical assemblies presented in Figure 4.12 deal with achiral tectons. Consequently, in all three cases, racemates are obtained. In order to generate optically pure helical strands, we used enantiomerically pure tectons such as **25** and **26** (Figure 4.13) [55]. Both tectons are based on isomannide backbone and bear two pyridyl units as neutral coordinating sites. The difference between the two is a consequence of the orientation of the N atom of the pyridyl ring. A combination of **25** with $HgCl_2$ leads to the formation of a single-stranded helical coordination network with *P* helicity. The latter is imposed by the stereochemistry of isomannide. The infinite helical arrangement is obtained upon bridging of consecutive tectons **25** by mercury cations (Figure 4.13a). It is interesting to note that the helical networks are packed in the *syn*-parallel fashion, thus leading to a polar crystal (Figure 4.13b).

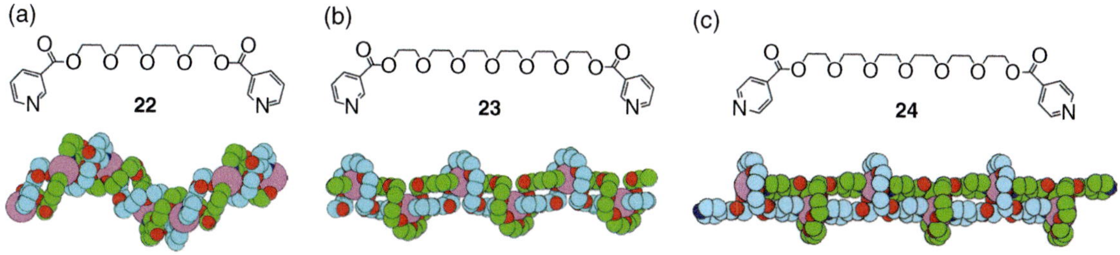

Figure 4.12 *Portions of X-ray structures of double-stranded helical coordination networks formed between silver cation and tectons* **22** *(a),* **23** *(b) and* **24** *(c). H atoms are omitted for clarity*

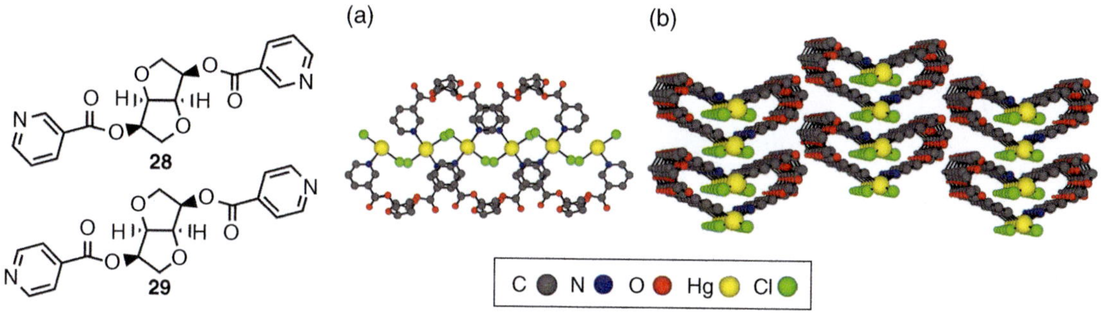

Figure 4.13 *Portions of the structure of the single-stranded helical coordination networks formed between* $HgCl_2$ *and tecton* **25** *(a) and the syn-parallel packing of 1D networks (b), affording a polar crystal*

The combination of **26** and $HgCl_2$ leads to the formation of an enantiomerically pure triple-stranded helical strand with *P* helicity. Again, the network is generated by bridging of consecutive tectons **26** by mercury cations (Figure 4.14). Interestingly, but unexpectedly, because of interactions between consecutive stands, a triple-stranded helical arrangement is obtained. The latter is not cylindrical but rather flat. The triple-stranded helices are again arranged in a parallel fashion.

The formation of a helical strand based on another combination using the exo-bischelate tecton **27** (Figure 4.15) and silver cation was also demonstrated [56]. These macrocyclic ligands, because of the spacer groups connecting the two 2,2′-bipyridyl units at positions 4 and 4′, adopt a roof-type conformation. For that reason, when tectons **27** are interconnected by cations such as silver adopting a tetrahedral coordination geometry, a single-stranded helical network is formed (Figure 4.15). Because of the achiral nature of the organic tecton, both *P* and *M* helices are present in the crystal, thus affording a racemate.

4.3.9 Directional 1D coordination networks

The design of directional networks is an important and challenging issue for exploiting directional physical properties. Here, we focus only on 1D networks. The strategy investigated was to combine dissymmetric organic tectons bearing monodentate and a tridentate coordination poles such as **28** and **29** (Figure 4.16) with a transition metal complex such as $CoCl_2$ in which the metal adopts an octahedral coordination geometry and thus, owing to the occupation of the two apical positions by chloride anions, offers four coordination sites in a square arrangement [57].

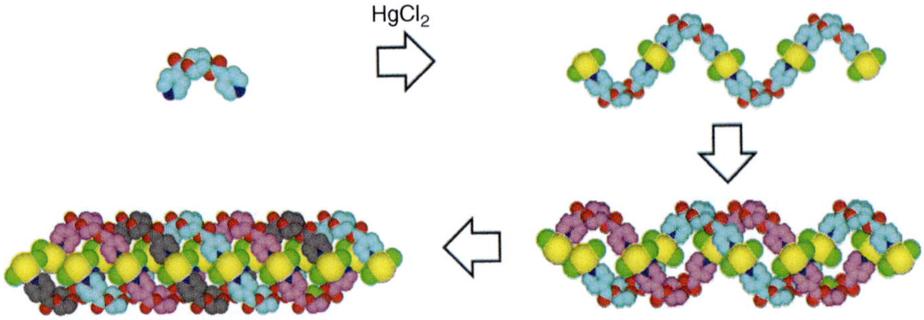

Figure 4.14 Generation of the triple-stranded helical coordination networks formed between $HgCl_2$ and tecton **26**

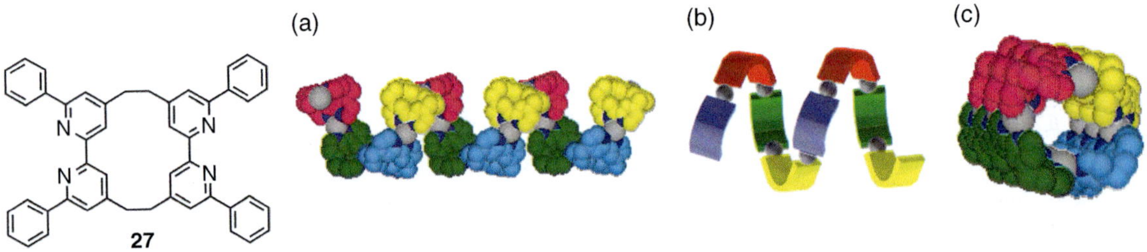

Figure 4.15 Portions of the X-ray structure of the single-stranded helical coordination networks formed between silver cation and tecton **27** (a); its schematic representation (b); and a lateral view (c). H atoms and phenyl groups are not represented for clarity

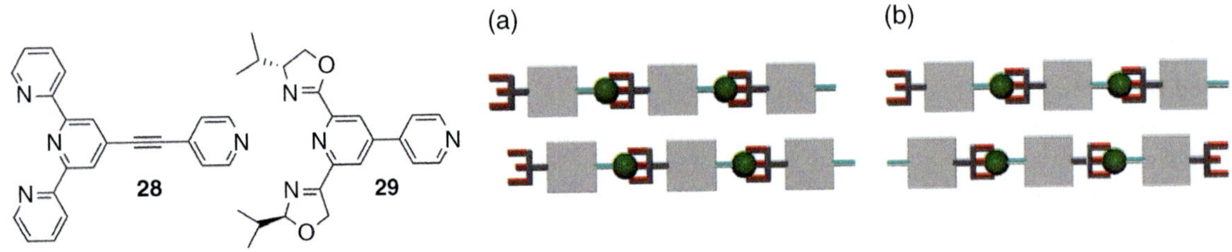

Figure 4.16 Directional tectons **28** and **29** and schematic representation of centric (a) and acentric (b) packing of two consecutive directional 1D networks

The strategy developed based on mutual bridging of the organic and metallic tectons leads, by construction, to a neutral directional 1D network (Figure 4.16). For such an organization, two packing possibilities (*anti*-parallel (Figure 4.16a) and *syn*-parallel (Figure 4.16b)) may be envisaged. In the case of the achiral tecton **28**, as expected, the centric packing (Figure 4.17a) leading to an apolar crystal is observed.

In order to investigate the possibility of avoiding *anti*-parallel packing, the introduction of chirality within the backbone of the organic tecton was explored. Thus, the neutral acentric and *c*2-chiral tecton **29** was designed. The latter is based on two coordination poles composed of a pyridyl unit connected at the 4-position to a pyridyl moiety bearing at the 2 and 6 positions two optically active oxazoline groups [58]. Upon combining the chiral tecton **29** with $CoCl_2$, a chiral and directional 1D coordination network is obtained (Figure 4.17b). As stated above, the directionality of the network is due to the acentric nature of the organic tecton **29**.

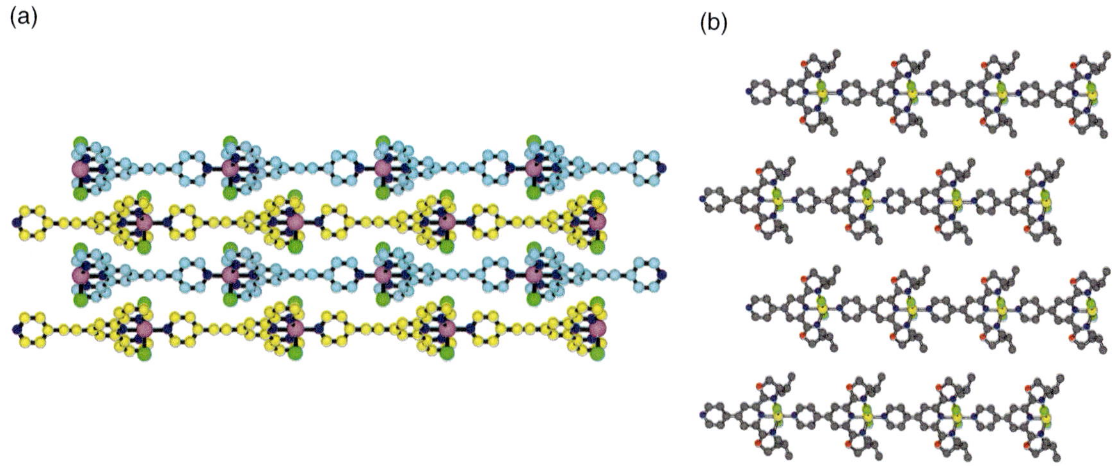

Figure 4.17 Portions of the X-ray structure of neutral network formed between $CoCl_2$ and tecton **28** (a) and **29** (b) showing the centric (a) and acentric (b) packing of consecutive directional 1D networks. In the case of tecton **28**, carbon atoms belonging to consecutive strands are differentiated by color. H atoms are omitted for clarity

In marked contrast with the case mentioned above, here, owing to the chiral nature of the tecton, a *syn*-parallel (Figure 4.16b) arrangement of consecutive chiral 1D networks is observed. This packing of 1D networks leads to a polar solid.

4.4 Summary and outlook

In this short contribution, we have tried to demonstrate that molecular tectonics is a viable approach for the design and generation of molecular networks in the crystalline phase. Indeed, by designing programmed and active molecular tectons capable of mutual interactions through molecular recognition events (structural node of the network) and allowing the translation of the recognition event (repetition of the association events generating periodicity) by self-assembly processes, one may predict, in some cases, the connectivity pattern between tectons and thus the geometry, topology and directionality of molecular networks. An important issue here is the reliability of the prediction. The latter is related to the quality of design of molecular tectons and the recognition patterns between them. This, obviously, depends on the nature and thus the energy associated with interactions used, the number of components and the structural flexibility of tectons. Our experience showed that, using strong interactions such as a combination of charge/charge electrostatic interactions and dihapto mode of H-bonding, or coordination bonds based on polyhapto mode of coordination, reliability in the prediction of the network is acceptable. This may be further increased by using tectons with rather restricted conformational space (rigid structures). In our opinion, this domain is in its infancy, far from fully explored and mastered. More systematic investigations are needed before reaching maturity. On the matter of the interesting and recurrent question, actively under discussion, of the relationship between molecular tectonics and crystal engineering, in our opinion, the response is that the two approaches, although related, differ in their aims. With our present level of understanding, we are not to predict the crystal structure of a set of tectons, in particular, the adopted conformation of flexible tectons, the presence of solvent molecules, polymorphs etc. However, using concepts and principles derived from supramolecular chemistry, the molecular tectonics approach allows one to predict some of the connectivity patterns between properly designed tectons and their assembling nodes and thus the formation of molecular networks.

4.5 Acknowledgements

The research carried out in our laboratory over the last 25 years has been supported by the University of Strasbourg, the CNRS, the Institut Universitaire de France (IUF), the International Center for Frontier Research in Chemistry (icFRC, Strasbourg), the Labex CSC (ANR-10-LABX- 0026 CSC) within the Investissement d'Avenir program ANR-10-IDEX-0002-02. I have had the great pleasure of working with talented and dedicated coworkers and students who contributed to my research. I thank them all. Their names are given in the cited publications.

References

1. (a) Dietrich, B., Hosseini, M.W., Lehn, J.-M., Sessions, R.B. (1981) Anion receptor molecules. Synthesis and anion binding properties of polyammonium macrocycles, *Journal of the American Chemical Society*, **103**, 1282–1283; (b) Hosseini, M.W., Lehn, J.-M. (1982) Anion receptor molecules. Chain length dependent selective binding of organic and biologic dicarboxylate anions by ditopic polyammonium macrocycles, *Journal of the American Chemical Society*, **104**, 3525–3527.
2. Hosseini, M.W., Lehn, J.-M., Mertes, M.P. (1983) Efficient molecular catalysis of ATP hydrolysis by protonated macrocyclic polyamines, *Helvetica Chimica Acta*, **66**, 2454–2466.
3. Hosseini, M.W., Lehn, J.-M. (1985) COCATALYSIS: acetylphosphate hydrolysis and pyrophosphate synthesis catalysed by a macrocyclic polyamine, *Journal of the Chemical Society, Chemical Communication*, 1155–1157.
4. (a) Hosseini, M.W., Lehn, J.-M. (1987) Supramolecular catalysis of phosphoryl transfer: pyrophosphate synthesis from acetylphosphate mediated by macrocyclic polyamines, *Journal of the American Chemical Society*, **109**, 7047–7058; (b) Hosseini, M.W., Lehn, J.-M., (1988) Supramolecular catalysis: substrate phosphorylation and adenosine triphosphate synthesis with acetylphosphate catalysed by a macrocyclic polyamine, *Journal of the Chemical Society, Chemical Communication*, 397–399.
5. Hosseini, M.W., Blacker, A.J., Lehn, J.-M. (1990) Multiple molecular recognition and catalysis. A multifunctional anion receptor bearing an anion binding site, an intercalant group and a catalytic site for nucleotide binding and catalysis, *Journal of the American Chemical Society*, **112**, 3896–3904.
6. Oddon, G., Hosseini, M.W. (1993) Silacrown ethers: synthesis of macrocyclic diphenylpolyethyleneglycol mono- and disilanes, *Tetrahedron Letters*, **34**, 7413–7416.
7. Graf, E., Hosseini, M.W., Ruppert, R., De Cian, A., Fischer, J. (1995) Simultaneous binding of boron and ammonium cation by a pseudocryptand: synthesis, X-ray analysis, and solution studies by NMR spectroscopy, *Journal of the Chemical Society, Chemical Communication*, 1505–1506.
8. Graf, E., Hosseini, M.W., Ruppert, R., Kyritsakas, N, De Cian, A., Fischer, J., Estounes, C., Taulelle, F. (1995) Cryptand[222] revisited: simultaneous binding of boron and alkaline cations by a pseudocryptand, *Angewandte Chemie International Edition*, **34**, 1115–1117.
9. Drexler, C., Hosseini, M.W., Pratviel, G., Meunier, B. (1998) Design, synthesis and cleaving activity of an abiotic nuclease based on a Mn^{III}-porphyrin complex bearing two acridine moieties, *Journal of the Chemical Society, Chemical Communication*, 1343–1344.
10. Delaigue, X., Harrowfield, J.McB., Hosseini, M.W., De Cian, A., Fischer, J., Kyritsakas, N. (1994) Exoditopic receptor I: synthesis and structural studies on *p-tert*-butyltetramercapto-calix[4]arene and its mercury complexes, *Journal of the Chemical Society, Chemical Communication*, 1579–1580.
11. Akdas, H., Bringel, L., Graf, E., Hosseini, M.W., Mislin, G., Pansanel, J., De Cian, A., Fischer, J. (1998) Thiacalixarenes: synthesis and structural analysis of thiacalix[4]arene and of *p-tert*-butylthiacalix[4]arene, *Tetrahedron Letters*, **39**, 2311–2314.
12. Rao, P., Hosseini, M.W., De Cian, A., Fischer, J. (1999) Synthesis and structural analysis of mercaptothiacalix[4]arene, *Journal of the Chemical Society, Chemical Communication*, 2169–2170.
13. Mislin, G., Graf, E., Hosseini, M.W., Bilyk, A., Hall, A.K., Harrowfield, J.McB., Skelton, B.W., White, A.H. (1999) Thiacalixarenes as cluster keepers: synthesis and structural analysis of a magnetically coupled tetracopper(II) square, *Journal of the Chemical Society, Chemical Communication*, 373–374.

14. Gehin, A., Ferlay, S., Harrowfield, J.McB., Fenske, D., Kyritsakas, N, Hosseini, M.W. (2012) Giant core-shell nanospheric clusters composed of 32Co or 32Ni held by 6 *p-tert*-butylthiacalix[4]arenes, *Inorganic Chemistry*, **51**, 5481–5486.
15. Rao, P., Enger, O., Graf, E., Hosseini, M.W., De Cian, A., Fischer, J. (2000) Mercaptocalixarenes as Mercury(II) extractants: synthesis, structural analysis and extraction properties of lipophilic dimercaptocalix[4]arenes, *European Journal of Inorganic Chemistry*, 1503–1508.
16. Eckes, F., Bulach, V., Guenet, A., Strassert, C., De Cola, L., Hosseini, M.W. (2010) Sensitization of the NIR emission of Nd(III) by the 4a atropoisomer of a *meso*-tetraphenyl porphyrin bearing four 8-hydroxyquinolinylamide chelates, *Chemical Communications*, **46**, 619–621.
17. Guenet, A., Graf, E., Kyritsakas, N., Allouche, L., Hosseini, M.W. (2007) En route to molecular machines: a porphyrin-based gate with a silver lock, *Chemical Communications*, 2935–2937.
18. Lang, T., Graf, E., Kyritsakas, N., Hosseini, M.W. (2011) An oscillating molecular turnstile, *Journal of the Chemical Society, Dalton Transactions*, **40**, 5244–5248.
19. Lang, T., Graf, E., Kyritsakas, N., Hosseini, M.W. (2012) Strapped porphyrin based molecular turnstiles, *Chemistry – A European Journal*, **33**, 10419–10426.
20. (a) Zigon, N., Guenet, A., Graf, E., Hosseini, M.W. (2013) A platinum based organometallic turnstile, *Chemical Communications*, **49**, 3637–3639; (b) Zigon, N., Guenet, A., Graf, E., Kyritsakas, N, Hosseini, M.W. (2013) A platinum turnstile with a palladium lock, *Journal of the Chemical Society, Dalton Transactions*, **42**, 9740–9745; (c) Zigon, N., Kyritsakas, N., Hosseini, M.W. (2014) Organometallic turnstiles: acid and base locking and unlocking, *Journal of the Chemical Society, Dalton Transaction*, **43**, 152–157; (d) Zigon, N., Hosseini, M.W. (2015) A bi-stable organometallic turnstile, *Chemical Communications*, **51**, 12486–12489.
21. (a) Zigon, N., Larpent, P., Jouaiti, A., Kyritsakas, N., Hosseini, M.W. (2014) Optical reading of the open and closed states of a molecular turnstile, *Chemical Communications*, **50**, 5040–5042; (b) Zigon, N., Larpent, P., Jouaiti, A., Kyritsakas, N., Hosseini, M.W. (2014) A luminescent molecular turnstile, *Journal of the Chemical Society, Dalton Transactions*, **43**, 15779–15784.
22. Dunitz, J.D. (1991) Phase transition in molecular crystals from a chemical point of view, *Pure and Applied Chemistry*, **63**, 177–185.
23. Lehn, J.-M. (1995) *Supramolecular Chemistry*, Concepts and Perspectives, VCH, Weinheim.
24. (a) Desiraju, G.D. (1989) *Crystal Engineering: The Design of Organic Solids*, Elsevier, New York; (b) Desiraju, G.R. (1995) Supramolecular synthons in crystal engineering-a new organic synthesis, *Angewandte Chemie International Edition*, **34**, 2311–2327.
25. Etter, M.C. (1990) Encoding and decoding hydrogen-bond patterns of organic compounds, *Accounts of Chemical Research*, **23**, 120–126.
26. Fowler, F.W., Lauher, J.W. (1993) An approach to the design of molecular solids. Strategies for controlling the assembly of molecules into two-dimensional layered structures, *Journal of the American Chemical Society*, **115**, 5991–6000.
27. Mann, S. (1993) Molecular tectonics in biomineralization and biomimetic materials chemistry, *Nature*, **365**, 499–505.
28. Hosseini, M.W. (2004) Molecular tectonics: from simple tectons to complex molecular networks, *Accounts of Chemical Research*, **38**, 313–323.
29. Brand, G., Hosseini, M.W., Félix, O., Schaeffer, P., Ruppert, R. (1996) *Molecular tectonics: self-assembly of charged molecular tectons into one- and two-dimensional solids*. NATO ASI Series, Kahn, O. (Ed), 484, 129.
30. Delaigue, X., Graf, E., Hajek, F., Hosseini, M.W., Planeix J.-M. (1996) *Molecular tectonics*. NATO ASI Series, Tsoucaris, G., Atwood, J.L., Lipkowski, J. (Eds), 480, 159.
31. Simard, M., Su, D., Wuest, J.D. (1991) Use of hydrogen bonds to control molecular aggregation. Self-assembly of three-dimensional networks with large chambers, *Journal of the American Chemical Society*, **113**, 4696–4698.
32. Whitesides, G.M.; Mathias, J.P., Seto, T. (1991) Molecular self-assembly and nanochemistry: a chemical strategy for the synthesis of nanostructures, *Science*, **254**, 1312–1319.
33. Gavezzotti, A. (1994) Are crystal structures predictable?, *Accounts of Chemical Research*, **27**, 309–314.
34. (a) Ermer, O. (1988) Five-fold diamond structure of adamantane-1,3,5,7-tetracarboxylic acid, *Journal of the American Chemical Society*, **110**, 3747–3754; (b) Aakeröy, C.B., Seddon K.R. (1993) The hydrogen bond and crystal engineering, *Chemical Society Reviews*, **22**, 397–407; (c) Subramanian, S., Zaworotko, M.J. (1994) Exploitation of the hydrogen bond: recent developments in the context of crystal engineering, *Coordination*

Chemistry Reviews, **137**, 357–401; (d) Lawrence, D.S., Jiang, T., Levett, M. (1995) Self-assembling supramolecular complexes, *Chemical Reviews*, **95**, 2229–2260; (e) Stoddart, J.F., Philip, D. (1996) Self-assembly in natural and unnatural systems, *Angewandte Chemie International Edition*, **35**, 1154–1196.

35. (a) Batten, S.R., Robson, R. (1998) Interpenetrating nets: ordered, periodic entanglement, *Angewandte Chemie International Edition*, **37**, 1460–1494; (b) Blake, A.J., Champness, N.R., Hubberstey, P., Li, W.-S., Withersby, M.A., Schröder, M. (1999) Inorganic crystal engineering using self-assembly of tailored building-blocks, *Coordination Chemistry Reviews*, **193**, 117–138; (c) Hosseini, M.W. (1999) NATO ASI Series, Braga, D., Grepiono, F., Orpen, G. (Eds), Serie C, Kluwer, Dordrecht, 538, 181; (d) Eddaoui, M., Moler, D.B., Li, H., Chen, B., Reineke, T.M., O'Keeffe, M., Yaghi, O.M. (2001) Modular chemistry: secondary building units as a basis for the design of highly porous and robust metal–organic carboxylate frameworks, *Accounts of Chemical Research*, **34**, 319–330; (e) Swiergers, G.F., Malefetse, T.J. (2000) New self-assembled structural motifs in coordination chemistry, *Chemical Reviews*, **100**, 3483–3538; (f) Janiak, C. (2003) Engineering coordination polymers towards applications, *Journal of the Chemical Society, Dalton Transactions*, 2781–2804; (g) Ferey, G. (2008) Hybrid porous solids: past, present, future, *Chemical Society Reviews*, **37**, 191–214; (h) *Chemical Society Reviews* (2009), **38** (5), themed issue on metal-organic frameworks,; (i) Janiak, C., Vieth, J.K. (2010) MOFs, MILs and more: concepts, properties and applications for porous coordination networks (PCNs), *New Journal of Chemistry*, **34**, 2366–2388 (j) Leong, W.L., Vittal, J.J. (2011) One-dimensional coordination polymers: complexity and diversity in structures, properties, and applications, *Chemical Reviews*, **111**, 688–764; (k) Kole, G.K., Vittal, J.J. (2013) Solid-state reactivity and structural transformations involving coordination polymers, *Chemical Society Reviews*, **42**, 1755–1775.

36. Hosseini, M.W., De Cian, A. (1998) Crystal engineering: molecular networks based on inclusion phenomena, *Journal of the Chemical Society, Chemical Communications*, 727–734.

37. Martz, J., Graf, E., De Cian, A., Hosseini, M.W. (2003) In *Perspectives in Supramolecular Chemistry*, Desiraju, G. (Ed), John Wiley, New York, 7, 177.

38. Delaigue, X., Hosseini, M.W., De Cian, A., Fischer, J., Leize, E., Kieffer, S., Van Dorsselaer, A. (1993) Multicavitands I: synthesis and X-ray crystal structure of a bis *p-tert*-butylcalix[4]arene fused by two silicon atoms, *Tetrahedron Letters*, **34**, 3285–3288.

39. Gutsche, C.D. (1989) *Calixarenes*, Monographs in Supramolecular Chemistry, Stoddart, J.F. (Ed), RSC, London.

40. Martz, J., Graf, E., Hosseini, M.W., De Cian, A., Fischer J. (1998) Molecular tectonics X: design, synthesis and structural analysis of a molecular network based on inclusion processes in the crystalline phase, *Journal of Materials Chemistry*, **8**, 2331–2333.

41. Hajek, F., Graf, E., Hosseini, M.W., Delaigue, X., De Cian, A., Fischer, J. (1996) Molecular tectonics I: the first synthesis and X-ray analysis of a linear koilate obtained by self-assembly of linear koilands and hexadiyne, *Tetrahedron Letters*, **37**, 1401–1404.

42. Martz, J., Graf, E., Hosseini, M.W., De Cian, A., Fischer, J. (2000) Directional 1-D inclusion networks: self-assembly of unsymmetrical koilands into directional koilates in the crystalline phase, *Journal of the Chemical Society, Dalton Transactions*, 3791–3795.

43. Taylor, R., Kennard, O. (1984) Hydrogen-bond geometry in organic crystals, *Accounts of Chemical Research*, **17**, 320–326.

44. Saladino, R., Hanessian S. (2003) In *Crystal Design and Function, Perspectives in Supramolecular Chemistry*, Desiraju, G.R. (Ed), John Wiley, New York.

45. Grosshans, P., Jouaiti, A., Bulach, V., Planeix, J.-M., Hosseini, M.W., Nicoud, J.-F. (2003) Molecular tectonics: design of enantiomerically pure single-stranded helical H-bonded molecular networks, *Crystal Engineering Communications*, **5**, 414–416.

46. Hosseini, M.W. (2003) Molecular tectonics: from molecular recognition of anions to molecular networks, *Coordination Chemistry Reviews*, **240**, 157–166.

47. (a) Holman, K.T., Pivovar, A.M., Swift, J.A., Ward, M.D. (2001) Metric engineering of soft molecular host frameworks, *Accounts of Chemical Research*, **34**, 107–118; (b) Ward, M.D. (2005) Design of crystalline molecular networks with charge-assisted hydrogen bonds, *Journal of the Chemical Society, Chemical Communincations*, 5838–5842; (c) Russell, A., Etter, M.C., Ward, M.D. (1994) Layered materials by molecular design: structural enforcement by hydrogen bonding in guanidinium alkane- and arenesulfonates, *Journal of the American Chemical Society*, **116**, 1941–1952; (d) Russell, V.A., Evans, C.C., Li, W., Ward, M.D. (1997) Nanoporous molecular sandwiches: pillared two-dimensional hydrogen-bonded networks with adjustable porosity, *Science*, **276**, 575–579.

48. (a) Hosseini, M.W., Ruppert, R., Schaeffer, P., De Cian, A., Kyritsakas, N., Fischer, J. (1994) A molecular approach to the solid state synthesis: prediction and synthesis of self-assembled infinite rods, *Journal of the Chemical Society, Chemical Communications*, 2135–2136; (b) Brand, G., Hosseini, M.W., Ruppert, R., Cian, A. De, Fischer, J., Kyritsakas, N. (1995) A molecular approach to organic solids: molecular recognition of arene sulfonate by a biscyclic amidinium in the solid state, *New Journal of Chemistry*, **19**, 9–13; (c) Félix, O., Hosseini, M.W., De Cian, A., Fischer, J. (1997) Molecular tectonics IV: molecular networks based on hydrogen bonding and elecrostatic interactions, *Tetrahedron Letters*, **38**, 1755–1758; (d) Félix, O., Hosseini, M.W., De Cian, A., Fischer, J. (1997) Molecular tectonics V: molecular recognition in the formation of molecular networks based on hydrogen bonding and electrostatic interaction, *Tetrahedron Letters*, **38**, 1933–1936; (e) Félix, O., Hosseini, M.W., De Cian, A., Fischer, J. (1997) Molecular tectonics III: the simultaneous use of h-bonding and charge-charge interactions for the self-assembly of fumaric acid and cyclic bisamidinium into one- and two-dimensional molecular networks, *Angewandte Chemie International Edition*, **36**, 102–104; (f) Félix, O., Hosseini, M.W., De Cian, A., Fischer, J. (2000) Crystal engineering of 2-D hydrogen bonded molecular networks based on the self-assembly of anionic and cationic modules, *Journal of the Chemical Society, Chemical Communications*, 281–282; (g) Félix, O., Hosseini, M.W., De Cian, A. (2001) Design of 2-D hydrogen bonded molecular networks using pyromellitate dianion and cyclic bisamidinium dication as complementary tectons, *Solid State Sciences*, **3**, 789–793.
49. Hosseini, M.W., Brand, G., Schaeffer, P., Ruppert, R., De Cian, A., Fischer, J. (1996) A molecular approach to solid structures II: prediction and synthesis of sheets through assembly of complementary molecular units, *Tetrahedron Letters*, **37**, 1405–1408.
50. Ferlay, S., Hosseini, M.W. (2011) Molecular tectonics: design of hybrid networks and crystals based on charge assisted hydrogen bonds, In *Functional Supramolecular Architectures for Organic Electronics and Nanotechnology*, Samori, P., Cacialli, F. (Eds), Wiley-VCH, Weinheim.
51. Ferlay, S., Félix, O., Hosseini, M.W., Planeix, J.-M., Kyritsakas, N. (2002) Second sphere supramolecular chirality: racemic hybrid H-bonded 2-D molecular networks, *Journal of the Chemical Society, Chemical Communications*, 702–703.
52. (a) Dechambenoit, P., Ferlay, S., Hosseini, M.W. (2005) From tectons to composite crystals, *Crystal Growth & Design*, **5**, 2310–2312; (b) Ferlay, S., Hosseini, M.W. (2004) Crystalline molecular alloys, *Journal of the Chemical Society, Chemical Communications*, 788–789.
53. (a) MacDonald, J.C., Dorrestein, P.C., Pilley, M.M., Foote, M.M., Lundburg, J.L., Henning, R.W., Schutlz, A.J., Manson, J.L. (2000) Design of layered crystalline materials using coordination chemistry and hydrogen bonds, *Journal of the American Chemical Society*, **122**, 11692–11702; (b) Luo, T.-J.M., MacDonald, J.C., Palmore, G.T.R. (2004) Fabrication of complex crystals using kinetic control, chemical additives, and epitaxial growth, *Chemistry of Materials*, **16**, 4916–4927; (c) Noveron, J.C., Lah, M.S., Del Sesto, R.E., Arif, A.M., Miller, J.S., Stang, P.J. (2002) Engineering the structure and magnetic properties of crystalline solids via the metal-directed self-assembly of a versatile molecular building unit, *Journal of the American Chemical Society*, **124**, 6613–6625.
54. (a) Schmaltz, B., Jouaiti, A.; Hosseini, M.W., De Cian, A. (2001) Double stranded interwound infinite linear silver coordination network, *Journal of the Chemical Society, Chemical Communications*, 1242–1243; (b) Jouaiti, A., Hosseini, M.W., Kyritsakas, N. (2003) Molecular tectonics: infinite cationic double stranded helical coordination networks, *Journal of the Chemical Society, Chemical Communications*, 472–473.
55. (a) Grosshans, P., Jouaiti, A., Bulach, V., Planeix, J.-M., Hosseini, M.W., Nicoud, J.-F. (2003) Molecular tectonics: from enantiomerically pure sugars to enantiomerically pure triple stranded helical coordination network, *Journal of the Chemical Society, Chemical Communications*, 1336–1337; (b) Grosshans, P., Jouaiti, A., Bulach, V., Planeix, J.-M., Hosseini, M.W., Nicoud, J.-F. (2004) Molecular tectonics: design and structural analysis of enantiomerically pure tectons and helical coordination networks, *Comptes Rendus Chimie*, **7**, 189–196.
56. Kaes, C., Hosseini, M.W., Rickard, C.E.F., Skelton, B.W., White, A.H. (1998) Synthesis and structural analysis of a helical coordination polymer formed by the self-assembly of a 2,2′-bipyridine based exoditopic macrocyclic ligand and silver cations, *Angewandte Chemie International Edition*, **37**, 920–922.
57. Jouaiti, A., Hosseini, M.W., De Cian, A. (2000) Design, synthesis and structural investigation of a 1-D directional coordination network based on the self-assembly of an unsymmetrical mono-tridentate ligand and cobalt cation, *Journal of the Chemical Society, Chemical Communications*, 1863–1864.
58. Jouaiti, A., Hosseini, M.W., Kyritsakas, N. (2002) Non-centrosymmetric packing of 1-D coordination networks based on chirality, *Chemical Communications*, 17, 1898–1899.

5

Three Tales of Supramolecular Analytical Chemistry

Margaret K. Meadows and Eric V. Anslyn

Department of Chemistry and Biochemistry, The University of Texas at Austin, Austin, TX, USA

5.1 Introduction

Eric Anslyn received the Izatt–Christensen Award in 2013 during the ISMSC conference that was organized by Lyle Issacs, Jeff Davis, and Amar Flood. It was a great honor to receive this award, particularly at a symposium wonderfully organized and executed by these three close friends. Receiving the I.C. award is a hallmark achievement in one's career, both due to the fact that it is the highest recognized award for chemists involved in the study of supramolecular and macrocyclic systems, and because the previous winners of this award are among inspirational leaders of these fields. Thus, an invitation to write a chapter as a perspective across his career was eagerly accepted.

Over the last decade and a half, our group has made a direct effort to show how supramolecular chemistry can be used in analytical protocols with real-life applications. As this chapter will reveal, many of the inspirations for these endeavors came serendipitously, or were initiated through conversations with colleagues who requested particular assays. Helping colleagues solve analytical problems with supramolecular chemistry still guides the vast majority of the work our group is pursuing. We have coined the term "Supramolecular Analytical Chemistry" to describe the specific endeavor of making sensors and generating assays using synthetic receptors and assemblies.

When considering how to structure this chapter, it seemed best to highlight precisely the same topics that were covered during the award lecture in 2013. Therefore, the chapter name is the same as that lecture, and follows the same course of scientific logic. The goal is to discuss three topics, not in great depth, but instead to show the initial inspiration and the way in which the topic has evolved to the current research that is still ongoing. Thus, herein we lay out "Three Tales of Supramolecular Analytical Chemistry."

Macrocyclic and Supramolecular Chemistry: How Izatt–Christensen Award Winners Shaped the Field, First Edition.
Edited by Reed M. Izatt.

5.2 Citrate sensing

5.2.1 Receptor design

In the process of designing bis-alkylguanidinium enzyme mimics for RNA hydrolysis, our research group accidentally discovered a sensor for citrate. We were developing artificial enzymes that could catalyze RNA cleavage and had the benefit of being metal- and peptide-free [1].

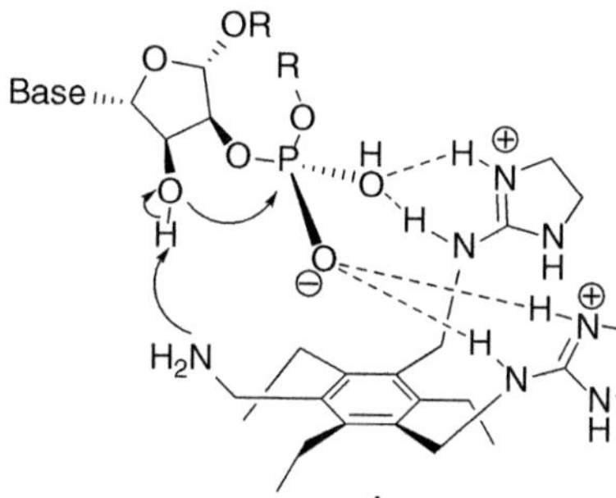

1

During the synthesis of the pinwheel-like receptor **1**, a major byproduct was the tris-alkylguanidinium **2**. In attempting to find uses for this tris-aminoimidazolinated benzene, a receptor for citrate was proposed [2]. A trianion at neutral pH, citrate is prevalent both in biological systems and in common beverages, but had proven difficult to sense in aqueous solutions, as hydrophilic substrates such as citrate are known to be difficult to bind in competitive media. Before our work, most reported citrate receptors did not demonstrate the sensitivity or selectivity necessary to function in solution with competitive contaminates.

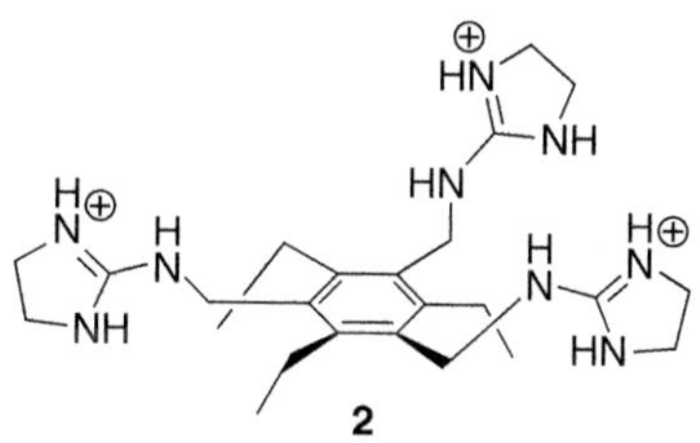

2

The structure of tris-alkylguanidinium **2** complements the structure of citrate through the respective distribution of charges and hydrogen bonds. The three protonated guanidinium groups bind the three carboxylates of citrate in a tridentate fashion (Figure 5.1). The binding properties of **2** were studied in different solvent/buffer

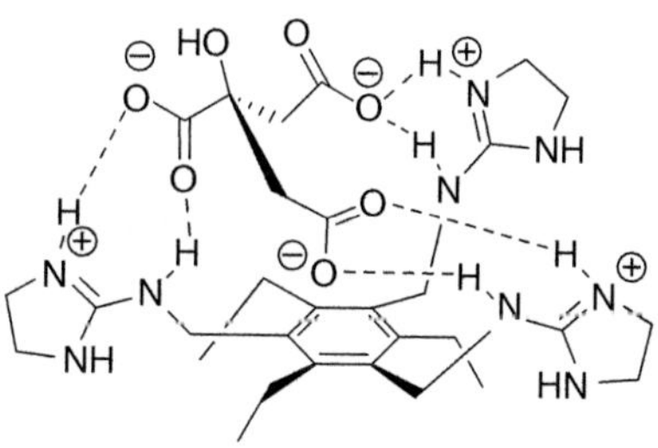

Figure 5.1 *Receptor **2** bound to citrate*

Table 5.1 Binding constants of the receptors **2, 3**, and **4** in D_2O at pH 7.4. Source: [2]. Adapted with permission of Wiley-VCH Verlag GmbH & Co. KGaA

Host	Guest	Binding constant [M^{-1}]
2	citrate	6.9×10^3
2	tricarballate	7.3×10^3
2	succinate	2.1×10^2
2	glutarate	2.2×10^2
2	acetate	<10
2	ATP^{4-}	1.2×10^3
2	3′-Up	2.1×10^2
2	cyclo-2,3-Up	<10
3	citrate	3.0×10^3
4	citrate	2.4×10^3

systems and with different guests. We found that the selectivity of the receptor for citrate was strongest in high dielectric solutions. Buffers with phosphate or sulfonate groups exhibited lower binding, as these groups are known to interact with guanidinium functionalities [3, 4]. Citrate and tricarballate, a citrate analog lacking the OH functionality, bound more strongly than other guests by an order of magnitude, including mono- and di-carboxylates and ATP (Table 5.1). The utility of both the guanidinium and the ethyl groups were verified through testing the binding of citrate to compounds **3** and **4**. Although **3** bound citrate in pure water, the binding was less than half that of **2** and a similar drop in binding affinity was found for **4**.

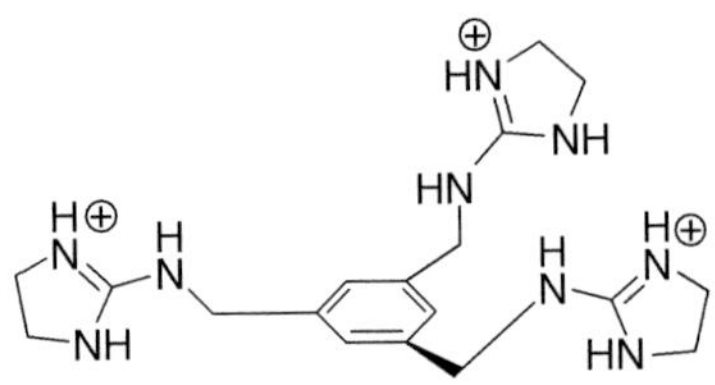

3

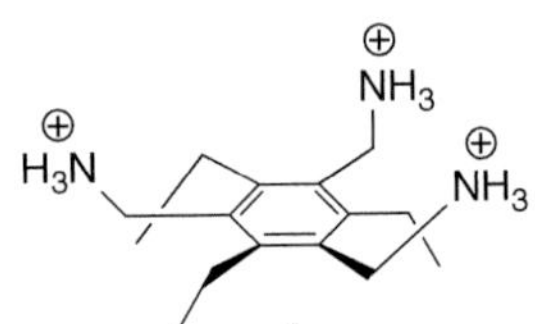

4

5.2.2 Indicator-displacement assays

In transitioning from a receptor for citrate to a sensor for citrate, we developed what became known as an indicator-displacement assay (IDA). Up to this point, most chemosensors used the receptor–spacer–reporter approach [5]. In this strategy, a chromophore or fluorophore is attached to a spacer that is attached to the receptor. Upon binding, there is a response from the reporter in the form of a change in absorbance or fluorescence. In contrast, indicator-displacement assays use an indicator that binds reversibly to the receptor and can be displaced by the substrate of interest. This displacement causes a change in the indicator's optical signal (Scheme 5.1). These assays are similar to immunoassays which use labeled and unlabeled antigens to measure binding [6].

Scheme 5.1 *Indicator-displacement assay*

There are several pathways through which the indicator's optical signal can be modulated in an IDA: photoinduced electron transfer (PET), fluorescence resonance energy transfer (FRET), electronic energy transfer (EET), or changes in local ionic strength or pH [7]. Another benefit of IDAs is that the indicator does not need to be covalently attached to the receptor, which simplifies the synthesis of the receptor. In addition, multiple indicators can be used with each receptor to provide the best optical signal.

5.2.3 Soda analysis

With our citrate receptor **2**, we used commercially available 5-carboxyfluorescein (CF) as an indicator [8]. Citrate was titrated into a solution of receptor **2** and indicator. These studies allowed us to plot calibration curves for the addition of citrate to our sensing assay from the fluorescence data. We also studied the addition of succinate and acetate, which gave smaller responses (Figure 5.2).

Once we had these calibration curves, we were able to measure the concentration of citrate in various beverages (Table 5.2). Monitoring either absorbance or fluorescence resulted in good accuracy of concentrations. In order to verify the concentrations, we used NMR and gravimetric analysis. We also analyzed sports drinks that are high in electrolyte content to ensure that other anions were not affecting our citrate assay.

5.2.4 Wine analysis

A similar assay was used to create a chemosensor for tartrate and malate [9]. Our citrate receptor was modified to include one boronic acid moiety for binding the diol of tartaric acid (**5**). Using alizarin complexone (**AC**) as a colorimetric indicator, the binding constant of receptor **5** and tartrate in a 75% methanol/water mixture was found to be 5.5×10^4 M^{-1} (Scheme 5.2). This receptor was also used in the threshold detection of tartrate and malate in wine and grape juices (Table 5.3).

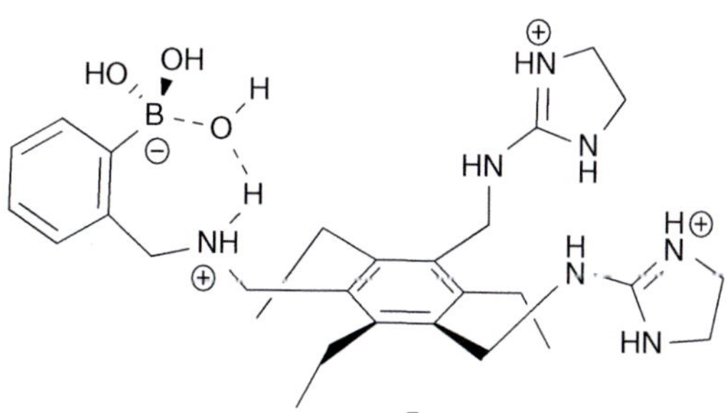

5

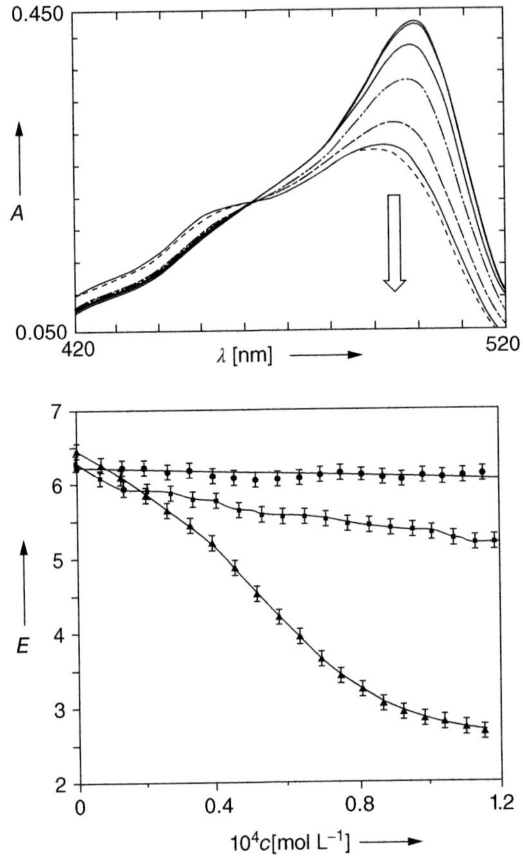

Figure 5.2 Top: Absorbance change when citrate is titrated into receptor **2** and 5-carboxyfluorescein. Bottom: Fluorescence-based calibration curves for citrate (▲), succinate (■), and acetate (●). Source: [8]. Reproduced with permission of Wiley-VCH Verlag GmbH & Co. KGaA

Table 5.2 Concentration of citrate [mM] in beverages determined by various methods. Source: [9]. Adapted with permission of the American Chemical Society

Solution	Gravimetric	NMR	**2** + **CF** absorption	**2** + **CF** emission
citrate model solution	30.1	–	30.3	29.9
calibration solution	1.33	–	1.205	1.39
orange juice	–	43.1	44.1	44.7
Gatorade	–	15.95	15.05	15.1
Powerade	–	12.4	11.1	11.3
All Sport	–	7.4	7.1	8.1
Mountain Dew	–	7.95	5.5	5.4
tonic water	–	21	21.12	20.8
Coca Cola	–	0	0	<0.5
Diet Coke	–	<0.2	<0.4	<0.7

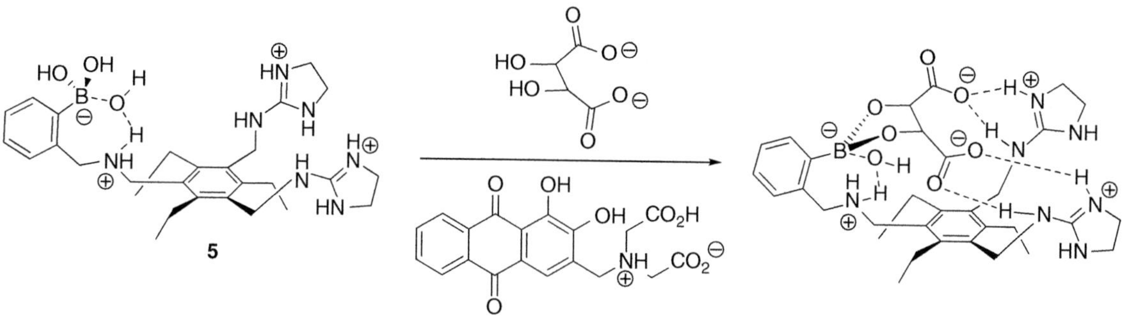

Scheme 5.2 *Receptor **5** binding tartrate. Indicator alizarin complexone (**AC**) shown*

Table 5.3 *Concentration of tartaric and malic acids [mM] in solutions determined by various methods. Source: [9] Adapted with permission of the American Chemical Society*

Solution	Gravimetric	NMR	**5** + **AC** absorption
tartaric acid model solution	50	51.2	50.2
calibration solution	50	–	47.9
Ernest & Julio Gallo Sauvignon Blanc	–	35.6	32.9
Ste. Genevieve Chardonnay	–	34.1	36.3
Henri Marchant Spumante	–	26.5	24.9
Talus Merlot	–	19.5	20.3
Santa Cruz organic white grape juice	–	43.6	42.3
Welch's grape juice	–	69.4	71.3

5.2.5 Scotch analysis

Phenolic acids, such as gallic acid, are present in Scotch whisky. During the production process, the whisky extracts phenolic acids from the oak barrels it is aged in. Gallic acid and similar compounds are indicators of the age of Scotch. Receptor **6** was used to bind gallate and gallate-like structures as a method of measuring the age of Scotch [10]. Two of the nitrogen groups of our original citrate receptor were capped with *ortho*-phenylboronic acids and the remaining amine was capped with a guanidinium group. Using an indicator displacement assay with pyrocatechol violet in a 75% methanol/water mixture, receptor **6** was found to bind gallate with a binding constant of $1.0 \times 10^4\ \mathrm{M}^{-1}$. UV/Vis analysis provided a "response number" via calibration curves for various phenolic acids. This approach correlated the response to more than just the concentration of gallate as previous sensing approaches had; instead, we were able to visualize a family of phenolic acids. The low selectivity of this sensor allowed for a more inclusive response than selective gallate sensors.

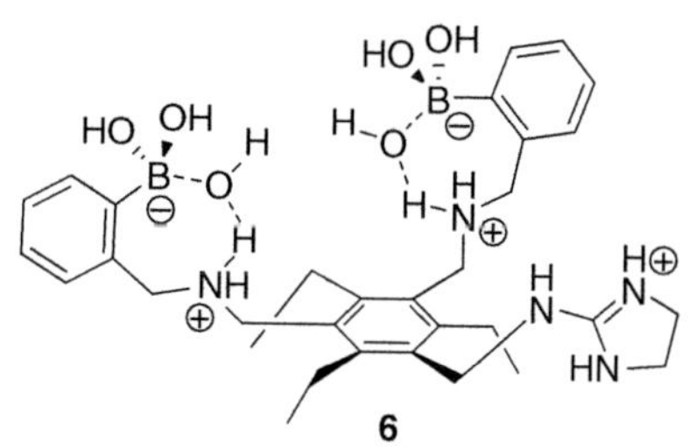

6

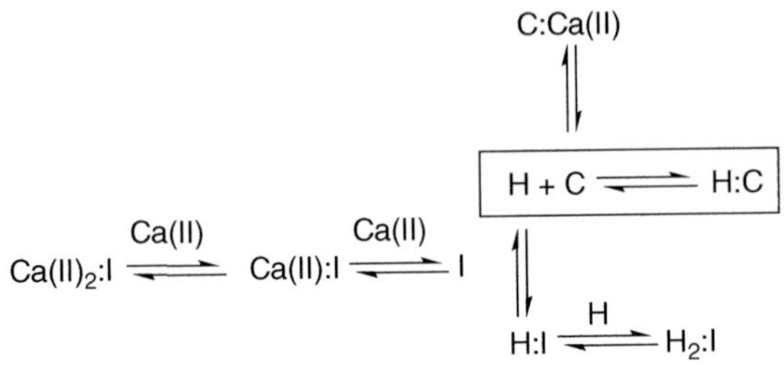

Scheme 5.3 Equilibria between host (H), citrate (C), and indicator (I). Source: [12]. Reproduced with permission of Elsevier

5.2.6 Calcium and citrate multianalyte sensing

To expand the versatility of our family of receptors, we developed multianalyte sensing Schemes, some of which used artificial neural networks (ANN) for analysis [11]. One of the ANNs was used in conjunction with our trisguanidinated citrate receptor **2** to develop an assay for simultaneously measuring the concentrations of citrate and calcium (Ca^{2+}). We selected xylenol orange (**XO**) as our indicator of choice due to its affinity for divalent cations like calcium. The benefit of this IDA is that citrate can displace xylenol orange, allowing it to bind calcium, or calcium can bind to xylenol orange which releases the indicator from the citrate receptor (Scheme 5.3).

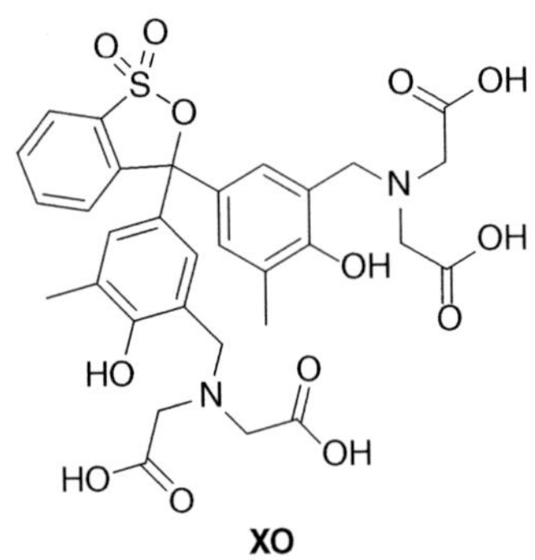

XO

We used this system to determine the concentrations of citrate and calcium in flavored vodkas. The optical data from our assay was analyzed using a multi-layer perceptron (MLP) artificial neural network (ANN) [12]. The input values were absorbance measured at several wavelengths in the UV–Vis spectrum, with the output being concentrations of calcium and citrate (Figure 5.3).We used a training set, one where we knew the concentrations of our analytes, to "teach" the ANN to convert absorbance data into measurements of concentration. We then used the ANN to analyze different flavored vodkas (Table 5.4). The concentrations of citrate in the samples were validated through NMR.

5.2.7 Dialysis

This method to simultaneously measure citrate and Ca^{2+} in vodka led to an application that we had never considered when we first studied citrate in soft drinks. A few years after publishing the paper about vodka,

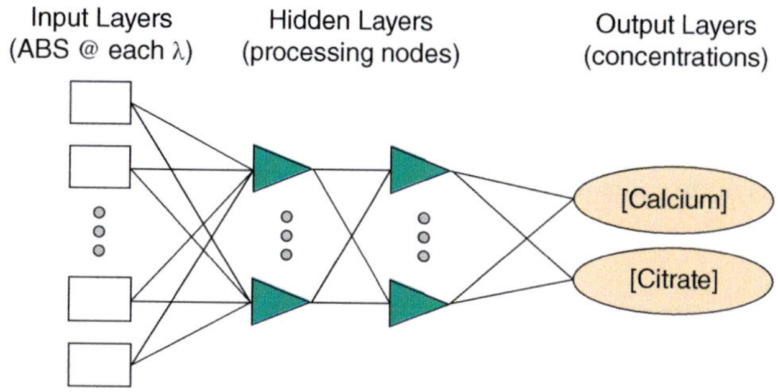

Figure 5.3 *Schematic of the ANN. Source: [12]. Reproduced with permission of Elsevier*

Table 5.4 *Concentration of citrate and calcium in solutions determined by various methods. Source: [12]. Adapted with permission of Elsevier*

Prepared Samples			
Real values		Predicted values	
[citrate] (mM)	[calcium] (mM)	[citrate] (mM)	[calcium] (mM)
0	0.1	0.03±0.01	0.12±0.04
0.2	0.4	0.15±0.03	0.37±0.01
0.4	0.15	0.35±0.005	0.12±0.002
0.6	0.2	0.56±0.01	0.18±0.001
0.8	0.35	0.87±0.05	0.30±0.01

Vodkas			
Flavor	Predicted values		NMR determined values
	[citrate] (mM)	[calcium] (mM)	[citrate] (mM) (15% error)
Citrus twist	0.98	0.72	0.9
Orange	0.31	0.21	0.44
Vanilla	1.18	0.74	1.16
Green apple	1.55	0.02	1.04
Raspberry	1.37	0.11	1.3

Dr. Anslyn received a phone call from a nephrologist, Dr. Balazs Salmosfalvi, from Henry Ford Hospital in Detroit. This physician had found our paper and described a problem that he encounters during dialysis therapy. In essence, he described that the standard anti-coagulant heparin [13] could be replaced with the safer anti-coagulant citrate, if both citrate and calcium were measureable in real time. Thus, we set out to extend our vodka work to dialysis fluid [14]. Hemodialysis is a renal replacement therapy for patients who have advanced or permanent kidney failure. Treatments such as these remove toxins from the blood, but clotting is a serious concern during this process. Citrate is used as an anticoagulant because it binds Ca^{2+}, a key cofactor in the initiation of coagulation; however, the concentration of citrate must be monitored because too much citrate can cause hypocalcemia and sudden death. Previous methods for determination of citrate and

Ca^{2+} levels required off-site lab analysis, which could take hours. Our sensor for calcium and citrate allowed us to develop an automated online method of analysis [15].

Using a sequential injection analysis (SIA) approach, we built a prototype instrument (Figure 5.4). We used our pinwheel receptor **5** and alizarin complexone for citrate and introduced Fura-2 as an independent Ca^{2+} receptor (Scheme 5.4). This allowed us to avoid the use of an ANN.

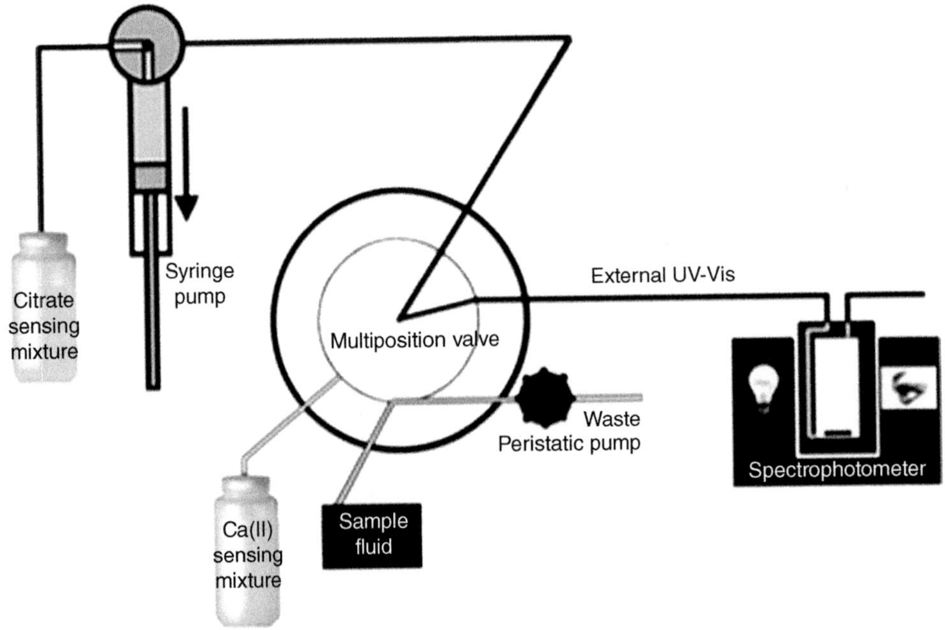

Figure 5.4 *Schematic of the SIA prototype instrument. Source: [15]. Reproduced with permission of Royal Society of Chemistry*

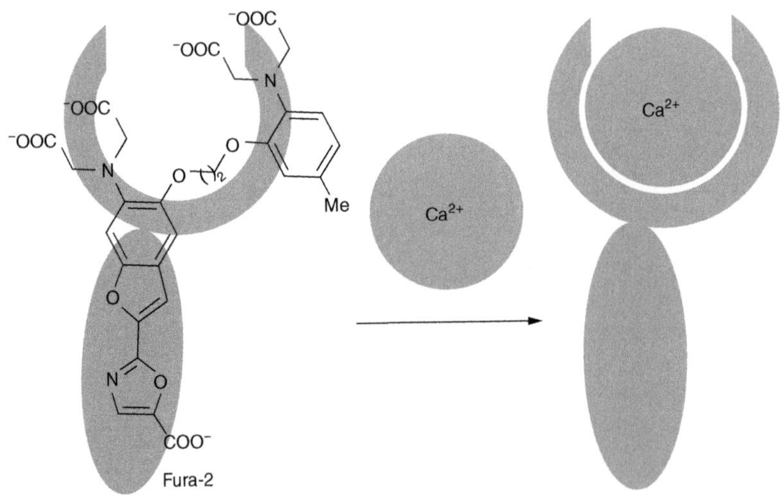

Scheme 5.4 *Fura-2 binding Ca^{2+}. Source: [15]. Reproduced with permission of Royal Society of Chemistry*

Samples from hemodialysis were obtained from Henry Ford Hospital, and these were analyzed through our system. Less than 15% error was observed in our analysis of these samples (results corroborated by NMR).

5.2.8 Conclusion

In this first tale we described how an initially unwanted side-product of a reaction challenged us to devise a use for that compound. Having recognized that **2** would bind citrate, we then needed to develop an optical method to signal the concentration of citrate. This led to the creation of Indicator Displacement Assays (IDAs), which have become generally used in many settings. We developed IDAs simply because we were forced to devise a spectroscopic technique to signal a guest lacking a chromophore with a host that possessed only a benzene ring, and we did not want to do any additional synthesis. Thus, the old adage, "necessity is the mother of invention", is apropos. Further, we had no foreknowledge that the study of vodka would lead ultimately to the analysis of dialysis fluids in a hospital setting. The overarching lesson is to take any serendipitous discovery as far as possible. Such studies demonstrate that one never knows where basic science breakthroughs can lead.

5.3 Rapid analysis of enantiomeric excess

This second tale describes an idea that was given to our group during a consulting visit to Merck Frost Montreal in 2003 where the seminar had a focus on the analysis of soft drinks, wines, and whiskies. Chemists at Merck specializing in synthetic methodology inquired if supramolecular assays could be used to create optical methods to measure enantiomeric excess (*ee*). These questions from Merck scientists became the inspiration for the work now described [16].

Asymmetric synthesis provides an avenue for synthesizing one enantiomer over the other, but the effectiveness of each method must be analyzed [17]. High-throughput screening (HTS) enables the determination of the method that yields the highest *ee*. HTS requires rapid analysis methods to determine these *ee* values; traditionally, this has been done through chiral HPLC, but this method is time-consuming [18]. When screening many samples, any additional time for analysis can result in a bottleneck in the HTS process. Optical spectroscopy provides rapid analysis of samples and omits chromatographic separation; however, these methods require derivatization of the analytes with a reporter unit [19]. Host–guest chemistry is ideal as a reversible derivatization process.

5.3.1 Enantioselective indicator-displacement assays

Our group developed enantioselective indicator displacement assays (eIDAs) that have been used to determine the *ee* values of α-hydroxycarboxylates, vicinal diols, and α-amino acids [20–22]. In eIDAs, the different enantiomers bind to a chiral receptor to form diastereomeric complexes. These diastereomers have differing energies that lead to different levels of indicator displacement and therefore different signals (Figure 5.5).

5.3.1.1 ee Analysis with boronic acids

With our first eIDA, we used the well-known binding of boronic acids with α-hydroxyacids and vicinal diols to form boronate esters [23]. Achiral boronic acid **7** binds α-hydroxycarboxylates and displaces indicators such as pyrocatechol violet well, but naturally does not differentiate between enantiomers. Chiral receptors with one stereogenic center, such as compound (*S*)-**8,** were found to be less effective than those with two,

Figure 5.5 An example of an enantioselective indicator displacement assay and the color change associated with each enantiomer. [21]. Reproduced with permission of American Chemical Society

such as (*S,S*)-**9**. The initial study measured *ee* values of α-hydroxyacids using boronic acid receptor (*S,S*)-**9** and indicator pyrocatechol violet with an average of ±15% error. Although this error is far larger than chiral HPLC, low analysis times can allow for a cursory screening of analytes and "hits" can be further characterized. We expanded this *ee*-sensing scheme to include 1,2-*cis*-diols, and various fluorescent indicators were investigated to see if they would improve the assay's sensitivity and accuracy. We found that 4-methylesculetin and receptor (*S,S*)-**9** could be used to correlate optical signal with *ee* and concentration (Figure 5.6). This approach gave an average error in *ee* of ±7%.

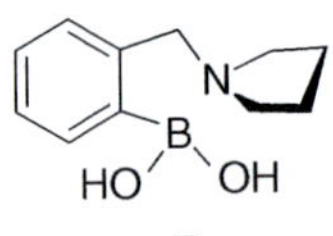

7

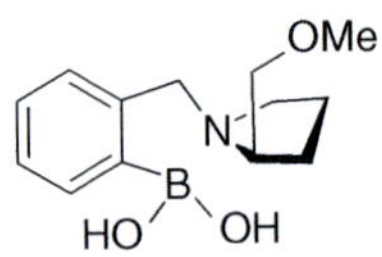

(*S*)-**8**

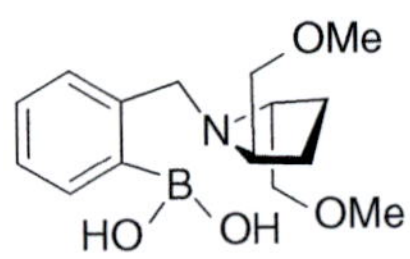

(*S,S*)-**9**

To demonstrate the effectiveness of our eIDA on HTS, we developed a stepwise protocol [24]. This protocol was used to analyze samples of hydrobenzoin with unknown *ee*. First, chiral boronic acid receptors and

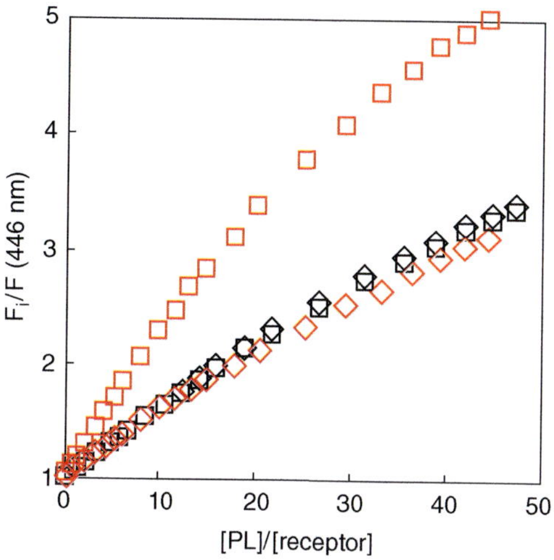

Figure 5.6 *Fluorescence quenching profiles of phenyllactic acid (PL) and receptors 7 and (S,S)-9. (◊=D-PL/7, □=L-PL/7, ◊=D-PL/(S,S)-10, □=L-PL/(S,S)-10). Source: [21]. Reproduced with permission of American Chemical Society*

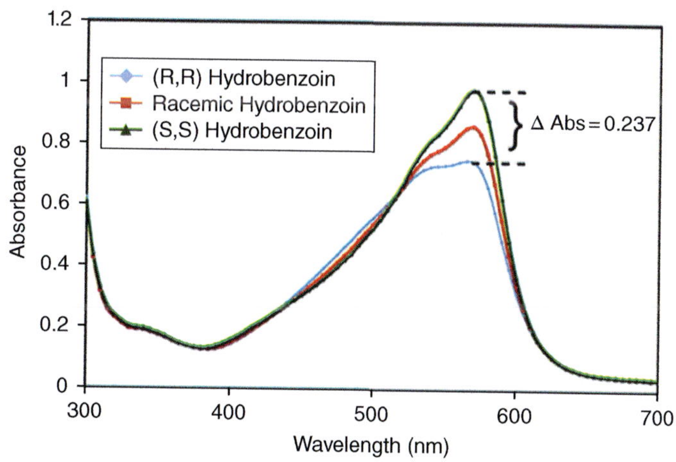

Figure 5.7 *Absorbance data of hydrobenzoin and receptor (S,S)-**9**. Source: [24]. Reproduced with permission of the National Academy of Sciences of the United States of America*

catechol-based indicators were screened for the best enantiodifferentiation of hydrobenzoin. We found (*S,S*)-**9** and bromopyrogallol red were the best combination for *ee* determination (Figure 5.7). We then set up three 96-well plates with three different hosts (**7**, (*S,S*)-**9**, and (*R,R*)-**9**) and bromopyrogallol red, and created a training set of known concentrations of hydrobenzoin enantiomers. This training set was used with an artificial neural network (ANN) for data analysis. Unknown samples of hydrobenzoin with varying *ee* were made and tested with good results. Following this, hydrobenzoin was synthesized using Sharpless asymmetric dihydroxylation chemistry and a variety of ligands [24]. Our assay was able to determine the best ligand for the reaction as reported in the literature.

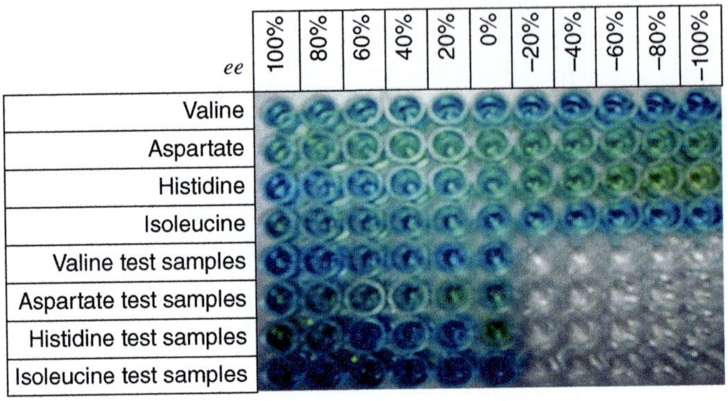

Figure 5.8 96-well plates of amino acids, **10**, and chromazurol S. Source: [26]. Reproduced with permission of the American Chemical Society

5.3.1.2 *ee* Analysis of amino acids

We also developed an eIDA for determining the *ee* of α-amino acids in protic media. This utilized a Cu^{II} complex with indicator chromazurol S to form the receptor complex (**10**). Complexation of chiral amino acids created diastereomeric compounds and displaced the indicator. An *ee* calibration curve showed an average of ±12% error in *ee* determination. This eIDA was also extended to a 96-well plate platform with four α-amino acids (Figure 5.8) [25].

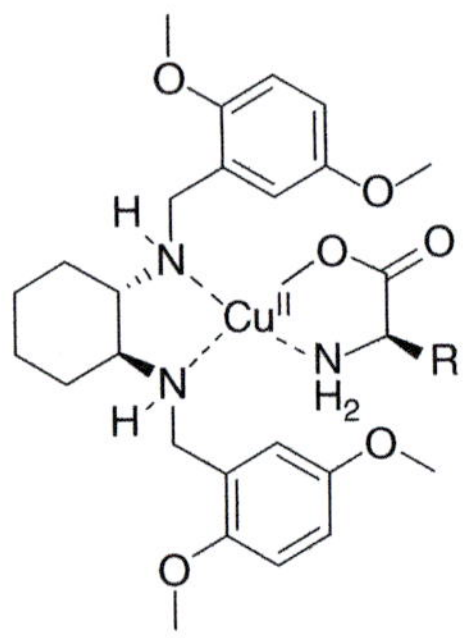

10

5.3.2 *ee* Determination using circular dichroism

While the above-mentioned colorimetric methods for determining *ee* values work, it must be conceded that the assays are complicated. One needs to synthesize a chiral host, pair it with an indicator for an IDA, and process the data with chemometrics, and even then, after all is said and done, the assays are not very accurate. Therefore, after taking stock of the limitations of our assays, we decided to change from UV/Vis absorbance spectroscopy to circular dichroism.

Circular dichroism is a spectroscopy method that is sensitive to chirality [26–29]. CD uses circularly polarized light that interacts with chiral chromophores, producing mirrored spectra for a pair of enantiomers. The use of CD also allows for simplification of host design. Although most chiral organic building blocks do not display strong CD signals, our goal is to amplify this signal with supramolecular assemblies. Metal complexes can have metal-to-ligand charge transfer (MLCT) which gives more intense CD signals [30]. A helical twist on chromophores leads to exciton-coupled circular dichroism (ECCD) which also amplifies the CD signal [31].

5.3.2.1 *CD analysis using metal-to-ligand charge transfer*

Our group's first CD assay involved a modulation of metal-to-ligand charge transfer bands (MLCT bands) [32]. CD spectroscopy of common organic molecules is difficult due to overlap in the 190–220 nm region. We sought to address this by using MLCT bands from the simple inorganic copper complex **11**. This complex was used with *R*- or *S*-BINAP to discriminate the chirality of various vicinal diamines (**12–15**). BINAP provides axial chirality due to limited rotation from steric hindrance of the naphthyl rings, which produces a large CD signal. With this assay, we were able to successfully distinguish enantiomers of the chiral analytes as well as the identity of the diamine in unknown samples. We were also able to quantitate concentration and *ee* for known analytes.

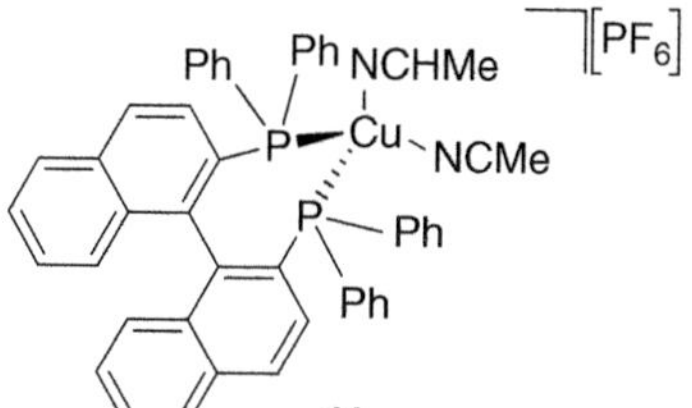

11

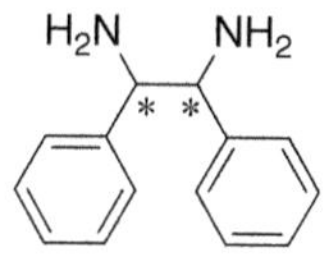

12

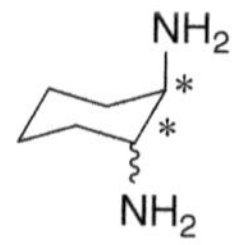

13

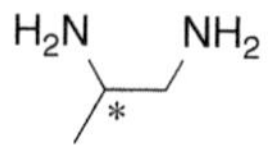

14

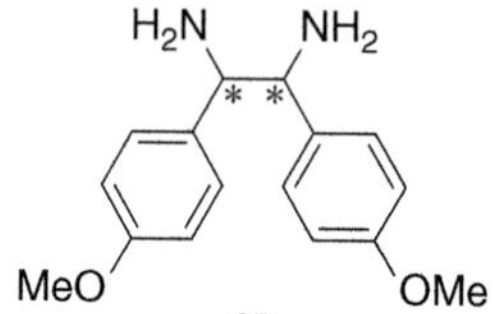

15

Complex **11** was also used for the analysis of α-chiral primary amines and α-chiral cyclohexanones [28, 33]. Both of these classes of compounds require derivatization prior to discrimination. The amines are derivatized though imine formation (Scheme 5.5). These chiral imines are then complexed with the copper–BINAP complex. We were able to determine *ee* with an average error of $\pm17\%$. For analysis of the α chiral cyclohexanones, the cyclohexanones were derivatized with 1-methyl-1-(2-pyridyl)hydrazine to form bidentate ligands for complexation (Scheme 5.6). Upon binding to the copper complex **11**, the modified cyclohexanones showed a CD signal which allowed for determination of *ee*. Once a calibration curve was created through analysis of known samples, unknown samples were tested and identified with good accuracy.

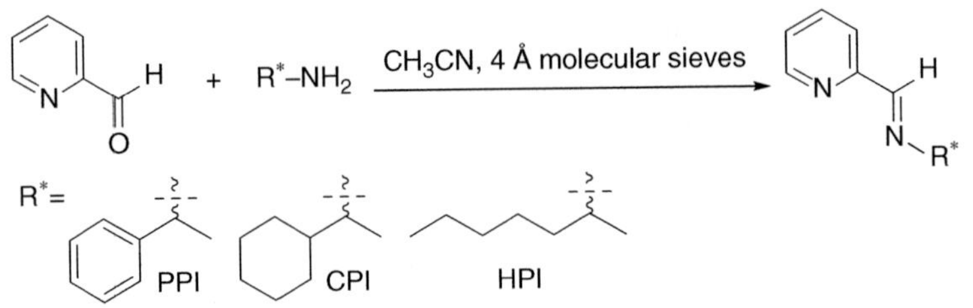

Scheme 5.5 Derivatization of primary amines to Schiff bases for CD analysis

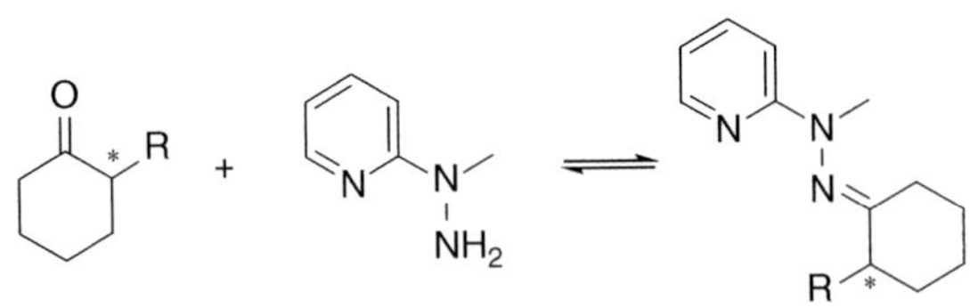

Scheme 5.6 Derivatization of cyclohexanones to hydrazones for CD analysis

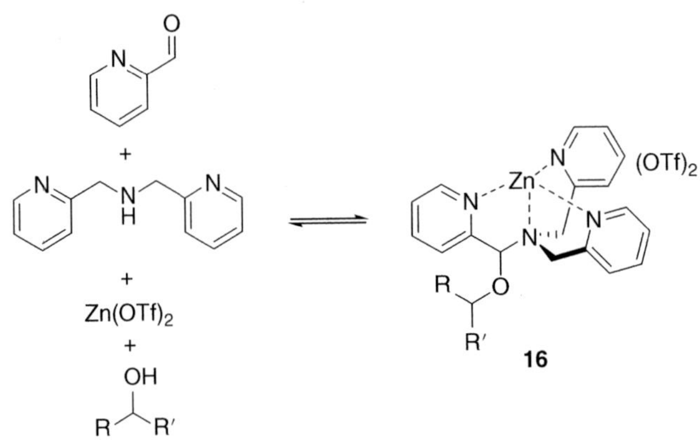

Scheme 5.7 Three component assembly for ee determination of chiral alcohols. Source: You 2012 [36]. Reproduced with permission of the American Chemical Society

5.3.2.2 *ECCD analysis of alcohols*

In order to measure *ee* of chiral secondary alcohols without needing prior derivatization, we developed a dynamic multicomponent assembly [34, 35]. Using this assembly (**16**), we were able to amplify the ECCD signal relative to the original alcohol (Scheme 5.7). The helicity of the assembly is dependent on the configuration of the stereocenter at the hemiaminal ether carbon, which is in turn dictated by the chirality of the alcohol (Figure 5.9). This assembly (**16**) gives a diastereomeric ratio which can be used to determine the original *ee* of the starting alcohol. ECCD spectra of various alcohols shown in Figure 5.10.

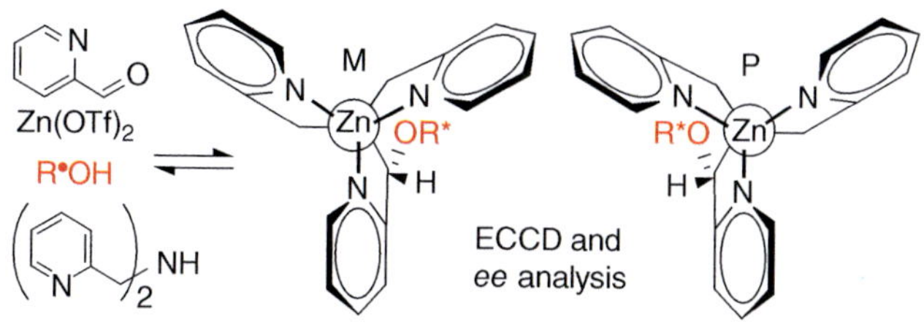

Figure 5.9 M and P helicity of the alcohol assembly. Source: [36] Reproduced with permission of the American Chemical Society

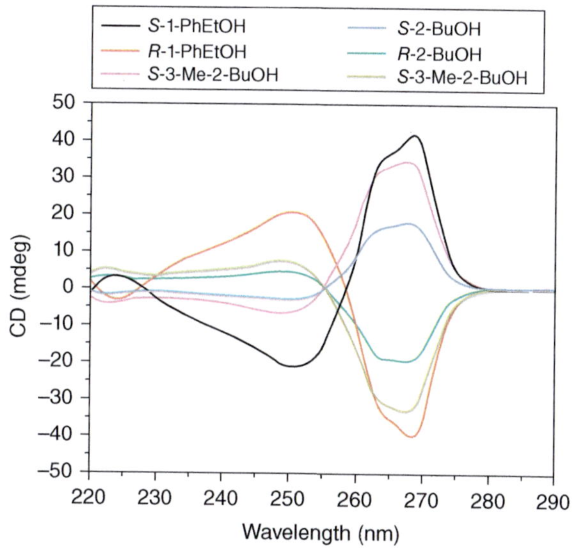

Figure 5.10 CD spectra of various alcohols. Source: [35]. Reproduced with the permission of Nature Publishing Group

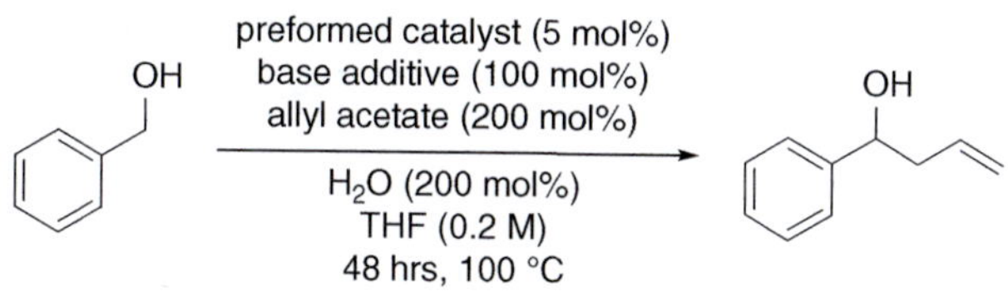

Scheme 5.8 Asymmetric alcohol allylation

In applying our assay to the determination of *ee* of chiral alcohols, we partnered with Michael Krische and his research group. The Krische group has developed a method for enantioselective allylation using iridium catalysis (Scheme 5.8) [36]. This method adds an allyl group to a carbon containing a hydroxyl group, resulting in an alcohol at the stereocenter. Our assay was used to measure the *ee* values of a variety of catalysts and reaction conditions in a high-throughput manner [38]. The allylation reactions were conducted in a 96-well plate and nearly four hundred carbonyl allylations were studied with varying conditions. After workup, CD analysis similar to the one described above was performed on each reaction to determine *ee* and find the ideal reaction conditions.

5.3.2.3 *ECCD analysis of amines*

Inspired by the success of our work with ECCD and chiral alcohols, we worked to develop an alternate method for analyzing the *ee* of chiral primary amines (Scheme 5.9) [39]. This method eliminates the need for prior derivatization of the amine and instead forms the imine *in situ*. The chiral amine reacts with the aldehyde to form the Schiff base which then forms an octahedral complex with the iron (**17**). The chirality of the complex is determined by the R groups on the amine (Figure 5.11). This complex is then analyzed via spectroscopy and the *ee* of the starting amine can be determined (Figure 5.12).

5.3.2.4 *Adaptation to high-throughput methods*

Due to our success with CD and *ee* determination, we sought to adapt this method to be more high-throughput [40]. In a collaboration with Bart Kahr at New York University, we constructed a CD spectropolarimeter with multiple photoelastic modulators that was capable of rapidly measuring CD in a 96-well glass-bottomed plate (Figure 5.13). By using four photoelastic modulators, we were able to reduce the interference from the well plate. We tested our prototype by running the same experiments we had previously tested with our assay and found good agreement.

5.3.3 Conclusion

In summary, this second tale describes how a need in the synthetic organic community led to the use of supramolecular assemblies to generate functional assays for enantiomeric excess. Further, not only did novel chemistry arise in this study, but it led to the creation of a new instrument: the first plate reader for CD spectropolarimetry. As in the first tale, the overarching lesson is not to stop at simply creating interesting supramolecular architectures, but instead to push the field's methods all the way to a final application in the real-world.

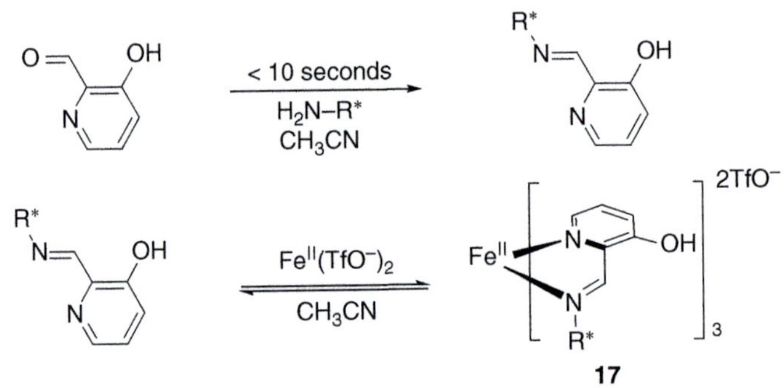

17

Scheme 5.9 *Formation of iron complex* ***17***

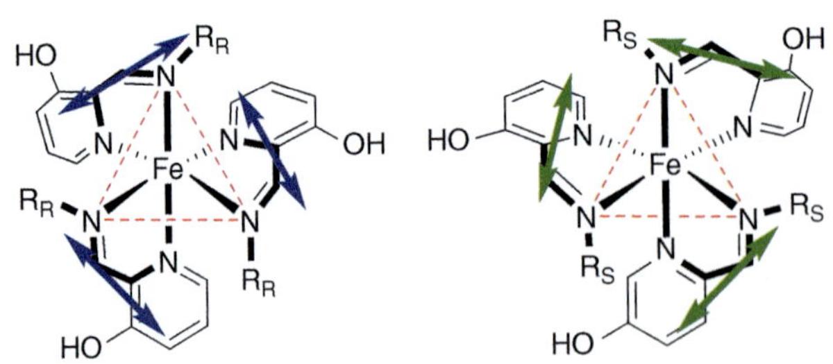

Figure 5.11 *Chirality of the iron complexes*

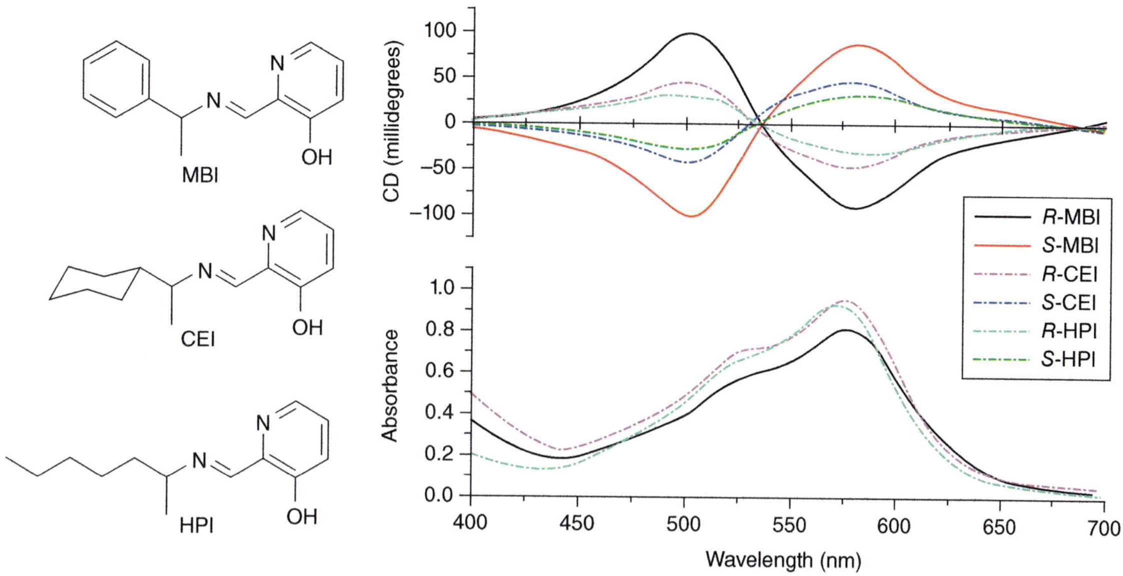

Figure 5.12 Some of the chiral amines analyzed and their CD and absorbance spectra. Source: [16]. Adapted with permission of the American Chemical Society

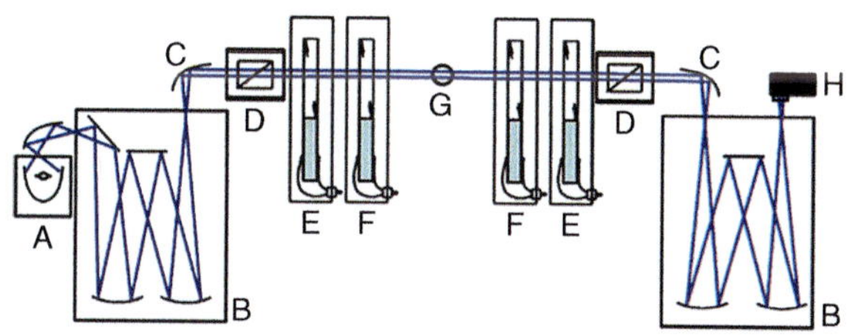

Figure 5.13 Four photoelastic modulator (4-PEM) spectropolarimeter. (A) Light source, (B) monochromators, (C) mirrors, (D) linear polarizers, (E and F) photoelastic modulators set at 0° (E) and 45° (F). (G) Sample holder, (H) photomultiplier tube. Source: [40]. Reproduced with permission of the Royal Society of Chemistry

5.4 Differential sensing

The assays described above have an element of selectivity that is necessary. For example, the citrate and Ca^{2+} assays need to be selective for these analytes in the complex mixtures inherent with soft drinks, vodka, or dialysis fluid. The *ee* assays for chiral functional groups need to be selective for just their specific functional group in a complex molecule. Our next tale approaches sensing from a different perspective.

During a casual lunch conversation with Dr. John McDevitt at the University of Texas at Austin, the topic of "electronic noses" arose [41]. These are analytical devices that classify gases and components of gases via pattern-recognition protocols using cross-reactive receptors. This chance conversation led to a new approach within supramolecular chemistry, for which we coined the term "differential sensing [42, 43]."

Historically, molecular recognition has mimicked biological receptors by designing a highly specific host for a single guest (Figure 5.14) [44]. More recently the field has started to evolve more diverse approaches. Differential sensing uses multiple receptors to create a fingerprint for analytes. These sensing arrays allow for discrimination of multiple analytes in a solution and remove the necessity of synthesizing specific receptors for each analyte and simplifying the design of the array (Figure 5.15). This type of analysis mimics the mammalian senses of taste and smell, which use a combination of specific and nonspecific receptors to create an overall interpretation.

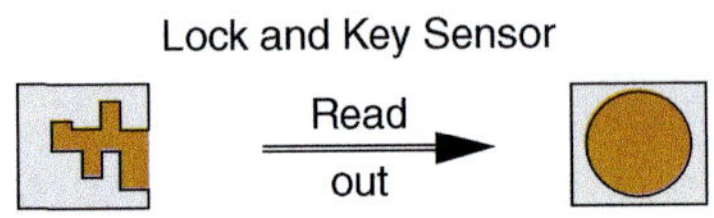

Figure 5.14 Lock and key design sensor for a specific analyte. Source: [43]. Reproduced with permission of Wiley-VCH Verlag GmbH & Co. KGaA

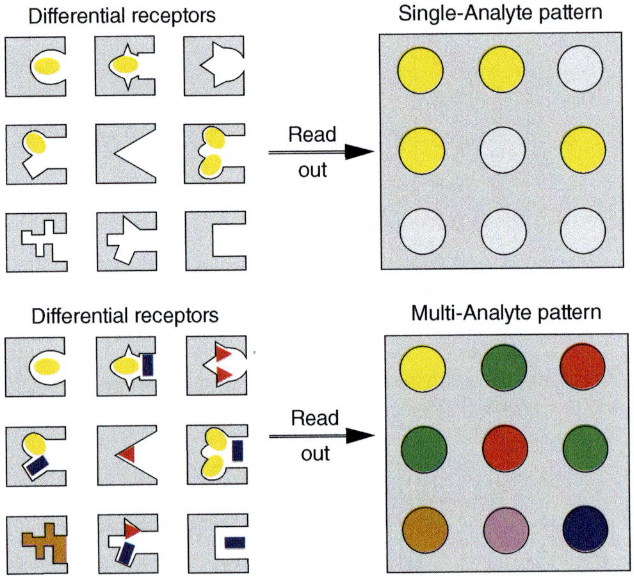

Figure 5.15 Differential sensor arrays for single- and multi-analyte patterns. Source: [43]. Reproduced with permission of Wiley-VCH Verlag GmbH & Co. KGaA

5.4.1 Chemometric analysis

The data provided by these arrays can be complex and difficult to interpret without computational methods. Our group has used the chemometric analysis methods of linear discriminant analysis (LDA) and principal component analysis (PCA) with success. PCA and LDA are statistical procedures that use eigenvectors to display the data along orthogonal axes with the most variance [45]. Via different algorithms, described as unsupervised and supervised, respectively, these methods generate score plots that classify data, and thereby analytes and/or complex mixtures. Classification arises from variance in the multivariate data which are displayed in a new multi-dimensional space. As will be shown below, when the method is successful, the score plots cluster like samples together and separate them from unlike samples. For example, samples of beef broth should be reliably differentiated from samples of chicken broth, and vice versa – just as our mammalian senses of taste and smell are able to make the distinction.

5.4.2 An electronic tongue

Our earliest foray into differential sensing was in 1998 when our group with several collaborators developed a sensor methodology called an "electronic tongue" [46]. Poly-(ethylene glycol)-polystyrene (PEG-PS) beads were functionalized with a variety of indicators to mimic taste buds (Figure 5.16). The indicators were chosen to be selective for individual analytes but not so selective that they were unreactive to other analytes.

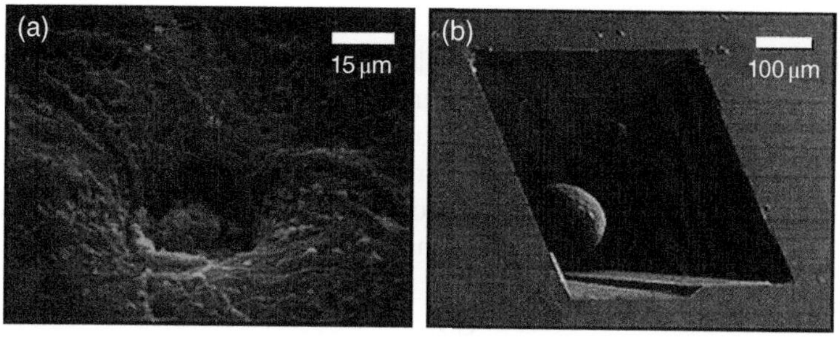

Figure 5.16 (a) scanning electron micrograph (SEM) image of a rat fungiform papilla (taste bud). (b) SEM image of a PEG-PS bead immobilized in a micromachined Si well. Source: [46]. Reproduced with permission of the American Chemical Society

These PEG-PS resins were positioned in micromachined wells formed in Si/SiN wafers meant to mimic natural taste buds. Four types of sensors were studied along with a control: fluorescein, *o*-cresolphthalein complexone, alizarin complexone, and a boronic ester of resorufin-derivatized galactose. Change in absorbance was measured upon addition of Ca^{2+}, Ce^{3+}, and mixtures of the two cations at varying pH.

5.4.3 Phosphopeptide classification

Protein phosphorylation and dephosphorylation are common methods through which cell processes are regulated [47]. Phosphorylation states can also be indicative of certain diseases, which makes identification of protein phosphorylation key to understanding the onset and progression of these diseases [48].

Historically, peptide phosphorylation has been studied through fluorescently modified peptide sequences that produce a change in signal upon phosphorylation [49–51]. These probes are limited in substrate scope due to the necessity of using modified peptides or synthetic analogues. We sought to develop a method that avoided modification of the analytes. Our copper-containing receptor **18** binds tetrahedral oxoanions such as inorganic phosphate and most monophosphoesters [52]. We decided to modify this receptor in order to study kinase activity by measuring phosphorylation of peptides. Although specific kinase activity can be monitored through selective molecular receptors for each phosphorylated substrate, this method requires the synthesis of a receptor for each substrate. Alternatively, an array of receptors could be used to pattern phosphorylation events in a fingerprint unique to each mixture of phosphopeptides.

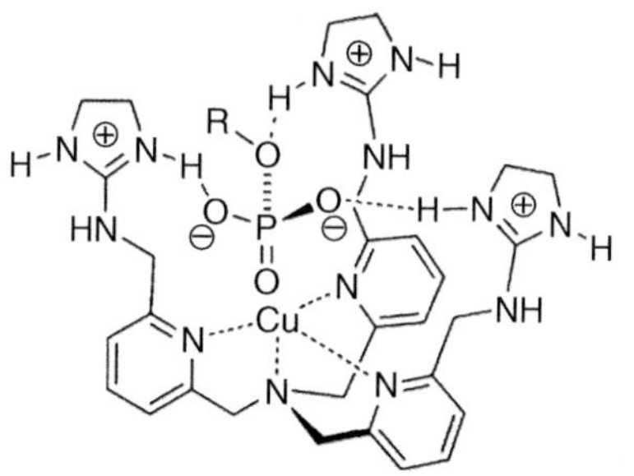

18

Source: *[52]. Reproduced with permission of the American Chemical Society*

We designed a differential sensing array of five receptors, three colorimetric indicators, and three metal ions. This array was used to differentiate a variety of short peptides and their phosphorylated analogues (Figure 5.17). The tripeptides that were used as analytes were selected from a sequence of peptides known to be involved in Parkinson's disease. Phosphorylation of the serine residue in this sequence has been found in cerebral protein aggregates in Parkinson's patients [53]. Receptors were based on **18** but modified with two tripeptide chains (**19**). Fourteen peptide sequences were screened against the phosphorylated peptides and five of these were selected. Indicators with metal coordination sites and first-row transition metals with two or three coordination sites were screened in a 96-well plate against phosphoserine. Based on the magnitude of change in absorbance, three indicators (pyrocatechol violet, celestine blue, and gallocyanine) and three metal ions (Ni^{II}, Co^{II}, and Cu^{II}) were selected for the final assay.

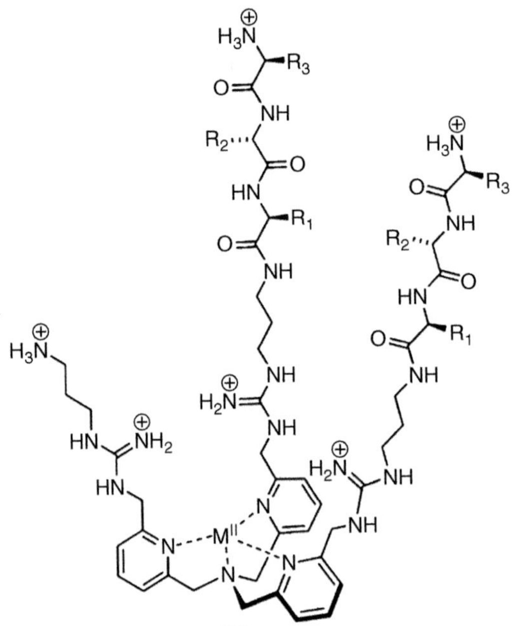

19

Source: *[52]. Reproduced with permission of the American Chemical Society*

After the components of the array had been selected, each combination of receptor, metal ion, indicator, and either phosphorylated or nonphosphorylated tripeptide was added into a 96-well plate for a total of ninety combinations. To analyze the data, we used linear discriminant analysis (LDA) to find correlations between the data (Figure 5.18). To test the validity of this method, nine replicates were conducted and the data was subjected to the resubstitution method with near 100% success. We studied various combinations of the array and found that the variety of peptidic arms on the receptors and the different indicators were critical, but the various metal cations were not. With both the full array and this truncated version, we were able to obtain good classification of phosphorylated and nonphosphorylated tripeptides.

5.4.4 Serum albumin receptors for terpenes

Serum albumins are blood proteins that bind a variety of hydrophobic molecules for transport [54]. This non-selective binding is appealing for use in cross-reactive arrays. Each species has a different albumin with variance in peptide sequences. To showcase the ability of serum albumins (SA) for differential sensing, we studied their binding to terpenes [55]. We developed an assay that used bovine serum albumin (BSA), human serum albumin (HSA), and rabbit serum albumin (RSA). To show a measurable change upon terpene binding,

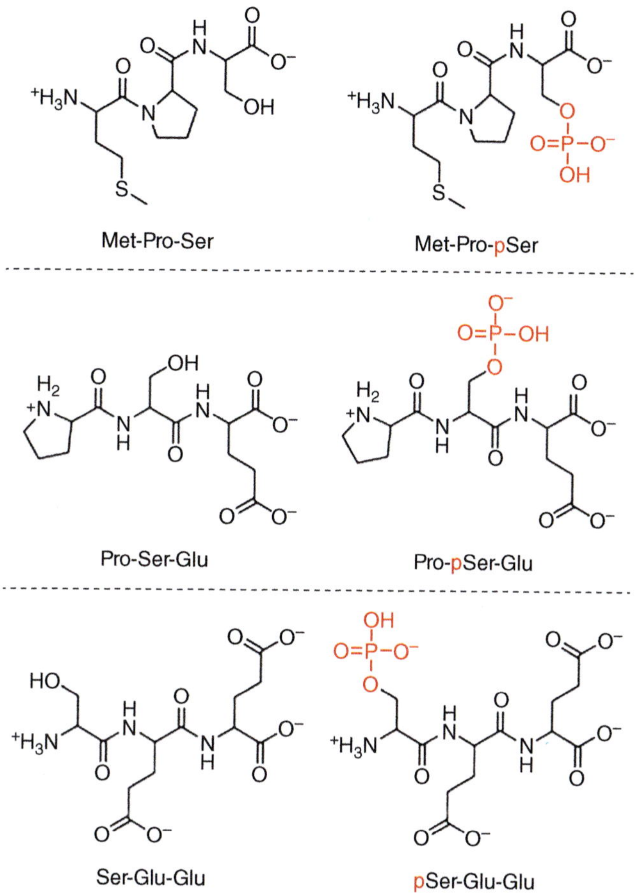

Figure 5.17 Tripeptides and their phosphorylated analogues used as analytes. Source: [52]. Reproduced with permission of the American Chemical Society

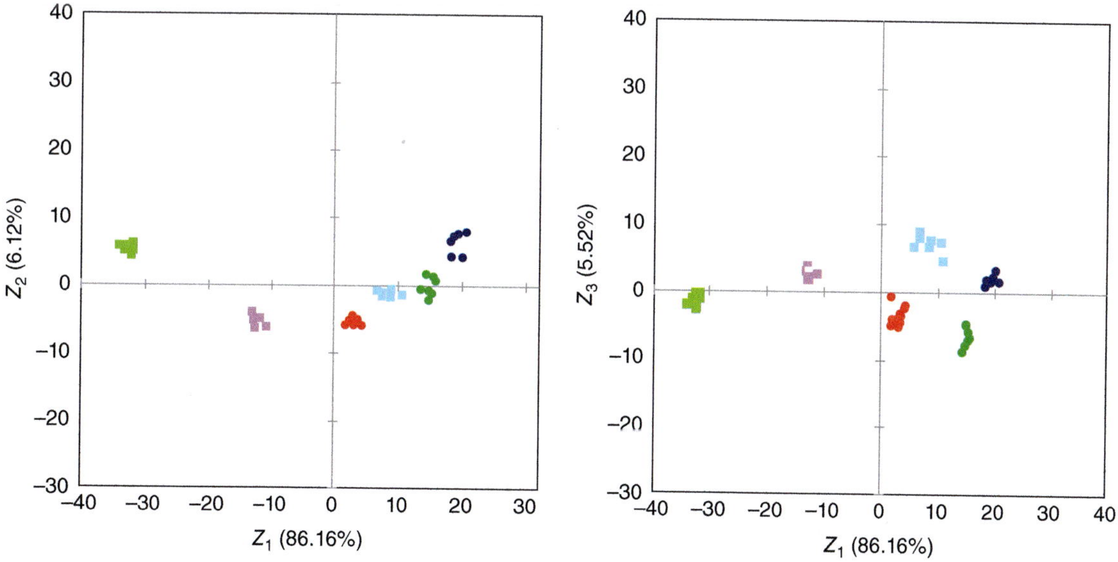

Figure 5.18 LDA plots of discriminant scores for all data. Source: [52]. Reproduced with permission of the American Chemical Society

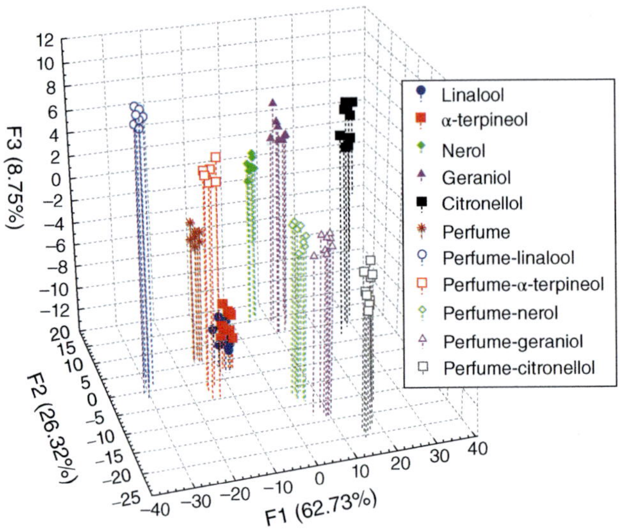

Figure 5.19 *LDA plot of terpene analysis. Source: [55]. Reproduced with permission of the American Chemical Society*

prodan (**20**) was used as an indicator. Prodan is known to show a hypsochromic shift in emission upon binding to SAs [56]. Additionally, similar to terpenes, prodan is hydrophobic and is neutrally charged, which allows it to bind to SAs in a similar manner. The bile acid deoxycholate was included in the assembly as another variable to help with differentiation. The five terpenes selected to study were linalool, α-terpineol, nerol, geraniol, and citronellol. To show the utility of this method in complex mixtures, these terpenes were also studied in perfume and the results were analyzed by LDA (Figure 5.19). Our data showed successful discrimination of the terpene solutions.

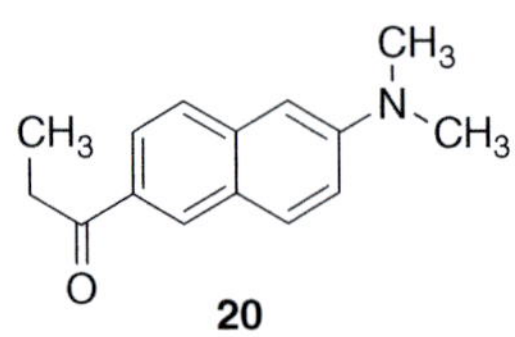

20

5.4.5 Serum albumin receptors for glycerides

We continued our work with SA cross-reactive arrays to study glycerides [57]. Glycerides are the main components of fat cells. They comprise mono-, di-, and tri-glycerides in varying mixtures and are important in understanding lipid metabolism [58]. Glycerides are difficult to analyze due to the similarities between fatty acids. The primary method for identification is mass spectrometry [59]. This approach is difficult in that glycerides do not ionize well, which prevents quantification, and regio- and stereoisomers are often indistinguishable. Chemical derivatization coupled with mass spectrometry has been more successful. We hypothesized that serum albumins could be used to help differentiate glycerides when used in a sensory array. SAs nonspecifically bind fatty acids for transport through the bloodstream, which suggests that they would bind glycerides. A variety of glycerides were examined, containing saturated, monounsaturated, and polyunsaturated mono-, di-, and triglycerides. We also included stereo- and regioisomers, which are not distinguishable by traditional methods.

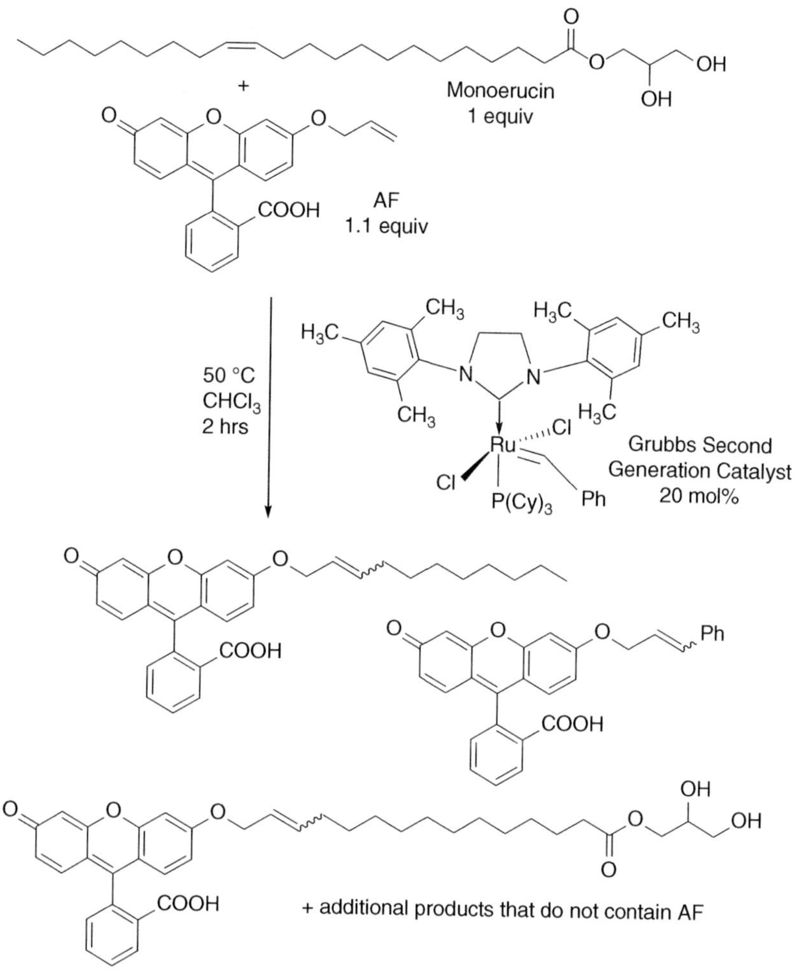

Scheme 5.10 *Allyl fluorescein (AF) derivatization of glycerides*

Similar to our terpene assay, fluorescent indicators were used to measure binding to different parts of SA. To add an additional level of analysis to our array, we subjected all fatty acids to an olefin metathesis reaction with allyl fluorescein (**AF**) prior to their inclusion in the array (Scheme 5.10). With this assay, we were able to successfully discriminate twenty glycerides and group them based on their structural characteristics (Figure 5.20).

5.4.6 Mitogen-activated protein kinases differentiation

Mitogen-activated protein kinases (MAPKs) are key regulators in cell processes such as mitosis, cell differentiation, and apoptosis [60]. These kinases are specific to serine, threonine, and tyrosine residues. There are three major types of MAPKs: extracellular signal regulated kinases (ERK1 and 2), c-jun N-terminal kinases (JNK1, 2, and 3), and p38 MAPKs (p38α, β, γ, and δ). Detection of these proteins and

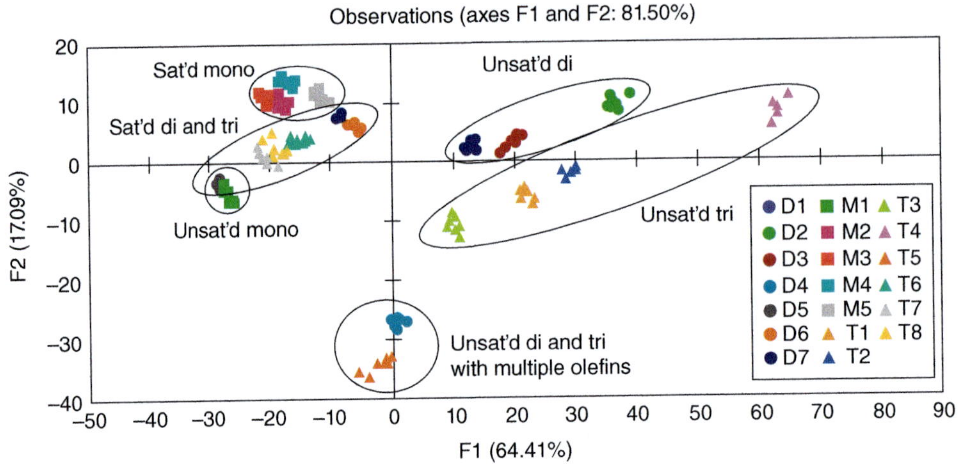

Figure 5.20 LDA plot of data collected from 96-well plates. Source: [57]. Reproduced with permission of the National Academy of Sciences of the United States of America

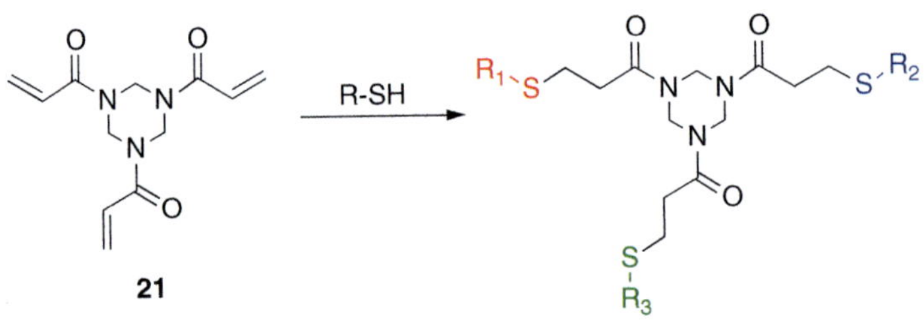

Scheme 5.11 Thiol addition to triazine conjugate acceptor (**21**). Source: [66]. Reproduced with permission of the American Chemical Society

their phosphorylation is done through radioassays, antibodies and, more recently, optical-based biosensors [61–63]. These biosensors are peptide-based and contain a sulfonamide-oxine (SOX) fluorophore that, upon phosphorylation of an adjacent residue, binds Mg^{2+} which enhances emission [64]. Varying peptide structure allows for selectivity between MAPK classifications. Some groups have worked to create supramolecular receptors. Hamachi *et al.* published a zinc-dipicolylamine [Zn(DPA)] chemosensor [65]. Both this system and the SOX fluorophore based system require specific receptors designed for each phosphorylation target.

We hypothesized that by using a cross-reactive array, we could simplify the analysis of these kinases [66]. The first generation of arrays was based on Hamachi's Zn(DPA) chemosensor. We developed a sensor array that utilized the ability of thiols to reversibly add via conjugate addition in water at room temperature. We used triazine **21** as a scaffold (Scheme 5.11) and included the Zn(DPA) unit modified with a nucleophilic thiol (**22**). We also included two peptides (**KALIC** and **PVDAC**) reported to bind MAPKs and dithiothreitol (**DTT**). DTT was hypothesized to bind to two of the triazine scaffolds, resulting in oligomerization and adding yet another level of complexity to the array. The other components of the array were added in three-fold excess of DTT to form the receptor array. The fluorescent indicator **23** was used to allow measurement of binding.

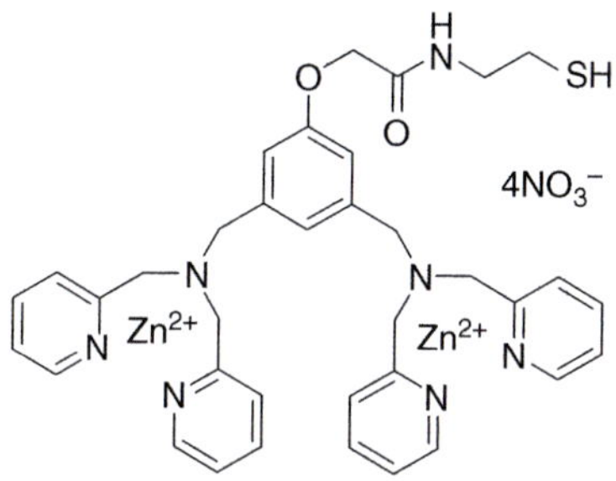

22

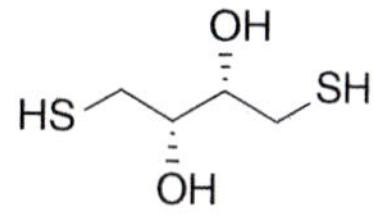

DTT

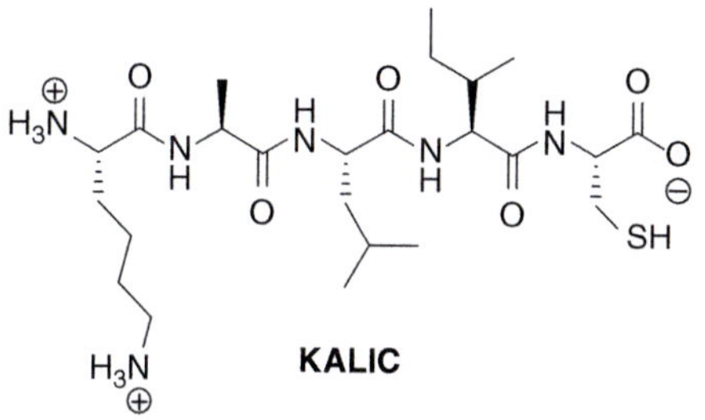

KALIC

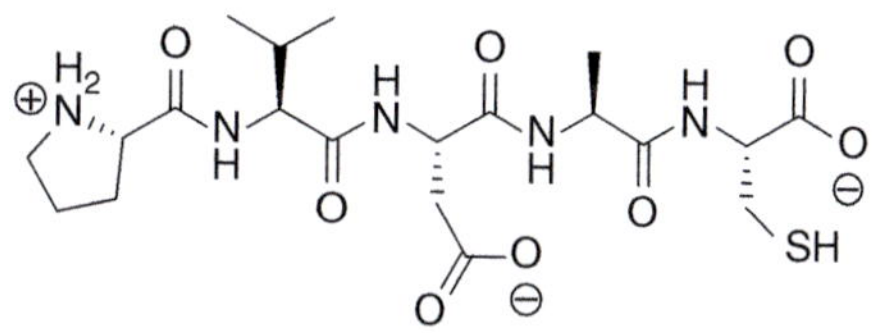

PVDAC

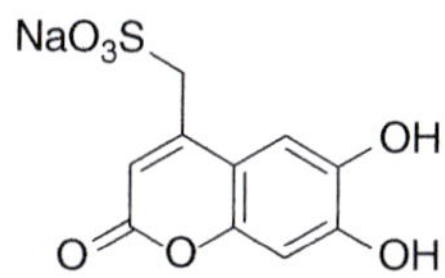

23

We tested our assay first with a phosphorylated peptide known to be a substrate of ERK-1, sub-F (YAEPLTPRILAKWEWPA) [67]. We compared the phosphorylated peptide with its non-phosphorylated counterpart; only the phosphorylated peptide displaced the indicator. After this preliminary test, we tested our array against phosphorylated and nonphosphorylated MAPKs with good differentiation (Figure 5.21).

We also lysed cells where MAPKs had been expressed and measured the response from the lysates (Figure 5.22). We used HEK293T cells with different kinase expression and activity. Western blot analyses

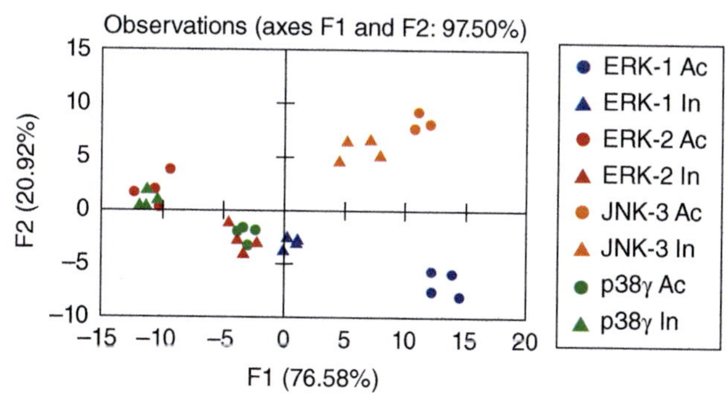

Figure 5.21 *LDA plot of the results of our array from phosphorylated and nonphosphorylated MAPKs. Source: [66]. Reproduced with permission of the American Chemical Society*

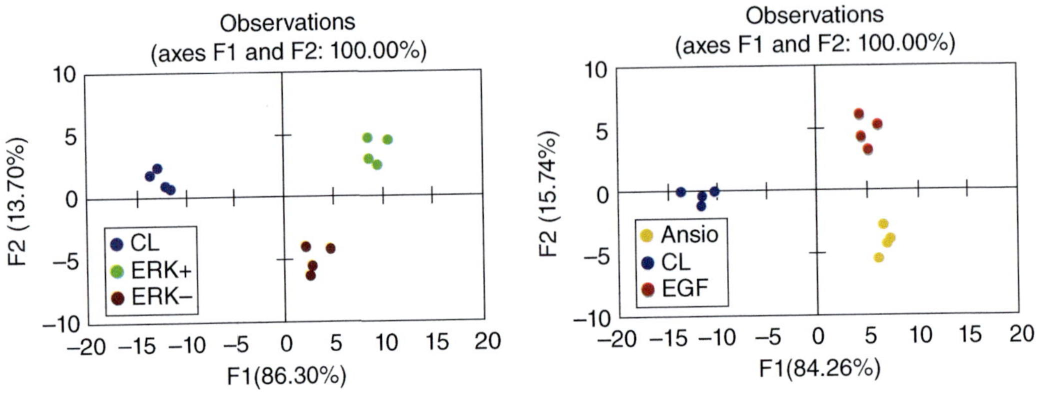

Figure 5.22 LDA plots of the results of our array from cell lysates. Source: [66]. Reproduced with permission of the American Chemical Society

of the lysates were performed to confirm that the cells had the expected different expressions of kinases. The LDA plots of these lysates showed very good differentiation.

Our second generation of differential sensing for MAPKs used sulfonamide-oxine (SOX) modified peptides [68]. We selected five peptides previously reported to bind different MAPKs and modified the cysteine residues in these peptides with the SOX fluorophore (Figure 5.23) [67, 69, 70]. The fluorescence of these SOX groups turns on upon phosphorylation of proximal amino acids and binding of Mg^{2+}.

We used the change in fluorescence and calculated rate constants as data for our cross-reactive array. We conducted four replicates of the array with five peptides and nine kinases and were successful in discriminating the different MAPKs (Figure 5.24).

We also explored whether our assay could quantitatively determine the concentration of one kinase in a mixture. We were able to successfully determine the concentrations of ERK1 in MAPK mixtures using support vector machines [71]. Through this, we sought to determine the ability of our array to analyze kinase inhibitor activity. We used the ERK inhibitor VX-11e (**24**) to investigate the inhibition of ERK1 in the presence of JNK3 and p38γ [72]. We found that VTX-11e activated release of the p38 kinase from an inhibitory heterodimer formed with ERK and we were able to determine the concentration of inhibitor after creating a calibration curve (Figure 5.25).

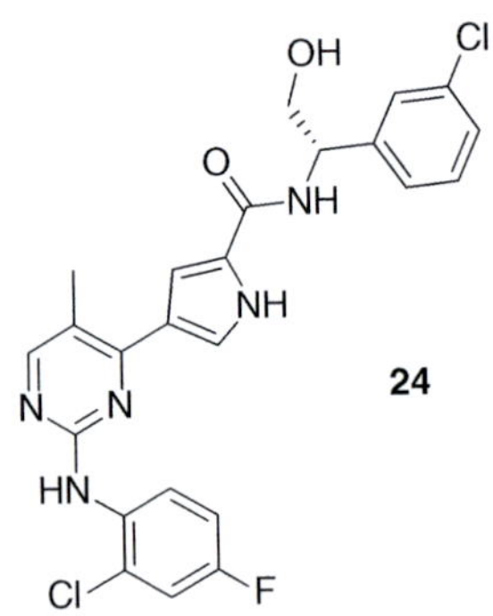

24

5.4.7 Wine

Analysis of specific organic molecules or classes of molecules within complex mixtures is difficult but has extensive application [73–75]. Classical techniques for this analysis include HPLC, mass spectrometry, and UV–Vis spectrometry [76]. More recently, array sensing has gained interest in differentiating complex mixtures [77]. Wine is essentially a complex aqueous mixture of organic molecules that varies based on the types of grapes used, the

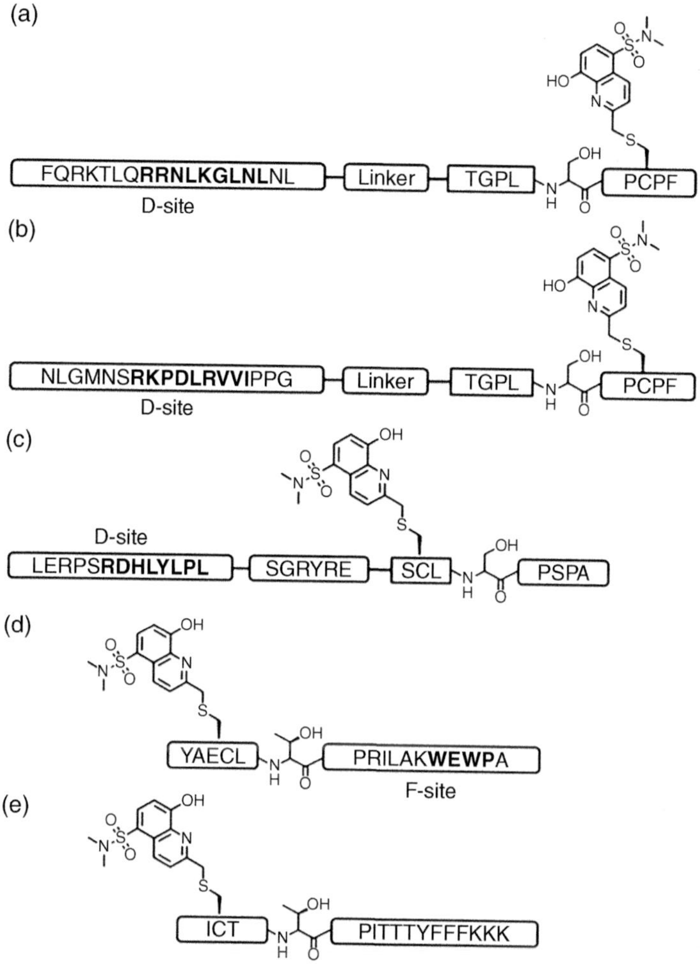

Figure 5.23 *SOX-peptides used in the assay. Source: [68]. Reproduced with permission of John Wiley & Sons, Ltd*

strain of yeast used in fermentation, the environment where the grapes are grown, and other factors [78]. Flavonoids are a class of compounds that vary in concentration according to the different types of grapes in each varietal and are ideal for differential sensing [79]. We designed an array using peptidic, colorimetric sensors that was used to differentiate wines [80]. Using histidine-rich peptides, metal ions, and colorimetric indicators, we were able to successfully differentiate flavonoids **25–29** and six different types of wines (varietals) (Figure 5.26).

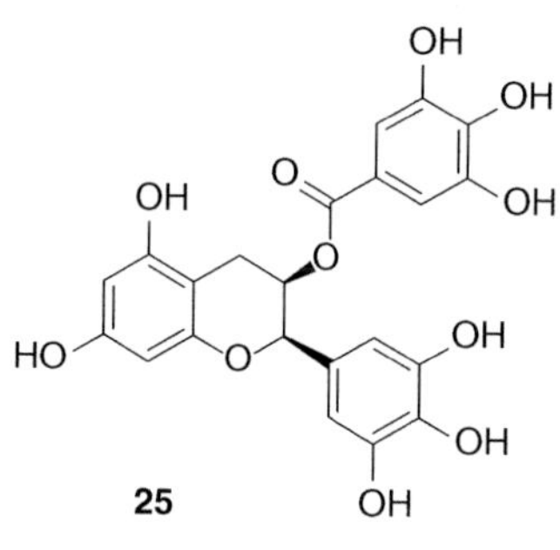

25

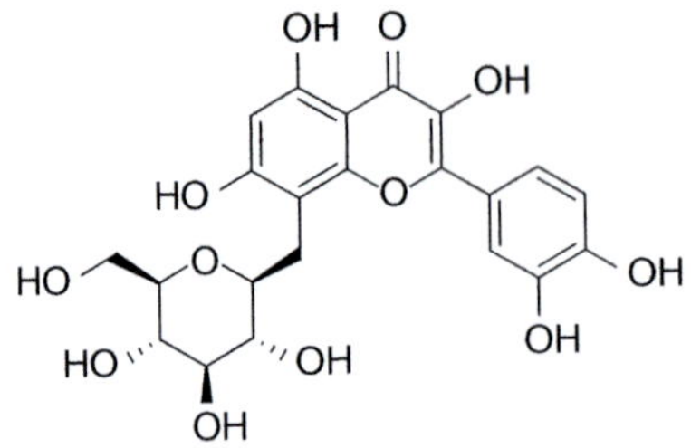

26

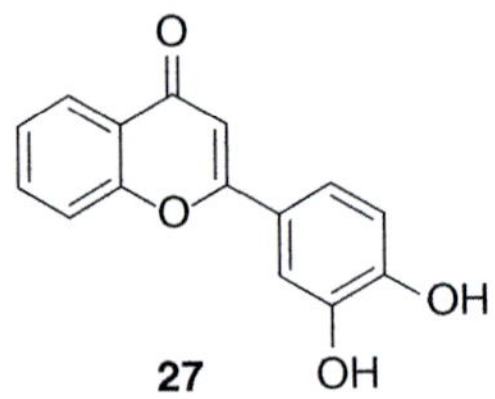

27

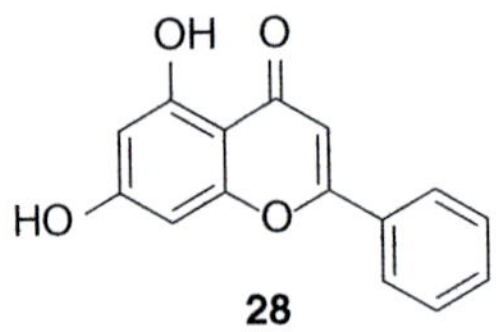

28

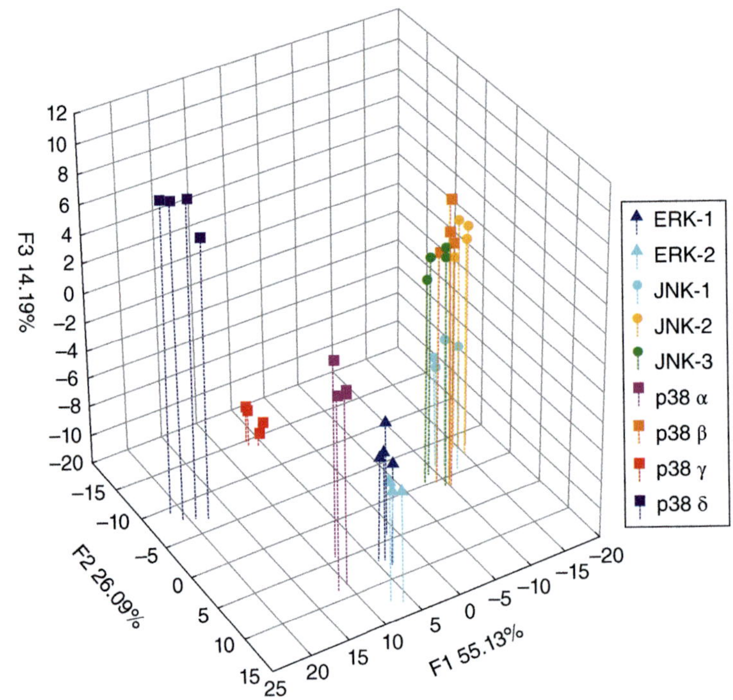

Figure 5.24 *Three dimensional LDA plot of the responses form the SOX-peptides to different MAPKs. Source: [68]. Reproduced with permission of John Wiley & Sons, Ltd*

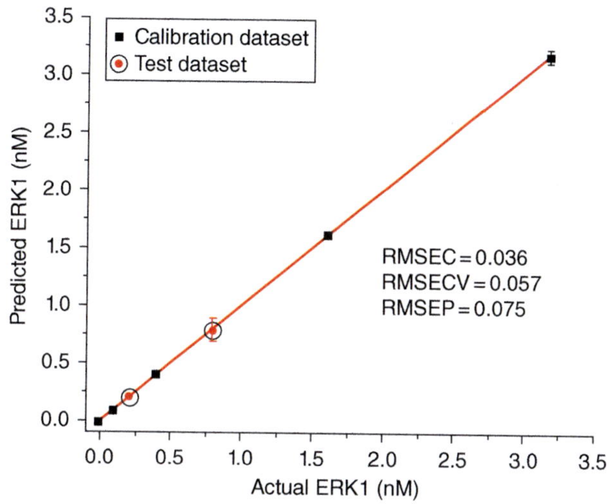

Figure 5.25 *Calibration curve and test validation of VX-11e inhibition. Source: [68]. Reproduced with permission of John Wiley & Sons, Ltd*

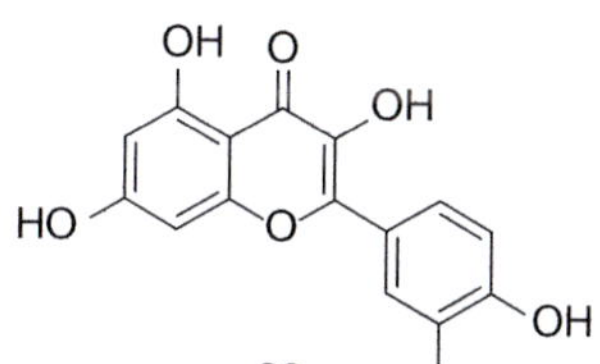

29

The organic acids malate, tartrate, and citrate can also be used to discriminate types, or varietals, of wine. In Section 5.2, we discussed our work with colorimetric sensors for these compounds in other solutions. Receptors **5**, **6**, and **30** were used to develop an array to differentiate these compounds in wine [81]. We created a 3 × 3 sensing array using these three receptors and three indicators (alizarone complexone, pyrocatechol violet, and bromopyrogallol red) for discriminating the acids in solution and in wines (Figure 5.27). This array was able to successfully discriminate six different types of wine.

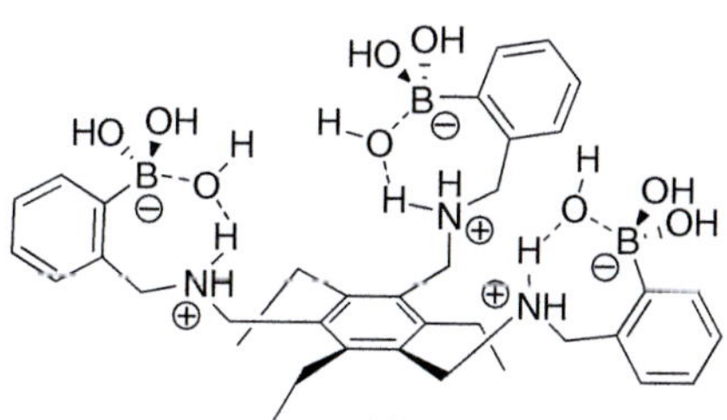

30

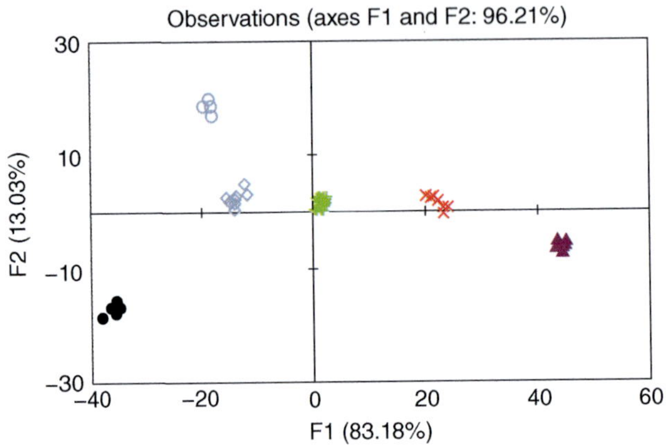

Figure 5.26 *LDA score plot of UV-vis responses from different wine varietals: Beaujolais (○), Zinfandel (*), Pinot Noir (◇), Shiraz (●), Merlot (■), Cabernet Sauvignon (▲). Source: [81]. Reproduced with permission of Taylor & Francis Group*

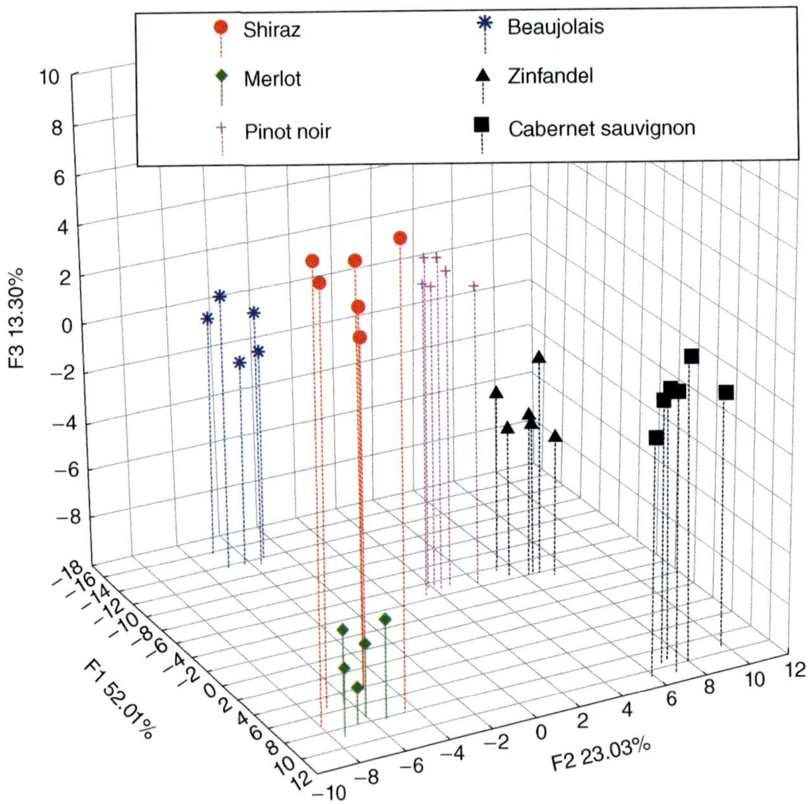

Figure 5.27 *LDA plot of the spectroscopic data from the evaluation of different wine varietals. Source: [80]. Reproduced with permission of the Royal Society of Chemistry*

Wine is an excellent proving ground for the field of differential sensing due to the fact that humans are able to distinguish very subtle differences, and our overall goal is to probe whether our method can be as sophisticated as the human senses. Here, much like the previous two tales, the basic science turned out to have a practical application we did not foresee. Recently, it has become evident that fraudulent labeling of wine varietals is more widespread than may be expected. In fact, our group was awarded a grant to continue our work to verify wine varietals from a class action lawsuit that involved the mislabeling of Gamay wines as Pinot Noir.

5.5 Conclusion

In conclusion, these three tales have had one unifying theme – take the basic science of supramolecular chemistry all the way to final applications. Admittedly, as described herein, much of our ability to do so was via chance events: phone calls, suggestions, and passing conversations. There is another old adage that seems appropriate, "we make our own luck." To the students reading this chapter the following advice may be useful: be open to suggestions, listen and respond, push the science to its limits via innovation and creativity, but, above all, take risks. Don't just be satisfied with another paper, instead test the final real-life goal. The worst that can happen is you have a failure, which will only inspire further rounds of innovation and creativity, ultimately yielding success. This was the main message of the Izatt–Christensen Award lecture in 2013.

References

1. Perreault, D.M., Cabell, L.A., Anslyn, E.V. (1997) Using guanidinium groups for the recognition of RNA and as catalysts for the hydrolysis of RNA. *Bioorganic & Medicinal Chemistry*, **5** (6), 1209–1220.
2. Metzger, A., Lynch, V.M., Anslyn, E.V. (1997) A synthetic receptor selective for citrate. *Angewandte Chemie (International Edition)*, **36** (8), 862–865.
3. Woods, A.S. (2004) The mighty arginine, the sTable quaternary amines, the powerful aromatics, and the aggressive phosphate: their role in the noncovalent minuet. *Journal of Proteome Research*, **3** (3), 478–484.
4. Berger, M., Schmidtchen, F.P. (1998) The binding of sulfate anions by guanidinium receptors is entropy-driven. *Angewandte Chemie International edition*, **37** (19), 2694–2696.
5. Roberts, S.M. (1989). *Molecular recognition: Chemical and Biochemical Problems*. Proceedings of an International Symposium, University of Exeter, April 1989, Volume **1**. Cambridge: Royal Society of Chemistry.
6. Smith, D.S., Eremin, S.A. (2008) Fluorescence polarization immunoassays and related methods for simple, high-throughput screening of small molecules. *Analytical and Bioanalytical Chemistry*, **391** (5), 1499–1507.
7. Nguyen, B.T., Anslyn, E.V. (2006) Indicator-displacement assays. *Coordination Chemistry Reviews*, **250** (23-24), 3118–3127.
8. Metzger, A., Anslyn, E.V. (1998) A chemosensor for citrate in beverages. *Angewandte Chemie International edition*, **37** (5), 649–652.
9. Wiskur, S.L., Ait-Haddou, H., Lavigne, J.J., Anslyn, E.V. (2001). Teaching old indicators new tricks. *Accounts of Chemical Research*, **34** (12), 963–972.
10. Lavigne, J., Anslyn, E. (1999) Teaching old indicators new tricks: a colorimetric chemosensing ensemble for tartrate/malate in beverages. *Angewandte Chemie International edition*, **38** (24), 3666–3669.
11. Wiskur, S.L., Anslyn, E.V. (2001) Using a synthetic receptor to create an optical-sensing ensemble for a class of analytes: a colorimetric assay for the aging of scotch. *Journal of the American Chemical Society*, **123** (41), 10109–10110.
12. McCleskey, S.C., Floriano, P.N., Wiskur, S.L., Anslyn, E.V, McDevitt, J.T. (2003) Citrate and calcium determination in flavored vodkas using artificial neural networks. *Tetrahedron*, **59** (50), 10089–10092.

13. Cirovic, D.A. (1997) Feed-forward artificial neural networks: applications to spectroscopy. *TrAC Trends in Analytical Chemistry*, **16** (3), 148–155.
14. Gray, E., Hogwood, J., Mulloy, B. (2012) The anticoagulant and antithrombotic mechanisms of heparin. *Handbook of Experimental Pharmacology*, **207**, 43–61.
15. Yang, Y., Szamosfalvi, B., Yee, J., Frinak, S., Anslyn, E.V. (2011) Development of an online citrate/Ca^{2+} sensing system for dialysis. *The Analyst*, **136** (2), 317–320.
16. Jo, H.H., Lin, C.-Y., Anslyn, E.V. (2014) Rapid optical methods for enantiomeric excess analysis: from enantioselective indicator displacement assays to exciton-coupled circular dichroism. *Accounts of Chemical Research*, **47** (7), 2212–2221.
17. Noyori, R. (2002) Asymmetric catalysis: science and opportunities (Nobel Lecture). *Angewandte Chemie International edition*, **41** (12), 2008.
18. Okamoto, Y., Ikai, T. (2008) Chiral HPLC for efficient resolution of enantiomers. *Chemical Society Reviews*, **37** (12), 2593–2608.
19. Finn, M.G. (2002) Emerging methods for the rapid determination of enantiomeric excess. *Chirality*, **14** (7), 534–540.
20. Zhu, L., Anslyn, E.V. (2004) Facile quantification of enantiomeric excess and concentration with indicator-displacement assays: an example in the analyses of alpha-hydroxyacids. *Journal of the American Chemical Society*, **126** (12), 3676–3677.
21. Zhu, L., Zhong, Z., Anslyn, E.V. (2005) Guidelines in implementing enantioselective indicator-displacement assays for alpha-hydroxycarboxylates and diols. *Journal of the American Chemical Society*, **127** (12), 4260–4269.
22. Folmer-Andersen, J.F., Lynch, V.M., Anslyn, E.V. (2005) Colorimetric enantiodiscrimination of alpha-amino acids in protic media. *Journal of the American Chemical Society*, **127** (22), 7986–7987.
23. Bull, S.D., Davidson, M.G., van den Elsen, J.M.H., *et al.* (2013) Exploiting the reversible covalent bonding of boronic acids: recognition, sensing, assembly. *Accounts of Chemical Research*, **46** (2), 312–326.
24. Shabbir, S.H., Regan, C.J., Anslyn, E.V. (2009) Molecular recognition and self-assembly special feature: A general protocol for creating high-throughput screening assays for reaction yield and enantiomeric excess applied to hydrobenzoin. *Proceedings of the National Academy of Sciences of the USA*, **106** (26), 10487–92.
25. Jacobsen, E.N., Marko, I., Mungall, W.S., Schroeder, G., Sharpless, K.B. (1988) Asymmetric dihydroxylation via ligand-accelerated catalysis. *Journal of the American Chemical Society*, **110** (6), 1968–1970.
26. Leung, D., Anslyn, E.V. (2008) Transitioning enantioselective indicator displacement assays for alpha-amino acids to protocols amenable to high-throughput screening. *Journal of the American Chemical Society*, **130** (37), 12328–12333.
27. Kelly, S.M., Price, N.C. (2000) The use of circular dichroism in the investigation of protein structure and function. *Current Protein & Peptide Science*, **1** (4), 349–384.
28. Gottarelli, G., Lena, S., Masiero, S., Pieraccini, S., Spada, G.P. (2008) The use of circular dichroism spectroscopy for studying the chiral molecular self-assembly: an overview. *Chirality*, **20** (3–4), 471–485.
29. Nieto, S., Dragna, J.M., Anslyn, E.V. (2010) A facile circular dichroism protocol for rapid determination of enantiomeric excess and concentration of chiral primary amines. *Chemistry – A European Journal*, **16** (1), 227–232.
30. Bringmann, G., Bruhn, T., Maksimenka, K., Hemberger, Y. (2009) The Assignment of absolute stereostructures through quantum chemical circular dichroism calculations. *European Journal of Organic Chemistry*, **2009** (17), 2717–2727.
31. Ziegler, M. (1998) Charge-transfer excited state properties of chiral transition metal coordination compounds studied by chiroptical spectroscopy. *Coordination Chemistry Reviews*, **177** (1), 257–300.
32. Gawronski, J., Skowronek, P. (2015) New Chromophores for organic stereochemical analysis by exciton- coupled circular dichroism. *Current Organic Chemistry*, **8** (1), 65–82.
33. Nieto, S., Lynch, V.M., Anslyn, E.V, Kim, H., Chin, J. (2008) High-throughput screening of identity, enantiomeric excess, and concentration using MLCT transitions in CD spectroscopy. *Journal of the American Chemical Society*, **130** (29), 9232–9233.
34. Leung, D., Anslyn, E.V. (2011) Rapid determination of enantiomeric excess of α-chiral cyclohexanones using circular dichroism spectroscopy. *Organic Letters*, **13** (9), 2298–22301.

35. You, L., Berman, J.S., Anslyn, E.V. (2011) Dynamic multi-component covalent assembly for the reversible binding of secondary alcohols and chirality sensing. *Nature Chemistry*, **3** (12), 943–948.
36. You, L., Pescitelli, G., Anslyn, E.V, Di Bari, L. (2012) An exciton-coupled circular dichroism protocol for the determination of identity, chirality, enantiomeric excess of chiral secondary alcohols. *Journal of the American Chemical Society*, **134** (16), 7117–7125.
37. Kim, I.S., Ngai, M.-Y., Krische, M.J. (2008) Enantioselective iridium-catalyzed carbonyl allylation from the alcohol or aldehyde oxidation level via transfer hydrogenative coupling of allyl acetate: departure from chirally modified allyl metal reagents in carbonyl addition. *Journal of the American Chemical Society*, **130** (44), 14891–14899.
38. Jo, H.H., Gao, 5., You, L., Anslyn, E.V., Krische, M.J. (2015) Application of a high-throughput enantiomeric excess optical assay involving a dynamic covalent assembly: parallel asymmetric allylation and *ee* sensing of homoallylic alcohols. *Chemical Science*, **6** (12), 6747–6753.
39. Dragna, J.M., Pescitelli, G., Tran, L., *et al.* (2012) In-situ assembly of octahedral Fe(II) complexes for the enantiomeric excess determination of chiral amines using circular dichroism spectroscopy. *Journal of the American Chemical Society*, **134** (9), 4398–4407.
40. Metola, P., Nichols, S.M., Kahr, B., Anslyn, E.V. (2014) Well Plate circular dichroism reader for the rapid determination of enantiomeric excess. *Chemical Science*, **5** (11), 4278–4282.
41. James, D., Scott, S.M., Ali, Z., O'Hare, W.T. (2004) Chemical sensors for electronic nose systems. *Microchimica Acta*, **149** (1-2), 1–17.
42. Anslyn, E.V. (2007) Supramolecular analytical chemistry. *The Journal of Organic Chemistry*, **72** (3), 687–699.
43. Lavigne, J.J., Anslyn, E.V. (2001) Sensing a paradigm shift in the field of molecular recognition: from selective to differential receptors. *Angewandte Chemie International edition*, **40** (17), 3118–3130.
44. Chen, B., Piletsky, S., Turner, A.P.F. (2002) High molecular recognition: design of "Keys". *Combinatorial Chemistry & High Throughput Screening*, **5** (6), 409–27.
45. Stewart, S., Ivy, M.A., Anslyn, E.V. (2014) The use of principal component analysis and discriminant analysis in differential sensing routines. *Chemical Society Reviews*, **43** (1), 70–84.
46. Lavigne, J.J., Savoy, S., Clevenger, M.B., *et al.* (1998) Solution-based analysis of multiple analytes by a sensor array: toward the development of an "electronic tongue." *Journal of the American Chemical Society*, **120** (25), 6429–6430.
47. Hunter, T. (1995) Protein kinases and phosphatases: The Yin and Yang of protein phosphorylation and signaling. *Cell*, **80** (2), 225–236.
48. Cohen, P. (2001) The role of protein phosphorylation in human health and disease. *European Journal of Biochemistry*, **268** (19), 5001–5010.
49. Chen, C.-A., Yeh, R.-H., Lawrence, D.S. (2002) Design and Synthesis of a fluorescent reporter of protein kinase activity. *Journal of the American Chemical Society*, **124** (15), 3840–3841.
50. McIlroy, B.K., Walters, J.D., Johnson, J.D. (1991) A continuous fluorescence assay for protein kinase C. *Analytical Biochemistry*, **195** (1), 148–152.
51. Yeh, R.-H., Yan, 5., Cammer, M., Bresnick, A.R., Lawrence, D.S. (2002) Real time visualization of protein kinase activity in living cells. *The Journal of Biological Chemistry*, **277** (13), 11527–11532.
52. Zhang, T., Edwards, N.Y., Bonizzoni, M., Anslyn, E.V. (2009) The use of differential receptors to pattern peptide phosphorylation. *Journal of the American Chemical Society*, **131** (33), 11976–11984.
53. Fink, A.L. (2006) The aggregation and fibrillation of alpha-synuclein. *Accounts of Chemical Research*, **39** (9), 628–634.
54. Peters, T. (1985) *Advances in Protein Chemistry Volume 37*. Elsevier, Amsterdam.
55. Adams, M.M., Anslyn, E.V. (2009) Differential sensing using proteins: exploiting the cross-reactivity of serum albumin to pattern individual terpenes and terpenes in perfume. *Journal of the American Chemical Society*, **131** (47), 17068–17069.
56. Moreno, F., Cortijo, M., González-Jiménez, J. (1999) The Fluorescent probe prodan characterizes the warfarin binding site on human serum albumin. *Photochemistry and Photobiology*, **69** (1), 8–15.
57. Diehl, K.L., Ivy, M.A., Rabidoux, S., *et al.* (2015) Differential sensing for the regio- and stereoselective identification and quantitation of glycerides. *Proceedings of the National Academy of Sciences of the USA*, **112** (30), E3977-E3986.

58. Voet, D., Voet, J.G., Pratt, C.W. (2013) *Fundamentals of Biochemistry: Life at the Molecular Level*, 3rd edition, Wiley, pp. 245–295.
59. Andrikopoulos, N.K. (2002) Chromatographic and spectroscopic methods in the analysis of triacylglycerol species and regiospecific isomers of oils and fats. *Critical Reviews in Food Science and Nutrition*, **42** (5), 473–505.
60. Raman, M., Chen, W., Cobb, M.H. (2007) Differential regulation and properties of MAPKs. *Oncogene*, **26** (22), 3100–3112.
61. Bonetta, L. (2005) Probing the kinome. *Nature Methods*, **2** (3), 225–232.
62. Kinoshita, E., Kinoshita-Kikuta, E., Koike, T. (2009) Separation and detection of large phosphoproteins using Phos-tag SDS-PAGE. *Nature Protocols*, **4** (10), 1513–1521.
63. Takahashi, M., Kawamura, A., Kato, N., *et al.* (2012) Phosphopeptide-dependent labeling of 14-3-3ζ proteins by fusicoccin-based fluorescent probes. *Angewandte Chemie International edition*, **51** (2), 509–512.
64. Shults, M.D., Janes, K.A., Lauffenburger, D.A., Imperiali, B. (2005) A multiplexed homogeneous fluorescence-based assay for protein kinase activity in cell lysates. *Nature Methods*, **2** (4), 277–283.
65. Ojida, A., Mito-oka, Y., Inoue, M., Hamachi, I. (2002) First artificial receptors and chemosensors toward phosphorylated peptide in aqueous solution. *Journal of the American Chemical Society*, **124** (22), 6256–6258.
66. Zamora-Olivares, D., Kaoud, T.S., Dalby, K.N., Anslyn, E.V. (2013) In-situ generation of differential sensors that fingerprint kinases and the cellular response to their expression. *Journal of the American Chemical Society*, **135** (39), 14814–14820.
67. Lee, S., Warthaka, M., Yan, C., *et al.* (2011) Examining docking interactions on ERK2 with modular peptide substrates. *Biochemistry*, **50** (44), 9500–9510.
68. Zamora-Olivares, D., Kaoud, T.S., Jose, J., *et al.* (2014) Differential sensing of MAP kinases using SOX-peptides. *Angewandte Chemie International edition*, **53** (51), 14064–14068.
69. Garai, Á., Zeke, A., Gógl, G., *et al.* (2012) Specificity of linear motifs that bind to a common mitogen-activated protein kinase docking groove. *Science Signaling*, **5** (245), ra74.
70. Chen, G., Porter, M.D., Bristol, J.R., Fitzgibbon, M.J., Pazhanisamy, S. (2000) Kinetic mechanism of the p38-α MAP Kinase: phosphoryl transfer to synthetic peptides. *Biochemistry*, **39** (8), 2079–2087.
71. Minami, T., Esipenko, N.A., Akdeniz, A., *et al.* (2013) Multianalyte sensing of addictive over-the-counter (OTC) drugs. *Journal of the American Chemical Society*, **135** (40), 15238–15243.
72. Aronov, A.M., Tang, Q., Martinez-Botella, G., *et al.* (2009) Structure-guided design of potent and selective pyrimidylpyrrole inhibitors of extracellular signal-regulated kinase (ERK) using conformational control. *Journal of Medicinal Chemistry*, **52**(20), 6362–6368.
73. Shi, S.-Y., Zhang, Y.-P., Jiang, 5.-Y.,*et al.* (2009) Coupling HPLC to on-line, post-column (bio)chemical assays for high-resolution screening of bioactive compounds from complex mixtures. *TrAC Trends in Analytical Chemistry*, **28** (7), 865–877.
74. Zygler, A., Wasik, A., Namieśnik, J. (2009) Analytical methodologies for determination of artificial sweeteners in foodstuffs. *TrAC Trends in Analytical Chemistry*, **28** (9), 1082–1102.
75. Ebbels, T.M.D., Keun, H.C., Beckonert, O.P. *et al.* (2007) Prediction and classification of drug toxicity using probabilistic modeling of temporal metabolic data: the consortium on metabonomic toxicology screening approach. *Journal of Proteome Research*, **6** (11), 4407–4422.
76. Ogrinc, N., Kosir, I.J., Spangenberg, J.E., Kidric, J. (2003. The application of NMR and MS methods for detection of adulteration of wine, fruit juices, and olive oil. A review. *Analytical and Bioanalytical Chemistry*, **376** (4), 424–430.
77. Adams, M.M., Joyce, L.A., Anslyn, E.V. (2012) Uses of differential sensing and arrays in chemical analysis, in *Supramolecular Chemistry*, P.A. Gale, J.W. Steed (Eds), John Wiley & Sons, Chichester, pp. 709–730.
78. Jackson, R.S. (2008) *Wine Science*. Elsevier.
79. Lund, S.T. (2006) The molecular basis for wine grape quality – a volatile subject. *Science*, **311** (5762), 804–805.
80. Umali, A.P., LeBoeuf, S.E., Newberry, R.W., *et al.* (2011) Discrimination of flavonoids and red wine varietals by arrays of differential peptidic sensors. *Chemical Science*, **2** (3), 439–445.
81. Gallagher, L.T., Heo, J.S., Lopez, M.A., *et al.* (2012) Pattern-based discrimination of organic acids and red wine varietals by arrays of synthetic receptors. *Supramolecular Chemistry*, **24** (2), 143–148.

6

Robust Host–Guest Chemistry of Cucurbit[*n*]uril: Fundamentals and Applications of the Synthetic Receptor Family

Kimoon Kim,[1,2] Dinesh Shetty,[1] and Kyeng Min Park[1,3]

[1] *Center for Self-assembly and Complexity, Institute for Basic Science (IBS), Pohang, Republic of Korea*
[2] *Department of Chemistry and Division of Advanced Materials Science, Pohang University of Science and Technology (POSTECH), Pohang, Republic of Korea*
[3] *Department of Nanomaterials Science and Engineering, University of Science and Technology (UST), Pohang, Republic of Korea*

6.1 Personal pathway to the discovery of cucurbit[*n*]uril and early day developments

On a fine spring day in 1991 I (KK) was sitting in a library trying to catch up with current literature. A macrocyclic molecule described in a review article by Lindsey [1] caught my attention, which gave a timely fillip for my research career. Although it was first reported by Behrend and coworkers in 1905 [2] as a condensation product of glycoluril and formaldehyde in acidic solution, the exact nature of the compound was not known until it was rediscovered and fully characterized by Mock and coworkers in 1981 [3]. The hexameric macrocyclic compound was named cucurbituril as it resembles a pumpkin, which belongs to the botanical family *cucurbitaceae*. Fascinated by the beautiful structure and easy synthesis, I started to dig into the details on this molecule and learned that Mock, who was the only player in this field at the time, had explored the supramolecular chemistry of cucurbituril by studying the thermodynamics and kinetics of its guest binding, observing the enzyme-like acceleration of reaction in the cavity of the molecule, and inventing a molecular switch based on a shuttling motion of cucurbituril threaded on an amine-based thread. I was intrigued by the exceptionally high binding affinity and selectivity of cucurbituril in guest binding. However, two questions arouse immediately

Macrocyclic and Supramolecular Chemistry: How Izatt–Christensen Award Winners Shaped the Field, First Edition.
Edited by Reed M. Izatt.

in my mind. First, why does the condensation reaction between glycoluril and formaldyde exclusively produce the hexameric species, namely cucurbit[6]uril (CB[6]) among a series of possible homologues, cucurbit[*n*]uril (CB[*n*])? Second, why did Mock use strongly acidic solutions such as a 1:1 mixture of formic acid and water to study the host–guest chemistry of CB[6]? The answer to the second question became apparent soon; CB[6] has extremely poor solubility in common solvents except strongly acidic solution. Having decided to work on cucurbituril, therefore, the first thing we tried to do was to synthesize alkylated CB[6] to improve the solubility by using alkylated glycolurils for the condensation reaction. This effort led us to the successful synthesis of decamethylcucurbit[5]uril; however, to a great disappointment the same result was published by Stoddart [4] by the time we finished our work. Adding to our frustration, efforts to synthesize other cucurbituril derivatives using different alkyl glycolurils failed to produce any cyclized products. Several years of our efforts led to more questions than answers. However, the success of decamethylcucurbit[5]uril synthesis reignited our assumption of the existence of other cucurbituril homologues. In pursuit of finding new homologues we spent several years without much success even though we eventually discovered and isolated other homologues.

During this frustrating period, one interesting discovery my students made was that CB[6] dissolves freely in neutral aqueous solution in the presence of alkali metal ions such as Na^+, which led us to study the reversible encapsulation and release of small neutral guest molecules such as THF inside the cavity of CB[6] in aqueous solution at a neutral pH [5]. Meanwhile, I asked my students to explore the synthesis of mechanically interlocked molecules such as catenanes and rotaxanes using CB[6] as a molecular bead as there was no such molecule reported then, except a pseudorotaxane reported by Mock [6], even though it was one of the hot topics in chemistry at the time. After a simple [2]rotaxane containing CB[6] had been synthesized [7], we decided to develop a general strategy to synthesize polyrotaxanes containing CB[6] by combining the host–guest chemistry and transition metal directed self-assembly. This effort led us to the successful synthesis of 1D, 2D, and 3D polyrotaxanes in the subsequent years [8–10]. Another important contribution to the field during this period was the synthesis of molecular necklaces, mechanically interlocked molecules in which a number of small rings are threaded onto a large ring [11, 12]. Extending our strategy for the synthesis of polyrotaxanes, we were able to synthesize [4]- and [5]-molecular necklaces in which three and four CB[6] molecules, respectively, are threaded on a triangular or square framework, respectively, made of transition metal complexes and organic linkers. We also explored the functional aspects of CB[6]-incorporated mechanically interlocked systems to design molecular machines and switches. Our work in this area was summarized in a review article [13].

While focusing on these projects, we continued to search for CB[6] homologues. After a long struggle, we finally observed small additional peaks besides those for CB[6] in the NMR spectra of the condensation product of glycoluril and formaldehyde, especially when the condensation reaction was carried out at a temperature below 95 °C. Having realized that these small peaks correspond to the elusive targets that we had been chasing – other members of the cucurbituril host family – we carried out a systematic study to optimize the synthetic conditions and isolate the individual CB homologues. After spending more than a year on the study, we successfully isolated and fully characterized the CB homologues, CB[*n*] (n=5, 7, and 8), and published the results in early 2000 [14]. Independently, Day and coworkers also discovered and isolated the CB homologues [15]. This breakthrough discovery led us to study the exciting host–guest chemistry of the CB homologues, especially CB[7] and CB[8]. The discovery and early studies on the host–guest chemistry of the CB homologues were documented in a review article [16].

One interesting observation made during this period was that an electron-donor and an electron-acceptor molecule form a charge-transfer (CT) complex inside the cavity of CB[8] to produce a stable ternary complex [17]. This discovery led us to design and synthesize a variety of supramolecular assemblies including molecular necklaces [18], supramolecular vesicles [19], and rotaxane dendrimers [20] through the intramolecular or intermolecular host-stabilized charge transfer complex formation. Furthermore, several molecular machines and switches were invented based on this new chemistry. A review article published in 2007 documented our work in this area [21].

After the discovery and successful isolation of the new members of the host family, we resumed our efforts to find suitable methods to functionalize CB[*n*]. Many years' efforts were finally rewarded with the discovery of a direct functionalization method to produce hydroxy-substituted CB[*n*] [22], which opened a new horizon in CB chemistry. With tailor-made CB derivatives, especially those of CB[6] and CB[7] in our hands, we have been exploring their applications in many areas including sensor, chromatography, polymers, drug delivery, tissue engineering, protein isolation, and synthetic ion channels. Earlier results in this area have been summarized in a review article [23]. In this chapter, we will discuss the structures, physical properties and host–guest chemistry of CB[*n*] with a special emphasis on the development of high-affinity host–guest chemistry of CB[6] and CB[7] with specific guests, and its applications

6.2 Structures and physical properties of CB[*n*]

CB[*n*] (n=5–8, 10) have a rigid structure [24] which makes the cavity parameters particularly informative. The most extensively studied members, CB[6], CB[7] and CB[8] have the same height (9.1 Å), but show the significant variation in cavity volume (164 Å^3, 279 Å^3, and 479 Å^3, respectively), which is close to those of α-, β-, and γ-CD (174 Å^3, 262 Å^3 and 400 Å^3, respectively). The mean diameter of the internal cavity of CBs increases progressively from CB[5] to CB[8] (~4.4 Å to ~8.8 Å). The portals of CBs are approximately ~2 Å narrower than the cavity, which results in constrictive binding that produces significant steric barriers to guest association and dissociation [25]. NMR spectra analysis of these homologues revealed that the chemical shift difference between the two sets of methylene protons progressively move downfield on going up in the homologous series.

In general, CB[*n*] show poor solubility ($<10^{-5}$ M) in common solvents except that CB[5] and CB[7] have a moderate solubility in water (20–30 mM). However, they are soluble in acidic aqueous solution or aqueous solution containing alkali metal ions. These CB homologues have high thermal stability in the solid states and no decomposition is observed up to 420 °C for CB[*n*] (n = 5, 6, and 8) although CB[7] starts decomposing at 370 °C. Although comprehensive toxicity studies are still under investigation, previous research works showed non-toxic nature of CBs. For example, a study *in vivo* in mice demonstrated that a dosage of CB[7] up to 250 mg kg^{-1} showed little toxic effect [26].

6.3 General host–guest chemistry of CB[*n*]

It has been shown that the cavity of CBs is best suited to assess the key-and-lock principle required for the formation of an inclusion complex, in which there should be a complementarity between the guest size and the volume of binding site. Even though CB homologues share characteristic features such as hydrophobic cavity and polar carbonyl groups surrounding the portals, their varying cavity and portal sizes lead to remarkable molecular recognition properties different to each other. In general, CB[6] forms very stable complexes with protonated diaminoalkanes whereas CB[7] forms 1:1 complexes with larger guest molecules that are not included in CB[6]. On the other hand, CB[8] has a cavity large enough to accommodate two guest molecules to form 1:2 complexes. Nau and coworkers successfully applied an optimal packing coefficient argument to explain the stability of these inclusion complexes [27]. This analysis for representative sets of known guests with well-defined hydrophobic binding motifs revealed average values of 58% for CB[6], 52% for CB[7], and 53% for CB[8], which well match the ideal packing coefficient (~55%) for other host–guest systems, first discovered by Rebek [28].

Although the cavity sizes are similar, the host–guest behavior of CBs is quite different from that of CDs. In general, CBs bind guest molecules with higher affinity and higher selectivity than CDs do. For example,

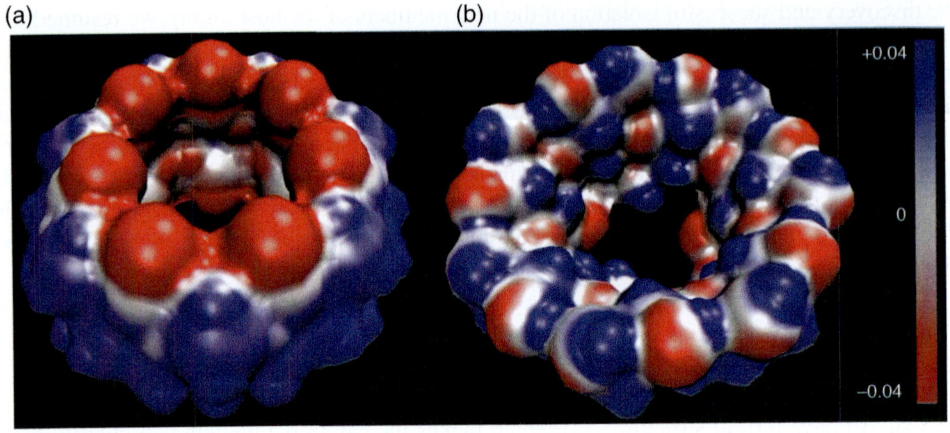

Figure 6.1 The surface electrostatic potential for (a) CB[7] and (b) β-CD (cavity view). The color ranges from negative (red) to positive (blue)

the binding constant (K_a) values for β-CD with different guests are rarely get over $10^6\,M^{-1}$ whereas the corresponding member of the CB family, CB[7] displays much higher binding affinity up to $10^{17}\,M^{-1}$. Furthermore, CBs show strong affinity toward cationic guests, moderate affinity toward neutral guests, and little affinity towards anionic guests [29], while CDs show moderate affinity to hydrophobic guests regardless of their charges, which can be explained by the electrostatic potential of the host families (Figure 6.1). In CBs, the regions around carbonyl oxygens are found to be significantly negative (red colored) and the inner surface of the cavity is also quite negative while the outer surface is somewhat positive. On the other hand, the portal and cavity of CDs are almost neutral. We should also note that the high structural rigidity of CBs in comparison to CDs allows highly selective recognition processes. Furthermore, the inner cavity of CBs has extremely low polarizability, which is closer to the gas-phase than to that of any other known solvents [30], and a very high quadruple moment [27], which allows higher-order electrostatic interactions.

6.4 High-affinity host–guest pairs

If we look into the behavior of most common biomolecules (e.g. proteins, antibodies), a key constituent in the inception of lifelike complexity is the availability of molecules that recognize other components through high-affinity recognition events. The biotin–avidin (or streptavidin) system is the strongest noncovalent interaction pair found in nature, with a binding constant in the order of 10^{13}–$10^{15}\,M^{-1}$ [31, 32]. Most attempts to achieve a synthetic host–guest pair which can mimic the ultrahigh binding affinity and specificity of the natural system have achieved limited success, except for the multivalency approach [33, 34]. In this regard, CB[n] family attracted tremendous interests among researchers, which is depicted in multiple reviews [16, 21, 35–38]. We and others reported that CBs can form stable host–guest complexes with different guest molecules [16, 38–40]. Especially, the unique, high-affinity properties of CB[6] and CB[7] make them attractive not only for their host–guest chemistry but also as a synthetic receptor in many challenging applications. These high-affinity inclusion complex formations are generally driven by cooperative and multiple intermolecular interactions. The hydrophobic effect coupled with ion–dipole interactions have long been known to contribute to high-affinity binding. Recently, it has been reported that the release of high energy water molecules from the host cavity is also important for tight binding of guest molecules [41]. In this chapter, we will discuss the development of high-affinity host–guest chemistry of CB[6] and CB[7] with different guests and highlight associated applications.

6.4.1 High-affinity CB[6]–guest complexes

In their pioneering work [3, 42–45], Mock and coworkers studied the complexation of CB[6] with alkylammonium and alkyldiammonium ions in formic acid/water (1:1) at 40 °C and determined the binding affinity by NMR spectroscopy, which ranged from 10^1 to $10^7\,\text{M}^{-1}$. Remarkably, CB[6] exhibits high selectivity in guest binding, depending on the size, shape and functional groups of guests [25]. Diammonium ions with a C_5 or C_6 chain at the middle bind to CB[6] more strongly than those with shorter or longer lengths. For 1,6-diammoniumhexane, the alkyl chain contracts to adopt a slightly folded conformation in order to optimize cavity filling and additionally to maximize ion–dipole interactions. We must note that because of the poor solubility of CB[6] in water, the binding constants were measured in aqueous solutions containing strong acid or alkali metal ions such as formic acid and sodium acetate [5, 42–47]. However, proton and alkali metal ions bind to the portal of CB[6] and behave as competitive binders, which hampered measurement of the "intrinsic" binding affinity of CB[6] toward guests. In 2001, our group synthesized six fused cyclohexane rings decorated CB[6] derivative (CB*[6]) which has essentially identical portal and cavity sizes, but remarkably enhanced solubility in water (~200 mM) [48, 49]. This finding prompted us to study the binding affinities of CB*[6] in pure water with different guest molecules and found two to three orders of magnitude higher binding affinity than those observed for CB[6] in an aqueous solution containing either proton or alkali metal ions [49]. In particular, the binding constants for spermine reached to $3.4 \times 10^{12}\,\text{M}^{-1}$, which is the highest binding affinity reported for CB[6]. The higher binding observation in pure water is attributed to a higher enthalpic gain upon complexation and the absence of interfering ions in the solution competing with guest molecules for binding.

Although kinetic studies on the guest binding to CB[6] are still limited compared to thermodynamic studies, initial investigation of kinetic studies by Mock on the moderately binding guest molecules to CB[6] found a slow rate of guest ingression, egression, and exchange on the NMR timescale [43, 45]. However, it is hard to correlate these observations to the kinetics of high-affinity guest binding to CB[6]. Further studies are needed to elucidate the kinetics of complexation and decomplexation of high-affinity guests to CB[6], which is important not only for fundamental studies but also for applications.

6.4.2 High-affinity CB[7]–guest complexes

CB[7] is one of the most interesting and important members of the CB[*n*] family, which forms exceptionally stable 1:1 inclusion complexes with several guests in aqueous media (Figure 6.2) [39, 40, 50–53]. The reported affinities for such guests approach or even exceed those of the strongest noncovalent interactions found in nature [54, 55]. Compared to CB[6], the cavity size of CB[7] is large enough to accommodate rigid and sphere-shaped guest molecules. In 2005, Isaacs group [40], and our group [39] in collaboration with Inoue and Kaifer independently measured that all neutral (**1**) and cationic ferrocene derivatives (**2** and **3**) form highly stable inclusion complexes with CB[7] with K_a values in the range of 10^9–$10^{12}\,\text{M}^{-1}$. Interestingly, the introduction of an ammonium unit on the guest leads to a significant increase in the binding affinity for **3** that suggests the importance of ion–dipole interaction between the host and guest molecules for the high-affinity binding process. Having seen the high-affinity of CB[7] toward **3**, we decided to extend this work to develop even tighter binding guests for CB[7] in collaboration with Inoue, Kaifer, Isaacs, and Gilson. We thought that addition of a second ammonium unit to ferrocene unit in **3**, which would interact with the other portal of CB[7], would further increase the affinity of the guest to CB[7]. Indeed, the binding affinity of CB[7] toward **4** having two ammonium units further increased to $3 \times 10^{15}\,\text{M}^{-1}$, as measured by a four-step competition ITC experiment [51]. Another interesting observation was the enthalpy and entropy changes associated with the

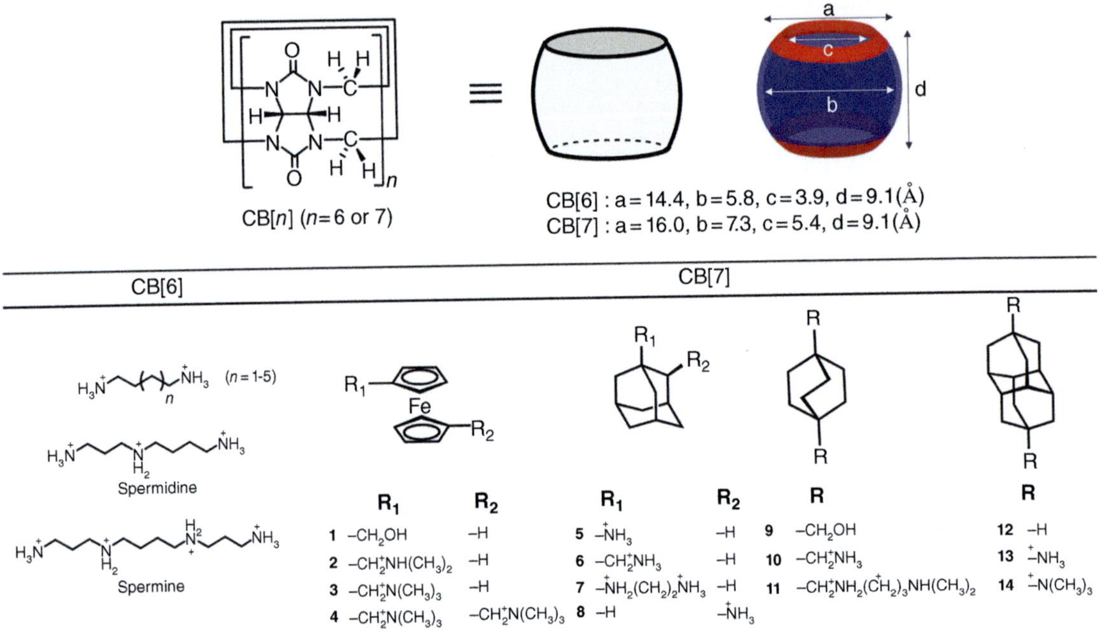

Figure 6.2 *Molecular structures of CB[6] and CB[7] and high-affinity guests (**1–14**)*

guest binding. Unexpectedly, the addition of ammonium groups onto the ferrocene core does not affect the enthalpy change (ΔH = ~–21 kcal mol^{-1}) associated with the complexation; instead, it causes an increase in the entropy change by 3.8–4.3 kcal mol^{-1}, which results in a thousandfold enhancement in the corresponding binding constant. The change in solvation entropy ($\Delta S°_{solv} = \Delta S° - \Delta S°_{config}$) upon complexation is more favorable for the strong binding dication **4** (18 kcal mol^{-1}) compared to monocationic or neutral guests (11 kcal mol^{-1} for **3** and 9 kcal mol^{-1} for **1**). This observation suggests that the enthalpy–entropy compensation effect, which is commonly observed in other molecular recognition systems [29, 50, 56], is not being obeyed in this case. Figure 6.3 presents the data points for different high-affinity guests (**1–14**) along with the data for the binding of CDs with a variety of guest molecules, which greatly deviate from the enthalpy–entropy compensation plot for CD–guest complexation.

In 2003, we observed the formation of a stable host–guest complex between CB[7] and protonated adamantylamine (**5**) [16] right after the successful isolation of CB[7]. In 2005, Isaacs and coworkers measured the binding constant quantitatively for the first time and found that the binding between CB[7] and **A2** is indeed very strong, with a binding constant as high as 4.2 × 10^{12} M^{-1} [40]. Recently, Inoue, Gilson, and coworkers designed a new series of guest molecules having a rigid, aliphatic (bicyclo[2.2.2]octane or adamantine) moiety, outfitted with cationic groups positioned to interact favorably with one or both CB[7] portals [57]. These metal-free guest molecules also show ultrahigh binding affinity toward CB[7], with a binding constant in the range 10^{14}–10^{15} M^{-1}. Remarkably, the binding affinity of mono-cationic adamantane derivatives (**5** and **6**) match those of **4** or **11** having two cationic substituents. However, in these series of guest molecules, the increase in overall charge of the molecules resulted in only a modest increment in binding affinities. Similar to ferrocene, the remarkably high affinities of these guests are attributed in part to the fact that they lose an unusually small amount of entropy on binding in proportion to their enthalpic changes. A very recent work by Gilson suggests that dispersion forces contribute strongly to the computed binding enthalpies of these complexes, partly due to the fact that their packing density is greater than that of water [27, 58].

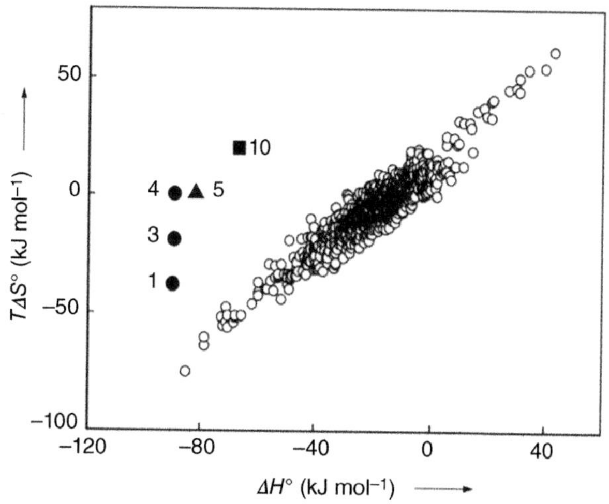

Figure 6.3 *The thermodynamic data obtained for complexation of ferrocene (**1**, **3**, **4**), adamantane (**5**), and bicyclo[2.2.2]octane (**10**) derivatives with CB[7] which are significantly deviated from the enthalpy–entropy compensation plot for cyclodextrin–guest complexation. (Reproduced from ref. [35]. Copyright © 2011 Wiley-VCH Verlag GmbH &Co. KGaA, Weinheim. Black open circles: data adapted from ref. [29]. Copyright © 1998 American Chemical Society.)*

Very recently, Isaacs and co-workers reported an attomolar-affinity host–guest complex between CB[7] and diamantane quaternary diammonium ion derivative **14** ($K_a = 7.2 \times 10^{17}\ \text{M}^{-1}$ in pure D_2O) [53, 59]. The X-ray crystal structure of this complex shows that the core of **14** is complementary to the cavity of CB[7] and the quaternary ammonium groups of **14** allow the guest to form 14 optimal ion–dipole interactions with CB[7]. The authors speculated that the highest binding constant is the outcome of additional solvating water molecules released from CB[7] upon the inclusion complex formation providing an additional entropic driving force.

Overall, the complementarity between the cavity of CB[7] with the size and shape of guest molecules, hydrophobic effect between host cavity and guests, ion–dipole and dipole–dipole interactions of guests with CB portals, have been addressed as the main driving forces for the high-affinity binding of different guests by CB[7]. Above all, recently, Nau and co-workers demonstrated that another major driving force for complexation of hydrophobic residues with CB[*n*] originates from the release of "high-energy" water molecules in the CB cavity [38, 41]. The calculated energy for the release of all high-energy water molecules is highest for CB[7], which provides perfect compromise between the number and energy of encapsulated water molecules in order to allow the maximum energetic gain upon guest binding. These findings can consolidate the reasons for the high-affinity host–guest binding in small synthetic macrocycles such as CB[*n*].

6.5 Functionalized CBs

Significant efforts have been devoted to the functionalization of CB[*n*] not only to improve their solubility but also to synthesize tailor-made derivatives for various applications [23]. The functionalization was achieved either by use of substituted glycoluril derivatives in CB[*n*]-forming reactions [4, 48, 60–62] or by direct substitution at the "equator" positions of CB[*n*] [22]. In a major breakthrough in the CB chemistry, we discovered a direct functionalization method to produce hydroxylated products of CB[*n*] through oxidation with $K_2S_2O_8$

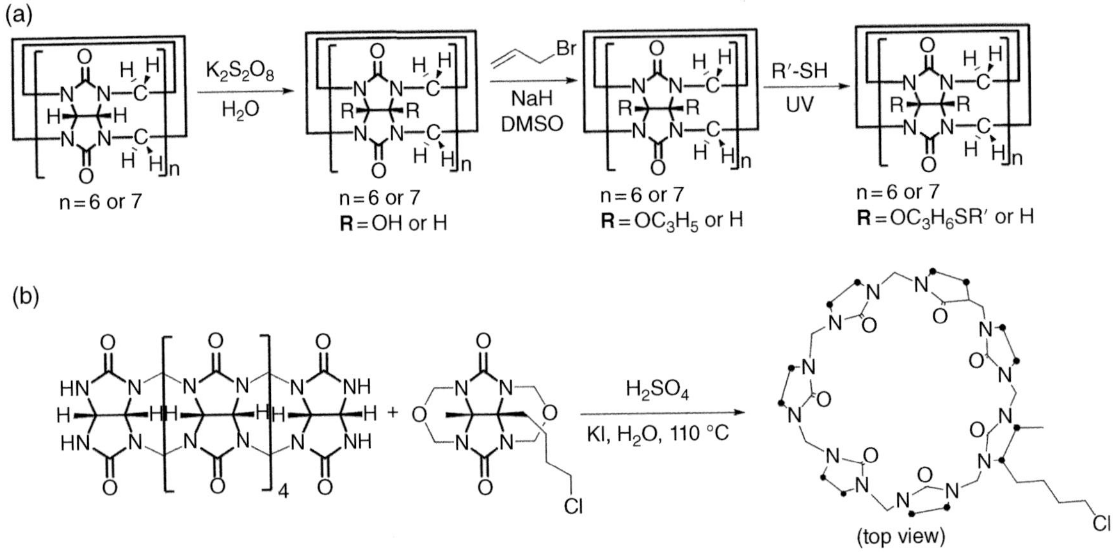

Scheme 6.1 *(a) Synthesis of various CB derivatives from mono- or multi-hydroxy CB[6] and CB[7]. (b) Recent report on the synthesis of a monofunctionalized CB[7] derivative; dots mark connection points to the second identical half of the symmetric structures*

in water [22]. This finding allowed us to synthesize a wide variety of tailor-made CB derivatives for various applications [23]. The hydroxy derivatives of CB[6] or CB[7] are soluble in organic solvents such as DMSO and DMF, which allow further functionalization via conventional organic transformation routes including acylation and alkylation (Scheme 6.1). More recently, Isaacs and co-workers reported a building-block approach to obtain monofunctionalized CB[7] derivatives (Scheme 6.1) [63]. This approach involves the condensation of acyclic glycoluril hexamer with glycoluril bis(cyclic ethers) to produce monofunctionalized CB[7] which can be further transformed into tailor-made CB[7] derivatives for various applications.

6.6 Applications of high-affinity CB[6] complexes

6.6.1 Pseudorotaxane formation and biological applications

By taking advantage of the stable complex formation between CB[6] and alkanediammonium ions, we synthesized various pseudorotaxanes, (pseudo)polyrotaxanes, and rotaxane dendrimers and explored their biological applications (Figure 6.4) [64, 65]. In collaboration with our group, Nakamura carried out DNA-binding studies of pseudorotaxanes comprising CB[6] and polyamines such as spermine, which showed that CB[6] can modulate the DNA-binding ability of polyamines (Figure 6.4a) [66, 67]. The high-affinity binding between spermine and CB[6] can induce the change in chemical reactivity of the amine, which in turn enhances the activity of topoisomerase (Topo 1) and reduces the activity of endonuclease (*Ban*II) [67]. We also synthesized pseudorotaxane-terminated dendrimers (rotaxane dendrimers), in which CB[6] beads are threaded on the diaminobutane units at the terminals of polypropylimine dendrimers (Figure 6.4b) [64]. The structures and properties of the dendrimers can be reversibly changed by threading and dethreding of CB[6]. We utilized these rotaxane dendrimers to develop non-covalently assembled ternary gene delivery vectors [68]. Interestingly, these delivery vectors transfect mammalian cells with high efficiency and the cytotoxicity of the vectors was relatively low compared to other polyamine-based delivery systems. Further, we demonstrated the targeted delivery of genes into cancer cells by introducing targeting moieties to CB[6], such as β-galactose, which forms a stable complex with a spermine-conjugated dextran polymer [69, 70]. This strategy enabled us to tailor the supramolecular delivery systems in terms of overall transfection efficiency and target specificity by varying the spermine derivatives conjugated to diverse platforms as well as by tagging different target specific ligands to CB[6].

By taking advantage of their stimuli responsive properties, Zink, Stoddart, and coworkers recently utilized CB[6]-based pseudorotaxanes as a new class of mechanical nanovalves which are attached on mesoporous silica nanoparticles (MSNs) to deliver and release interesting cargos to target sites (Figure 6.4c) [71–75]. In response to pH or temperature changes, the nanovalves can be regulated to release the entrapped cargos from the pores of MSNs. In their first report, this concept was demonstrated with a pH-sensitive movement of CB[6] on alkanediammonium molecule (DA) conjugated on MSNs [71]. A model cargo such as rhodamine B loaded in the MSNs at neutral pH starts to be released when the pHis raised above 10, as a result of the decomplexation of CB[6] from the deprotonated DA leading to opening of the pores. Later, a more sophisticated design of thread molecules led to development of another nanovalve system working in a physiological pH condition. The thread molecule comprises two different stations, on one of which CB[6] stays depending on pH conditions [72]. At neutral pH, CB[6] is placed at a station in the thread molecule close to the pores thereby blocking the pores. At a pH below 5, CB[6] moves toward the other station far from the pores, which leads to open the pores and release the entrapped cargos. The pH at which these systems respond can be

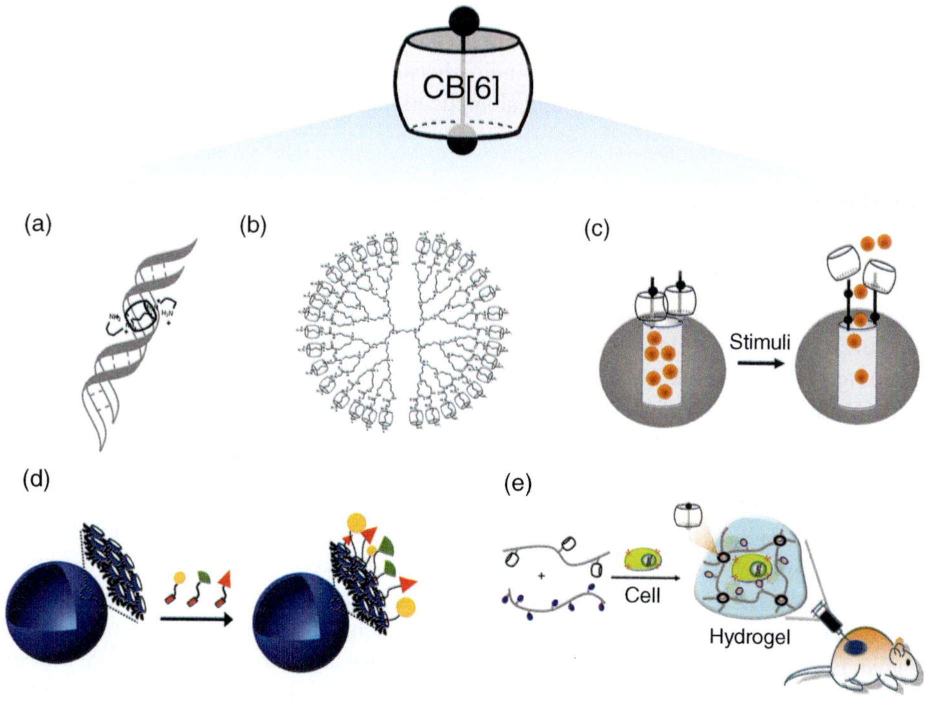

Figure 6.4 *Supramolecular assemblies based on high-affinity CB[6] complexes and their applications. (a) Complexation of spermine@CB[6] pseudorotaxane to DNA; (b) pseudoroxane-terminated dendrimer (rotaxane dendrimer); (c) pseudorataxane-based nanovalves; (d) non-covalent surface modification of CB[6]-based nanomaterials; and (e) hydrogel formation using stable non-covalent cross-linking*

fine-tuned for use in a wide range of different tissues with different pH values. In addition, temperature-sensitive nanovalve systems attached to magnetic nanopoarticles and gold nanoparticles have also been studied and utilized as a remotely controllable heating nanomaterials [73, 74].

6.6.2 Modular multi-functionalities through non-covalent modification

Surface modification of nanomaterials mostly formed by metals or polymers often requires laborious and time-consuming processes. Our group demonstrated an easy and modular surface modification technique by utilizing the high-affinity host–guest chemistry between CB[6] and DA (Figure 6.4d). We found that thiol-ene photoreaction of perallyloxylated CB[6] with various organic thiols produced various molecular building blocks bearing an accessible cavity of CB[6] [69, 76–79]. In our observation, tri- or hexa-ethylene glycol thiol-conjugated CB[6] derivatives act as amphiphiles in water and spontaneously form stable nanosized vesicles (~200 nm in diameter) [76, 80]. When more hydrophobic moieties such as 1-hydroxyhexane thiols were conjugated to perallyloxylated CB[6], the resulting CB[6] derivative self-assembles into a spherical nanoparticle (~190 nm in diameter) in water [77]. Another important development is the template-free synthesis of polymer nanocapsules and 2D polymer films with molecular-scale thickness by photoreaction between perallyloxylated CB[6] and ditholated linear molecules [69, 79, 81]. These CB[6]-based nanomaterials prepared by either non-covalent or covalent synthesis can be used as multifunctional delivery vehicles by taking advantage of their non-covalently-tailorable surfaces which can be easily and modularly modified with various functional tags conjugated with CB[6]-binding moieties such as spermine or spermidine. Carbohydrates, folic acid, and fluorescein isothiocyanate (FITC) were already introduced on the surface of these materials as functional tags for targeting or imaging to demonstrate the multifunctionalities for efficient intracellular targeted delivery and imaging. The high-affinity binding between CB[6] and its guests is a crucial factor for the utility of these materials, especially in biological applications.

6.6.3 Hydrogelation through non-covalent crosslinking

Synthetic hydrogels have emerged as novel scaffolds to provide artificial 3D environments that mimic the extracellular matrix (ECM) [82–85]. These hydrogels are mostly fabricated by crosslinking of polymers with additional chemical reagents or by spontaneous physical crosslinking through noncovalent interactions [86]. However, such hydrogels lack required physical properties mostly due to the weak interactions among gel forming molecules. Taking advantage of the strong host–guest interaction of alkylammonium molecule to CB[6] under physiological condition, our group demonstrated the formation of robust supramolecular hydrogels by simple mixing of hyaluronic acid (HA)-conjugated CB[6] derivatives and another HA conjugated hexamethylenediamine (HMDA) (Figure 6.4e) [87]. The resulting hydrogels are mechanically stable enough to hold their shape even under in-vivo conditions for more than a few weeks and soft enough for cell growth and differentiation without cytotoxicity [88]. Later, we found that engineered mesenchymal stem cells (eMSCs) entrapped in the hydrogels well kept their antitumoral therapeutic functions, such as expression of a tumor growth-inhibition protein, mutant interleukin-12, even in mice, for more than 60 days [89], which was the first demonstration of such a long-term assessment of eMSCs within hydrogels *in vivo*. Encouraged by the successful results *in vivo*, we and our collaborators are now expending the utilities of the supramolecular hydrogel to regeneration of tissues such as blood vessels and osteochondral tissues *in vivo* using vascular endothelial growth factor (VEGF)-transduced MSCs and human tubal mesenchymal stem cells (htMSCs) entrapped hydrogels, respectively. These findings show the feasibility of CB[6]-based hydrogels as a practical artificial ECM mimicking material for cell therapies and regenerative tissue engineering applications.

6.7 Applications of high-affinity CB[7] complexes

6.7.1 Immobilization of biomolecules on a solid surface

Convenient ways to link biomolecules to predefined locations in a site-specific manner on reactive surfaces can provide precise control over immobilizing biomolecules and yield improved performance of biochips [90, 91]. The biotin–streptavidin pair is one of the most widely used tools for surface immobilization of biomolecules because of its strong and specific interaction. However, this system suffers some shortcomings, including denaturation by organic solvents or elevated temperatures, and high cost. Considering the high stability of ferrocene–CB[7] complex, we thought that these ultrastable host–guest complexes could be used for the immobilization of biomolecules on solid surfaces.

With this idea in mind, we first decided to develop a novel method to immobilize proteins on gold using the ultrastable ferrocenemethylammonium–CB[7] (FA–CB[7]) complexes, and explore their applications as a biosensor [92]. The surface immobilization of CB[7] was achieved by olefin metathesis reaction between allyloxyCB[7] and a vinyl-terminated mixed self-assembled monolayer (SAM) on gold. Ferrocene conjugated proteins can directly interact with the CB[7] moieties on the surface (Figure 6.5a). We chose glucose oxidase (GOx), one of the most well-known redox proteins extensively studied for glucose sensor applications, as a model protein and the "ferrocenylated" GOx (FA–GOx) was prepared by simple coupling reaction. The protein was immobilized on the CB[7]-anchored gold substrate by direct immersion of the substrate in a solution containing FA–GOx. The surface plasmon resonance (SPR) technique confirmed that FA–GOx was anchored with a surface density of $1.1 \times 10^{12}\,\mathrm{mol\,cm^{-2}}$, which is approximately 5.5 times higher than that of GOx non-specifically bound on gold. The protein immobilized on the surface was utilized as a glucose sensor with millimolar sensitivity as the catalytic current increases linearly with increasing analyte concentration.

More recently, we extended this platform utility to immobilize aptamers for sensor applications [93]. By taking advantage of the fact that CB[7] readily forms a self-assembled monolayer (CB[7]SAM) on gold [94], we developed a new SPR-based aptasensor chip which can be easily prepared by simple addition of ferrocenlyated aptamers (FA-aptamers) to CB[7]SAM. (Figure 6.5b) [93]. A concentration-dependent sensor response was observed when a target protein (thrombin) was injected into a channel of the sensor chip; a nanomolar (sub-ppm) level of sensitivity to thrombin was achieved. Brunsveld and others utilized the same strategy to immobilize different proteins and tested their applications [95–97]. Their method allowed to print stable protein monolayers in well-defined formats with controlled protein orientation and its release from the surface by treating an excess amount of a competitive guest such as (ferrocenylmethyl)trimethyl-ammonium iodide. In general, these reports suggest that CB[7]–FA pairs can be applied to reversible immobilization of different biomolecules on solid surfaces.

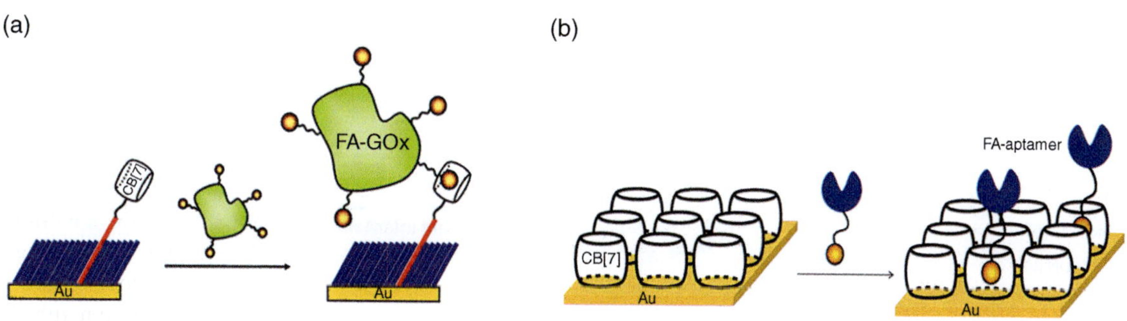

Figure 6.5 Immobilization of (a) ferrocenylated proteins (Adapted from ref. [94]. Copyright © 2007, American Chemical Society); and (b) ferrocenylated aptamers on a gold surface using an ultrastable synthetic binding pair system. (Reproduced from ref. [95] with permission from The Royal Society of Chemistry.)

6.7.2 Plasma membrane protein fishing

The isolation of plasma membrane proteins (PMPs) is highly challenging because of their inherent hydrophobicity and heterogeneity [98]. A well-known method for isolation of PMPs involves biotinylation of PMPs and capture of the biotinylated PMPs with streptavidin-immobilized beads, exploiting the strong interaction between streptavidin and biotin. However, it has some intrinsic drawbacks including low yield, contamination with endogenously biotinylated proteins, and interference by streptavidin detached from beads [99–102]. To address the issues, we developed a new supramolecular fishing method utilizing an ultrastable and highly selective synthetic binding pair, CB[7]–FA system (Figure 6.6) [103]. CB[7]-conjugated sepharose beads (CB[7]-beads) were utilized to capture FA-labelled proteins selectively from heterogeneous protein mixtures of various mammalian cells including rat fibroblast (Rat-1) and HEK 293 cells. The capture efficiency for plasma membrane proteins such as EGFR, is much higher compared to cytoplasmic proteins. In addition, the captured proteins were easily recovered from the beads at room temperature by treating with a strong competitor such as 1′,1-bis(trimethylammoniomethyl)ferrocene (BAF), which cannot be achieved by the streptavidin–biotin-based method. These results demonstrated another promising biological application of the ultrastable host–guest pair system – efficient isolation and purification of proteins. This work may be extended to selective isolation of cytoplasmic proteins, which is currently under investigation.

6.7.3 Supramolecular velcro system

Another important application of high-affinity host–guest pairs is the development of underwater adhesive system. A serious problem in developing underwater adhesive surfaces is how to expel the water molecules from the region between the two interacting surfaces. Since the formation of extremely stable complexes between CB[7] and hydrophobic guests is strongly associated with the transfer of water molecules to the bulk solution [41], we utilize the ultrastable host–guest system to develop functionalized surfaces with strong underwater adhesion properties, which work like a hook and loop fastener,Velcro®. Our system comprises two complementary silicon wafer surfaces: a "loop" surface which is functionalized with CB[7]; and a "hook" surface which is covered with FA molecules (Figure 6.7) [104]. When these complementary surfaces are

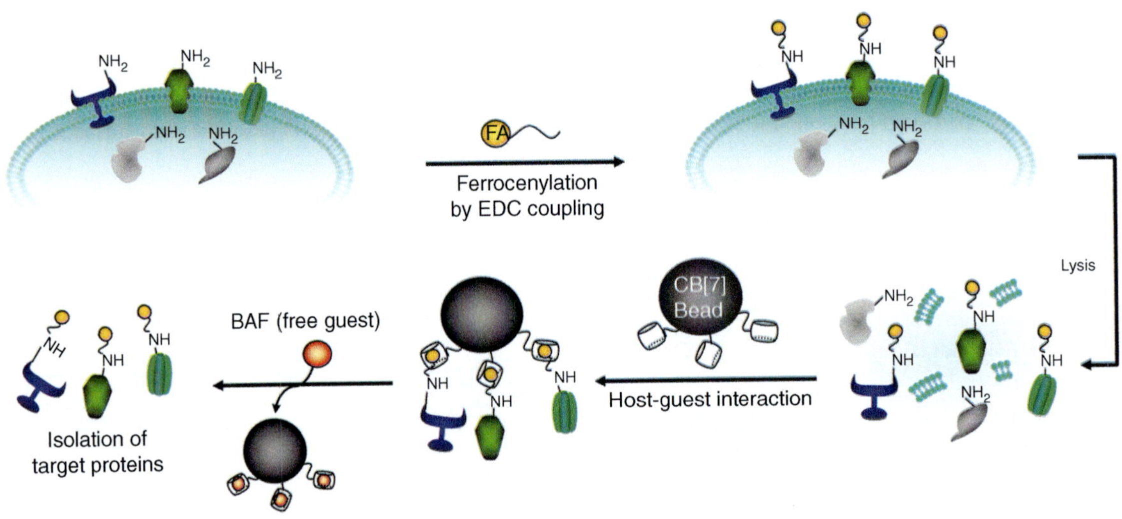

Figure 6.6 Strategy for isolation of plasma membrane proteins using an ultrastable synthetic host–guest binding pair system. (Reproduced from ref. [105]. Copyright © 2010, Nature Publishing Group.)

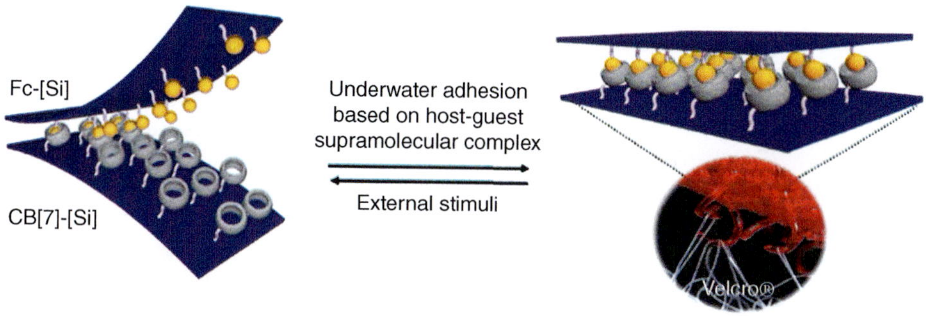

Figure 6.7 Supramolecular Velcro or hook-and-loop strategy for underwater adhesion based on CB[7]- and FA-modified surfaces. (Adapted from ref. [106]. Copyright © 2013 Wiley-VCH Verlag GmbH &Co. KGaA, Weinheim.)

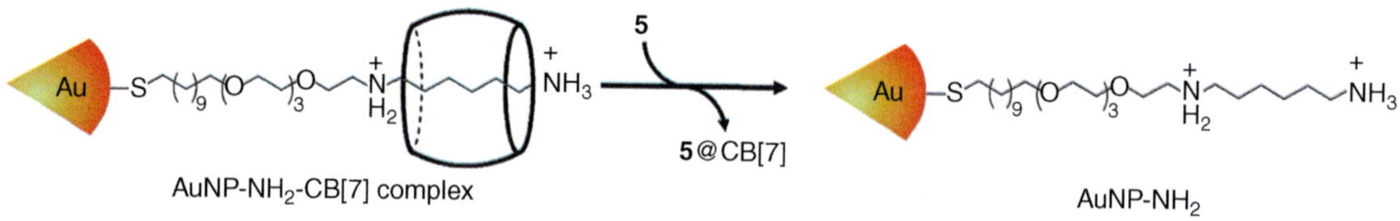

Figure 6.8 Schematic illustration depicting the use of intracellular host–guest complexation to trigger gold nanoparticle (AuNP–NH2) cytotoxicity

pushed together under water, the FA moiety binds tightly inside the CB[7] cavity which is sufficient to "glue" them together with enough strength to lift a 2 kg weight. The lap-shear adhesion strength of this system in air is much higher than that of commercially available adhesives, including double-sided tapes and traditional Velcro. Importantly, our supramolecular Velcro is not only mechanically reversible but also chemically switchable, as the binding affinity of the FA@CB[7] complex can be modulated by oxidation of the ferrocenyl group [105]. The redox properties of the FA@CB[7] complex afford an interesting method to modulate the strength of adhesion, which is not available with any other adhesive systems. The non-cytotoxic nature of these host and guest molecules may permit potential biological applications.

6.7.4 Other applications

Apart from the utility of high-affinity host–guest pairs in the separation of proteins and surface immobilization, Rotello and co-workers have exploited it as a trigger in therapeutic systems. Their system consists of complementary diaminohexane-terminated gold nanoparticles (AuNP–NH_2) and CB[7] that form a non-toxic nanoparticle assembly, which is readily taken up by cells (Figure 6.8) [106]. Intracellular removal of CB[7] from the nanoparticle surface using orthogonal competitor guest **5** results in endosomal escape of the AuNP–NH_2, which activates the cytotoxicity of the amine tethered gold nanoparticles and thereby causes cell death. This strategy for triggering therapeutic systems has some advantages in dosage control and potential utility in dual-targeting therapies. This result demonstrates the use of synthetic host–guest chemistry to provide triggered activation of a therapeutic system by competitive complexation, and provides a potential strategy for the construction of such systems using CB[7], which has an ability to form a strong complex within living cells. Based on the same principle, the authors have recently reported the regulation of exocytosis of nanoparticles using CB[7] host–guest chemistry [107]. This approach provides a potential strategy for prolonged retention of drug carriers within endosomes, enabling sustained therapeutic effect of the carriers.

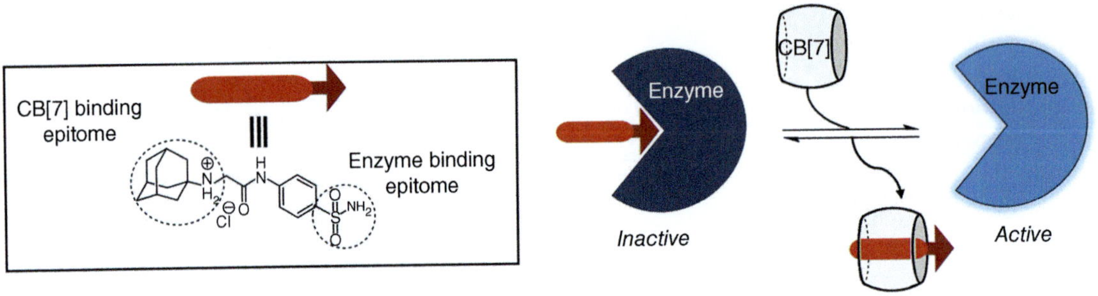

Figure 6.9 *Control of enzymatic activity using CB[7] and adamantine derivative*

By applying the quality of high-affinity and selectivity of CB[7] toward specific guests, Isaacs and co-workers demonstrated the influence of CB[7] on the behavior of biological receptors toward their guests, thereby changing in their biological activity (Figure 6.9) [80]. They have explored the self-sorting capability of two-faced guests in the regulation of the catalytic activity of bovine carbonic anhydrase (BCA). The active site of the enzyme was initially occupied by a two-faced inhibitor that contains benzenesulfonamide (enzyme inhibitor unit) and trimethylsilylmethylammonium group (CB[7] binding unit). Initial enzymatic activity was recovered by addition of CB[7], which forms a 1:1 host–guest complex with the trimethylsilylmethylammonium moiety ($K_a \sim 10^9 M^{-1}$) of the two-faced guest to release the inhibitor from the enzyme. However, the enzyme activity was inhibited again by addition of **5**, which triggered the dissociation of the CB[7]•enzyme inhibitor, resulting in the release of free two-faced guest, which then re-occupied the catalytic site of the enzyme. One of the key controlling factors in this process is the relative binding affinity of the two-faced inhibitor toward the enzyme versus CB[7].

6.8 Conclusions

Thirty five years ago, the structure of CB[6] was unveiled, and it was noticed that this fascinating macrocycle could encapsulate guest molecules with high-affinity. However, its poor solubility and its stubbornness in resisting chemical modification kept the field rather sluggish until the discovery of other CB homologues and the functionalization of these macrocycles. This later discovery, to which we contributed significantly, triggered a dramatic increase in the number of articles, reviews and patents published every year. In this chapter we discussed a number of important issues related to high-affinity host–guest chemistry of CBs which are directly relevant to the development of synthetic systems that can meet the functions of their natural counterparts. The popularity of CB[*n*]s is largely due to their outstanding recognition properties, and to the high-affinity interactions with specific guests. Our findings not only changed the notion on the properties of CBs but also helped greatly to broaden the horizon of its applications. Even today, wherever CBs are involved, unexpected and sometimes very unusual results emerge. This has been the story on several occasions during our past two decades of involvement in the CB[*n*] chemistry.

This chapter offered a brief history of CB[*n*] chemistry and highlighted one of the exciting developments, i.e the high-affinity host–guest interactions of CB[6] and CB[7] with various guest molecules and their applications. These high-affinity pairs offer numerous advantages over other conventional macrocycles: attomolar affinities and tunable complexation kinetics. These host molecules show remarkable binding characteristics based on guest size, shape, length, and chemical functionality. The positive entropy change from the extensive dehydration of the host and guest upon complexation, thereby overcoming enthalpy–entropy compensation, critically assists the achievement of high binding affinity. The ultrastable host–guest complexes are unique because they rely on monovalency to achieve such high binding affinity. While there have been continuing efforts to develop new high-affinity guest molecules, their implementation in challenging applications has rapidly unfolded during the past decade. Besides those we have discussed here, there should be plenty of other applications of these ultrahigh-affinity host–guest pairs. Because of its high affinity and selectivity, CB[7] is most useful to deal with challenging applications in materials science, biology, and medicine. Moreover, the high-affinity features of CB[7] straddle the line between designed synthetic and natural biomolecular receptors and hence promise the development of synthetic systems whose functions match their biological counterparts.

6.9 Acknowledgements

This work was supported by the Institute for Basic Science (IBS; CA1303) in Korea. We thank all the present and past group members who were involved in the development of CB[*n*] chemistry for their great contributions.

References

1. Lindsey, J.S. (1991) Self-assembly in synthetic routes to devices. biological principles and chemical perspectives: a review. *New Journal of Chemistry*, **15** (2–3), 153–180.
2. Behrend, R., Meyer, E., Rusche, F. (1905) I. ueber condensationsproducte aus glycoluril und formaldehyd. *Justus Liebigs Annalen der Chemie*, **339** (1), 1–37.
3. Freeman, W.A., Mock, W.L., Shih, N.Y. (1981) Cucurbituril. *Journal of the American Chemical Society*, **103** (24), 7367–7368.
4. Flinn, A., Hough, G.C., Stoddart, J.F., Williams, D.J. (1992) Decamethylcucurbit[5]uril. *Angewandte Chemie International Edition*, **31** (11), 1475–1477.
5. Jeon, Y.-M., Kim, J., Whang, D., Kim, K. (1996) Molecular container assembly capable of controlling binding and release of its guest molecules: reversible encapsulation of organic molecules in sodium ion complexed cucurbituril. *Journal of the American Chemical Society*, **118** (40), 9790–9791.
6. Mock, W.L., Pierpont, J. (1990) A cucurbituril-based molecular switch. *Journal of the Chemical Society, Chemical Communications*, (**21**), 1509–1511.
7. Jeon, Y.-M., Whang, D., Kim, J., Kim, K. (1996) A simple construction of a rotaxane and pseudorotaxane: syntheses and X-ray crystal structures of cucurbituril threaded on substituted spermine. *Chemistry Letters*, **25** (7), 503–504.
8. Whang, D., Jeon, Y.-M., Heo, J., Kim, K. (1996) Self-assembly of a polyrotaxane containing a cyclic "bead" in every structural unit in the solid state: cucurbituril molecules threaded on a one-dimensional coordination polymer. *Journal of the American Chemical Society*, **118** (45), 11333–11334.
9. Whang, D., Kim, K. (1997) Polycatenated two-dimensional polyrotaxane net. *Journal of the American Chemical Society*, **119** (2), 451–452.
10. Lee, E., Heo, J., Kim, K. (2000) A three-dimensional polyrotaxane network. *Angewandte Chemie International Edition*, **39** (15), 2699–2701.
11. Whang, D.M., Park, K.M., Heo, J., *et al.* (1998) Molecular necklace: quantitative self-assembly of a cyclic oligorotaxane from nine molecules. *Journal of the American Chemical Society*, **120** (19), 4899–4900.
12. Roh, S.G., Park, K.M., Park, G.J., *et al.* (1999) Synthesis of a five-membered molecular necklace: a 2+2 approach. *Angewandte Chemie International Edition*, **38** (5), 638–641.
13. Kim, K. (2002) Mechanically interlocked molecules incorporating cucurbituril and their supramolecular assemblies. *Chemical Society Reviews*, **31** (2), 96–107.
14. Kim, J., Jung, I.S., Kim, S.Y., *et al.* (2000) New cucurbituril homologues: syntheses, isolation, characterization, and X-ray crystal structures of cucurbit[*n*]uril (n=5, 7, and 8). *Journal of the American Chemical Society*, **122** (3), 540–541.
15. Day, A., Arnold, A.P., Blanch, R.J., Snushall B. (2001) Controlling factors in the synthesis of cucurbituril and its homologues. *The Journal of Organic Chemistry*, **66** (24), 8094–8100.

16. Lee, J.W., Samal, S., Selvapalam, N., *et al.* (2003) Cucurbituril homologues and derivatives: new opportunities in supramolecular chemistry. *Accounts of Chemical Research*, **36** (8), 621–630.
17. Kim, H.J., Heo, J., Jeon, W.S., *et al.* (2001) Selective inclusion of a hetero–guest pair in a molecular host: formation of stable charge-transfer complexes in cucurbit[8]uril. *Angewandte Chemie International Edition*, **40** (8), 1526–1529.
18. Ko, Y.H., Kim, K., Kang, J.K., *et al.* (2004) Designed self-assembly of molecular necklaces using host-stabilized charge-transfer interactions. *Journal of the American Chemical Society*, **126** (7), 1932–1933.
19. Jeon, Y.J., Bharadwaj, P.K., Choi, S.W., *et al.* (2002) Supramolecular amphiphiles: spontaneous formation of vesicles triggered by formation of a charge-transfer complex in a host. *Angewandte Chemie International Edition*, **41** (23), 4474–4476.
20. Kim, S.Y., Ko, Y.H., Lee, J.W., *et al.* (2007) Toward high-generation rotaxane dendrimers that incorporate a ring component on every branch: noncovalent synthesis of a dendritic[10]pseudorotaxane with 13 molecular components. *Chemistry – An Asian Journal*, **2** (6), 747–754.
21. Ko, Y.H., Kim, E., Hwang, I., Kim, K. (2007) Supramolecular assemblies built with host-stabilized charge-transfer interactions. *Chemical Communications*, (**13**), 1305–1315.
22. Jon, S.Y., Selvapalam, N., Oh, D.H., *et al.* (2003) Facile synthesis of cucurbit[*n*]uril derivatives via direct functionalization: expanding utilization of cucurbit[*n*]uril. *Journal of the American Chemical Society*, **125** (34), 10186–10187.
23. Kim, K., Selvapalam, N., Ko, Y.H., *et al.* (2007) Functionalized cucurbiturils and their applications. *Chemical Society Reviews*, **36** (2), 267–279.
24. Samsonenko, D.G., Virovets, A.V., Lipkowski, J., *et al.* (2002) Distortion of the cucurbituril molecule by an included 4-methylpyridinum cation. *Journal of Structural Chemistry*, **43** (4), 664–668.
25. Marquez, C., Hudgins, R.R., Nau, W.M. (2004) Mechanism of host–guest complexation by cucurbituril. *Journal of the American Chemical Society*, **126** (18), 5806–5816.
26. Uzunova, V.D., Cullinane, C., Brix, K., *et al.* (2010) Toxicity of cucurbit[7]uril and cucurbit[8]uril: an exploratory in vitro and in vivo study. *Organic & Biomolecular Chemistry*, **8** (9), 2037–2042.
27. Nau, W.M., Florea, M., Assaf, K.I. (2011) Deep inside cucurbiturils: physical properties and volumes of their inner cavity determine the hydrophobic driving force for host–guest complexation. *Israel Journal of Chemistry*, **51** (5–6), 559–577.
28. Mecozzi, S., Rebek, J. (1998) The 55% solution: a formula for molecular recognition in the liquid state. *Chemistry – A European Journal*, **4** (6), 1016–1022.
29. Rekharsky, M.V., Inoue, Y. (1998) Complexation thermodynamics of cyclodextrins. *Chemical Reviews*, **98** (5), 1875–1918.
30. Marquez, C., Nau, W.M. (2001) Polarizabilities inside molecular containers. *Angewandte Chemie International Edition*, **40** (23), 4387–4390.
31. Swamy, M.J. (1995) Thermodynamic analysis of biotin binding to avidin. A high sensitivity titration calorimetric study. *Biochemistry & Molecular Biology International*, **36** (1), 219–225.
32. Kuhn, B., Kollman, P.A. (2000) A ligand that is predicted to bind better to avidin than biotin: insights from computational fluorine scanning. *Journal of the American Chemical Society*, **122** (16), 3909–3916.
33. Rao, J.H., Lahiri, J., Isaacs, L., *et al.* (1998) A trivalent system from vancomycin·D-Ala-D-Ala with higher affinity than avidin·biotin. *Science*, **280** (5364), 708–711.
34. Ludden, M.J.W., Reinhoudt, D.N., Huskens, J. (2006) Molecular printboards: versatile platforms for the creation and positioning of supramolecular assemblies and materials. *Chemical Society Reviews*, **35** (11), 1122–1134.
35. Ko, Y.H., Hwang, I., Lee, D.W., Kim, K. (2011) Ultrastable host–guest complexes and their applications. *Israel Journal of Chemistry*, **51** (5–6), 506–514.
36. Isaacs, L. (2014) Stimuli responsive systems constructed using cucurbit[*n*]uril-type molecular containers. *Accounts of Chemical Research*, **47** (7), 2052–2062.
37. Yang, H., Yuan, B., Zhang, X., Scherman, O.A. (2014) Supramolecular chemistry at interfaces: host–guest interactions for fabricating multifunctional biointerfaces. *Accounts of Chemical Research*, **47** (7), 2106–2115.
38. Assaf, K.I., Nau, W.M. (2015) Cucurbiturils: from synthesis to high-affinity binding and catalysis. *Chemical Society Reviews*, **44** (2), 394–418.

39. Jeon, W.S., Moon, K., Park, S.H., *et al.* (2005) Complexation of ferrocene derivatives by the cucurbit[7]uril host: a comparative study of the cucurbituril and cyclodextrin host families. *Journal of the American Chemical Society*, **127** (37), 12984–12989.
40. Liu, S., Ruspic, C., Mukhopadhyay, P., *et al.* (2005) The cucurbit[*n*]uril family: prime components for self-sorting systems. *Journal of the American Chemical Society*, **127** (45), 15959–15967.
41. Biedermann, F., Uzunova, V.D., Scherman, O.A., *et al.* (2012) Release of high-energy water as an essential driving force for the high-affinity binding of cucurbit[*n*]urils. *Journal of the American Chemical Society*, **134** (37), 15318–15323.
42. Mock, W.L., Shih, N.Y. (1983) Host–guest binding capacity of cucurbituril. *The Journal of Organic Chemistry*, **48** (20), 3618–3619.
43. Mock, W.L., Shih, N.Y. (1986) Structure and selectivity in host–guest complexes of cucurbituril. *The Journal of Organic Chemistry*, **51** (23), 4440–4446.
44. Mock, W.L., Shih, N.Y. (1988) Organic ligand–receptor interactions between cucurbituril and alkylammonium ions. *Journal of the American Chemical Society*, **110** (14), 4706–4710.
45. Mock, W.L., Shih, N.Y. (1989) Dynamics of molecular recognition involving cucurbituril. *Journal of the American Chemical Society*, **111** (7), 2697–2699.
46. Whang, D., Heo, J., Park, J.H., Kim, K. (1998) A molecular bowl with metal ion as bottom: reversible inclusion of organic molecules in cesium ion complexed cucurbituril. *Angewandte Chemie International Edition*, **37** (1–2), 78–80.
47. Rekharsky, M.V., Ko, Y.H., Selvapalam, N., *et al.* (2007) Complexation thermodynamics of cucurbit[6]uril with aliphatic alcohols, amines, and diamines. *Supramolecular Chemistry*, **19** (1–2), 39–46.
48. Zhao, J.Z., Kim, H.J., Oh, J., *et al.* (2001) Cucurbit[*n*]uril derivatives soluble in water and organic solvents. *Angewandte Chemie International Edition*, **40** (22), 4233–4235.
49. Kim, Y., Kim, H., Ko, Y.H., *et al.* (2009) Complexation of aliphatic ammonium ions with a water-soluble cucurbit[6]uril derivative in pure water: isothermal calorimetric, NMR, and X-ray crystallographic study. *Chemistry – A European Journal*, **15** (25), 6143–6151.
50. Chang, C.-E., Gilson, M.K. (2004) Free energy, entropy, and induced fit in host–guest recognition: calculations with the second-generation mining minima algorithm. *Journal of the American Chemical Society*, **126** (40), 13156–13164.
51. Rekharsky, M.V., Mori, T., Yang, C., *et al.* (2007) A synthetic host–guest system achieves avidin–biotin affinity by overcoming enthalpy–entropy compensation. *Proceedings of the National Academy of Sciences of the United States of America*, **104** (52), 20737–20742.
52. Moghaddam, S., Yang, C., Rekharsky, M., *et al.* (2011) New ultrahigh affinity host–guest complexes of cucurbit[7]uril with bicyclo[2.2.2]octane and adamantane guests: thermodynamic analysis and evaluation of M2 affinity calculations. *Journal of the American Chemical Society*, **133** (10), 3570–3581.
53. Cao, L., Sekutor, M., Zavalij, P.Y., *et al.* (2014) Cucurbit[7]uril–guest pair with an attomolar dissociation constant. *Angewandte Chemie International Edition*, **53** (4), 988–993.
54. Green, N.M. (1990) Avidin and streptavidin. *Methods in Enzymology*, **184**, 51–67.
55. Pazy, Y., Kulik, T., Bayer, E.A., *et al.* (2002) Ligand exchange between proteins. Exchange of biotin and biotin derivatives between avidin and streptavidin. *The Journal of Biological Chemistry*, **277** (34), 30892–30900.
56. Rekharsky, M., Inoue, Y. (2000) Chiral recognition thermodynamics of β-cyclodextrin: the thermodynamic origin of enantioselectivity and the enthalpy–entropy compensation effect. *Journal of the American Chemical Society*, **122** (18), 4418–4435.
57. Moghaddam, S., Inoue, Y., Gilson, M.K. (2009) Host–guest complexes with protein-ligand-like affinities: computational analysis and design. *Journal of the American Chemical Society*, **131** (11), 4012–4021.
58. Yang, L., Adam, C., Nichol, G.S., Cockroft, S.L. (2013) How much do van der Waals dispersion forces contribute to molecular recognition in solution? *Nature Chemistry*, **5** (12), 1006–1010.
59. Šekutor, M., Molčanov, K., Cao, L., *et al.* (2014) Design, synthesis, and X-ray structural analyses of diamantane diammonium salts: guests for cucurbit[n]uril (CB[n]) hosts. *European Journal of Organic Chemistry*, **2014** (12), 2533–2542.
60. Isobe, H., Sato, S., Nakamura, E. (2002) Synthesis of disubstituted cucurbit[6]uril and its rotaxane derivative. *Organic letters*, **4** (8), 1287–1289.

61. Burnett, C.A., Witt, D., Fettinger, J.C., Isaacs L. (2003) Acyclic congener of cucurbituril: synthesis and recognition properties. *The Journal of Organic Chemistry*, **68** (16), 6184–6191.
62. Jansen K., Wego, A., Buschmann, H.-J., *et al.* (2003) Glycoluril derivatives as precursors in the preparation of dubstituted cucurbit[*n*]urils. *Designed monomers and polymers*, **6** (1), 43–55.
63. Vinciguerra, B., Cao, L., Cannon, J.R., *et al.* (2012) Synthesis and self-assembly processes of monofunctionalized cucurbit[7]uril. *Journal of the American Chemical Society*, **134** (31), 13133–13140.
64. Lee, J.W., Ko, Y.H., Park, S.H., *et al.* (2001) Novel pseudorotaxane-terminated dendrimers: supramolecular modification of dendrimer periphery. *Angewandte Chemie International Edition*, **40** (4), 746–749.
65. Tan, Y.B., Choi, S.W., Lee, J.W., *et al.* (2002) Synthesis and characterization of novel side-chain pseudopolyrotaxanes containing cucurbituril. *Macromolecules*, **35** (18), 7161–7165.
66. Isobe, H., Tomita, N., Lee, J.W., *et al.* (2000) Self-assembled ternary complex of cationic dendrimer, cucurbituril, and DNA: noncovalent strategy in developing a gene delivery carrier. *Angewandte Chemie International Edition*, **39** (23), 4257–4260.
67. Isobe, H., Sato, S., Lee, J.W., *et al.* (2005) Supramolecular modulation of action of polyamine on enzyme/DNA interactions. *Chemical Communications*, (12), 1549–1551.
68. Lim, Y.B., Kim, T., Lee, J.W., *et al.* (2002) Self-assembled ternary complex of cationic dendrimer, cucurbituril, and DNA: noncovalent strategy in developing a gene delivery carrier. *Bioconjugate Chemistry*, **13** (6), 1181–1185.
69. Kim, J., Ahn, Y., Park, K.M., *et al.* (2007) Carbohydrate wheels: cucurbituril-based carbohydrate clusters. *Angewandte Chemie International Edition*, **46** (39), 7393–7395.
70. Kim, S.K., Park, K.M., Singha, K., *et al.* (2010) Galactosylated cucurbituril-inclusion polyplex for hepatocyte-targeted gene delivery. *Chemical Communications*, **46** (5), 692–694.
71. Angelos, S., Yang, Y.W., Patel, K., *et al.* (2008) pH-responsive supramolecular nanovalves based on cucurbit[6]uril pseudorotaxanes. *Angewandte Chemie International Edition*, **47** (12), 2222–2226.
72. Angelos, S., Khashab, N.M., Yang, Y.W., *et al.* (2009) pH clock-operated mechanized nanoparticles. *Journal of the American Chemical Society*, **131** (36), 12912–12914.
73. Thomas, C.R., Ferris, D.P., Lee, J.H., *et al.* (2010) Noninvasive remote-controlled release of drug molecules in vitro using magnetic actuation of mechanized nanoparticles. *Journal of the American Chemical Society*, **132** (31), 10623–10625.
74. Croissant, J., Zink, J.I. (2012) Nanovalve-controlled cargo release activated by plasmonic heating. *Journal of the American Chemical Society*, **134** (18), 7628–7631.
75. Ambrogio, M.W., Thomas, C.R., Zhao, Y.-L., *et al.* (2011) Mechanized silica nanoparticles: a new frontier in theranostic nanomedicine. *Accounts of Chemical Research*, **44** (10), 903–913.
76. Lee, H.K., Park, K.M., Jeon, Y.J., *et al.* (2005) Vesicle formed by amphiphilc cucurbit[6]uril: versatile, noncovalent modification of the vesicle surface, and multivalent binding of sugar-decorated vesicles to lectin. *Journal of the American Chemical Society*, **127** (14), 5006–5007.
77. Park, K.M., Suh, K., Jung, H., *et al.* (2009) Cucurbituril-based nanoparticles: a new efficient vehicle for targeted intracellular delivery of hydrophobic drugs. *Chemical Communications*, (**1**), 71–73.
78. Park, K.M., Lee, D.W., Sarkar, B., *et al.* (2010) Reduction-sensitive, robust vesicles with a non-covalently modifiable surface as a multifunctional drug-delivery platform. *Small*, **6** (13), 1430–1441.
79. Baek, K., Yun, G., Kim, Y., *et al.* (2013) Free-standing, single-monomer-thick two-dimensional polymers through covalent self-assembly in solution. *Journal of the American Chemical Society*, **135** (17), 6523–6528.
80. Ghosh, S., Isaacs, L. (2010) Biological catalysis regulated by cucurbit[7]uril molecular containers. *Journal of the American Chemical Society*, **132** (12), 4445–4454.
81. Baek, K., Hwang, I., Roy, I., *et al.* (2015) Self-assembly of nanostructured materials through irreversible covalent bond formation. *Accounts of Chemical Research*, **48** (8), 2221–2229.
82. Zhang, S. (2004) Beyond the petri dish. *Nature Biotechnology*, **22** (2), 151–152.
83. Lutolf, M.P., Hubbell, J.A. (2005) Synthetic biomaterials as instructive extracellular microenvironments for morphogenesis in tissue engineering. *Nature Biotechnology*, **23** (1), 47–55.
84. Cushing, M.C., Anseth, K.S. (2007) Hydrogel cell cultures. *Science*, **316** (5828), 1133–1134.
85. Dvir, T., Timko, B.P., Kohane, D.S., Langer, R. (2011) Nanotechnological strategies for engineering complex tissues. *Nature Nanotechnology*, **6** (1), 13–22.

86. Pek, Y.S., Wan, A.C.A., Shekaran, A., *et al.* (2008) A thixotropic nanocomposite gel for three-dimensional cell culture. *Nature Nanotechnology*, **3** (11), 671–675.
87. Park, K.M., Yang, J.A., Jung, H., *et al.* (2012) In situ supramolecular assembly and modular modification of hyaluronic acid hydrogels for 3D cellular engineering. *ACS Nano*, **6** (4), 2960–2968.
88. Jung, H., Park, J.S., Yeom, J., *et al.* (2014) 3D tissue engineered supramolecular hydrogels for controlled chondrogenesis of human mesenchymal stem cells. *Biomacromolecules*, **15** (3), 707–714.
89. Yeom, J., Kim, S.J., Jung, H., *et al.* (2015) Supramolecular hydrogels for long-term bioengineered stem cell therapy. *Advanced Healthcare Materials*, **4** (2), 237–244.
90. Wong, L.S., Khan, F., Micklefield, J. (2009) Selective covalent protein immobilization: strategies and applications. *Chemical Reviews*, **109** (9), 4025–4053.
91. Chen, Y.X., Triola, G., Waldmann, H. (2011) Bioorthogonal chemistry for site-specific labeling and surface immobilization of proteins. *Accounts of Chemical Research*, **44** (9), 762–773.
92. Hwang, I., Baek, K., Jung, M., *et al.* (2007) Noncovalent immobilization of proteins on a solid surface by cucurbit[7]uril–ferrocenemethylammonium pair, a potential replacement of biotin–avidin pair. *Journal of the American Chemical Society*, **129** (14), 4170–4171.
93. Lee, D.-W., Park, K.M., Gong, B., *et al.* (2015) A simple modular aptasensor platform utilizing cucurbit[7]uril and a ferrocene derivative as an ultrastable supramolecular linker. *Chemical Communications*, **51**, 3098–3101.
94. An, Q., Li, G.T., Tao, C.G., *et al.* (2008) A general and efficient method to form self-assembled cucurbit[*n*]uril monolayers on gold surfaces. *Chemical Communications*, (**17**), 1989–1991.
95. Young, J.F., Nguyen, H.D., Yang, L., *et al.* (2010) Strong and reversible monovalent supramolecular protein immobilization. *ChemBioChem*, **11** (2), 180–183.
96. Neirynck, P., Brinkmann, J., An Q., *et al.* (2013) Supramolecular control of cell adhesion via ferrocene–cucurbit[7]uril host–guest binding on gold surfaces. *Chemical Communications*, **49** (35), 3679–3681.
97. Wasserberg, D., Uhlenheuer, D.A., Neirynck, P., *et al.* (2013) Immobilization of ferrocene-modified SNAP-fusion proteins. *International Journal of Molecular Sciences*, **14** (2), 4066–4080.
98. Josic, D., Clifton, J.G. (2007) Mammalian plasma membrane proteomics. *Proteomics*, **7** (16), 3010–3029.
99. Jaffrey, S.R., Erdjument-Bromage H., Ferris C.D., *et al.* (2001) Protein S-nitrosylation: a physiological signal for neuronal nitric oxide. *Nature Cell Biology*, **3** (2), 193–197.
100. Zhang, W., Zhou, G., Zhao, Y., *et al.* (2003) Affinity enrichment of plasma membrane for proteomics analysis. *Electrophoresis*, **24** (16), 2855–2863.
101. Zhang, L., Wang, X.E., Peng, X., *et al.* (2007) Immunoaffinity purification of plasma membrane with secondary antibody superparamagnetic beads for proteomic analysis. *Journal of Proteome Research*, **6** (1), 34–43.
102. Takechi, R., Taniguchi, A., Ebara, S., *et al.* (2008) Biotin deficiency affects the proliferation of human embryonic palatal mesenchymal cells in culture. *Journal of Nutrition*, **138** (4), 680–684.
103. Lee, D.W., Park, K.M., Banerjee, M., *et al.* (2011) Supramolecular fishing for plasma membrane proteins using an ultrastable synthetic host–guest binding pair. *Nature Chemistry*, **3** (2), 154–159.
104. Ahn, Y., Jang, Y., Selvapalam, N., *et al.* (2013) Supramolecular velcro for reversible underwater adhesion. *Angewandte Chemie International Edition*, **52** (11), 3140–3144.
105. Li, W., Kaifer, A.E. (2012) Combining proton and electron transfer to modulate the stability of cucurbit[7]uril complexes. *Langmuir*, **28** (42), 15075–15079.
106. Kim, C., Agasti, S.S., Zhu, Z., *et al.* (2010) Recognition-mediated activation of therapeutic gold nanoparticles inside living cells. *Nature Chemistry*, **2** (11), 962–966.
107. Kim, C., Tonga, G.Y., Yan, B., *et al.* (2015) Regulating exocytosis of nanoparticles via host–guest chemistry. *Organic & Biomolecular Chemistry*, **13** (8), 2474–2479.

7

Molecular Recognition in Biomimetic Receptors

Peter C. Knipe, Sam Thompson, and Andrew D. Hamilton

Chemistry Research Laboratory, Department of Chemistry, University of Oxford, Oxford, UK

7.1 Molecular recognition in biological systems

The selective recognition of one molecule by another is of fundamental importance in biological systems. This includes the binding of small-molecule substrates and drug molecules to the active site of enzymes and their allosteric regulation, to the vast surfaces whose selective binding lies at the heart of the protein–protein interactions (PPIs) that regulate cell signaling and metabolism. Taken as a whole, these constitute the interactome [1], the complex interconnected network of cellular interactions evolved by nature over eons. While the majority of these interactions are desirable, those of an aberrant type have been implicated in the progression of disease. It has been a long-term goal within our laboratory to develop a deeper understanding of these interactions and to devise new tools to probe and mediate them.

7.2 Model systems to investigate fundamental forces

Although the recognition of one molecule by another is often complex and seemingly intractable, there are a small number of fundamental forces at play. We envisaged that we could interrogate and control these underlying forces within the confines of small, simple and structurally well-defined systems, to gain a greater understanding of their role within the ensemble of interactions present in larger molecular recognition events [2].

Macrocyclic and Supramolecular Chemistry: How Izatt–Christensen Award Winners Shaped the Field, First Edition.
Edited by Reed M. Izatt.

7.2.1 Hydrogen bonding

Hydrogen bonding is a crucial non-covalent interaction in biological systems: backbone hydrogen bonds stabilize the key elements of secondary protein structure, while complementary Watson–Crick hydrogen-bonding gives rise to the critical selectivity of base-pairing within oligonucleic acids. Therefore, harnessing the power of hydrogen bonding to develop new receptors [3], catalysts [4], and supramolecular assemblies [5] remains a vibrant area of research.

To investigate the use of directed hydrogen bonding in binding small molecules, we developed an artificial receptor for barbiturates, exploiting the triple hydrogen bond complementarity between 2,6-diaminopyridines and imides [6–8]. By incorporating a pair of strategically positioned 2,6-aminopyridines into the binding pocket of the receptor, an array of up to six hydrogen bonds was formed with a range of barbiturates and related substrates (Figure 7.1). Binding constants were determined by UV/vis and fluorescence spectroscopy, with K_a values of up to 6×10^5 M^{-1} in dichloromethane. These values are extremely high for the binding of a *neutral* small molecule, and demonstrate the power of complementary hydrogen-bonded networks for molecular recognition. The importance of multiple hydrogen bonds in favoring complexation was illustrated by selective deletion of hydrogen bonding groups. For example, replacing the alkyl amides with an ether linkage incapable of making the lower pair of hydrogen bonds led to a 2000-fold decrease in association constant. Similarly, alkylation of one imide nitrogen in the substrate prevents the formation of 3–4 hydrogen bonding interactions, and gave a corresponding 1000-fold decrease in affinity. X-Ray crystallographic studies revealed the precise atomic structure of the complex, with all six possible hydrogen-bonding interactions satisfied. In addition, a 2.21 Å distance between the isophthaloyl 2-hydrogen and the central barbital C=O was suggestive of the formation of a weak CH···O hydrogen bond in the solid state.

7.2.2 Aromatic stacking

Whilst hydrogen bonding is a key primary interaction involved in molecular recognition – particularly within biological systems – others are crucial. We sought to examine the geometry of stacking interactions that occur between adjacent aromatic rings [9–11], and their electronic characteristics. A two-site binding strategy was

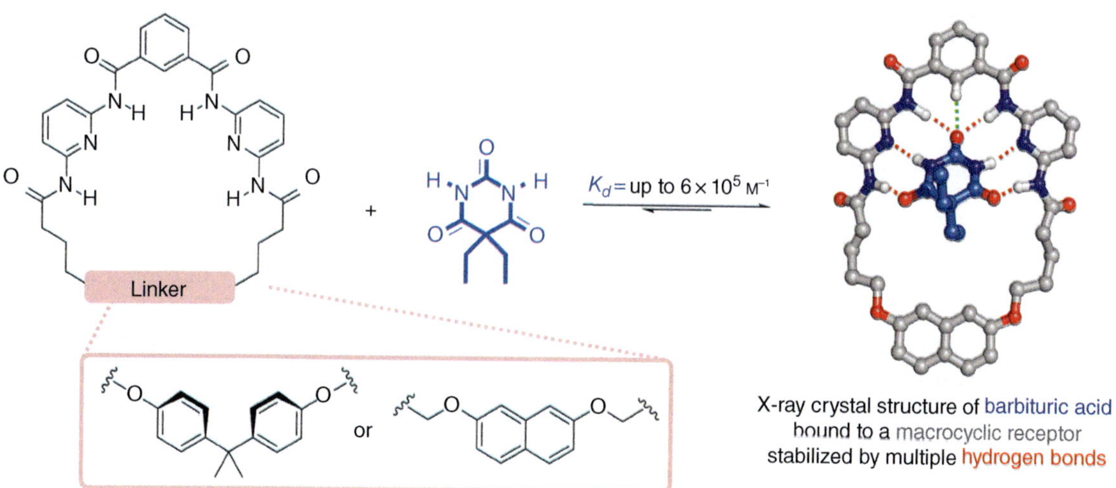

Figure 7.1 Strong binding of barbiturates by a synthetic receptor mediated by an array of NH···O hydrogen bonds (red). A putative CH···O hydrogen bond is also indicated (green)

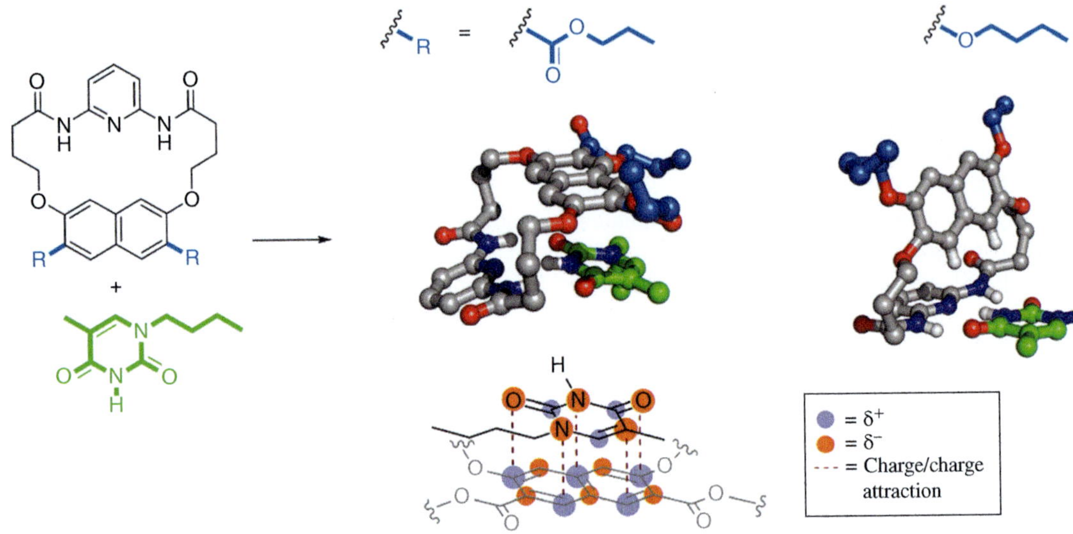

Figure 7.2 *A thymine-binding molecular hinge for the examination of aromatic stacking interactions*

employed to examine the interactions between a thymine derivative and a series of synthetic receptors (Figure 7.2) [12, 13]. The formation of a substrate–receptor pair was assured by the formation of three strong intermolecular hydrogen bonds, allowing examination of the weaker aromatic stacking interaction which otherwise would be insufficient in strength to lead to their association. When the stacking component of the receptor (a macrocyclic naphthalene linker) was functionalized with electron-withdrawing ester groups at the 3- and 6-positions, a face-to-face stacked arrangement was observed by ^{1}H NMR and X-ray crystallography. Quantum mechanical MNDO calculation of charge distribution in the receptor and thymine revealed that this interaction is a consequence of electrostatic pairing of partial charges (Figure 7.2, bottom) [14]. Conversely, when the electronic structure of the naphthalene is altered by replacing the ester groups with ethers the face-to-face stacking is no longer favorable since this electrostatic interaction is no longer attractive. Rather, an edge-to-face complex is formed, as demonstrated in solution by upfield shifting of the deshielded naphthalene 1,8-protons, and in the solid state by X-ray crystallography, with Foster–Fife NMR analysis indicating that this interaction provides a small attractive stabilization (0.26 kcal·mol^{-1}) [15].

7.2.3 A first foray into mimicry – minimalist vancomycin analogues

Vancomycin is a potent antibiotic used in the treatment of drug-resistant gram-positive bacterial infections [16]. It prevents cell wall peptidoglycan synthesis by selectively binding the *C*-terminal D-Ala-D-Ala-OH of peptidoglycan precursor UDP-MurNAc-pentapeptide [17]. This occurs predominantly through a network of hydrogen bond donors and acceptors distributed around the macrocyclic core, with five of the six key hydrogen bonds at the carboxylate terminus directed to the eastern ether macrocycle (Figure 7.3, highlighted). In principle these hydrogen-bonding patterns are accessible to all *C*-terminal dipeptide units; substrate specificity is achieved through hydrophobic interactions between the dipeptide side-chains and the antibiotic [18]. We developed a minimalist mimic of this key domain that was effective in binding small-molecule carboxylates in a similar manner to the parent antibiotic (Figure 7.3, inset) [19, 20]. The mimic exhibited a binding constant of 580 M^{-1} in $CDCl_3$ for the formation of a 1:1 complex with cyanoacetic acid as a model substrate. With the emergence of multi-drug-resistant bacteria and fears of an "antibiotic apocalypse," [21] research into vancomycin analogues and their recognition properties continues to be extremely active [22].

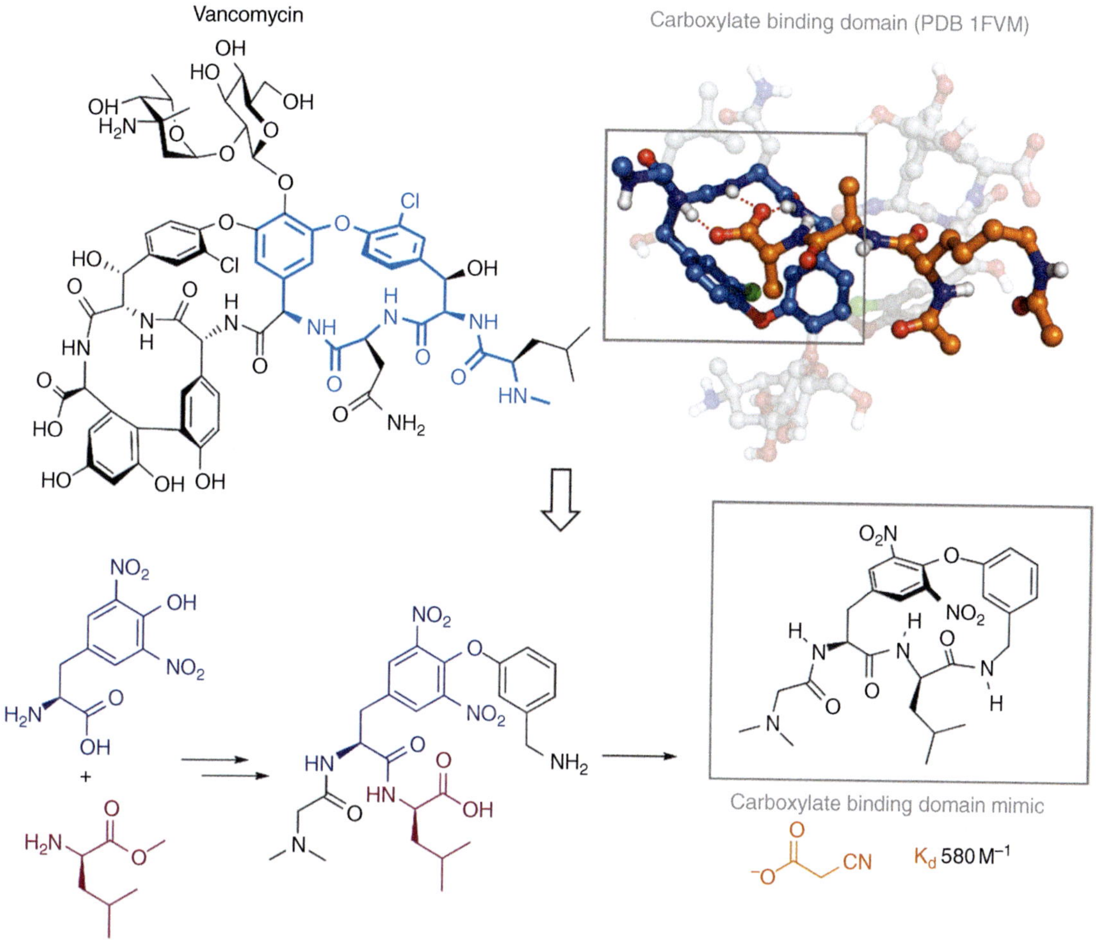

Figure 7.3 *A minimalist mimic of the key macrocyclic binding domain of last-resort antibiotic vancomycin*

7.3 Recognition of more complex systems – into the realm of peptides

7.3.1 Carboxylic acid binding

The binding of a single carboxylic acid is inherently challenging: the entropic penalty paid for uniting the receptor and acid in the correct orientation is large, and the enthalpic gains from forming a maximum of two hydrogen bonds may not provide adequate compensatory stabilization. We have investigated a series of receptors designed to bind small molecules containing multiple carboxylic acid groups, where the multiple hydrogen bonds formed begin to significantly outweigh the entropic penalty of bringing the molecules together. Initially a dipyridylamide was developed which bound a variety of diacids with K_a values of up to $>10^5$ M^{-1} in $CDCl_3$ (Figure 7.4) [23]. This motif was also applied to the *cis–trans* isomerization of acylprolines, where it induced a 32-fold change in the rotamer equilibrium by selectively binding the *s-cis* form [24]. However, binding of diacids to this receptor was extremely sensitive to solvent: when competitive hydrogen-bonding solvents such as DMSO or water were added, the binding was almost completely lost.

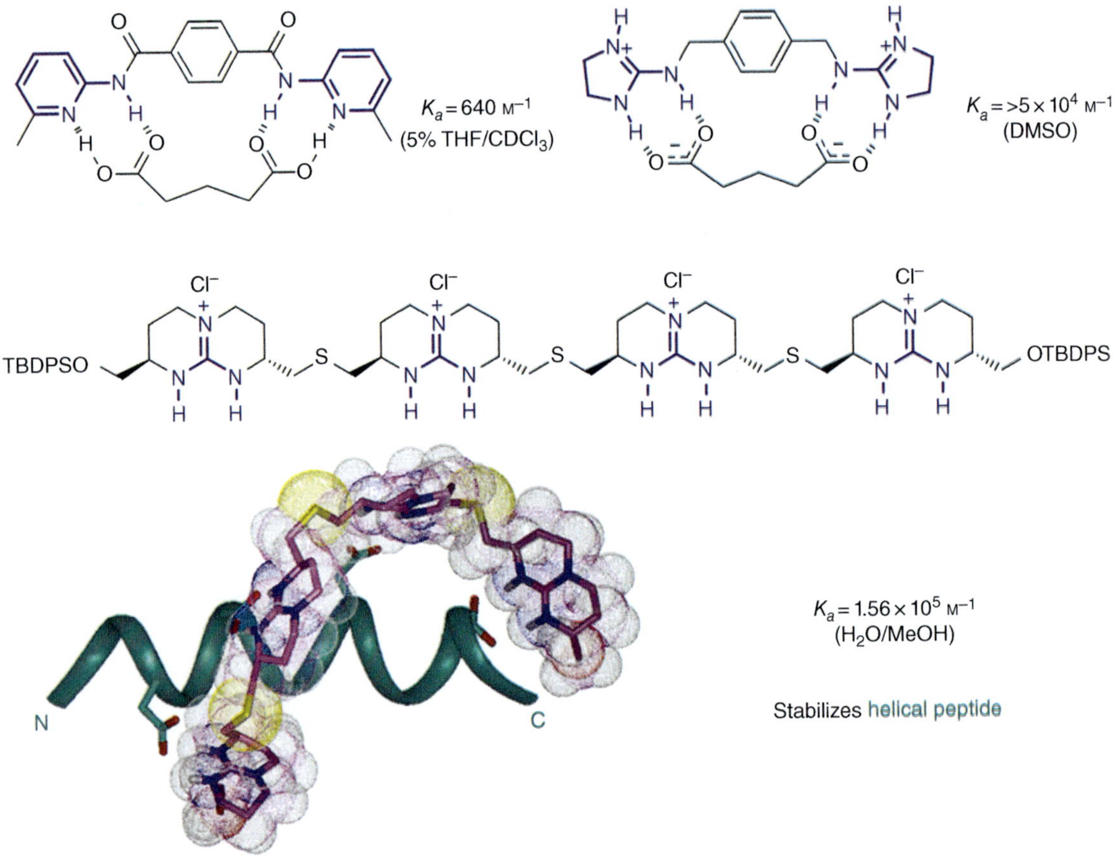

Figure 7.4 A series of receptors designed to bind substrates bearing multiple carboxylic acid groups. Molecular dynamics calculations demonstrate the stabilization of the helical conformation of $(Asp)_4$ by a tetrameric guanidinium receptor. Adapted with permission from ref. [29]. Copyright © 1997 American Chemical Society

This spurred the development of a second-generation of receptor where: (i) *both* donor hydrogens were situated on the receptor, an arrangement that gives rise to stronger hydrogen bonds [25, 26]; and (ii) both donor and acceptor are formally charged, adding an electrostatic attraction to further stabilize the complex. A bis-alkylguanidinium receptor gave extremely strong recognition in DMSO (K_a >5 × 10^5 M^{-1}), and was impressive in mixed DMSO/water (K_a 480 M^{-1} in 25% D_2O/d_6-DMSO) [27]. This improved binding allowed the use of these synthetic receptors in applications more directly biological in nature. We reasoned that receptors containing multiple guanidinium groups could be used as stabilizers of secondary structure in peptides containing carboxylate side-chains such as aspartate. A *cis*-fused bicylo[3.3.0]octane bis(guanidinium) gave strong binding (K_a 2200 M^{-1} in 10% water/methanol) and modest stabilization of helicity in a polypeptide containing Asp residues at the *i* and *i*+3 positions [28]. The concept was extended to a tetrameric guanidinium receptor, which binds tetrapeptide $(Asp)_4$ with a K_a of 1.56 × 10^5 M^{-1} in 10% water/methanol, and increases the equilibrium helicity (as measured by CD spectroscopy) by almost 150% [29]. It is also possible to combine these ion-pairing and hydrogen-bonding interactions synergistically with the aromatic stacking described previously (see Figure 7.2).

7.3.2 Peptide mimicry as a strategy for the inhibition of a key protein–protein interaction: prenyltransferases

Prenyltransferases (PTases) are the family of enzymes that catalyze the post-translational addition of various prenyl groups onto appropriately tagged proteins. These include farnesyltransferases (FTases) and geranylgeranyltransferases (GGTases), both of which have been implicated in the oncogenic behavior of many proteins, including the RAS family [30]. Prenylation leads to localization of the target protein to the cell membrane, and a subsequent methylation of the C-terminal acid further anchors the modified protein in this position. The downstream effects of these modifications are increased cell proliferation, cell cycle progression and cell survival – the hallmarks of cancer (Figure 7.5, left). Intriguingly, both FTase and GGTase-1 recognize specific four-residue CAAX sequences [31, 32], which tag the pendant protein for post-translational modification (PTM), where C is the cysteine undergoing modification, A is an aliphatic amino acid and X is a residue which determines whether FTase or GGTase-1 will act (X = Met, Ser, Glu or Cys favors FTase; X = Leu or Ile favors GGTase). This unique reactivity provides an opportunity to selectively inhibit PTases by mimicry of these four-residue sequences.

Through an extended collaborative effort [30, 33–44]. We and the Sebti research group have developed a series of semi- and non-peptidic inhibitors of PTases. For example, an increase in IC_{50} of two orders of magnitude was achieved over the natural FTase tetrapeptide ligand CVIM by replacing the central AA dipeptide unit with an alternative lipophilic *para*-substituted 2-phenylbenzamide linker (Figure 7.5, right) [40]. The mimic also exhibited high selectivity for FTase over GGTase, which gave an IC_{50} of 50 nM. A confounding factor in the progression of these inhibitors to the clinic is the propensity of some oncogenic proteins (e.g. KRAS) to be both farnesylated *and* geranylgeranylated *in vivo*. Since both modifications are associated with cancer development, this necessitates the inhibition of *both* FTase and GGTase for the therapy to be

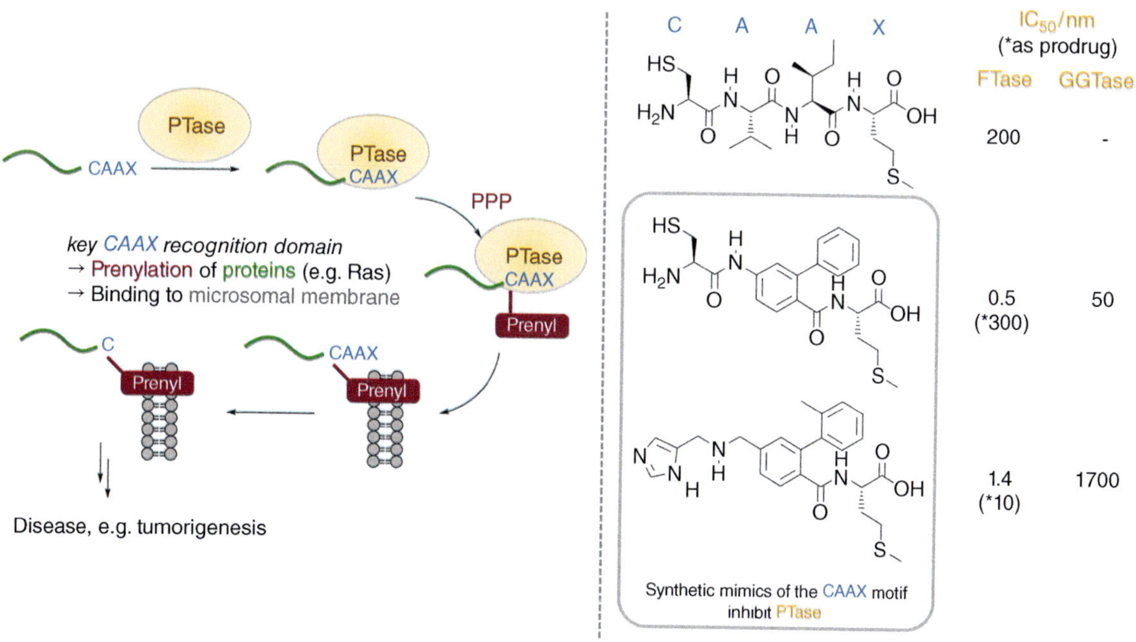

Figure 7.5 Left: The PTase-mediated prenylation pathway leading to protein membrane localization and downstream carcinogenic effects (PPP = generic prenyl diphosphate); right: the development of prenyltransferase (PTase) inhibitors mimicking the key CAAX recognition domain

effective. However, we have shown that FTase and GGTases can both act in isolation to suppress the growth of human tumor tissue in mouse models [37]. Removal of the thiol group in the inhibitor is also tolerated, and leads to much greater selectivity for FTase over GGTase (1200-fold selectivity) [38]. When administered as a methyl ester pro-drug the imidazole gave a 30-fold increase in potency over the thiol analogue. The successful mimicry of primary structure, such as CAAX above, led us to examine more complex therapeutically relevant targets – many mediated by large interfacial protein surfaces.

7.4 A general approach to peptide mimicry – targeting secondary structure

7.4.1 Terphenyls: the first class of α-helix mimic

Protein–protein interactions (PPIs) have been referred to as "undruggable," owing to the large surface areas and weak binding constants involved in their formation [45]. However, the presence of "hot-spot" regions on protein surfaces – areas that contribute disproportionately to binding – means that elements of these large surfaces may be targeted [46], bringing the inhibition of PPIs into the realm of synthetic molecules. α-Helices are present at the interface of 62% of PPIs reported in the PDB [47, 48], hence their mimicry would provide a general approach to target a large proportion of protein-protein interactions.

We recognized that the majority of α-helices involved in PPIs interact through the side-chains with little involvement of the backbone polypeptide chain; the development of a non-peptidic scaffold which projects

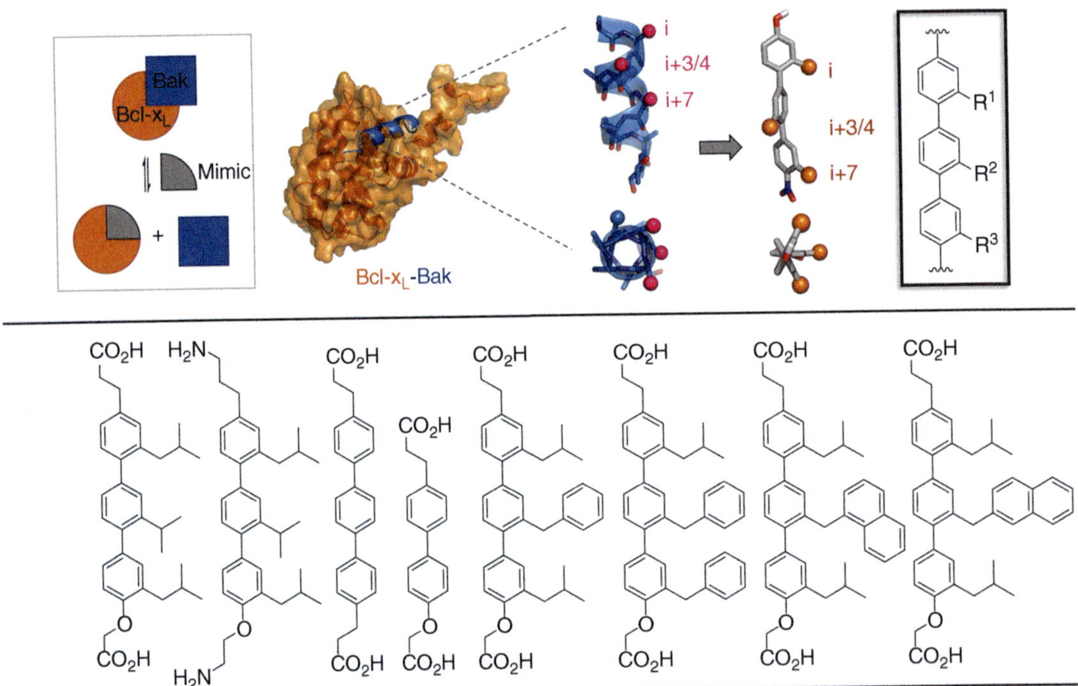

K_i /μM	Bcl-x$_L$	2.1	13.7	27.4	28.3	1.8	2.7	0.11	2.5
	hDM2							25.7	0.18

Figure 7.6 The terphenyl scaffold as a general platform for α-helix mimicry. Judicious choice of substituents leads to selective, sub-micromolar binding of Bcl-x$_L$ or hDM2

natural or unnatural side-chains along the same vectors would therefore be an effective mimic of the protein surface, and a viable therapeutic for intercepting PPIs (Figure 7.6) [49]. Deviation from the naturally occurring proteinogenic amino acids also provides an opportunity to incorporate a vast variety of side-chains and functional groups, and to circumvent the oral bioavailability and proteolysis issues that limit the use of peptides as therapeutics [50]. The terphenyl motif was developed as such a scaffold, where staggering along the aryl–aryl axis leads to substituents being presented with remarkable angular and spatial similarity to the projection of the *i*, *i*+3/4 and *i*+7 positions of a natural α-helix. The synthetic approach allowed the use of both natural and unnatural side-chain mimics. This motif has been successfully applied to the inhibition of a variety of PPIs including gp1 formation [51], p53/*h*DM2 [52], and Bcl-x_L/Bak (Figure 7.6) [53, 54]. In the latter case, sub-micromolar activity was observed for a naphthyl-substituted terphenyl, and through simple isomeric substitution of the central side-chain it was possible to achieve >10-fold selectivity for binding of Bcl-x_L or *h*DM2.

7.4.2 The development of next-generation α-helix mimetics

The field of secondary structural – in particular α-helix – mimicry as a general approach to targeting PPIs has since expanded rapidly with seminal contributions from Boger [55], Arora [56], and others [47], while Burgess has pursued an alternative strategy termed "universal peptidomimetics" [57, 58]. Within our laboratory, a variety of novel aromatic scaffolds have been developed to improve the ease with which mimetics can be synthesized, as well as enhancing their water solubility and drug-like properties (Figure 7.7) [59]. For example, the arylamide scaffold may be rapidly constructed via an iterative sequence of amide bond formations and nitroarene reduction reactions, and has been demonstrated as a viable platform for the mediation of the Bcl-x_L/Bak PPI [60, 61]. More recently we, in collaboration with Andrew Miranker, have employed this scaffold as a potential therapeutic for type II diabetes, where aggregation of islet amyloid polypeptide (IAPP) is believed to be involved in β-cell toxicity. When substituted with appropriate anionic side-chains, the helix

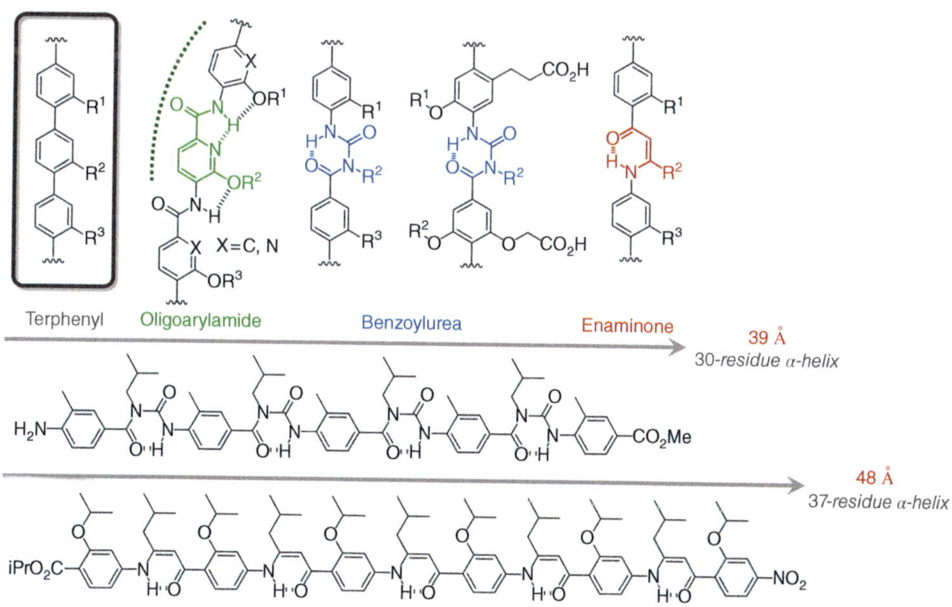

Figure 7.7 The terphenyl motif as an inspiration for the development of new helix mimetic scaffolds

mimetics markedly decreased the rate of IAPP aggregation in the presence of phospholipids, an effect which, intriguingly, is reversed when the phospholipid is removed [62–66]. The oligoarylamides offer an additional degree of freedom in scaffold design: incorporation of pyridines introduces backbone hydrogen bonding and leads to in-plane curvature of the scaffold, allowing further exploration of conformational space and enabling fine-tuning of the structure of the inhibitor [60, 67]. Recently we have developed highly functionalized oligobenzamides where both edges of the scaffold can act as recognition domains [68]. Others have extended the scope of oligoarylamides: Wilson recently revealed an elegant mixed α-amino acid/oligoarylamide approach to selectively target the p53/*h*DM2 PPI, along with greatly improved synthetic throughput, having developed solid-phase synthesis [69, 70].

The subsequent development of the benzoylurea and enaminone scaffolds – where a non-covalent hydrogen-bond-stabilized ring acts as a surrogate for the central arene of the terphenyl backbone – has enabled the synthesis of extended mimetics with increased aqueous solubilities [71–73]. In the case of the enaminones, this has extended to an oligomer 48 Å in length, that projects 11 side-chains and mimics an α-helix of 37 residues. The synthetic flexibility of these scaffolds has also allowed the development of amphiphilic variants, where sections of the scaffold are functionalized with hydrophobic and hydrophilic side-chains, as is frequently encountered on protein surfaces where the former is buried and the latter is solvent exposed [65, 74].

7.4.3 Approaches to β-strand mimicry

In comparison to α-helices, the mimicry of β-sheets and strands is less well explored. Despite Hirschmann and Smith's seminal contribution to the field [75], there have been a limited number of subsequent designs. There are two distinct recognition domains that may be replicated by the putative mimic: (i) the hydrogen-bonding edges, which form the primary attractive interaction between strands within a sheet; and (ii) the surface of side-chains presented in an alternating and perpendicular manner from the plane of the backbone. In 2009 we reported a novel scaffold for strand side-chain mimicry, where a strong N-H···O hydrogen bond formed across a diphenylacetylene linkage is used to limit rotation about the C_{Ar}–C_{sp} bond (Figure 7.8) [76–78]. This favors the conformation in which the side-chains are presented uniformly on the same side of the molecule – as illustrated in the solid phase by X-ray crystallography – and the persistence of the H-bond in solution was verified by ^{1}H NMR spectroscopy. More recently, a system based on alternating *meta*-substituted arenes and amino acid-derived homo-chiral imidazolin-2-ones was developed [79–81]. It was demonstrated by ^{1}H NMR and X-ray crystallography that when the arene subunit is phenyl, free rotation about the N_{urea}–C_{Ar} bond leads to conformational flexibility, and non-uniform presentation of side-chains. However, when 2,6-disubtituted pyridines are employed a strong dipole repulsion effect is observed, leading to mimics almost exclusively adopting the desired conformation, with side-chains projected from a single face of the strand. This motif has subsequently been modified to the six-membered ring urea homologue, which leads to reduced curvature and greater agreement with the natural strand side-chain vectors [82].

7.4.4 β-sheet templation

While the mimicry of β-strands necessitates the defined positioning of side-chains on a linear template, β-sheet mimicry is a more challenging problem since side-chains must instead be projected in a controlled manner from the face of a surface. Kemp demonstrated the utility of the diphenylalkyne linkage as a β-turn inducer, where the covalent link reduces the entropic penalty for associating strands, and positions the donor–acceptor pairs in a favorable orientation for intramolecular hydrogen-bonding [76, 77]. We adapted this design to stabilize bi-directional β-sheet formation from a tetrasubstituted diphenylalkyne (Figure 7.9) [83]. Solution- and solid-state data confirmed that the sheet conformation was adopted for a variety of lipophilic (Ala), cationic (Lys) and anionic (Glu) side-chains, even in challenging d_6-DMSO/D_2O solvent mixtures.

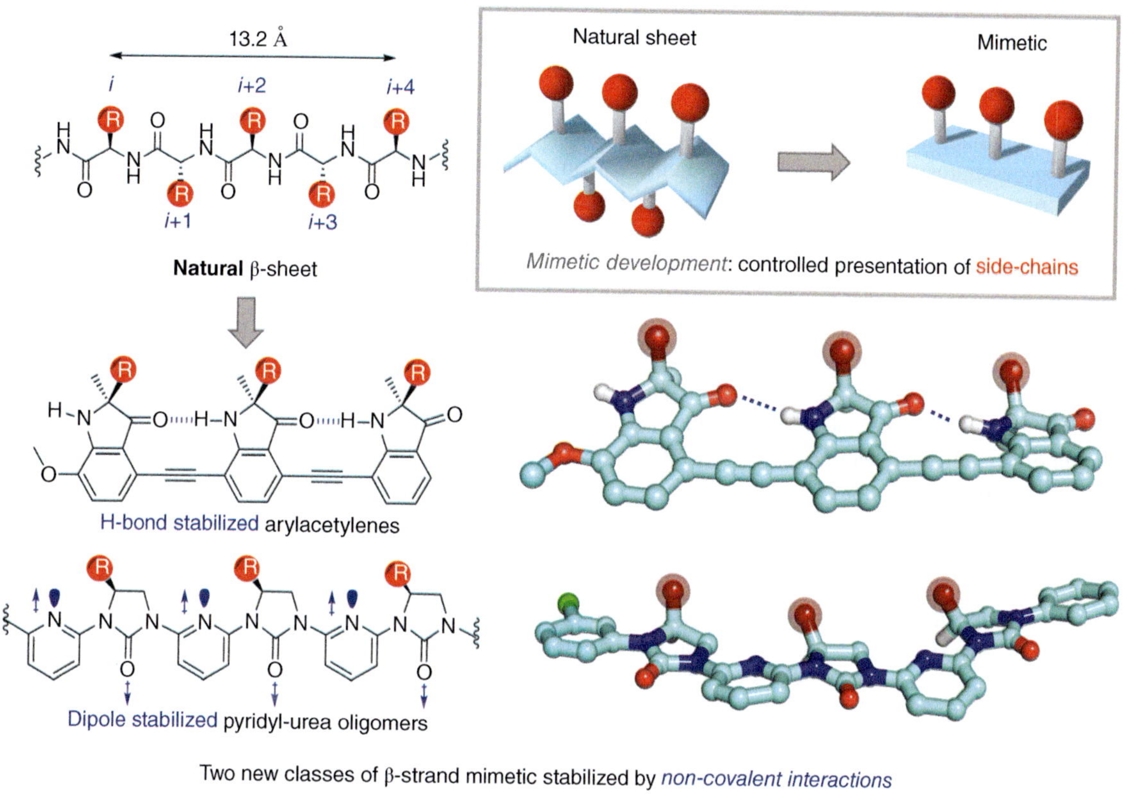

Figure 7.8 Non-peptidic structural mimics of single β-strands. Side-chain residues are highlighted in red, and are presented on a common face of each mimetic [80]. Adapted with permission of John Wiley & Sons, Ltd

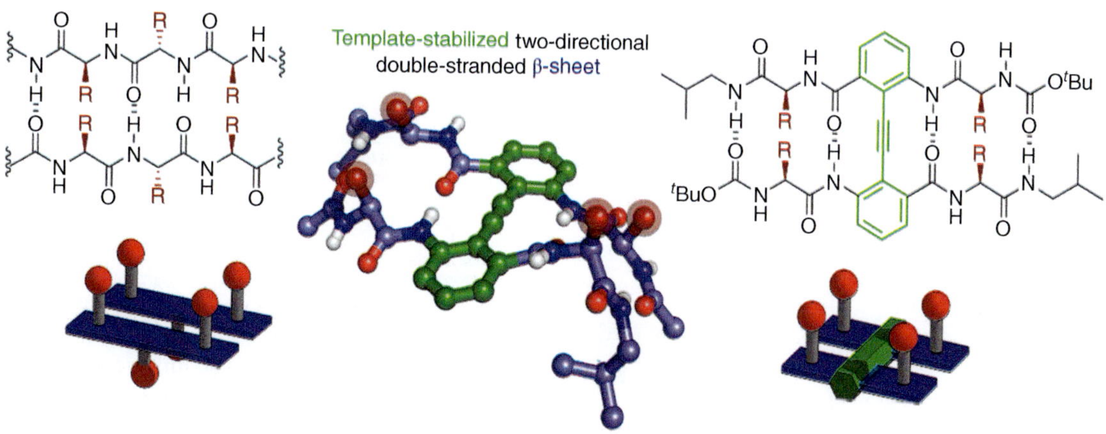

Figure 7.9 Covalent stabilization of β-sheets by incorporation of a diphenylalkyne linker. Side-chains are highlighted in red, and are uniformly projected from a single face of the sheet surface. Cartoon schematic representations of a natural sheet (left) and the templated sheet mimetic (right) are provided

7.5 Super-secondary structures and beyond

7.5.1 The β-meander

While the stabilization and mimicry of single secondary structural elements such as α-helices, β-strands and β-sheets necessitates the projection of side-chains from a single, well-defined scaffold, mimicry of larger areas of proteins or multiple elements of secondary structure requires the linking of these elements in a controlled manner. For example, the β-meander is a super-secondary structural motif in which a single contiguous polypeptide chain adopts an antiparallel β-sheet arrangement through an alternating pattern of β-strands and turns. Side chains are projected from both faces of the extended sheet structure, and any mimic must replicate this arrangement. We have developed a semi-peptidic mimic of this motif by a templation strategy: incorporation of diphenylalkyne linkers into a polypeptide backbone leads to a foldamer containing periodic artificial β-turns (Figure 7.10) [84]. The role of the diphenylalkynes is to lower the entropic penalty for the formation of the meander by reducing rotational degrees-of-freedom. A series of NMR, CD and crystallographic experiments were consistent with this hypothesis, indicating that the major conformer is that of a multistranded sheet.

7.5.2 Mimicking α-helix bundles

The clinical importance of multi-helix bundles is exemplified by the success of therapies targeting G-protein coupled receptors (GPCRs) – trans-membrane helix heptamers involved in signal transduction – which constitute 30–40% of marketed drugs [85]. Leucine zippers have received less attention, but are an impressive example of a system in which multiple secondary structural elements act in unison. These structural motifs arise in certain transcription factors, and involve a pair of leucine-rich helical domains which combine in a two-helix bundle that binds certain DNA sequences in a "scissors-grip" manner [86]. We sought to replicate the structural properties of these functional helix bundles with a pair of oligobenzamide α-helix mimetics (see Figure 7.7), covalently linked with an appropriate head group (Figure 7.11) [87]. A variety of head groups give different inter-helix angles varying between 22° and 111°, demonstrating this approach as a broad and flexible approach to the mimicry of a variety of helical bundles. It is necessary that the head group be rigid in order for the helix mimetic strands to be held in a defined orientation, but this rigidity may be induced *in situ* by chelation of a flexible, Lewis basic linker to an appropriate metal template [88].

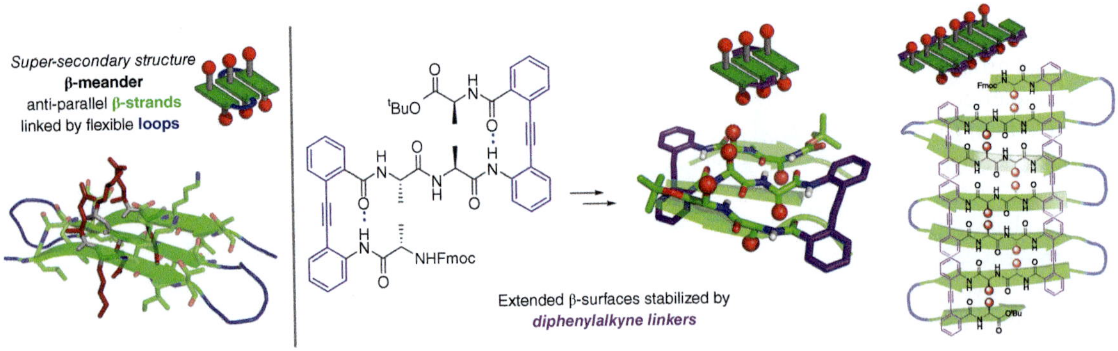

Figure 7.10 A diphenylalkyne foldamer templating the mimicry of a natural β-meander. Cartoon schematic representations of the natural meander (left) and three- and seven-stranded mimics (right) are provided [84]. Adapted with permission of John Wiley & Sons, Ltd

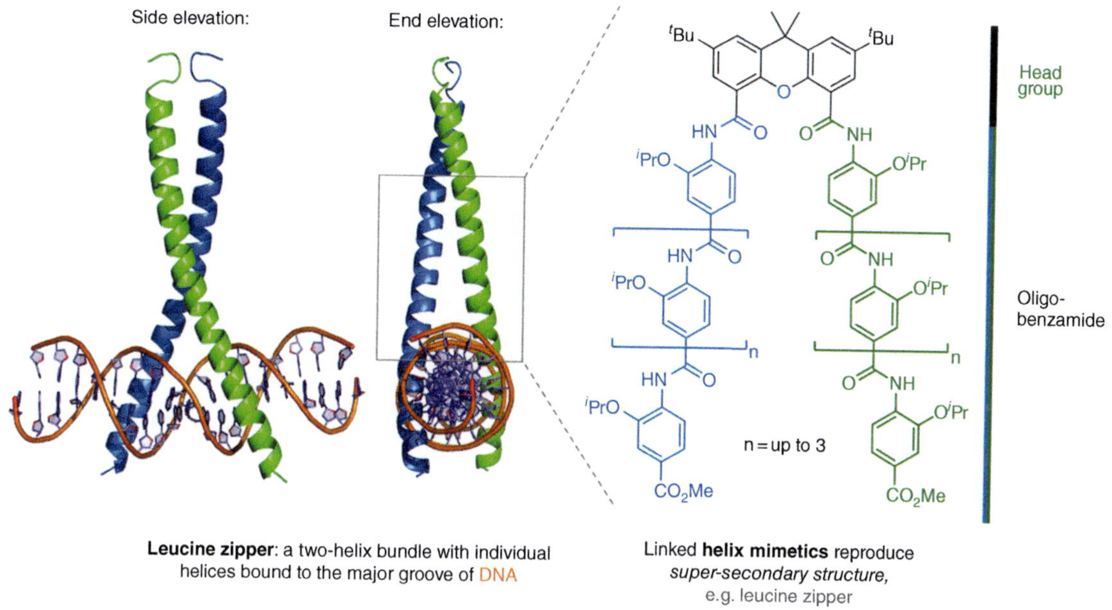

Figure 7.11 *Terminally-linked helix mimetics replicate the structure of helical bundles including the dimeric leucine zipper motif*

7.5.3 Binding multiple sites through multivalent receptors

The choice of head group in the linking of subsections of mimetics or receptors is of critical importance, as it determines the number of interactions between the receptor and the substrate. In cytochrome c, a lipophilic portion of the surface is flanked by multiple cationic residues, occupying four distinct regions. We envisaged that complementary anionic receptor domains could be appended to an appropriate multipodal head group, to achieve selective binding of multiple discrete areas of the protein surface. Both C_4-symmetrical calix[4]arenes [89–91] and D_{4h}-symmetrical porphyrins [92, 93] were amenable to this application, binding cytochrome c with micromolar and nanomolar affinities respectively (Figure 7.12). Control experiments where isolated peptide strands and loops were added to cytochrome c in the absence of the head group linker exhibited much weaker binding, demonstrating the importance of the multipodal approach to effective recognition. The calix[4]arene scaffold has also been used as a binder of serine protease α-chymotrypsin, and effectively inhibited its catalytic activity in the hydrolysis of *N*-benzoyltyrosine-*p*-nitroanilide, with an inhibition constant (K_i) of 0.81 μM in aqueous buffer [94]. The interaction between cytochrome c and a series of metallo-porphyrins was also examined: binding induced large-scale conformational change in the protein, catalyzing a 50-fold increase in the rate of its proteolysis [95, 96]. The porphyrin scaffold has also been applied to the development of a diagnostic test for the presence of proteins [97]. We synthesized a variety of porphyrins differing in size, charge, hydrophobicity and symmetry (by appending different R-groups). Since binding of these porphyrins to proteins induces a change in fluorescence properties characteristic of both the binding partner and porphyrin substituents, we reasoned that when displayed in an array and exposed to a particular protein, a representative "fingerprint" of fluorescent spots would result. An eight-porphyrin variant successfully distinguished between cytochrome c, ferredoxin, cytochrome c551 and myoglobin.

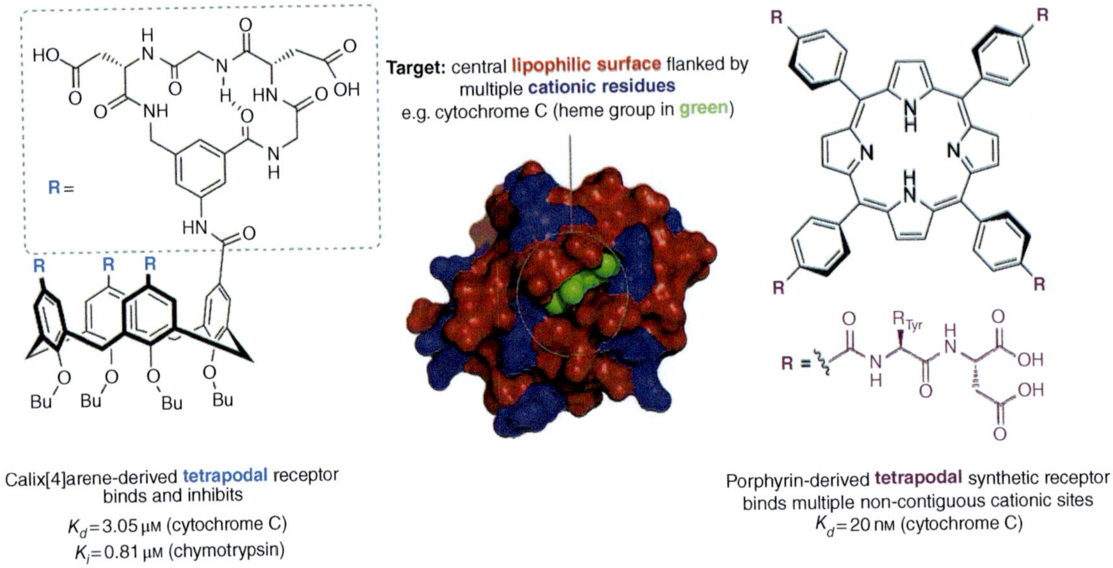

Figure 7.12 Porphyrins and calix[4]arenes as templates for multipodal binding

7.5.4 Self-assembly of receptors – binding quaternary structures

As the target protein surfaces increase in size, so must the artificial receptors. This presents a practical problem: the synthesis of such large mimetics becomes prohibitive and the synthesis of analogues unfeasible. We reasoned that a potential solution would be to generate smaller, monomeric protein surface binders programmed to self-assemble *in situ*. The synthetic problem would then be reduced to the synthesis of these monomers, while the self-assembly provides the large surface areas and multiple points of binding necessary to recognise extended protein surfaces. Guanine-rich regions of nucleic acid oligomers self-assemble via Hoogsteen hydrogen-bonding to form multiplet strands, with a central templating cation occupying the center of the C_n-axis [98, 99]. The most common of these is the G-quadruplex, which could, we reasoned, provide a viable platform for the self-assembly of a polycarboxylate binder of cytochrome c (Figure 7.13). Binding of receptors to cytochrome c increases its rate of proteolysis (*vide supra*) by inducing conformational change [95]. G-quadruplex-based receptors formed *in situ* by self-assembly[100] demonstrated the same behavior, as well as accelerating thermal denaturation of the enzyme [101]. The same approach was successful in the inhibition of chymotrypsin [102], suggesting it is a generally applicable strategy, in array-based protein detection assays [103, 104], and in the templation of four-helix bundles [105].

We have developed strategies to bind multiple, non-contiguous regions on the surface of a number of globular proteins. However, in all cases these areas have been components of the same protein molecule. A particularly lofty goal in molecular recognition within biological systems is to bind multiple proteins, interacting with and potentially influencing *quaternary* structure. C-reactive protein (CRP) is a homopentameric protein expressed as part of the natural inflammatory response, and is an important biomarker of disease [106]. A shallow pocket on each monomeric protein binds phosphorylcholine (PC) via two protein-bound calcium ions [107]. We postulated that a self-assembled DNA pentaplex appended – through a tether of appropriate length – with PC binding domains would simultaneously bind all five protein monomers of the homopentamer, selectively targeting protein quaternary structure. The use of a templating cesium cation led to selective formation of the G-pentaplex, and the resulting receptor complex bound CRP with a K_d of

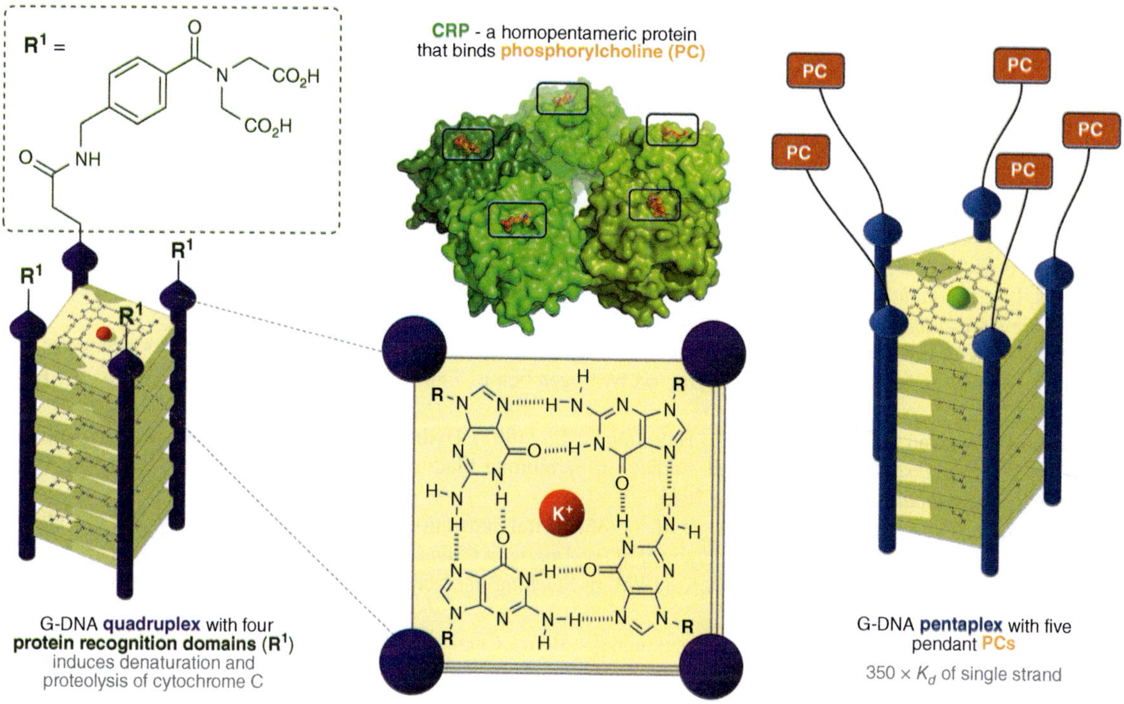

Figure 7.13 Multiplex DNA as a template for binding non-contiguous sites of protein surface, and binding quaternary structures

141 nM in aqueous buffer. This constitutes a three orders-of-magnitude increase in potency over the monomer, confirming a cooperative mode of binding since if the contribution of each PC subunit was simply additive a K_d of 10 μM would be expected [108, 109].

7.6 Outlook

Over a period spanning almost thirty years we have developed a series of tools and strategies to achieve molecular recognition within biological systems. A key theme has been the use of multiple points of binding between receptor and substrate in a cooperative manner, a strategy employed from binding small-molecules such as barbiturates to large multidomain protein assemblies such as CRP.

When applied to the problem of protein–protein interactions, the synergy between molecular recognition and protein mimicry is clear. This led to the development of a number of scaffolds mimicking canonical components of secondary structure including the α-helix, β-strand and β-sheet. Such a general approach to protein mimicry allows a broad range of PPIs to be rationally targeted by the selection of appropriate side-chains.

Given the central importance of molecular recognition in all aspects of chemistry and biology it is likely that interest in the field will continue to grow. It is our hope that the advances made within our laboratory will be built upon to further our understanding of biological networks, and to inform the design of novel therapies for currently intractable diseases.

References

1. Sanchez, C., Lachaize, C., Janody, F., *et al.* (1999) Grasping at molecular interactions and genetic networks in *Drosophila melanogaster* using FlyNets, an internet database. *Nucleic Acids Research*, **27**, 89–94.
2. Hamilton, A.D., Pant, N., Muehldorf, A. (1988) Artificial receptors for biologically active molecules. *Pure and Applied Chemistry*, **60**, 533–538.
3. Amendola, V., Fabbrizzi, L., Mosca, L. (2010) Anion recognition by hydrogen bonding: urea-based receptors. *Chemical Society Reviews*, **39**, 3889–3915.
4. Pikho, P. (Ed) (2009) *Hydrogen Bonding in Organic Synthesis*, Wiley-VCH Verlag, Weinheim.
5. Prins, L.J., Reinhoudt, D.N., Timmerman, P. (2001) Noncovalent Synthesis using hydrogen bonding. *Angewandte Chemie International Edition*, **40**, 2382–2426.
6. Chang, S.-K., Hamilton, A.D. (1988) Molecular recognition of biologically interesting substrates: synthesis of an artificial receptor for barbiturates employing six hydrogen bonds. *Journal of the American Chemical Society*, **110**, 1318–1319.
7. Chang, S.-K., Van Engen, D., Fan, E., Hamilton, A.D. (1991) Hydrogen bonding and molecular recognition: synthetic, complexation, and structural studies on barbiturate binding to an artificial receptor. *Journal of the American Chemical Society*, **113**, 7640–7645.
8. Feibush, B., Figueroa, A., Charles, R., *et al.* (1986) Chiral separation of heterocyclic drugs by HPLC: solute-stationary phase base-pair interactions. *Journal of the American Chemical Society*, **108**, 3310–3318.
9. Mcgaughey, G.B., Gagne, M., Rappe, A.K. (1998) pi-Stacking interactions. *Journal of Biological Chemistry*, **273**, 15458–15463.
10. Waters, M.L. (2002) Aromatic interactions in model systems. *Current Opinion in Chemical Biology*, **6**, 736–741.
11. Hunter, C.A., Sanders, J.K.M. (1990) The Nature of pi–pi Interactions. *Journal of the American Chemical Society*, **112**, 5525–5534.
12. Hamilton, A.D., Van Engen, D. (1987) Induced fit in synthetic receptors: nucleotide base recognition by a molecular hinge. *Journal of the American Chemical Society*, **109**, 5035–5036.
13. Muehldorf, A., Engen, D. Van, Warner, J.C., Hamilton, A.D. (1988) Aromatic–aromatic interactions in molecular recognition: a family of artificial receptors for thymine that shows both face-to-face and edge-to-face orientations. *Journal of the American Chemical Society*, **110**, 6561–6562.
14. Dewar, M.J.S., Thiel, W. (1977) Ground states of molecules. 38. The MNDO method. Approximations and parameters. *Journal of the American Chemical Society*, **99**, 4899–4907.
15. Foster, R., Fyfe, C.A. (1969) Chapter 1 Nuclear magnetic resonance of organic charge-transfer complexes. *Progress in Nuclear Magnetic Resonance Spectroscopy*, **4**, 1–89.
16. Kahne, D., Leimkuhler, C., Lu, W., Walsh, C. (2005) Glycopeptide and lipoglycopeptide antibiotics. *Chemical Reviews*, **105**, 425–448.
17. Reynolds, P.E. (1989) Structure, biochemistry and mechanism of action of glycopeptide antibiotics. *European Journal of Clinical Microbiology and Infectious Diseases*, **8**, 943–950.
18. Williamson, M.P., William, D.H., Hammond, S.J. (1984) Interactions of vancomycin and ristocetin with peptides as a model for protein binding. *Tetrahedron*, **40**, 569–577.
19. Mann, M.J., Pant, N., Hamilton, A.D. (1986) A synthetic analogue of the carboxylate-binding pocket of vancomycin. *Journal of the Chemical Society, Chemical Communications*, 158.
20. Pant, N., Hamilton, A. (1988) Carboxylic acid complexation by a synthetic analog of the "carboxylate-binding pocket" of vancomycin. *Journal of the American Chemical Society*, **110**, 2002–2003.
21. MacKenzie, D. (2015) Are we facing an antibiotic apocalypse? *New Scientist*, **225**, 22–23.
22. O'Connell, K.M.G., Hodgkinson, J.T., Sore, H.F., *et al.* (2013) Combating multidrug-resistant bacteria: Current strategies for the discovery of novel antibacterials. *Angewandte Chemie International Edition*, **52**, 10706–10733.
23. Garcia-Tellado, F., Goswami, S., Chang, S.-K., *et al.* (1990) Molecular recognition: a remarkably simple receptor for the selective complexation of dicarboxylic acids. *Journal of the American Chemical Society*, **112**, 7393–7394.
24. Vicent, C., Hirst, S., Garcia-Tellado, F., Hamilton, A.D. (1991) Conformational selectivity in molecular recognition: the influence of artificial receptors on the cis-trans isomerization of acylprolines. *Journal of the American Chemical Society*, **113**, 5466–5467.

25. Jorgensen, W.L., Pranata, J. (1990) Importance of secondary interactions in triply hydrogen bonded complexes: Guanine–cytosine vs uracil–2,6-diaminopyridine. *Journal of the American Chemical Society*, **112**, 2008–2010.
26. Murray, T.J., Zimmerman, S.C. (1992) New triply hydrogen bonded complexes with highly variable stabilities. *Journal of the American Chemical Society*, **114**, 4010–4011.
27. Fan, E., Van Arman, S.A., Kincaid, S., Hamilton, A.D. (1993) Molecular recognition: hydrogen-bonding receptors that function in highly competitive solvents. *Journal of the American Chemical Society*, **115**, 369–370.
28. Albert, J.S., Goodman, M.S., Hamilton, A.D. (1995) Molecular recognition of proteins: sequence-selective binding of aspartate pairs in helical peptides. *Journal of the American Chemical Society*, **117**, 1143–1144.
29. Peczuh, M.W., Hamilton, A.D., Sánchez-Quesada, J., *et al.* (1997) Recognition and stabilization of an α-helical peptide by a synthetic receptor. *Journal of the American Chemical Society*, **119**, 9327–9328.
30. Berndt, N., Hamilton, A.D., Sebti, S.M. (2011) Targeting protein prenylation for cancer therapy. *Nature Reviews Cancer*, **11**, 775–791.
31. Gutierrez, L., Magee, A.I., Marshall, C.J., Hancock, J.F. (1989) Post-translational processing of p21ras is two-step and involves carboxyl-methylation and carboxy-terminal proteolysis. *The EMBO journal*, **8**, 1093–1098.
32. Vogt, A., Qian, Y., Blaskovich, M.A., *et al.* (1995) A Non-peptide mimetic of Ras-CAAX: Selective Inhibition of farnesyltransferase and Ras processing. *Journal of Biological Chemistry*, **270**, 660–664.
33. Lerner, E.C., Qian, Y., Blaskovich, M.A., *et al.* (1995) Ras CAAX peptidomimetic FTI-277 selectively blocks oncogenic Ras signaling by inducing cytoplasmic accumulation of inactive Ras–Raf complexes. *Journal of Biological Chemistry*, **270**, 26802–26806.
34. Qian, Y., Blaskovich, M.A., Saleem, M., *et al.* (1994) Design and structural requirements of potent peptidomimetic inhibitors of p21ras farnesyltransferase. *Journal of Biological Chemistry*, **269**, 12410–12413.
35. Vogt, A., Qian, Y., Blaskovich, M.A., *et al.* (1995) A non-peptide mimetic of Ras-CAAX: Selective inhibition of farnesyltransferase and Ras processing. *Journal of Biological Chemistry*, **270**, 660–664.
36. Sun, J., Qian, Y., Hamilton, A.D., Sebti, S.M. (1995) Ras CAAX peptidomimetic fti 276 selectively blocks tumor growth in nude mice of a human lung carcinoma with K-Ras Mutation and p53 deletion. *Cancer Research*, **55**, 4243–4247.
37. Sun, J., Qian, Y., Hamilton, A.D., Sebti, S.M. (1998) Both farnesyltransferase and geranylgeranyltransferase I inhibitors are required for inhibition of oncogenic K-Ras prenylation but each alone is sufficient to suppress human tumor growth in nude mouse xenografts. *Oncogene*, **16**, 1467–1473.
38. Sun, J., Blaskovich, M.A., Knowles, D., *et al.* (1999) Non-thiol-containing peptidomimetic inhibitors of farnesyltransferase and geranylgeranyltransferase i combination therapy with the cytotoxic agents Cisplatin, Taxol. *Cancer research*, **59**, 4919–4926.
39. Sebti, S.M., Hamilton, A.D. (1997) Inhibition of Ras prenylation: A novel approach to cancer chemotherapy. *Pharmacology and Therapeutics*, **74**, 103–114.
40. Lernert, E.C., Qian, Y., Hamilton, A.D., Sebti, S.M. (1995) Disruption of oncogenic K-Ras4B processing and signaling by a potent geranylgeranyltransferase I inhibitor. *Journal of Biological Chemistry*, **270**, 26770–26773.
41. Sebti, S.M., Hamilton, A.D. (1997) Anticancer activity of farnesyltransferase and geranylgeranyltransferase I inhibitors: prospects for drug development. *Expert Opinion on Investigational Drugs*, **6**, 1711–1714.
42. Kazi, A., Carie, A., Blaskovich, M.A., *et al.* (2009) Blockade of protein geranylgeranylation inhibits Cdk2-dependent p27Kip1 phosphorylation on Thr187 and accumulates p27Kip1 in the nucleus: implications for breast cancer therapy. *Molecular and Cellular Biology*, **29**, 2254–2263.
43. Nigam, M., Seong, C.-M., Qian, Y., *et al.* (1993) Potent Inhibition of Human Tumor p21ras Farnesyltransferase by A1A2-lacking p21ras CA1A2X Peptidomimetics. *Journal of Biological Chemistry*, **268**, 20695–20698.
44. Sebti, S.M., Hamilton, A.D. (2000) Farnesyltransferase and geranylgeranyltransferase I inhibitors in cancer therapy: important mechanistic and bench to bedside issues. *Expert Opinion on Investigational Drugs*, **9**, 2767–2782.
45. Verdine, G.L., Walensky, L.D. (2007) The challenge of drugging undruggable targets in cancer: Lessons learned from targeting BCL-2 family members. *Clinical Cancer Research*, **13**, 7264–7270.
46. Clackson, T., Wells, J.A. (1995) A hot spot of binding energy in a hormone–receptor interface. *Science*, **267**, 383–386.
47. Jayatunga, M.K.P., Thompson, S., Hamilton, A.D. (2014) α-Helix mimetics: outwards and upwards. *Bioorganic & medicinal Chemistry Letters*, **24**, 717–724.

48. Bullock, B.N., Jochim, A.L., Arora, P.S. (2011) Assessing helical protein interfaces for inhibitor design. *Journal of the American Chemical Society*, **133**, 14220–14223.
49. Orner, B.P., Ernst, J.T., Hamilton, A.D. (2001) Toward Proteomimetics: terphenyl derivatives as structural and functional mimics of extended regions of an α-helix. *Journal of the American Chemical Society*, **123**, 5382–5383.
50. Morishita, M., Peppas, N.A. (2006) Is the oral route possible for peptide and protein drug delivery? *Drug Discovery Today*, **11**, 905–910.
51. Ernst, J.T., Kutzki, O., Debnath, A.K., *et al.* (2002) design of a protein surface antagonist based on alpha-helix mimicry : inhibition of gp41 assembly and viral fusion. *Angewandte Chemie International Edition*, **41**, 278–281.
52. Yin, H., Lee, G., Park, H.S., *et al.* (2005) Terphenyl-based helical mimetics that disrupt the p53/HDM2 interaction. *Angewandte Chemie International Edition*, **44**, 2704–2707.
53. Kutzki, O., Park, H.S., Ernst, J.T., *et al.* (2002) Development of a potent Bcl-xL Antagonist based on α-helix mimicry. *Journal of the American Chemical Society*, **124**, 11838–11839.
54. Yin, H., Lee, G., Sedey, K.A., *et al.* (2005) Terphenyl-based Bak BH3 alpha-helical proteomimetics as low-molecular-weight antagonists of Bcl-xL. *Journal of the American Chemical Society*, **127**, 10191–10196.
55. Shaginian, A., Whitby, L.R., Hong, S., *et al.* (2009) Design, synthesis, and evaluation of an alpha-helix mimetic library targeting protein–protein interactions. *Journal of the American Chemical Society*, **131**, 5564–5572.
56. Liu, J., Wang, D., Zheng, Q., *et al.* (2008) Atomic structure of a short α-helix stabilized by a main chain hydrogen-bond surrogate. *Journal of the American Chemical Society*, **130**, 4334–4337.
57. Ko, E., Liu, J., Burgess, K. (2011) Minimalist and universal peptidomimetics. *Chemical Society Reviews*, **40**, 4411–4421.
58. Ko, E., Liu, J., Perez, L.M., *et al.* (2011) Universal peptidomimetics. *Journal of the American Chemical Society*, **133**, 462–477.
59. Saraogi, I., Hamilton, A.D. (2009) Recent advances in the development of aryl-based foldamers. *Chemical Society Reviews*, **38**, 1726–1743.
60. Saraogi, I., Incarvito, C.D., Hamilton, A.D. (2008) Controlling curvature in a family of oligoamide alpha-helix mimetics. *Angewandte Chemie International Edition*, **47**, 9691–9694.
61. Ernst, J.T., Becerril, J., Park, H.S., *et al.* (2003) Design and application of an alpha-helix-mimetic scaffold based on an oligoamide-foldamer strategy: antagonism of the Bak BH3/Bcl-xL complex. *Angewandte Chemie International Edition*, **42**, 535–539.
62. Hebda, J.A., Saraogi, I., Magzoub, M., *et al.* (2009) A peptidomimetic approach to targeting pre-amyloidogenic states in type II diabetes. *Chemistry & Biology*, **16**, 943–950.
63. Saraogi, I., Hebda, J.A., Becerril, J., *et al.* (2010) Synthetic alpha-helix mimetics as agonists and antagonists of islet amyloid polypeptide aggregation. *Angewandte Chemie International Edition*, **49**, 736–739.
64. Kumar, S., Miranker, A.D. (2013) A foldamer approach to targeting membrane-bound helical states of islet amyloid polypeptide. *Chemical Communications*, **49**, 4749–4751.
65. Kulikov, O. V., Kumar, S., Magzoub, M., *et al.* (2015) Amphiphilic oligoamide α-helix peptidomimetics inhibit islet amyloid polypeptide aggregation. *Tetrahedron Letters*, **56**, 3670–3673.
66. Kumar, S., Schlamadinger, D.E., Brown, M.A., *et al.* (2015) Islet amyloid-induced cell death and bilayer integrity loss share a molecular origin targetable with oligopyridylamide-based α-helical mimetics. *Chemistry & Biology*, **22**, 369–378.
67. Azzarito, V., Prabhakaran, P., Bartlett, A.I., *et al.* (2012) 2-O-alkylated para-benzamide α-helix mimetics: the role of scaffold curvature. *Organic & Biomolecular Chemistry*, **10**, 6469–6472.
68. Thompson, S., Vallinayagam, R., Adler, M.J., *et al.* (2012) Double-sided α-helix mimetics. *Tetrahedron*, **68**, 4501–4505.
69. Azzarito, V., Long, K., Murphy, N.S., Wilson, A.J. (2013) Inhibition of α-helix-mediated protein-protein interactions using designed molecules. *Nature Chemistry*, **5**, 161–173.
70. Azzarito, V., Miles, J.A., Fisher, J., *et al.* (2015) Stereocontrolled protein surface recognition using chiral oligoamide proteomimetic foldamers. *Chemical Science*, **6**, 2434–2443.
71. Adler, M.J., Scott, R.T.W., Hamilton, A.D. (2012) Enaminone-based mimics of extended and hydrophilic α-helices. *Chemistry – A European Journal*, **18**, 12974–12977.

72. Adler, M.J., Hamilton, A.D. (2011) Oligophenylenaminones as scaffolds for α-helix mimicry. *The Journal of Organic Chemistry*, **76**, 7040–7047.
73. Rodriguez, J.M., Hamilton, A.D. (2007) Benzoylurea oligomers: synthetic foldamers that mimic extended alpha helices. *Angewandte Chemie International Edition*, **46**, 8614–8617.
74. Thompson, S., Hamilton, A.D. (2012) Amphiphilic α-helix mimetics based on a benzoylurea scaffold. *Organic & Biomolecular Chemistry*, **10**, 5780.
75. Smith, A.B., Guzman, M.C., Sprengeler, P.A., *et al.* (1994) De novo design, synthesis, X-ray crystal structures of pyrrolinone-based beta-strand peptidomimetics. *Journal of the American Chemical Society*, **116**, 9947–9962.
76. Kemp, D.S., Li, Z.Q. (1995) 2-Amino-2′-carboxydiphenylacetylenes as β-turn mimetics. synthesis and conformational properties. *Tetrahedron Letters*, **36**, 4175–4178.
77. Kemp, D.S., Li, Z.Q. (1995) A short β-sheet containing proline nucleated by a 2,2′-substituted tolan β-turn mimetic. *Tetrahedron Letters*, **36**, 4179–4180.
78. Wyrembak, P.N., Hamilton, A.D. (2009) Alkyne-linked 2,2-disubstituted-indolin-3-one oligomers as extended beta-strand mimetics. *Journal of the American Chemical Society*, **131**, 4566–4567.
79. Sutherell, C.L., Thompson, S., Scott, R.T.W., Hamilton, A.D. (2012) Aryl-linked imidazolidin-2-ones as non-peptidic β-strand mimetics. *Chemical Communications*, **48**, 9834.
80. German, E.A., Ross, J.E., Knipe, P.C., *et al.* (2015) β-Strand Mimetic foldamers rigidified through dipolar repulsion. *Angewandte Chemie International Edition*, **54**, 2649–2652.
81. Jamieson, A.G., Russell, D., Hamilton, A.D. (2012) A 1,3-phenyl-linked hydantoin oligomer scaffold as a β-strand mimetic. *Chemical Communications*, **48**, 3709.
82. Yamashita, T., Knipe, P.C., Busschaert, N., *et al.* (2015) A modular synthesis of conformationally preorganised extended beta-strand peptidomimetics. *Chemistry – A European Journal*, **21**, 14699–14702.
83. Lingard, H., Han, J.T., Thompson, A.L., *et al.* (2014) Diphenylacetylene-linked peptide strands induce bidirectional β-sheet formation. *Angewandte Chemie International Edition*, **53**, 3650–3653.
84. Ross, J.E., Knipe, P.C., Thompson, S., Hamilton, A.D. (2015) Hybrid diphenylalkyne–dipeptide oligomers induce multi-strand β-sheet formation. *Chemistry – A European Journal*, **21**, 13518–13521.
85. Wise, A., Gearing, K., Rees, S. (2002) Target validation of G-protein coupled receptors. *Drug Discovery Today*, **7**, 235–246.
86. Vinson, C.R., Sigler, P.B., McKnight, S.L. (1989) Scissors-grip model for DNA recognition by a family of leucine zipper proteins. *Science*, **246**, 911–916.
87. Kulikov, O. V., Thompson, S., Xu, H., *et al.* (2013) Design and synthesis of oligoamide-based double α-helix mimetics. *European Journal of Organic Chemistry*, **2013**, 3433–3445.
88. Nevola, L., Rodriguez, J.M., Thompson, S., Hamilton, A.D. (2013) Super-secondary structure peptidomimetics: design and synthesis of an α–α hairpin analogue. *Supramolecular Chemistry*, **25**, 586–590.
89. Hamuro, Y., Calama, M.C., Park, H.S., Hamilton, A.D. (1997) A calixarene with four peptide loops: an antibody mimic for recognition of protein surfaces. *Angewandte Chemie International Edition*, **36**, 2680–2683.
90. Lin, Q., Park, H.S., Hamuro, Y., *et al.* (1998) Protein surface recognition by synthetic agents: Design and structural requirements of a family of artificial receptors that bind to cytochrome c. *Peptide Science*, **47**, 285–297.
91. Lin, Q., Hamilton, A.D. (2002) Design and synthesis of multiple-loop receptors based on a calix[4] arene scaffold for protein surface recognition. *Comptes Rendus Chimie*, **5**, 441–450.
92. Jain, R.K., Hamilton, A.D. (2000) Protein surface recognition by synthetic receptors based on a tetraphenylporphyrin scaffold. *Organic Letters*, **2**, 1721–1723.
93. Aya, T., Hamilton, A.D. (2003) Tetrabiphenylporphyrin-based receptors for protein surfaces show sub-nanomolar affinity and enhance unfolding. *Bioorganic and Medicinal Chemistry Letters*, **13**, 2651–2654.
94. Park, H.S., Lin, Q., Hamilton, A.D. (2002) Modulation of protein-protein interactions by synthetic receptors: design of molecules that disrupt serine protease-proteinaceous inhibitor interaction. *Proceedings of the National Academy of Sciences of the United States of America*, **99**, 5105–5109.
95. Wilson, A.J., Groves, K., Jain, R.K., *et al.* (2003) Directed denaturation: Room temperature and stoichiometric unfolding of cytochrome c by a metalloporphyrin dimer. *Journal of the American Chemical Society*, **125**, 4420–4421.
96. Groves, K., Wilson, A.J., Hamilton, A.D. (2004) Catalytic unfolding and proteolysis of cytochrome C induced by synthetic binding agents. *Journal of the American Chemical Society*, **126**, 12833–12842.

97. Baldini, L., Wilson, A.J., Hong, J., Hamilton, A.D. (2004) Pattern-based detection of different proteins using an array of fluorescent protein surface receptors. *Journal of the American Chemical Society*, **126**, 5656–5657.
98. Diezmann, F., Seitz, O. (2011) DNA-guided display of proteins and protein ligands for the interrogation of biology. *Chemical Society Reviews*, **40**, 5789–5801.
99. Davis, J.T. (2004) G-Quartets 40 Years Later: From 5'-GMP to Molecular Biology and Supramolecular Chemistry. *Angewandte Chemie International Edition*, **43**, 668–698.
100. Sprinz, K.I., Tagore, D.M., Hamilton, A.D. (2005) Self-assembly of bivalent protein-binding agents based on oligonucleotide-linked organic fragments. *Bioorganic and Medicinal Chemistry Letters*, **15**, 3908–3911.
101. Tagore, D.M., Sprinz, K.I., Fletcher, S., *et al.* (2007) Protein recognition and denaturation by self-assembling fragments on a DNA quadruplex scaffold. *Angewandte Chemie International Edition*, **46**, 223–225.
102. Cai, J., Rosenzweig, B.A., Hamilton, A.D. (2009) Inhibition of chymotrypsin by a self-assembled DNA quadruplex functionalized with cyclic peptide binding fragments. *Chemistry - A European Journal*, **15**, 328–332.
103. Margulies, D., Hamilton, A.D. (2009) Protein recognition by an ensemble of fluorescent DNA G-quadruplexes. *Angewandte Chemie International Edition*, **48**, 1771–1774.
104. Margulies, D., Hamilton, A.D. (2009) Digital analysis of protein properties by an ensemble of DNA quadruplexes. *Journal of the American Chemical Society*, **131**, 9142–9143.
105. Rosenzweig, B.A., Hamilton, A.D. (2009) Self-assembly of a four-helix bundle on a DNA quadruplex. *Angewandte Chemie International Edition*, **48**, 2749–2751.
106. Pepys, M.B., Hirschfield, G.M. (2003) C-reactive protein: A critical update. *Journal of Clinical Investigation*, **111**, 1805–1812.
107. Thompson, D., Pepys, M.B., Wood, S.P. (1999) The physiological structure of human C-reactive protein and its complex with phosphocholine. *Structure*, **7**, 169–177.
108. Rosenzweig, B.A., Ross, N.T., Adler, M.J., Hamilton, A.D. (2010) Altered binding of a multimeric protein by changing the self-assembling properties of its substrate. *Journal of the American Chemical Society*, **132**, 6749–6754.
109. Rosenzweig, B.A., Ross, N.T., Tagore, D.M., *et al.* (2009) Multivalent protein binding and precipitation by self-assembling molecules on a DNA pentaplex scaffold. *Journal of the American Chemical Society*, **131**, 5020–5021.

8
A Lifetime Walk in the Realm of Cyclam

Luigi Fabbrizzi

Dipartimento di Chimica, Università di Pavia, Pavia, Italy

8.1 Synthesis and development of cyclam and related macrocycles

In 1936, a distinguished Dutch chemist, Jan van Alphen, University of Leiden, published a series of papers on the synthesis of aliphatic polyamines. Among these substances was 1,4,8,11-tetra-aza-undecane (**1**, which in the following will be indicated as 2·3·2-tet, where tet is the abbreviation of tetramine, and the 2·3·2 figures indicate the number of $-CH_2-$ groups in the linking aliphatic chains) [1]. 2·3·2-tet was obtained from the reaction of 1,3-dibromopropane with an excess of 1,2-diaminoethane (en) in absolute ethanol in the presence of KOH (see Scheme 8.1). The wanted tetramine distilled at 157 °C (12 mm Hg). The author was surprised to observe that, on addition of a copper(II) salt, an aqueous solution of the tetramine took a reddish-violet color ("closely resembling that of biuret reaction"), in contrast to what observed with other polyamines, e.g. 2·2·2-tet, for which the characteristic blue color had been observed. A significant fraction distilled at 316 °C (14 mm Hg) and was collected as a yellow viscous oil [2]. The greatest part of this oil did not react with benzaldehyde, which indicated the presence as a major component of a tetramine constituted by all secondary amine groups: this suggested the formation of a cyclic compound. The elemental analysis (nitrogen) of the tetra-hydrochloride and tetra-nitrate derivatives was consistent with the cyclic tetramine 1,4,7,11-tetra-aza-cyclotetradecane (**2**), which formed as illustrated in Scheme 8.1. Also, tetramine **2** gave a biuret-like color on reaction with Cu^{II}.

24 years later, Stetter intentionally synthesized **2** through the more elaborated route illustrated in Scheme 8.2 [3]. The 14-membered macrocycle was obtained as a crystalline solid, which could be further purified by sublimation and was characterized by the IR spectrum and by elemental analysis (C, H, N) of the tetramine and of its tetra-tosyl derivative. Stetter synthesized a variety of tetramine macrocycles, from the 12- to the 21-membered, and noticed that, in contrast to all the other compounds, hygroscopic, oily liquids or low-melting-point solids, cyclam was a solid melting at 185 °C, sublimating over 120 °C, poorly soluble in

---

Macrocyclic and Supramolecular Chemistry: How Izatt–Christensen Award Winners Shaped the Field, First Edition.
Edited by Reed M. Izatt.

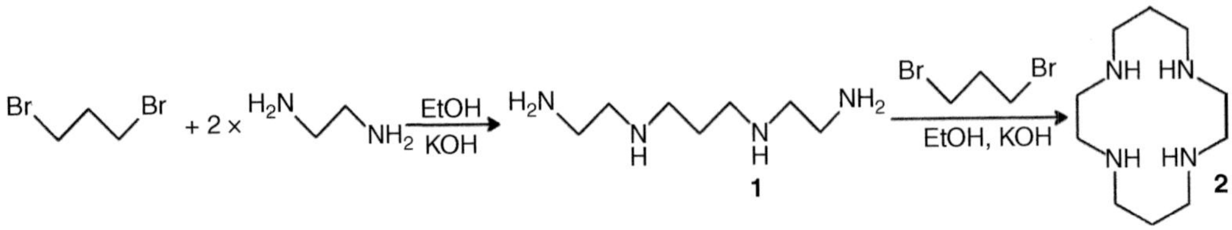

Scheme 8.1 *Van Alphen's synthetic route to 2·3·2 tet, 1, [1] and to cyclam, 2 [2]*

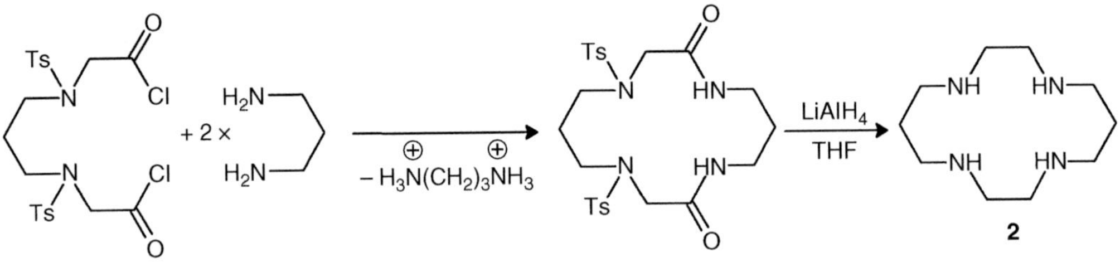

Scheme 8.2 *Stetter's synthetic route to cyclam, 2 [3]*

water, and insensitive to CO_2. He suggested that such unusual stability had to be ascribed to the presence of a network of four intramolecular hydrogen bonds, as illustrated in Figure 8.1.

Stereochemical considerations and, in particular, the crystal and molecular structure reported about three decades later [4], showed that things are not exactly so: indeed, four intramolecular bonds are present, but they involve only the two equatorial N–H fragments, whereas the other two axial N–H fragments cannot be engaged in intramolecular interactions. In particular, each equatorial N–H fragment interacts with the lone pairs of the two adjacent nitrogen atoms: N–H (a) with nitrogens (b) and (d); N–H (c) with nitrogens (b) and (d) (see Figure 8.1, right). In any case, Stetter's intuition that something special is present in the cyclam arrangement was totally correct.

Coordination chemistry had developed as a well-defined discipline thanks to the pioneering research by Alfred Werner on ammonia complexes with inert metal ions (Co^{III}, Cr^{III}, Pt^{IV}), a monumental work rewarded by the Nobel Prize in 1913 [5]. Werner's chemistry was extended in the following decades to complexes of polyamines with inert and labile metal ions. Investigated polyamines were linear or branched, not cyclic. Thanks to the contributions by van Alphen and Stetter, in the early 1960s, cyclic polyamines were available on the shelf, probably anxious to be tested as ligands for transition metals.

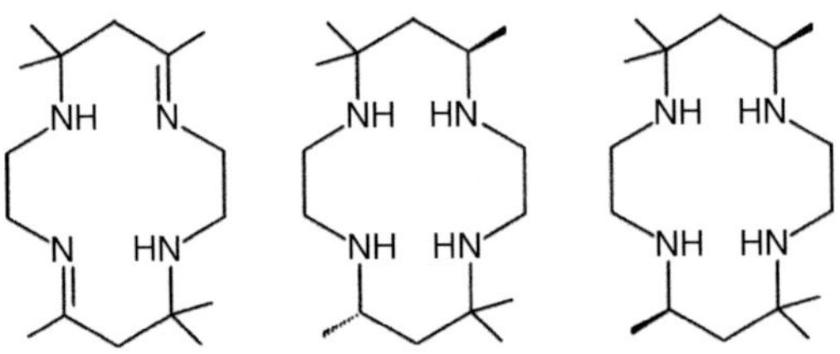

3 CT

4 tet-a *ms*-CTH

5 tet-b *rac*-CTH

However, in these years, the still waters of polyamine coordination chemistry were vigorously shaken by the serendipitous discovery by Neil F. Curtis, Victoria University, Wellington, New Zealand, that, on reaction

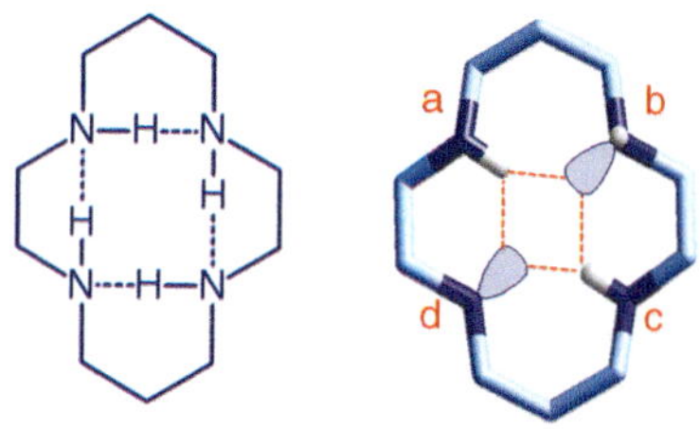

Figure 8.1 *The hypothesized structure of cyclam, with a network of four intramolecular hydrogen bonds, left [3] and its crystal and molecular structure [4], right, which shows that only the two equatorial N–H fragments can establish H bonds with the lone pairs of the adjacent nitrogen atoms*

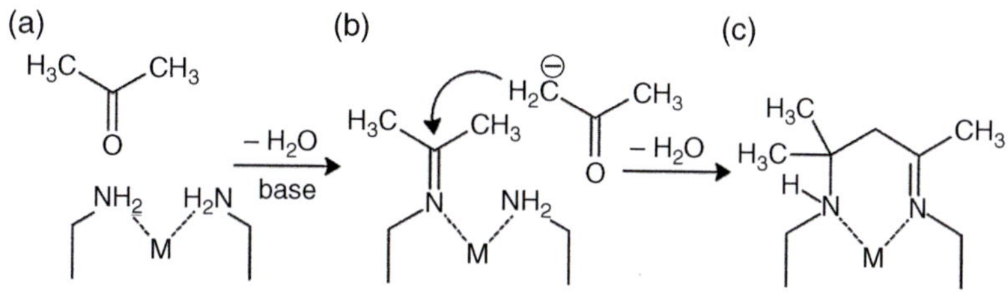

Scheme 8.3 *The mechanism of cyclization of metal bis-ethylenediamine complexes in the presence of acetone (M = Ni^{II}, Cu^{II}); only half of the metal complex has been drawn*

of $[Ni^{II}(en)_3]^{2+}$ with acetone, the nickel(II) complex of the 14-membered diene macrocycle **3** (later called CT, for Cyclic Tetramine) could be obtained [6, 7]. Interest increased due to the special properties displayed by the $[Ni^{II}(CT)]^{2+}$ complex, for instance its extreme inertness to demetalation, even in strongly acidic solutions. Moreover, the synthesis of $[Ni^{II}(CT)]^{2+}$ was the first example of a reaction in which a metal ion of defined geometric preferences promoted the reactivity of a coordinated ligand towards exotic reactants and addressed the formation of a non-trivial structure, in the present case cyclic. As stated at the time by Daryle H. Busch, of Ohio State University, Columbus, [8] the metal acts as a *template* (which, in this particular case, is a square).

The reactive metal complex is $[Ni^{II}(en)_2]^{2+}$ (half-complex drawn in Scheme 8.3**a**), in which one $-NH_2$ group undergoes Schiff base condensation with the C=O fragment of one acetone molecule. Then, an acetone carbanion, formed by deprotonation of a methyl group in the presence of excess ethylenediamine, makes a nucleophilic attack to the C=N double bond of **b**. Thus, Schiff base condensation at the facing $-NH_2$ group operates, linking the two nitrogen atoms, to give **c**, and, overall, the cyclization process yields the $Ni^{II}(CT)]^{2+}$ complex. The formation of a bidimensional square structure requires a template effect by a metal inclined to square or tetragonal coordination geometry. In this sense, the Cu^{II} ion gives, under the same conditions, an analogous reaction to form $[Cu^{II}(CT)]^{2+}$.

The two C=N bonds of the $[Ni^{II}(CT)]^{2+}$ complex may be reduced by catalytic hydrogenation or by reaction with sodium borohydride in aqueous solution, to give the couple of *meso* and *racemic* diastereoisomers $[Ni^{II}(ms\text{-}CTH)]^{2+}$ and $[Ni^{II}(rac\text{-}CTH)]^{2+}$, which were easily separated due to their markedly different solubilities [9]. Diastereoisomerism was generated by the orientation of the methyl substituents at carbon atoms 5 and 7 of the 5,7,7,12,14,14-hexamethyl-1,4,8,11-tetra-azacyclotetradecane ligand. Hydrogenation of the corresponding $[Cu^{II}(CT)]^{2+}$ did not work well, due to the simultaneous reduction of Cu^{II} to metal. Moreover, the two diastereoisomeric complexes of nickel(II) could be demetaled by boiling in a water/ethanol solution containing an excess of NaCN, and the two diastereoisomers, *ms*-CTH (**4**) and *rac*-CTH (**5**), could be obtained as stable hydrated crystalline solids. The simple and inexpensive synthesis of the two macrocycles opened the route to the preparation of hundreds of complex salts with a variety of metal ions and anions (in particular with *ms*-CTH) and to the investigations of their properties in solution, a topic still vivid, 50 years later.

In 1965, Bosnich, Poon and Tobe, of University College, London, reported the first complexes of 1,4,7,11-tetra-aza-cyclo-tetradecane, which they named cyclam (from cycl[ic] [tetr]am[ine]) [10]. They were interested in the kinetics of substitution (e.g. aquation) in octahedral complexes of type *trans*-$[Co(N_4)X_2]^+$, where N4 is the donor set constituted by two diamine molecules (e.g. en) and X^- an inorganic anion (e.g. Cl^-). They were looking in particular at the nature of the five-coordinate intermediate and were seeking an amine donor set, which, unlike two en molecules, could avoid any geometrical change. Due to its cyclic structure and rigidity, cyclam seemed the ideal candidate ligand. They chose cyclam, instead of the more easily accessible *ms*-CTH macrocycle, because it "did not contain superfluous organic side chains." They used an improved version of the van Alphen procedure and obtained cyclam as white needle crystals (2.5 g from 40 g of 2·3·2-tet), melting at 175 °C (in a sealed tube) and subliming at 125 °C. They prepared a number of *trans*-$[Co(cyclam)XY]ClO_4$ complex salts and studied their aquation equilibria. Then, they took the view that each nitrogen in the cyclam molecule, when coordinated, represents an asymmetric center, thus giving rise to five distinct configurational isomers, which they indicated as *trans*-I –*trans*-V. Then, on the basis of a qualitative analysis made with Dreiding stereomodels, they concluded that the most stable configurations should be *trans*-I and *trans*-III, whose arrangements are shown in Figure 8.2.

Indeed, the first structure of a cyclam complex, $[Ni^{II}(cyclam)Cl_2]$ [11] (see Figure 8.3), showing a *trans*-dichloro-octahedral geometry, exhibited the *trans*-III configuration, which hundreds of X-ray determined crystal and molecular structures have later demonstrated to be the most common arrangement in metal cyclam complexes. It has been also shown, through crystallographic studies, that the *trans*-I configuration prevails in metal complexes of tetra-*N*-alkyl substituted cyclam derivatives [12].

The purely organic lengthy procedure, the recommended high-dilution technique (to minimize polymerization), the low overall yield (24% Stetter, [3] 5% Bosnich *et al.* [10]) discouraged from the synthesis of cyclam by inorganic chemists, who preferred to investigate macrocyclic properties of metal complexes of tet a (*ms*-CTH). Things drastically changed in 1972, when E. Kent Barefield, of University of Illinois, Urbana, a former student of Busch, reported the very convenient metal template synthesis of cyclam, illustrated in Scheme 8.4 [13].

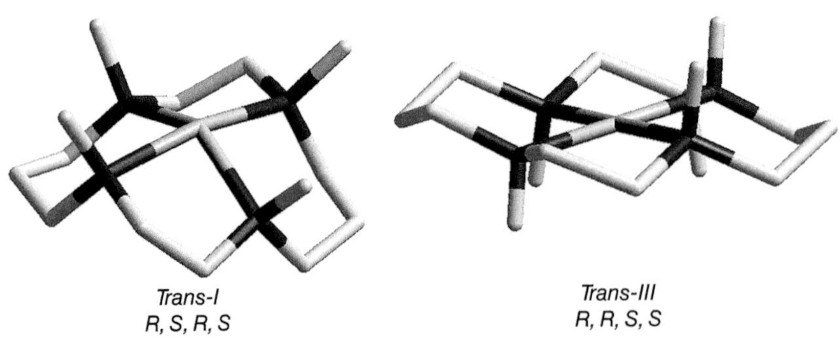

Figure 8.2 *The* trans-I *configuration of a metallo-cyclam complex, in which all the substituents (H, CH_3, etc.) point on the same side of the plane of nitrogen atoms, to which R and S descriptors have been assigned according to CIP rules. In the* trans-III *isomers, the two N–H fragments linked by one trimethylenic chain point above the plane, the other two point below*

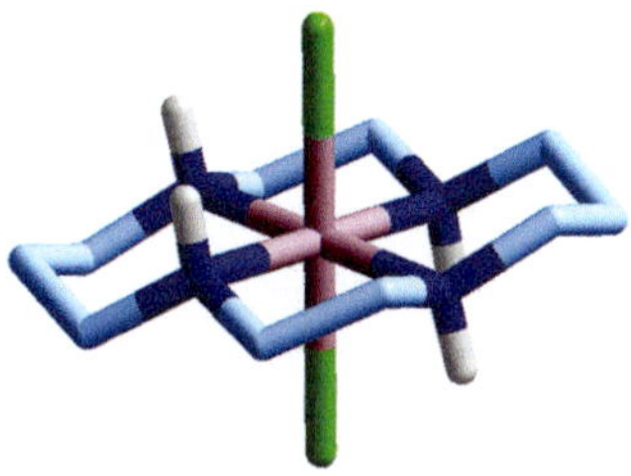

Figure 8.3 *The first crystal and molecular structure of a cyclam metal complex: [Ni^{II}(cyclam)Cl_2] [11]. The complex is octahedral, paramagnetic for two unpaired electrons and exhibits a* trans-*III configuration*

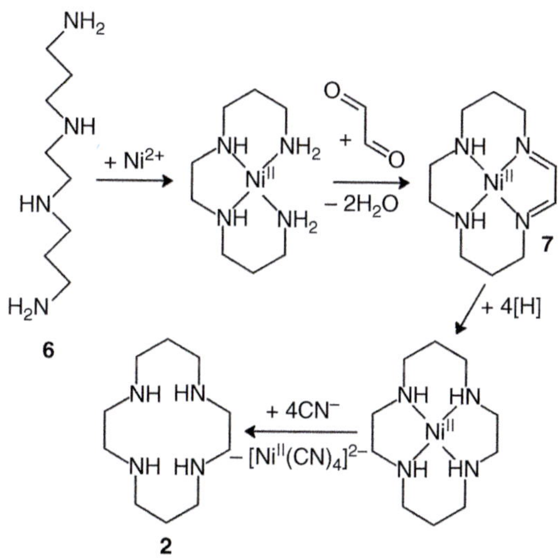

Scheme 8.4 *The Ni^{II} template synthesis of cyclam [13]*

The templating role was exerted by Ni^{II}, which, in an aqueous solution, formed a stable complex with the open-chain tetramine **3** (3·2·3-tet), positioning the four nitrogen atoms at the corners of a square. In this way, the two primary amine groups of **3** were in the right positions to undergo Schiff base condensation with one molecule of glyoxal, thus giving the diene macrocyclic complex **4**. The two C=N double bonds were reduced by catalytic hydrogenation on nickel Raney, the metal was removed from the [Ni^{II}(cyclam)]$^{2+}$ complex by treatment with boiling cyanide in the presence of NaOH and the ligand was extracted in $CHCl_3$ from the aqueous solution. All the process was carried out in the same solution at a very high concentration (0.25 M). The overall yield was quite low (20%), but the procedure was one-pot and involved cheap reagents with water as a solvent. Yield was dramatically improved a few years later by performing C=N hydrogenation with $NaBH_4$, still in the same aqueous solution [14]. Since then, cyclam has become commercially available and now can be bought at a reasonable price: 1 g ~50€; 5 g ~ 200€). The synthesis of cyclam is now matter of laboratory exercise for undergraduates (a good introduction to metal template reactions). In our university, students produce some kg of Ni(cyclam)$(ClO_4)_2$ per year. Demetalation with cyanide is a task left to the graduate students interested in using cyclam as an intermediate on the route to more sophisticated systems.

8.2 Macrocyclic effects and the importance of being 14-membered

In 1969, Dale W. Margerum, of Purdue University, West Lafayette, reported the first investigation of the thermodynamic stability of metal macrocyclic complexes, a study made difficult by their extreme inertness [15]. In particular, Margerum determined the constant of the decomposition equilibrium of the $[Cu^{II}L]^{2+}$ complex (L = tet a) in a 0.1 M HCl solution at varying temperatures, by measuring spectrophotometrically the concentration of the remaining complex in the equilibrated solution after 47 days at 50 °C and 75 °C, and after 14 days at 100 °C.

$$\left[Cu^{II}L\right]^{2+} + 4H^{+} \rightleftarrows Cu^{2+} + LH_4^{+} \quad (1)$$

The constant $K^{(1)}$ of eq. (1) in 0.1 M HCl at 25 °C was obtained by extrapolation from a plot of log $K^{(1)}$ vs the reciprocal of absolute temperature. The constant of the complex formation equilibrium (2) at 25 °C was obtained through the combination of $K^{(1)}$ and the four protonation constants of L.

$$Cu^{2+} + L \rightleftarrows \left[Cu^{II}L\right]^{2+} \quad (2)$$

The determined constant ($K = 10^{28}$) was four orders of magnitude higher than that of the most stable complex of open-chain tetramines, $[Cu^{II}(2{\cdot}3{\cdot}2\text{-tet})]^{2+}$, $K = 10^{24}$. Such an extra stability could not be considered as an extended *chelate effect*, [16] and was a ascribed to an effect specifically related to the cyclic structure of the ligand (*thermodynamic macrocyclic effect*).

Five years later, Margerum, in order to determine the constant of the formation of the $[Ni^{II}(\text{cyclam})]^{2+}$ complex, investigated the equilibrium (3) involving the slow decomposition by cyanide of the same complex in 0.1 M NaOH [17, 18].

$$\left[Ni^{II}(\text{cyclam})\right]^{2+} + 4CN^{-} \rightleftarrows \left[Ni^{II}(CN)_4\right]^{2-} + \text{cyclam} \quad (3)$$

The constants of eq. (3) were determined at 10, 25 and 40 °C and from the van't Hoff plot, ln K vs $1/T$(K), $\Delta H°$ and $\Delta S°$ values were determined. On combination of these values with those pertaining to the formation equilibrium of $[Ni^{II}(CN)_4]^{2-}$, the thermodynamic quantities for equilibrium (4) at 25 °C were calculated: $\Delta G° = -30.3$, $\Delta H° = -31.0\,\text{kcal mol}^{-1}$, $T\Delta S° = 0.7\,\text{kcal mol}^{-1}$.

$$Ni^{2+} + \text{cyclam} \rightleftarrows \left[Ni^{II}(\text{cyclam})\right]^{2+} \quad (4)$$

Corresponding values for the formation equilibrium of the $[Ni^{II}(2{\cdot}3{\cdot}2\text{-tet})]^{2+}$ complex are: $\Delta G° = -21.5$, $\Delta H° = -19.4\,\text{kcal mol}^{-1}$, $T\Delta S° = -2.1\,\text{kcal mol}^{-1}$.Then, the thermodynamic quantities of the ligand displacement equilibrium (5), defining the macrocyclic effect, can be calculated:

$$\left[Ni^{II}(2{\cdot}3{\cdot}2\text{-tet})\right]^{2+} + \text{cyclam} \rightleftarrows \left[Ni^{II}(\text{cyclam})\right]^{2+} + 2{\cdot}3{\cdot}2\text{-tet} \quad (5)$$

$\Delta G° = -8.8$, $\Delta H° = -11.6\,\text{kcal mol}^{-1}$, $T\Delta S° = -2.8\,\text{kcal mol}^{-1}$. These data demonstrate that the macrocyclic effect is totally driven by an extremely favorable enthalpy change and moderately disfavored by the entropy contribution. The interpretation suggested by Margerum was based exclusively on the different hydration of the uncomplexed ligands 2·3·2-tet and cyclam. In particular, it was considered that the open-chain tetramine in the primary hydration shell interacted with four water molecules, whereas the macrocycle, due to its closed arrangement, could establish hydrogen bonds at most with two H_2O molecules. Thus, the macrocyclic effect should be solely ascribed to the fact that the cyclic ligand, on complexation, suffers from dehydration to a lesser extent than its non-cyclic counterpart (enthalpy term); the open-chain ligand has an advantage, because on complexation releases a higher number of water molecules, to which a more positive contribution of translational entropy corresponds, but this term reduces the enthalpy gain only by one third, looking at the thermodynamic quantities of eq. (5).

At that time, we were interested in the thermochemistry of metal–polyamine complexes and we had found that a linear correlation existed between the enthalpy of complex formation in aqueous solution (determined calorimetrically) and the energy of the d–d absorption band for a series of copper(II) tetramine and bis-diamine complexes [19], illustrated in Figure 8.4. The d–d band of Cu^{II} tetramine complexes, showing a tetragonal geometry, is the envelope of three transitions ($x^2-y^2 \rightarrow z^2$, $\rightarrow xy$, $\rightarrow xz, yz$). The energy of the band can be reasonably assumed to coincide with the energy of the $x^2-y^2 \rightarrow xy$ transition, which expresses the intensity of the in-plane interaction, thus giving a measure of the energy of the bonds established by Cu^{II} with the four amine nitrogen atoms. It seems therefore reasonable to assume a correlation of the wavenumber of the absorption band with the enthalpy of complex formation, which includes as a main contribution the energy of the $Cu^{II}-N$ coordinative interaction.

As the energy of the d–d band increases, the color changes to blue, to violet, to pink red. This accounts for the observation by van Alphen that $[Cu^{II}(2{\cdot}2{\cdot}2)]^{2+}$ is blue and $[Cu^{II}(2{\cdot}3{\cdot}2)]^{2+}$ pink-violet (the biuret color). Noticeably, a solution of $[Cu^{II}(cyclam)]^{2+}$ showed a red color, to which a wavenumber of 19.8 kK corresponded and, according to the linear relationship in Figure 8.3, a value of $\Delta H° = -30.9\,kcal\,mol^{-1}$ should be associated. The calculated value of $\Delta H°$ does not necessarily correspond to real value: it reflects the sole contribution of the energy of the metal–ligand interaction, but does not include solvational and configurational terms.

This observation prompted us to plan a calorimetric determination of the enthalpy of metal complexation by cyclam and related tetramine macrocycles. The design of the experiment was not straightforward in view

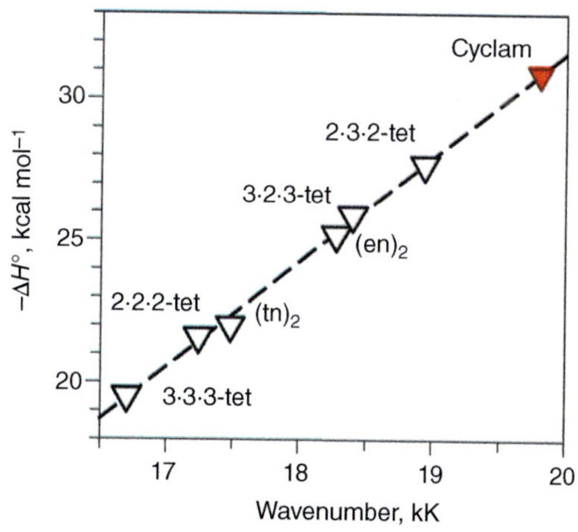

Figure 8.4 *A linear relationship between the energy of the d–d absorption band (in kK, 1 k = 1000 cm^{-1}) and the enthalpy of complex formation in aqueous solution at 25 °C for copper(II) tetramine and bis-diamine complexes (tn = 1,3-diaminopropane) [19]. The equation of the least-squares straight line is: ΔH° (kcal mol^{-1}) = −42.2 + 3.69 × wavenumber (kK). The red triangle refers to the value calculated for [Cu^{II}(cyclam)]$^{2+}$ from the energy of its absorption band: 30.9 kcal mol^{-1}*

of the sluggishness of tetra-aza-macrocyclic complexes towards both formation and demetalation. This is in particular the case of cyclam, which exists in the diprotonated form, LH_2^{2+} over a large pH interval, 3–9. As a consequence, the complexation process is made slow by the electrostatic repulsion by the ligand dication and the doubly positively charged metal ion. The problem could be circumnavigated by carrying out the complexation reaction in basic solution, e.g. in 1 M NaOH. This was possible in the case of copper(II), which in a strongly alkaline solution does not precipitate as a hydroxide, but forms the hydroxo-complex $[Cu^{II}(OH)_3]^-$. Thus, the following equilibrium was investigated by flow-microcalorimetry, by mixing aqueous solutions of cyclam and of $Cu^{II}(ClO_4)_2$, both made 1 M in NaOH [20].

$$\left[Cu^{II}(OH)_3\right]^- + L \rightleftarrows \left[CuL\right]^{2+} + 3OH^- \qquad (6)$$

The heat effect was corrected for the enthalpy of formation of $[Cu^{II}(OH)_3]^-$, calorimetrically determined, to give the $\Delta H°$ value for the complexation equilibrium of the type represented by eq. (2): – 32.2 kcal mol^{-1}. This term is slightly more exothermic than that estimated from the linear relationship of Figure 8.3: –30.9 kcal mol^{-1}. This means that all the other energy terms associated to the complexation either favorable (hydration of the $[Cu^{II}(cyclam)]^{2+}$complex) or unfavorable (dehydration of $[Cu^{II}(H_2O)_4]^{2+}$ and of cyclam) roughly compensate.

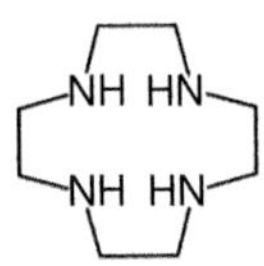
[12]aneN$_4$

[13]aneN$_4$

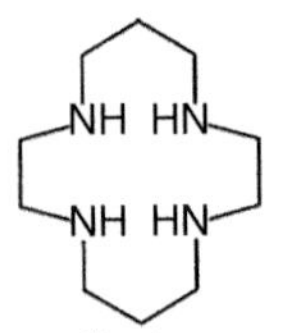
Cyclam

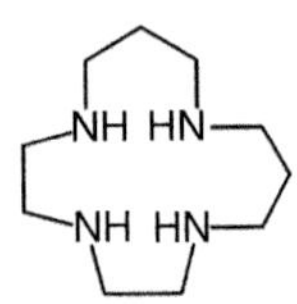
Isocyclam

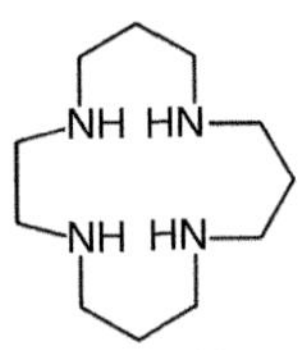
[15]aneN$_4$

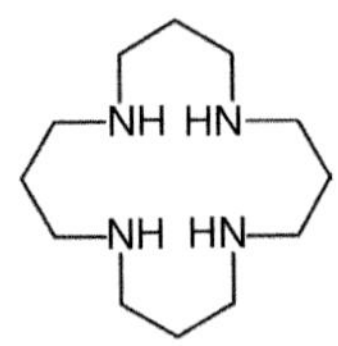
[16]aneN$_4$

The study was extended to macrocycles of varying atomicity, from 12 to 15 [21]: the enthalpy changes associated to the formation of the corresponding copper(II) complexes, in aqueous solution, at 25 °C, are displayed in Figure 8.5.

A sharp peak selectivity exists, in favor of the 14-membered macrocycle (cyclam): either expansion or contraction of the macrocyclic cavity induces a drastic reduction of the heat of complex formation. Noticeably, the same ring-size dependence is observed for the energy of the d–d band (wavenumber, kK, open circles in Figure 8.5), which suggests that the enthalpy selectivity is strictly related to the capability of the macrocycle to establish more or less intense in-plane metal–ligand interaction. Ring size selectivity had been observed in the complexation of alkaline metal ions by crown ethers and cryptands. However, enthalpic selectivity of copper(II) tetra-aza-macrocyclic complexes is not a simple matter of ring size. This is demonstrated by the behavior of the "other" 14-membered macrocycle in addition to cyclam: the so-called isocyclam [22]. Cyclam possesses a sequence of alternating aliphatic chains 2,3,2,3 and on complexation gives a 5,6,5,6 series of chelate rings, whereas isocyclam shows a 2,2,3,3 sequence of aliphatic chains and a 5,5,6,6 series of chelate rings. $\Delta H°$ for the formation of $[Cu^{II}(isocyclam)]^{2+}$ is – 27.8 kcal mol^{-1}, i.e. 4.6 kcal less exothermic than for

$[Cu^{II}(cyclam)]^{2+}$. Figure 8.6 shows the structure of the two complexes $[Cu^{II}(cyclam)]^{2+}$ [23], and $[Cu^{II}(isocyclam)]^{2+}$ [24] (both crystallized as diperchlorate salts).

The nitrogen atoms of each macrocycle have been linked with a red dotted straight segment, to define a quadrilateral. The quadrilateral of $[Cu^{II}(cyclam)]^{2+}$ is an almost regular square with four right angles, a circumstance which allows a substantial overlap of the nitrogen lone pairs with the empty $d_{x^2-y^2}$ metal orbital and affords a strong binding interaction. On the other hand, $[Cu^{II}(isocyclam)]^{2+}$ gives a kite-like quadrilateral, to which a poor lone pair/orbital overlap should correspond. This accounts for the distinctly less negative enthalpy change and for the lower energy of the d–d band (18.18 kK). Noticeably, the uncomplexed cyclam, whose structure is shown in Figure 8.6, [4] shows a quite regular square of nitrogen atoms, whose edges are close to those observed in the $[Cu^{II}(cyclam)]^{2+}$ complex. This may provide a sound interpretation of the

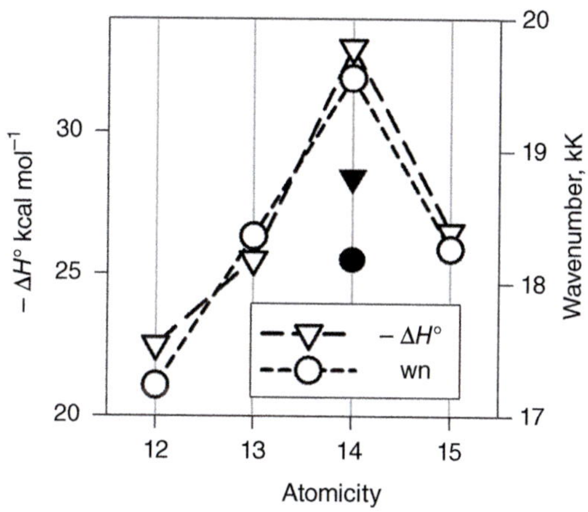

Figure 8.5 *Complexation of Cu^II by tetramine macrocyles of varying atomicity [21]. Triangles: ΔH° in aqueous solution at 25 °C (kcal mol⁻¹); circles: energy of the d–d band (kK). Filled symbols refer to the copper(II) complex of isocyclam [22]*

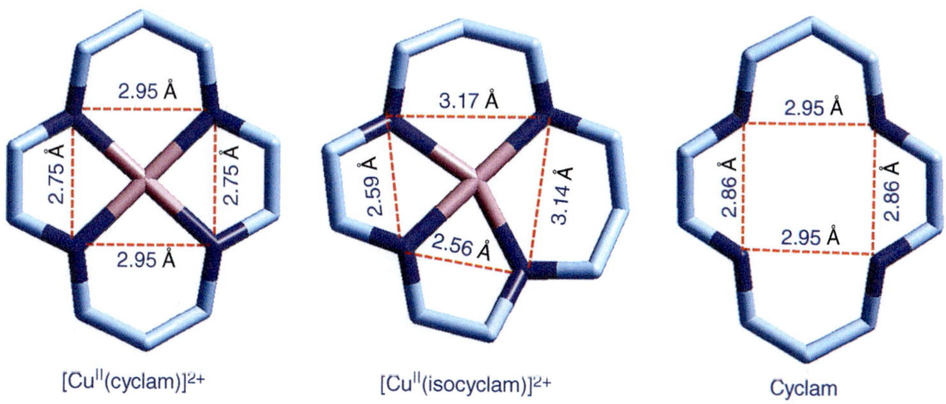

Figure 8.6 *The molecular structure of the two complexes $[Cu^{II}(cyclam)]^{2+}$ [23], and $[Cu^{II}(isocyclam)]^{2+}$ [24] (both crystallized as diperchlorate salts, counteranions omitted in the picture for clarity) and of the uncomplexed cyclam. Red dotted lines define a quadrilateral whose vertices are the nitrogen atoms. The structure of uncomplexed cyclam [4] is also reported for comparative purposes*

macrocyclic effect: cyclam forms especially stable complexes because it is perfectly *preorganized and preoriented* for including and establishing strong interactions with those 3d metals that experience the highest ligand field stabilization from a tetragonal coordination geometry. Such a macrocyclic advantage diminishes or vanishes when the ligand's structure deviates from that of cyclam. Obviously, macrocyclic preorganization should provide in addition a substantial entropy advantage with respect to a non-cyclic ligand (e.g. 2·3·2-tet), which on complexation loses most of its conformational freedom and flexibility.

In the case of nickel(II), the complexation with cyclam could not be calorimetrically investigated through the direct [metal ion + ligand] reaction in alkaline solution, due to insolubility of $Ni(OH)_2$. The problem was circumnavigated by studying the decomposition of the $[Ni^{II}(macrocycle)]^{2+}$ complex by excess cyanide, in alkaline solution (see eq. (3)) [25]. The reaction, due to the typical sluggishness of nickel(II) macrocyclic complexes, is not instantaneous, but, at 25 °C, takes hours to go to completion. Thus, it was studied with a "batch" microcalorimeter, which allowed the precise determination of the heat released over a period of several hours. In the case of $[Ni^{II}(cyclam)]^{2+}$, the destruction of the complex was complete in 5 hours (no more heat effect detected after that time). The heat effect was corrected for that associated to the formation of $[Ni^{II}(CN)_4]^{2-}$, to give the enthalpy of macrocyclic complexation. $\Delta H°$ values for complexes with ligands of 12-to-15 atomicity are displayed in Figure 8.7.

It must be noted that complexes of [12]aneN_4 and of [15]aneN_4 exist in aqueous solution at 25 °C as 100% of the octahedral form, blue color, two unpaired electrons, high-spin state (HS). On the other hand, complexes of macrocycles [13]aneN_4 and of cyclam give an equilibrium mixture of the octahedral HS form and of the square, yellow color, diamagnetic low-spin form (LS) [27]. The blue-to-yellow equilibrium is pictorially illustrated in Figure 8.8. For [13]aneN_4 in aqueous solution at 25 °C there is 13% of the blue, octahedral HS form and 87% of the yellow, square, LS form; for cyclam 21% of blue and 79% of yellow coexist. The blue-to-yellow conversion is endothermic ($\Delta H° = 7.5\,kcal\,mol^{-1}$ for [13]aneN_4 and $5.4\,kcal\,mol^{-1}$ for cyclam). Thus, enthalpy changes for the formation of both HS and LS complexes could be calculated and they are shown in the diagram in Figure 8.7. In particular, the plot of HS complexes (open triangles) shows the same peak selectivity observed for copper(II), with the highest heat effect occurring for cyclam, while either reduction or increase of the macrocyclic ring size makes complexation less exothermic. Also in the present case, complexation by cyclam, to give both HS and LS complexes, is more exothermic than for the isomeric macrocycle

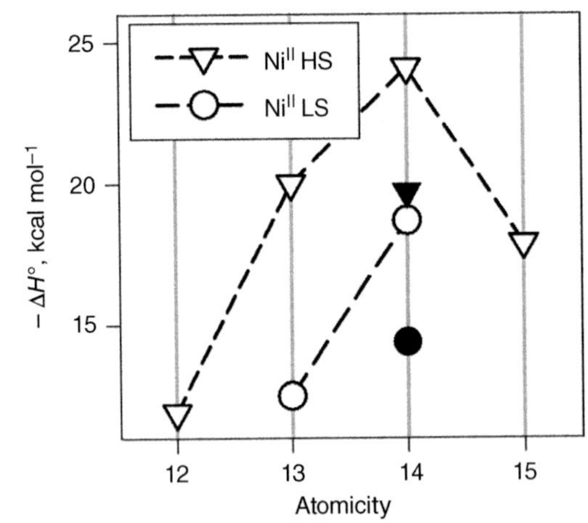

Figure 8.7 *Enthalpy changes associated to the complexation of Ni^{II} by tetramine macrocyles of varying atomicity, in aqueous solution at 25 °C [25]: open triangles refer to the formation of Ni^{II} high-spin (HS) complexes; open circles to low-spin (LS) complexes. Filled symbols refer to isocyclam complexes, either HS (triangle) or LS (circle) [26]*

isocyclam [26]. Crystal and molecular structures of the LS [Ni^{II}(cyclam)](ClO_4)$_2$ [29], and [Ni^{II}(isocyclam)](ClO_4)$_2$ complex salts [30], shown in Figure 8.9, indicate that the cyclam complex exhibits a coordination geometry (square) much more favorable than isocyclam (kite-like) to the establishing of strong in-plane interactions.

It has to be noticed that low-spin Ni^{II}, in view of its more compact electronic configuration, has a ionic radius distinctly smaller than high-spin Ni^{II}, as demonstrated by the smaller Ni^{II}–N distances (1.93–2.00 Å in the cyclam and isocyclam low-spin complexes vs 2.07 Å in the high-spin cyclam complex considered in Figure 8.9). Spin-pairing induces a further contraction of the roughly square quadrilateral obtained by joining the nitrogen atoms, whose area is 7.6 Å^2 to be compared to the area of the uncomplexed macrocycle 8.4 Å^2. On the other hand, HS Ni^{II} (area: 8.5 Å^2) seems to have an almost perfect size to fit the "square" provided by the strain-free uncomplexed ligand. It is worthwhile noting that Busch in 1974 [31], on the basis of molecular mechanics calculations, was able to predict that best fitting of the cavity of cyclam should be provided by a metal ion M giving an M–N distance of 2.07 Å.

Curtis had realized immediately the most astonishing property of nickel(II) and copper(II) complexes with tet a, i.e. the extreme resistance to demetalation in strongly acidic or alkaline solution, an effect to be ascribed to the cyclic structure of the ligand [7]. Some years later, Margerum quantified such an effect comparing the kinetic stability of [Cu^{II}(tet a)]$^{2+}$ and [Cu^{II}(2·3·2-tet)]$^{2+}$ in 6.1 M HCl at 25 °C, for which following lifetimes, τ, were determined by spectrophotometry: 32.4 days and 0.24 s, respectively [32]. This defined the *kinetic macrocyclic effect.*

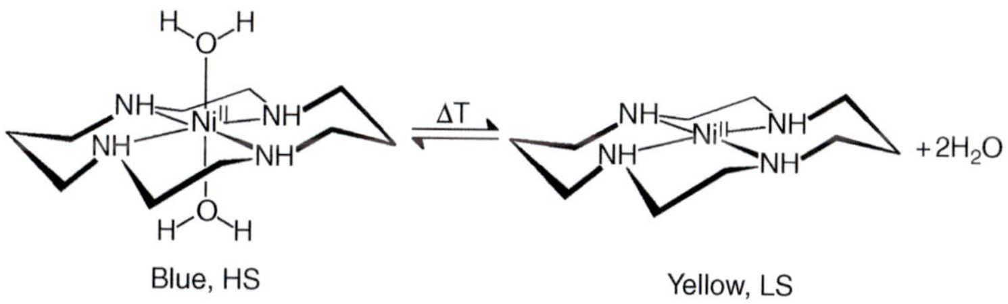

Figure 8.8 *The blue-to-yellow equilibrium of [Ni^{II}(cyclam)]$^{2+}$. In aqueous solution at 25 °C there is 21% of the blue, octahedral, high-spin form and 79% of yellow, square, low-spin form [28]*

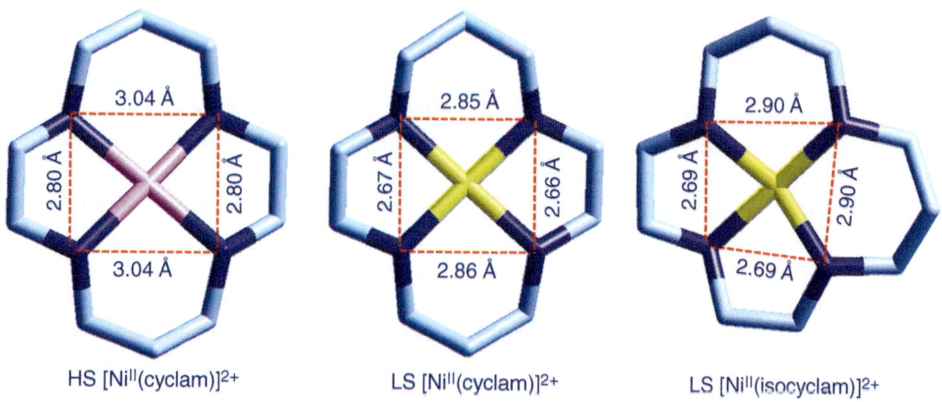

Figure 8.9 *The molecular structure of the complexes: high-spin [Ni^{II}(cyclam)]$^{2+}$ (dichloro-trans-octahedral structure shown in Figure 8.3 [11], chloride ions not reported; low-spin [Ni^{II}(cyclam)]$^{2+}$ [29], and low-spin [Ni^{II}(isocyclam)]$^{2+}$ [30] (both crystallized as diperchlorate salts, ClO_4^- counterions not reported). Hydrogen atoms have been omitted for clarity. Red dotted lines define a quadrilateral whose vertices are the nitrogen atoms*

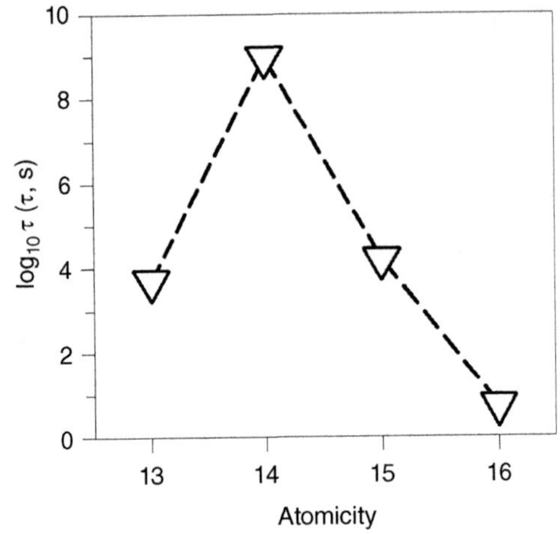

Figure 8.10 *Lifetimes, τ, s, reported in a logarithmic scale, for the process: $[Ni^{II}(L)]^{2+} + 4H^+ \leftrightarrows Ni^{2+} + LH_4^{4+}$ in 0.3 M HCl at 25 °C [33] (L = tetramine macrocycle of varying ring atomicity). The value for the 14-membered macrocycle (cyclam, τ = 30 years) refers to a 1 M $HClO_4$ solution [34]*

However, as observed in the case of thermodynamic stability, the magnitude of the kinetic macrocyclic effect is strongly dependent upon ring size. This is shown, for instance, by the lifetimes, τ, of the demetalation in acidic solution of nickel(II) complexes of 13-to-16 membered saturated tetra-aza macrocycles [33], which ranges from 5.3 s of $[Ni^{II}([16]aneN_4)]^{2+}$ to 30 years of $[Ni^{II}(cyclam)]^{2+}$ [34], a difference of more than nine orders of magnitude. A peak selectivity in favor of cyclam is illustrated by the diagram of τ values vs ring atomicity in Figure 8.10.

Thus, it appears that being cyclic is a necessary, but not sufficient condition to profit from especially high thermodynamic and kinetic stabilities. The most important point is to be 14-membered and symmetric: to be cyclam.

8.3 Cyclam promotes the redox activity of the encircled metal ion

Curtis had in his hands the first metal complexes of 14-membered tetra-aza-macrocycles, tet a and tet b, and, thanks to a well-developed chemical ingenuity, he discovered two of their distinctive properties: (i) the extreme kinetic stability of 3d metal complexes, discussed in the previous Section, (ii) the ability to promote the redox activity of the encircled metal ion. As far as point (ii) is concerned, Curtis observed that, when treated with nitric acid, a yellow solution of the $[Ni^{II}(\text{tet a})](ClO_4)_2$ complex salt turned green, from which the olive-green salt $[Ni^{III}(\text{tet a})(NO_3)_2]ClO_4$ could be precipitated [35]. The magnetic moment of this compound pointed towards the presence of one unpaired electron, as expected for the formation of a low-spin d^7 complex (Ni^{III}) of octahedral geometry (unpaired electron in the d_{z^2} orbital). Later, Olson and Vasilevskis carried out accurate electrochemical investigations on the same complex and demonstrated the formation of Ni^{III} and Ni^{I} species, which were isolated as solid and characterized [36]. Busch extended this study to the nickel(II) complexes with a variety of 14-membered tetra-aza macrocycles (including cyclam) and unambiguously determined the electronic configuration of the Ni^{III} (d^7, low-spin) and of Ni^{I} complex (d^9), through ESR studies on MeCN solutions frozen at 77 K [37]. It was also demonstrated in those years that coordination by cyclam and by cyclam-like macrocycles allowed the access to otherwise elusive oxidation states of the encircled metal ions: Cu^{III} (d^8) [38, 39], Ag^{II} (d^9) [40, 41], Ag^{III} (d^8) [42] and the transient Hg^{III} (d^9) [43]. The notion emerged that, if you have a metal ion immobilized in a tetra-aza macrocycle, you can either abstract from it or put on it electrons and it will not undergo further redox reactions because it has been made inert and unreactive by the closed arrangement of the ligand.

We demonstrated that such a rich and versatile redox activity is not to be generally ascribed to the macrocyclic structure, but to the coordination by the symmetric 14-membered ligand, cyclam. This was documented, for instance, by the electrochemical investigation we carried out on the Ni^{II}/Ni^{III} oxidation process in MeCN on complexes of tetramine macrocycles whose ring atomicity ranged from 12 to 16 [44]. $E_{1/2}$ values measured at a platinum working electrode are shown in the diagram in Figure 8.11. Again, a peak selectivity in favor of the symmetric 14-membered macrocycle (cyclam) is observed. Coordination by cyclam affords the easiest attainment of the trivalent state (expressed by the lowest value of $E_{1/2}(Ni^{III}/Ni^{II})$). Either decreasing or increasing of the ring atomicity makes the half-wave potential become more and more positive and the oxidation process more and more difficult. The explanation is also in this case related to the especially favorable spatial arrangement of cyclam. Strong in-plane metal–ligand interactions raise the energy of the highest occupied metal centered orbital, antibonding in nature, from which one electron is abstracted, thus making oxidation easier. It has to be noticed that oxidation of $[Ni^{II}(isocyclam)]^{2+}$ takes place at a potential well 340 mV more positive than for $[Ni^{II}(cyclam)]^{2+}$.

The $[Ni^{III,II}(cyclam)]^{3+/2+}$ couple also operates in water, provided that the solution is acidic, in order to prevent the oxidative degradation of the macrocycle promoted by Ni^{III} [45, 46]. The $E°$ value for the $[Ni^{III,II}(cyclam)]^{3+/2+}$ couple in 1 M HCl at 25 °C is 0.71 V vs NHE, to be compared, for instance, to the Fe^{3+}/Fe^{2+} couple in 1 M HCl, 0.77 V vs NHE, which means that $[Ni^{III}(cyclam)]^{3+}$ in aqueous solution is a moderate oxidizing agent, like Fe^{3+}, which oxidizes I^-, while $Ni^{II}(cyclam)]^{2+}$ is a moderate reducing agent, like Fe^{2+}, which reduces MnO_4^-. On these bases, we considered the $[Ni^{III,II}(cyclam)]^{3+/2+}$ couple as a reasonable candidate to act as an electron carrier in a liquid organic membrane **c** interfacing an aqueous layer **a** (electron source phase, ESP) containing a reducing agent Red, and an aqueous layer **b** (electron receiving phase, ERP), containing an oxidizing agent Ox, according to the mechanism illustrated in Figure 8.12. However, to act as

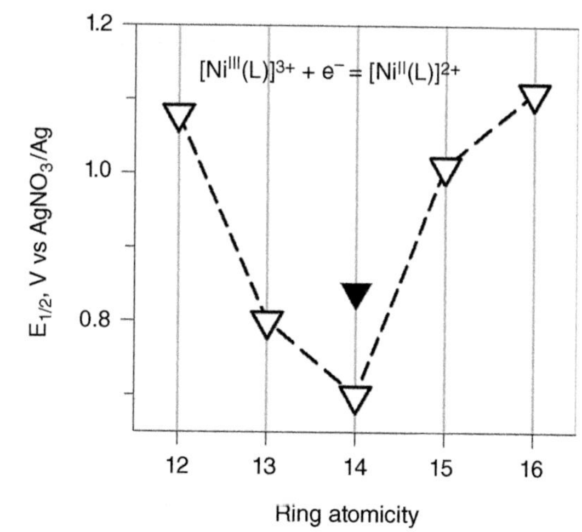

Figure 8.11 *Half-wave potentials, $E_{1/2}$, associated to the half-reaction $[Ni^{III}(L)]^{3+} + e^- \leftrightarrows [Ni^{II}(L)]^{2+}$ in MeCN solution at 25 °C (L=[12]aneN_4→[16]aneN_4) [44]). The filled symbol refers to the asymmetric 14-membered macrocycle isocyclam. The less positive the potential, the easier the access to the Ni^{III} state*

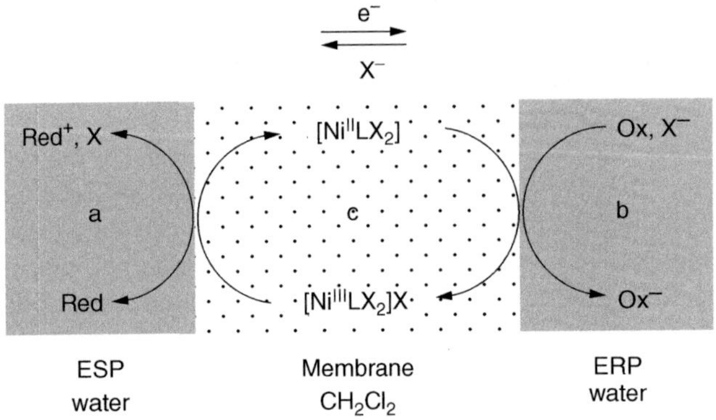

Figure 8.12 *The mechanism of electron transport from the reducing aqueous layer* **a** *(electron source phase, ESP) to an aqueous oxidizing layer* **b** *(electron receiving phase, ERP), mediated by the electron carrier* $[Ni^{II}(L)]Cl_2]$*, confined in the interfacing* CH_2Cl_2 *layer* **c** *(the liquid organic membrane; L = N-cetylcyclam,* **8***). The* **a**-to-**b** *transport of electrons is coupled to a* **b**-to-**a** *countertransport of anions* X^-

a carrier the macrocyclic complex has to be confined in the membrane. This was made by appending a long aliphatic chain to a nitrogen atom of cyclam, to give *N*-cetylcyclam [47] (**8**).

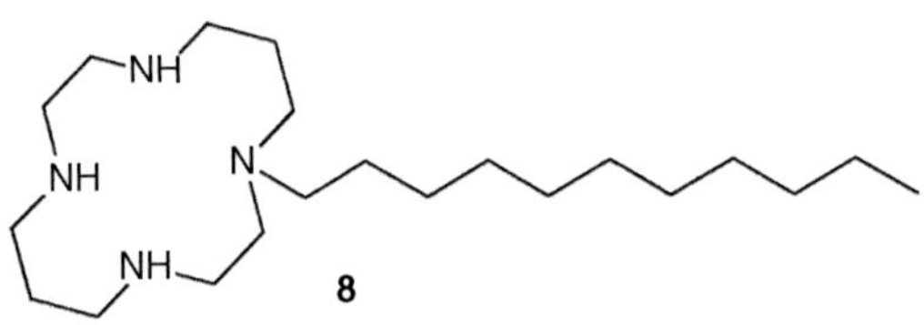

8

The three-phase experiment was carried out in the V-shaped tube sketched in Figure 8.13. ERP (**a**) contains the oxidizing agent $Na_2S_2O_8$ and is 1 M in an NaX salt. The liquid membrane **c** was made 10^{-3} M in $[Ni^{II}(L)X_2]$. The choice of the anion X^- is crucial: when the X^- is ClO_4^-, the Ni^{II}-to-Ni^{III} oxidation process at the ERP/membrane interfaces does not take place. It does take place when $X^- = Cl^-$, both as counterion of the metal complex and as a background electrolyte of the aqueous ERP. This was visually demonstrated by the fact that the membrane solution near to the interface with ERP takes the lemon yellow color of $[Ni^{III}(L)Cl_2]Cl)$. Cl^-/ClO_4^- selectivity can be explained on the basis of the diagram in Figure 8.14, in which the electrochemical scale in water has been juxtaposed to that in CH_2Cl_2. According to the empirical matching of the two scales, the potential of the $S_2O_8^{2-}/SO_4^{2-}$ couple in water (2.1 V vs NHE) is lower than that of the $[Ni^{III/II}L(ClO_4)_2]^{+/0}$ couple in CH_2Cl_2 (0.920 V vs Fc^+/Fc), but higher than that of the $[Ni^{III/II}LCl_2]^{+/0}$ couple in CH_2Cl_2 (0.175 V vs Fc^+/Fc). The marked difference of the two $E_{1/2}$ values of nickel complexes depends upon the fact that the Ni^{III} oxidation state (d^7, low-spin) is strongly stabilized by the two highly coordinating Cl^- anions occupying the two apical positions of an elongated octahedron. Such a stabilization effect is not exerted by the poorly coordinating ClO_4^-.

The ESP aqueous layer (**b**) is 1 M in HCl and 10^{-2} M in one of the following one-electron reducing agents: Ti^{III} (from $TiCl_3$), Cr^{II}, Fe^{II} and $[Co^{II}(diamsarH_2)]^{4+}$. All the reducing agents are oxidized under the three-phase conditions and the mechanism illustrated in Figure 8.12. Two electrodes (indicator, platinum band, and

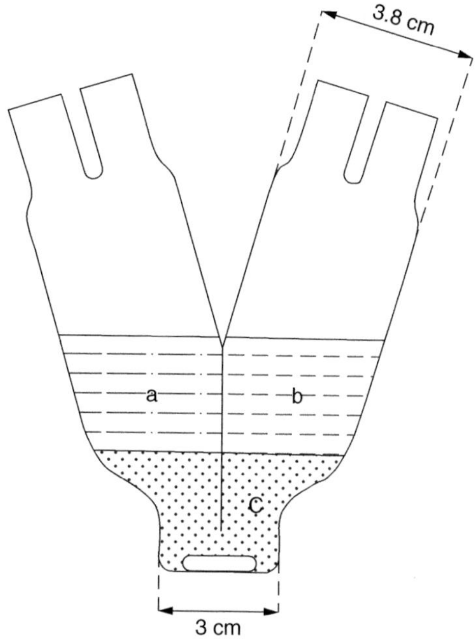

Figure 8.13 *The V-shaped tube used in the e$^-$/X$^-$ cross transport experiments. The ESP aqueous layer (**a**) was 10^{-2}m in Red and 1 M HCl. The water-immiscible liquid membrane (**c**, CH_2Cl_2) was 10^{-3}M in [Ni^{II}(L)X_2] (L = N-cetylcyclam, **8**). The ERP aqueous layer (**b**) was 1 M in NaCl and 0.1 M in $Na_2S_2O_8$. The membrane was magnetically stirred; ESP and ERP were stirred by rotating glass rods, driven by twin electric motors, at a constant speed. The concentration of Red$^+$ in ESP was monitored through the potential of a platinum electrode vs a calomel electrode, both dipped in the aqueous layer [47]*

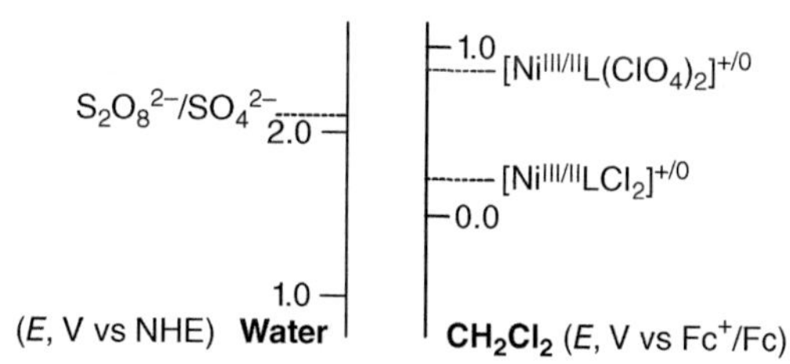

Figure 8.14 *Empirical juxtaposition of the electrochemical scales in water (ERP, left) and in CH_2Cl_2 (membrane, right). Potentials of nickel complexes are $E_{1/2}$ values determined by cyclic voltammetry studies in a CH_2Cl_2 solution made 0.1 M in [Bu_4N]X (X$^-$ = ClO_4^-, Cl$^-$), at 25 °C*

reference, calomel)) were dipped in ESP (**b**) to monitor the progress of the e$^-$/X$^-$ cross-transport experiment, through the potential of the aqueous Red$^+$/Red couple. From the measured potential, the concentration of Red$^+$ was calculated and corresponding values for all the investigated reducing agents are plotted vs time (minutes) in the diagram in Figure 8.15. The rate of cross transport depends dramatically upon the nature of the reductant, the transport being complete in a range varying from 10 minutes to 12 h. End of the transport is signaled by the reaching of a plateau in the transport profile as well as by the color change of the CH_2Cl_2

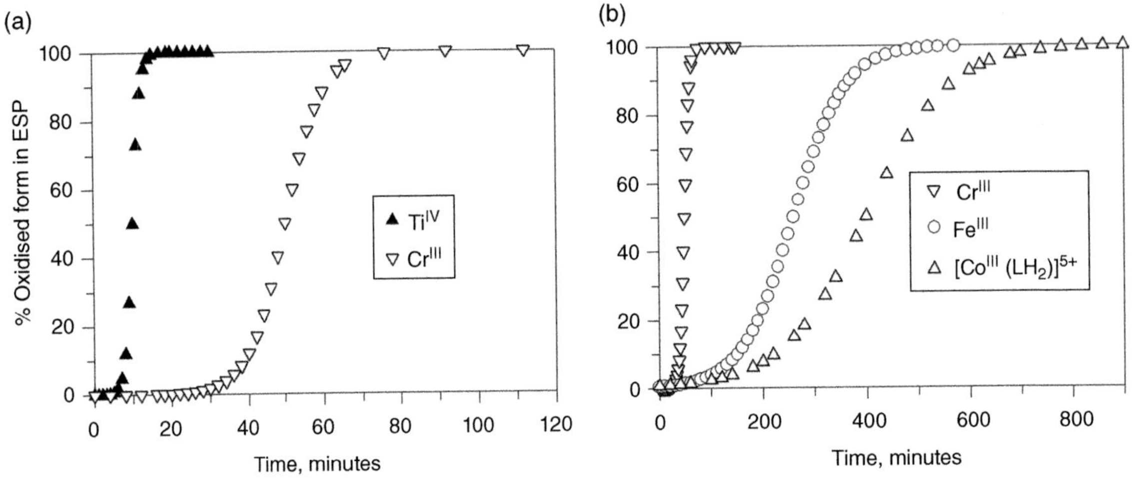

Figure 8.15 Profiles of the concentration of the oxidized species Red^+ in ESP over the course of the e^-/X^- cross transport. L=diamsarH_2^{2+}, i.e. the doubly protonated form of diaminosarcophagine

layer, from lemon yellow ($[Ni^{III}(L)Cl_2]Cl$) to pale violet ($[Ni^{II}(L)Cl_2]$). Transport profiles show a sigmoidal shape, whose initial induction period may be related to the complicated reduction kinetics of peroxydisulfate. The sequence of the transport rate ($Ti^{III}>Cr^{II}>Fe^{III}>$ $[Co^{II}(diamsarH_2^{2+})]^{4+}$ does not correlate with either the electrode potential or the self-exchange rate of the Red^+/Red couples. It is probable that in the complex environment at the water/CH_2Cl_2 interface the electron is transferred from the reductants to the Ni^{III} center through a chloride ion bridging the two metal centers [47]. This could account for the lowest rate observed with the cage complex $[Co^{II}(diamsarH_2^{2+})]^{4+}$, for which an inner-sphere electron transfer process is prevented.

We know two main procedures to carry out redox processes in the lab: (i) through direct mixing of the solutions of the oxidizing agent Ox and of the reducing agent Red (short circuit, electrons are transferred from Red to Ox, which are in close contact), and (ii) by keeping the beaker containing Ox (A) and the beaker containing Red (B) separated and connected through a salt bridge: the electrons flow from B to A through metal wire linking two electrodes each dipped in one of the two beakers, while electrical charges are balanced by the flow of cations and anions from the salt bridge (Volta's cell). The three-phase system described above still keeps the solutions of Ox and Red separated, and electrons and anions counter-flow from an aqueous layer to the other thanks to a carrier confined in the interfacing organic layer, present in a catalytic amount and going back and forth, cyclically, from one interface to the other. In the examples described above, the carrier performed ten cycles.

8.4 Scorpionands: cyclam derivatives with an aggressive tail, biting a chelated metal from the top

In 1977 Thomas Kaden, University of Basel, Switzerland, reported the quinquidentate ligand **9** [48] in which a $-CH_2CH_2N(CH_3)_2$ side-chain had been covalently linked to an amine nitrogen atom of the Busch's macrocycle *ms*-CRH [49].

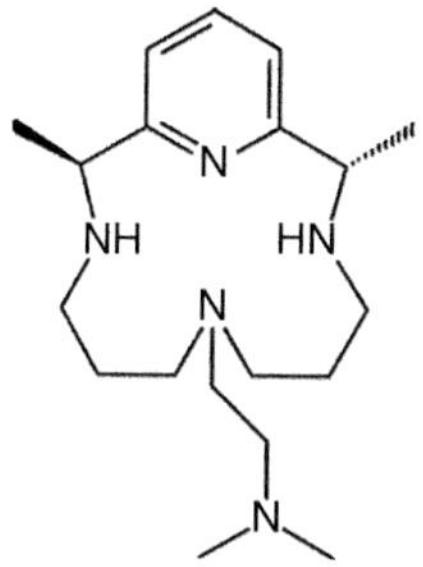

9

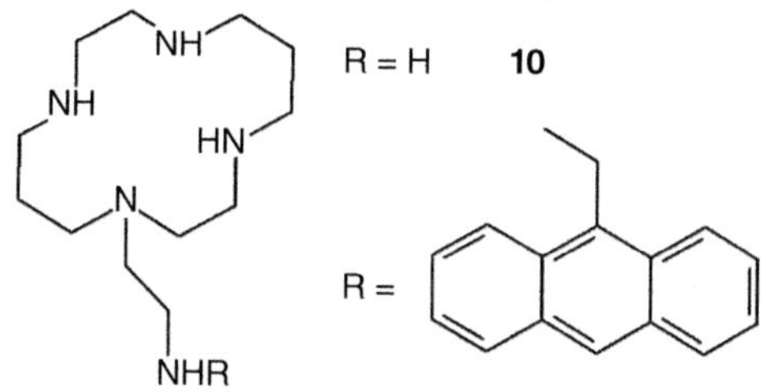

10

11

The nickel(II) complex of **9** shows a unique combination of the rigid and kinetically stable arrangement of the tetra-aza macrocycle and the flexible and kinetically labile group of the side-chain [48]. This determines a spectacular change of the structure and of the electronic properties of the Ni^{II} complex on consecutive addition of acid and base, as illustrated in the general scheme of Figure 8.16.

In an acidic solution the species **I** is present, with the dimethylamino group of the side-chain protonated, the chain being stiffened and pointing far away from the complex in order to minimize the electrostatic repulsion between the secondary ammonium fragment and the metal ion. On the other hand, the 14-membered macrocycle resists protonation and forms a square complex with Ni^{II}, low-spin and yellow color. If pH is increased, the dimethylammonium group deprotonates and, as an amino group, goes to bind the Ni^{II} center from the top. A water molecule completes six-coordination, according to a distorted octahedral geometry (complex **II** in Figure 8.16). Axial binding by the $-N(CH_3)_2$ fragment and H_2O stabilizes the high-spin state, which makes color turn blue-violet. The pK_a of the $[Ni(L\cap BH)]^{3+}$ acid is 6.29 at 25 °C and $I = 0.5$ M.

Penta-aza ligands of this type are called *scorpionands*, because they possess an aggressive tail (the side-chain), which bites from the top a firmly chelated individual (the metal ion). The prototype of scorpionands is aminoethylyclam (AEC, **10**) [50], whose nickel(II) complex undergoes the equilibrium described in Figure 8.14, with a $pK_a = 2.8$. It should be noticed that the ethylammonium fragment in the AEC complex is ten orders of magnitude more acidic than plain ethylammonium ($pK_a = 10.8$). A contribution to this extra-acidity may be given by the formation of a new $Ni^{II}-N$ bond plus the $Ni^{II}-OH_2$ bond. However, this contribution (an enthalpy term) must be moderate, if one considers that the log K_5 value associated to the binding of the fifth ammonia molecule to Ni^{2+} is 0.8. Moreover, one can observe that ammonia neither binds the metal center of $[Ni^{II}(cyclam)]^{2+}$ in aqueous solution, even if added in a large excess, nor modifies its spectral features. Thus, the determining contribution to the strong acidity of the ethylammonium side-chain

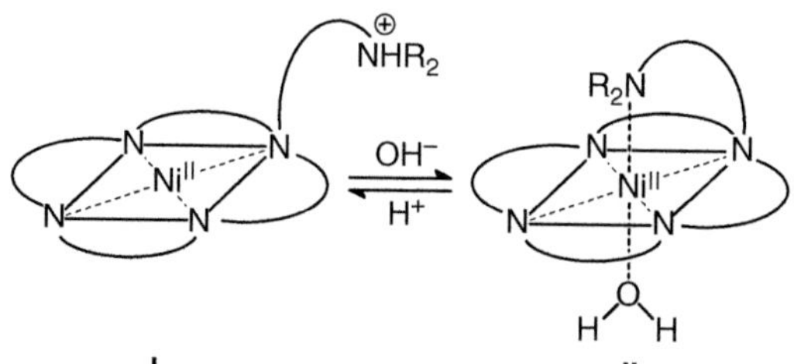

I

II

Figure 8.16 *The pH dependent equilibrium of nickel(II) complexes with tetra-aza macrocycles bearing an alkylamino pendant arm (scorpionands, e.g.* ***9, 10, 11****)*

essentially results from the high probability of the corresponding ethylamine pendant arm to collide and to bind the close metal center, according to a sort of chelate or *scorpionate effect*. Such an effect should be entropic in nature. Indeed, the scorpionate effect, in this particular case, profits from a further moderate, still significant entropy contribution: that associated to the spin expansion of Ni^{II} from singlet to triplet: R ln 3 − R ln 1 = 2.18 cal mol^{-1} K^{-1}.

In the two examples discussed above, the location of the pendant arm with respect to the tetra-aza macrocycle, whether pointing far away from it or into it, was signaled by the color of the solution, bright yellow or pale blue-violet, respectively. However, pendant arm position can be revealed by a more powerful signal, thus detectable even in very dilute solution: fluorescence emission. Such a goal could be achieved with the nickel(II) complex of scorpionand **11**, L∩B, in which an anthracenyl substituent was covalently linked to the amine group of the side-chain [51]. On potentiometric titration on an MeCN/H_2O solution (4:1 v/v) of the complex, the three following complex species were found to be present at the equilibrium over the 2–12 pH range: $[Ni^{II}(L\cap BH)]^{3+}$, $[Ni^{II}(L\cap B)H_2O]^{2+}$, $[Ni^{II}(L\cap B)OH]^{+}$. The concentration profiles of the three species are shown in Figure 8.17 (lines).

Emission spectra due to the anthracene subunit were measured at varying pH and the emission intensity at 400 nm measured over the course of the titration (symbols) is shown in Figure 8.17. Three distinct levels of emission are observed: (i) full, corresponding to the protonated complex $[Ni^{II}(L\cap BH)]^{3+}$; (ii) 60%, associated with the complex $[Ni^{II}(L\cap B)H_2O]^{2+}$, and (iii) completely quenched, pertaining to the deprotonated complex $[Ni^{II}(L\cap B)OH]^{+}$. The hypothesized structural arrangements of the three species are sketched in Figure 8.18. In the protonated complex $[Ni^{II}(L\cap BH)]^{3+}$, square geometry, low-spin, structure **a** in Figure 8.18, the fluorogenic fragment is positioned far away from the metal center and displays the most intense emission. On increasing pH, the antracenylammonium fragment deprotonates and the complex species $[Ni^{II}(L\cap B)H_2O]^{2+}$ forms, in which the secondary amine group of the side-chain goes to bind the metal from the top, while one water molecule completes six coordination according to an octahedral geometry, high-spin state (**b**). In this structural arrangement the anthracene subunit is close enough to the macrocycle to establish an intramolecular energy transfer process (ET), which quenches fluorescence at 60%. Then, on a further pH increase, deprotonation of the axially bound water molecule takes place (**c**), which promotes the occurrence of an electron

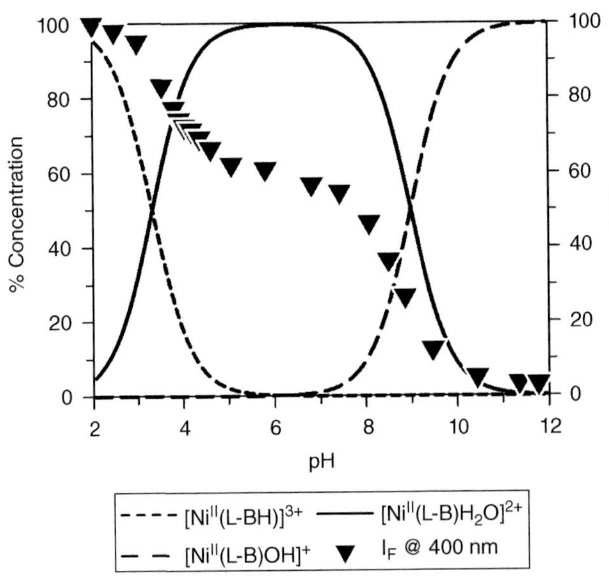

Figure 8.17 *Lines: concentration profiles of the species at the equilibrium in a solution of the complex of nickel(II) with* ***11*** *(= L-B) in MeCN/H_2O (4:1 v/v) over the pH interval 2-12; symbols: fluorescence intensity, I_F, a.u., of the solution at 400 nm*

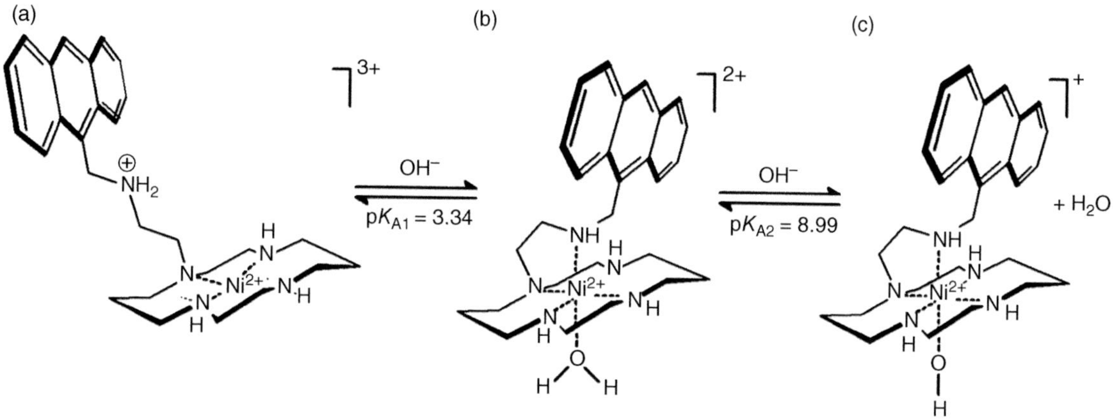

*Figure 8.18 Structural arrangements of the three species present at the equilibrium in an MeCN/H_2O solution (4:1 v/v) of the [Ni^{II}(**11**)](ClO_4)$_2$ complex salt*

transfer process (eT) from the metal to the excited anthracene fragment. Simultaneously operating ET and eT processes quench completely fluorescence. The nickel(II) complex of **11** appears as an unusual high/low/off fluorescence switch, whose control has a mechanical nature (the oscillation of the pendant arm).

The scorpionate effect, with its powerful entropy advantage, can be used in principle to favor metal binding by poorly coordinating fragments. This could be the case of the nitro group. Nitromethane and nitrobenzene are classical non-coordinating solvents, typically used to handle in solution undissociated metal complex salts in which anions either coordinate the metal ion or interact electrostatically with the complex according to an outer-sphere mode. In the perspective of favoring coordination by the nitro group, we considered scorpionand **12**, in which a 2-nitrobenzene subunit had been linked through a –CH_2– group to a nitrogen atom of cyclam [52].

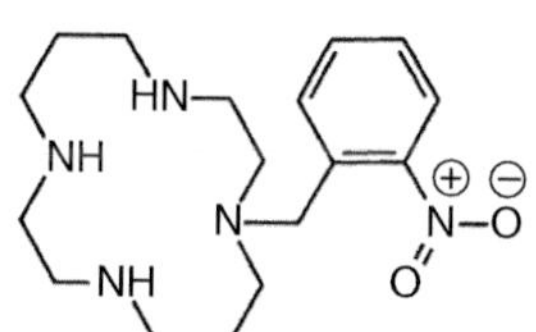

12

On reaction of $Ni^{II}(ClO_4)_2$ with **12**, yellow crystals of the complex salt [Ni^{II}(LH)](ClO_4)$_2$ were obtained (LH = **12**). The corresponding crystal and molecular structure is shown in Figure 8.19. The low-spin nickel(II) complex shows a square geometry with the nitrobenzyl side-chain pointing to the opposite side with respect to the macrocycle. The distance between the Ni^{II} ion and the closest oxygen atom of the nitro group is 4.8 Å, which excludes any interaction. When dissolved in water, the metal complex shows an intense absorption band at 330 nm, due to the nitrobenzene chromophore, with a shoulder at 450 nm, which is ascribed to the d–d transition of the low-spin square-planar Ni^{II} tetramine complex (see solid line in Figure 8.20a). This confirms the non-coordination of the nitrobenzyl pendant arm in solution. On increasing pH, through addition of standard NaOH, no modification of the spectrum was observed until pH 10, when the shoulder at 450 nm began to decrease and a new, less intense, band developed at 550 nm (Figure 8.20a).

Such a spectral pattern is typically observed in the presence of a low-spin to high-spin conversion, which accompanies a change of coordination geometry from square-planar to octahedral. It is suggested that under strongly alkaline conditions the –CH_2– group linking the cyclam ring to the 2-nitrobenzene fragment undergoes deprotonation, which simultaneously promotes the following events: (i) electronic rearrangement to

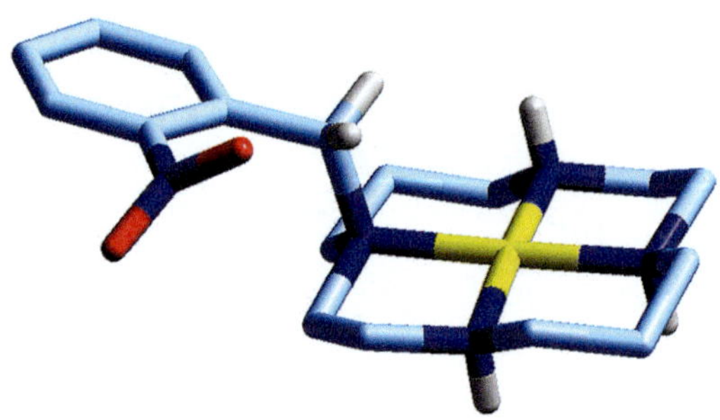

Figure 8.19 The crystal and molecular structure of the complex salt $[Ni^{II}(\mathbf{12})](CF_3SO_3)_2{\cdot}CH_3OH$ [52]. Counteranions, solvational molecules and most of the hydrogen atoms have been omitted for clarity. Only hydrogen atoms of the benzylic $-CH_2-$ group and of the secondary amine nitrogen atoms are shown

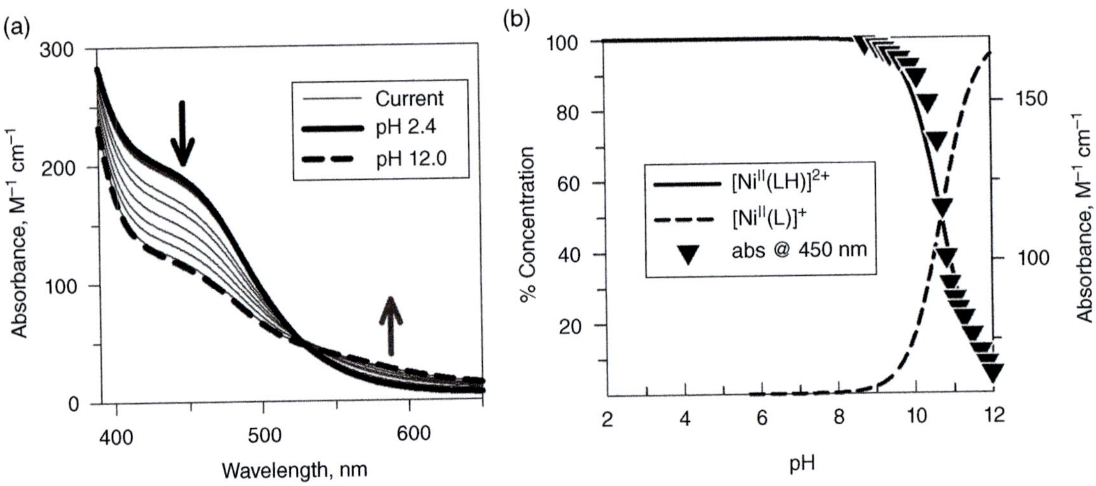

Figure 8.20 (a) visible spectra recorded over the course of the titration of an aqueous solution of $[Ni^{II}(\mathbf{12})](CF_3SO_3)_2$ with standard base; (b) symbols: change of the absorbance at 450 nm with pH; lines: concentration profiles of the complex species present at the equilibrium: $[Ni^{II}(LH)]^{2+} \leftrightarrows [Ni^{II}(L)]^{+} + H^{+}$ (LH = **12**) [52]

give the anion of the nitronic acid; (ii) intramolecular motion of the pendant arm towards the metallo-cyclam subunit; (iii) establishing of a coordinative interaction between one oxygen atom of the $-NO_2^-$ group and the Ni^{II} ion; (iv) metal coordination of a water molecule at the remaining site of the octahedron. The overall process is illustrated in Figure 8.21. The plot of the absorbance at 450 nm vs. pH in Figure 8.20b exhibits a sigmoidal profile, to which a $pK_a = 10.71 \pm 0.01$ for the acid dissociation equilibrium: $[Ni^{II}(LH)]^{2+} \leftrightarrows [Ni^{II}(L)]^{+} + H^{+}$ corresponded. On addition of standard acid, the band at 450 nm was fully restored, demonstrating the reversibility of the process.

The study described above can be considered in the framework of the nitro-*aci*nitro tautomerism of 2-nitrotoluene, illustrated by the thermodynamic cycle in Figure 8.22 [53]. The cycle shows that the tautomeric equilibrium nitro–*aci*nitro is poorly displaced to the right ($K_T = 10^{-17}$) due to the very low acidity of the nitro form ($pK_a = 20.6$) compared to the *aci*nitro form ($pK_a = 3.6$). It is observed that appending a $Ni^{II}(cyclam)^{2+}$ moiety to the methyl(ene) fragment increases the acidity of the 2-nitrotoluene subunit of nearly ten orders of magnitude (20.6 – 10.7 = 9.9), a further demonstration of the power of the scorpionate effect.

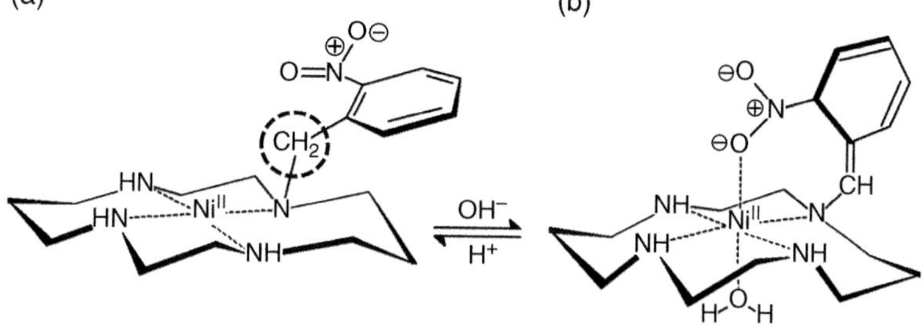

Figure 8.21 *The acidic behavior of the low-spin[$Ni^{II}(LH)]^{2+}$ complex in aqueous solution (**a**, LH = **12**). On deprotonation of the benzylic –CH_2– group (dashed circle in structure (**a**)) the electronic charge is delocalized onto the nitro group, which changes to nitronate. The pendant arm bearing the nitronate fragment moves to bind the metal and the high-spin octahedral [$Ni^{II}(L)H_2O]^+$ complex forms (structure (**b**). The pK_a of the [$Ni^{II}(LH)]^{2+}$ monoprotic acid is 10.71 ± 0.01 [52]*

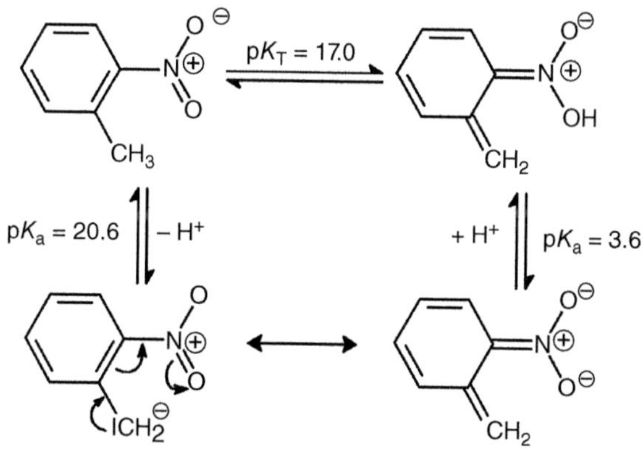

Figure 8.22 *Thermodynamic cycle at the basis of the nitro-acinitro tautomerism of 2-nitrotoluene [53]*

Direct coordination of the neutral nitro group has been recently observed in copper(II) complexes with macrocycle **13**, in which a 2,4-nitrobenzyl group is directly linked to a nitrogen atom of the cyclam subunit [54]. Molecular models suggest that the –NO_2 fragment can coordinate a metal encircled by cyclam.

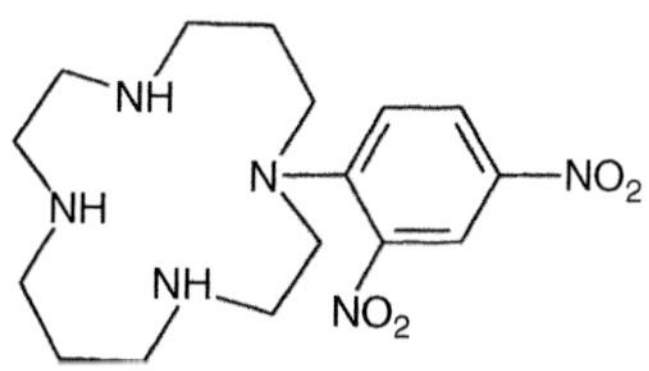

13

On addition of an MeOH solution of copper(II) perchlorate to an MeOH solution of **13** a red precipitate formed, the molecular structure of which is shown in Figure 8.23a.

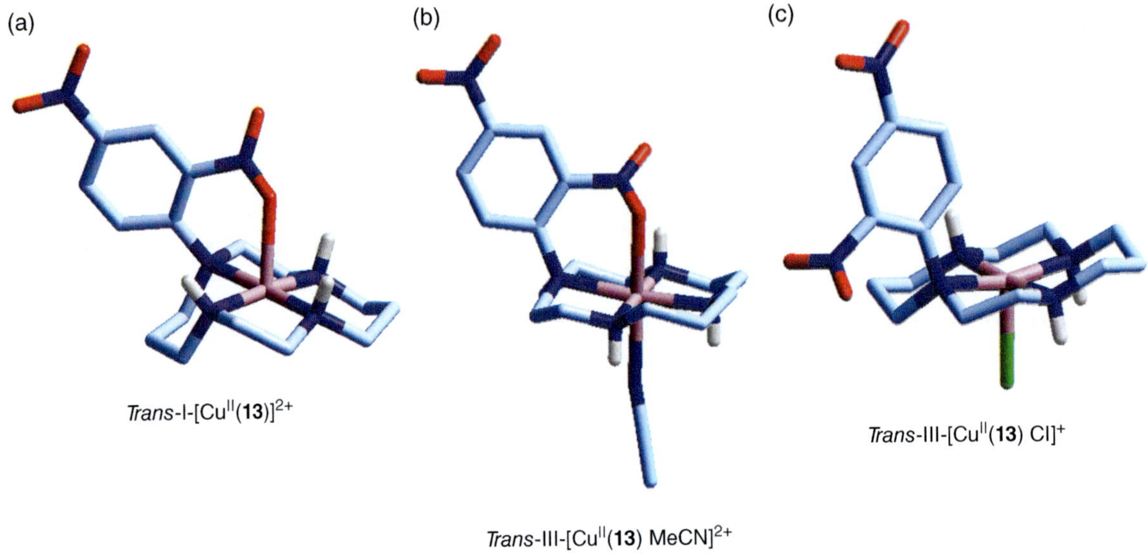

Figure 8.23 *The crystal and molecular structures of copper(II) complex salts with macrocycle* ***13****: (a)* trans-*I-[Cu^{II}(**13**)](ClO$_4$)$_2$, red color; (b)* trans-*III-[Cu^{II}(**13**)(ClO$_4$)$_2$·MeCN], violet color; (c)* trans-*III-[Cu^{II}(**13**)Cl] Cl·H$_2$O.½(C$_2$H$_5$)$_2$O, blue-violet color [54]*

One of the oxygen atoms of the nitro group in *ortho* is bound to the Cu^{II} ion, forming a six-membered chelate ring. The complex is five-coordinate, according to a regular square-pyramidal geometry and adopts the *trans*-I configuration. When dissolved in MeCN, it gives a red solution and the absorption spectrum displayed in Figure 8.24a (solid line). Over a period of weeks, the red solution turns violet, the spectrum of which (dashed line) is shown in Figure 8.24a. On diethyl ether diffusion of this solution, violet crystals were obtained, the structure of which is shown in Figure 8.23b. The scorpionate binding mode is maintained, an MeCN molecule has entered into the coordination sphere to complete an octahedral geometry and the configuration has rearranged to *trans*-III. On the other hand, if diethylether vapor is allowed to diffuse on an MeCN solution of the red complex salt [Cu^{II}(**13**)](ClO$_4$)$_2$, in the presence of [Bu$_4$N]Cl, blue-violet crystals of the salt Cu^{II}(**13**)Cl]Cl·H$_2$O·½Et$_2$O form. The corresponding crystal and molecular structure is shown in Figure 8.23c. The nitro group is no longer bound to the metal, a chloride ion coordinates Cu^{II} from the opposite side, the metallo-cyclam subunit has assumed a *trans*-III configuration. Thus, it appears that: (i) the kinetically less complicated pathway leads to the formation of a scorpionate complex in a *trans*-I configuration, of square-pyramidal geometry. This is not the thermodynamically stable species and the *trans*-I scorpionate species (red) slowly converts in MeCN to the octahedral scorpionate *trans*-III form (violet, one MeCN molecule coordinated). If a coordinating anion is present (e.g. Cl$^-$), the *trans*-III complex prefers to bind the anion according to a square pyramidal geometry, diverting the –NO$_2$ oxygen from the axial site and releasing the bound MeCN molecule to the solution.

Spectrophotometric investigations showed that the red *trans*-I complex converts to the violet *trans*-III complex according to first-order kinetics with $t_{1/2}$=290 h (see the decay profile in Figure 8.24b). However, *trans*-I→*trans*-III conversion can be made faster in the presence of catalytic amounts of anions added as [Bu$_4$N]X salts. In the presence of Cl$^-$ the half-time is 5.5 h, in the presence of NCO$^-$ 0.5 h. It has been suggested that the configurational conversion may take place through a dissociative mechanism (detachment of an amine

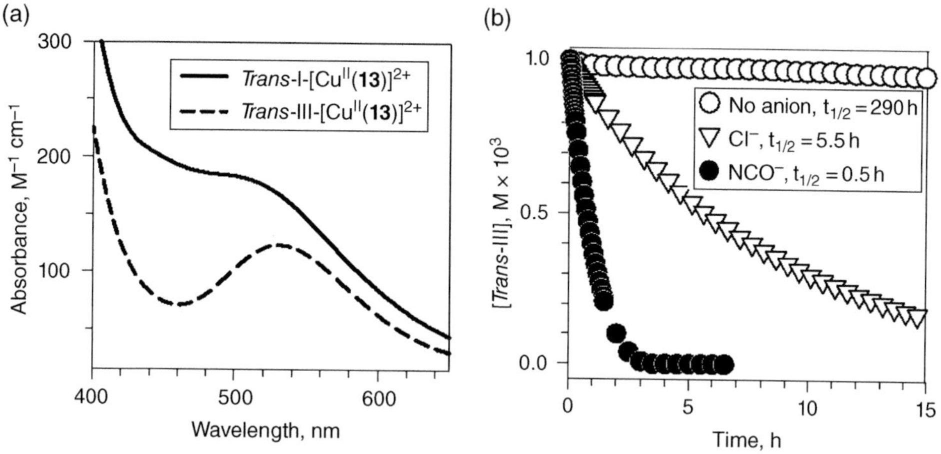

Figure 8.24 *(a) Absorption spectra of: (1) a freshly prepared MeCN solution of* trans-*I-[Cu^{II}(**13**)](ClO$_4$)$_2$ (solid line); (ii) spectrum of the same solution after several weeks (dashed line); (b)* trans-*I-to-*trans-*III conversion in the absence (open circles) and in the presence of catalytic amounts of chloride (triangles) and azide (filled circles), as monitored through the decay of the concentration of the* trans-*I form*

nitrogen atom, as a preliminary step to the configurational reorganization). The coordinating anion is assumed to temporarily occupy the empty coordination site, thus decreasing the energy of the transition state. The higher the coordinating tendencies of the anion ($NCO^- > Cl^-$), the faster the conversion process.

8.5 Azacyclams: cyclam-like macrocycles with built-in functionalization

Functionalization of tetramine macrocycles (e.g. with a coordinating pendant arm, to give scorpionands) is typically carried out through a time-consuming and low-yield multistep procedure, which involves: (i) protection of all but one of the amine nitrogen atoms; (ii) reaction of the unprotected–NH group with a R–X synthon (X, reactive fragment; R, function to be appended); (iii) deprotection of the functionalized macrocycle. However, there exists a one-pot metal template procedure that gives 14-membered cyclam-like macrocycles possessing a desired functionalization R [55]. The reaction scheme is illustrated in Figure 8.25.

A metal ion prone to square coordination (Ni^{II}, Cu^{II}) properly orientates 2·3·2-tet (**a**); then, the two terminal primary amine groups give Schiff base condensation with two molecules of formaldehyde, to give **b**. At this stage, the $-NH_2$ fragment of an amide (either carboxyamide or sulfonamide), in the presence of base, undergoes stepwise deprotonation and the $-NH^-$ fragment first, then the $-N^-$ fragment make consecutive nucleophilic attack at the two C=N bonds, to give **c** and **d** in sequence, eventually performing cyclization. Metal template ring-closure of a tetramine involving nucleophilic attack on the terminal C=N bonds by a diprotic acid had been first obtained by Sargeson [56], using nitroethane (CH_3**CH**$_2NO_2$, the 2D version of the 3D template synthesis of [Co^{III}(sarcophagine)]$^{3+}$ involving the triprotic acid nitromethane, **CH**$_3NO_2$, [57]). Also, aliphatic primary amines (CH_3**NH**$_2$, C_2H_5**NH**$_2$) operate as diprotic acids in the synthesis of cyclam-like macrocycles [58]. Amides are especially effective locking fragments in this type of synthesis [59–62], in view of their relatively high acidity (and the more acidic sulfonamides ensure higher cyclization yields than carboxyamides). The synthetic procedure is very simple: an ethanolic solution containing equimolar amounts of $NiCl_2{\cdot}6H_2O$, of 2·3·2-tet and of the chosen amide plus excess of triethylamine and formaldehyde (40% aqueous solution) is

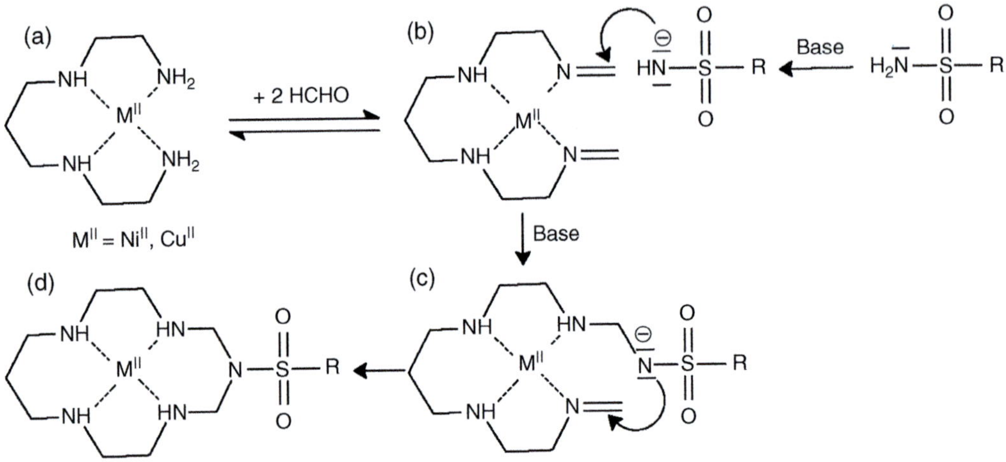

Figure 8.25 *Synthesis of metal azacyclam complexes. Cyclization takes place through stepwise nucleophilic attack to the C=N bonds by the deprotonated diprotic acid, either carboxyamide or sulfonamide (a sulphonamide in the Figure). Each primary amide brings a functionalization R [55]*

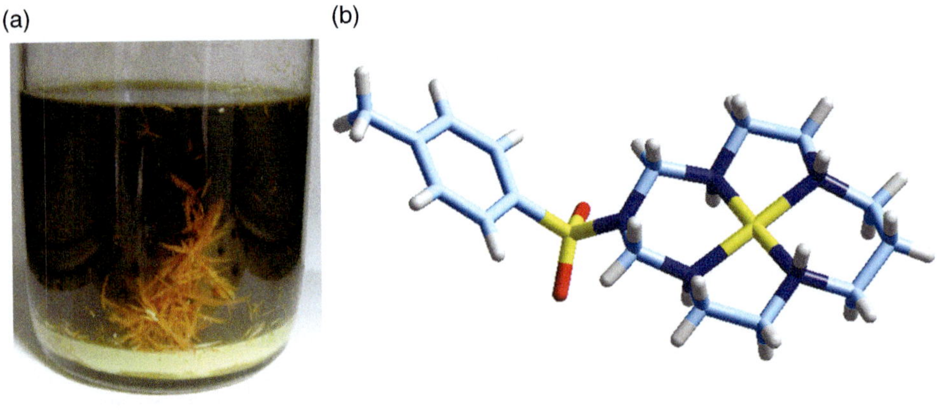

Figure 8.26 *(a) Crystals of the low-spin [3-(4-tolylsulfonyl)-1,3,5,8,12-penta-azacyclotetradecane]nickel(II) perchlorate salt, obtained through the metal template reaction of Figure 8.25, in which the diprotic acid acting as a locking fragment is tosylamide; (b) molecular structure of the corresponding complex cation [62]*

kept in a stoppered bottle which was heated to 50 °C in a thermostated bath for one night. Typically an orange-yellow precipitate forms. If not, precipitation can be induced through addition of an aqueous solution of 6 M $NaClO_4$ or conc. $HClO_4$. Figure 8.26a shows the crystals formed from the reaction involving tosylamide as a locking fragment. Figure 8.26b displays the crystal and molecular structure of the corresponding complex salt $[Ni^{II}(L)](ClO_4)_2$. (L = [3-(4-tolylsulfonyl)-1,3,5,8,12-pentaazacyclotetradecane]). The low-spin complex exhibits a square coordination geometry, in a *trans*-III configurational arrangement [62]. Macrocycles of this type are called *azacyclams*.

Azacyclams are penta-aza macrocycles, but, as shown by the complex in Figure 8.26b, behave as quadridentate ligands: the amide nitrogen atom is never involved in metal coordination and plays only a structural role. Azacyclam complexes are structurally similar to the corresponding metal cyclam complexes, with some small, though significant, differences. In Figure 8.27 some distances (in Å) of structurally investigated low-spin nickel(II) azacyclam complexes, containing carboxyamide[61] and sulfonamide nitrogen atoms [62], are compared to those of a nickel(II) complex of plain cyclam [29]. In all three complexes the $Ni^{II}-N$ distances range from 1.93 to 1.96 Å, values expected for low-spin nickel(II) polyamine complexes. However, looking at the squares whose vertices are the four amine nitrogen atoms, it appears that the most regular one is that of cyclam (the opposite edges show the same distance), whereas the squares of CO-azacyclam and SO_2-azacyclam show a significant asymmetry (opposite edges show a different length). This may result from (i) the tendency of the amide nitrogen atoms to planarization; (ii) to π–π interaction between a filled dπ orbital of Ni^{II} (low-spin d^8), donor, and the LUMO of the carboxyamide/sulfonamide group (acceptor). The existence of the latter interaction is suggested by the distance between the metal center and the amide nitrogen atom (CO-azacyclam: 3.25 Å; SO_2-azacyclam: 3.22 Å), distinctly smaller than the distance observed in the cyclam complex between Ni^{II} and the carbon atom in the middle of each trimethylenic chain ((3.33 Å). The more extended π-delocalization in the sulfonamide subunit may explain the smaller distance with respect to the carboxyamide counterpart.

In any case, azacyclam complexes display the same properties as cyclam complexes, including extreme resistance to demetalation/decomposition in strongly acidic media. Nickel(II) azacyclam complexes undergo one-electron oxidation (to Ni^{III}) and one-electron reduction (to Ni^{I}) in MeCN solution. Some relevant parameters referring to azacyclam complexes are compared to those for cyclam in Table 8.1. The energy of the d–d band in square low-spin nickel(II) tetramine complexes is a direct measure of the intensity of the metal–ligand interaction and for low-spin $[Ni^{II}(cyclam)]^{2+}$ it is considerably higher than for $[Ni^{II}(CO\text{-azacyclam})]^{2+}$ and for $Ni^{II}(SO_2\text{-azacyclam})]^{2+}$. This should be ascribed to the capability of cyclam to position its donor atoms at the vertices of a square and to establish strong in-plane interactions. These interactions may be also responsible for the different stability of the Ni^{III} state, which is expressed by the values of $E_{1/2}$ (Ni^{III}/Ni^{II}). Values decrease along the sequence: cyclam < CO-azacyclam < SO_2-azacyclam.

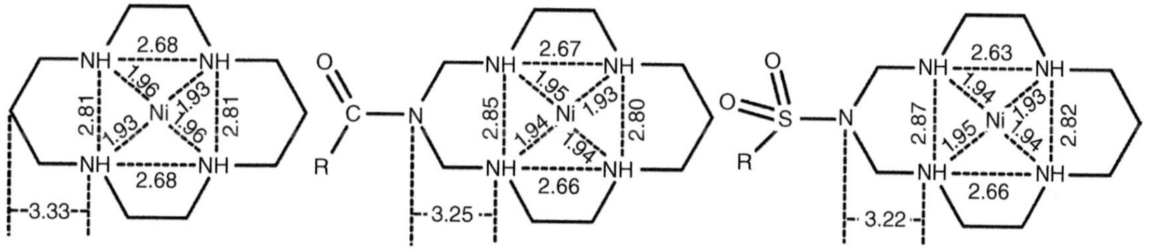

Figure 8.27 *Some distances (Å) observed in the low-spin nickel(II) complexes of cyclam [29], of an azacyclam containing a carboxyamide nitrogen atom (CO-azacyclam = 3-(4-pyridiniumylcarbonyl)-1,3,5,8,12-penta-azacyclotetradecane) [61], and of an azacyclam containing a sulfonamide nitrogen atom (SO_2-azacyclam = [3-(4-tolylsulfonyl)-1,3,5,8,12-penta-azacyclotetradecane]) [62], structure shown in Figure 8.26*

Table 8.1 *Spectral and electrochemical parameters for nickel(II) complexes of cyclam and selected azacyclam macrocycles in MeCN solution [62]*

	$[Ni^{II}(cyclam)]^{2+}$	$[Ni^{II}(CO\text{-azacyclam})]^{2+}$	$Ni^{II}(SO_2\text{-azacyclam})]^{2+}$
λ_{max} (nm)	464	470	482
$E_{1/2}$ (Ni^{III}/Ni^{II}) V vs Fc^+/Fc	0.59	0.72	0.78
$E_{1/2}$ (Ni^{II}/Ni^{I}) V vs Fc^+/Fc	−1.82	−1.74	−1.76

However, unlike $[Ni^{II}(cyclam)]^{2+}$ which can be demetaled (e.g. in boiling aqueous NaCN and NaOH) to give cyclam, demetalation of $[Ni^{II}(azacyclam)]^{2+}$ complexes causes decomposition, due to the intrinsic instability of the azacyclam molecule, which contains aminal groups. This restricts the coordination chemistry of azacyclams to the complexes of the metals which can operate as templating agents, i.e. Ni^{II} and Cu^{II}. In any case, the valuable opportunity remains of synthesizing metal complexes of 14-membered macrocycles with an appended desired functionality. In fact, chosen a given molecular subunit, it is always possible to prepare its carboxyamide or sulfonamide derivative.

An example is provided by the one-pot synthesis of redox switches of fluorescence (RSFs). The design of a RSF involves the covalent linking of a fluorescent fragment (Fl) and of a control unit, as illustrated in Figure 8.28 [63]. The control unit consists of a redox active moiety (the real "switch"), whose oxidation states, the reduced form Red and the oxidized form Ox, have a comparable stability. The principle is that one of the two oxidation states, either Ox or Red, quenches the excited fluorogenic subunit Fl* and the other does not. Then, it should happen that sequential oxidation and reduction processes cause consecutive quenching and revival of the fluorescent emission. Communication between Red/Ox and Fl* may take place through an electron Transfer mechanism (eT, linker permeable to electrons) or Energy Transfer mechanism (ET, Dexter type, insulating linker, but flexible enough to allow a close contact between the redox center and the excited fluorogenic subunit).

Complex **14**, shown in Figure 8.29, is a possible candidate for the role of RSF [64]: it contains a redox active subunit, the Ni–azacyclam complex, whose two oxidation states are separated by a fast and reversible one-electron process ($E_{1/2}$=0.18 V vs Fc^+/Fc in MeCN, 0.1 M in $[Bu_3Bn]Cl$ at 25 °C), and a fluorogenic

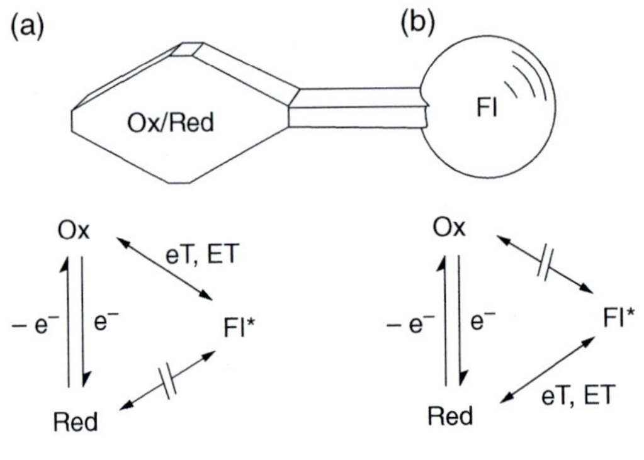

Figure 8.28 *The design of a redox switch of fluorescence (RSF). Switching requires that: (a) the oxidized form Ox quenches the excited fluorophore Fl* through either an electron transfer (eT) or an energy transfer (ET), to give an OFF/ON switch; (b) Red quenches the excited fluorophore Fl*, to give an ON/OFF switch [63]*

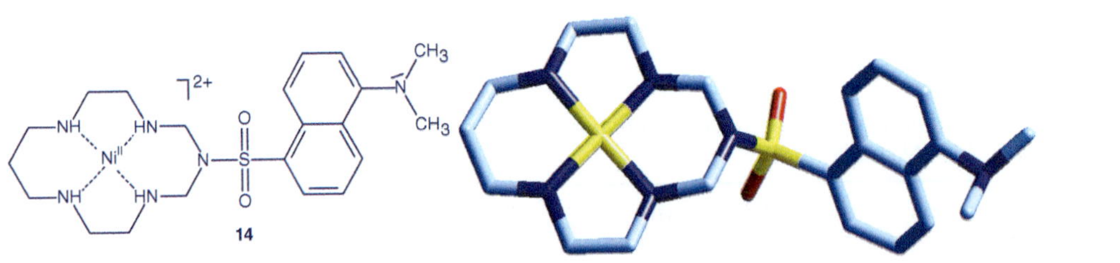

Figure 8.29 *The crystal and molecular structure of the complex salt low-spin $[Ni^{II}(LH)](ClO_4)_3$ (L=**14**). The dimethylaniline group is protonated [64]. Hydrogen atoms and counterions have been omitted for clarity*

subunit, the dansylamide fragment. Complex **14** was obtained through the metal template procedure described above, by using dansylamide as a molecular padlock. Figure 8.29 shows the crystal and molecular structure of the low-spin Ni^{II} complex, which was crystallized from an acidified solution and in which the dimethylaniline group is protonated. The distance between the metal center and the closest aromatic carbon atom is 5.9 Å. The sulfonamide spacer is too rigid to allow molecular folding and to put in close contact the redox center and the fluorophore. However, low-energy empty π* orbitals extended on the sulfonamido bridge may allow electron communication between the metal and the dansyl moiety, thus making possible an eT process.

The complex **14** emits an orange fluorescence in solution. Figure 8.30 shows (solid line) the emission spectrum of an MeCN solution of **14** at 25 °C: it consists of a low-energy, non-structured emission band (Φ = 0.40), which originates from a charge-transfer excited state and is nearly coincident with the band observed for the plain dansylamide molecule. On controlled potential electrolysis at a platinum gauze electrode, set at 0.23 V vs Fc^+/Fc, the yellow-orange solution turns green ($[Ni^{III}(L)Cl_2]^+$ complex) and fluorescence is lost. The corresponding spectrum (dashed line) is shown in Figure 8.30. On setting the potential of the platinum gauze electrode at – 0.07 V vs Fc^+/Fc, the solution turns yellow-orange again and the fluorescence emission is fully restored (see the emission spectrum (dotted line) in Figure 8.30). The nature of the quenching process, whether eT or ET, was ascertained through the measure of the emission spectrum of an EtOH/water solution (95:5 v/v) of the Ni^{III} complex, obtained through oxidation with $S_2O_8^{2-}$, frozen at 77 K. On freezing, fluorescence is fully restored, which demonstrates the eT nature of the quenching process at room temperature. In fact, vitrification at 77 K immobilizes solvent molecules, thus preventing their reorganization as a preliminary event to the occurrence of charge redistribution; as a consequence, the eT process cannot take place (whereas the ET process, which does not require any charge separation or redistribution, could).

The thermodynamic bases of the ON-OFF behavior of the RSF **14** are illustrated in Figure 8.31. Occurrence of the eT process from the excited dansyl subunit, *Dns, to Ni^{III}, responsible for quenching, is justified by the largely negative value of ΔG°_{eT} (−1.9 V). An examination of the different energy contributions in the thermodynamic cycle in the left of Figure 8.31 indicates that the driving force of the process is the feeble resistance to oxidation of the dansylamide subunit. On the other hand (thermodynamic cycle on

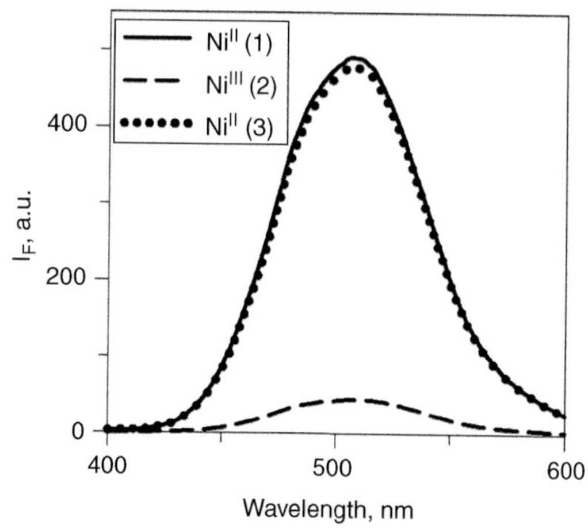

Figure 8.30 *Emission spectra at 25 °C in MeCN, 0.1 M in $[Bu_3Bn]Cl$ of: (1) original solution of **14** (Ni^{II}, solid line); (2) after electrolysis at 0.23 V vs Fc^+/Fc (Ni^{III}, dashed line); (3) after electrolysis at – 0.07 V vs Fc^+/Fc (Ni^{II} again, dotted line) [64]*

the right), the non-occurrence of the Ni^{II}-to-*Dns eT process (and persistence of the fluorescent emission) has to be ascribed to the extremely high resistance to the reduction of the Dns fragment, which makes $\Delta G°_{eT}$ substantially positive (>0.7 V). Consecutive processes of oxidation and reduction, with concurring quenching and revival of fluorescent emission can be carried out also chemically in an ethanol/water solution (95:5 v/v) by successive additions of $Na_2S_2O_8$ and $NaNO_2$ for several cycles, with a moderate fatigue.

The same switch can operate different light bulbs: the template reaction involving equimolar amounts of Ni^{II}, 2·3·2-tet and 2-naphthalensulfonamide, plus an excess of formaldehyde and triethylamine, gives the RSF **15**, whose "light bulb" emits a bluish fluorescence. An MeCN solution of the low-spin complex **15** displays the emission spectrum of the locally excited naphthalene fragment, with its typical vibrational structure (λ_{max} = 360 nm, see solid line in Figure 8.32). On setting the working electrode at 0.33 V vs Fc^+/Fc, Ni^{II} is oxidized to Ni^{III} and fluorescence is completely quenched (dashed line). When the potential of the platinum gauze electrode is adjusted to 0.03 V, Ni^{III} is reduced to Ni^{II} and naphthalene fluorescence revives (dotted line).

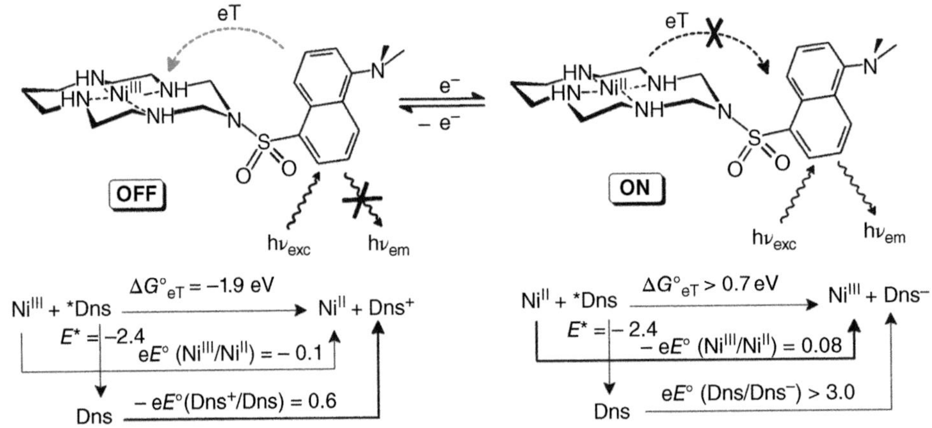

Figure 8.31 *Thermodynamic cycles accounting for the occurrence of the eT process from *Dns to Ni^{III} (Dns = dansylamide fluorophore) and the non-occurrence of the eT process from *Dns to Ni^{II} [64]. E* is the photonic energy, E° the electrode potential in MeCN solution. All quantities are expressed in eV*

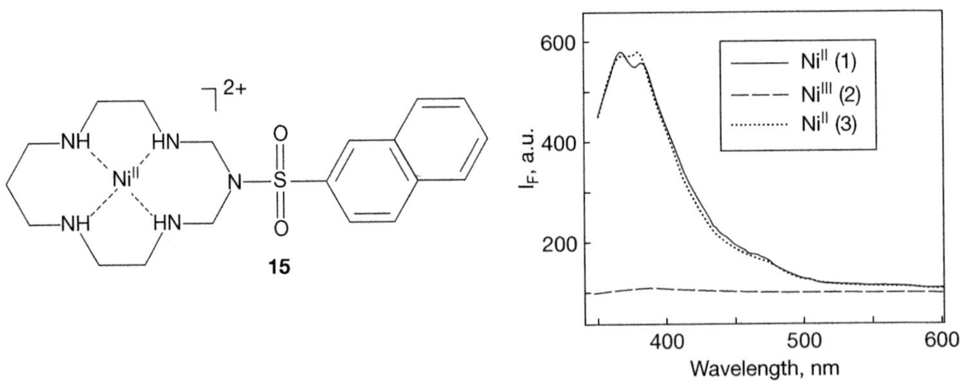

Figure 8.32 *Emission spectra at 25 °C in MeCN, 0.1 M in $[Bu_3Bn]Cl$ of: (1) original solution of **15** (Ni^{II}, solid line); (2) after electrolysis at 0.33 V vs Fc^+/Fc (Ni^{III}, dashed line); (3) after electrolysis at 0.03 V vs Fc^+/Fc (Ni^{II} again, dotted line) [64]*

In principle, any desired fluorescent amide can be "screwed in" to the $[Ni^{II}(azacyclam)]^{2+}$ device. Changing of the metal centered switch is possible, but it requires more sophisticated synthetic procedures [65].

8.6 Conclusion

In the present chapter only some selected aspects of the rich and versatile coordination chemistry of cyclam have been considered, aspects which were of particular interest for the author or to which the author and his students made a contribution. Other important topics of the coordination chemistry of cyclam have not been considered, including: (i) the use of $[Ni^{II}(cyclam)]^{2+}$ as an electron-transfer catalyst (through the Ni^{II}/Ni^{I} couple) in the electrochemical and photochemical reduction of CO_2 [66, 67] (ii) ethylene cross-bridged cyclam derivatives, e.g. **16**, which impart further inertness to their metal complexes, thus favoring the formation of otherwise unstable tetramine complexes (e.g. those of Mn^{II}) and promoting an unusually rich redox activity ($Mn^{II} \rightarrow Mn^{III}$, $Mn^{III} \rightarrow Mn^{IV}$);[68] (iii) *N,N′,N′′,N′′′*-tetra-alkylcyclam derivatives (e.g. Me_4cyclam, **17**), which give complexes of *trans*-I configuration, in which the metal ion is significantly displaced above the N_4 plane and presents an unusual reactivity [12].

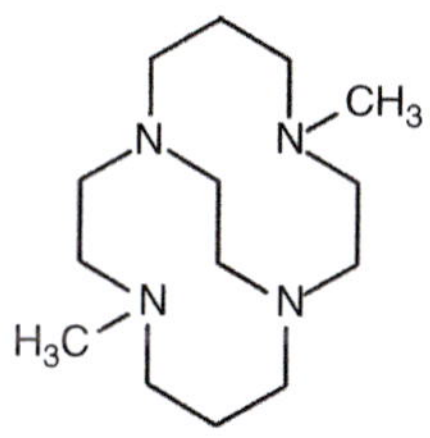

16

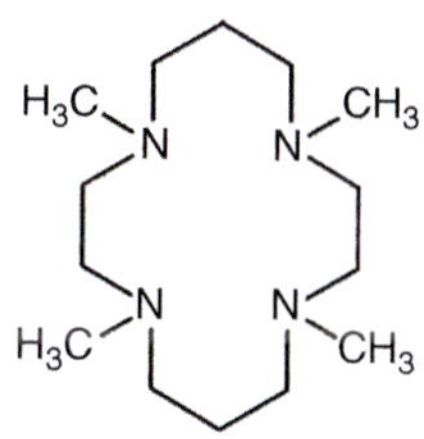

17

In particular, $[Mn^{II}(Me_4cyclam)]^{2+}$ [69] and $[Fe^{II}(Me_4cyclam)]^{2+}$ [70] react with H_2O_2, to give the side-on peroxo complexes $[Mn^{III}(Me_4cyclam)O_2]^+$ and $[Fe^{III}(Me_4cyclam)O_2]^+$, whose crystal and molecular structures are reported in Figure 8.33. In the two *trans*-I complexes, Mn^{III} and Fe^{III} ions are displaced above the mean N4 plane by 0.71 and 0.64 Å respectively, a circumstance which may explain their marked affinity for anions, including peroxide. On the other hand, methyl substituents pointing on the side of O_2^{2-} may create a protective hydrophobic cavity for the coordinated peroxide anion. *Trans*-I-$[Fe^{III}(Me_4cyclam)\eta^2\text{-}O_2]^+$ is the most relevant iron(III) peroxo complex with a non-heme ligand.

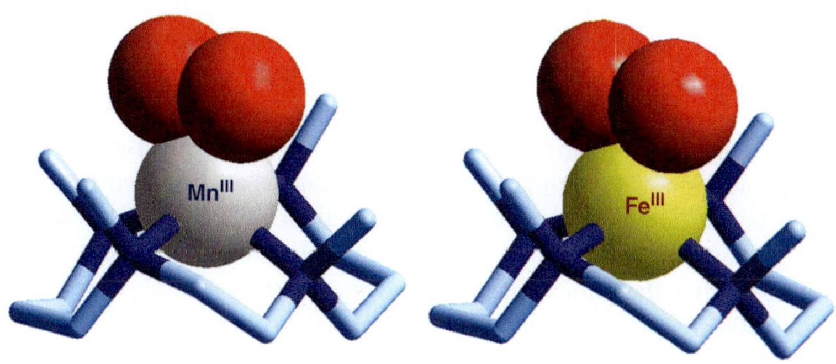

Figure 8.33 *Crystal and molecular structure of the complexes* trans-*I*-$[M^{III}(Me_4cyclam)\eta^2\text{-}O_2]^+$ *(M = Mn [69], Fe [70]). Hydrogen atoms of the ligand and counteranions omitted for clarity*

Chemistry is probably the scientific discipline which contributes most to the improvement of life quality of humans, in particular in its medical and pharmaceutical aspects. So the question is: have cyclam and its derivatives played any role in medicine and pharmacology? Indeed, derivatized ethylene cross-bridged cyclams work well as ligands and carriers of the radionuclide ^{64}Cu ($t_{1/2}$ = 12.7 h), in its +2 oxidation state, the ideal isotope for Positron Emission Tomography (PET) imaging and radiotherapy [71]. However, the by far most significant impact on medicine has been made by the bicyclam molecule **18**, in which two cyclam subunits have been linked by a 1,4-xylyl spacer at the nitrogen atoms.

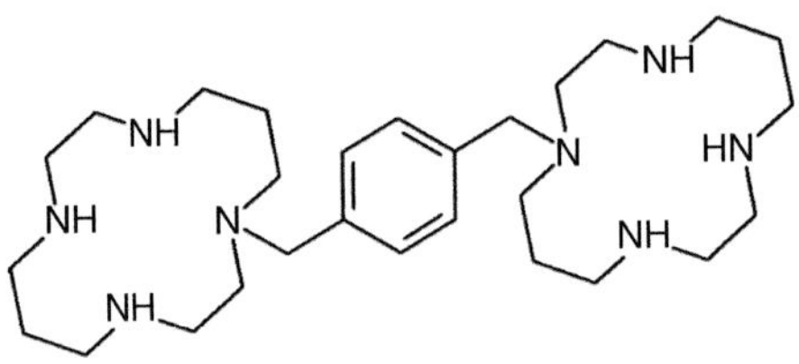

18

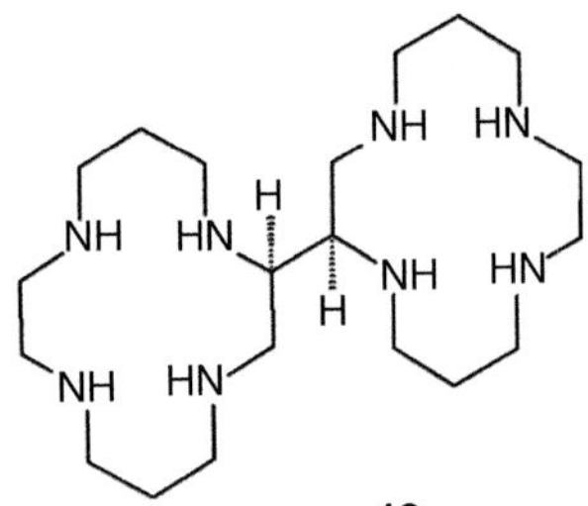

19

This molecule was first synthesized in our laboratory in 1987 along with a series of bicyclam analogs of different spacers ($-CH_2CH_2-$, $-CH_2CH_2CH_2-$, $-CH_2CH_2CH_2CH_2-$, 1,3-xylyl), whose dinickel(II) complexes were prepared and the two-step Ni^{II}-to-Ni^{III} oxidation processes voltammetrically investigated in order to evaluate mutual electrostatic effects [72]. The interest of this study was typically restricted in the area of the coordination chemistry of synthetic macrocycles. A few years later, De Clercq tested for Johnson-Matthey the antiviral activity of $[Pt^{II}(cyclam)]^{2+}$ and he realized that the ligand itself, not its platinum(II) complex, exhibited inhibitory effects on the replication of human immunodeficiency virus (HIV) [73]. However, such an effect was present only with cyclam obtained through the nickel(II) template synthesis illustrated in Scheme 8.4; it was totally absent for cyclam obtained with purely organic nonmetal template procedures. It was then recognized that the antiviral activity was exerted by bicyclam **19** (named JM1657), which forms as a by-product in the Ni^{II} template synthesis of cyclam [74], and was present in traces in the commercial product. As the synthesis of **19** appeared hardly feasible and chancy, a variety of *N,N*-linked bicyclams were tested and the highest activity was found for **18**, isolated as octahydrobromide (JM2987) and octahydrochloride (JM 3100), which displayed the highest inhibitory effect. When the ongoing research was transferred to AnorMED, JM3100 took the name of AMD3100. In particular, it was ascertained that the inhibitory mechanism is related to the capability of the drug xylylbicyclam **18** to interact with coreceptor CXCR4, a protein membrane which favors the entry of HIV into the cell. Xylylbicyclam/CXCR4 interaction prevents the entry. The bicyclam drug reached phase II clinical trials, but its development for treatment of HIV was terminated because of its lack of oral availability and cardiac disturbances. Among its many important functions, the membrane protein CXCR4 is also responsible for anchoring stem cells in the bone marrow. Thus, it happens that xylylbicyclam **18** interacting with CRCX4 induces the mobilization of hematopoietic stem cells from the bone marrow and their release into the circulating blood [75]. In particular, a patient affected by hematological malignancies such as non-Hodgkin's lymphoma or multiple myeloma first takes the appropriate dose of drug, then is connected to an apheresis machine, which collects stem cells and returns to the body the remaining blood components. Stem cells are processed, frozen and stored. After the patient has received the prescribed high-dose chemotherapy and/or radiation-therapy, stem cells are infused back into his bloodstream. In conclusion, xylylbicyclam **18**, marketed as Mozobil™ by Genzyme, has outdated and eliminated risky and invasive surgical intervention. From a strictly chemical point of view, it has to be noted that at a physiological pH Mozobil™ exists as a

Figure 8.34 *Sandro Botticelli,* Venus and Mars, ca. *1483. Tempera on panel, 69 cm × 173 cm. National Gallery, London*

partly protonated form and is believed to act as a hydrogen-bonding donor and to bridge two aspartate residues of the CXCR4 co-receptor, facing each other in the main pocket of the protein. Recent equilibrium studies have shown that the pentaprotonated form of Mozobil gives 1:1 complexes with linear dicarboxylates, whose stability strongly depends upon the length of the spacer linking the two $-COO^-$ groups [76]. This study may help the understanding of the interaction mode of Mozobil with CXCR4.

At the end of the story, one could ask what is the ultimate reason for the fortune of cyclam in coordination chemistry and related disciplines. Perhaps, a stimulating suggestion can come from a painting of the Florentine Renaissance of the second half of 15th century: *Mars and Venus* by Sandro Botticelli (Figure 8.34). The picture, complex and intriguing, was probably commissioned in the occasion of the marriage of a member of the Vespuccis, Botticelli's patron family, (as suggested by the wasps buzzing around Mars, in Italian *vespa* = *wasp*). The painting may appear provocative, if intended to celebrate a wedding, as the male partner, Mars, is lost in sleep, while Venus, the gentle and beautiful bride, is waiting for the promised love, perplexed. However, jokes and provocations were part of the Florentine spirit of that time. In any case, the viewer realizes that the painting carries an enormous potential energy, and he may have the feeling that something may happen at any moment. But nothing takes place. Why? Nothing takes place because Mars is totally inert, a prisoner of sleep, and, because he is inert, young satyrs are able to make fun of the sleeping god: one wears Mars' helmet, another steals his lance, a third blows through a shell into the god's ear in an unsuccessful attempt to wake him. In the same way, moving to chemistry, the metal ion is sleeping, prisoner of the tight cyclam framework, and, because it is inert, anyone, either a student or a professor, like a playful satyr, can make fun of it, taking away or putting on electrons, spilling strong acids or bases on it, with no reaction. There is no problem in admitting that the chemical allegory is far-fetched. Nevertheless, Botticelli's painting is fine and captivating.

8.7 Acknowledgements

I am indebted to the many students and coworkers who accompanied me in the neverending exploration of the world of macrocyclic polyamine complexes, since mid 1970' to date. Such an interest had the beneficial effect of making me join the community of macrocyclic and supramolecular chemistry and meet so many people, who have enriched my life from both the scientific and human side.

References

1. van Alphen, J. (1936) On aliphatic polyamines III. *Recueil des Travaux Chimiques des Pays-Bas*, **55** (10), 835–840.
2. van Alphen, J. (1937) On aliphatic polyamines IV. *Recueil des Travaux Chimiques des Pays-Bas*, **56** (4), 343–349.
3. Stetter, H., Mayer. K.-H. (1961) Zur Kenntis der makrocyclischen Ringsysteme, VII. Herstellung und Eigenschaften makrociclischer Tetramine, *Chemische Berichte*, **94** (6), 1410–1416.
4. Airey, S., Drljaca, A., Hardie, M.J., Raston, C.L. (1999) Cyclam as a supramolecular synthon: infinite stacked arrays to encapsulation in superanions. *Chemical Communications*, 1137–1138.
5. Werner, A. (1893) Beitrag zur Konstitution anorganischer Verbindungen. *Zeitschrift für anorganische und allgemeine Chemie*, **3** (1), 267–330.
6. Curtis, N.F. (1960) Transition-metal complexes with aliphatic Schiff bases. Part I.Nickel(I1) complexes with *N*-isopropylidene-ethylenediamine Schiff bases. *Journal of the Chemical Society,* 4409–4413.
7. Blight, M.M., Curtis, N.F. (1962) Transition-metal complexes with aliphatic Schiff bases. Part III. Compounds formed by reaction of some 1,2-diamine complexes of Copper(II) with some ketones. *Journal of the Chemical Society,* 3016–3020.
8. Thompson, M.C., Busch, D.H. (1964) Reactions of coordinated ligands. I8. Utilization of the template hypothesis to synthesize macrocyclic ligands in situ. *Journal of the American Chemical Society*, **86** (18), 3651–3656.
9. Curtis, N.F. (1964). Some cyclic tetra-amines and their metal-ion complexes. Part I. Two isomeric hexamethyltetra-azacyclotetradecanes and their Copper(II) and Nickel(II) complexes *Journal of the Chemical Society,* 2644–2650.
10. Bosnich, B., Poon, C.K., Tobe, M.L. (1965) complexes of Cobalt(III) with a cyclic tetradentate secondary amine. *Inorganic Chemistry*, **4** (8), 1102–1108.
11. Bosnich, B., Mason, R., Pauling, P.J., Robertson, G.B., Tobe, M. L. (1965) The molecular structure of dichloro-1,4,8,11-tetra-azacyclotetradecanenickel(II). *Chemical Communications (London)*, 97–98.
12. Barefield, E.K. (2011) Coordination chemistry of *N*-tetraalkylated cyclam ligands – A status report. *Coordination Chemistry Reviews*, **254** (15–16), 1607–1627.
13. Barefield, E.K. (1972) New synthesis of 1,4,8,11-tetraazacyclotetradecane (Cyclam) via the nickel(II) complex. *Inorganic Chemistry*, **11** (9), 2273–2274.
14. Barefield, E.K., Wagner, F., Herlinger, A.W., Dahl, A.R. (1976) [Preparation of] (1,4,8,11- tetraazacyclotetradecane) nickel(II) perchlorate and 1,4,8,11-tetraazacyclotetradecane. *Inorganic Syntheses*, **16**, 220–225.
15. Cabbiness, D.K., Margerum, D.W. (1969) Macrocyclic effect on the stability of copper(II) tetramine complexes. *Journal of the American Chemical Society*, **91** (23), 6540–6541.
16. Schwarzenbach, G. (1952) Der Chelateffect. *Helvetica Chimica Acta*, **35** (7), 2344–2359.
17. Hinz, F.P., Margerum, D.W. (1974) Effect of ligand solvation on the stability of metal complexes in solution. Explanation of the macrocyclic effect. *Journal of the American Chemical Society*, **96** (15), 4993–4994.
18. Hinz, F.P., Margerum, D.W. (1974) Ligand solvation and the macrocyclic effect. Nickel(II)-tetramine complexes. *Inorganic Chemistry*, **13** (12), 2941–2949.
19. Paoletti, P., Fabbrizzi, L., Barbucci (1973) A linear relationship between the enthalpy of formation and the frequency of the maximum in the electronic absorption spectra of the Copper(I1)-tetramine complexes. *Inorganic Chemistry*, **12** (8), 1961–1962.
20. Anichini, A., Fabbrizzi, L. Paoletti, P., Clay, R. (1977) Calorimetric determination of macrocyclic enthalpy. Copper(II) and Zinc(II) complexes with 1,4,8,11-tetra-azacyclotetradecane. *Journal of the Chemical Society, Chemical Communications*, 244–245.
21. Anichini, A., Fabbrizzi, L. Paoletti, P., Clay, R. (1978) A microcalorimetric study of the macrocyclic effect. Enthalpies of formation of Copper(II) and Zinc(II) complexes with some tetra-aza macrocyclic ligands in aqueous solution. *Journal of the Chemical Society, Dalton*, 577–583.
22. Fabbrizzi, L., Micheloni, M., Paoletti, P. (1978) Enthalpy of formation of Copper(II) complexes with 14-membered tetra-aza macrocycles. *Journal of the Chemical Society, Chemical Communications*, 833–834.
23. Tasker, P.A., Sklar, L. (1975) Crystal and molecular structure of di(perchlorato)(1,4,8,11-tetraazacyclotetradecane) copper(II). Cu(cyclam)$(ClO_4)_2$. *Journal of Crystal and Molecular Structure*, **5** (5), 329–344.
24. Panneerselvam, K., Lu, T.-H., Tung, S.-F. *et al.* (1998) Crystal structure of (diperchlorato-*O,O′*)(1,4,7,11-tetraazac yclotetradecane-*N,N′,N′′,N′′′*)copper(II). *Analytical Sciences*, **14** (5), 1029–1030.

25. Fabbrizzi, L., Micheloni, M., Paoletti, P. (1980) A microcalorimetric determination of the enthalpies of formation in solution of Nickel(II) complexes with tetraaza macrocyclic ligands of varying size. *Inorganic Chemistry*, **19** (2), 535–538.
26. Fabbrizzi, L., Micheloni, M., Paoletti, P. (1980) Encircling of metal ions by tetra-aza macrocycles. Enthalpies of formation of high- and low-spin Nickel(II) complexes of 14-membered ligands. *Journal of the Chemical Society, Dalton*, 134–136.
27. Fabbrizzi, L. (1979) Selectivity of tetra-ata macrocycles towards metal spin states. The high-spin–low-spin equilibrium for Nickel(II) complexes with 13- and 14-membered ligands. *Journal of the Chemical Society, Dalton Transactions,* 1857–1861.
28. Sabatini, L. Fabbrizzi, L. (1979) Fitting of Nickel(II) ion into the two 14-membered tetraaza macrocycles. Blue-to-yellow conversion and the oxidation and reduction behavior. *Inorganic Chemistry*, **18** (2), 438–444.
29. Prasad, L., Nyburg, S.C., McAuley, A. (1987) The structure of Ni(cyclam)$(ClO_4)_2$. *Acta Crystallographica*, **C43**, 1038–1042.
30. Boeyens, J.C.A. (1983) Analysis of disorder by force-field calculation: the structure of 1,4,7,11-tetraazacyclotetradecane (isocyclam)-nickel perchlorate, $C_{10}H_{24}N_4$-Ni$(ClO_4)_2$. *Acta Crystallographica*, **C39**, 846–849.
31. Martin, L.Y., DeHayes, L.J., Zompa, L.J., Busch, D.H. (1974) The relationship between metal–donor distance and ring size in macrocyclic complexes. *Journal of the American Chemical Society*, **96** (12), 4046–4048.
32. Cabbiness, D.K., Margerum, D.W. (1970) Effect of macrocyclic structures on the rate of formation and dissociation of Copper(II) complexes. *Journal of the American Chemical Society*, **92** (7), 2151–2153.
33. Busch, D.H. (1978) Distinctive coordination chemistry and biological significance of complexes with macrocyclic ligands. *Accounts of Chemical Research*, **11** (10), 392–400.
34. Billo, E.J. (1984) Equilibria and kinetics of complexation of bidentate ligands with the macrocyclic complex Ni([14]aneN_4)$^{2+}$, *Inorganic Chemistry*, **23** (15), 2223–2227.
35. Curtis, N.F., Cook, D.F. (1967) Nickel(III) complexes with cyclic tetrarnines and nitrato- or sulphato-ligands. *Chemical Communications*, 962–963.
36. Olson, D.C., Vasilevskis, J. (1968) Cyclic amine complexes of nickel(I), -(II), and -(III). Electrochemistry, preparation, and properties. *Inorganic Chemistry*, **8** (8), 1611–1621.
37. LoVecchio, F.V., Gore, E.S., Busch, D.H. (1974) Oxidation and reduction behavior of macrocyclic complexes of nickel. Electrochemical and electron spin resonance studies. *Journal of the American Chemical Society*, **96** (10), 3109–3118.
38. Olson, D.C., Vasilevskis, J. (1971) Cyclic amine complexes of copper(I), -(II), and -(III). Electrochemistry, preparation, and properties. *Inorganic Chemistry*, **10** (3), 463–470.
39. Bisi Castellani, C., Fabbrizzi, L., Licchelli, M., Perotti, A., Poggi, A. (1984) Stabilization by a strongly acidic medium of trivalent copper tetra-aza macrocyclic complexes. *Journal of the Chemical Society, Chemical Communications*, 806–808.
40. Kestner, M.O., Allred, A.L. (1972) Ligand-induced disproportionation of Silver(I). *Journal of the American Chemical Society*, **94** (20), 7189–7189.
41. Barefield, E.K., Mocella, M.T. (1973) Complexes of Silver(II) and Silver(III) with macrocyclic tetraaza ligands. *Inorganic Chemistry*, **12** (12), 2829–2832.
42. Pesavento, M., Profumo, A., Soldi, T., Fabbrizzi, L. (1985) Formation of a trivalent silver tetraaza macrocyclic complex in aqueous solution: hydrolytic tendencies and interaction with the sulfate ion. *Inorganic Chemistry*, **24** (23), 3873–3875.
43. Deming, R.L., Allred, A.L., Dahl, A.R., *et al.* (1976) Tripositive mercury. Low temperature electrochemical oxidation of 1,4,8,11-tetraazacyclotetradecanemercury(II) tetrafluoroborate. *Journal of the American Chemical Society*, **98** (14), 4132–4137.
44. Bencini, A., Fabbrizzi, L., Poggi A. (1981) Formation of Nickel(III) complexes with n-dentate amine macrocycles (n=4, 5, 6). ESR and electrochemical studies. *Inorganic Chemistry*, **20** (8), 2544–2549.
45. Barefield, E.K., Mocella, M.T. (1975) Mechanism of base promoted reduction of Nickel(III) complexes of macrocyclic amines. Coordinated ligand radical intermediate. *Journal of the American Chemical Society*, **97** (15), 4238–4246.
46. De Santis, G., Fabbrizzi, L., Poggi, A., Taglietti, A. (1994) Nickel(III)-promoted deprotonation of an amine group of cyclam. Characterization of the violet transient through stopped-flow spectrophotometric techniques and determination of the pK_a value. *Inorganic Chemistry*, **33** (1), 134–139.

47. De Santis, G., Di Casa, M., Mariani, M. *et al.* (1989). Design of redox systems for the selective transport of electrons across liquid membranes: Nickel(II,III) tetraaza macrocyclic complexes. *Journal of the American Chemical Society*, **111** (7), 2422–2427.
48. Lotz, T., Kaden, T.A. (1977) pH-induced Co-ordination Geometry Change in a Macrocyclic Nickel(I1) Complex. *Journal of the Chemical Society, Chemical Communications*, 15–16.
49. Karn, J.K.; Busch, D.H. (1966) Nickel (II) complexes of the tetradentate macrocycle 2,12-dimethyl-3,7,11,17-tetraazabi*cyclo* (11.3.1) heptadeca-1 (17),2,11,13,15-pentaene. *Nature*, **211**, 160–162.
50. Pallavicini, P., Perotti, A., Poggi, A. *et al.* (1987) *N*-(aminoethyl)cyclam: a tetraaza macrocycle with a coordinating tail (scorpiand). Acidity controlled coordination of the side chain to nickel(II) and nickel(III) cations. *Journal of the American Chemical Society*, **109** (17), 5139–5144.
51. Fabbrizzi, L., Licchelli, M., Pallavicini, P., Parodi, L. (1998) controllable intramolecular motions that generate fluorescent signals for a metal scorpionate complex. *Angewandte Chemie International Edition*, **37** (6), 800–802.
52. Boiocchi, M., Fabbrizzi, L., Foti, F. *et al.* (2005) Dramatically enhanced carbon acidity of the nitrobenzyl fragment in a Nickel(II) scorpionate complex. *Organic Letters*, **7** (16), 3417–3420.
53. Schwörer, M., Wirz, J. (2001) Photochemical reaction mechanisms of 2-nitrobenzyl compounds in solution, I. 2-Nitrotoluene: thermodynamic and kinetic parameters of the *aci*-nitro tautomer. *Helvetica Chimica Acta*, **84** (6), 1441–1458.
54. Boiocchi, M., Ciarrocchi, C., Fabbrizzi, L. *et al.* (2015) Copper(II) complexes of cyclams containing nitrophenyl substituents: push–pull behaviour and scorpionate coordination of the nitro group, *Inorganic Chemistry*, **54** (21), 10197–10207.
55. Fabbrizzi, L., Licchelli, M., Mosca, L., Poggi, A. (2010) Template synthesis of azacyclam metal complexes using primary amides as locking fragments. *Coordination Chemistry Reviews*, **254** (15–16), 1628–1636.
56. Geue, R.J., Hambley, T.W., Harrowfield, J.M. *et al.* (1984) Metal ion encapsulation: cobalt cages derived from polyamines, formaldehyde, and nitromethane. *Journal of the American Chemical Society*, **106** (19), 5478–5488.
57. Comba, P, Curtis, N.F., Lawrance, G.A. *et al.* (1986) Template syntheses involving carbon acids. Synthesis and characterization of (3,10-dimethyl-3,10-dinitro-1,4,8,11-tetra-azacyclotetradecane)Copper(II) and (1,9-diamino-5-methyl-5-nitro-3,7-diazanonane)copper(II) cations and nitro group reduction products. *Inorganic Chemistry*, **25** (23), 4260–4267.
58. Suh, M.P., Kang, S.G. (1988) Synthesis and properties of Nickel(II) and Copper(II) complexes of 14-membered hexaaza macrocycles, 1,8-dimethyl- and 1,8-diethyl-1,3,6,8,10,13-hexaazacyclotetradecane. *Inorganic Chemistry*, **27** (14), 2544–2546.
59. De Blas, A., De Santis, G. Fabbrizzi, L. (1993) Novel routes to functionalized cyclam-like macrocycles. *Pure & Applied Chemistry*, **65** (3), 455–459.
60. De Blas, A., De Santis, G., Fabbrizzi L. *et al.* (1993) Amides and sulfonamides: efficient molecular padlocks for the template synthesis of azacyclam (1,3,5,8,12-pentaazacyclotetradecane) macrocycles. *Journal of the Chemical Society, Dalton Transactions*, 1411–1416.
61. De Blas, A., De Santis, G., Fabbrizzi, L. *et al.* (1993) Pyridines with an appended metallocyclam subunit. Versatile building blocks to supramolecular multielectron redox systems. *Inorganic Chemistry*, **32** (1), 106–113.
62. Abbà, F., De Santis, G., Fabbrizzi, L. et al. (1994) Nickel(II) complexes of azacyclams: oxidation and reduction behavior and catalytic effects in the electroreduction of carbon dioxide. *Inorganic Chemistry*, **33** (7), 1366–1375.
63. Fabbrizzi, L., Licchelli, M., Pallavicini, P. (1999) transition metals as switches. *Accounts of Chemical Research*, **32** (10), 846–853.
64. De Santis, G., Fabbrizzi, L., Licchelli, M. (1996) fluorescence redox switching systems operating through metal centres: the NI^{III}/NI^{II} couple. *Chemistry – A European Journal*, **2** (10), 1243–1250.
65. Fabbrizzi, L., Poggi, A. (1995) Sensors and switches from supramolecular chemistry. *Chemical Society Reviews*, **24**, 197–202.
66. Beley, M., Collin, J.-P., Ruppert, R., Sauvage, J.-P (1984) Electrocatalytic reduction of carbon dioxide by nickel $cyclam^{2+}$ in water: study of the factors affecting the efficiency and the selectivity of the process. *Journal of the American Chemical Society*, **108** (24), 7461–7467.

67. Grant, J.L., Goswami, K., Spreer, L.O. *et al.* (1987) Photochemical reduction of carbon dioxide to carbon monoxide in water using a nickel(II) tetra-azamacrocycle complex as catalyst. *Journal of the Chemical Society, Dalton Transactions*, 2105–2109.
68. Hubin, T.J., McCormick, J.M., Collinson, S.R. *et al.* (2000) New Iron(II) and Manganese(II) complexes of two ultra-rigid, cross-bridged tetraazamacrocycles for catalysis and biomimicry. *Journal of the American Chemical Society*, **122** (11), 2512–2522.
69. Seo, M.S., Kim, J.Y., Annaraj, J. *et al.* (2007) [Mn(tmc)(O_2)]$^+$: A Side-on peroxido Manganese(III) complex bearing a non-heme ligand. *Angewandte Chemie International Edition*, **46** (3), 377–380.
70. Cho, J., Jeon, S., Wilson, S.A. *et al.* (2011) Structure and reactivity of a mononuclear non-haem iron(III)–peroxo complex. *Nature*, **478**, 502–505.
71. Shokeen, M., Anderson, C.J. (2009) Molecular imaging of cancer with Copper-64 radiopharmaceuticals and positron emission tomography (PET). *Accounts of Chemical Research*, **42** (7), 832–841.
72. Ciampolini, M., Fabbrizzi, L., Perotti, A. *et al.* (1987) Dinickel and dicopper complexes with *N,N*-linked bis(cyclam) ligands. An ideal system for the investigation of electrostatic effects on the redox behavior of pairs of metal ions. *Inorganic Chemistry*, **26** (21), 3527–3533.
73. De Clercq, E., Yamamoto N., Pauwels, R. *et al.* (1992) Potent and selective inhibition of human immunodeficiency virus (HIV)-1 and HIV-2 replication by a class of bicyclams interacting with a viral uncoating event. *Proceedings of the National Academy of Science of the USA*, **89** (12), 5286–90.
74. Barefield, E.K., Chueng, D., Van Derveer, D.G., Wagner, F. (1981) Characterization of 2,2′-bi-(1,4,8,11-tetra-azacyclotetradecane): X-ray structure and properties of the dinuclear complex [$Ni_2(C_{20}H_{46}N_8)$][ClO_4]$_4$. *Journal of the Chemical Society, Chemical Communications*, 302–304.
75. Liles, W.C., Broxmeyer, H.E., Rodger, E. *et al.* (2003) Mobilization of hematopoietic progenitor cells in healthy volunteers by AMD3100, a CXCR4 antagonist. *Blood*, **102** (8), 2728–2730.
76. Amendola, V., Bergamaschi, G., Fabbrizzi, L. *et al.* (2016) The interaction of Mozobil™ with carboxylates. *Organic & Biomolecular Chemistry*, **14** (3), 905–912.

9

Porosity in Metal–Organic Compounds

Alexander Schoedel[1,2,3] and Omar M. Yaghi[1,2,3,4]

[1] *Department of Chemistry, University of California, Berkeley, CA, USA*
[2] *Materials Sciences Division, Lawrence Berkeley National Laboratory, Berkeley, CA, USA*
[3] *Kavli Energy Nanoscience Institute, Berkeley, CA, USA*
[4] *King Fahd University of Petroleum and Minerals, Dhahran, Saudi Arabia*

9.1 Introduction

Inorganic zeolites [1], mesoporous silica [2], and porous carbon [3] are useful materials because of their permanent porosity imparted by their architectural stability in the absence of guest molecules. The chemical functionalization and structure control of these classes of porous materials have been long-standing challenges. Indeed, the incorporation of organic functionality and transition metal ions within the structures of these porous materials remain objectives highly sought after. Inorganic and organic molecules when linked into extended structures provide vast opportunities for making porous metal–organic crystals by design and for translating molecular organic and inorganic functionality and reactivity into the solid state. Metal–organic compounds having crystal structures with open space have been known since Werner's work on coordination complexes [4, 5]. Since that time many different classes of porous molecular crystals and extended solids have been studied, but none of these had been shown to have permanent porosity until metal–organic frameworks (MOFs) were reported in the 1990s [6–11]. This contribution highlights the structures and porous properties of important crystalline molecular complexes and extended structures dating from Werner compounds up to and including MOFs.

Macrocyclic and Supramolecular Chemistry: How Izatt–Christensen Award Winners Shaped the Field, First Edition.
Edited by Reed M. Izatt.

9.2 Werner complexes

At the end of the 19th century, little was known about what has since been termed coordination compounds, and conclusions were mainly based on phenomenological observations. Assigned empirical formulas were entirely speculative. However, coordination compounds were in widespread practical use, such as in pigments (e.g. Prussian blue) and dyes (e.g. alizarin) [12]. No theory existed to rationalize the behavior of these materials, when Alfred Werner started his research in 1890 [13]. In the following decades he conceptualized coordination chemistry, work which was finally recognized with the award of the 1913 Nobel Prize in Chemistry. Before Werner's groundbreaking work, the formulation of metal complexes strongly relied on findings from valence theory in organic and inorganic compounds [14]. Kekulé suggested that each element could only exist in one valence and formulas were made accordingly, leading to the dot notation, e.g. $CoCl_3{\cdot}6NH_3$. This meant that cobalt shows a valence of +3 and contains six equivalents of ammonia, which described the composition of matter correctly, but provided no information concerning the actual structure. In contrast, Werner's concept postulated that there is a fixed coordination number for a given metal center, with a defined arrangement of the ligands around it. He also made a distinction between charged, e.g. Cl^- and neutral ligands, e.g. NH_3. The Werner complex could then be formulated as $[Co(NH_3)_6]Cl_3$ which clearly indicates six ammonia ligands surrounding the Co^{3+} central ion [15]. In order to elucidate the binding mode at the metal center, Werner performed experiments with complexes of formula $[M(NH_3)_4Cl_2]$ which could principally show three distinct geometries: hexagonal planar, trigonal prismatic or octahedral. Hexagonal planar and trigonal prismatic geometries would produce three isomers, whereas the octahedral geometry could only produce two isomers (Figure 9.1).

Werner confirmed the latter by isolation of only two geometrical isomers and subsequently validated that model by introduction of chelating amine ligands such as ethylenediamine. We should emphasize here that, in contrast to today's technical capabilities, Werner and his coworkers had to rely on experimental techniques, e.g. crystallization and solubility tests as well as the use of very basic characterization tools such as polarimetry and conductivity [4]. This work enabled the synthesis and characterization of an important class of Werner complexes, β-$[M(PIC)_4(NCS)_2]$ ($M=Ni^{2+}$, Co^{2+}, NCS = thiocyanate) [16] whose crystal structure showed open spaces between the complexes (Figure 9.2) [17]. The work of Barrer in 1969 first showed that these complexes could take up gases, such as nitrogen, oxygen, noble gases, and hydrocarbons in a reversible manner [18]. Heats of adsorption were calculated and the authors finally concluded a zeolite-like sorption behavior of these molecular materials. However, these compounds lost porosity upon cycling due to their molecular structures collapsing into non-porous solids.

9.3 Hofmann clathrates

Hofmann clathrates represent an important class of materials, since they can be considered as the first coordination networks. In this context a network of formula $[Ni(CN)_2(NH_3)]{\cdot}C_6H_6$ was prepared by Hofmann as early as 1897; unfortunately, structural characterization was not possible at that time [19]. The structure of the network therefore remained unclear for almost 60 years, until Powell and coworkers performed a successful X-ray analysis in 1952 [20]. It was discovered to be principally composed of two-dimensional layers of alternating octahedral and square planar Ni^{2+}, linearly connected through 4 CN^- moieties. The terminal NH_3 ligand is axially bound and points towards the adjacent layer, thus forming a cavity that can encapsulate benzene molecules (Figure 9.3).

Later, in the 1960s, Iwamoto and coworkers focused on a more systematic concept in synthesizing Hofmann complexes and utilized two charged components to build these networks, in addition to having neutral, encapsulated guest molecules [21, 22]. Their clathrates with general formula $[M^{I}(NH_3)_2M^{II}(CN)_4]{\cdot}G$ (G=Guest)

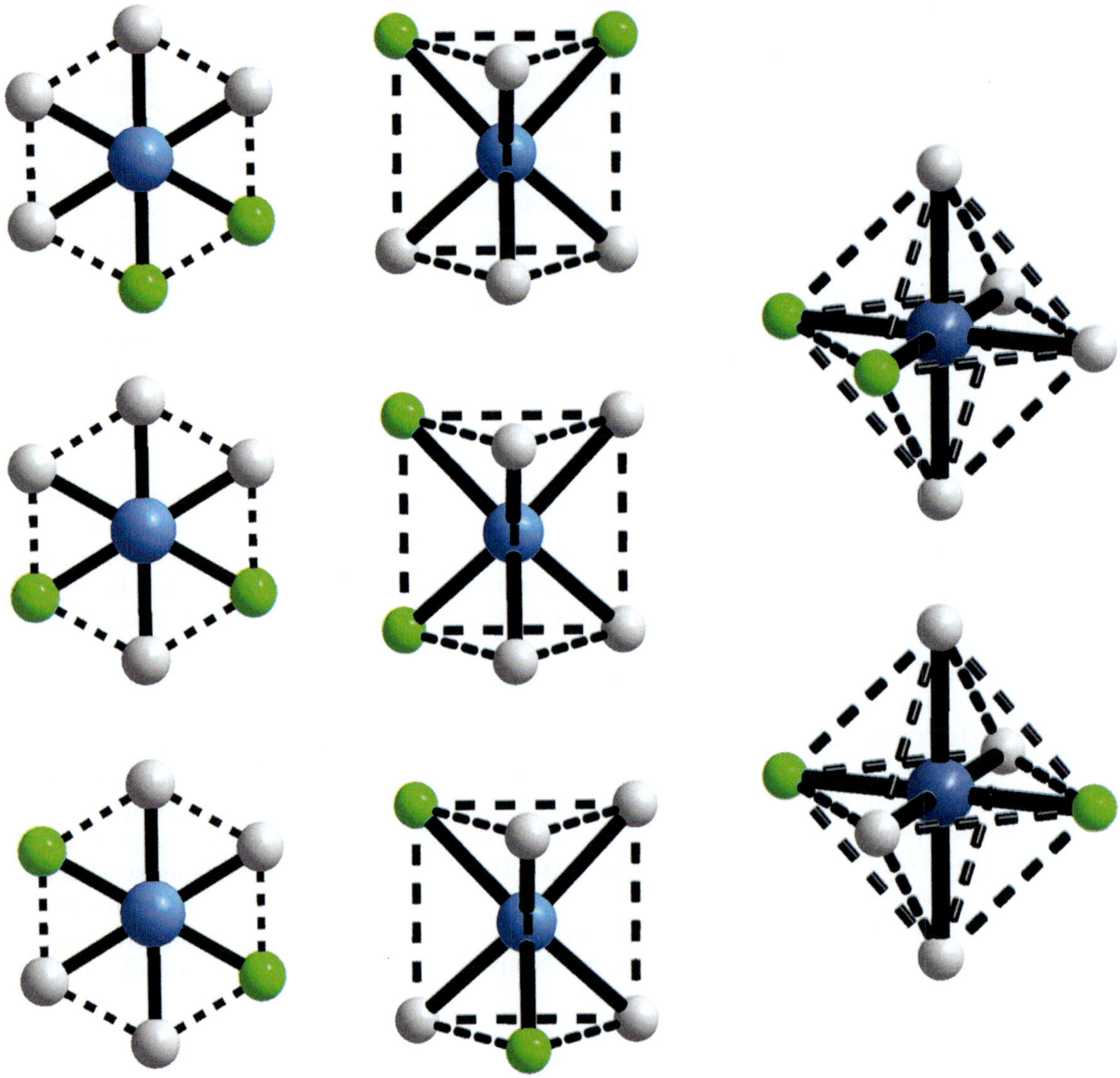

Figure 9.1 *Possible geometries of the $M(NH_3)_4Cl_2$ complex. Three isomers of hexagonal planar, three isomers of trigonal prismatic and two isomers of octahedral geometry are theoretically feasible (from left to right). Color code: blue, M; green, Cl; light grey, NH_3*

were obtained from cationic $[M^1(NH_3)_2]^{2+}$ complexes together with anionic square planar $[M^2(CN)_4]^{2-}$ complexes and two molecules of benzene, aniline, pyrrole or thiophene. The structural characterization was performed by refinement from X-ray powder diffraction data. In 1967, Walker and Hawthorne replaced the ammonia ligand by an *n*-alkylamine and synthesized nickel cyanide complexes [23]. The next logical step forward was again reported by Iwamoto et al., who first introduced ethylenediamine into these types of networks [24]. The use of this bifunctional linker enabled the formation of an organic bridge between the square $M(CN)_2$ layers. Encapsulation of aromatic guests such as aniline, benzene, thiophene, or pyrrole was

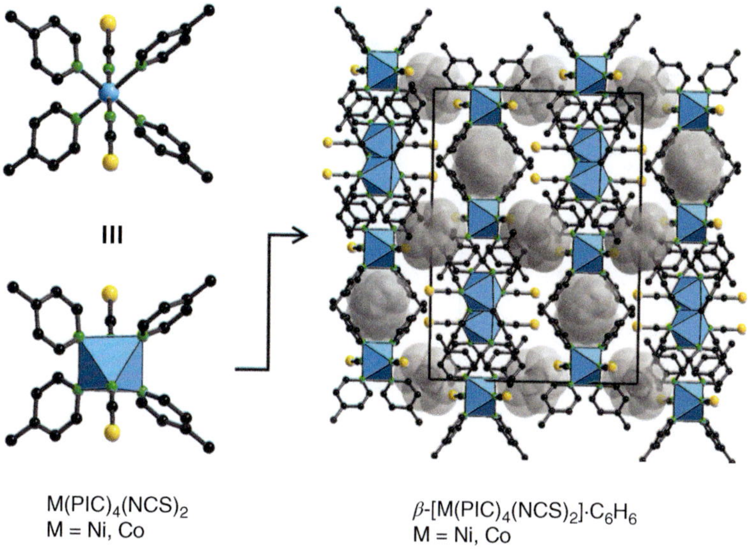

Figure 9.2 *Werner complexes of formula β-[M(PIC)$_4$(NCS)$_2$] (M = Ni^{2+}, Co^{2+}, PIC = γ-picoline) (left). The crystal packing, depicted along the [010] axis, shows the encapsulation of benzene molecules (shown in space-filling representation) occupying the zigzag channels between the complexes. Color code: black, C; green, N; yellow, S; blue polyhedra, M*

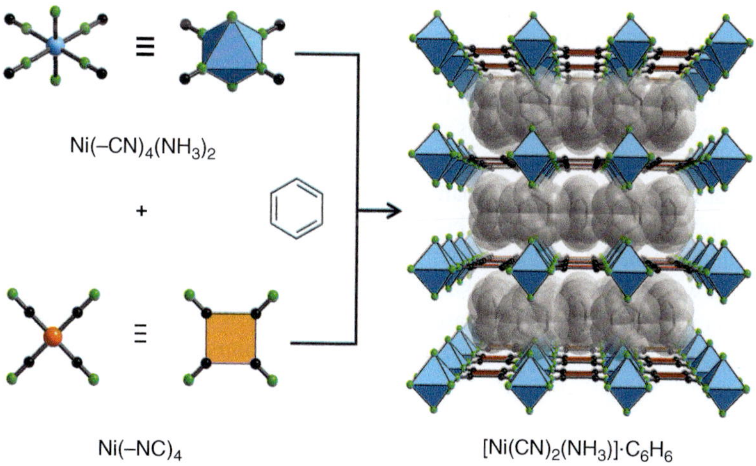

Figure 9.3 *Single-crystal X-ray structure of the original Hofmann clathrate [Ni(CN)$_2$(NH$_3$)]·C$_6$H$_6$ as analyzed by Powell and coworkers [20]. The structure is reproduced from given atomic coordinates; benzene guest molecules are shown in the space-filling model between the two-dimensional [Ni(CN)$_2$(NH$_3$)] layers. Hydrogen atoms are omitted for clarity. Color code: black, C; green, N; blue and orange polyhedra, Ni*

successfully demonstrated. Mathey and coworkers then further expanded this concept in 1977 to show that variation of linker length provides the opportunity for selective rather than random encapsulation of aromatic guests and solvents [25, 26]. Later, Iwamoto et al. followed up and synthesized a series of α,ω-diaminoalkane complexes; each deliberately designed to encapsulate specific molecules (Figure 9.4) [27, 28].

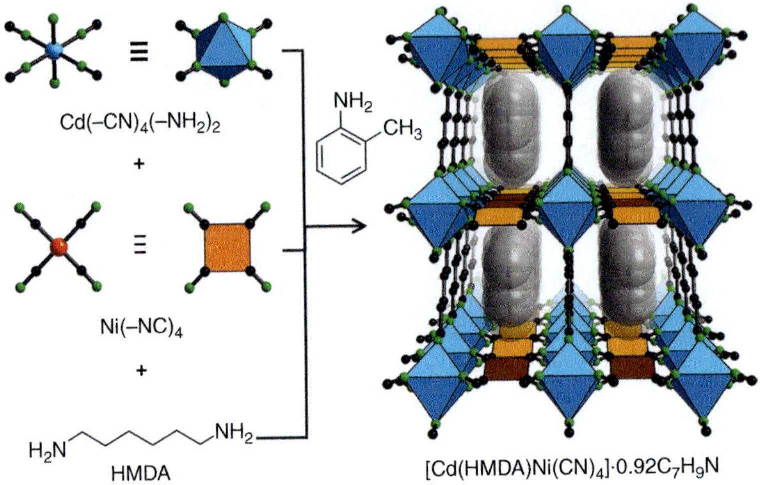

Figure 9.4 Single-crystal X-ray structure of the Hofmann clathrate [Cd(HMDA)Ni(CN)$_4$]·0.92C$_7$H$_9$N (HMDA = hexamethylene-1,6-diamine) sustained by α,ω-diaminoalkanes. The encapsulated o-toluidine (shown in space-filling model) appears disordered. Hydrogen atoms are omitted for clarity. Color code: black, C; green, N; orange polyhedra, Ni; blue polyhedra, Cd

All these valuable contributions established early principles of materials design, and clearly showed how linker length can control material properties, enabling selectivity at the molecular level. The concept introduced at that time provided the foundation for the synthesis of modular coordination polymers, a term first used by John Bailar in 1963 to describe Beryllium-bis-(β-diketone) complexes [29]. Although these are not polymers but extended crystals, the term is still in use today. We note that although Hofmann clathrates and related compounds showed absorption of guests from liquid phase, no examination by gas adsorption at low-temperature (the golden standard of measuring permanent porosity) was performed until 1995 when Ramprasad et al. demonstrated that lithium pentacyanocobaltate coordination networks of formula $Li_3[Co(CN)_6]$·4DMF (DMF = *N,N*-dimethylformamide) show reversible oxygen sorption at room temperature [30]. The type I isotherm, fitted to the Langmuir model, was explained by a reversible chemisorption of O_2; however, the material did not exhibit pores. It was rather suggested that the oxygen uptake was occurring via a solid state diffusion mechanism. A point to make about the cyano-based frameworks is that the cyanide ion is not functionalizable and that the pillaring organic units may serve as functionalization points to some limited degree.

9.4 Coordination polymers

Coordination polymers, even if the term had not been used at that time, can be dated back to 1959, when Saito et al. crystallized the first bis(alkylnitrilo)copper(I) structures in an attempt to gain insights to the dyeing of polyacrylonitrile fibers [31–33]. The obtained structures were characterized by means of single-crystal X-ray diffraction. They are composed of tetrahedral Cu(–CN)$_4$ units that have been known to coordination chemists since the early 20th century [34]. These building units are connected through organic moieties with different length and thus propagate into extended networks of varying dimensions. Depending on the length and conformation of the dinitrile linker, one-dimensional chains, in the case of succinonitrile (SUC), two-dimensional grid structures, in the case of glutaronitrile (GLU) or three-dimensional networks, in the case of adiponitrile (ADI) were obtained (Figure 9.5).

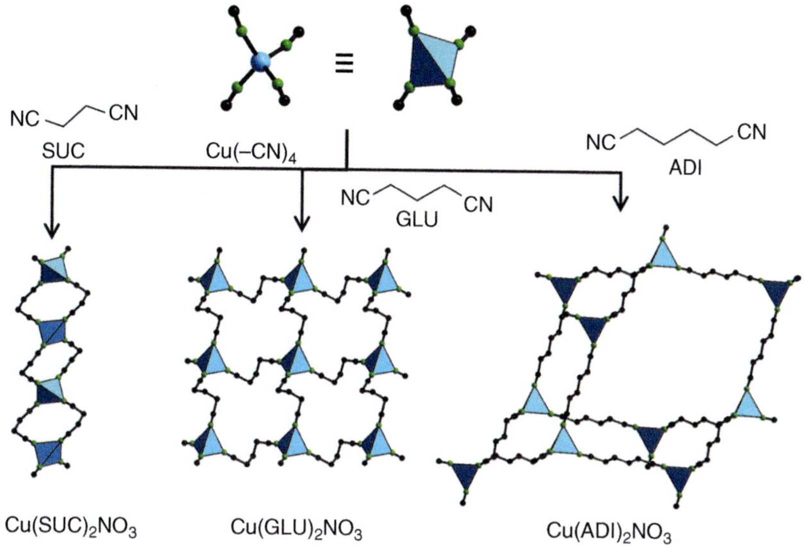

Figure 9.5 *Diverse bis(alkylnitrilo)copper(I) structures: one-dimensional chains sustained by succinonitrile (SUC) [31], two-dimensional grid structures with glutaronitrile (GLU) linkers [32], and three-dimensional networks made of adiponitrile (ADI) linkers [33]. Hydrogen atoms, anions, and interpenetrating networks are omitted for clarity. Color code: black, C; green, N; blue polyhedra, Cu*

In this context, bis(adiponitrilo)copper(I) nitrate deserves special attention since it represents the first three-dimensional, four-connected network reported. The authors classified it correctly as a diamond lattice according to the geometry principles of crystal chemistry established by A. F. Wells in 1954 [35]. The relative length of the organic linker and the resulting large open architecture led to the formation of six interpenetrated nets[36] to fill the void space of the adamantine cage.

Besides several other one- or two- dimensional nets, a diamond-like coordination polymer was synthesized in 1986 by Hünig and co-workers in an attempt to synthesize a radical anion salt of 2,5-dimethyl-*N,N′*-dicyanoquinonediimine with high electrical conductivity [37]. However, the structural outcome was not discussed in detail, though it shows a seven-fold interpenetrated diamond net. This particular compound was later derivatized by Kobayashi et al., using methoxy-, chloro- and bromo-substituted linkers with retention of the overall structure [38].

In 1989, Hoskins and Robson provided an important rationale for the deliberate design and synthesis of extended nets. In their contribution they utilized the tetrahedral single-metal building unit $Cu(-CN)_4$ together with the tetrahedral organic linker 4,4′,4′′,4′′′-tetracyanotetraphenylmethane (TCTPM) to afford an infinite, non-interpenetrated, cationic framework of formula $[Cu(TCTPM)]^+$ (Figure 9.6) [39]. This framework contains charge balancing BF_4^- counteranions as well as neutral nitrobenzene guests in an adamantine cavity that was estimated to be around 700 Å^3 in volume. It was further proved that the BF_4^- anions are indeed exchangeable through PF_6^- without disturbing the framework's integrity.

Many diamond-like frameworks have subsequently evolved based on the simple design principles from tetrahedral building units. The shortest moieties in this context were CN^-, to yield isostructural, two-fold interpenetrated $M(CN)_2$ nets, where M = Zn, Cd [40, 41]. The introduction of longer linkers, such as 1,4-dicyanobenzene, 4,4′-dipyrdiyl or 2,5-dimethylpyrazine together with Cu^I afforded five-fold, four-fold and one-fold interpenetrated diamond networks. In all these cases, the anions (BF_4^- or PF_6^-) reside in the channels and do not interact with the metal cations [42].

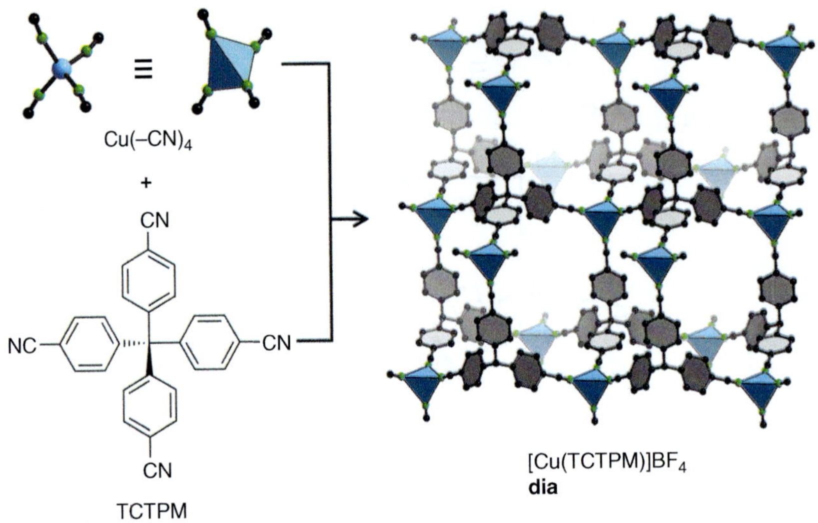

Figure 9.6 Diamond-like framework composed of tetrahedral single-metal building units $Cu(-CN)_4$ and tetrahedral organic linkers. Anions and solvent molecules were found heavily disordered and were not refined. Hydrogen atoms are omitted for clarity. Color code: black, C; green, N; light blue polyhedra, Cu

Although diamond-type frameworks were predominant at that time, other nets structurally related to platinum(II) sulfide (**pts**) were synthesized by replacing the tetrahedral (S^{2-}) and the square planar (Pt^{2+}) through Cu^+ and $[Pt(CN)_4]^{2-}$ respectively [43]. This strategy resulted in an anionic open framework structure, containing hexagonal channels partially filled with charge balancing $[NMe_4]^+$ cations. The contribution provided further evidence that geometric design principles were feasible for the synthesis of frameworks; however, a general concept had to be developed.

Following this initial strategy, in 1994, a tetracyano-functionalized porphyrin (TCP) linker was introduced as a square planar building unit of the PtS-type framework $Cu^{II}(TCP)Cu^{I}BF_4$ (Figure 9.7, left) [44]. This net shows two-fold interpenetration which could be avoided through structure fine-tuning based on pyridine-derived linkers. We note here that Cu^{II} resides in the center of the porphyrin linker, whereas Cu^{I} serves as a tetrahedral single-metal building unit. The tetrapyridyl-functionalized porphyrin (TPP) network $Cu^{II}(TPP)Cu^{I}BF_4$ (Figure 9.7, right) relies on a single net without interpenetration, which the authors attribute to the steric bulk of the pyridyl- in comparison to the cyano-functionality.

Later, in 1995, this building block approach was utilized to combine tritopic cyano-based linkers with Ag^{I} cations to afford networks that exhibit either AlB_2 (**alb**) or $ThSi_2$ (**ths**) topology [45]. The latter contains channels of inner diameter 15 Å and thus enables guest exchange by retaining framework crystallinity.

In summary, cyano-based structures have long been well known to coordination and supramolecular chemists. The initial results, which appeared rather serendipitous throughout the 1960s to 1980s, were conceptualized in the 1990s according to simple geometric design principles. The use of the molecular building block approach [46] rendered these coordination polymers into fine-tunable materials, though with strong limitations at that time. Nonetheless, the results paved the way for the design of materials with superior properties and levels of synthetic control hitherto unknown to materials chemists.

Although there had already been many reports on pyridine based structures, it took until 1994 for the first two-dimensional square grid network with catalytic activity to be reported [47]. It is composed of $Cd(BIPY)_2(NO_3)_2$ (BIPY = 4,4′bipyridine) and contains large guest-filled voids that can render catalytic processes with respect to cyanosilylation. The framework was derived from a closely related molecular material

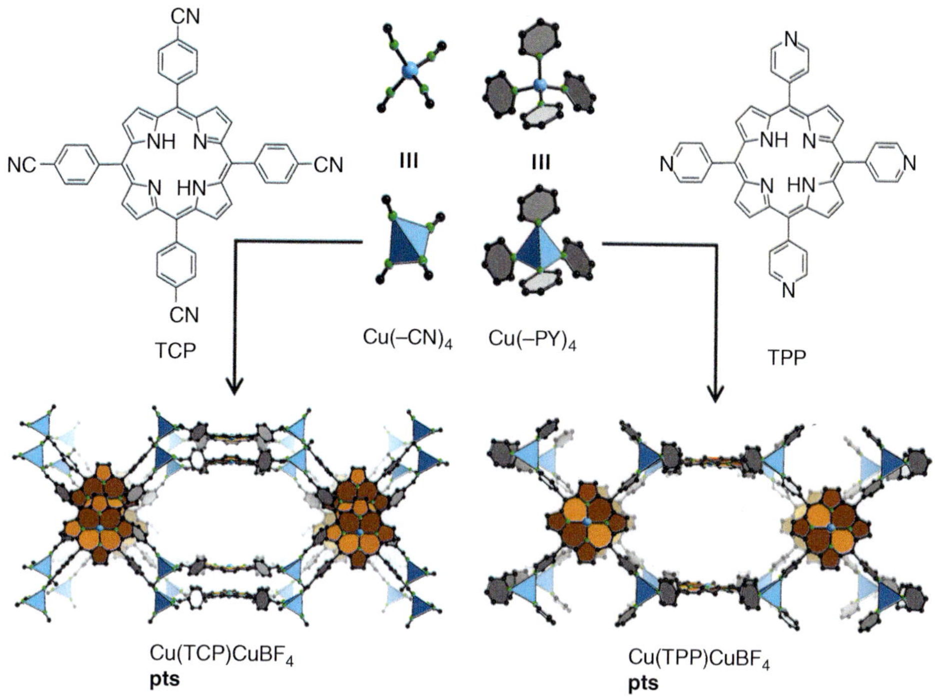

Figure 9.7 Comparison of the two-fold interpenetrated framework $Cu^{II}(TCP)Cu^{I}BF_4$ and the single net $Cu^{II}(TPP)Cu^{I}BF_4$. Hydrogen atoms, solvent molecules and anions are omitted for clarity. Color code: black, C; green, N; light blue, Cu^{II}; light blue polyhedra, Cu^{I}

where square planar Pd^{II} ions are capped in a *cis*-fashion by ethylenediamine [48]. The two remaining coordination sites then connect through BIPY to yield a molecular square. Based on this structure, the extended two-dimensional $Cd(BIPY)_2(NO_3)_2$ was synthesized. Single-crystal X-ray diffraction provided evidence that the structure forms a clathrate with *o*-dibromobenzene that sits inside the square microchannels of the framework (Figure 9.8).

More importantly, the first catalytic process within a coordination polymer was addressed through cyanosilylation of benzaldehyde and imines. Although compounds of formula $[M(BIPY)_2(H_2O)]X$ ($M = Zn^{2+}$, Cd^{2+}, Cu^{2+}; $X = PF_6^-$, SiF_6^{2-}) and $[Cd(BIPY)_2(H_2O)_2]PF_6 \cdot 2(BIPY) \cdot 4H_2O$ have previously been reported, they were not investigated for guest encapsulation or catalytic performance [49, 50]. In such cases interpenetration has precluded the formation of cavities and thus hampered clathrate formation or catalysis. We therefore believe, that $Cd(BIPY)_2(NO_3)_2$ can be considered special since it inspired researchers to explore these particular types of networks, which has led to numerous elegant examples of $M(BIPY)_2$ square grids [51–55]. Some of them show interesting properties, especially in terms of gas storage and separation and still represent a very active branch of MOF research up to date. We consider certain contributions on $M(BIPY)_2$ and related materials that emerged through the mid-1990s as essential for the development of MOFs, and will discuss them here.

The term metal–organic frameworks, abbreviated to MOFs, was first coined in 1995 initially referring to a compound with formula $Cu(BIPY)_{1.5}(NO_3)$ that shows a diamond-like structure composed of Cu^I and 4,4′-bipyridine (Figure 9.9) [56]. Since then, thousands of MOFs, sometimes also termed porous coordination polymers (PCPs), have appeared in the open literature [9].

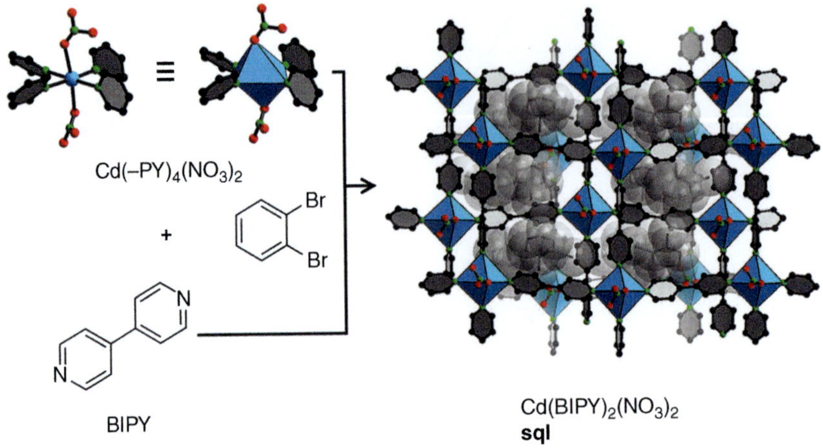

Figure 9.8 *Square grid net of formula Cd(BIPY)$_2$(NO$_3$)$_2$ shown approximately along the crystallographic [001] axis. It forms a clathrate with o-dibromobenzene molecules (depicted in space-filling representation) that reside inside the square microchannels. Hydrogen atoms are omitted for clarity. Color code: black, C; red, O; green, N; blue polyhedra, Cd*

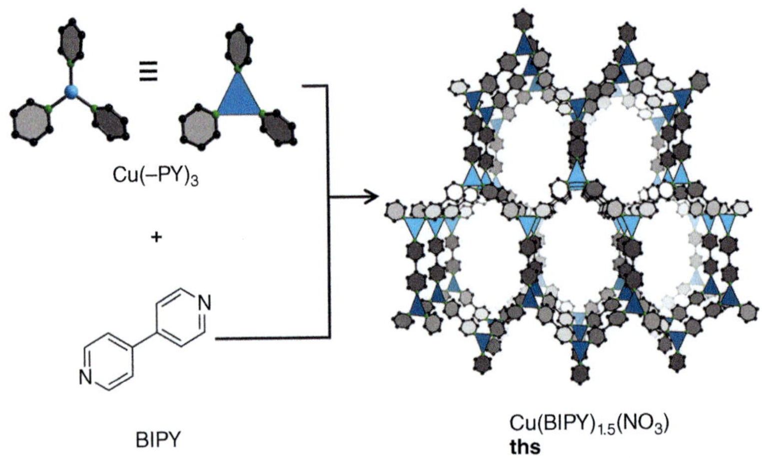

Figure 9.9 *Left: triangular Cu^{I} building units connected through 4,4′-bipyridine; right: the overall cationic three-dimensional ThSi$_2$ (**ths**) topology framework is based on linearly connected triangular units. Hydrogen atoms, solvent molecules, charge balancing anions, and interpenetrating networks are omitted for clarity. Color code: black, C; green, N; blue polyhedra, Cu*

Cu(BIPY)$_{1.5}$(NO$_3$) is composed of a triangular Cu(-PY)$_3$ building unit connected by 4,4′-bipyridine into an extended framework with ThSi$_2$ (**ths**) topology. The six interpenetrated nets contain charge-balancing anions that reside in the rectangular channels with dimensions of 3×6 and 6×6 Å. The nitrate ions were demonstrated to be exchangeable with other simple inorganic anions such as BF_4^- or SO_4^{2-}. The contribution also demonstrated that hydrothermal synthetic conditions represent a viable route to such frameworks, resembling a route that has proved fruitful in zeolite chemistry.

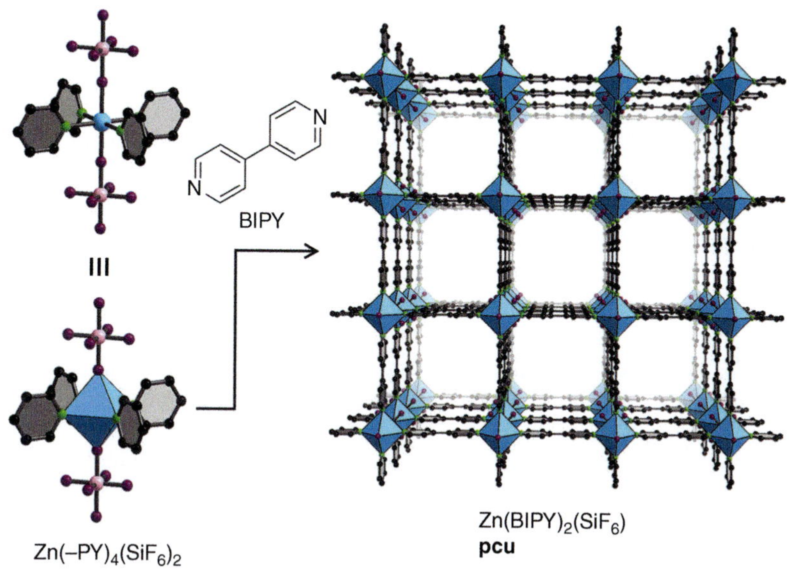

Figure 9.10 *Single-crystal X-ray structure of Zn(BIPY)$_2$(SiF$_6$); the two-dimensional net is connected through SiF$_6^{2-}$ anions to give a 3D framework with rectangular channels of 8×8Å along the crystallographic [001] direction. Color code: black, C; green, N; purple, F; rose, Si; blue polyhedra, Zn*

In the same year, another three-dimensional net was reported, based on octahedral Zn(–PY)$_4$(–SiF$_6$)$_2$ building units [57]. The resulting framework consists of an octahedral Zn^{2+} cation coordinated to four BIPY linkers into a square grid net that is in turn axially pillared by anionic SiF_6^{2-} moieties. Preparation of Zn(BIPY)$_2$(SiF$_6$) under certain synthetic conditions then led to the formation of a non-interpenetrated open framework with a primitive cubic (**pcu**) topology showing large square channels along the crystallographic [001] axis (Figure 9.10).

These channels have dimensions of 8×8Å, and the empty space generated comprises about 50% of the total unit cell volume. We note here that the Zn-version of this material was reported to be unstable upon loss of guest molecules under vacuum, and that the Cu-analogue shows a surface area of 1337 m^2g^{-1} [54]. This material was reported in 2000 and is, today, one of the best sorbents for methane in terms of volumetric uptake at room temperature.

After these seminal contributions it took more than ten years for it to be discovered that materials of composition Cu(BIPY-*n*)$_2$(SiF$_6$) (BIPY-*n* generally represents 4,4′-bipyridine or its derivatives, i.e. linear linkers with two terminal aromatic nitrogen donor moieties) are exceptional with respect to CO_2 uptake and separation [58, 59]. These materials were shown as amenable to structure fine-tuning and to be stable under humid conditions, which makes them one of the best-performing materials for this particular application to date.

9.5 Porosity in metal–organic frameworks

Previous experiments on reversible sorption of small molecules in Werner complexes, such as β-M(PIC)$_4$(NCS)$_2$ (M = Ni^{2+}, Co^{2+}), Prussian Blue of general formula $Fe_4[Fe(CN)_6]$ and Hofmann clathrates have already been established and well-documented. Many crystal structures of coordination polymers following the structure classification of "node" and "linker" put forward by Wells have been reported all throughout the last 70 years or so. One of us (O.Y.) recalls in the early 1990s chemists commenting about these structures that they were "sculptures rather than useful materials." This underlined the need for developing strategies for making robust frameworks that exhibit permanent porosity so that they can be utilized in many applications such as separations and catalysis. A strategy reported in 1998 outlined the use of carboxyl functionality to lock the metal–ion positions in place and allow for robust metal–carboxyl clusters to be used as nodes in making extended structures. The other advantage of these clusters was that they had one kind of an overall coordination mode and therefore there was no ambiguity as to the coordination geometry to be adopted upon formation of the network; a factor coordination polymers suffered from, and one that prevented the making of structures by design.

Thus in 1998, MOFs gained traction due to the first report establishing microporosity by surface area and pore volume determination. In particular, a material, named MOF-2, with formula $Zn(BDC)(H_2O)$ (BDC = 1,4-benzenedicarboxylate) was synthesized and shown to contain potential voids filled with DMF and H_2O guest molecules, respectively (Figure 9.11) [60]. The structure is composed of $Zn_2(-COO)_4(H_2O)_2$ paddlewheel-like secondary building units (SBUs) that are in turn connected by BDC linkers. The resulting two-dimensional grid network is held together by strong hydrogen-bonding interactions among the axial water molecules as well as the carboxylate moieties on the paddlewheel units.

The guests can be removed under vacuum with heating to yield the desolvated structure Zn(BDC). Gas adsorption studies demonstrated that this material is permanently microporous, as exemplified by reversible type I gas sorption isotherms using nitrogen at 77 K (Figure 9.12) or carbon dioxide at 195 K as sorbates. These experiments enabled the estimation of Langmuir apparent surface areas between 270 and 310 m^2g^{-1}, as well as micropore volumes between 0.094 and 0.086 $cm^3 g^{-1}$.

Other materials that validated the existence of permanent porosity in MOFs are represented by structures later termed as MOF-3 and MOF-4, $Zn_3(BDC)_3$ and $Zn_2(BTC)(NO_3)$ (BTC = 1,3,5-benzenetricarboxylate) respectively [7, 61, 62], MOF-3 was initially reported in 1998 and is composed of a trimeric zinc SBU that is in turn coordinated to six carboxylate groups of BDC [61]. This affords an overall three-dimensional framework with rhombic shaped pores, filled with methanol guest molecules. In addition, the remaining four coordination sites at the SBU were occupied by methanol, giving rise to potential open metal sites (OMS). The apparent surface area of this material was estimated in a later contribution to be 140 m^2g^{-1}. MOF-4 was first reported in 1997 by reacting zinc nitrate together with H_3BTC at room temperature. The resulting framework

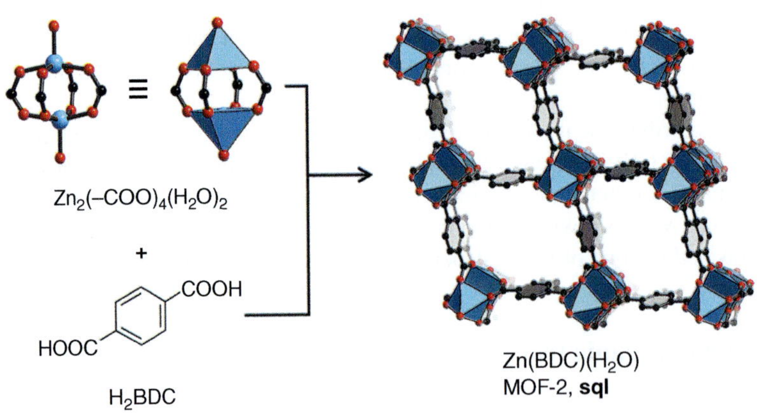

Figure 9.11 *View of MOF-2 along the crystallographic [001] direction, the rectangular channels are clearly visible. Hydrogen atoms, except on terminal water ligands, and solvent molecules, are omitted for clarity. Color code: black, C; red, O; blue polyhedra, Zn*

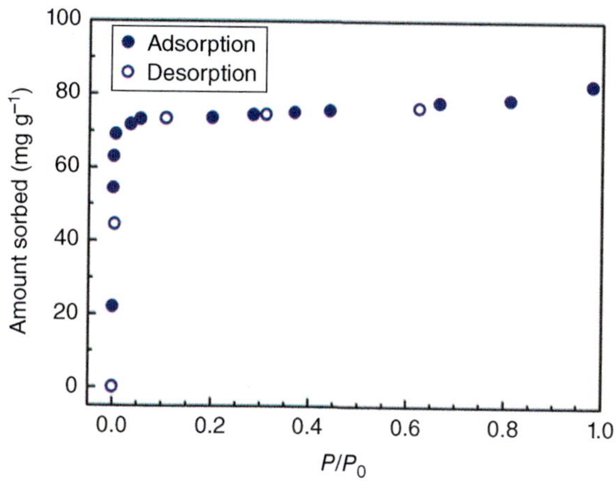

Figure 9.12 *Gas adsorption isotherm for MOF-2 with N_2 at 77 K. This served as the first example of microporosity reported in a metal–organic framework*

is composed of a trimeric $Zn_2(–COO)_3(NO_3)$ SBU that is connected to two other SBUs through a BTC linker to afford a three-dimensional framework [62]. Remaining coordination sites are occupied by ethanol solvent molecules. It was later proven by ethanol-sorption experiments, that MOF-4 is also a permanently porous structure.

These seminal contributions, especially the discovery of microporosity in MOF-2, imparted momentum to the development of structures based on paddlewheel SBUs and spurred interest in exploitation of new carboxylate based clusters as building blocks for MOFs. We note here that other MOFs have previously appeared in the literature together with reported gas sorption isotherms. Herein, the T-shaped $Co_2(BIPY)_3(NO_3)_4$ readily adsorbs gases such as methane, nitrogen, and oxygen at high pressures (up to 36 atm) [63]. However, since the experiments were performed at room temperature and not at the boiling point of the particular gas, an estimation of the apparent surface area and pore volume was not possible. The gas adsorption isotherm at 77 K is the gold standard for proof of porosity.

9.6 The discovery of MOF-5: the golden age of metal–organic frameworks

MOF-5 is recognized as the most prominent example of a metal–organic framework by many researchers in material science but also from other fields of chemistry (Figure 9.13) [64].

The enormous popularity of this structure, in our view, has three main causes: the advantage of a simple network accessible through rational design principles, a facile route of preparation, and a very high porosity, in fact the highest porosity of any material at the time. We will therefore provide an overview of MOF-5 and its features as well as discussing other materials that helped to popularize the field at this early stage.

In 1999, little was known about porosity of metal–organic frameworks, since MOF-2 and its associated microporosity had just been discovered. This shifted the interest of many researchers from pyridine- toward carboxylate-based frameworks and thus represented a turning point in the field. At that time, the first MOF based on linking of octahedral basic zinc acetate building units, $Zn_4O(–COO)_6$, was reported. The resulting framework shows primitive cubic (**pcu**) topology and was termed MOF-5. The name was coined in reminiscence of the famous zeolite ZSM-5.

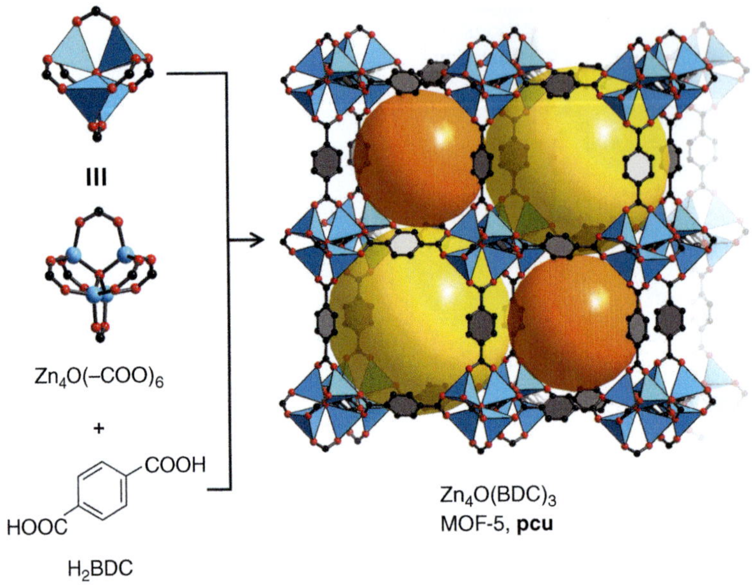

Figure 9.13 *Single-crystal X-ray structure of MOF-5 constructed from $Zn_4O(-COO)_6$ SBUs, connected through BDC. The overall structure is composed of a large cavity (15.1 Å, yellow sphere) together with a smaller cavity (11.0 Å in diameter, orange sphere) in an alternating fashion. Hydrogen atoms are omitted for clarity. Color code: black, C; red, O; light blue polyhedra, Zn*

The SBU of MOF-5 is composed of a tetrahedral central oxide surrounded by four tetrahedrally coordinated Zn^{2+} ions which are in turn coordinated by a total of six bridging carboxylate groups (–COO). This SBU represents a well-known cluster entity that has previously been obtained as discrete complexes with monocarboxylates such as acetate, benzoate, and pivalate [65].

MOF-5 was synthesized by solvothermal reaction of zinc nitrate tetrahydrate and H_2BDC in a mixture of DMF and chlorobenzene. The authors were able to show that the resulting framework contains voids filled with DMF and chlorobenzene guests, which are readily exchangeable with chloroform that could later be removed under vacuum at room temperature. The desolvated crystals kept their integrity, and nitrogen gas adsorption experiments at 77 K revealed a type I isotherm with an estimated Langmuir surface area of 2,900 m^2g^{-1} and a pore volume of 1.04 cm^3g^{-1} (Figure 9.14). These values exceeded by far all known conventional porous materials such as zeolites, silicates, or porous carbon.

It was later shown that MOF-5 is a modular structure and several isoreticular (based on the same net or topology, here a **pcu** net) frameworks could be obtained by variation of length and functionality of the linker [66].

At around the same time, another prototypal MOF, of formula $Cu_3(BTC)_2(H_2O)_3$, was synthesized and termed HKUST-1 (HKUST=Hong Kong University of Science and Technology) [67]. This structure is composed of a $Cu_2(-COO)_4$ paddlewheel SBU, similar to the one that sustains MOF-2. The SBU is connected to triangular BTC to form an overall 3,4-c twisted boracite (**tbo**) network (Figure 9.15). The $Cu_2(-COO)_4(H_2O)_2$ paddlewheel cluster has also been known for quite some time since it sustains the structure of molecular copper(II)-acetate hydrate and other transition metal analogues such as Mo or Rh [68, 69]. Removal of the terminal water ligands is possible and generates the anhydrous MOF $Cu_3(BTC)_2$, with a color change of the material from blue to purple. Surface area analysis was performed and yielded a Brunauer-Emmett-Teller (BET) area of around 700 m^2g^{-1}, which indicates a not fully desolvated structure. More recent BET measurements of this material have yielded around 1800 m^2g^{-1} after full activation [70]. In the following years, the modularity of the **tbo** net was

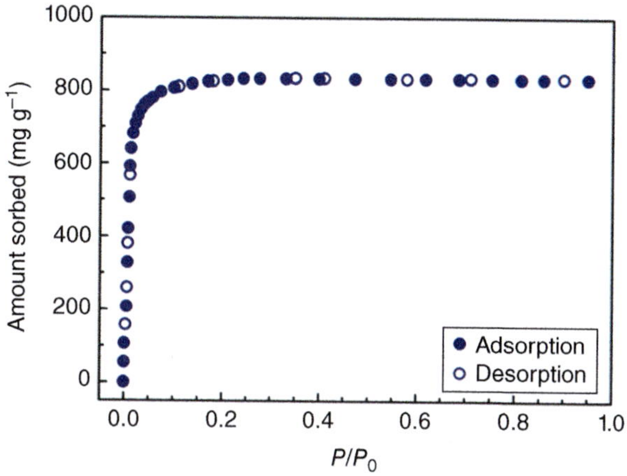

Figure 9.14 *Nitrogen gas adsorption isotherm for MOF-5 at 77 K. The microporous nature of the material is indicated by a type I isotherm*

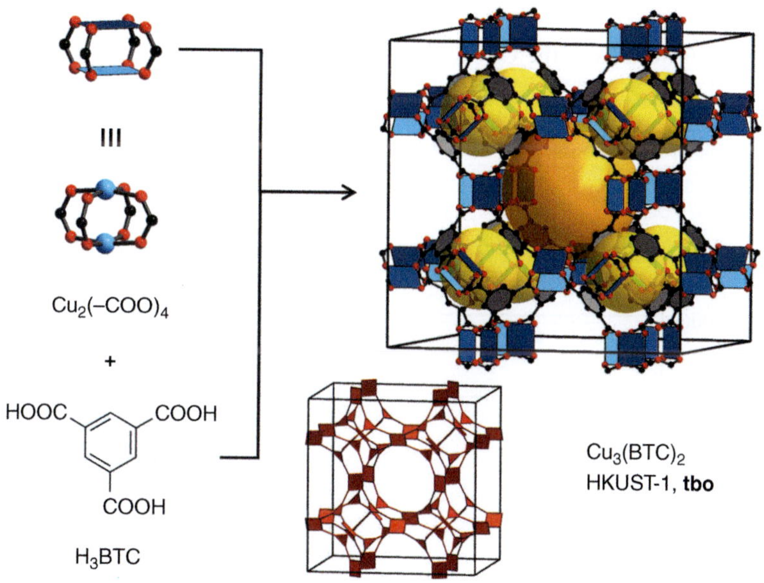

Figure 9.15 *HKUST-1 synthesized by a combination of $Cu_2(-COO)_4$ and BTC. Hydrogen atoms and solvent molecules are omitted for clarity. Color code: black, C; red, O; blue, Cu. The yellow and orange balls represent the empty space in the framework*

further explored by the synthesis of isoreticular variants [71]. with MOF-399 holding the record for the lowest density MOF (0.126 g/cm^3) known to date [72].

In summary, by end of the 1990s, MOFs started to attract considerable scientific interest due to the discovery of ways to design high surface area materials, unparalleled for conventional inorganic and organic porous materials. The materials display a large structural diversity based on different SBUs and linkers. Therefore, MOF-5 and HKUST-1 are prototypical for this class of material and are still under investigation by many research groups around the world, even 15 years after their initial discovery.

9.7 The Cambridge Structural Database – an essential tool for MOF chemists

In the context of more than 20,000 metal–organic frameworks made during the past decade, an essential resource for researchers is a database where crystal structures and related literature of existing materials can be retrieved in a facile and systematic manner. The Cambridge Structural Database (CSD), which is a part of the Cambridge Crystallographic Data Centre (CCDC), is ideally suited for this purpose since it represents a resource for small molecules containing a wide range of organic, metal–organic and organometallic crystal structures [73, 74]. We introduce discussion of this particular database here in order to give an overview of history, and, more importantly, describe the basic features of this powerful research tool.

The CCDC as a part of the Department of Chemistry at the University of Cambridge was principally established in 1965 by Dr. Olga Kennard, together with a handful of coworkers. They compiled, organized, and gradually included the relatively small number of crystal structures published at that time into a database, the CSD. Through constant development in the late 1970s and 1980s, systems for search, retrieval, analysis, and visualization of CSD information began to emerge and are still under constant improvement today. Having these systematic search capabilities, the CSD became a useful tool for what has later been termed "data mining" in fundamental research [75, 76]. The CCDC started to serve academia as well as major pharmaceutical companies during the 1980s, and it finally became an independent self-financing non-profit institution in 1989. On October 19, 2015 the CSD recorded its 800,000th structure, and is showing exponential growth, with more than 40,000 structures currently added per year, thus leading to a doubling time of around 9 years. A statistic that can be retrieved from the CSD clearly shows that the number of MOF structures increases exponentially, with today over 6000 newly reported structures per year (Figure 9.16). These enormous research efforts are mainly related to carboxylates, which have exceeded all other linker types by far (Figure 9.16 inset). This in turn means that two-thirds of all newly reported MOF structures nowadays are based on robust carboxylate linkers.

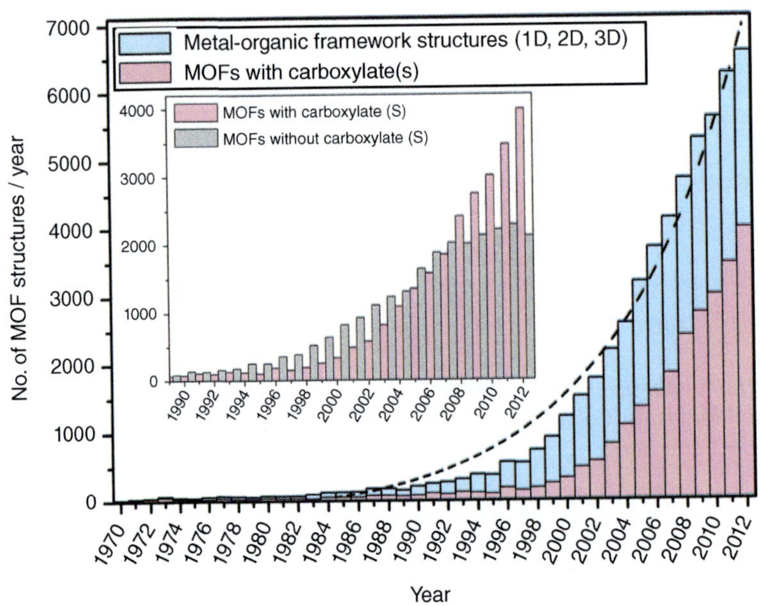

Figure 9.16 A statistic obtained from the CSD revealed an exponential increase of reported metal–organic framework structures (1D, 2D and 3D) over the decade to 2012. Notably, the utilization of carboxylate moieties represents a turning-point in the late 1990s, and now exceeds all other types of linkers by far (see inset)

New structural data are acquired by the CCDC in form of crystallographic information files (.cif), standardized text files according to the International Union of Crystallography (UICr) [77, 78], These are almost exclusively used throughout the scientific universe for small-molecule crystallographic data. Each individual crystal structure determination represents one entry in the CSD, denoted by a unique six letter code, the reference code (e.g. SAHYIK for the original MOF-5 structure). Two supplementary digits that may be added identify additional determinations of the same structure and are commonly included for data from different research groups, when done under different experimental conditions or if there is an improved refinement. In the early 2000s, with the improvement of computational capabilities and the conversion of data into connectivity tables, substructure searching at the molecular and the supramolecular level became possible. The resulting structures are now displayed as a two-dimensional diagram as well as in a 3D visualizer. In addition, plentiful crystallographic information together with the literature reference is made accessible.

Hence, the CSD represents an essential tool for researchers in the MOF field, enabling rapid access to data on published structures and related literature. This information greatly facilitates the quest for and design of new materials and the application of reticular chemistry that would otherwise be very difficult.

9.8 Concluding remarks

In summary, we have provided a short introduction to metal–organic compounds, starting from fundamental examples of coordination chemistry dated back to the end of the 19th century. The historical overview covered Werner and Hofmann complexes that enabled fundamental studies based on molecular porous crystals and multidimensional porous networks. Coordination polymers sustained by metal ions connected through organic linkers attracted considerable interest in the 1990s and have led to numerous structures with interesting properties. The utilization of strong bonds, i.e. metal–carboxylate bonds, at the end of the 1990s, represented a turning-point in the field, since it enabled the synthesis of framework materials with structural integrity and permanent porosity. Subsequently, very high surface area MOFs were synthesized, with a capacity that exceeded all other known porous inorganic and organic materials. With respect to the ever increasing interest in metal–organic frameworks throughout the last 15 years, we anticipate a promising and long-lasting future for this relatively new class of material.

9.9 Acknowledgement

Funding of MOF research in the Yaghi group is supported by BASF SE (Ludwigshafen, Germany), the US Department of Defense, Defense Threat Reduction Agency (HDTRA 1-12-1-0053) and the US Department of Energy, Office of Science, Office of Basic Energy Sciences, Energy Frontier Research Center grant (DE-SC0001015). A. S. gratefully acknowledges financial assistance from the Deutsche Forschungsgemeinschaft (DFG, SCHO1639/1-1).

References

1. Baerlocher, C., McCusker, L.B., Olson, D.H. (2007) *Atlas of Zeolite Framework Types*, 6th edn, Elsevier Science, Amsterdam.
2. Zhao, D., Feng, J., Huo, Q. et al. (1998) Triblock copolymer syntheses of mesoporous silica with periodic 50 to 300 Angstrom pores, *Science*, **279**, 548–552.
3. Zakhidov, A.A., Baughman, R.H., Iqbal, Z. et al. (1998) Carbon structures with three-dimensional periodicity at optical wavelengths, *Science*, **282**, 897–901.

4. Constable, E.C., Housecroft, C.E. (2013) Coordination chemistry: the scientific legacy of Alfred Werner, *Chemical Society Reviews*, **42**, 1429–1439.
5. Bowman-James, K. (2005) Alfred Werner revisited: The coordination chemistry of anions, *Accounts of Chemical Research*, **38**, 671–678.
6. Furukawa, H., Cordova, K.E., O'Keeffe, M., Yaghi, O.M. (2013) The chemistry and applications of metal–organic frameworks, *Science*, **341**, 1230444.
7. Eddaoudi, M., Moler, D.B., Li, H. et al. (2001) Modular chemistry: Secondary building units as a basis for the design of highly porous and robust metal – organic carboxylate frameworks, *Accounts of Chemical Research*, **34**, 319–330.
8. Moulton, B., Zaworotko, M.J. (2001) From molecules to crystal engineering: Supramolecular isomerism and polymorphism in network solids, *Chemical Reviews*, **101**, 1629–1658.
9. Kitagawa, S., Kitaura, R., Noro, S.-i. (2004) Functional porous coordination polymers, *Angewandte Chemie International Edition*, **43**, 2334–2375.
10. James, S.L. (2003) Metal–organic frameworks, *Chemical Society Reviews*, **32**, 276–288.
11. Janiak, C. (2003) Engineering coordination polymers towards applications, *Dalton Transactions*, 2781–2804.
12. Wunderlich, C.-H., Bergerhoff, G. (1994) Konstitution und Farbe von Alizarin- und Purpurin-Farblacken, *Chemische Berichte*, **127**, 1185–1190.
13. Kauffmann, G.B. (1966) *Alfred Werner. Founder of Coordination Chemistry*, Springer Verlag, Berlin.
14. Ramberg, P.J. (2003) *Chemical Structure, Spatial Arrangement: The Early History of Stereochemistry, 1874–1914*, Ashgate, Aldershot.
15. Werner, A. (1893) Beitrag zur Konstitution anorganischer Verbindungen, *Zeitschrift für anorganische Chemie*, **3**, 267–330.
16. Schaeffer, W.D., Dorsey, W.S., Skinner, D.A., Christian, C.G. (1957) Separation of xylenes, cymenes, methylnaphthalenes and other isomers by clathration with inorganic complexes, *Journal of the American Chemical Society*, **79**, 5870–5876.
17. Soldatov, D.V., Enright, G.D., Ripmeester, J.A. (2004) Polymorphism and pseudopolymorphism of the [Ni(4-methylpyridine)$_4$(NCS)$_2$] Werner complex, the compound that led to the concept of "organic zeolites", *Crystal Growth & Design*, **4**, 1185–1194.
18. Allison, S.A., Barrer, R.M. (1969) Sorption in the β-phases of transition metal(II) tetra-(4-methylpyridine) thiocyanates and related compounds, *Journal of the Chemical Society A: Inorganic, Physical, Theoretical*, 1717–1723.
19. Hofmann, K.A., Küspert, F. (1897) Verbindungen von Kohlenwasserstoffen mit Metallsalzen, *Zeitschrift für anorganische Chemie*, **15**, 204–207.
20. Rayner, J.H., Powell, H.M. (1952) 67. Structure of molecular compounds. Part 9. Crystal structure of the compound of benzene with an ammonia–nickel cyanide complex, *Journal of the Chemical Society (Resumed)*, 319–328.
21. Iwamoto, T., Miyoshi, T., Miyamoto, T. et al. (1967) The metal ammine cyanide aromatics clathrates. I. The preparation and stoichiometry of the diamminemetal(II) tetracyano-niccolate(II) dibenzene and dianiline, *Bulletin of the Chemical Society of Japan*, **40**, 1174–1178.
22. Iwamoto, T., Nakano, T., Morita, M. et al. (1968) The Hofman-type clathrate: $M(NH_3)_2M'(CN)_4 \cdot 2G$, *Inorganica Chimica Acta*, **2**, 313–316.
23. Walker, G.F., Hawthorne, D.G. (1967) Complexes between n-alkylamines and nickel cyanide, *Transactions of the Faraday Society*, **63**, 166–174.
24. Miyoshi, T., Iwamoto, T., Sasaki, Y. (1972) The structure of catena-μ-ethylenediaminecadmium(II) tetracyanoniccolate(II) dibenzene clathrate: Cd(en)Ni(CN)$_4$.2C$_6$H$_6$, *Inorganica Chimica Acta*, **6**, 59–64.
25. Mathey, Y., Mazières, C., Setton, R. (1977) Formation of 3-dimensional structures using bifunctional ligand bridges between nickel cyanide planes, *Inorganic and Nuclear Chemistry Letters*, **13**, 1–3.
26. Mathey, Y., Setton, R., Mazieres, C. (1977) Molecular inclusions in the paraffin layers of the nickel cyanide n-alkylamine systems, *Canadian Journal of Chemistry*, **55**, 17–20.
27. Hasegawa, T., Nishikiori, S.-i., Iwamoto, T. (1984) The crystal structures of α,ω-diaminoalkanecadmium(II) tetracyanonickelate(II)–aromatic molecule inclusion compounds. II. 1,6-diaminohexanecadmium(II) tetracyanonickelate(II)-*o*-toluidine (1/1): Cd(NH$_2$(CH$_2$)$_6$NH$_2$)Ni(CN)$_4$.*o*-CH$_3$C$_6$H$_4$NH$_2$, *Journal of Inclusion Phenomena*, **2**, 351–357.

28. Hasegawa, T., Nishikiori, S.-i., Iwamoto, T. (1985) Isomer selection of 1,6-diaminohexanecadmium(II) tetracyanonickelate(II) for m- and p-toluidine. Formation of 1,6-diaminohexanecadmium(II) tetracyanonickelate(II)-m-toluidine(1/1) inclusion compound and bis(p-toluidine)-1,6-diaminohexanecadmium(II) tetracyanonickelate(II) complex, *Chemistry Letters*, **14**, 1659–1662.
29. Klein, R.M., Bailar, J.C. (1963) Reactions of coordination compounds. Polymers from 3-substituted bis-(β-diketone)-beryllium complexes, *Inorganic Chemistry*, **2**, 1190–1194.
30. Ramprasad, D., Pez, G.P., Toby, B.H. et al. (1995) Solid state lithium cyanocobaltates with a high capacity for reversible dioxygen binding: Synthesis, reactivity, and structures, *Journal of the American Chemical Society*, **117**, 10694–10701.
31. Kinoshita, Y., Matsubara, I., Saito, Y. (1959) The crystal structure of bis(succinonitrilo)copper(I) nitrate, *Bulletin of the Chemical Society of Japan*, **32**, 741–747.
32. Kinoshita, Y., Matsubara, I., Saito, Y. (1959) The crystal structure of bis(glutaronitrilo)copper(I) nitrate, *Bulletin of the Chemical Society of Japan*, **32**, 1216–1221.
33. Kinoshita, Y., Matsubara, I., Higuchi, T., Saito, Y. (1959) The crystal structure of bis(adiponitrilo)copper(I) nitrate, *Bulletin of the Chemical Society of Japan*, **32**, 1221–1226.
34. Morgan, H.H. (1923) CCCXLII.–Preparation and stability of cuprous nitrate and other cuprous salts in presence of nitriles, *Journal of the Chemical Society, Transactions*, **123**, 2901–2907.
35. Wells, A. (1954) The geometrical basis of crystal chemistry. Part 1, *Acta Crystallographica*, **7**, 535–544.
36. Carlucci, L., Ciani, G., Proserpio, D.M. (2003) Polycatenation, polythreading and polyknotting in coordination network chemistry, *Coordination Chemistry Reviews*, **246**, 247–289.
37. Aumüller, A., Erk, P., Klebe, G. et al. (1986) A radical anion salt of 2,5-dimethyl-N,N′-dicyanoquinonediimine with extremely high electrical conductivity, *Angewandte Chemie International EditionEnglish*, **25**, 740–741.
38. Kato, R., Kobayashi, H., Kobayashi, A. (1989) Crystal and electronic structures of conductive anion–radical salts, (2,5-R_1,R_2-DCNQI)$_2$Cu (DCNQI = *N,N′*-dicyanoquinonediimine; R_1, R_2 = CH_3, CH_3O, Cl, Br), *Journal of the American Chemical Society*, **111**, 5224–5232.
39. Hoskins, B.F., Robson, R. (1989) Infinite polymeric frameworks consisting of three dimensionally linked rod-like segments, *Journal of the American Chemical Society*, **111**, 5962–5964.
40. Hoskins, B.F., Robson, R. (1990) Design and construction of a new class of scaffolding-like materials comprising infinite polymeric frameworks of 3D-linked molecular rods. A reappraisal of the $Zn(CN)_2$ and $Cd(CN)_2$ structures and the synthesis and structure of the diamond-related frameworks $[N(CH_3)_4][Cu^IZn^{II}(CN)_4]$ and Cu^I[4,4′,4′′,4′′′-tetracyanotetraphenylmethane]$BF_4{\cdot}xC_6H_5NO_2$, *Journal of the American Chemical Society*, **112**, 1546–1554.
41. Kitazawa, T., Nishikiori, S.-i., Kuroda, R., Iwamoto, T. (1988) Novel clathrate compound of cadmium cyanide host with an adamantane-like cavity. cadmium cyanide–carbon tetrachloride(1/1), *Chemistry Letters*, **17**, 1729–1732.
42. Zaworotko, M.J. (1994) Crystal engineering of diamondoid networks, *Chemical Society Reviews*, **23**, 283–288.
43. Gable, R.W., Hoskins, B.F., Robson, R. (1990) Synthesis and structure of [NMe4][CuPt$(CN)_4$]: an infinite three-dimensional framework related to PtS which generates intersecting hexagonal channels of large cross section, *Journal of the Chemical Society, Chemical Communications*, 762–763.
44. Abrahams, B.F., Hoskins, B.F., Michail, D.M., Robson, R. (1994) Assembly of porphyrin building blocks into network structures with large channels, *Nature*, **369**, 727–729.
45. Gardner, G.B., Venkataraman, D., Moore, J.S., Lee, S. (1995) Spontaneous assembly of a hinged coordination network, *Nature*, **374**, 792–795.
46. Kaszynski, P., Friedli, A.C., Michl, J. (1992) Toward a molecular-size tinkertoy construction set. Preparation of terminally functionalized [*n*]staffanes from [1.1.1]propellane. *Journal of the American Chemical Society*, **114**, 601–620.
47. Fujita, M., Kwon, Y.J., Washizu, S., Ogura, K. (1994) Preparation, clathration ability, and catalysis of a two-dimensional square network material composed of cadmium(II) and 4,4′-bipyridine, *Journal of the American Chemical Society*, **116**, 1151–1152.
48. Fujita, M., Yazaki, J., Ogura, K. (1990) Preparation of a macrocyclic polynuclear complex, $[(en)Pd(4,4'\text{-bpy})]_4(NO_3)_8$ (en = ethylenediamine, bpy = bipyridine), which recognizes an organic molecule in aqueous media, *Journal of the American Chemical Society*, **112**, 5645–5647.

49. Gable, R.W., Hoskins, B.F., Robson, R. (1990) A new type of interpenetration involving enmeshed independent square grid sheets. The structure of diaquabis-(4,4′-bipyridine)zinc hexafluorosilicate, *Journal of the Chemical Society, Chemical Communications*, 1677–1678.
50. Robson, R., Abrahams, B.F., Batten, S.R., Gable, R.W. Hoskins, B.F. and Liu, J. (1992) Crystal engineering of novel materials composed of infinite two- and three-dimensional frameworks, In *Supramolecular Architecture*, vol. **499**, T. Bein (Ed), American Chemical Society, Washington, DC, pp. 256–273.
51. Real, J.A., Andrés, E., Muñoz, M.C. et al. (1995) Spin crossover in a catenane supramolecular system, *Science*, **268**, 265–267.
52. Halder, G.J., Chapman, K.W., Neville, S.M. et al. (2008) Elucidating the mechanism of a two-step spin transition in a nanoporous metal – organic framework, *Journal of the American Chemical Society*, **130**, 17552–17562.
53. Neville, S.M., Halder, G.J., Chapman, K.W. et al. (2009) Guest tunable structure and spin crossover properties in a nanoporous coordination framework material, *Journal of the American Chemical Society*, **131**, 12106–12108.
54. Noro, S.-i., Kitaura, R., Kondo, M. et al. (2002) Framework engineering by anions and porous functionalities of Cu(II)/4,4′-bpy coordination polymers, *Journal of the American Chemical Society*, **124**, 2568–2583.
55. Carlucci, L., Ciani, G., Proserpio, D.M. et al. (2014) Entangled two-dimensional coordination networks: A general survey, *Chemical Reviews*, **114**, 7557–7580.
56. Yaghi, O.M., Li, H. (1995) Hydrothermal synthesis of a metal–organic framework containing large rectangular channels, *Journal of the American Chemical Society*, **117**, 10401–10402.
57. Subramanian, S., Zaworotko, M.J. (1995) Porous solids by design: $[Zn(4,4'\text{-bpy})_2(SiF_6)]_n \cdot xDMF$, a single framework octahedral coordination polymer with large square channels, *Angewandte Chemie International EditionEnglish*, **34**, 2127–2129.
58. Nugent, P., Belmabkhout, Y., Burd, S.D. et al. (2013) Porous materials with optimal adsorption thermodynamics and kinetics for CO_2 separation, *Nature*, **495**, 80–84.
59. Shekhah, O., Belmabkhout, Y., Chen, Z. et al. (2014) Made-to-order metal–organic frameworks for trace carbon dioxide removal and air capture, *Nature Communications*, **5**, 4228.
60. Li, H., Eddaoudi, M., Groy, T.L., Yaghi, O.M. (1998) Establishing microporosity in open metal–organic frameworks: Gas sorption isotherms for Zn(BDC) (BDC = 1,4-benzenedicarboxylate), *Journal of the American Chemical Society*, **120**, 8571–8572.
61. Eddaoudi, M., Li, H., Reineke, T. et al. (1999) Design and synthesis of metal–carboxylate frameworks with permanent microporosity, *Topics in Catalysis*, **9**, 105–111.
62. Yaghi, O.M., Davis, C.E., Li, G., Li, H. (1997) Selective guest binding by tailored channels in a 3-D porous zinc(II) – benzenetricarboxylate network, *Journal of the American Chemical Society*, **119**, 2861–2868.
63. Kondo, M., Yoshitomi, T., Matsuzaka, H. et al. (1997) Three-dimensional framework with channeling cavities for small molecules: $\{[M_2(4,4'\text{-bpy})_3(NO_3)_4] \cdot xH_2O\}_n$ (M=Co, Ni, Zn), *Angewandte Chemie International Edition English*, **36**, 1725–1727.
64. Li, H., Eddaoudi, M., O'Keeffe, M., Yaghi, O.M. (1999) Design and synthesis of an exceptionally stable and highly porous metal–organic framework, *Nature*, **402**, 276–279.
65. Clegg, W., Harbron, D.R., Homan, C.D. et al. (1991) Crystal structures of three basic zinc carboxylates together with infrared and FAB mass spectrometry studies in solution, *Inorganica Chimica Acta*, **186**, 51–60.
66. Eddaoudi, M., Kim, J., Rosi, N. et al. (2002) Systematic design of pore size and functionality in isoreticular MOFs and their application in methane storage, *Science*, **295**, 469–472.
67. Chui, S.S.Y., Lo, S.M.F., Charmant, J.P.H. et al. (1999) A chemically functionalizable nanoporous material $[Cu_3(TMA)_2(H2O)_3]_n$, *Science*, **283**, 1148–1150.
68. Cotton, F.A., Mester, Z.C., Webb, T.R. (1974) Dimolybdenum tetraacetate, *Acta Crystallographica Section B*, **30**, 2768–2770.
69. Ceccherelli, P., Curini, M., Marcotullio, M.C., Rosati, O. (1991) Regiocontrol by the carbon–carbon double bond in the $Rh_2(OAc)_4$ mediated carbon–hydrogen insertion of α-diazo-ketones, *Tetrahedron*, **47**, 7403–7408.
70. Millward, A.R., Yaghi, O.M. (2005) Metal – organic frameworks with exceptionally high capacity for storage of carbon dioxide at room temperature, *Journal of the American Chemical Society*, **127**, 17998–17999.
71. Ma, S., Sun, D., Ambrogio, M. et al. (2007) Framework–catenation isomerism in metal – organic frameworks and its impact on hydrogen uptake, *Journal of the American Chemical Society*, **129**, 1858–1859.

72. Furukawa, H., Go, Y.B., Ko, N. et al. (2011) Isoreticular expansion of metal–organic frameworks with triangular and square building units and the lowest calculated density for porous crystals, *Inorganic Chemistry*, **50**, 9147–9152.
73. Allen, F. (2002) The Cambridge Structural Database: A quarter of a million crystal structures and rising, *Acta Crystallographica Section B*, **58**, 380–388.
74. Groom, C.R., Allen, F.H. (2014) The Cambridge Structural Database in retrospect and prospect, *Angewandte Chemie International Edition*, **53**, 662–671.
75. Buergi, H.B., Dunitz, J.D. (1983) From crystal statics to chemical dynamics, *Accounts of Chemical Research*, **16**, 153–161.
76. Allen, F.H., Kennard, O., Taylor, R. (1983) Systematic analysis of structural data as a research technique in organic chemistry, *Accounts of Chemical Research*, **16**, 146–153.
77. Hall, S.R., Allen, F.H., Brown, I.D. (1991) The crystallographic information file (CIF): a new standard archive file for crystallography, *Acta Crystallographica Section A*, **47**, 655–685.
78. Brown, I.D., McMahon, B. (2002) CIF: the computer language of crystallography, *Acta Crystallographica Section B*, **58**, 317–324.

10

Cyclodextrin-based Supramolecular Systems

Akira Harada

Graduate School of Science, Osaka University, Osaka, Japan

10.1 Introduction

Molecular recognition [1], host–guest chemistry [2], and supramolecular chemistry [3] have received much attention in the past few decades. Crown ethers, cryptands, cyclophanes, and cucurbiturils have been extensively used as host molecules. Pedersen, Cram, and Lehn received the Nobel Prize in Chemistry in 1987 for the development and use of highly selective and structurally sensitive molecules.

The molecular recognition selectivity of these molecules is much poorer than that of biological systems. Our research strives to achieve specific molecular recognition similar to biological systems for larger and more complicated molecules using simple artificial organic systems. We have focused on cyclodextrins (CDs) as host molecules for studies on specific molecular recognition and the construction of the supramolecular architecture. CDs are cyclic molecules consisting of six to eight glucose units and are known to form inclusion complexes with various molecules (see Figure 10.1) [4]. They are produced from starch, which can be synthesized using carbon dioxide, water, and solar energy. CDs are renewable, environmentally benign, and safe in our body.

10.2 Cyclodextrin-containing polymers

10.2.1 Preparation

Using cyclodextrin-containing monomers, we have synthesized various homopolymers and copolymers [5]. CD acrylates were prepared using ester-exchange reactions of *m*-nitrophenyl esters with CDs via host–guest interactions. Because the reaction forms an inclusion complex, mono-functional derivatives are predominantly realized. Pure monomers containing CDs obtained by column chromatography can be polymerized by radical polymerization to give water-soluble homopolymers and various copolymers (see Figure 10.2).

Macrocyclic and Supramolecular Chemistry: How Izatt–Christensen Award Winners Shaped the Field, First Edition.
Edited by Reed M. Izatt.

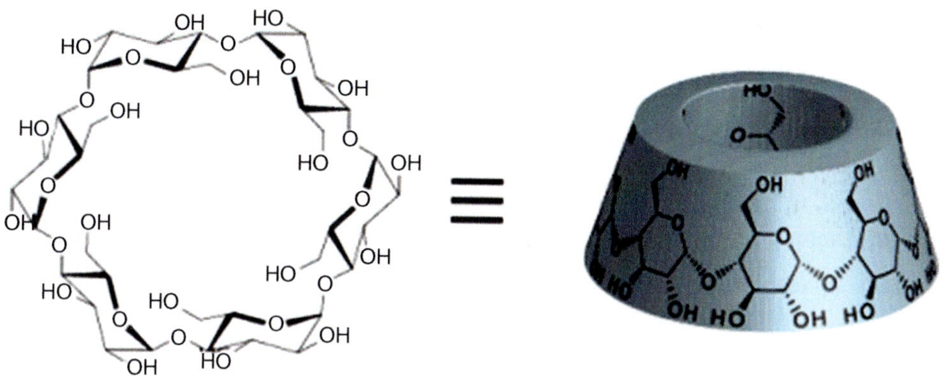

Figure 10.1 *Cyclodextrin*

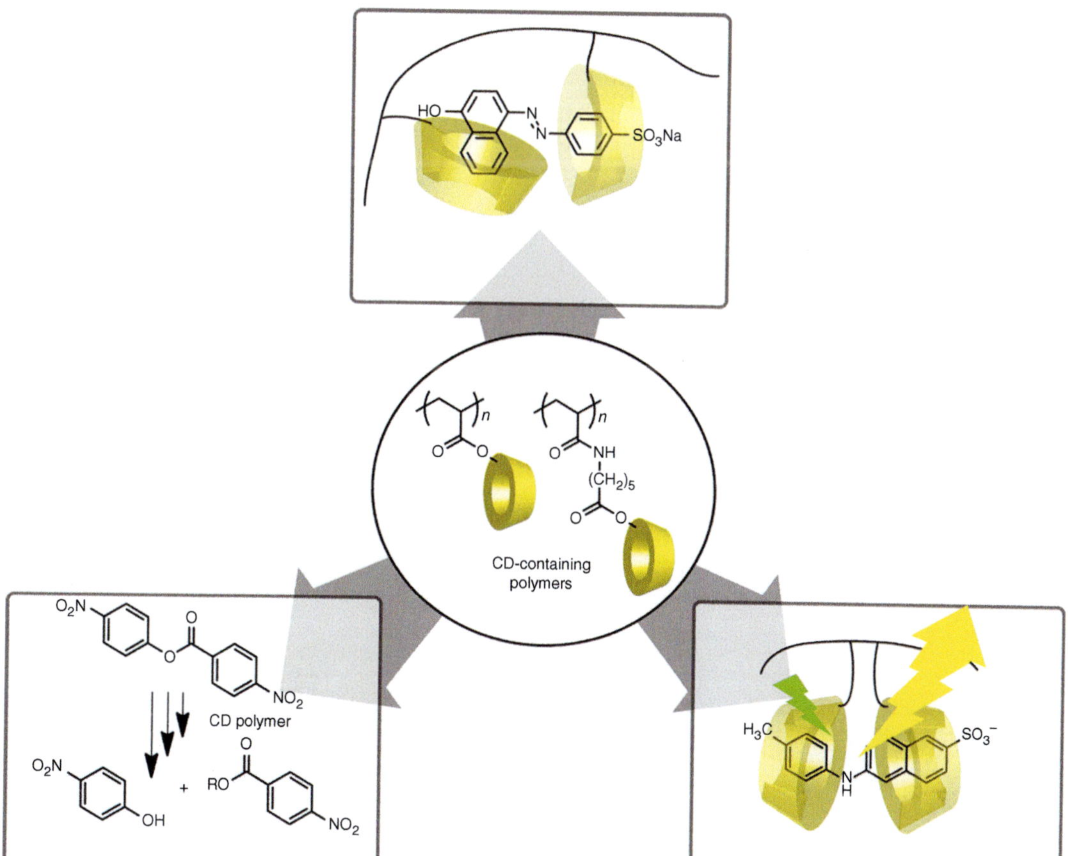

Figure 10.2 *Cyclodextrin-containing polymers*

10.2.2 Cooperativity

These polymers include long molecules more efficiently than short, small molecules [6] due to cooperative effects between the CD moieties. In addition, we found that the CD polymers accelerate the ester-hydrolysis of some activated esters. In this case, the effects are also selective [7]. Furthermore, some fluorescent probe molecules, ANS or TNS, may be included cooperatively in the CD cavities in the polymers, yielding highly fluorescent species.

10.2.3 Optical resolution by CD

We carried out the optical resolution of mandelic acid and its derivatives by column chromatography using CDs [8] where the stationary phase was a hydrogel obtained by crosslinking CD with epichlorohydrin. The configuration realizes optically pure mandelic acid derivatives.

10.3 CD-organometallic complexes

CDs form inclusion complexes with some organometallic complexes with a high selectivity [9]. For example, we found that ferrocenes form inclusion complexes with CDs [10]. Although α-CD realizes 2:1 complexes with ferrocene, β- and γ-CD form 1:1 complexes. The structure of the β-CD–ferrocene complex was elucidated by single crystal X-ray studies (see Figure 10.3) [11]. Our studies also revealed that racemic ferrocene

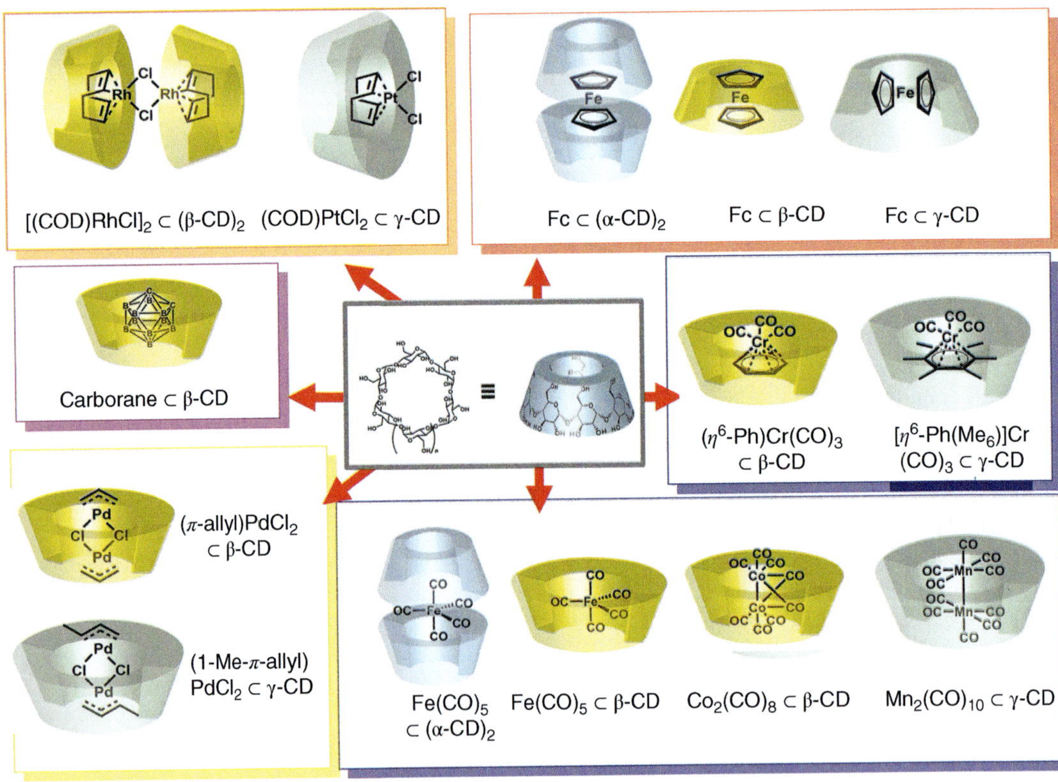

Figure 10.3 Cyclodextrin–organometallic complexes

derivatives can be optically resolved using CD as a mobile phase in column chromatography. Our research was the first demonstration of CDs completely resolving racemic compounds [12].

We found that arene–chromcarbonyl derivatives form inclusion complexes with CDs selectively [13]. Benzene–chromcarbonyl forms a 1:1 complex with β-CD but not with α-CD. Hexamethylbenzene-chromcarbonyl compounds form a 1:1 complex only with γ-CD.

In addition, a η^4-cyclooctadiene Rh or Pt complex forms inclusion complexes with β-CD and γ-CD [14], but not α-CD. Using β-CD and γ-CD, we found that cyclooctadiene ligands can be observed separately in the ^{1}H NMR spectra of the inclusion complex and the free complex. Stoddart et al. also reported complexation between CDs and Pt complexes (*cis*-platin) [15].

Some other complexes between CDs and organometallics such as η^3-allyl Pd complexes [16] or carboranes [17] have been reported. CDs have even been reported to form inclusion complexes with metal (Cr, Fe, Mn)–carbonyl complexes, usually with a 1:1 stoichiometry. We found that Wacker reactions, which oxidize alkenes by palladium, are selectively accelerated by CDs in reversed phase transfer reactions [18].

10.4 Complex formation of cyclodextrin with polymers

10.4.1 Macromolecular recognition

In biological systems, macromolecular recognition where macromolecules recognize macromolecules plays an important role in maintaining life. Enzymes recognize substrates such as proteins and polysaccharides. Antibodies specifically recognize antigens, while DNA recognizes each strand to give double strands.

Our research has focused on realizing macromolecular recognition using artificial synthetic systems. Macromolecular recognition can be classified either as main-chain recognition or side-chain recognition [19]. Main-chain recognition may be involved in the formation of *pseudo*-poly rotaxanes. Each unit of the polymers is included and recognized by a cyclic host molecule. In our research, we chose cyclodextrins (CDs) as the host molecule.

Adding polyethylene glycol (PEG) to an aqueous solution of α-CD causes the solution to become turbid and a crystalline precipitate to form [20], while adding polypropylene glycol (PPG) to a α-CD solution has no effect. In contrast, adding PEG to a β-CD solution does not induce a change, but adding PPG to a β-CD solution causes the solution to become turbid and complexes form. These observations confirm that α-CD and β-CD exhibit completely different behaviors [21].

In addition, poly(methyl vinyl ether) (PMVE), which has the same composition as PPG, forms complexes only with the largest CD, γ-CD [22] but not α-CD or β-CD. Although PMVE and PPG are isomers, the PPG oxygen atom is in the main chain, whereas the PMVE oxygen atom is in the side-chain. In other words, PMVE has a larger cross-sectional area than PPG. In this case, only γ-CD forms complexes with PMVE. There is a good correlation between the polymer thickness and CD cavity size.

We also found that CDs form complexes with hydrophobic polymers with a high selectivity. α-CD, for example, forms complexes only with poly(ethylene) (PE) [23], and not with polypropylene (PP) [24, 25]. On the other hand, β-CD realizes complexes with PP but not with PE. γ-CD forms complexes only with poly(isobutylene) efficiently [25]. Again, polymer thickness and polymer size are well correlated. These results indicate that the polymer chain is included in the CD cavities.

We tested complexation of CDs with PEG with bulky groups at the end of the polymer chain. α-CD does not form inclusion complexes with PEG when there are bulky substituents at both ends of the polymer chain. However, α-CD does form inclusion compounds with PEG that has a bulky group at one end of the chain and a small group at the other. Additionally, α-CD does not form inclusion complexes with large cyclic PEGs [26],

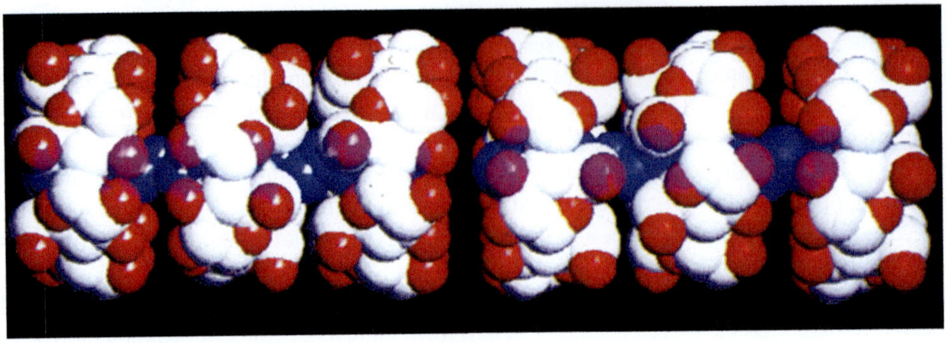

Figure 10.4 Single-crystal X-ray studies on a-CD-hexa(ethylene glycol) complex. Reproduced from ref. [29] with permission of Royal Society of Chemistry

suggesting that small end-groups are required for complexation. These results demonstrate that the α-CD ring threads from the small end-groups.

Interestingly, α-CD and PEG form stoichiometric complexes regardless of the mixing ratio. The Job plots for complex formation between α-CD and PEG (MW = 1600) show that CD includes two PEG units. We confirmed the composition of CD and PEG in the complexes using ^{1}H NMR. All the complexes have the same 2:1 composition (two ethylene glycol units per single CD). Molecular model studies show that the α-CD ring (6.7 Å) can accommodate two ethylene glycol units (7 Å). In addition, the PEG chain is slim enough to be included in the α-CD ring.

Years later, we prepared single crystals of complexes with α-CD and hexa(ethylene glycol) (HEG). Our X-ray crystallographic studies of a single crystal of a complex between α-CD and HEG revealed that the HEG chain is included in a tunnel formed by α-CD (see Figure 10.4) [27].

10.4.2 Polyrotaxanes

As mentioned above, α-CD forms inclusion complexes with PEG except when PEG possesses bulky substituents at both ends of the polymer chain. Therefore, we hypothesized that if complexes of α-CD with PEG are initially isolated and both ends of the polymer chain are subsequently capped by bulky groups, polyrotaxanes in which many CDs are entrapped along the polymer chain could be realized [28]. Actually, we prepared inclusion complexes of α-CDs with PEG bisamine and subsequently capped both ends using 2,4-dinitro fluorobenzene to give polyrotaxanes (see Figure 10.5). This is a unique molecule because many cyclic molecules (CDs) are imprisoned along the polymer chain. Although the ring molecules and the polymer chain do not have covalent bonds, the CD rings cannot escape [29], because each component is mechanically interlocked. In polyrotaxanes, the ring component can move and rotate along the polymer chain.

10.4.3 Tubular polymers

Tubular polymers have been prepared starting from polyrotaxanes. We used epichlorohydrin as a binding agent to bind neighboring CDs by short cross-linking reagents. Although tubular polymers are formed, the polymer chain remains in the CD cavity. To remove the polymer chain, bulky groups at the chain ends were removed and the polymer chain was threaded out, yielding tubular polymers with diameters less than 1 nm (see Figure 10.6) [30].

Tubular polymers are water soluble and flexible. Because they are derived from glucose, tubular polymers are safe in the human body. In addition, tubular polymers are renewable and environmentally friendly.

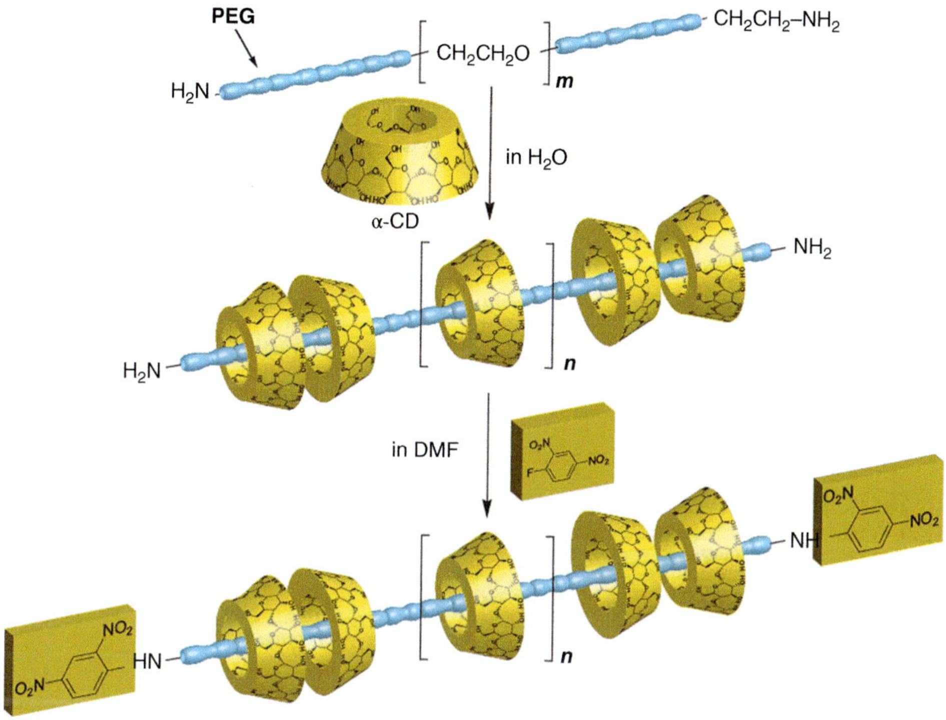

Figure 10.5 Polyrotaxane

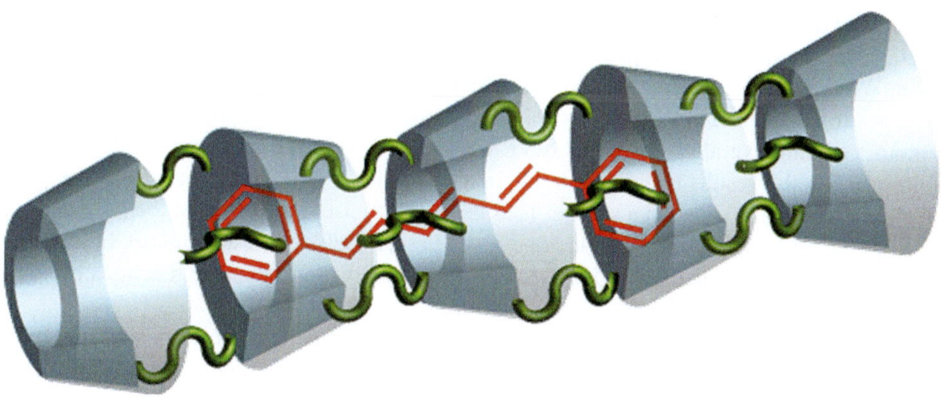

Figure 10.6 Molecular tube

10.5 Polymerization by CDs

We found that CDs form inclusion complexes with some polyesters such as poly(ε-caprolactone) [31], poly(valerolactone) (PVL), and poly(alkylene adipate) [32] to give *pseudo*-polyrotaxanes. PVL forms inclusion complexes with α-CD in a 1:1 (CD:monomer unit) ratio. Adding PVL to an aqueous solution α-CD followed by heating and sonicating gives complexes as white solids. Moreover, we discovered that CDs

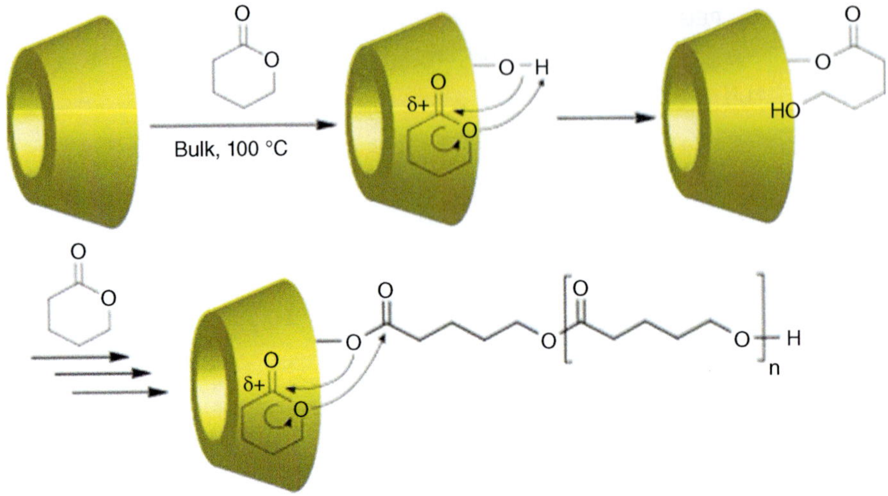

Figure 10.7 Polymerization mechanism 1

selectively form inclusion complexes with some lactones (the starting materials of the polyesters) and that the hydrolysis of select lactones is promoted or suppressed according to the CD size [33].

While preparing inclusion complexes of PVL with α-CD, we found that the polymer is partially hydrolyzed to give polymers with lower molecular weights, even though the polymers are not easily hydrolyzed under the same conditions without CD. In addition, their monomers, valerolactone(δ-VL) and caprolatone(ε-CL), are hydrolyzed in water in the presence of CDs. These results indicate that lactones and polymers can be hydrolyzed by CDs like hydrolyzing enzymes. Therefore, we speculated that if CDs and lactones (cyclic esters) are mixed and heated without water, they could form polymers because hydrolysis does not occur. We used α-, β-, and γ-CDs and butyrolactone (β-BL), δ-VL, and ε-CL, as monomers.

Mixing and heating β-CD and δ-VL without solvent produces poly δ-VL in high yields [34]. Although the reaction is almost quantitative after 48 hours, δ-VL does not produce any polymers without CD under the same conditions. γ-CD also results in a polymer, but in a lower yield. α-CD does not give any polymers under the same conditions. In contrast, when a smaller lactone, β-BL, is used, the order of the polymer yields with CDs is α-CD, β-CD, and γ-CD. A smaller lactone gives higher yields with a smaller CD (α-CD), while a larger lactone produces higher yields with a larger CD (γ-CD), indicating that the reactions occur via inclusion of cyclic esters in the CD cavities.

An induction period occurs when using β-CD. In contrast, the polymerization reactions start smoothly and efficiently when an inclusion complex of δ-VL with β-CD is used as the initiator. These results indicate that the formation of inclusion complexes is the rate-determining step for the polymerization reaction. When an inclusion complex of adamantane carboxylic acid (AdCxacid) with β-CD is used as an initiator, polymerization does not occur, indicating that lactones are inaccessible to the CD cavities because AdCxacid is strongly included in the β-CD cavity.

The MALDI-TOF mass spectrum of poly(δ-VL) obtained from a mixture of β-CD and δ-VL shows that each polymer has a single β-CD at the end of the polymer chain. A possible mechanism for the polymerization of lactones is shown in Figure 10.7. The first step is complexation of lactones with CD to give a 1:1 inclusion complex. Then the secondary –OH group of CD should attack the carbonyl carbon of the included cyclic esters, cleaving the C–O bond to form a CD-ester.

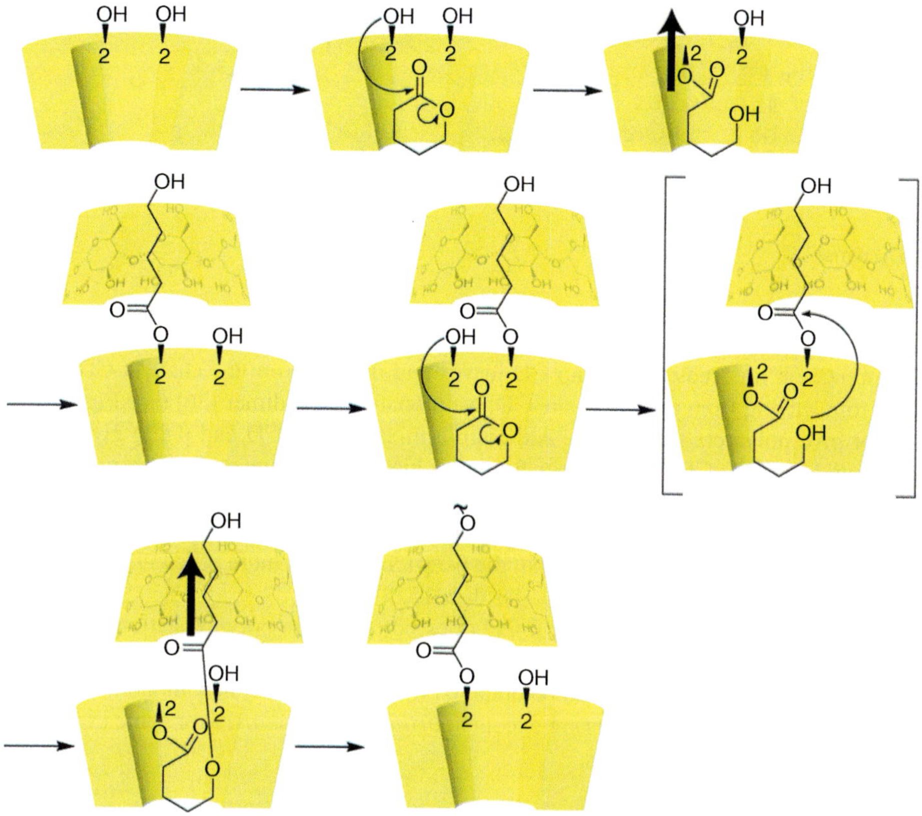

Figure 10.8 *Polymerization mechanism 2*

After the monomers are consumed by the polymerization reactions, new monomers can be added into the reaction mixture. Further polymerization is observed, but the purified and isolated polymer does not re-initiate polymerization. Even if the isolated polymer is used, when the polymer chain is included by CDs, polymerization starts again. In this case, the polyester chain is included in the β-CDs to give *pseudo*-polyrotaxane. Consequently, the polymer chain is extended and the CD cavities open for the new monomers, reinitiating polymerization.

Although α-CD cannot initiate polymerization of δ-VL when α-CD is added onto the isolated product, the product reinitiates polymerization of lactone, indicating that β-CD with a single polyester chain is included by α-CD and can reinitiate the polymerization. The *pseudo*-rotaxane structure plays an important role in the polymerization. CD has two roles: inclusion of the monomers to initiate and propagate the polyester and inclusion of the polyester to support polymerization similar to Chaperon proteins [35].

Figure 10.8 shows a more detailed mechanism. The hydroxyl group of β-CD attacks the C–O bond of the included lactone (initiation). Then another CD extends the opened chain by pulling it out from the cavity. Next the –OH group of the CD attacks the second lactone to open the cyclic ester. Soon after ring cleavage, the O anion attacks the growing ester to insert the second lactone between the CD and the growing chain.

The proposed mechanism is similar to the clamp part of DNA polymerase, which is shaped like a doughnut and encircles double-helix DNA. The DNA chain is elongated in the enzyme. Although the clamp part does not have a catalytic site, polymerization does not proceed without it. Its structure is similar to that of the CD system.

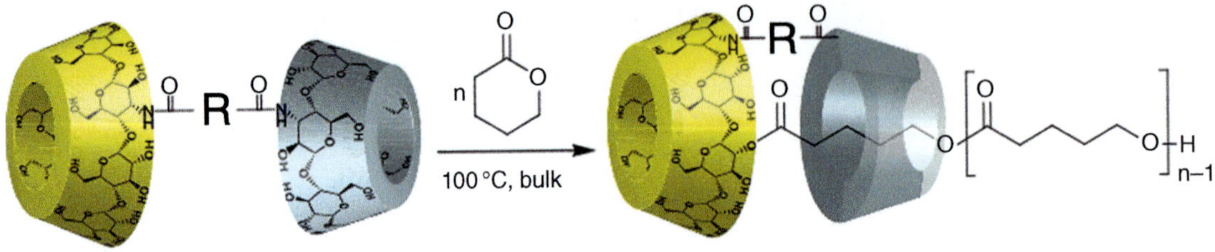

Figure 10.9 Polymerization of lactones by cyclodextrin dimer

As the polymerization proceeds, the clamp CD leaves and departs from the catalyst CD. To prevent the clamp part CD from leaving from the active site CD, we prepared a CD dimer [36] (see Figure 10.9).

α-CD does not give polymers from δ-VL. Although a mixture of α-CD and β-CD gives only low MW polymers in low yields, the α-CD–β-CD dimer produces high MW polymers in high yields, suggesting a cooperative effect on the polymerization reactions.

The combination of CDs in the dimers is also important. The α-CD dimer does not give polymers. The α – β-CD dimer is effective for polymerization, but the β-, γ-CD dimer is more efficient. The polymer yield is 96% using the β-, γ-CD dimer and the MW is larger than 15,000.

Although the isolated polymer obtained by β-CD does not reinitiate polymerization, the CD dimer can initiate polymerization of new monomers to give polymers with higher molecular weights. For example, poly(δ-VL) with MW higher than 80,000 can be obtained by the addition of new monomers.

As this section demonstrates, CD dimers are excellent *artificial polymerases* for some cyclic esters, yielding polyesters with high molecular weights.

10.6 Supramolecular polymers

In biological systems, the structures of virus, phages, microtubules, filaments, and flagella are composed through host–guest interactions. Dimers, oligomers, helices, rings, and filaments are formed.

We chose CD as the host and a benzene ring as the guest because the benzene ring can fit in the CD cavity. However, benzoyl CD does not form supramolecular polymers, suggesting that a spacer group is required to efficiently form intermolecular complexes. Thus, we tested a longer group as the guest moiety [37].

Flexible hydrocinnamoyl β-CD forms intramolecular complexes, and a hydrocinnamoyl α-CD derivative gives weak intermolecular complexes. Rigid cinnamoyl β-CD is almost insoluble in water, but the X-ray data of the single crystal of this compound shows that the guest part is included in another CD cavity to form a supramolecular dimer. In contrast, the rigid cinnamoyl α-CD derivative is water-soluble and the molecular weight is about 3000, which is three times larger than the monomer, indicating that the rigid α-CD forms a trimer. Attaching a bulky group to the cinnamoyl α-CD derivative in water gives supramolecular oligomers. The mass spectra and ^{1}H NMR spectra of the product show that the product is a cyclic trimer (a cyclic daisy chain) [38].

Although cinnamoyl β-CD has a limited solubility in water, adding AdCrbx to a suspension of this compound can solubilize the rigid β-CD derivative to give a clear solution, indicating that AdCrX acid is included in the β-CD cavity in lieu of the benzene group to give soluble complex. Adding α-CD to this solution allows the benzene group to be included in the α-CD cavity. Treating the complex with bulky groups produces a unique [2]rotaxane. The rotaxane forms supramolecular polymers to give an α – β – α – β alternating structure [39]. Figure 10.10, shows an AFM image of the supramolecular polymer. Part of the supramolecule can be cut off by an STM tip.

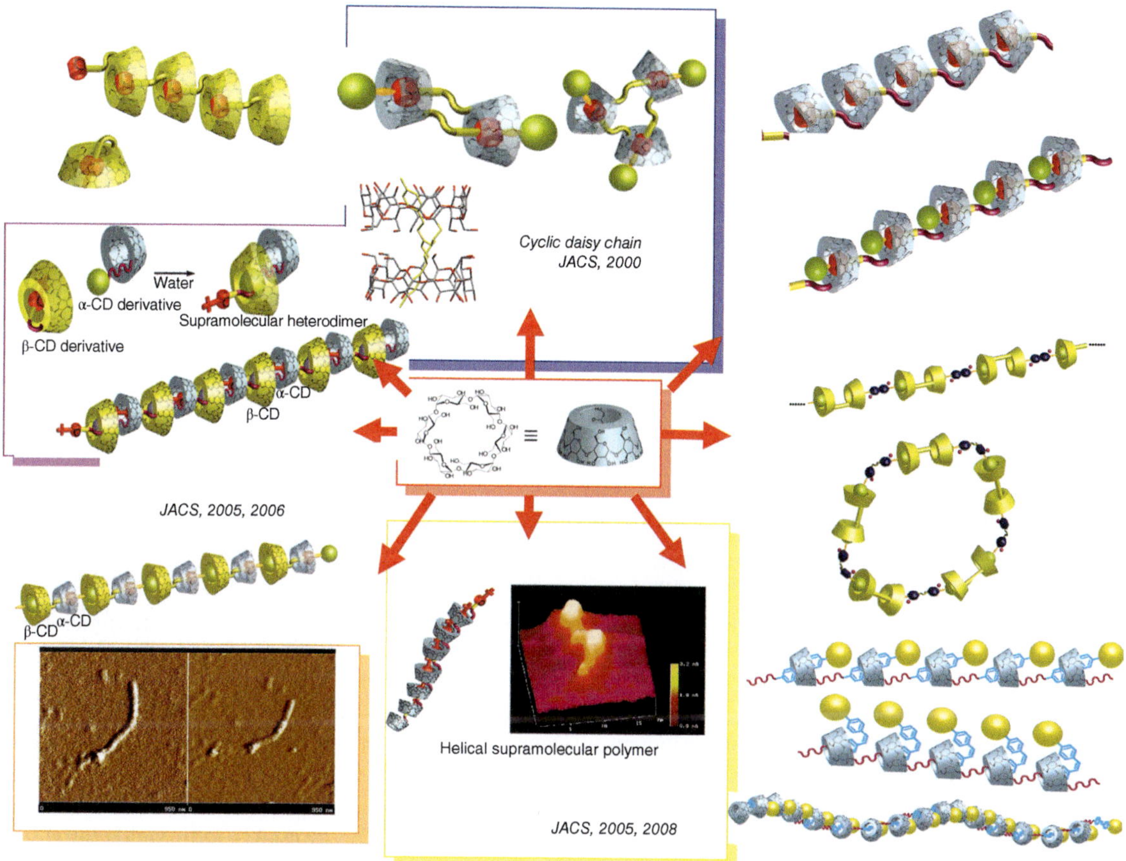

Figure 10.10 *Supramolecular polymers*

When a guest group is attached by an amide bond, a cyclic dimer is formed. Attaching a guest group on the other side (the secondary OH group side) of CD with an amide bond forms a longer chain. The MW increases as the concentration increases and approaches a MW of 20,000. The circular dichroism spectrum of the complex shows negative bands, indicating that the guest moiety is included in the α-CD cavity in a slantwise (tilted) orientation [40].

When a *t*-butyl group is attached to the cinnamoyl group, more stable supramolecular polymers are formed because the *t*-butyl group is tightly included in the CD cavity. The *t*-butyl derivative forms longer supramolecular polymers even at much lower concentrations than cinnamoyl derivatives without *t*-butyl groups. The circular dichroism spectrum of the *t*-butyl derivative shows that the guest parts can be oriented in a left-handed anti-configuration. These results indicate that helical supramolecular polymers are formed. Figure 10.10 shows an STM image of the helical supramolecular polymers formed by *tert*-Boc cinnamoyl CD.

When *t*-Boc cinnamate-β-CD and Ad-appended-α-CD (1:1 ratio) are mixed, an αβ−αβ-alternating supramolecular copolymer is formed even though each CD derivative does not form intermolecular complexes [41]. This is because when AdCx acid appended α-CD is added onto *t*-ButCin–β-CD, the AdCa group ejects the *t*-ButylCin group, exposing the cinnamoyl group to water. Then the *t*-ButylCin group is included in a α-CD cavity to give a supramolecular polymer.

In summary, when a guest part is covalently attached to a CD host, supramolecular polymers are formed via host–guest-alternating copolymers. Not only linear supramolecular polymers but helical, cyclic, and α – β-alternating copolymers can be obtained.

10.7 Side-chain recognition by CDs

CDs form inclusion complexes with guest parts in the polymer side-chains. α-CD, for example, forms inclusion complexes with linear alkyl parts (e.g., *n*-butyl, *n*-hexyl, and *n*-octyl groups) but not with *t*-butyl, cyclohexyl, and *iso*-octyl groups as side-chains of polyacrylamides. In contrast, β-CD forms inclusion complexes with *iso*-alkyl, *t*-butyl chains, and cyclohexyl groups but not with *n*-alkyl chains [42]. CDs have a higher selectivity for polymer side-chains than for low–molecular weight guest molecules.

10.8 CD-based molecular machines

Although α-CD forms inclusion complexes with poly(ethylene glycol), it is difficult to observe how α-CD threads over the end-groups and moves along the polymer chain because both CD and PEG lack chromophores. Therefore, we prepared CD derivatives with a PEG chain via a benzene group as a probe. This compound forms an intramolecular complex (poly[1]rotaxane) at room temperature. However, adding an equimolar amount of adamantane in this solution causes the adamantane group to be included in the CD cavity and the PEG part to be ejected. When a half molar adamantine solution is added to the solution, fast exchanges between inclusion and exclusion of the PEG chain can be observed on a ^{1}H NMR time-scale. The exchange rates decrease as the chain length increases. Depending on the chain length, the exchange can vary from about 1000 times to several times a second [43].

When an azobenzene (AB) group is attached to the end of the PEG chain, we can observe the exchange between inclusion and exclusion of a PEG chain by 2D ROESY NMR. The *trans*-AB-form yields intramolecular complexes at low temperatures where the cinnamoyl group is included in the CD cavity. In contrast, the AB part is included at higher temperature and shuttling between the two occurs at room temperature. Moreover, photoisomerization causes a reversible shuttling between them at low temperature (see Figure 10.11) [44].

To observe the translational movement of CD in detail, we chose a polymethylene chain as the combining site and a cationic part as the potential surface [45]. CDs formed complexes with such cationic polymers. In this case, the complexes are soluble in water and the ^{1}H NMR spectrum of the polymer in D_2O does not change upon adding the largest CD (γ-CD). In contrast, the spectrum of the polymer splits upon adding α-CD and the peaks due to β-CD broaden. These results indicate that α-CD forms stable complexes with this polymer while β-CD gives dynamic complexes. Upon measuring the ^{1}H NMR spectra of the cationic polymer in the presence of CD, the spectrum changes with time. Thus, we observed the threading process in real time.

We prepared a molecular shuttle [46] in which the α-CD ring moves back and forth along a linear chain [47] (see Figure 10.12) where a polymethylene chain and bipyridinium groups were used as the station and the potential surface, respectively. The CD ring moves back and forth 80 times a second at 130 °C and 0.9 times per minute at 30 °C. These results are consistent with the observation that α-CD threads on a polymer chain on the minute to hour timescale.

Our experiments revealed that the cationic groups serve as a potential barrier for threading of the CD ring. We used this knowledge to successfully entrap a CD ring on an axle molecule by ionization of the groups [48]. Thus, both steric effects and electronic effects are important for controlling the threading over the end-groups.

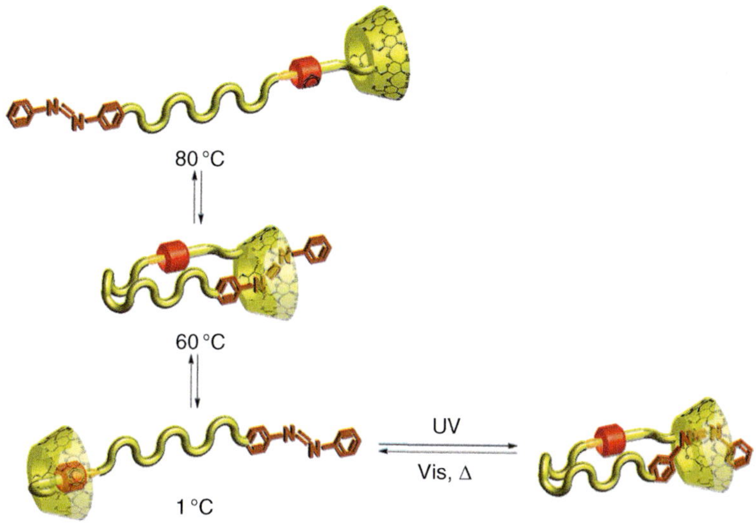

Figure 10.11 *Stimuli(temperature and photoirradiation)-responsive polymers*

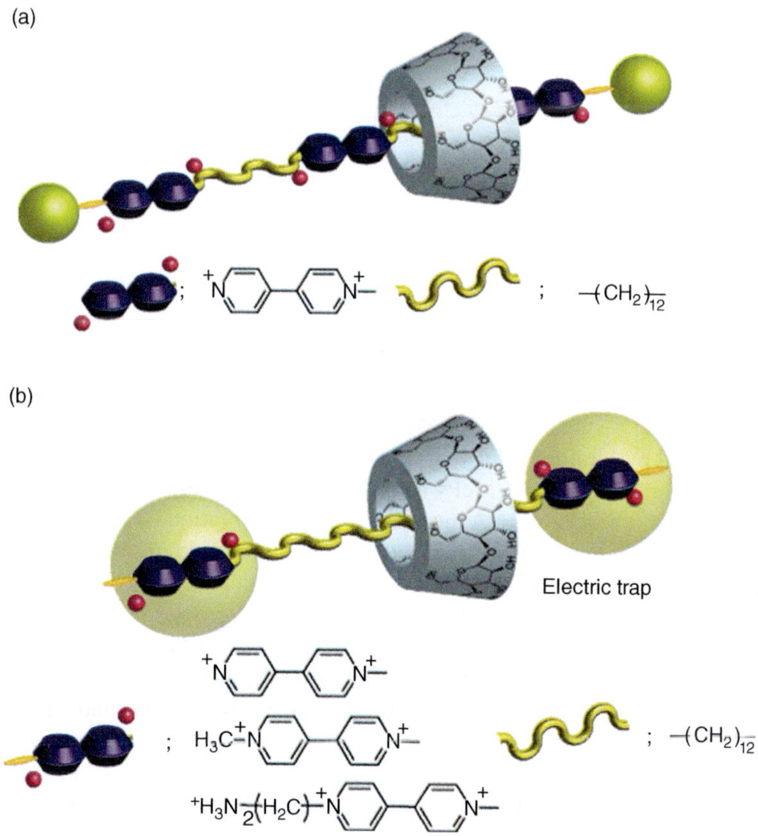

Figure 10.12 *Molecular shuttle (a) and electrical trap (b). (Reproduced from ref. [29] with permission of Royal Society of Chemistry.)*

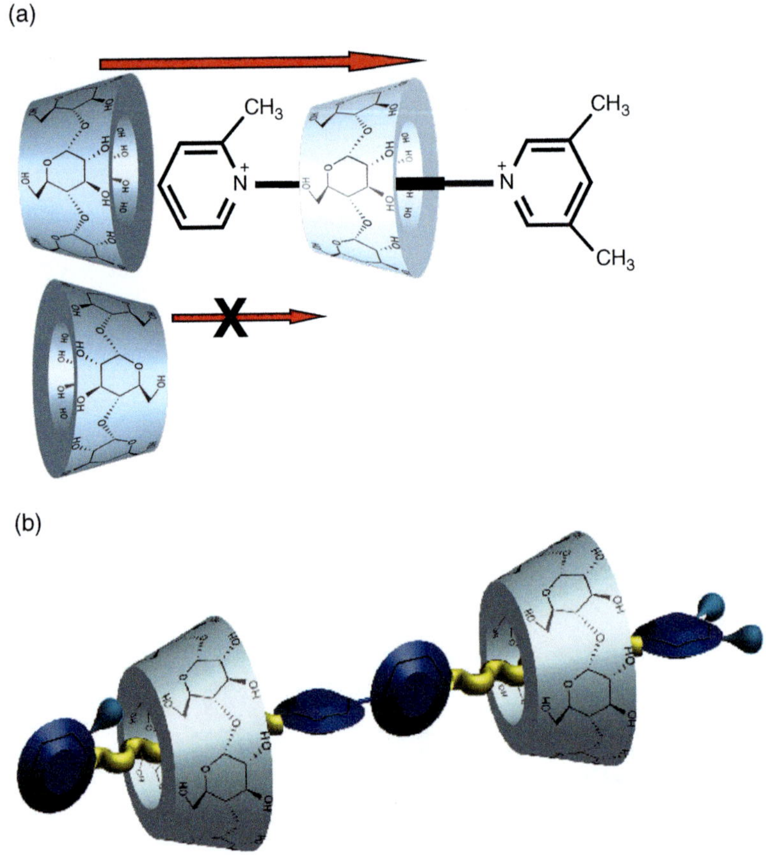

Figure 10.13 Unidirectional threading (a) and unidirectional arrangement of CD rings (b)

Next we studied the steric effects of the end-groups on threading in detail. The α-CD ring can be easily threaded onto an axle through a pyridinium group, whereas a CD ring cannot get past a 3,5-dimethyl-pyridinium(3,5-diMePy) group because the 3,5-diMePy group is too large for α-CD to pass. Thus, we examined whether a mono-methyl pyridinium (MePy) group can be attached to the end of the axle molecule.

Interestingly, a CD ring can pass over a 3-MePy (or 4-MePy) group with ease to give a rotaxane. Although the α-CD ring passes 2-MePy, it is very slow. A more interesting observation is that the CD ring can pass over the 2-MePy group only from the wider side (the secondary –OH-group side) of the CD but not from the other side (the primary hydroxyl side) at room temperature. Accordingly, the product is a (2)-rotaxane with a CD ring where the secondary hydroxyl group directs the diMePy stopper group [49]. This is quite an interesting result because the diameter of the narrowest part of CD is the same regardless of which side is the threading side. Hence, we achieved perfect unidirectional threading at room temperature (see Figure 10.13).

The 2-MePy group controls the direction of the CD ring. Usually, when CDs and a polymer are mixed in an aqueous solution, the CDs are arranged in a head-to-head, tail-to-tail, or random distribution on a polymer chain. When a 2-MePy group is attached to the end of the polymer chain, α-CDs are arranged in one direction to give a *pseudo*-polyrotaxane. This result can be explained by the transition states; because the 2-MPy ring is asymmetric but CD is not, the transition states are not identical for both directions (see Figure 10.14).

We designed and prepared a compound where the 2-MePy group is incorporated in an axle molecule. In contrast to the previous threading case, the potentials of both sides are almost the same. The CD ring passes

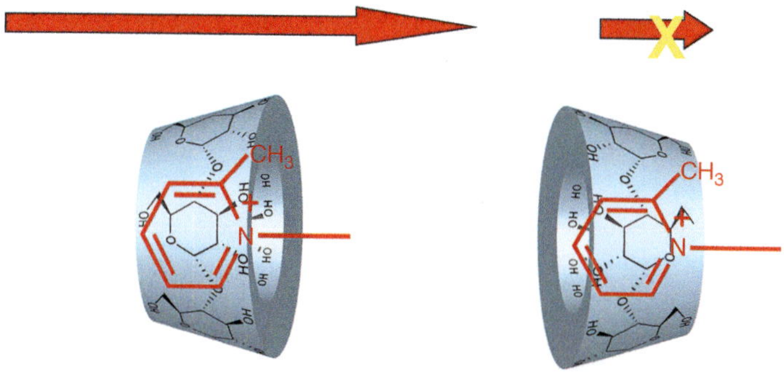

Figure 10.14 *Transition state for threading*

over the 2-MePy ring from the wider side much faster than from the narrower side. The ratio of the rate constant for the CD translation from the wider side of CD to that from the narrower side is 35 [50]. Although the potentials of the first and second stations are similar, the translation rates are not.

In summary, we achieved perfect *unidirectional threading* of α-CD using a 2-MePy group. The translation rate and direction of translation of the CD ring can be controlled incorporating this group into the polymer chain.

10.9 Macroscopic self-assembly through molecular recognition

10.9.1 Macroscopic self-assembly

Is it possible to observe a molecular recognition event with the naked eye? While investigating molecular recognition using different techniques, we were able to answer this simple question. This type of event can be observed using large visible molecules. Because polymers and supramolecular polymers are too small to see directly, we thought that a gel, which is a giant macromolecule linked by short crosslinking agents, might be suitable. We selected a polyacrylamide gel as a scaffold because it does not interact with DNA, RNA, or proteins, which is why it is used as the base for electrophoresis. Additionally, the gel has a negligible interaction with polysaccharides or small molecules, as exemplified by the fact that it is used for gel chromatography. Thus, CDs were added as the host part in the gel by copolymerization of a CD monomer, acrylamide, and bisacrylamide. Guest gels containing an adamantane group, *n*-, or *t*-butyl group were prepared in the same manner.

Blank gels (without hosts or guests) do not bind each other. Similarly, host gels containing β-CD (guest gels) do not bind each other; however, bringing an adamantane (guest) gel into contact with the host gel containing β-CD, the two adhere firmly. The interaction between β-CD gel and Ad gel is so strong that separating the gel assembly is difficult. Pulling both sides of a bound gel containing a guest gel (green) and a host gel (red) using a creep meter broke a gel piece, the Ad gel (green) without damaging the contact interfaces. It is noteworthy that the interface is intact (without damage) despite the fact that the gels are bound solely by non-covalent (host–guest) interactions [51]. Figure 10.15 schematically depicts the macroscopic self-assembly between the β-CD host gel and the Ad guest gel in which both gels are bound solely by host–guest interactions.

When pieces of β-CD gel and Ad gel are mixed and shaken in water, β-CD gel and Ad gel adhere to each other to form an assembly. These observations indicate that molecular recognition plays an important role on both the molecular and macroscopic level.

Lifting a gel aggregate with tweezers results in an alternating host–guest pattern in accordance with the high complementarity of the host and guest gels.

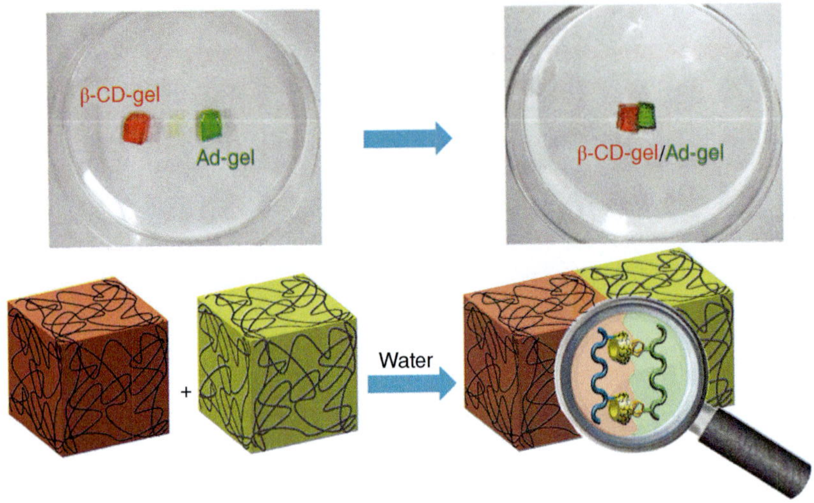

Figure 10.15 Macroscopic self-assembly by molecular recognition. (Reproduced from ref. [37] with permission. Copyright © Nature Publishing Group and the American Chemical Society

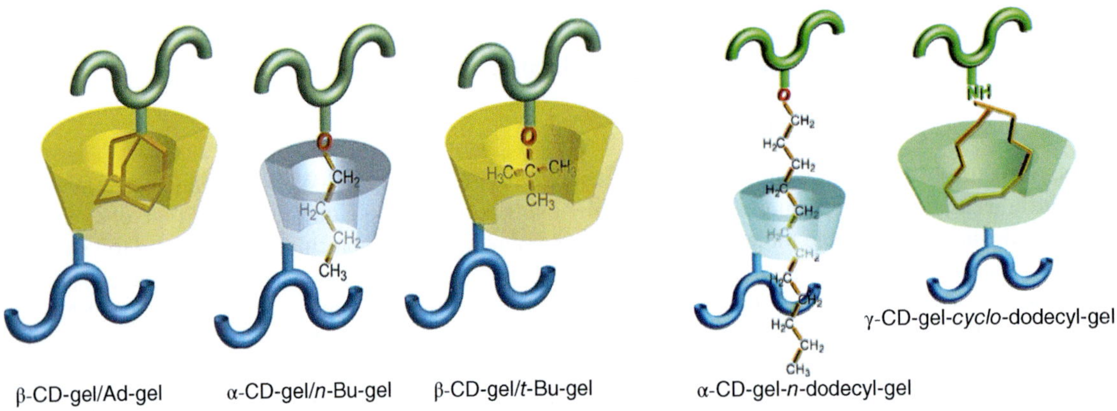

Figure 10.16 Interaction modes. (Reproduced from ref. [37] with permission. Copyright © 2011 Nature Publishing Group and the American Chemical Society.)

10.9.2 Selectivity

When pieces of α-CD gel (blue), β-CD gel (red), *n*-Bu gel (yellow), and *t*-Bu gel (green) are mixed and shaken in water, only the α-CD gel binds to *n*-Bu gel, while only β-CD gel sticks to *t*-Bu gel to give alternating or checkered structures. These results clearly indicate that a host gel recognizes the corresponding guest gel through molecular recognition to form macroscopic structures [51].

Figure 10.16 shows the host–guest interaction modes. The slim *n*-Bu group fits well in the small α-CD cavity, while the bulky *t*-Bu group fits well in the larger β-CD cavity: the host gels can differentiate such small variations in the guest parts. This system can be applied to differentiate macroscopic materials through molecular recognition of various guest parts.

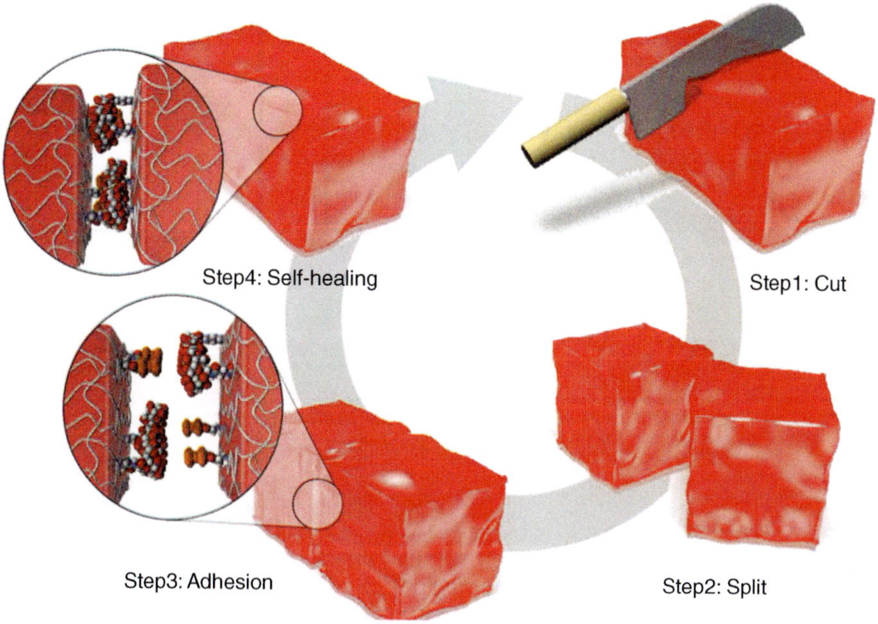

Figure 10.17 *Self-healing of CD–ferrocene gel*

When a cyclohexyl group is used as the guest group, only the β-CD gel and not the α-CD gel binds to the cyclohexyl gel. Adding *t*-butanol in water as a competitive guest causes the gel to disassemble, giving each component. After disassembling the gel aggregate and removing the competitive guest, the solvent changes to pure water and the gel begins to reassemble, yielding another (new) assembly [52].

This is the first instance of a molecular recognition event that has been used to assemble large objects. Researchers working with molecular systems use well-defined reversible interactions to generate assembled structures such as well-defined H-bond donors and acceptors or the formation of ligand–metal bonds. In addition, research on macroscopic systems typically use macroscopic interactions (capillary forces, electrostatics, magnetics, etc.) to directly assemble objects.

10.10 Self-healing by molecular recognition

A gel is formed upon mixing an aqueous solution of poly(acrylic acid: PAA) containing β-CD with that of PAA containing ferrocene. Even if the gel is cut in half by a knife, the gels bind and heal into a single gel upon contact; the contact surface disappears after several hours, and after 24 hours, the gel strength is almost 85% recovered (see Figure 10.17) [53].

The adamantane monomer is insoluble in water. However, mixing the guest monomer with the β-CD monomer in water solubilizes the guest monomer to provide a clear solution. β-CD includes the adamantane monomer to give a host–guest complex. A gel is formed when the acrylamide monomer is polymerized with host–guest complex monomers. There are no free CDs or adamantane.

In this case, the gel is formed by crosslinking between β-CD and the adamantane groups. Cutting the gel breaks the host–guest interactions, but bringing the cut ends of the gel pieces together realizes a single gel

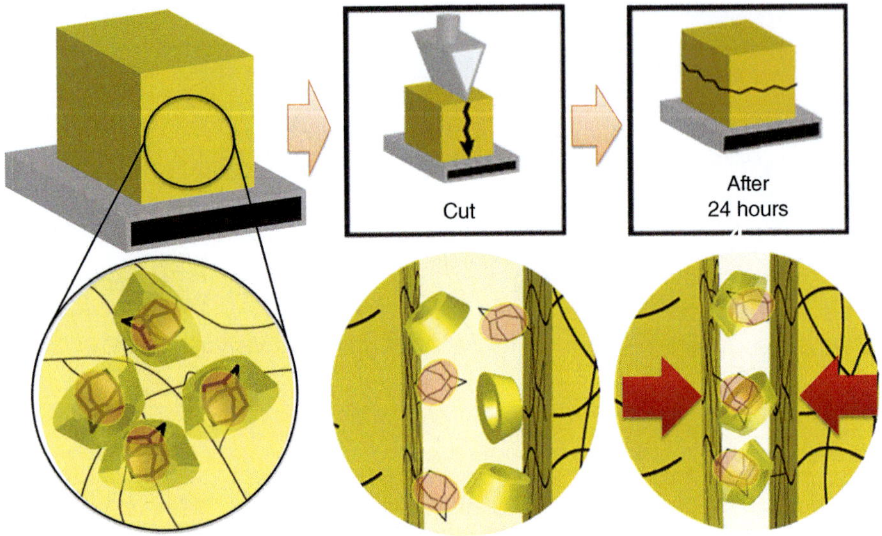

Figure 10.18 Self-healing of CD–adamantane gel

because the CDs bind adamantane to restore a single gel. In this case, the gel strength recovered 100%, achieving perfect self-healing (see Figure 10.18) [54].

10.11 Stimuli-responsive polymers

10.11.1 Photoresponsive polymers

A gel is realized when a polymer containing α-CD and a polymer containing azobenzene (AZ) group are mixed in water. α-CD on the polymer includes azobenzene, giving the crosslinked part. Irradiating this gel with UV light turns the gel into a sol: *trans*-azobenzene changes to *cis*-form upon UV irradiation, then the *cis*-AB leaves α-CD. This process is reversible. However, if the gel is partially cross-linked by covalent bonds, irradiating with UV light turns the *trans*-AB to *cis*-AB, which leaves α-CD, but this time the chemical cross-linking prevents conversion into the sol. Consequently, the gel structure is maintained.

In fact, we prepared such a gel on cloth via a polyacrylamide gel. Irradiating the gel surface with UV light causes the gel to bend, but the visible light restores the original shape. This process can be repeated many times (see Figure 10.19) [55].

We used azobenzene as a guest gel. The α-CD gel and *trans*-AB gel were bound to each other to give an assembly. Irradiating the gel assembly with UV light in the presence of β-CD gels causes the AB gels to leave the α-CD gels, switching to β-CD gels [56].

10.11.2 Redox responsive gel

A gel forms when polymers containing β-CD and others containing ferrocene are mixed in water. Treating the cut ends with an oxidizing agent prevents binding of both gel pieces because the ferrocene is oxidized into the ferrocenium cation, which cannot be bound in the cavity. However, treating the cut edges with a reducing

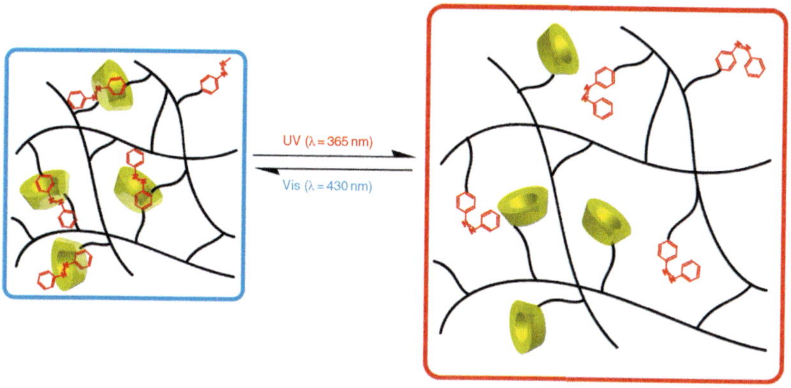

Figure 10.19 *Photoresponsive host–guest gel*

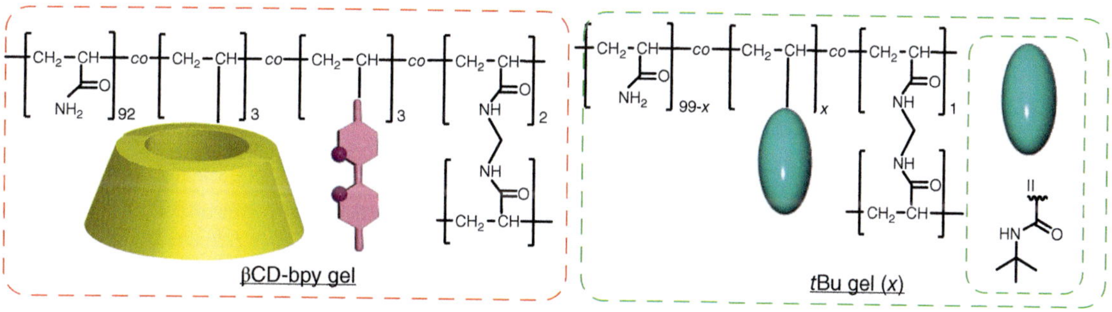

Figure 10.20 *Metal-ion-responsive gel*

agent allows the gel pieces to bind into a single gel [53]. Ferrocene has also been used as a guest to generate redox-generated mechanical motion [57], shape-memory [58], and macroscopic self-assembly [59].

10.11.3 Metal-ion responsive system

When a host gel containing β-CD and bipyridyl (BP) groups is placed into contact with a *t*-butyl-(*t*-Bu) guest gel, the gel pieces do not bind, whereas immersing the host gel into a copper solution results in firm adhesion between the host and the guest realizing an assembly (see Figure 10.20) [60]. However, immersing into an EDTA solution dissociates the gel.

The CD-BP gel does not bind a guest gel containing *t*-Bu groups because the BP groups are tightly included in β-CD cavities. This strong inclusion prevents the *t*-Bu groups from being included in the CD cavities. Consequently, the CD-BP gel does not bind to the guest gel. However, treating the host gel with a copper ion solution allows the host and guest gel to bind because the copper ions bind to the BP groups outside of the CD cavities. Hence, the CD cavities are free for new guests. Bringing the *t*-Bu gel in contact with this host gel realizes strong adhesion.

The adhesion of the gel pieces is selective. The strength of the gel adhesion is in the order Zn > Co > Cu > Ni. Fe, Mn, and Mg do not cause adhesion. In the case of Fe, three bipyridyl groups are involved in the coordination to Fe, causing the gel to shrink and harden, inhibiting binding.

10.12 Conclusion

We have studied molecular recognition using cyclodextrins, and used the elucidated phenomena for more specific systems. We have extended these events to larger and more complicated systems, even macromolecular recognition. Furthermore, we have successfully constructed supramolecular materials, macroscopic self-assemblies, and self-healing materials.

References

1. Pedersen, C.J. (1988) The discovery of crown ethers (Nobel Lecture), *Angewandte Chemie, International Edition*, **27**, 1021–1027.
2. Cram, D.J. (1988) The design of molecular hosts, guests, and their complexes (Nobel Lecture), *Angewandte Chemie, International Edition*, **27**, 1009–1112.
3. Lehn, J.-M. (1988) Supramolecular chemistry – Scope and perspectives molecules, supermolecules, and molecular devices (Nobel Lecture), *Angewandte Chemie, International Edition*, **27**, 89–112.
4. Lehn, J.-M., Atwood, J.L., Davies, J.E.D., MacNicol, D.D., Vogtle, F.(1996) *Comprehensive Supramolecular Chemistry, Vol.* **3**, *Cyclodextrins*, Elsevier (Pergamon), Amsterdam.
5. Harada, A., Furue, M., Nozakura, S. (1976) Cyclodextrin-containing polymers, 1. Preparation of polymers, *Macromolecules*, **9**, 701–704.
6. Harada, A., Furue, M., Nozakura, S. (1976) Cyclodextrin-containing polymers. 2. Cooperative effects in catalysis and binding, *Macromolecules*, **9**, 705–710.
7. Harada, A., Furue, M., Nozakura, S., (1977) Interaction of cyclodextrin-containing polymers with fluorescent compounds, *Macromolecules*, **10**, 676–681.
8. Harada, A., Furue, M., Nozakura, S., (1978) Optical resolution of mandelic acid derivatives by column chromatography on crosslinked cyclodextrin-gels, *Journal of Polymer Science, Polymer Chemistry Edition*, **16**, 189–196.
9. Zanello, P. (Ed) (1994) *Chains, Clusters, Inclusion Compounds, Paramagnetic Labels, and Organic Rings*. Elsevier, Amsterdam.
10. Harada, A., Takahashi, S. (1984) Preparation and properties of cyclodextrin–ferrocene inclusion complexes, *Journal of the Chemical Society, Chemical Communications*, **1984**, 645–646.
11. Odagaki, Y., Hirotsu, K., Higuchi, T., Harada, A., Takahashi, S. (1990) X-ray structure of the α-cyclodextrin–ferrocene (2:1) inclusion compound. *Journal of the Chemical Society, Perkin Transactions*, **1**, 1990, 1230–1231.
12. Harada, A., Saeki, K., Takahashi, S. (1989) Optical resolution of ferrocene derivatives by liquid chromatography using aqueous cyclomaltohexaose as the mobile phase and polyamide as the stationary phase, *Carbohydrate Research*, **192**, 1–7.
13. Harada, A., Saeki, K., Takahashi, S. (1989) Preparation and properties of inclusion compounds of (η6-arene) tricarbonylchromiumu(0) complexes with cyclodextrins, *Organometallics*, **8**, 730–733.
14. Harada, A., Yamamoto, S., Takahashi, S. (1989) Preparation and properties of inclusion compounds of transition metal complexes of cycloocta-1,5-diene and norbornadiene with cyclodextrins, *Organometallics*, **8**, 2560–2563.
15. Alston, E., Slawin, A.Z., Stoddart, J.F., Williams, D.J. (1985) Cyclodextrin as second sphere ligands for transition metal complexes – The X-ray crystal structure of $[Rh(cod)(NH_3)_2$-α-cyclodextrin$][PF_6]6H_2O$, *Angewandte Chemie, International Edition*, **24**, 786–787.
16. Harada, A., Takeuchi, M, Takahashi, S. (1988) Preparation and properties of inclusion compounds of (η^3-allyl) palladium complexes with cyclodextrins, *Bulletin of the Chemical Society of Japan.*, **61**, 4367–4370.
17. Harada, A., Takahashi, S. (1988) Preparation and properties of inclusion complexes of 1,2-dicarbadodecaborane(12) with cyclodextrin, *Journal of the Chemical Society, Chemical Communications*, **1988**, 1352–1353.
18. Harada, A., Hu, Y., Takahashi, S. (1986) Cyclodextrin–palladium chloride. New catalytic system for selective oxidation of olefins to ketones, *Chemistry Letters*, **15**, 2083–2084.
19. Harada, A., Hashidzume, A., Yamaguchi, H., Takashima, Y. (2009) Polymeric rotaxanes, *Chemical Reviews*, **109**, 5974–6023.

20. Harada, A., Kamachi, M. (1990) Complex formation between poly(ethylene glycol) and α-cyclodextrins, *Macromolecules* **23**, 2821–2823.
21. Harada, A., Kamachi, M. (1990) Complex formation between cyclodextrin and poly(propylene glycol), *Journal of the Chemical Society, Chemical Communications*, **1990**, 1322–1323.
22. Harada, A., Li, J., Takahashi, S. (1993) Complex formation between poly(methyl vinyl ether) and γ-cyclodextrin. *Chemistry Letters*, **22**, 237–240.
23. Harada, A., Okada, M. (1999) Complex formation between hydrophobic polymers and methylated cyclodextrins, oligo(ethylene) and poly(propylene), *Polymer Journal*, **31**, 1095–1098.
24. Harada, A., Li, J., Suzuki, S., Kamachi, M. (1993) Complex formation between polyisobutylene and cyclodextrins: Inversion of chain-length selectivity between β-cyclodextrin and γ-cyclodextrin, *Macromolecules* **26**, 5267–5268
25. Harada, A., Suzuki, S., Okada, M., Kamachi, M. (1996) Preparation and characterization of inclusion complexes of polyisobutylene with cyclodextrins, *Macromolecules* **29**, 5611–5614.
26. Harada, A., Li, J., Kamachi, M. (1993) Preparation and properties of inclusion complexes of polyethylene glycol with α-cyclodextrins, *Macromolecules* **26**, 5698–5703.
27. Harada, A., Li, J., Kamachi, M., Kitagawa Y., Katsube, Y. (1997) Structure of polyrotaxane models, *Carbohydrate Research.* **305**, 127–129.
28. Harada, A., Li, J., Kamachi, M. (1992) The molecular necklace: A rotaxane-containing many-threaded α-cyclodextrin, *Nature*, **356**, 325– 327.
29. Harada, A., Takashima, Y., Yamaguchi, H. (2009) Cyclodextrin-based supramolecular polymers, *Chemical Society Reviews*, **38**, 875–882.
30. Harada, A., Li, J., Kamachi, M. (1993) Synthesis of a tubular polymer from threaded cyclodextrins, *Nature*, **364**, 516–518.
31. Kawaguchi, Y., Nishiyama, T., Okada, M., Kamachi, M., Harada, A. (2000) Complex formation of poly (ε-caprolactone) with cyclodextrins, *Macromolecules* **33**, 4472–4477.
32. Harada, A., Nishiyama, T., Kawaguchi, Y., Okada, M., Kamachi, M. (1997) Preparation and properties of inclusion complexes of aliphatic polyesters with cyclodextrins, *Macromolecules* **30**, 7115–7118.
33. Takashima, Y., Kawaguchi, Y., Nakagawa, S., Harada, A. (2003) Inclusion complex formation and hydrolysis of lactones by cyclodextrins, *Chemistry Letters*, **32**, 1122–1123.
34. Harada, A., Osakai, M., Takashima, Y., Yamaguchi, H. (2008) Ring-opening polymerization of cyclic esters by cyclodextrins *Accounts of Chemical Research*, **41**, 1143–1152.
35. Osaki, M., Takashima, Y., Harada, A. (2007) An artificial molecular chaperone: Poly-*pseudo*-rotaxane with an extensible axle, *Journal of the American Chemical Society*, **129**, 14452–14457.
36. Takashima, Y., Osaki, M., Ishimaru, Y., Yamaguchi, H., Harada, A. (2011) Artificial molecular clamp: A novel device for synthetic polymerases, *Angewandte Chemie, International Edition*, **50**, 7524–7528.
37. Harada, A. (Ed) (2011) *Supramolecular Polymer Chemistry*, Wiley–VCH, Weinheim.
38. Hoshino,T., Miyauchi, M., Kawaguchi, K., Yamaguchi, H., Harada, A. (2000) Daisy chain necklace: Tri[2]rotaxane containing cyclodextrins, *Journal of the American Chemical Society*, **122**, 9876–9877.
39. Miyauchi, M., Hoshino,T., Yamaguchi, H., Kamitori, S., Harada, A. (2005) A [2]rotaxane capped by a cyclodextrin and a guest: Formation of a supramolecular [2]rotaxane polymer, *Journal of the American Chemical Society*, **127**, 2034–2035.
40. Miyauchi, M., Takashima, Y., Yamaguchi, H., Harada, A. (2005) Chiral supramolecular polymers formed by host–guest interactions, *Journal of the American Chemical Society*, **127**, 2984–2989.
41. Miyauchi, M., Harada, A. (2004) Construction of supramolecular polymers with alternating α-, β-cyclodextrin units using conformational change induced by competitive guests, *Journal of the American Chemical Society*, **126**, 11418–11419.
42. Harada, A., Adachi, H., Kawaguchi, Y., Kamachi, M. (1997) Recognition of alkyl groups on a polymer chain by cyclodextrins, *Macromolecules* **30**, 5181–5182.
43. Inoue, Y., Miyauchi, M., Nakajima, H. et al. (2006) Self-threading of a poly(ethylene glycol) chain in a cyclodextrin-ring: Control of the exchange dynamics by chain length, *Journal of the American Chemical Society*, **128**, 8994–8995.
44. Inoue, Y., Kuad, P., Okumura Y. et al. (2007) Thermal and photochemical switching of conformation of poly(ethylene glycol)-substituted cyclodextrin with an azobenzene group at the chain end, *Journal of the American Chemical Society*, **129**, 6396–6397.

45. Harada, A., Adachi, H., Kawaguchi, Y., Okada, M., Kamachi, M. (1996) Complex formation of cyclodextrins with cationic polymers, *Polymer Journal*, **28**, 159–163.
46. Harada, A. (2001) Cyclodextrin-based molecular machines, *Accounts of Chemical Research*, **34**, 456–464.
47. Kawaguchi, Y., Harada, A. (2000) A cyclodextrin-based molecular shuttle containing energetically favored and disfavored portions in its dumbbell component, *Organic Letters*, **2**, 1353–1356.
48. Kawaguchi, Y., Harada, A. (2000) An electric trap: A new method for entrapping cyclodextrin in a rotaxane structure, *Journal of the American Chemical Society*, **122**, 3797–3798.
49. Oshikiri, Y., Takashima, Y., Yamaguchi, H., Harada, A. (2007) Face-selective [2]-and [3]rotaxanes: kinetic control of the threading direction of cyclodextrins, *Chemistry – A European Journal*, **13**, 7091–7098.
50. Oshikiri, Y., Yamaguchi, H., Takashima, Y., Harada, A. (2009) Face-selective translation of a cyclodextrin ring along an axle, *Chemical Communications*, **2009**, 5515–5517.
51. Harada, A., Kobayashi, R., Takashima, Y., Hashidzume, A., Yamaguchi, H. (2011) Macroscopic self-assembly through molecular recognition, *Nature Chemistry*, **3**, 34–37.
52. Yamaguchi, H., Kobayashi, R., Takashima, Y., Hashidzume, A., Harada, A. (2011) A self-assembly of gels through molecular recognition of cyclodextrins: Shape selectivity for linear and cyclic guest molecules, *Macromolecules*, **44**, 2395–2399.
53. Nakahata, M., Takashima, Y., Yamaguchi, H., Harada, A. (2011) Redox-responsive self-healing materials formed from host–guest polymers, *Nature Communications*, **2**, 511
54. Kakuta, T., Takashima, Y., Nakahata, M. et al. (2013) A preorganized hydrogel: Self-healing properties of supramolecular hydrogels formed by polymerization of host–guest-monomers that contain cyclodextrins and hydrophobic guest groups, *Advanced Materials*, **25**, 2849–2853.
55. Takashima, Y., Hatanaka, S., Otsubo, M. et al. (2012) Expansion–contraction of photoresponsive artificial muscle regulated by host–guest interactions, *Nature Communications*, **3**, 1270.
56. Yamaguchi, H., Kobayashi, Y., Kobayashi, R. et al. (2012) Photoswitchable gel assembly based on molecular recognition, *Nature Communications*, **3**, 603.
57. Nakahata, M., Takashima, Y., Hashidzume, A., Harada, A. (2013) Redox-generated mechanical motion of a supramolecular polymeric actuator based on host–guest interactions, *Angewandte Chemie. International Edition*, **52**, 5731–5735.
58. Miyamae, K., Nakahata, M., Takashima, Y., Harada, A. (2015) Redox-generated mechanical motion of a supramolecular polymeric actuator: Self-healing, expansion–contraction, and shape-memory properties of a preorganized supramolecular hydrogel through host–guest interactions, *Angewandte Chemie, International Edition*, **54**, 8984–8987.
59. Nakahata, M., Takashima, Y., Harada, A. (2014) Redox-responsive macroscopic gel assembly based on discrete dual interactions, *Angewandte Chemie, International Edition*, **53**, 3617–3621.
60. Nakamura, T., Takashima, Y., Hashidzume, A., Yamaguchi, H., Harada, A. (2014) A metal ion-responsive adhesive material via switching of molecular recognition properties, *Nature Communications*, **5**. 4622.

11

Making the Tiniest Machines

David A. Leigh

School of Chemistry, The University of Manchester, Manchester, UK

11.1 Introduction

The widespread use of molecular machines in biology has long suggested that great rewards could come from bridging the gap between synthetic molecular systems and the machines of the macroscopic world [1, 2]. Perhaps the best way to appreciate the technological potential of controlled molecular-level motion is to recognize that nanomotors and molecular-level machines lie at the heart of every significant biological process. Over billions of years of evolution nature has not repeatedly chosen this solution for achieving complex task performance without good reason. When mankind learns how to build artificial structures that can control and exploit molecular-level motion, and interface their effects directly with other molecular-level substructures and the outside world, it will potentially impact on every aspect of functional molecule and materials design. An improved understanding of physics and biology will surely follow. This chapter highlights some of our group's contributions to this effort, from the serendipitous discovery of the benzylic amide catenane system in the mid-1990s to the design and synthesis of some relatively complex artificial molecular machines.

11.1.1 From a macrocyclic receptor for CO_2 to a [2]catenane

Our foray into the area of molecular machines began with a chance discovery in the laboratory. Given the tremendous interest shown in catenanes (molecular links) and rotaxanes (molecular rings mechanically locked onto a molecular axle by bulky "stoppers") in recent years, it is easy to forget that until relatively recently the synthesis of any mechanically interlocked structure was regarded as something very special indeed. Up to the early-1990s the only practical methods that had been developed to such molecules were the tetrahedral Cu(I) template method introduced by Sauvage in 1983 [3] and the π-rich-π-poor aromatic stacking system introduced by Stoddart in 1989 [4]. Around this time chemists recognized that mechanically

Macrocyclic and Supramolecular Chemistry: How Izatt–Christensen Award Winners Shaped the Field, First Edition.
Edited by Reed M. Izatt.

interlocked architectures could potentially undergo well-defined, large-amplitude changes in the positions of the components and, if those movements could be controlled, they could form the basis for molecular machines. This gave a compelling reason to make catenanes and rotaxanes; ours was one of several groups involved in this endeavor throughout the 1990s and 2000s [1].

Amide-based catenanes, in which hydrogen bonds template interlocking of the two rings, were first reported in 1992 by Hunter [5]. In Hunter's catenanes the steric bulk of cyclohexyl groups prevents circumrotation of the macrocyclic rings, a feature that allowed him to deduce the mechanically interlocked architecture through analysis of the complex 1D and 2D ^{1}H NMR data. Hunter's discovery prompted the publication of similar compounds by Vögtle [6]. Despite the accumulation of a large body of work over the subsequent decade, the fundamental problem that these catenane systems exist in a rigidly locked co-conformation was never sufficiently overcome to make them usable in molecular machine systems.

At around the same time our group were working on isophthalamide macrocycles as potential receptors for neutral molecules. Macrocycle **1** was identified as being the right size and shape to bind a CO_2 molecule within its cavity and the amide groups at each end meant it should be able to hydrogen bond to both oxygen atoms of the guest molecule. However, when isophthaloyl dichloride and xylylene diamine were reacted together in $CHCl_3$ in the presence of Et_3N we got a surprise: after the reaction was complete the only compound present in solution was a [2]catenane, consisting of two of the intended product macrocycles mechanically linked together (see Figure 11.1; **2**) [7].

The hydrogen-bonding patterns in the X-ray crystal structures of the benzylic amide catenanes suggested that modification of the synthetic procedure might allow rotaxanes to be synthesized. Sure enough, we discovered that if the catenane-forming reaction was carried out in the presence of a suitable stoppered benzylic 1,3-diamide thread (e.g. **3**) rotaxanes such as **4** were also formed (see Figure 11.2) [8].

Rotaxane **4** illustrates how an interlocked architecture can bring about significant changes in the chemical and physical properties of the components: rotaxane **4** is 10^5× more soluble in $CHCl_3$ than the parent macrocycle, **1** [8].

The wealth of structural data obtained on the benzylic amide catenane and rotaxane system showed that the mechanism for the formation of the interlocked products was primarily the templated assembly of the macrocycle around two amide bonds. Divergent hydrogen bonding sites in a similar spatial arrangement occur in adjacent amino acid residues in peptide chains and, indeed, it was found that simple peptides are also able to

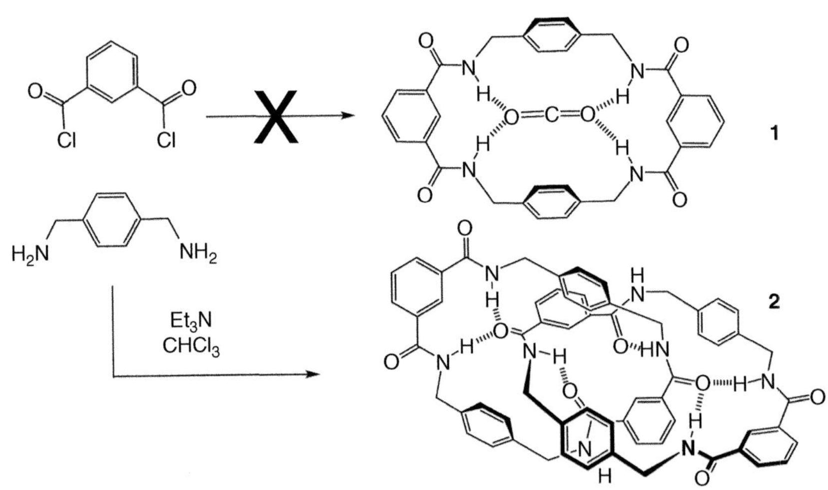

Figure 11.1 Eight-component assembly of a [2]catenane (**2**) [7]

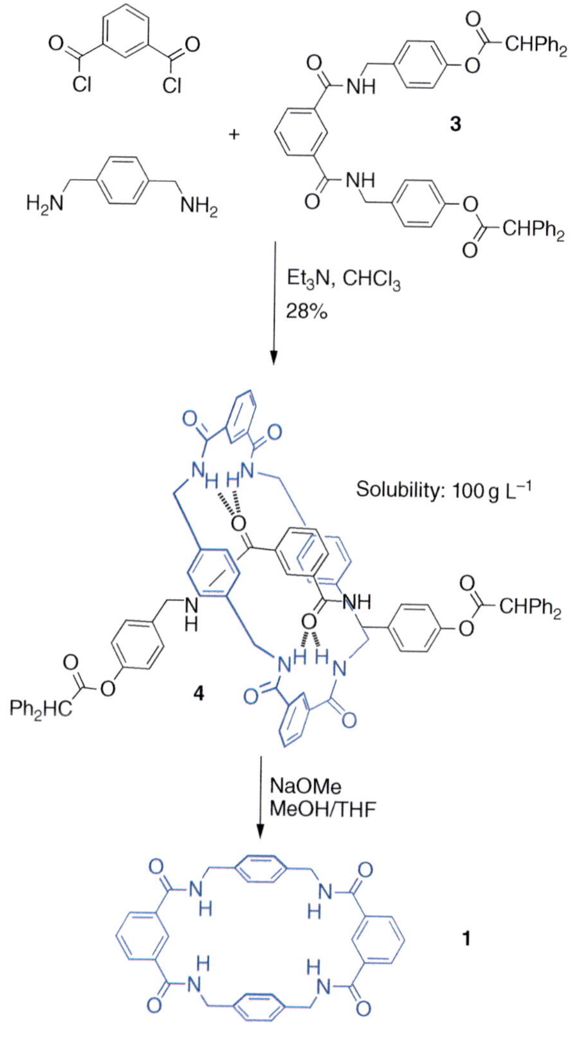

Figure 11.2 *Five-component assembly of a [2]rotaxane (**4**) [8]*

template the cyclization of benzylic amide macrocycles to give peptide rotaxanes [9]. However, the inherent flexibility of the peptide backbone is not optimal for rotaxane formation (Figure 11.3). Replacing the peptide motif with a fumaramide unit maintains the required hydrogen-bonding groups but with a rigidified backbone. This results in a remarkable 97% yield for a rotaxane (**5**), from a five-component assembly process in just two steps from commercial starting materials (Figure 11.3) [10].

11.1.2 From rotaxanes to molecular shuttles

Shuttling is the movement of a macrocycle back and forth along the thread component of a rotaxane. This motion takes the form of a Brownian motion-powered random walk, constrained to one dimension by the thread and to translational displacement boundaries by the bulky stoppers. In rotaxanes **6** and **7** (Figure 11.4) the macrocycle

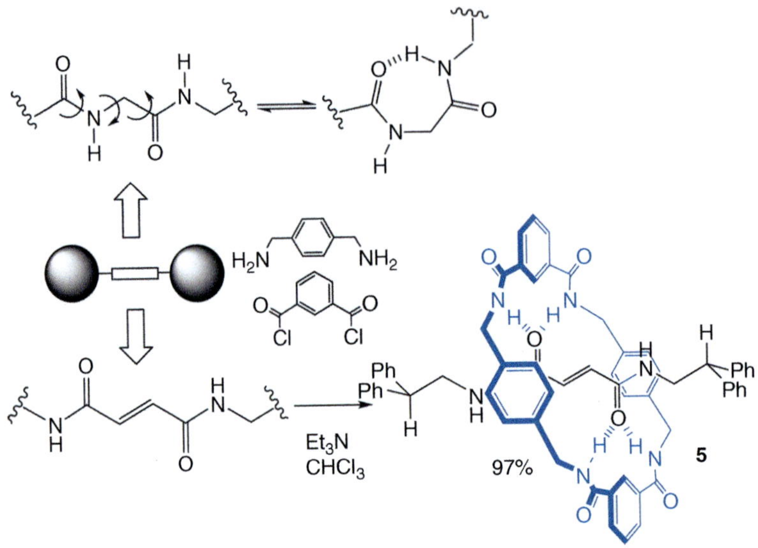

Figure 11.3 *High-yielding five-component assembly of a [2]rotaxane (**5**) about a fumaramide template [10]*

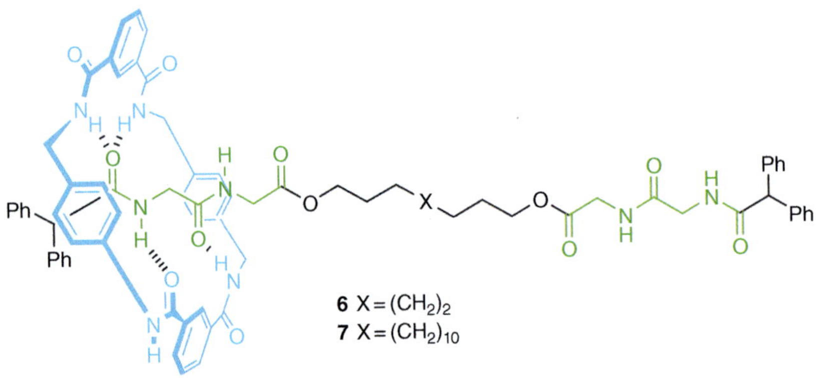

Figure 11.4 *Peptide-based molecular shuttles [11]*

shuttles between the two peptide stations rapidly at room temperature in $CDCl_3$, evidenced by the single set of signals for the two peptide stations in the ^{1}H NMR spectrum, which is resolved into two sets (for the occupied and unoccupied stations) on cooling the sample and freezing out the motion on the NMR timescale [11].

11.1.3 From shuttles to switches: stimuli-responsive molecular shuttles

Rotaxanes in which the macrocycle can be translocated between two or more well-separated stations in response to an external signal can, in principle, act as molecular-level mechanical switches. In a rotaxane the macrocycle typically distributes itself between the available binding sites according to the difference in the macrocycle binding energies and the temperature. If a suitably large difference in macrocycle affinity between two stations exists, the macrocycle resides overwhelmingly in one positional isomer or co-conformation. In stimuli-responsive molecular shuttles, an external trigger is used to chemically modify the system and alter the non-covalent intercomponent interactions such that the second macrocycle binding site becomes

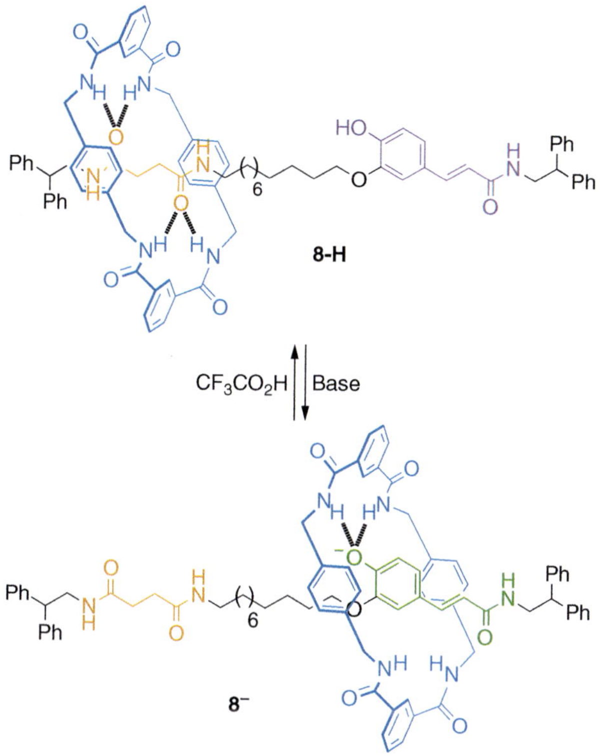

Figure 11.5 *A rotaxane switchable by NH···anion hydrogen-bonding interactions [12]*

energetically more favored, causing translocation of the macrocycle along the thread to the second station. From 1997 to 2010 we developed many effective switching methods for molecular shuttles using a broad range of stimuli (light, electrochemistry, pH, reversible covalent bond formation, redox processes, anion and cation binding, allosteric and competitive binding, changes in the polarity of the environment, temperature etc.), resulting in a "toolbox" that could be exploited in different types of functional molecular machine systems [1].

A typical example of the type of switchable molecular shuttles developed during this period is one that exploited N–H···anion hydrogen-bonding interactions [12]. In [2]rotaxane **8**·H (Figure 11.5), formation of the benzylic amide macrocycle about the thread is directed by a succinamide template. However, the thread also contains a cinnamate group. In the neutral form, the cinnamate phenol is a relatively poor hydrogen-bonding group and the macrocycle resides on the succinamide station >95% of the time in most solvents at room temperature. Deprotonation to give $\mathbf{8}^-$ in [D_7]DMF results in the macrocycle binding to the phenolate anion.

11.2 Property effects using molecular shuttles

Whatever the future application of synthetic molecular machines may be, it is clear practical devices will require that the machine and its components are able to interact with the macroscopic world, either directly or through further interactions with other molecular-scale devices. Stimuli-responsive molecular shuttles offer a generic approach which could be taken to create mechanical molecular switches for a variety of distance-dependent properties. Suitable functionalization of the macrocycle and one end of the thread can lead to molecular switches that can change a variety of properties in response to a particular switching stimulus (Pérez 2004) (Figure 11.6).

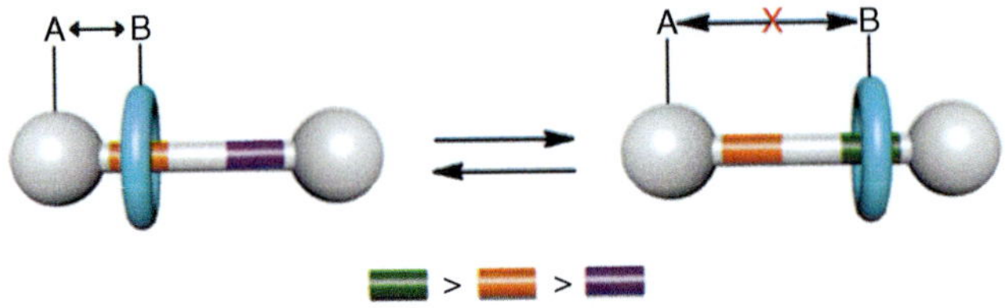

Figure 11.6 Exploiting a well-defined, large-amplitude positional change to trigger property changes. A and B interact to produce a physical response (fluorescence quenching, specific dipole or magnetic moment, NLO properties, color, creation/concealment of a binding site or reactive/catalytic group, hydrophobic/hydrophilic region, etc.). Moving A and B far apart mechanically switches off the interaction and the corresponding property effect

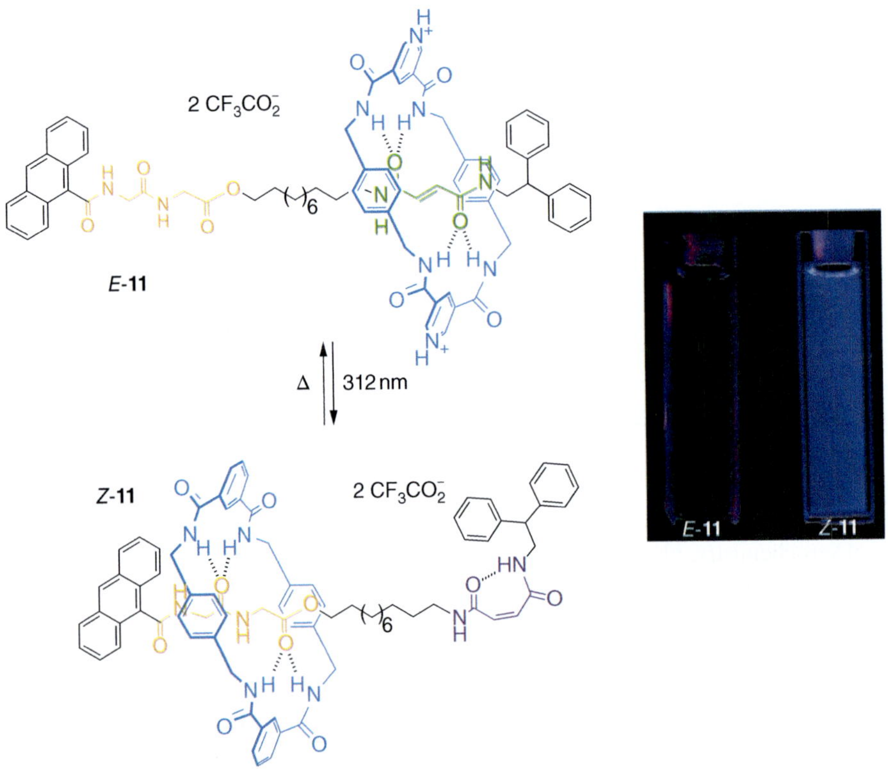

Figure 11.7 A light-switchable fluorescent [2]rotaxane E/Z-**11** [13]. Interconversion between fluorescent E-**11** and non-fluorescent Z-**11** and images of cuvettes containing solutions of Z-**11** and E-**11** respectively (0.8 μM, CH_2Cl_2) illustrating the clearly visible difference in fluorescence intensity. The photograph was taken while illuminating with UV light (365 nm)

11.2.1 Switching "on" and "off" fluorescence with a molecular shuttle

To illustrate the kinds of tasks these machines can perform we designed molecular shuttle *E/Z*-**11** as a fluorescence switch (Figure 11.7) [13]. The system relies on the photoswitchable fumaramide/maleamide binding site; attached to the intermediate-affinity dipeptide station is an anthracene fluorophore. The macrocycle contains pyridinium units able to quench anthracene fluorescence by electron transfer. In both the free thread and *E*-**11**, strong fluorescence (λ_{exc} = 365 nm) is observed, while shuttling of the macrocycle

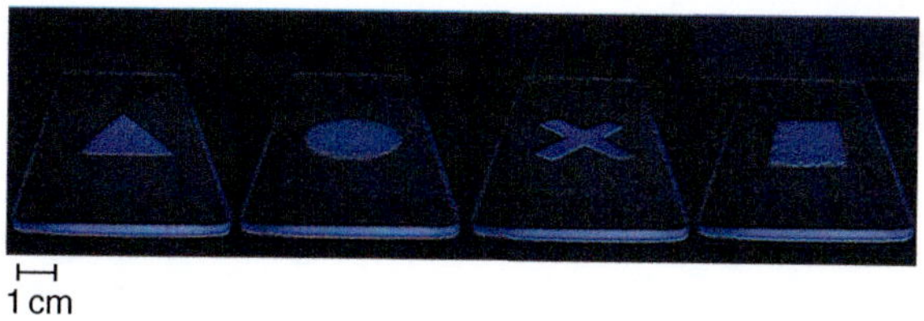

Figure 11.8 Images obtained by casting films of a polarity-switchable fluorescent rotaxane polymer on quartz slides, covering them with an aluminum mask and exposing the unmasked area to dimethyl sulfoxide vapors for 5 minutes [14]. The photograph was taken while illuminating with UV light (254–350 nm)

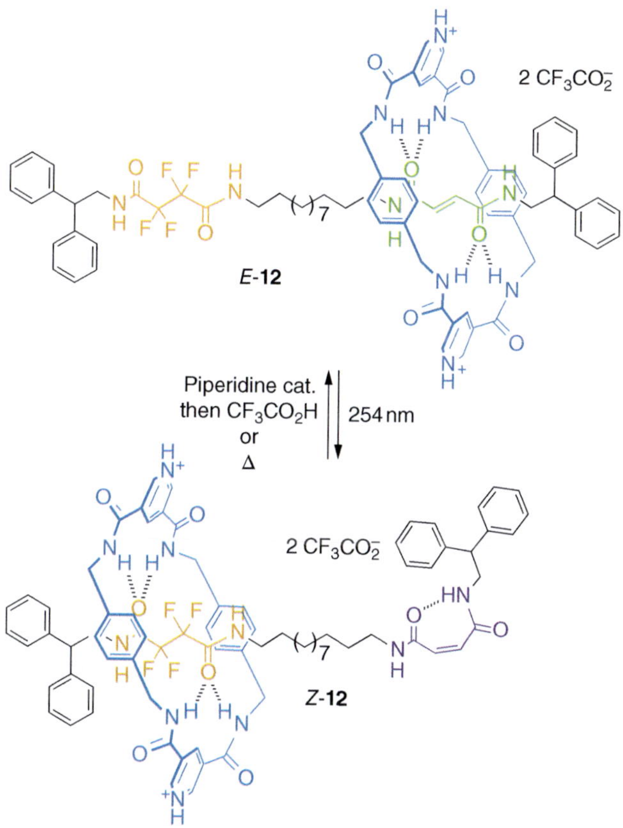

Figure 11.9 Stimuli-induced positional change of the macrocycle in 12 – a fluorinated molecular shuttle [15]

onto the glycylglycine station in *Z*-**11** quenches the emission almost completely. At the maximum of *E*-**11** emission (λ_{max} = 417 nm) there is a remarkable 200:1 difference in intensity between the two states – strikingly visible to the naked eye (Figure 11.7).

The same principles were used to create rotaxane switches that function in polymer films [14]. Patterns visible to the naked eye were generated using a polarity-switchable shuttle covalently derivatized with poly(methyl methacrylate) (PMMA) (Figure 11.8). A polymer film INHIBIT logic gate based on a combination of control of submolecular positioning and chemical modification (protonation) was also demonstrated.

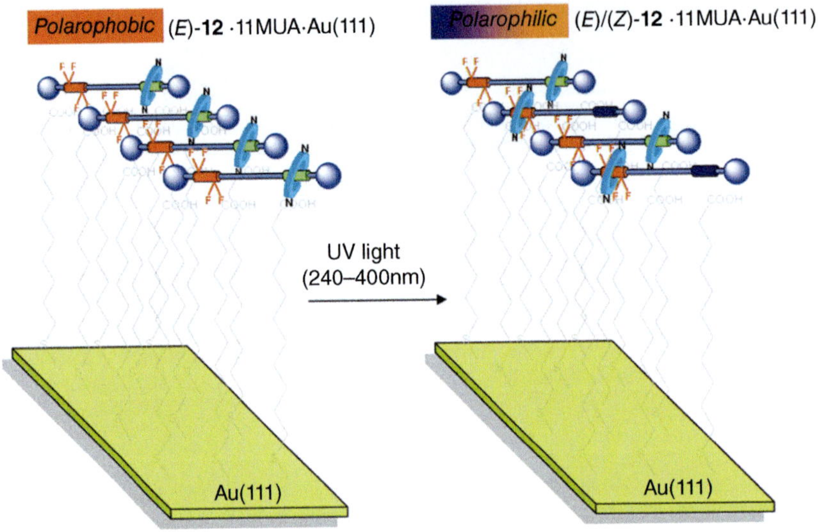

Figure 11.10 A photo-responsive surface based on switchable fluorinated molecular shuttles [15]. Light-switchable rotaxanes with the fluoroalkane region (orange) exposed (E-**12**) were physisorbed onto a SAM of 11-MUA on Au(111) deposited onto either glass or mica to create a polarophobic surface, E-**12**·11-MUA·Au(111). Illumination with 240–400 nm light isomerizes some of the E olefins to Z causing a nanometer displacement of the rotaxane threads in the Z-shuttles which encapsulates the fluoroalkane units leaving a more polarophilic surface, E-**12**·11-MUA·Au(111)

11.2.2 Rotaxane-based photoresponsive surfaces and macroscopic transport by molecular machines

A further illustration of the potential of controlled molecular level motion in this type of system is the creation and utility of photoresponsive surfaces based on rotaxanes (Figures 11.9–11.11). The millimeter scale directional transport of diiodomethane across a surface (Figure 11.11) was achieved using the biased Brownian motion of the components of a stimuli-responsive rotaxane **12** (Figure 11.9) to expose or conceal fluoroalkane residues and thereby modify surface tension [15]. The collective operation of a monolayer of the molecular shuttles attached to a self-assembled monolayer of 11-mercaptoundecanoic acid (11-MUA) on Au(111) (Figure 11.10) was sufficient to power the movement of a microliter droplet of diiodomethane up a 12 degree incline (Figure 11.11e–h).

Recently, we have begun to use rotaxane molecular switches to control catalysis [16]. One goal is to use molecular machines to control sequences of transformations, producing different products from a pool of building blocks according to the order and type of stimulus applied [17]. We see this as reminiscent of the way that allostery and other trigger-induced events are used to control enzymatic synthesis in biology.

11.3 Molecular motors and ratchet mechanisms

Although it was not appreciated by chemists for many years, it turns out that the difference between a "switch" and a "motor" is very important in molecular machine terms [1, 2]. A "switch" – e.g. the change in position of the ring on the thread of the rotaxanes in Section 2 – only influences a system as a function of state. Returning the components of a molecular switch to their original position undoes any mechanical effect it has

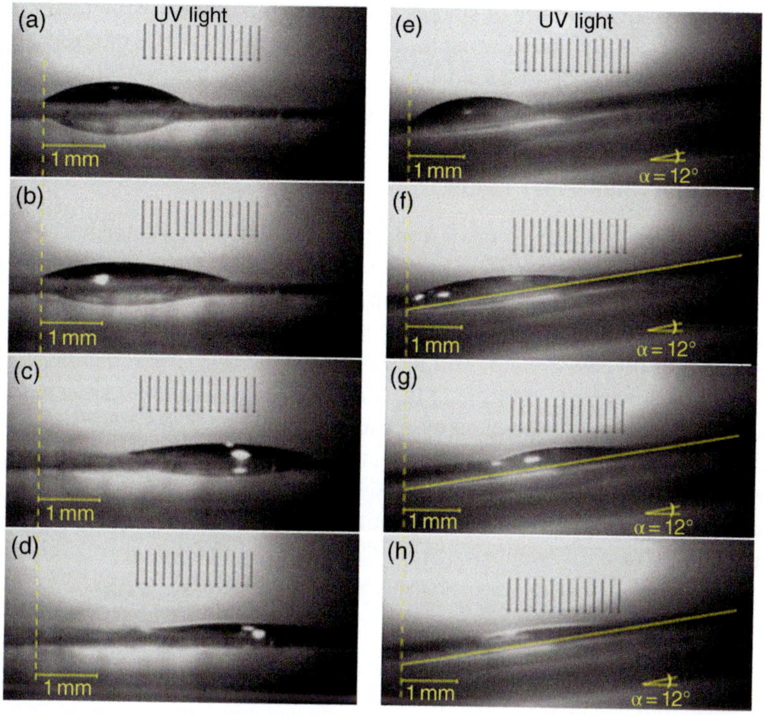

Figure 11.11 *Lateral photographs of light-driven directional transport of a 1.25 μL diiodomethane drop across the surface of a E-**12**·11-MUA·Au(111) substrate on mica arranged flat (a)–(d) and up a 12° incline (e)–(h) [15]. (a) Before irradiation (pristine E-**12**). (b) After 215 s of irradiation (20 s prior to transport) with UV light in the position shown (the right edge of the droplet and the adjacent surface). (c) After 370 s of irradiation (just after transport). (d) After 580 s of irradiation (at the photostationary state). (e) Before irradiation (pristine E-**12**). (f) After 160 s of irradiation (just prior to transport) with UV light in the position shown (the right edge of the droplet and the adjacent surface). (g) After 245 s of irradiation (just after transport). (h) After 640 s irradiation (at the photostationary state). For clarity, on photographs (f)–(h) a yellow line is used to indicate the surface of the substrate*

made; hence, a switch cannot be used to repetitively and progressively do work. A "motor," on the other hand, influences a system as a function of trajectory, meaning that when the components of a molecular motor arrive back at their original positions (after a 360° directional rotation, for example) any mechanical effect (e.g. work) that has been done by the movement of the components is not undone (unless the motor is rotated another 360° in reverse).

This difference in behavior is significant; none of the molecular machines described in Section 2 – which are all switches – can be used to progressively perform work in the way that biological motors can, such as those from the kinesin, myosin and dynein superfamilies.

11.3.1 From molecular switches to molecular motors

In order to make motors using molecular components that move through random thermal motions it is necessary to employ ratchet mechanisms. The quest for ways to move Brownian particles directionally in physical terms dates back to the Maxwell's demon (1874), Smoluchowski's trapdoor (1912) and Feynman's ratchet-and-pawl (1963) thought experiments (for a discussion of their relevance to artificial molecular machine design, see reference [1]). A huge number of studies on (mostly theoretical) Brownian ratchet mechanisms built up in the theoretical physics literature during the 1990s and 2000s. The use of ratchet mechanisms in the design of artificial molecular motors was first used in the design of catenane-based rotary motors (Figures 11.12 and 11.13) and developed over the next few years in rotaxanes.

11.3.2 A [3]catenane rotary molecular motor

In order to bias the direction the macrocycle takes from station-to-station in a catenane temporary barriers are required at each stage to restrict Brownian motion in one particular direction and bias the path taken by the macrocycle. Such a situation is intrinsically present in [3]catenane **13** (Figure 11.12) [18]. Irradiation at 350 nm of *E,E*-**13** causes counter-clockwise (as drawn) rotation of the light blue macrocycle to the succinic amide ester (orange) station to give *Z,E*-**13**. Isomerization (254 nm) of the remaining fumaramide group causes the other (purple) macrocycle to relocate to the single amide (dark green) station (*Z,Z*-**13**) and, again, this occurs counter-clockwise because the clockwise route is blocked by the other (light blue) macrocycle. This "follow-the-leader" process, each macrocycle in turn moving and then blocking a direction of passage for the other macrocycle, is repeated throughout the sequence of transformations shown in Figure 11.12. After three diastereomer interconversions, *E,E*-**13** is again formed but 360° rotation of each of the small rings has not yet occurred, they have only swapped places. Complete unidirectional rotation of both small rings occurs only after the synthetic sequence (i)–(iii) has been completed twice.

11.3.3 Controlled rotation in either direction: a [2]catenane reversible rotary molecular motor

The key to visualizing how the principles of a Brownian ratchet mechanism can be applied to a catenane architecture is not to consider the whole catenane as a molecular machine, but rather to think of one macrocycle as a motor that transports a substrate – the other ring – directionally around itself. In its simplest form this results in a [2]catenane such as **14** (Figure 11.13) which is able to directionally rotate the smaller ring about the larger one in either direction response to a series of chemical reactions – a reversible molecular rotary motor [19].

11.3.4 Molecular information ratchets

An information ratchet, such as Maxwell's pressure demon, is a different sort of ratchet mechanism that uses information regarding the position of a particle to lower a kinetic barrier when the particle approaches the barrier from a particular direction. This positional sorting generates net directional motion of the particle across the barrier. We have employed information ratchet mechanisms in the designs of several rotaxane machine systems. Molecular machine **15** consists of a dibenzo-24-crown-8-based macrocycle mechanically locked on a linear thread by bulky 3,5-di-*t*-butylphenyl stoppers (Figure 11.14) [20]. An α-methyl stilbene unit divides the thread into two compartments, both with ammonium binding sites for the crown ether macrocycle. When the stilbene unit is in the *E* form the macrocycle can move along the full length of the thread by Brownian motion, while in the *Z* form the macrocycle is trapped in one or the other of the two compartments. Selective "gate opening" was achieved by photosensitized energy transfer from the macrocycle (which has a benzophenone moiety attached) to the stilbene unit. As energy transfer is distance-dependent the intercomponent photosensitized energy transfer is much more efficient when the macrocycle is in the left-hand compartment (Figure 11.14a). Once the gate is open and the macrocycle moves to the ammonium binding site that is far from the gate, the gate is closed by isomerization by intermolecular energy transfer from another photosensitizer, benzil. It is the interplay between the two photosensitized isomerization processes that leads to ratcheting: when the macrocycle is in one

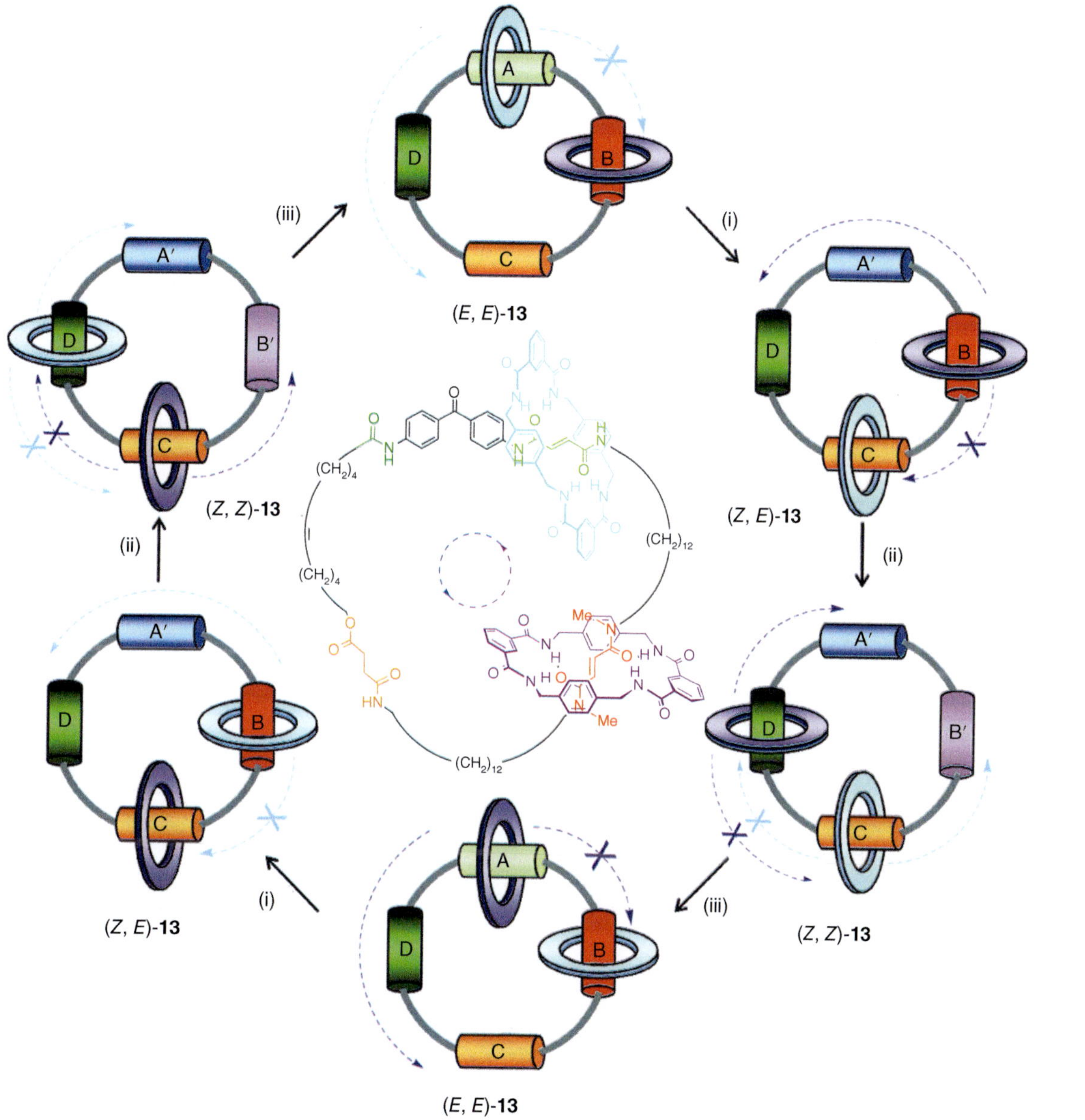

Figure 11.12 *A [3]catenane rotary molecular motor [18]. (i) hv 350 nm; (ii) hv 254 nm; (iii) Br_2 (cat), hv 400-670 nm*

compartment intramolecular, intercomponent energy transfer from the macrocycle dominates; when it is in the other intermolecular energy transfer from benzil dominates. This is an example of using the positional information of a Brownian particle to break detailed balance, driving the macrocycle distribution away from its equilibrium value.

We later explored an alternative approach to a molecular information ratchet using a chemical fuel to directionally transport a macrocycle in a rotaxane (Figure 11.15) [21]. In rotaxane **16** the macrocycle

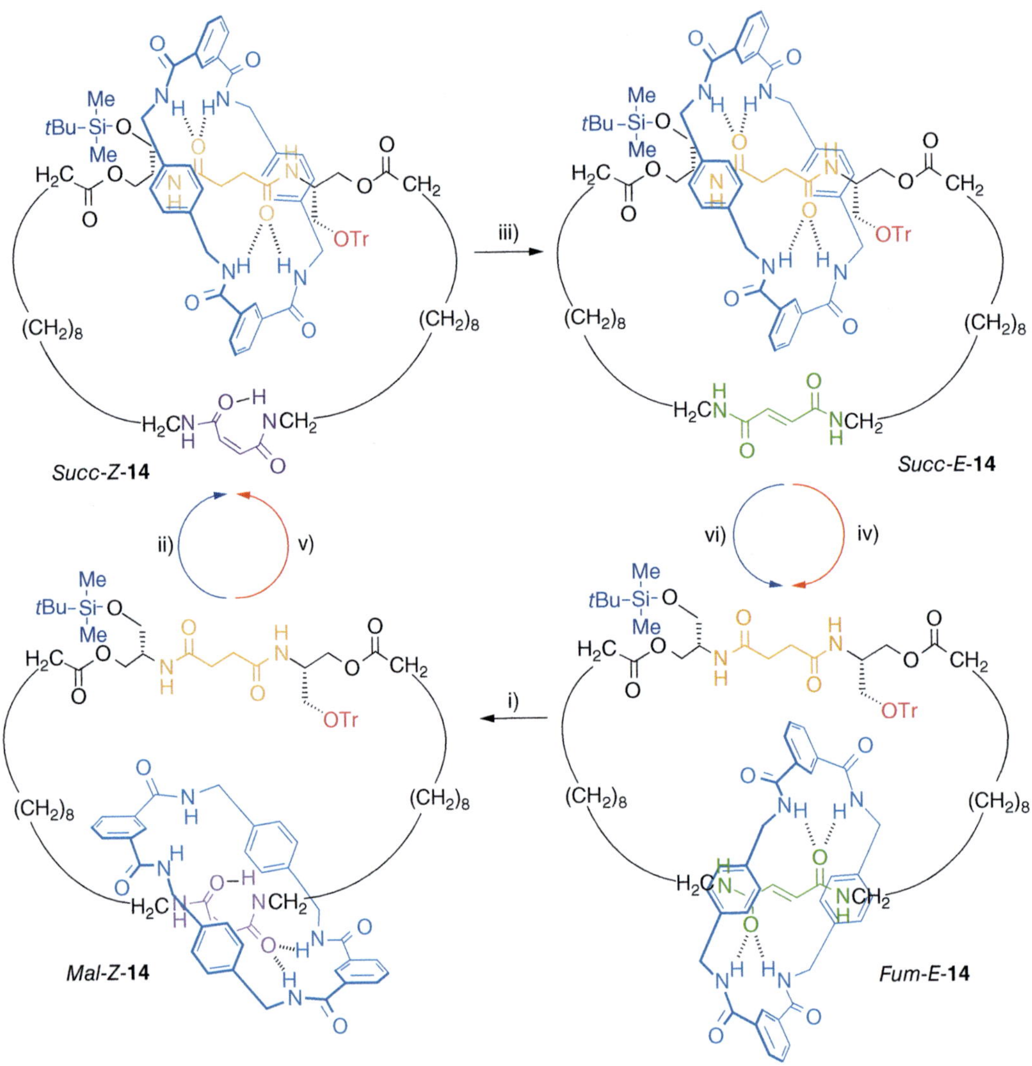

Figure 11.13 *A reversible [2]catenane rotary motor [19]. (i) hv 254 nm; (ii) (a) Bu_4NF, (b) $^tBuMe_2SiOTf$, 2,4,6-collidine; (iii) piperidine; (iv) (a) $Me_2S.BCl_3$, (b) Ph_3COTf, 2,4,6-collidine; (v) (a) $Me_2S.BCl_3$, (b) Ph_3COTf, 2,4,6-collidine; (vi) (a) Bu_4NF, (b) $^tBuMe_2SiOTf$, 2,4,6-collidine*

spends an equal amount of time on the two fumaramide-binding sites. When a barrier was introduced by benzoylation with the chiral DMAP catalyst (*S*)-**17**, a 33:67 distribution of the macrocycle in favor of the pro-(*S*)-fumaramide (*S*)-**18** station was obtained. Using the other hand of the catalyst ((*R*)-**17**) the opposite bias for the position of the macrocycle results. The origin of the product distribution bias is that the chiral catalyst discriminates between the macrocycle being on the different fumaramide sites, benzoylating the prochiral hydroxyl group faster with the ring on one particular site. In other words the *position* of the macrocycle determines how quickly the barrier to shuttling is introduced; a molecular information ratchet mechanism that drives the macrocycle distribution directionally away from its equilibrium value.

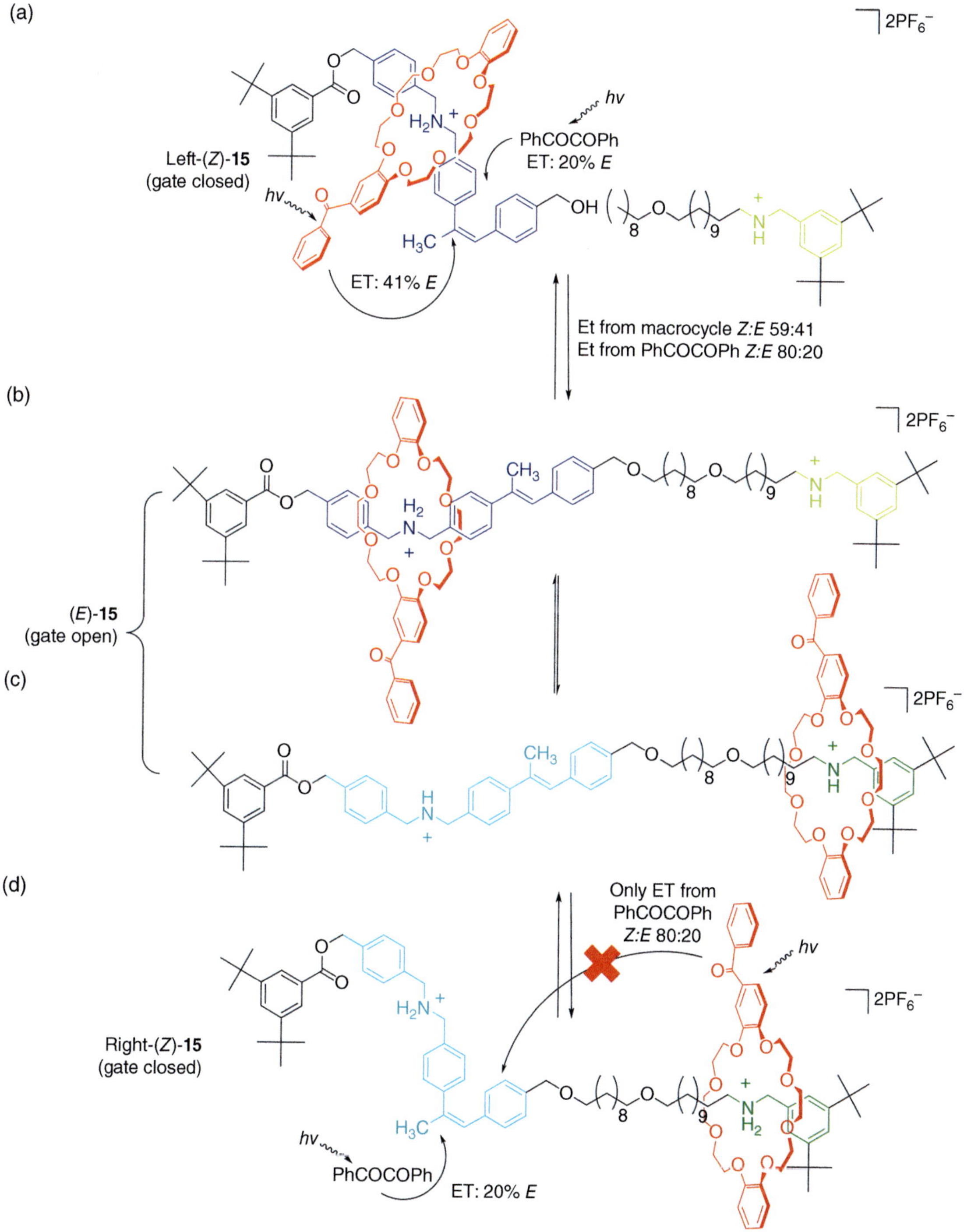

Figure 11.14 *A molecular information ratchet (**15**) [20]. (a) Gate closed, energy transfer from the macrocycle to the stilbene gate is efficient. (b & c) Gate is open, allowing shuttling of the macrocycle between the two ammonium sites. (d) The macrocycle is held far from the gate and intramolecular energy transfer (ET) from the macrocycle is inefficient; intermolecular energy transfer from PhCOCOPh dominates (closing the gate)*

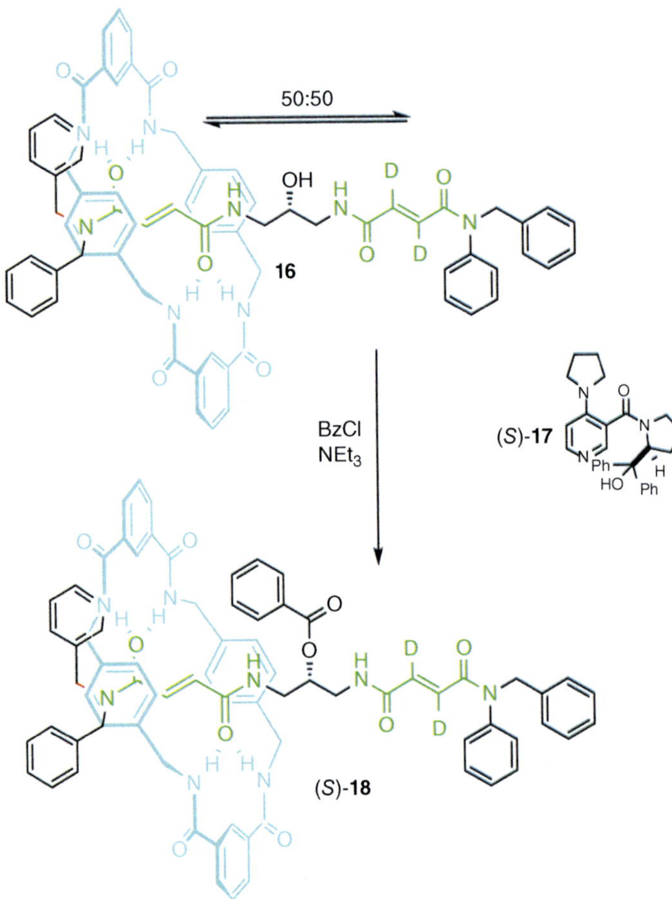

Figure 11.15 A chemically-driven molecular information ratchet [21]

This concept was further developed in a three-compartment, chemically driven, molecular information ratchet **19** (Figure 11.16) [22]. Here, the macrocycle could be efficiently driven to either end of the track by benzoylation in the presence of a chiral catalyst, resulting in a 1:21:79 distribution of the macrocycle between the three compartments. Again, the distributional bias was reversed using the catalyst enantiomer.

11.4 Small molecules that can "walk" along molecular tracks

Linear molecular motors are widely used in biology, to transport cargoes (e.g. kinesin I), generate mechanical force at the molecular (actin polymerases) and macroscopic levels (myosin in muscles), to carry out sequential synthesis (ribosome, DNA polymerases) and numerous other intracellular tasks. The first small molecules able to "walk" along tracks (reminiscent of the mode of transport of kinesin and other motor proteins) were introduced in 2010 [23], with directional versions employing ratchet mechanisms developed shortly afterwards (Figure 11.17) [24]. Unlike rotaxanes, these walkers move along their tracks progressively, and such systems should be capable of transporting cargoes directionally. However, current systems

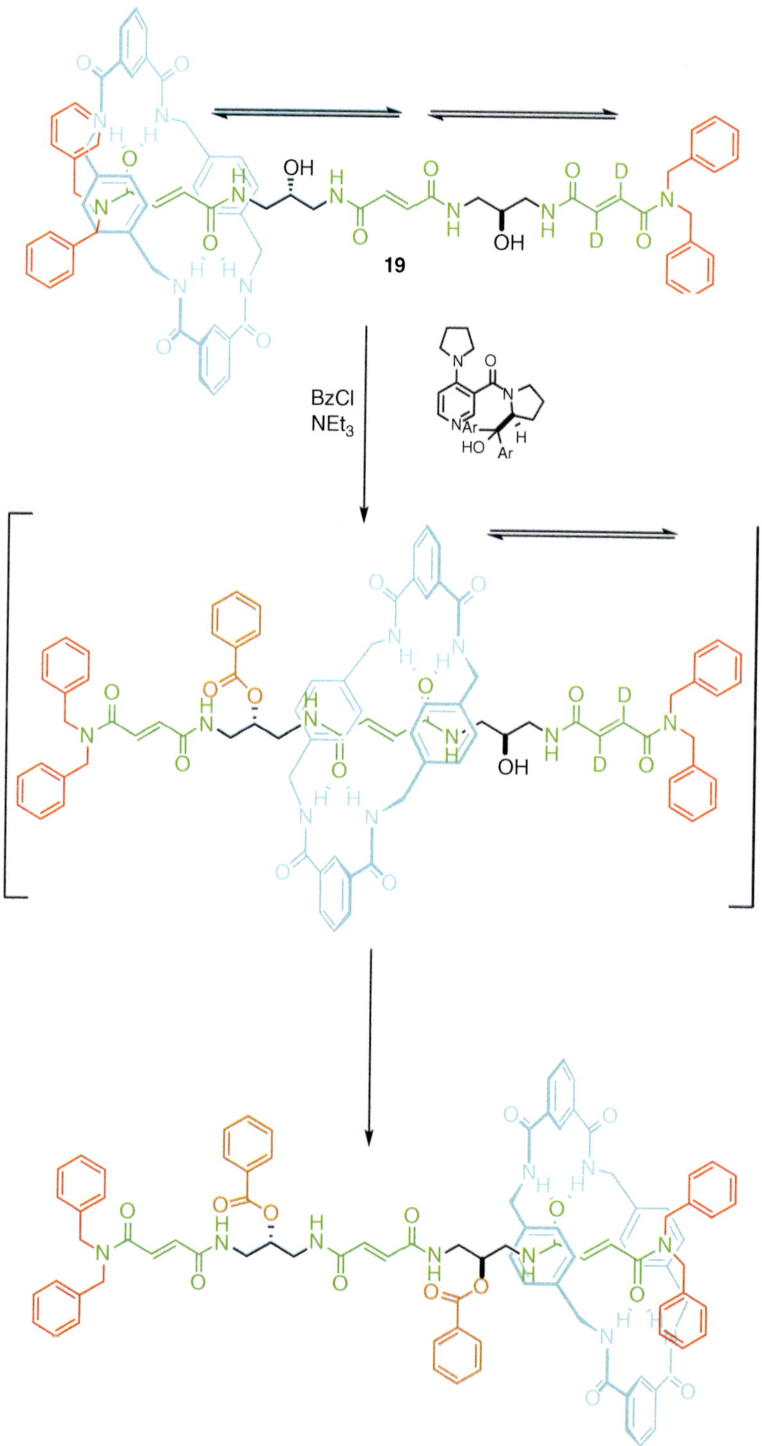

Figure 11.16 *Directional transport of a macrocycle in a [2]rotaxane three-compartment chemical information ratchet (**19**) [22]*

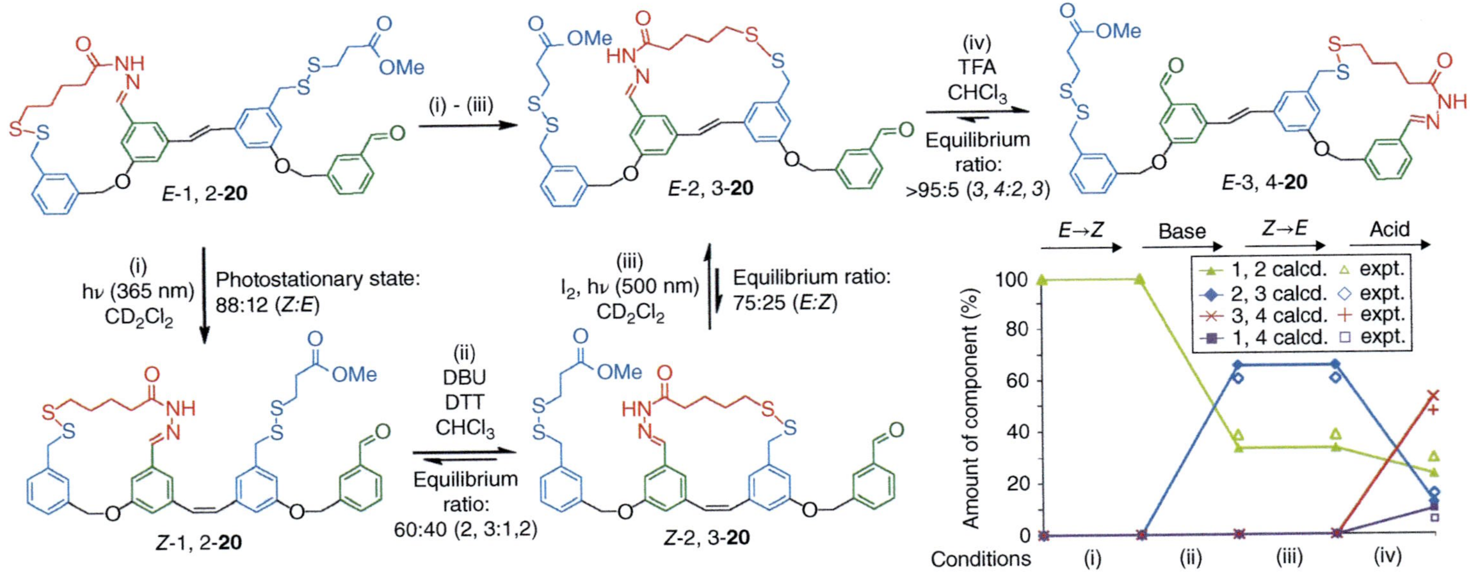

Figure 11.17 A small molecule that "walks" directionally along a molecular track using a light-fueled energy ratchet mechanism [24]

are only efficient enough to take a few steps along short tracks and are not yet robust enough to walk across surfaces, or along polymers, or to perform useful tasks.

One approach we have developed utilizes molecular walkers with two different "feet" that form dynamic linkages with the track under orthogonal sets of conditions. The rationale is to allow the feet to be able to step along the track without both feet ever becoming detached at the same time (which would result in the walker exchanging with others in the bulk; i.e. a loss of processivity. In molecule **20** this is accomplished using hydrazone formation (dynamic in acid; kinetically locked in base) for one foot-track interaction, and disulfide bond formation (labile in base; inert in acid) for the other [24]. In this system directionality is conferred by using *cis–trans*-isomerization of a stilbene unit in the track to put strain into the walkers when the double bond is part of the macrocycle formed between the walker and the track (i.e. *2,3*-**20**). Release of the strain by labilizing one of the foot-track linkages drives the walker in a particular direction (Figure 11.17).

An alternative system features molecular walkers with two identical feet, in the form of α-methylene-4-nitrostyrene bound to polyamine tracks and the dynamics of its subsequent migration from an amine group to amine group, without fully detaching, via a sequence of intramolecular Michael and retro-Michael reactions, generating the most thermodynamically favored distribution of walkers on the oligoamine track [25]. Remarkably, these Michael-retro-Walkers are highly processive; each α-methylene-4-nitrostyrene unit takes on average ~530 "steps" between amine groups before completely detaching from its track.

11.5 Making molecules that make molecules

From polyketide synthase to DNA polymerases and the ribosome, one of the key uses of molecular machines in biology is for the construction of other molecules. We recently introduced an artificial small-molecule machine that assembled a tripeptide of specific sequence by traveling along a track loaded with amino acid building (Figure 11.18) [26]. This is a very(!) primitive analog of the task performed by the ribosome in cells but is, nonetheless, one of the most sophisticated tasks to be performed by an artificial molecular machine to date. It is the first synthetic molecular machine to have a truly complex mechanism of operation, requiring the integrated interaction of several functional component parts (a reversibly attached reactive "arm" with regenerable catalytic site and a peptide-elongation site, a ring that threads the track catalytically with no residual ring-track interactions to retard the machine's action, a track with amino acid building blocks in a predetermined sequence separated by rigid spacers). The invention of molecular machines capable of complex task performance marks the start of what is likely to be an era of increasingly ambitious and useful applications of synthetic molecular machines.

11.6 Outlook

The future for the field of artificial molecular machines is extremely bright. There is already a working nanotechnology based on molecular machines that perform a myriad of useful tasks: it is called "biology." So we know that it is possible to do extraordinary things with molecular machines. In terms of artificial systems, advances over the past 25 years mean that chemists now have the knowhow and synthetic tools available to enable them to make appropriate machine architectures (e.g. catenanes, rotaxanes, overcrowded alkenes, molecules that walk upon tracks, etc.). They can switch the position of components (often by clever manipulation of non-covalent interactions between the various parts), they understand about how to use ratchet mechanisms to create motor-mechanisms and are learning how to introduce them into molecular machine systems. However, although there have been some notable successes in using artificial molecular machines to bring about property changes, they have not yet been shown to perform any useful task that cannot be

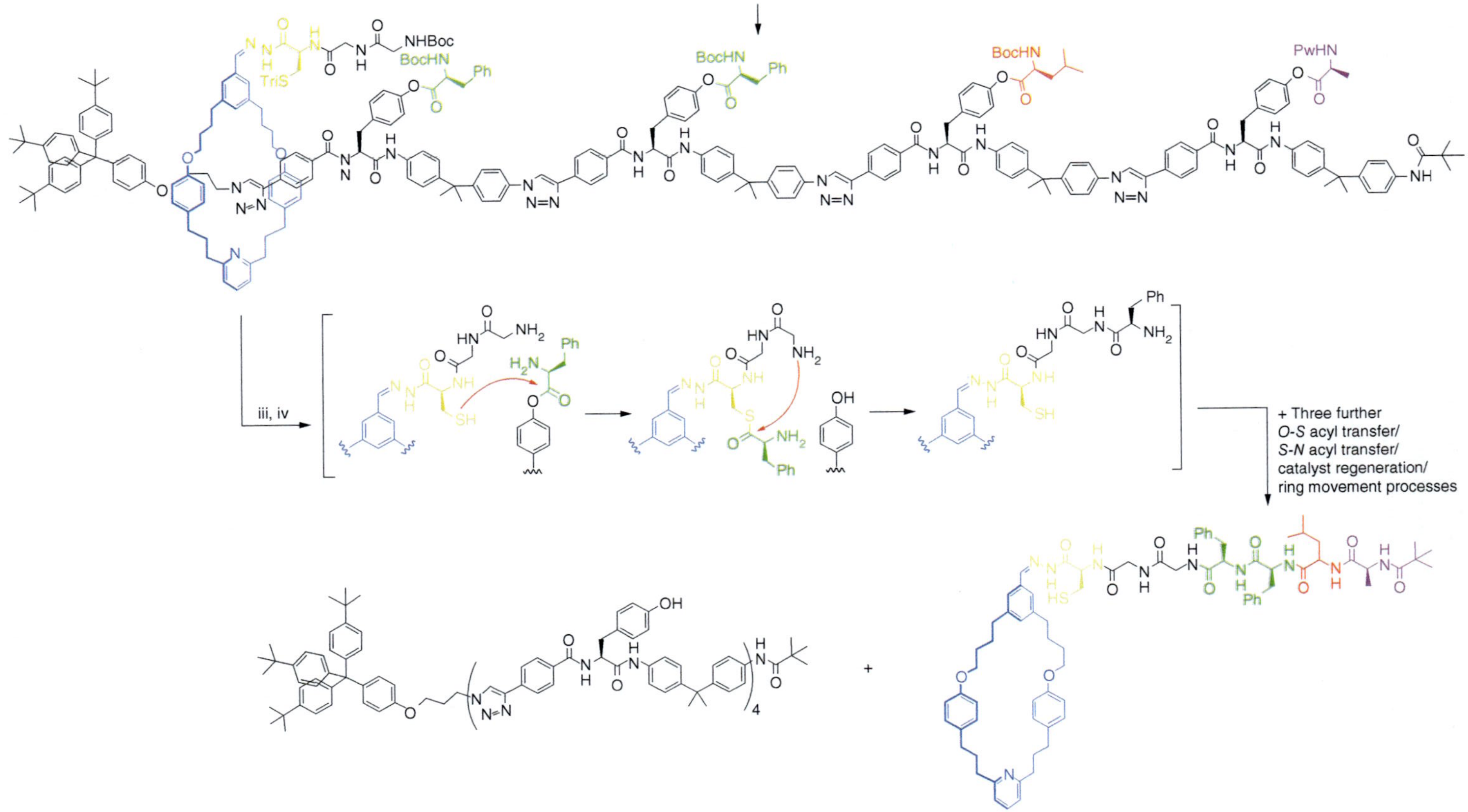

Figure 11.18 An artificial molecular peptide synthesizer [26]

accomplished by conventional chemical means. This contrasts with the crucial roles played by biological machines in numerous cellular processes. When this last step happens, and very significant advances have happened in this regard in the last few years, then artificial molecular machines will have started to fulfil their almost limitless potential.

11.7 Acknowledgements

I am hugely indebted to the many extremely talented, enthusiastic and creative students and postdoctoral researchers who have contributed over the years to our group's efforts on catenanes, rotaxanes, knots and molecular machines.

A journey of a thousand miles begins with a single step (Lao Tzu).

References

1. Kay, E.R., Leigh, D.A., Zerbetto, F. (2007) Synthetic molecular motors and mechanical machines, *Angewandte Chemie, International Edition*, **46**, 72–191.
2. Kay, E.R., Leigh, D.A. (2015) Rise of the molecular machines, *Angewandte Chemie, International Edition*, **54**, DOI: 10.1002/anie.201503375.
3. Dietrich-Buchecker, C.O., Sauvage, J.-P., Kintzinger, J.P. (1983) Une nouvelle famille de molecules: les metallo-catenanes, *Tetrahedron Letters*, **24**, 5095–5098.
4. Ashton, P.R., Goodnow, T.T., Kaifer, A.E., Reddington, M.V., Slawin, A.M.Z., Spencer, N., Stoddart, J.F., Vicent, C., Williams, D.J. (1989) A [2]catenane made to order, *Angewandte Chemie, International Edition English*, **28**, 1396–1399.
5. Hunter, C.A. (1992) Synthesis and structure elucidation of a new [2]-catenane, *Journal of the American Chemical Society*, **114**, 5303–5311.
6. Vögtle, F., Meier, S., Hoss, R. (1992) One-step synthesis of a fourfold functionalized catenane, *Angewandte Chemie, International Edition English*, **31**, 1619–1622.
7. Johnston, A.G., Leigh, D.A., Pritchard, R.J., Deegan, M.D. (1995) Facile synthesis and solid state structure of a benzylic amide [2]catenane, *Angewandte Chemie, International Edition English*, **34**, 1209–1212.
8. Johnston, A.G., Leigh, D.A., Murphy, A., Smart, J.P., Deegan, M.D. (1996) The synthesis and solubilization of amide macrocycles via rotaxane formation, *Journal of the American Chemical Society*, **118**, 10662–10663.
9. Leigh, D.A., Murphy, A., Smart, J.P., Slawin, A.M.Z. (1997) Glycylglycine rotaxanes - the hydrogen bond-directed assembly of synthetic peptide rotaxanes, *Angewandte Chemie, International Edition English*, **36**, 728–732.
10. Gatti, F.G., Leigh, D.A., Nepogodiev, S.A., Slawin, A.M.Z., Teat, S.J., Wong, J.K.Y. (2001) Stiff, and sticky in the right places, The dramatic influence of preorganizing guest binding sites on the hydrogen bond-directed assembly of rotaxanes, *Journal of the American Chemical Society*, **123**, 5983–5989.
11. Lane, A.S., Leigh, D.A., Murphy, A. (1997) Peptide-based molecular shuttles, *Journal of the American Chemical Society*, **119**, 11092–11093.
12. Keaveney, C.M., Leigh, D.A. (2004) Shuttling through anion recognition, *Angewandte Chemie, International Edition*, **43**, 1222–1224.
13. Pérez, E.M., Dryden, D.T.F., Leigh, D.A., Teobaldi, G., Zerbetto, F. (2004) A generic basis for some simple light-operated mechanical molecular machines, *Journal of the American Chemical Society* **126**, 12210–12211.
14. Leigh, D.A., Morales, M.Á.F., Pérez, E.M., Wong, J.K.Y., Saiz, C.G., Slawin, A.M.Z., Carmichael, A.J., Haddleton, D.M., Brouwer, A.M., Buma, W.J., Wurpel, G.W.H., León, S., Zerbetto, F. (2005) Patterning through controlled submolecular motion: Rotaxane-based switches and logic gates that function in solution and polymer films, *Angewandte Chemie, International Edition*, **44**, 3062–3067.
15. Berná, J., Leigh, D.A., Lubomska, M., Mendoza, S.M., Pérez, E.M., Rudolf, P., Teobaldi, G., Zerbetto, F. (2005) Macroscopic transport by synthetic molecular machines, *Nature Materials*, **4**, 704–710.
16. Blanco, V., Carlone, A., Hänni, K.D., Leigh, D.A., Lewandowski, B. (2012) A rotaxane-based switchable organocatalyst, *Angewandte Chemie, International Edition*, **51**, 5166–5169.

17. Beswick, J., Blanco, V., De Bo, G., Leigh, D.A., Lewandowska, U., Lewandowski, B., Mishiro, K. (2015) Selecting reactions and reactants using a switchable rotaxane organocatalyst with two different active sites, *Chemical Science*, **6**, 140–143.
18. Leigh, D.A., Wong, J.K.Y., Dehez, F., Zerbetto, F. (2003) Unidirectional rotation in a mechanically interlocked molecular rotor, *Nature*, **424**, 174–179.
19. Hernandez, J.V., Kay, E.R., Leigh, D.A. (2004) A reversible synthetic rotary molecular motor, *Science*, **306**, 1532–1537.
20. Serreli, V., Lee, C.-F., Kay, E.R. Leigh, D.A. (2007) A molecular information ratchet, *Nature*, **445**, 523–527.
21. Alvarez-Pérez, M., Goldup, S.M., Leigh, D.A., Slawin, A.M.Z. (2008) A chemically-driven molecular information ratchet, *Journal of the American Chemical Society*, **130**, 1836–1838.
22. Carlone, A., Goldup, S.M., Lebrasseur, N., Leigh, D.A., Wilson, A. (2012) A three-compartment chemically-driven molecular information ratchet, *Journal of the American Chemical Society*, **134**, 8321–8323.
23. von Delius, M., Geertsema, E.M., Leigh, D.A. (2010) A synthetic small molecule that can walk down a track, *Nature Chemistry*, **2**, 96–101.
24. Barrell, M.J., Campaña, A.G., von Delius, M., Geertsema, E.M., Leigh, D.A. (2011) Light-driven transport of a molecular walker in either direction along a molecular track, *Angewandte Chemie, International Edition*, **50**, 285–290.
25. Campaña, A.G., Carlone, A., Chen, K., Dryden, D.T.F., Leigh, D.A., Lewandowska, U., Mullen, K.M. (2012) A small molecule that walks non-directionally along a track without external intervention, *Angewandte Chemie, International Edition* **51**, 5480–5483.
26. Lewandowski, B., De Bo, G., Ward, J.W., Papmeyer, M., Kuschel, S., Aldegunde, M.J., Gramlich, P.M.E., Heckmann, D., Goldup, S.M., D'Souza, D.M., Fernandes, A.E., Leigh, D.A. (2013) Sequence-specific peptide synthesis by an artificial small-molecule machine, *Science*, **339**, 189–193.

12

Clipping an Angel's Wings

Roeland J.M. Nolte,[1] Alan E. Rowan,[2] and Johannes A.A.W. Elemans[1]

[1]Radboud University, Institute for Molecules and Materials, Nijmegen, The Netherlands
[2]Australian Institute for Bioengineering and Nanotechnology (AIBN), The University of Queensland, Brisbane, Australia

12.1 Introduction

> Do not all charms fly
> At the mere touch of cold philosophy?
> There was an awful rainbow once in heaven:
> We know her woof, her texture; she is given
> In the dull catalogue of common things.
> *Philosophy will clip an Angel's wings,*
> Conquer all mysteries by rule and line,
> Empty the haunted air, and gnomed mine –
> Unweave a rainbow, as it erewhile made
> The tender-person'd Lamia melt into a shade.

Newton's physical and mathematical analysis of natural phenomena (at that time described as a natural philosopher's view) dramatically changed our way of thinking and had a great impact on society. Not to the pleasure of everyone, though, as can be concluded from the quote of the romantic poet John Keats (1795–1821) shown above. As he complains in his poem, insight in how the colors of a rainbow are produced makes this phenomenon a dull event. The same can be said about the important discoveries of Antoine Lavoisier (1743–1794), which heralded the end of alchemy and opened the field of chemical science. Not something to please all his contemporaries. With the review below on molecular clips derived from glycoluril we hope to show that Keats' vision was too pessimistic, indeed wrong. Instead of destroying dreams, chemistry and physics create them: clipping an angels' wings has a different meaning here.

Macrocyclic and Supramolecular Chemistry: How Izatt–Christensen Award Winners Shaped the Field, First Edition.
Edited by Reed M. Izatt.

The serendipitous discovery of the class of crown ethers by Charles J. Pedersen in the 1960s and the observation that these compounds can bind alkali metal ions [1] is generally accepted as the start of a new field of research, i.e. that of macrocyclic compounds and molecular recognition, later named "host–guest chemistry" by Donald J. Cram and, in a more broad sense, "supramolecular chemistry" by Jean-Marie Lehn. In 1981 one of us (RN) worked in Cram's laboratory at the University of California, Los Angeles as a visiting scientist. Cram told him the story that after reading Pedersen's paper he had become so excited that he had made the decision to completely change his research program. After having seen the potential of host–guest chemistry and the way it was approached by Cram, i.e. by designing compounds with the help of space-filling (CPK) models, we became fascinated and concluded that we should start a line of research in the Netherlands in which this new type of chemistry was incorporated. We decided to focus on glycoluril as a building block for the construction of functional macrocyclic compounds. The motivation for this decision was two-fold: firstly, while working with Cram on cyclic urea containing cavitands [2] RN had become interested in cyclic urea molecules as components of macrocyclic rings. Secondly, during RN's time at UCLA a paper appeared in the literature by Mock and co-workers on the crystal structure of an until that time unknown compound (**1**, Figure 12.1) [3]. It involved a solid material that had already been prepared by Behrend and coworkers at the beginning of 1900 by a one-pot procedure from urea, glyoxal, and formaldehyde [4], but because of its insolubility they could not obtain any information about its structure.

Mock showed that it was a cyclic hexamer of glycoluril molecules linked by methylene bridges. The new material was extremely stable and formed crystalline complexes with a great variety of metal and alkyl ammonium salts. He proposed the name cucurbituril for this compound, because it looked like a pumpkin (*cucurbita* is the Latin name for pumpkin). Our initial efforts were concentrated on making cucurbituril a more functional compound; while trying to increase its solubility in various ways, e.g. by attaching substituents, we serendipitously discovered that the reaction of diphenylglycoluril with urea in benzene gave the clip-shaped compound **2** [5]. An X-ray study of the tetramethoxy derivative **2a** [6] confirmed this structure and revealed that the clip had an electron-rich bottom formed by the two fused five-membered rings of glycoluril, and carbonyl groups that were pointing upwards, offering the possibility of functioning as hydrogen-bond acceptor sites, see 2a. The aromatic side-walls were not parallel but placed at an angle, defining a tapering cavity suitable for π–π stacking interactions with guest molecules (Figure 12.1). In subsequent studies we prepared a large series of different clip molecules, which were found to display interesting host–guest binding properties, and we used them to develop supramolecular catalytic systems and self-assembled materials (vide infra) [7–9]. Somewhat later, the groups of Rebek and de Mendoza prepared larger clip-shaped glycoluril molecules and self-assembled these to construct cage compounds (called capsules) of varying sizes, for which they coined the names tennis ball,

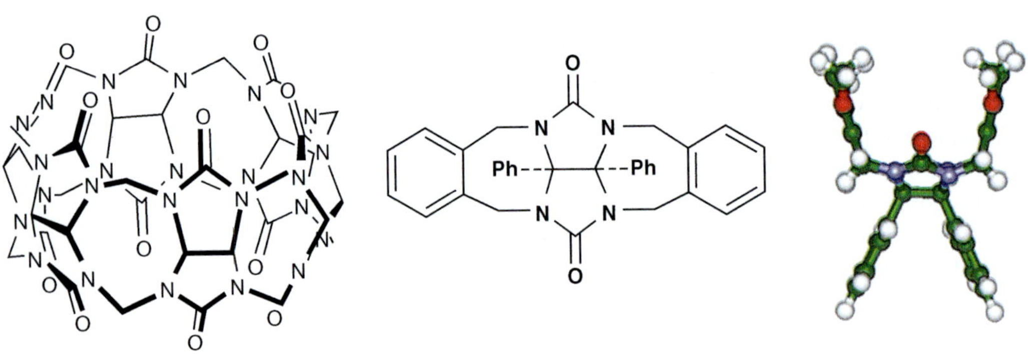

1 **2** **2a**

Figure 12.1 *Chemical structures of cucurbituril **1**, diphenylglycoluril clip **2**, and X-ray structure of **2a***

soft ball, etc. [10, 11]. These were shown to be capable of holding different types of guest molecules and to act as nanocontainers in which reactions can be carried out. These studies are described later in this chapter.

Since the clip compounds formed an interesting topic for further research we did not continue our investigations on cucurbiturils, also because the great insolubility of these compounds turned out to be a barrier to the making of functional derivatives. The solubility issue, however, was successfully addressed in later years, in particular by the groups of Kim and Isaacs. Once made soluble and accessible, cucurbituril turned out to be a versatile building block in the field of host–guest chemistry, allowing it to be used for many applications in a similar way to cyclodextrin molecules. A great variety of cucurbituril derivatives, abbreviated as CB[*n*], containing cavities of different sizes has now been prepared, including the series CB[5]–CB[10], and some of these compounds are commercially available [12]. The cucurbiturils are characterized by a hydrophobic cavity that can be entered via two portals lined by ureido carbonyl groups. This property causes them to have a pronounced affinity for metal ions and guest molecules, in particular alkyl ammonium ions, which are bound by hydrophobic interactions, hydrogen bonds, and ion–dipole interactions leading to association constants for CB[6] as high as $K_a = 10^7\,M^{-1}$ (water/HCOOH, 1:1 v/v, 40 °C). CB[7] is more voluminous and can bind a wider range of guest molecules than CB[6], e.g. adamantanes, naphthalenes, viologens, o-carboranes, ferrocene, and metal complexes. Cucurbiturils have been applied as molecular switches, as nanoreaction vessels, e.g. for cycloaddition reactions, as building blocks for the synthesis of supramolecular polymers, as containers for the delivery of drugs *in vivo*, and as complexants of heavy metal colorants and organic dyes in the removal of side products in textile waste streams. Furthermore, they have been used to prepare a great variety of supramolecular structures, such as molecular necklaces, rotaxanes and pseudorotaxanes, and also molecular clips, which will be discussed later in this review. Very recently, cucurbiturils were found to be useful tools to modulate protein assembly and to fabricate multifunctional biointerfaces [12–14]. The synthesis, physical properties, and applications of the cucurbituril family of compounds have been reviewed by Isaacs [15, 16] and are described by Kim in Chapter 6 of this book.

12.2 Molecular clips

12.2.1 Synthesis

Glycoluril-based molecular clips are synthesized by the condensation of a diketone with urea, followed by reaction with paraformaldehyde and base to give a cyclic ring ether (3), which is subsequently treated with an aromatic compound and acid [17]. The reaction is a straightforward electrophilic aromatic substitution reaction, but in the case of less reactive aromatic compounds and in order to be more flexible with respect to the choice of side-walls, routes via the tetrachloro derivative **4b** and the dichloro derivative **5b** (which are accessible from the corresponding acetates **4a** and **5a**) are preferred (Figure 12.2) [6]. Compound **5b** allows the preparation of clips with different side-walls, e.g. the metal-binding derivative **6** and porphyrin clips.

Isaacs and coworkers have synthesized various extended clip molecules (e.g. **7**) that contain aromatic side-walls and up to four glycoluril units following a building-block approach in which ring ether **3** is reacted with a C-shaped glycoluril dimer and finally with hydroquinone (Figure 12.3) [18]. Glycoluril clips can also be prepared by a direct nucleophilic substitution reaction between glycoluril and a bromomethylated aromatic compound, as was shown for the synthesis of clips containing two glycoluril units, e.g. **8** [19, 20] and clips with crown ether side-walls [21].

The glycoluril clips are usually isolated and purified by precipitation or column chromatography. Recently, an efficient purification procedure based on affinity chromatography using a phloroglucinol-functionalized Merrifield resin was published. The active component is a resorcinol guest molecule that binds efficiently in the clips (vide infra) and retards the passing of the latter compounds through the column, leading to improved separation from by-products [22].

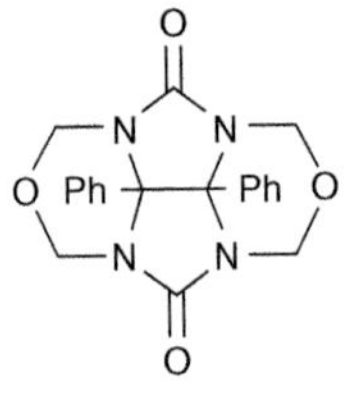

3

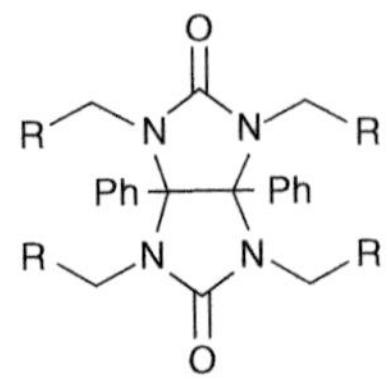

4a, R = OAc
4b, R = Cl

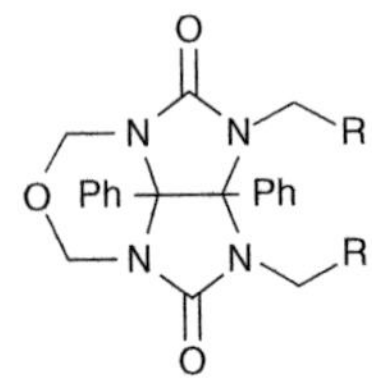

5a, R = OAc
5b, R = Cl

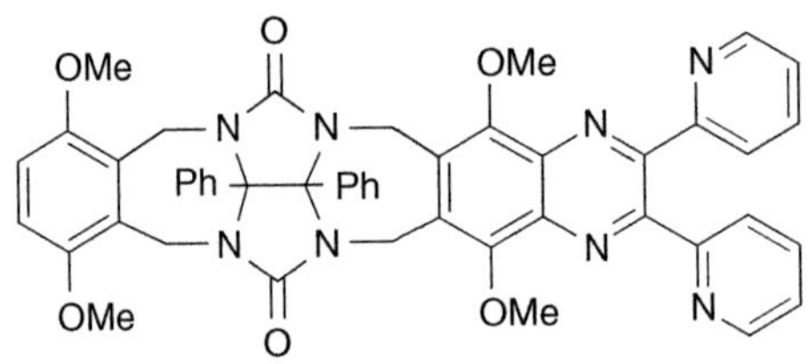

6

Figure 12.2 *Structures of ring ether **3**, tetraacetato and tetrachloro derivatives **4a** and **4b**, diacetato and dichloro derivatives **5a** and **5b**, and ligand clip **6***

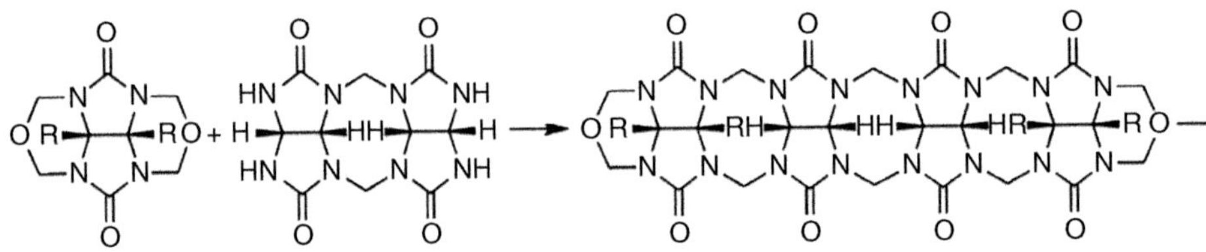

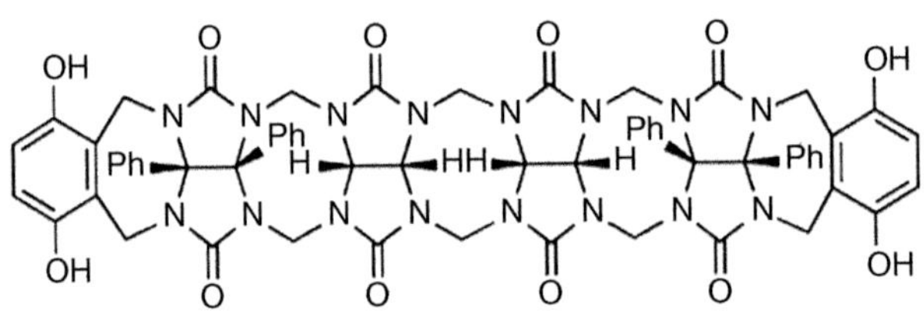

7

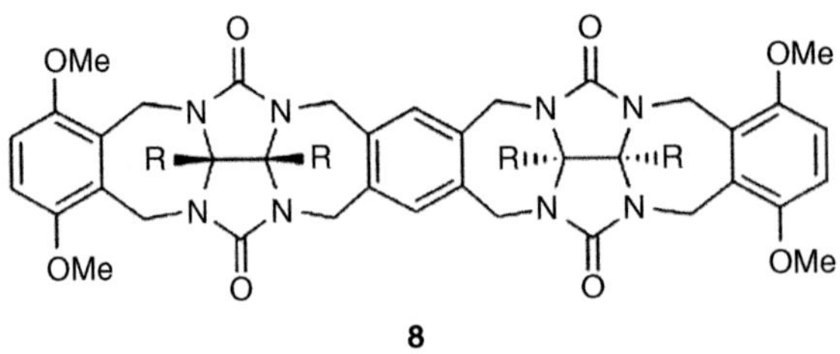

8

Figure 12.3 *Extended clip molecules **7** and **8** synthesized by Isaacs and co-workers [18, 19]*

12.2.2 Binding properties

Despite their relatively simple structure, glycoluril clips are excellent hosts for flat aromatic guest molecules, with association constants that can reach values up to $K_a = 10^6\,M^{-1}$ ($CDCl_3$, RT) [8]. Binding occurs by a combination of non-covalent interactions, as was concluded from extensive binding studies involving clips of different sizes, clips that lacked one side wall, and clips in which the glycoluril oxygen atoms were replaced by non-hydrogen bonding sulfur atoms or guanidine functions. The first important non-covalent interaction is π–π stacking between the aromatic ring of the guest and the two side-walls of the clip. Furthermore, if the guest has suitable hydrogen-bonding groups (e.g. OH groups) binding is strengthened by hydrogen bonding interactions with the π-systems of the carbonyl groups. A third contribution to the stability of the host–guest complex is a cavity effect, which is the entropically favorable filling of the empty cavity by the guest. From the binding measurements the latter effect was calculated to be $6\,kJ\,mol^{-1}$ [23].

Sometimes a conflict occurs between the different non-covalent interactions if the geometric requirements for π–π stacking and hydrogen-bond formation do not match. This was, for instance, found to occur for clips with naphthalene side-walls that were attached at the 2,3 positions (Figure 12.4: **9**) leading to considerably lower binding constants with aromatic guest molecules. This problem can be overcome by attaching the naphthalene side-walls in a different way, i.e. via their 1,8-positions (**10**). Interestingly, **10** turns out to exist as a mixture of three slowly interconverting conformers, in which the naphthalene side-walls are down–down (*syn–syn*, *ss*), down–up (*syn–anti*, *sa*), or up–up (*ant–anti*, *aa*) (Figure 12.4).The *sa* conformation is the one that is present in excess. When a guest binds to **10** the latter conformation changes to *aa*, leading to an induced-fit type of binding mechanism [24]. Also, silver ions can bind between the walls of **10** and induce a conformational change from *ss* and *sa* to *aa* [25].

Recently, the structure of a glycoluril clip with benzene walls and interactions with hydroxybenzene guests have been determined by DFT calculations [26]. In line with experiment, binding was found to take place inside the clip; for a series of hydroxybenzene guests the strongest host–guest interactions were found for 1,3-dihydroxybenzene, again in line with experimental results. Binding of guests (amines) in the extended clip **7** is very strong, with K_a values up to $3.5 \times 10^9\,M^{-1}$ (water/HCOOH, 1:1 v/v, RT), whereas the association

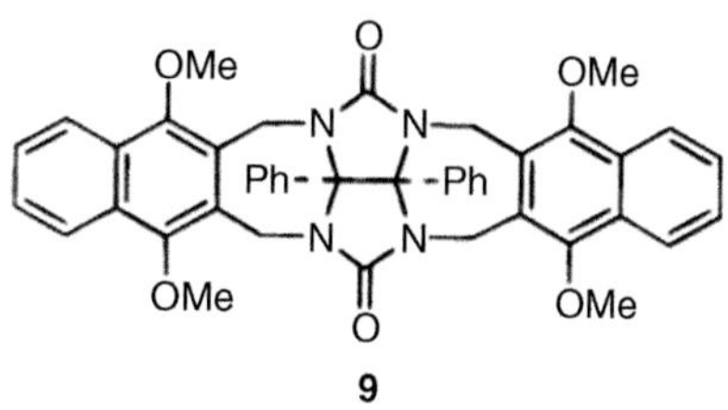

9

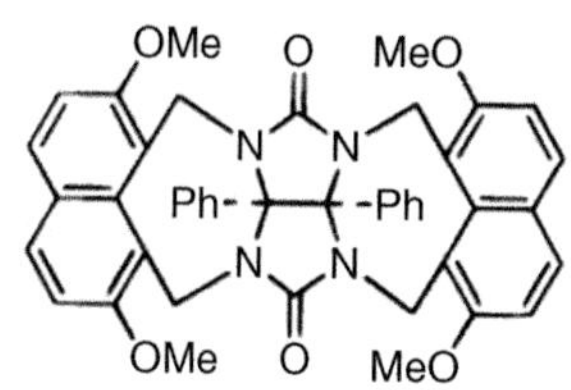

10

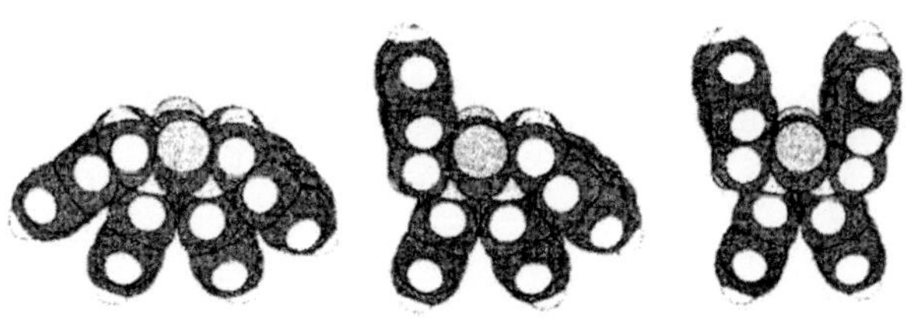

ss *sa* *aa*

Figure 12.4 *Chemical structures of naphthalene clips **9** and **10** and the three conformers of **10***

constants are smaller ($K_{ass} < 1.5 \times 10^4\ \text{M}^{-1}$ for the same guests in water/HCOOH, 1:1 v/v, RT) for extended clip **8**. A methylene bridged glycoluril tetramer model compound of **7** lacking the aromatic side-walls, binds guests with a much lower K_a value than **7**, which establishes the importance of these walls in the overall recognition process. Electroactive glycoluril clips with tetrathiafulvalene side-walls have been reported; their affinities towards a tetracyanoquinodimethane (TNCQ) derivative and *m*-dinitrobenzene were measured and found to be relatively high given the fact that only π–π stacking interactions are involved ($K_a = 2.5 \times 10^4\ \text{M}^{-1}$ for TNCQ in CH_2Cl_2, RT) [26, 27]. In another report the synthesis of a glycoluril clip with two zinc porphyrin walls has been described. It was found to bind amino- and diamino alkanes as well as bipyridine between the zinc centers in a mono-or bidentate fashion, with association constants ranging from K_a 10^4 to $10^7\ \text{M}^{-1}$ (CH_2Cl_2, RT). NMR studies and modelling studies indicated that the size of the cavity of the porphyrin clip adjusts itself to the size of the guest molecule [28]. Binding of a guest can also affect its reactivity, as was shown for the methylation of phloroglucinol with dimethyl sulfate in the presence of a glycoluril clip, which was found to give more monomethylated than dimethylated product as a result of the binding of the guest in the clip [29].

12.2.3 Self-assembly

When no guests are available glycoluril clips tend to self-assemble in order to efficiently fill space in the solid state or to prevent their cavities from becoming filled with solvent molecules that do not match well with the dimensions of the empty spaces in the clips. Compound **2** forms a dimer in the solid state and also other glycoluril clips display the same behavior, e.g. porphyrin clip **11** (Figure 12.5) [30]. In both cases the side-wall of one clip fills the cavity of the other clip.

In the solid state structure of **2** the aromatic walls of the two clips are shifted with respect to each other in line with calculations, which predict a similar offset geometry. ^{1}H NMR studies at different concentrations in $CDCl_3$ reveal that compound **2** also dimerizes in solution. On the basis of the chemically induced shift (CIS) values of the protons, the geometry of the dimer was found to be the same as in the solid state. The dimerization constants for a series of different clip molecules were determined in $CDCl_3$ solution and were found to be $K_{dim} = 16\ \text{M}^{-1}$ for **2** and $5\ \text{M}^{-1}$ for **11** [31]. A derivative of **2** in which the phenyl rings on the convex side of the clip are replaced by benzoate groups was found to dimerize in water as well, the dimerization constant being higher than in chloroform, i.e. $K_{dim} = 630\ \text{M}^{-1}$. Replacing the 1,4-dimethoxy walls in this derivative of **2** by 1,4-dimethoxynaphthalene walls increased the dimerization constant even further, to $K_{dim} = 1.45 \times 10^4\ \text{M}^{-1}$, which is the result of the large hydrophobic cleft

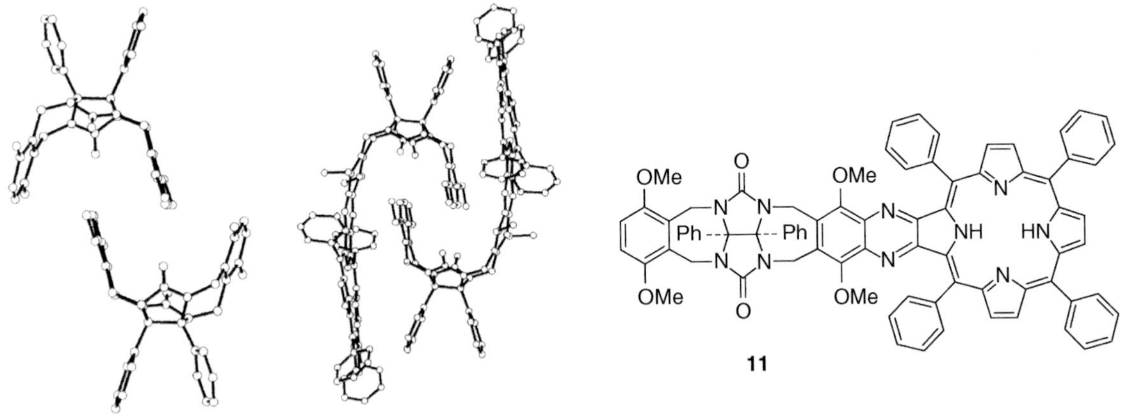

Figure 12.5 *X-ray structures of compounds* ***2*** *(left) and* ***11*** *(middle), and chemical structure of* ***11***

present in this compound. The measured CIS values indicated that in water the side-walls of the clips are inserted more deeply into the cavities of their dimeric partners than in chloroform [31].

Isaacs and coworkers have studied the X-ray structures of a series of glycoluril clip molecules with substituted o-xylylene side-walls. A large number of clips (8 out of 11) turned out to form dimeric structures in the solid state by penetration of the aromatic side-walls into the clefts. The dimeric structures are stabilized by π–π stacking and H-bond interactions between the methoxy groups on the side-walls and the urea carbonyl groups (C-H·O interactions) of the glycoluril frameworks. Three clips in which such interactions were not possible formed tapes [32]. In a separate paper the dimerization of chiral clips was studied and it was found that these compounds preferentially form heterochiral dimers by a highly diastereoselective recognition process [33].

Increasing the number of glycoluril units in the clip to two increases the propensity to form dimeric structures. For instance, clip **12** (Figure 12.6) forms dimeric aggregates in a wide range of solvents (from benzene to water) with dimerization constants as large as $K_{dim} > 9 \times 10^5\,\text{M}^{-1}$ ($CDCl_3$, RT) [34]. Compound **8**, on the other hand, which contains alternating substituted benzene and glycoluril units displays a much lower dimerization constant ($K_{dim} = 1127\,\text{M}^{-1}$, D_2O, RT) [20] and compound **7** does not dimerize at all [18]. An X-ray structure of **7** shows that the latter compound has an almost completely closed C-structure, very similar to cucurbituril, allowing no dimerization to take place. Glycoluril clips with tetrathiofulvalene side-walls have been reported to be monomeric in their neutral state, but, upon gradual oxidation, dimerization takes place first to mixed-valence dimers and then to radical cation–radical cation dimers. Complete oxidation leads to dissociation and again the formation of a monomeric species [35].

The remarkable selective dimerization of clip molecules is also evident from self-sorting experiments that have been described in the literature. Mixtures of different types of cucurbiturils, cyclodextrins, and glycoluril clip molecules were mixed with guest molecules in water and shown by NMR to undergo self-sorting, meaning that the hosts selected the most appropriate guests, which in the case of the clips were the clips themselves [36]. Similar results were obtained for self-sorting experiments in organic solvents using either clips [37] or mixtures of clips and other host–guest systems [38].

An unusual higher-order self-assembly process in the solid state was observed for the differentially substituted *o*-xylylene clip **13**. It forms dimers that further self-assemble into cyclododecamers, which in turn stack to form nanotubes. Interestingly, the rings of dimeric clip molecules hold cyclic clusters of methanol molecules, which adopt a crown-like geometry. The methyl groups of the methanol clusters are directed towards the inner walls of the cyclododecamers, in this way stabilizing the assembly by hydrophobic interactions (see Figure 12.6) [39]. Higher-order structures can also be formed in solution, e.g. by the water-soluble clip **14**. At relatively low concentrations in water (0.1–3 mM) the former compound self-assembles in both a head-to-head and a head-to-tail fashion (Figure 12.7). When the concentration in water is increased, well-defined, "razor-blade" shaped nanoaggregates are formed (Figure 12.7). These structures are generated by a process of hierarchical self-assembly, in which head-to-head dimers act as nuclei for further growth by binding new clip molecules in a head-to-tail arrangement, eventually leading to aggregates that have a hydrophilic surface. Interestingly, apart from having the same shape, all aggregates had approximately the same size, and were built up from a limited number of layers (*ca.* 50) [40]. This suggests that the assembly process is well-controlled and stops if the balance between the enthalpy gain by molecular recognition (π–π stacking) and the entropy loss upon aggregation switches over in favor of the latter. A similar higher order self-assembly behavior is displayed by the water-soluble clip **15**, which contains on its convex side a ruthenium bipyridine complex [41]. In this case head-to-head dimerization followed by head-to tail-binding involving the cleft of the clip and a bipyridine ligand of the ruthenium complex leads to "cigar"-like aggregates displaying a very narrow distribution in size, i.e. an aspect ratio (length/width) of 11 ± 2 (Figure 12.7). Also in this case growth stops if the enthalpy–entropy balance is no longer favoring aggregate formation.

Another route to self-assembled structures was demonstrated with glycoluril clip molecules that had been provided with crown ether and azacrown ether moieties to give basket-shaped compounds, e.g. **16** and **17**.

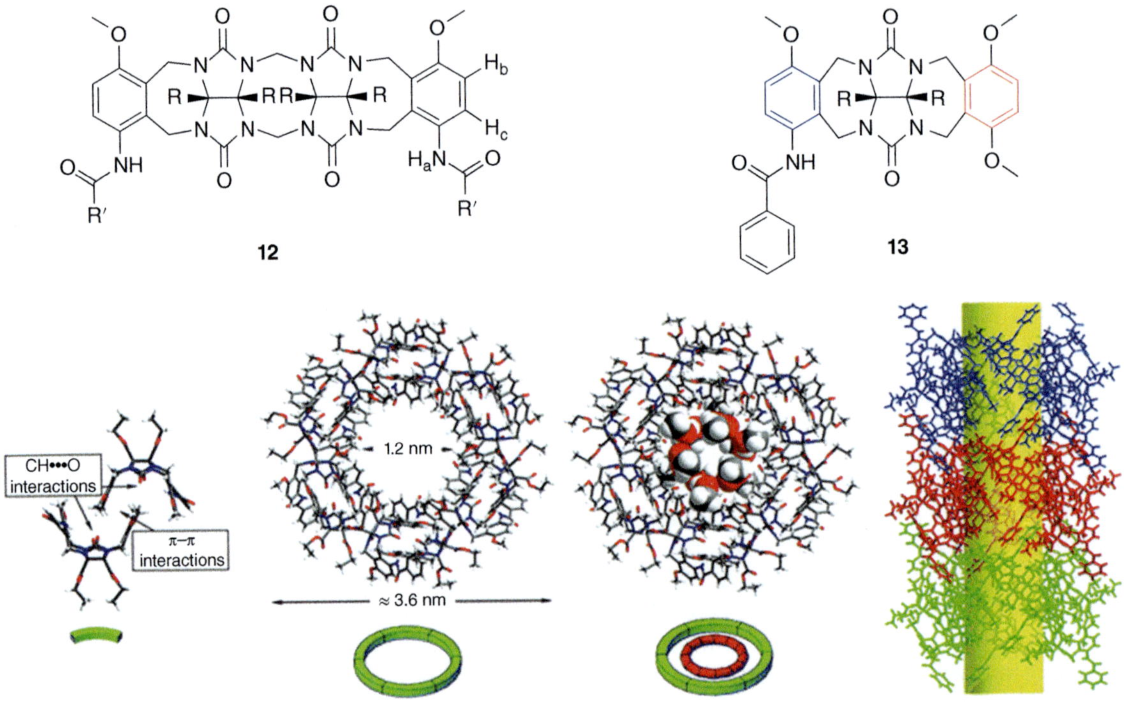

*Figure 12.6 Top: chemical structures of clips **12** and **13**. Bottom from left to right: dimer of clip **13** stabilized by CH·O and π–π interactions, cyclododecamer formed by **13**, assembly of methanol molecules within the cavity of the cyclododecamer, and nanotube formed by stacking of the cyclododecamer molecules. (Reproduced with permission from ref. [39] with permission of the Royal Society of Chemistry)*

The crown ether derivative **16** was found to strongly bind alkali metal ions ($K_a \sim 10^8 \text{M}^{-1}$ in water-saturated $CHCl_3$), diammonium salts of type $^+H_3N(CH_2)_nNH_3^+$ ($K_a \sim 10^9 \text{M}^{-1}$ in water-saturated $CHCl_3$), and paraquat (**18**, $K_a \sim 22{,}000 \text{M}^{-1}$ in water/$CHCl_3$, 1:1 v/v, RT; see X-ray structure in Figure 12.8) [8, 42, 43]. Basket compound **17**, which has long aliphatic tails, acts as an amphiphile and forms vesicles on dispersal in water with a critical aggregation constant of 2×10^{-5} M (Figure 12.8). The vesicles act as "golf-balls" and can bind in water organic guest molecules such as 4-(4-nitrophenylazo)resorcinol in their dimples with $K_a = 4 \times 10^5 \text{M}^{-1}$ (RT) [44]. The structure of the aggregates can be tuned by changing the pH or by adding alkali metal ions: tube-like aggregates are formed in aqueous HCl and vesicles in the presence of K^+ ions. Larger alkali metal ions (Rb^+ and Cs^+) also generated tubes. Physical studies showed that the protonated baskets had a relatively open geometry, leading to tubes, whereas the baskets with bound K^+ had a more compact, closed structure, resulting in vesicles. This more open structure was also present in the case of the baskets with bound Rb^+ and Cs^+ ions [45].

12.2.4 Functional clips

Glycoluril clips with fluorescent side-walls have been reported to be excellent sensors for Fe^{3+} ions, to the extent that this type of ion can be detected even in the presence of other metal ions (e.g. of Hg^{2+}, Cd^{2+}, Zn^{2+}, Co^{2+}, Ni^{2+}, Cu^{2+}, Pb^{2+}, and Cr^{3+}). It is proposed that the Fe^{3+} ions bind at the bottom of the concave side of the clip by coordinative bonds with the ureidyl carbonyl groups and by cation-π interactions with the aromatic side-walls of the clip. The capture of Fe^{3+} may cause electron or energy transfer from the excited aromatic

14

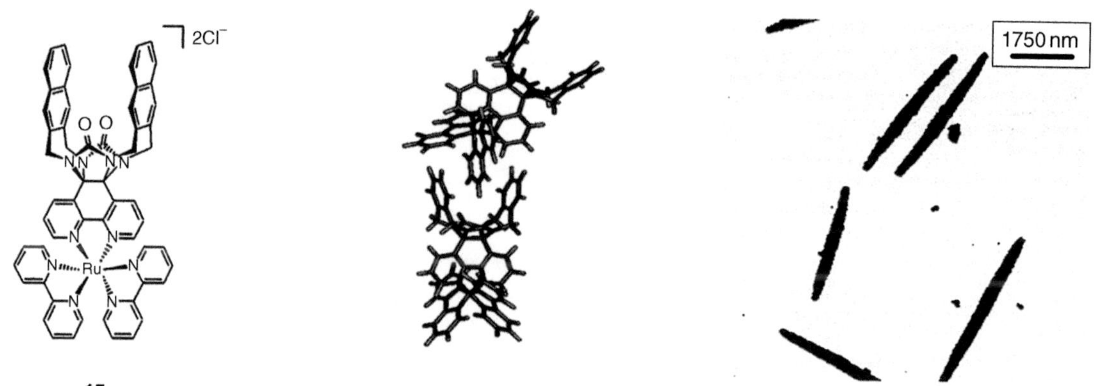

15

Figure 12.7 *Top: chemical structure of **14** (left), head-to-head and head-to-tail assembly of molecules of **14** (middle), and Transmission Electron Microscopic (TEM) pictures of the"razor blades" formed from **14** (right). Bottom: chemical structure of **15** (left), 1H-NMR-based and computer-generated structure of the head-to-tail binding of two molecules of **15** (middle), and TEM picture of "cigar"-like aggregates of **15** (right)*

side-walls to the bound metal ion [46, 47]. A related clip with fluorescent aromatic side-walls was found to be a selective sensor for nitrophenol, which is a mutagenic compound often used in active ester coupling reactions. It was possible to detect this compound in a mixture of nine other phenols [48]. Selective detection of fluoride ions has been described for a glycoluril clip with aromatic walls functionalized with two thiourea groups. These ions gave a visually noticeable color change with this clip, whereas other ions (Cl^-, Br^-, I^-, ClO_4^-, PF_6^-, $H_2PO_4^-$, and OAc^- ions) did not. The fluoride ions were found to be complexed via hydrogen-bond interactions with the thiourea NH functions and the binding was found to be strong ($K_a = 5 \times 10^4\,M^{-1}$, acetonitrile/DMSO, 9:1 v/v, RT) [49]. A clip with TTF side-walls has been used to construct a two-input NOR functioning logic gate. It complexes a pyridinium macrocycle yielding a colored complex (**19**) that can be reversibly dissociated and re-associated by the addition and removal of K^+ or NH_4^+ ions, which bind in the crown ether moiety of the macrocycle (Figure 12.9) [50].

An interesting application of glycoluril clips has been reported in the field of gene therapy. Ruthenium clip **20** (Figure 12.9), with N–N representing different ligands, e.g. dipyrido[3,2-*a*:2′,3′-*c*]phenazine (dppz), was found to bind to DNA with association constants ranging from $K_a = 2.5 \times 10^3$ to $1.0 \times 10^5\,M^{-1}$ (aqueous phosphate buffer, pH 7.2, RT). Interestingly, on binding to DNA the luminescence activity of the **20–dppz**

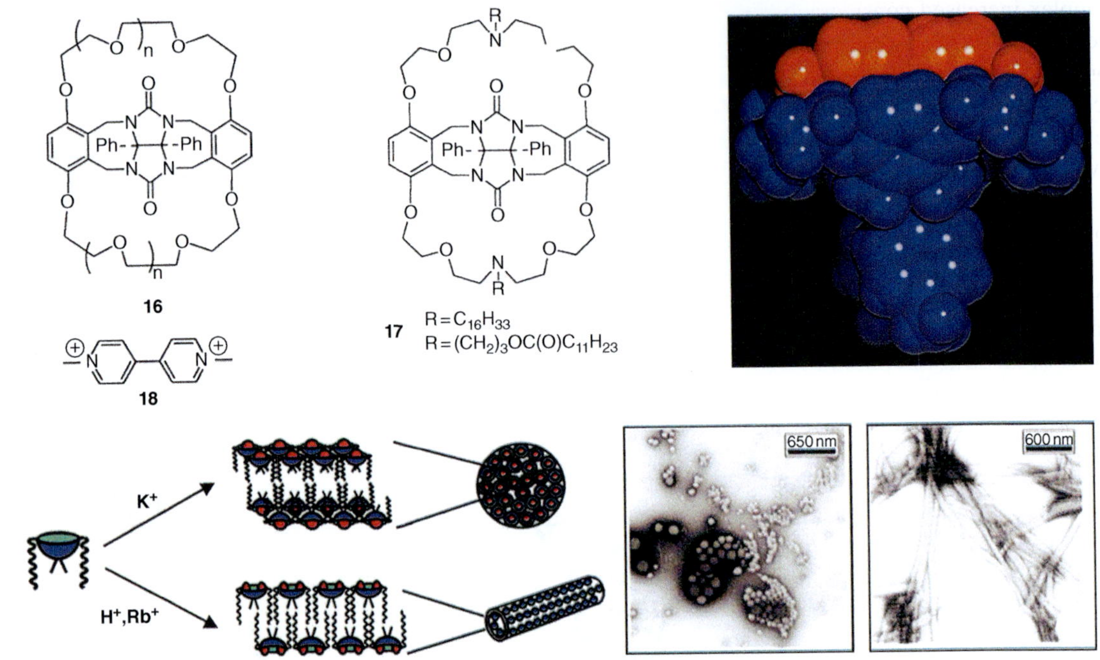

Figure 12.8 *Top: chemical structures of compounds* ***16–18*** *(left and middle) and X-ray structure of the complex between* ***16*** *and* ***18****. Bottom: schematic representation of the self-assembly of* ***17*** *in the presence of acid or alkali metal ions (left) and electron micrographs of the vesicle and tube-like structures that are formed (middle and right) (Reproduced from ref. [45] with permission. Copyright © 1994 the American Chemical Society.)*

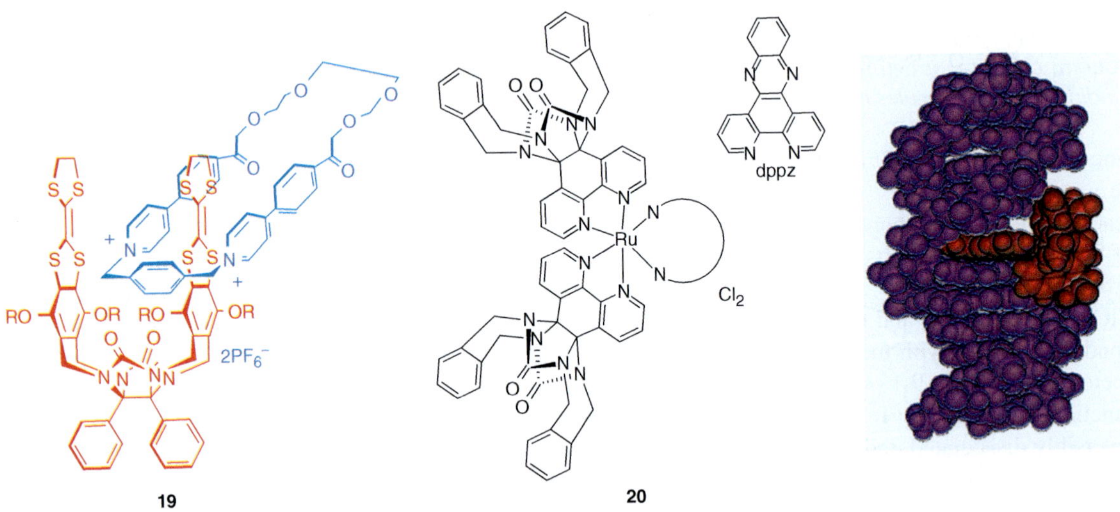

Figure 12.9 *Chemical structures of clip–macrocyclic complex* ***19****, ruthenium–clip complex* ***20****, and ligand* ***dppz****, and the mode of binding of* ***20*** *to DNA (right)(reprinted from ref. [52] with permission. Copyright © 2009 the American Chemical Society.)*

complex increased considerably (molecular light-switch effect), suggesting an interaction mode in which the ligands or walls of the clip stack with the base pairs of DNA (see Figure 12.9). The ruthenium clips were found to induce condensation of DNA to particles with average diameters of 1300–1600 nm, and these particles could be taken up by cells to act as gene delivery vehicles. Separate studies showed that for gene delivery only submicromolar concentrations of the complexes were needed, and that they had low toxicities towards cells [51].

Apart from condensing DNA, the ruthenium clip complexes are also capable of cleaving DNA, [52] which, however, did not play a major role under applied physiological conditions [51].

One of the challenges of supramolecular chemistry is to change the properties and functions of materials by non-covalent interactions. Clip compound **21** was designed with the objective of clipping it to polymers to make them liquid-crystalline (Figure 12.10). Just like compound **10**, clip **21** exists in three different conformations (*ss*, *sa*, *aa*), with the *sa* conformation being the dominant one. Surprisingly, **21** was not liquid-crystalline despite the presence of twelve long aliphatic tails. However, when the guest methyl 3,5-dihydroxybenzoate was added, the *sa* conformation of **21** changed into *aa*, the guest was bound, and the resulting complex was found to display liquid-crystalline properties. Different liquid-crystalline phases (two smectic phases and one nematic phase) were observed depending on the temperature and the ratio of host and guest. In a similar way, clipping **21** to a porphyrin which was modified at the *meso* positions with 3,5-dihydroxyphenyl groups led to a liquid-crystalline material (two smectic phases). A porphyrin lacking the host-binding functions did not give mesophases, indicating that the host–guest complex is the mesogenic species. Two different polystyrene derivatives **22** (MW 35,500, $M_w/M_n = 2.26$) and **23** (MW 24,000, $M_w/M_n = 2.38$) were prepared containing pendant 1,3-dihydroxy functions for binding to **21** (Figure 12.10). Clipping of the host to these polymeric guests

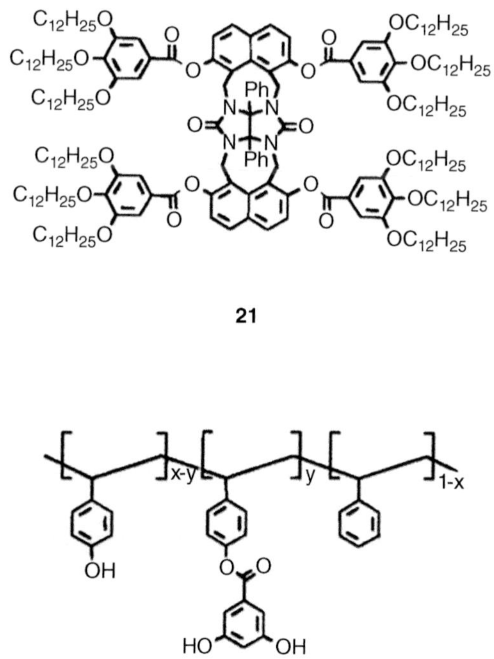

21

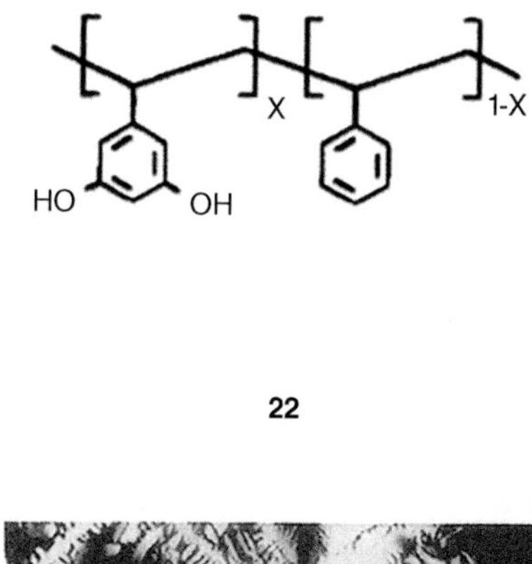

22

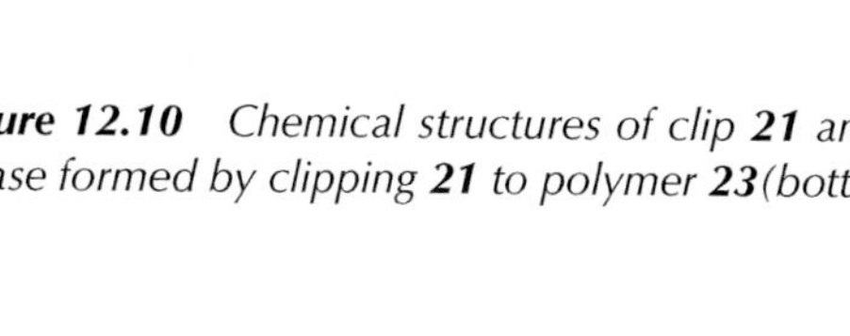

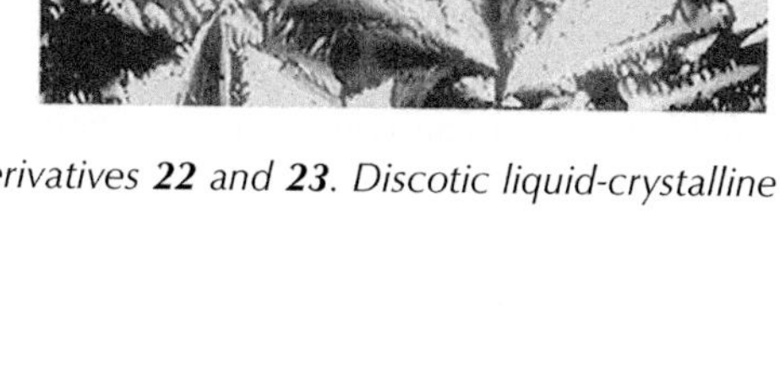

Figure 12.10 *Chemical structures of clip **21** and polystyrene derivatives **22** and **23**. Discotic liquid-crystalline phase formed by clipping **21** to polymer **23**(bottom right)*

resulted in liquid-crystalline phases, two discotic phases for **22** (21–86 °C and 86–131 °C) and a single discotic phase for **23** (30–141 °C). The formation of discotic phases for the polymeric guests and smectic/nematic phases for the low molecular weight ones is somewhat surprising. Probably, the binding of the clip molecules to the polymers leads to a conformational change such that the trialkoxybenzoyl groups can adopt a more spread-out arrangement, making it possible for the taper-shaped clips to surround the polystyrene chains in a cylindrical fashion [53].

12.2.5 Biomimetic systems

The clips and baskets derived from glycoluril are interesting building blocks for the development of biomimetic systems that display certain aspects of enzymatic catalysis, such as substrate binding and conversion at a nearby catalytic center, induced fit, and allosteric binding. An important class of proteins are the ferrodoxins, which catalyze electron-transfer reactions and are involved in a large number of biochemical pathways. They contain in their active centers a 4Fe–4S cluster the redox properties of which can be tuned by changes to the surrounding protein mantle. A glycoluril clip molecule provided with two aza-crown ether rings, each containing a sulfur ligand has been synthesized and used to construct a ferrodoxin mimic. In the model (**24**) the 4Fe–4S cluster is completely encapsulated by two clip molecules (Figure 12.11), which completely changed its redox properties, just as in nature. These properties could be further fine-tuned by the addition of metal ions, which bind to the crown ether functions of the ligand [54–56].

A derivative of porphyrin clip **11** (vide supra), in which the dimethoxybenzene wall is replaced by an electron accepting quinone function (**25**) has been used to construct a model that mimics certain aspects of the complex processes that take place in the photosynthetic reaction system after the absorption of a photon and the subsequent transfer of an electron. An unresolved question is the role of intervening aromatic amino acids in this electron-transfer process. In the X-ray structures of the photosynthetic reaction centers of *Rhodopseudomonas viridis* and *Rhodobacter sphaeroides* the aromatic ring of a tryptophan molecule is in van der Waals contact with the primary electron-acceptor (a bacteriochlorophyll) and the quinone, and may mediate electron transfer between the two partners. In clip **25**, a bound dihydroxybenzene molecule is also in van der Waals contact with the porphyrin and the quinone, and this supramolecular system may, therefore, serve as a model to study this particular mechanistic aspect of the photosynthetic system (Figure 12.11). Furthermore, the edge-to-edge and center-to-center distances between the porphyrins and the quinones in the photosynthetic reaction centers and in clip **25** are roughly the same, making the model even more reliable. Fluorescence studies were carried out with **11** and **25** in different solvents and in the presence of the aromatic guest hexyl 3,5-dihydroxybenzoate. It turned out that in a non-polar solvent the fluorescence quantum yields of **11** and **25** were the same, but in a polar solvent they were different, i.e. much smaller for **25** than for **11**, because of solvent-mediated electron transfer between the porphyrin and the quinone in the former clip. When in an apolar solvent the aromatic guest was added, the fluorescence of **25** was quenched to the same extent (80%) as in an apolar solvent, indicating that now the aromatic ring of the guest took the role of facilitating electron transfer. These studies indicate that in the photosynthetic reaction centers, too, aromatic amino acids may play an important function, i.e. in making the electron-transfer process easier [57].

Haemocyanine and tyrosinase are di-copper proteins that bind molecular oxygen. The former is involved in oxygen transport in molluscs and arthropods, and the latter uses molecular oxygen to oxidize phenols, e.g. tyrosine and dopamine, in plants, animals, and humans. The two copper centers are held by six histidine residues. Synthetic models mimicking the active sites of these proteins have been developed based on glycoluril clips that were provided with two azacrown ether rings. Each ring contained a pyridine or pyrazole ligand set, which can coordinate to a copper center, see **26** and **27** [58–63]. Just as the natural systems compound **26** was found to bind one molecule of oxygen in a bent μ–η^2: η^2 mode. Unfortunately, experiments to oxidize substrate molecules complexed in the cavity of **26** were unsuccessful, because the activated oxygen

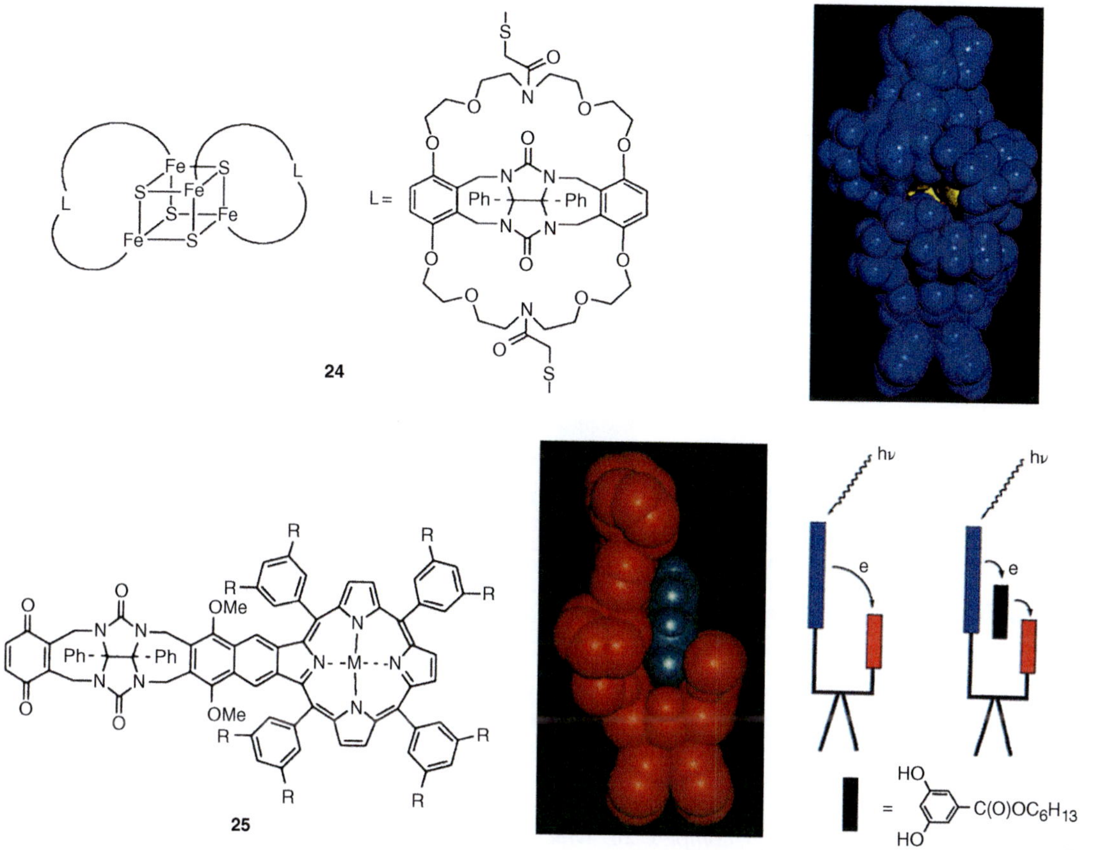

Figure 12.11 *Top: chemical structure of the 4Fe–4S cluster complex **24** (left and middle) and a computer generated model of this complex showing that the 4Fe–4S core is completely encapsulated by the ligand (right). Bottom: chemical structure of porphyrin–quinone clip **25** (left), a computer generated model, based on NMR studies, of the complex between hexyl 3,5-dihydroxybenzoate and **25** (middle), and schematic representation of the electron-transfer processes in **25** with and without guest present (right)*

preferentially attacked the benzylic carbon atoms of the ligand system, which according to modeling, are positioned directly next to the bound oxygen molecule. The pyrazole derivative **27**, however, could be used to selectively oxidize substrates (benzyl alcohols) bound in its cavity, albeit in a stoichiometric fashion. In this reaction the two Cu(II) centers are reduced to Cu(I), the electrons being taken from the alcohol, which is oxidized to an aldehyde. A Hammett plot of the rates of oxidation of a series of substituted benzyl alcohols clearly revealed that benzyl alcohols possessing hydroxy functions are oxidized with much higher rates (up to >50,000×) (Figure 12.12) [61].

Apart from oxidation catalysts, hydrogenation and isomerization catalysts have been attached to the glycoluril clips with azacrown ether rings [62–64]. An example is the tetrakis(triphenylphosphite)rhodium(I) hydride metallohost **28** shown in Figure 12.13. It is capable of selectively hydrogenating allylbenzene substrates that bind in its cavity. Apart from substrate binding reminiscent of enzymatic catalysis, the system also displays other enzymatic features, such as Michaelis–Menten kinetics and inhibition (by alkali metal ions). Furthermore, it was found that the cooperative binding of two molecules, i.e. the substrate allylcatechol and resorcinol, in the cavity of **28** substantially increases the rate of hydrogenation of the former compound

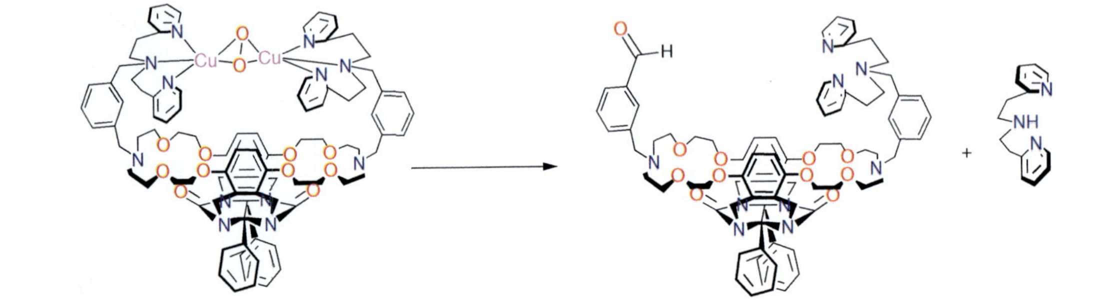

26

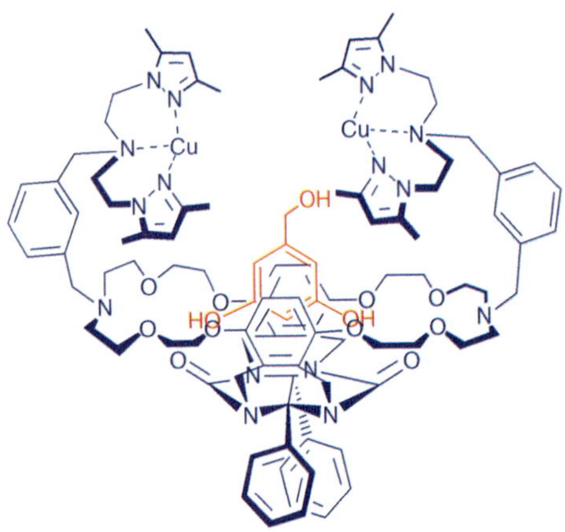

27

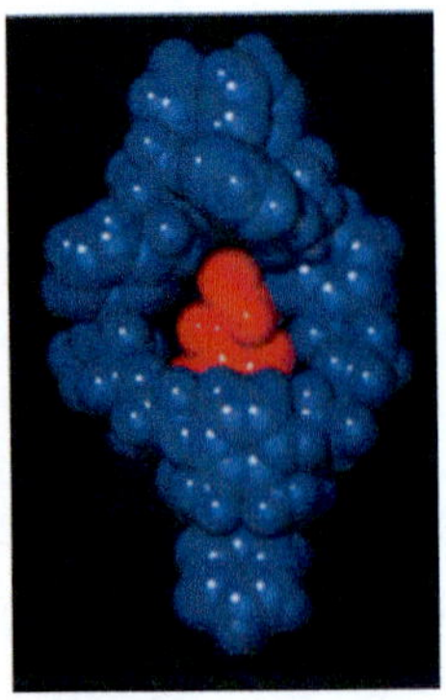

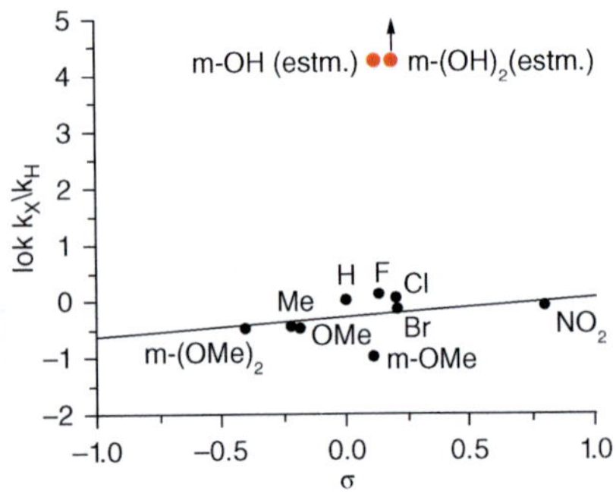

Figure 12.12 *Top: copper–oxygen complex* ***26****, which is catalytically decomposed by the bound oxygen molecule. Bottom: chemical structure of the di-copper pyrazole complex* ***27*** *with a bound 3,5-dihydroxybenzyl alcohol (left), computer generated model of this complex (middle), and Hammett plot of the rate of oxidation of substituted benzyl alcohols by* ***27***

(almost tenfold). With the help of resorcinol, allylcatechol can more favorably bridge the distance between the hydrogen-bond acceptor sites (carbonyl groups of the clip) while keeping its internal hydrogen bond (Figure 12.13) [63]. (For a general recent overview of substrate-selective catalysis, see ref. [65].)

12.2.6 Porphyrin clips: processive catalysis

In this section a recent example of a biomimetic system based on clips of glycoluril will be discussed in more detail. It involves a synthetic system that acts as a processive catalyst [66], i.e. a catalyst that remains bound to a substrate molecule, often a polymer, while performing several rounds of catalysis. Processive catalysts are important systems in nature because they are involved in the replication of a species, for instance DNA polymerase III, which makes copies of DNA using a single DNA strand as a template [67, 68]. An important element of the synthetic processive system is a clip with a porphyrin roof (**30**). It was designed as metalloenzyme mimic containing a substrate-binding cavity and a manganese(III) porphyrin, which catalyzes oxidation reactions. The cavity of **30** strongly binds paraquat (N,N-dimethylviologen) guests and ligands, e.g. pyridine (Figure 12.14), and also combinations of guests and ligands, the latter in a strongly cooperative

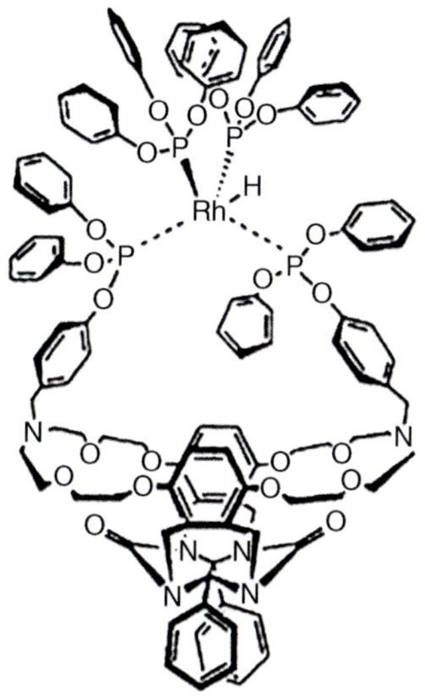

28

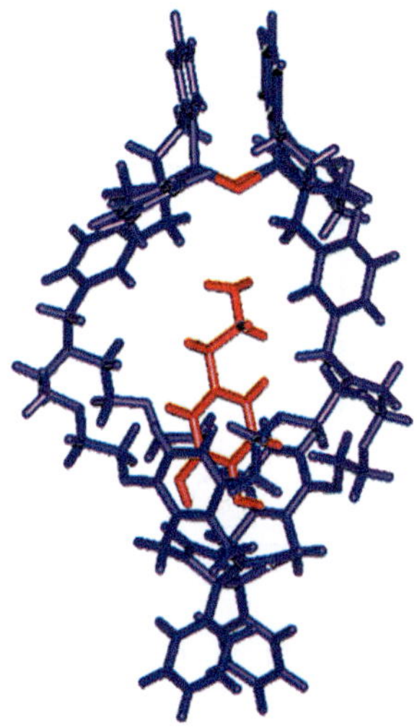

29

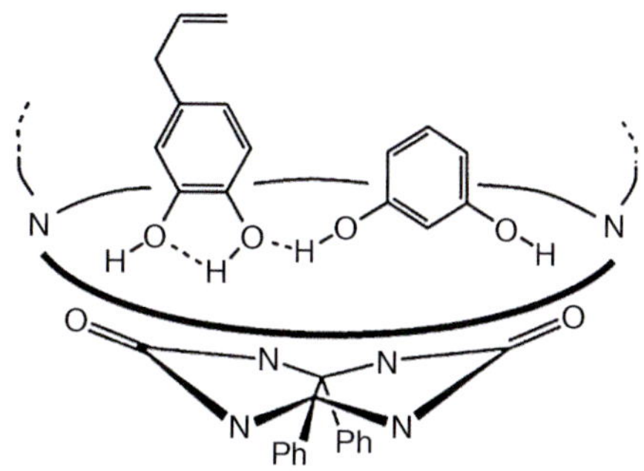

*Figure 12.13 Chemical structure of compound **28** (left) and calculated structure of this compound (**29**) with a bound allylcatechol substrate molecule (middle), and cooperative binding of allylcatechol and resorcinol in the cavity of **28** (right)*

fashion [69–73]. With bidentate ligands such as DABCO (1,4-diazabicyclo[2.2.2]octane) allosterically driven multicomponent assemblies are formed [74].

In the presence of sodium hypochlorite or iodosylbenzene as oxidants, porphyrin clip **30** was found to epoxidize low molecular weight alkenes (styrene, *cis*-and *trans*-stilbene, α-pinene) in two ways: either on the outside of the porphyrin ring or inside its cavity, depending on whether a small (pyridine) or bulky (tert-butylpyridine) ligand was used to activate the catalyst (Figure 12.14) [75, 76]. This behavior allowed the design of a catalytic system that could thread onto a polybutadiene polymer and move along it while performing oxidation reactions (see blueprint in Figure 12.14). With iodosylbenzene as oxidant and a 500-fold excess of tert-butylpyridine as ligand, the alkene double bonds of polybutadiene (>98% *cis*, molecular weight 300,000) could be completely epoxidized to the polymeric epoxide [77]. The rate of epoxidation was slower (factor 2.2) than that of a reference Mn(III) porphyrin catalyst containing o-methoxyphenyl groups, which is understandable because in the case of **30** the reaction has to take place inside the cavity. When the cavity was blocked by the addition of 1 equiv. of paraquat as an inhibitor the reaction rate dropped significantly. This is in contrast to the control experiments with the reference porphyrin catalyst, where the addition of paraquat caused a rate enhancement. The epoxidation with **30** produced 80% *trans*- and 20% *cis*-epoxide polymer from polybutadiene, whereas the reference catalyst gave predominantly *cis* product (78% *cis*-epoxide and 22% *trans*-epoxide). All this is in line with a rotaxane-like epoxidation mechanism, which was further confirmed by comparing the catalytic activity of **30** with that of the separately synthesized rotaxane catalyst **31**. The latter displays a two-fold higher turnover because the initial binding of the catalyst to the polymer is already achieved (Figure 12.14). A disadvantage of **30** is that an excess of a bulky pyridine ligand is needed to prevent the oxidation reaction from taking place on the outside of the catalyst. This problem can be overcome by shielding the outside with four attached ureaalkoxy groups, one of which binds to the manganese center and another forms a hydrogen bonding network, leading to both a more stable and a more active catalytic system [78].

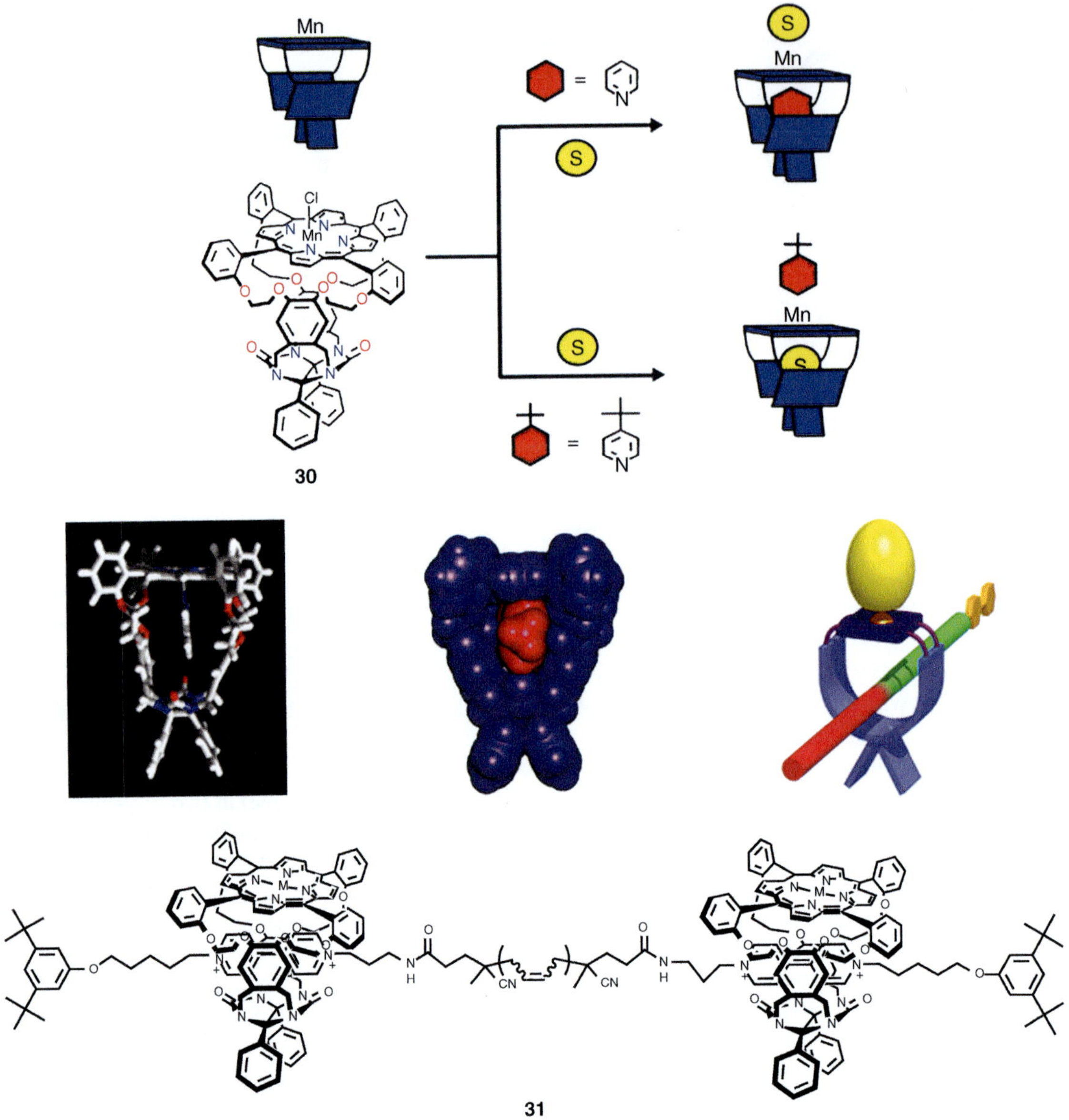

Figure 12.14 Top: porphyrin clip **30**, which can act as an epoxidation catalyst and convert a substrate S on the outside of the porphyrin ring or inside its cavity. Middle: X-ray structures of **30** with a bound pyridine ligand (left) and a bound paraquat guest (middle), and blueprint of the synthetic processive catalyst (right). Bottom: chemical structure of rotaxane catalyst **31**

The results above indirectly indicate that catalyst **30** first threads onto the polymer chain before carrying out the oxidation reactions. In order to firmly prove this threading, and to obtain more information about the precise mechanism of movement along the polymer chain, a large series of polymers (polytetrahydrofurans and polyalkenes), open at one end and blocked at the other end by a bulky 3,5-di(tert-butyl)phenyl group, was prepared (Figure 12.15). A paraquat molecule was located close to the blocking site, acting as a trap for the porphyrin cage compound. This trapping can be followed by fluorescence spectroscopy if, instead of the Mn complex, the fluorescent zinc or free-base derivatives of **30** are used. The threading of both compounds on the polymer chains followed a second-order process, and the dethreading a first-order process [79]. From the kinetic data it was concluded that after the initial binding (a second-order process involving a so-called entron effect, i.e. the filling of the cavity), the movement along the chain involves a series of first-order hopping steps, in which the cage compound travels randomly from one local energy minimum on the chain to the next (Figure 12.15). The rates of threading and dethreading became slower, and the free energy of activation increased, as the lengths of the polymers increased. From a detailed analysis of the threading kinetics as a function of the polymer length it was concluded that two mechanisms play a role. The first is a direct, intermolecular

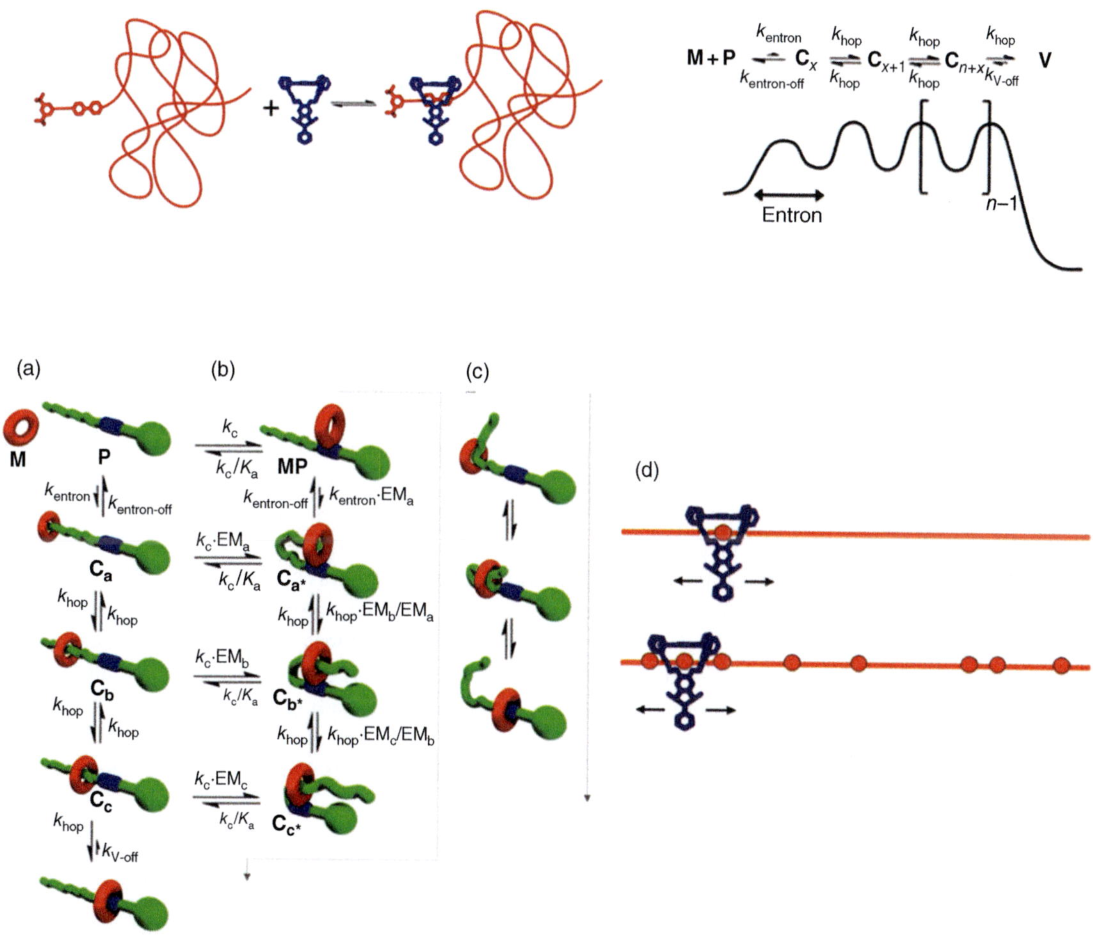

Figure 12.15 *Top: schematic representation of the threading of a porphyrin clip onto a polymer chain, which contains a trapping site and a blocking group (left). Consecutive hopping process with entron effect included (right). Bottom: intermolecular threading path (a), intramolecular looping path (b), and alternative path of movement over a folded chain (c). Oxidation of polybutadiene by a random gliding process (d). (Partly reproduced from ref. [80] with permission. Copyright © 2008 the American Association for the Advancement of Science)*

threading route, as shown in Figure 12.15a. The second route involves an intramolecular process in which the porphyrin clip first binds to the polymer followed by a looping of the end of the polymer chain inside the cavity, like a dog biting its own tail (Figure 12.15b) [80]. An alternative third route in which a porphyrin clip directly moves over the folded polymer chain (Figure 12.15c) was found to be unlikely [81]. The kinetic data also revealed that after binding there is unidirectional motion of the porphyrin clip to the paraquat trapping site. The Zn-porphyrin derivative of **30** appeared to thread more slowly than the free-base-porphyrin clip, which was found to be the result of solvent molecules (acetonitrile) blocking the entrance of the cavity by binding from the inside to the zinc center. A similar blocking could be achieved by adding pyridine [82].

From the above described catalytic oxidation of polybutadiene it was not possible to determine whether this reaction proceeded according to a stepwise processive or a random sliding mechanism. More information can be obtained, however, by comparing the threading rates with the rates of oxidation. From the former studies on a large series of polymers a lower limit for the movement of the catalyst along the polymer chain can be derived. For the combination of the free base derivative of **30** and the polytetrahydrofuran derivatives, this lower limit speed of movement is ~750 pm s^{-1} and for the combination with polybutadiene ~14 pm s^{-1}. These translocation speeds are much higher than the translocation speed that can be calculated from the epoxidation of polybutadiene, i.e. ~1 pm s^{-1}, assuming a sequential processive process. From this comparison it may be concluded that the oxidation of polybutadiene by **30** most likely follows a random sliding process, in which the catalyst hops from site to site eventually oxidizing all polymer double bonds (Figure 15d). In order to make this process sequential processive, the rate of threading and movement should be slowed down and made more similar to the rate of oxidation. This can be achieved by using a more flexible porphyrin cage compound that allows it to wrap itself around the polymer chain [83].

Porphyrin clip molecules are currently being used in a challenging project aimed at the development of a so-called molecular Turing machine. The basic ideas about the working principles of a computer were developed by Alan Turing in 1936. He proposed a (hypothetical) device, now carrying his name, that manipulates symbols on a strip of tape according to instructions provided by a tape head. A great challenge for the future will be the development of Turing computers that can perform calculations on the molecular level. The blueprint of a molecular machine that may be a first step in this direction is shown in Figure 12.16. It is based on two linked porphyrin clip molecules that move along a polymer chain, while writing (oxidizing or not oxidizing double bonds, which are the binary codes 1 and 0) and reading data on this chain using information from an attached ring. A porphyrin clip with such an attached ring, which can switch between two states (zero and one) might serve as the information-carrying component. Such a compound has been synthesized by an olefin metathesis reaction from **30** [84]. An important aspect will be the information transfer from the first cage in the dimer to the second one, which is attached to the polymer chain and glides along it. This may be realized by allosteric interactions. It was shown that two porphyrin clips can be connected by a bidentate ligand, e.g. DABCO (1,4 diazabicyclo[2.2.2]octane) to form a dimer in which the binding of a guest (e.g. paraquat) in one cavity increases the binding of another guest in the second cavity. If the porphyrin clip additionally contains four hydrogen bonding hydroxy groups on its roof, the interactions in the dimer become stronger, leading in the case of binding of paraquat to an allosteric effect $\alpha = 4K_2/K_1$ as high as 17 (Figure 12.16) [74]. A similar value was obtained from threading studies using polymers blocked on one side [85]. Whether such allosteric interactions can also be applied in oxidation reactions has to be determined.

12.3 Molecular capsules

Glycoluril's curved structure, and the fact that it possesses both hydrogen-bond donor and acceptor sites are features that have been elegantly exploited by Rebek's group in the construction of molecular architectures with the shape of a macroscopic object, i.e. tennis ball, doughnut, softball, and football.

[9–11, 86–88] C-shaped compound **32** is composed of a central benzene ring to which two diphenylglycoluril units are attached via methylene spacers. In solution it forms a dimer (molecular capsule or molecular tennis ball **32–32**), which is held together by a seam of eight hydrogen bonds (Figure 12.17) [89]. Its cavity is large enough to hold a guest molecule. The design of larger self-assembled capsules turned out to be more complicated because it required more hydrogen bond interactions and a differently shaped concave glycoluril framework. The molecular doughnut could be generated by self-assembling a concave building block synthesized by attaching three glycoluril units via methylene spacers to a central triphenylene ring [90]. For the construction of the softball, the group designed a series of very long curved compounds (**33**) in which two terminal glycoluril units are linked by as many as eight or nine rings, the central one of which was provided with an additional ethylene bridge in order to create the necessary curvature (Figure 12.18). Strategically placed –OH functions for extra hydrogen-bonding interactions cause such molecules (e.g. **33a**) to preferentially form capsules rather than supramolecular polymers [91]. A chiral softball could be made by

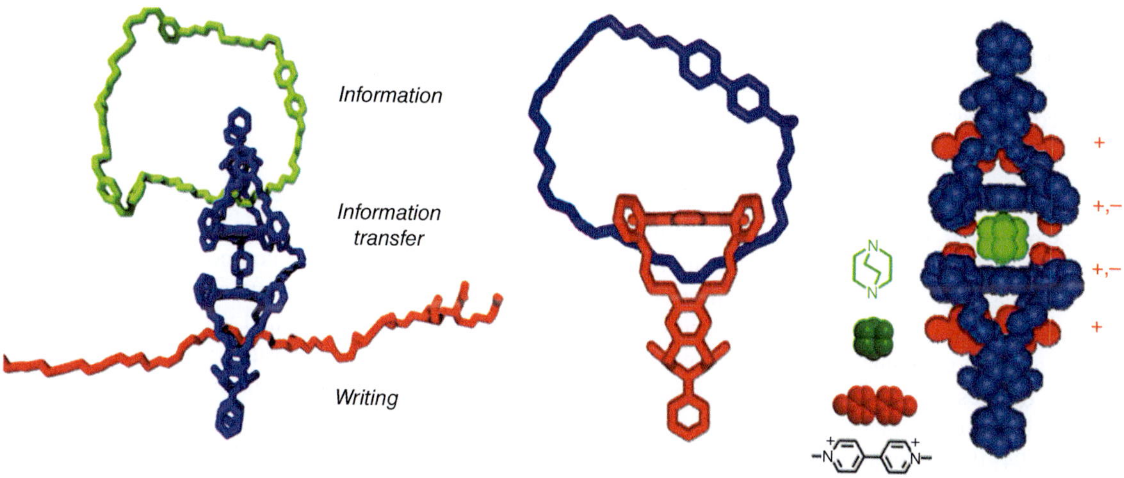

Figure 12.16 Blueprint of a molecular Turing machine (left), a porphyrin clip with attached switchable ring (middle), and a tetrahydroxyporphyrin clip dimer held together by coordinative and hydrogen bonds, which displays a strong allosteric effect ($\alpha = 0$ 17) in the binding of paraquat (right)

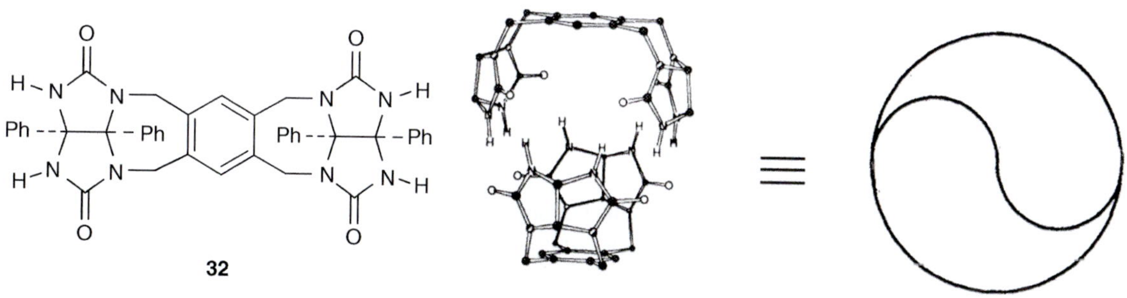

32

Figure 12.17 C-shaped glycoluril compound **32**, which self-assembles to form a molecular capsule (**32–32**) with the shape of a tennis ball. (Reprinted from ref. [9] with permission. Copyright © 2000 the American Chemical Society.)

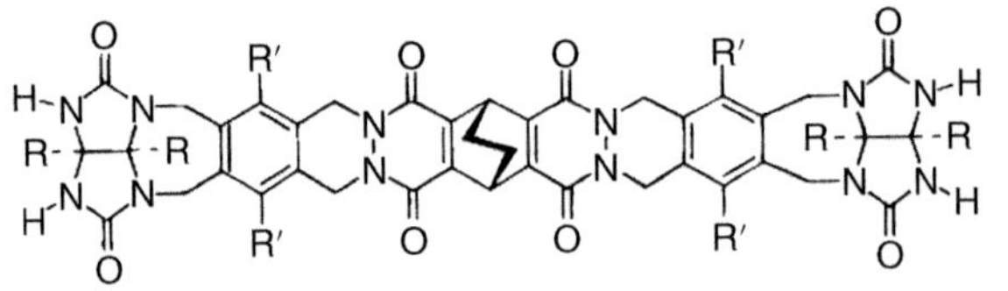

33a

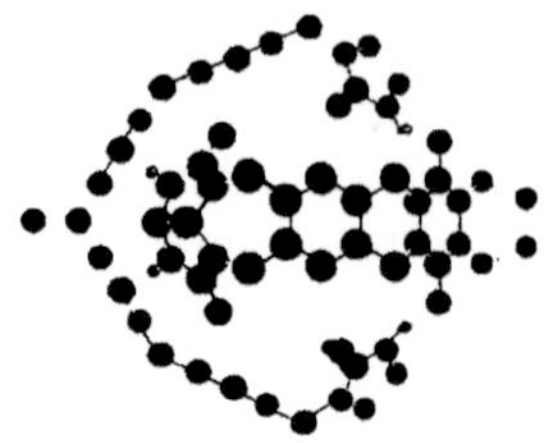

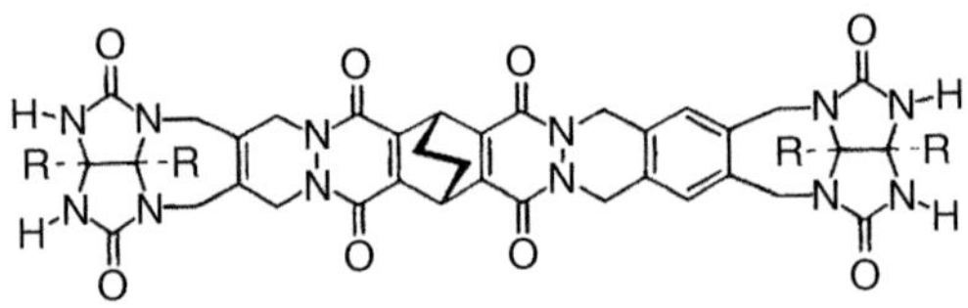

33b

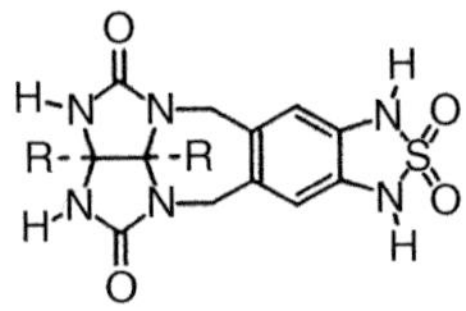

34

Figure 12.18 *Glycoluril derivatives **33a** and **33b**, which generate molecular soft-balls (**33–33**) by dimerization (see drawing top right). Compounds **33a** and **33b**: R=4-n-heptylphenyl, R′=OH. Compound **34**: R=4-n-heptylphenyl. (Reprinted from ref. [9] with permission. Copyright © 2000 the American Chemical Society.)*

self-assembling a derivative (**33b**) that had its terminal glycoluril moieties linked by an unsymmetrical spacer. Compound **33b** is achiral, but upon dimerization a racemate of an enantiomeric capsule with a chiral inner space is formed [92]. The self-assembly of four molecules of a glycoluril compound (**34**) that contained on one side an *ortho*-xylylene ring to which a cyclic sulfamide group was attached, gave a capsule with the shape of a football. It is stabilized by a pattern of 16 hydrogen bonds, and was only formed if a suitable guest (e.g. adamantane) was present as a template [93]. The various capsules were found to bind all kinds of guest molecules in solution: for instance, the molecular tennis ball can include methane and xenon, [89] the molecular doughnut can hold flat aromatic guest molecules but also a cyclohexane molecule, [90, 94] the molecular softball has a high affinity for ferrocene [91] and, if chiral, can preferentially bind one enantiomer of a chiral guest [92]. The football was shown to encapsulate adamantane (see above). Interestingly, the host–guest binding can be both enthalpy-driven (as in the case of the doughnut) [94] or entropy-driven (as in the case of the molecular softball) [91]. In separate papers Rebek's group demonstrated that the cavities of the capsules can be used as nanoreactors, for instance facilitating Diels–Alder reactions [95, 96].

The rate of the reaction between *p*-benzoquinone and cyclohexadiene, when both are bound inside the capsule **33a–33a**, was found to be increased 200-fold and, due to space limitations, the endo-isomer was the exclusively formed product. Product inhibition may be a problem for these systems and limit the realization of catalytic turnovers, but this feature could be overcome by selecting a reaction in which the product is too large to fit inside the capsule, allowing it to be replaced by new reactants [96].

In more recent years Rebek's group has employed glycoluril derivatives as spacers that can be reversibly incorporated in self-assembled cylindrical cages. Resorcinarene derivative **35** self-assembles in organic solvents into a dimeric capsule **35–35** in the presence of suitable guests that can fill its cavity, e.g. n-alkanes (Figure 12.19) [97]. Interestingly, n-$C_{15}H_{32}$ was not accommodated until the addition of an excess of compound **36** to the solution, which produced the characteristic ^{1}H NMR upfield chemical shifts of an encapsulated alkane species. Integration of the spectrum revealed the presence of four molecules of **36**, indicating expansion of the capsule as a result of co-assembly of the glycoluril derivatives, which form a hydrogen-bonded belt. The increase of the capsule's length by 7 Å and the volume by 200 Å^3 now allowed the accommodation of alkanes with lengths up to 18 carbon atoms. When longer alkanes were used as guests, even more extended capsules

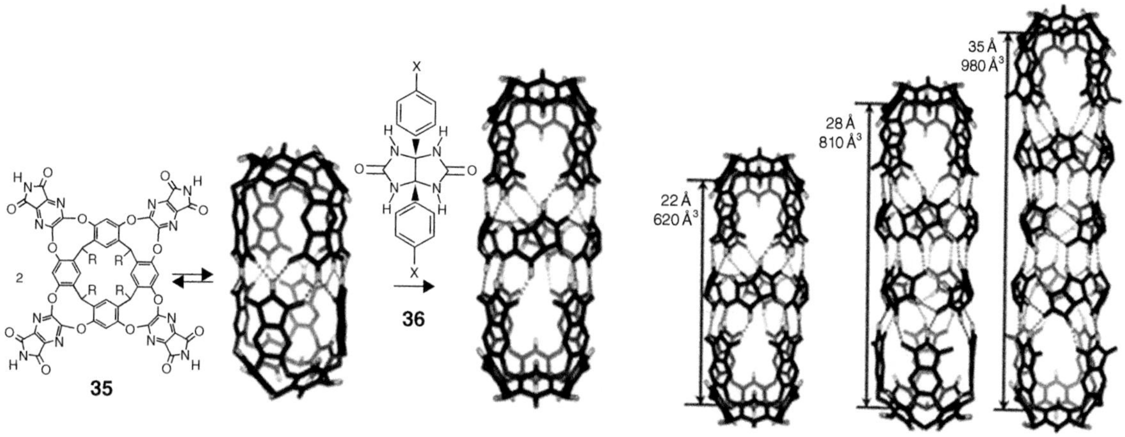

Figure 12.19 *Resorcinarene derivative **35** ($R = C_{11}H_{23}$) reversibly self-assembles into a homodimeric capsule, which can be extended through the progressive incorporation of multiple moieties of glycoluril derivative **36** ($X = OC_8H_{17}$, $C_{12}H_{25}$, or NBu_2)*

were formed: eight (with guests n-$C_{19}H_{40}$ to n-$C_{23}H_{48}$) or 12 molecules of **36** (Figure 12.19) [98, 99]. The belts of the co-assembled glycoluril species resemble the interior environment of the cucurbiturils, which are known to favorably accommodate linear alkanes. Interestingly, the well-defined rotation of the glycoluril moieties within a belt with respect to the axes of the capsule causes the expanded capsule to be chiral and, therefore, exist as a mixture of mirror-image enantiomers [100].

When glycoluril **36** is functionalized at its convex side with two p-dimethylaminophenyl groups (Figure 12.19, $X = NBu_2$) it becomes slightly basic. Upon addition of HCl gas to the solution containing capsule **35–36$_4$–35**, the dihydrochloric acid salt of the glycoluril precipitates and the contracted capsule **35–35** remains. This process is reversible, since deprotonation of the glycoluril salts by an amine resulted in full recovery of the extended capsule [101]. The fluctuations in capsule size also had an impact on the binding geometry of an n-$C_{14}H_{30}$ guest molecule, which is fully stretched in the extended capsule and coiled, like a "loaded spring," in the contracted one. The coiling of the n-alkane guests exerts pressure on the interior of the capsule, causing the hydrogen-bonded seams between the components to be loose and rotate. As a result, racemization of the mirror image capsule assemblies can occur. Longer n-alkanes, which exert more pressure, increased the racemization rate, while the perfectly fitting and non-coiled n-$C_{16}H_{34}$ guest did not result in racemization [100].

Apart from binding long n-alkanes, the extended capsules can host a variety of other guests. Because of their large volume they can bind multiple guests, for example small gas molecules such as ethane and cyclopropane [102], ion pairs [103], and hydrogen-bonded dimers of carboxylic [104] and boronic acids [105]. In analogy to the earlier described work with the molecular softball, the extended capsules were also used as dynamic nanoreactors in which the covalent connection of two simultaneously bound substrates can be accelerated. In the presence of **35–36$_4$–35**, n-hexyl isocyanate reacts with water to produce N,N-dihexylurea, which fits perfectly in the capsule, while in the absence of the capsule the reaction does not take place [106]. Interestingly, when the larger substrate n-octylisocyanate undergoes the same reaction, the product N, N-dioctylurea appeared in the more extended capsule **35–36$_8$–35**. The capsule assembly turned out to be highly substrate-selective: n-pentylisocyanate was not bound in the capsule and produced no urea product, nor did t-butylisocyanate, despite its efficient encapsulation. Because the reaction occurs relatively slowly, it could be followed in real time by ^{1}H NMR. These studies indicated that the amphiphilic interior of the capsule assembly, with the aromatic ends being hydrophobic and the hydrogen-bond-rich center hydrophilic, directed the substrates such that the carbamate intermediate, and probably also polar transition states, are stabilized.

12.4 Outlook

The field of (chemical) science spans a long period of time, starting more than 2000 years ago with Aristotle trying to describe and explain natural phenomena. Unfortunately, his works also created a fertile ground for alchemy, which was only abandoned after the important studies by Lavoisier at the end of the 18th century. Molecular chemistry started with Woehler's synthesis of urea (1826) and developed rapidly, later also expanding to include macromolecular chemistry (Staudinger's seminal work in the 1920s). Looking back in history it is evident that also the discoveries of Pedersen, Cram, and Lehn constituted a major development in the field of chemical sciences. Due to them the toolbox of the molecular chemist became enriched with new methodologies enabling the construction of more complex chemical architectures by using both covalent and non-covalent (supramolecular) interactions and by applying techniques such as self-assembly and self-organization. More than before, scientists became aware of the fact that this combination of interactions and techniques is also used by nature and forms the basis of the exceptional properties of biological materials and the highly sophisticated functions of biological machines. One of the great challenges for the future is to synthesize such advanced materials and functional machines *de novo* by using nature as a guide, and to study them with the help of the new physical methods that have become available in the last few decades, such as scanning probe microscopy and single molecule spectroscopy. The objective would be to unravel the mechanisms behind their formation, which is needed to develop new facile routes to their syntheses. Our studies on molecular clips and the works of the other Izatt–Christensen Award winners described in this book nicely show how the field of supramolecular chemistry has expanded over the years and how imaginative and creative it is. It is remarkable that with relatively simple building blocks such a great variety of chemical structures, displaying diverse properties and functions can be made. And this is probably only the beginning. More complex architectures are already on the drawing board, designed to be made not only from molecules and macromolecules, but also from natural building blocks, e.g. proteins, DNA and even viruses [107, 108]. In this connection it is worth mentioning that biological processes take place in nature under non-equilibrium conditions, i.e. under a constant supply of energy. Nearly all examples described in our chapter and in this book are based on chemical systems that are close to thermodynamic equilibrium. A great future challenge, therefore, will be to design and construct artificial systems that operate out of equilibrium. This will require a higher level of control than is currently possible in supramolecular chemistry, but great inspiration can be taken from nature, which during the process of evolution has made similar steps to reach the current state of structural and functional complexity. In order to achieve this (supramolecular) chemists increasingly will have to work together with scientists from other disciplines, in particular from biology and physics, and also from mathematics. However, more than any other discipline, chemistry has the ability to create structures, allowing it to solve a problem not only by mathematical and physical analysis but also by synthesis, which, after all, is the final proof.

John Keats' pessimistic vision about the physical and chemical sciences has proven to be wrong. The works of many scientists all over the world, including those presented in this book, clearly demonstrate that science is extremely imaginative and does not destroy but, rather, creates dreams. If he had lived 200 years later, he certainly would have admitted this, as we can conclude from a quote taken from a letter he wrote to a friend: "I am certain of nothing but the holiness of the heart's affections and the truth of imagination. What the imagination seizes as beauty must be truth, whether it existed before or not."[109]

12.5 Acknowledgements

This research was supported by the European Research Council in the form of an ERC Advanced grant to R.J.M.N. (ALPROS-290886). Further financial support was obtained from the Ministry of Education, Culture and Science (Gravity program 024.001.035).

References

1. Pedersen, C.J. (1967) Cyclic polyethers and their complexes with metal salts, *Journal of the American Chemical Society*, **89** (10), 2495–2496.
2. Nolte, R.J.M., Cram, D.J. (1984) Host–guest complexation. 27. Hosts containing only cyclic urea binding sites, *Journal of the American Chemical Society*, **106** (5), 1416–1420.
3. Freeman, W.A., Mock, W.L., Shih, N.Y. (1981) Cucurbituril, *Journal of the American Chemical Society*, **103** (24), 7367–7368.
4. Behrend, R., Meyer, E., Rusche, F. (1905) I. Ueber condensationsproducte aus glycoluril und formaldehyd, *Justus Liebigs Annalen der Chemie*, **339** (1), 1–37.
5. Smeets, J.W.H., Sijbesma, R.P., Niele, F.G.M., et al. (1987) Novel concave building block for the synthesis of organic hosts. *Journal of the American Chemical Society*, **109** (3), 928–929.
6. Reek, J.N.H., Elemans, J.A.A.W., Nolte, R.J.M. (1997) Synthesis, conformational analysis, and binding properties of molecular clips with two different side walls, *Journal of Organic Chemistry*, **62** (7), 2234–2243.
7. Sijbesma, R.P., Nolte, R.J.M. (1995) Molecular clips and cages derived from glycoluril, In *Supramolecular Chemistry II – Host Design and Molecular Recognition*, Weber, P.D.E. (Ed), Springer, Berlin Heidelberg, pp. 25–56.
8. Rowan, A.E., Elemans, J.A.A.W., Nolte, R.J.M. (1999) Molecular and supramolecular objects from glycoluril, *Accounts of Chemical Research*, **32** (12), 995–1006.
9. Elemans, J.A.A.W., Rowan, A.E., Nolte, R.J.M. (2000) Self-assembled architectures from glycoluril, *Industrial & Engineering Chemical Research*, **39** (10), 3419–3428.
10. Wyler, R., de Mendoza, J., Rebek, J. (1993) A synthetic cavity assembles through self-complementary hydrogen bonds, *Angewandte Chemie International Edition English*, **32** (12), 1699–1701.
11. Meissner, R., Garcias, X., Mecozzi, S., Rebek, J. (1997) Synthesis and assembly of new molecular hosts: solvation and the energetics of encapsulation, *Journal of the American Chemical Society*, **119** (1), 77–85.
12. Dang, D.T., Bosmans, R.P.G., Moitzi, C., et al. (2014) Solution structure of a cucurbit[8]uril induced compact supramolecular protein dimer. *Organic & Biomolecular Chemistry*, **12** (46), 9341–9344.
13. Ramaekers, M., Wijnands, S.P.W., Dongen, J.L.J. van, et al. (2015) Cucurbit[8]uril templated supramolecular ring structure formation and protein assembly modulation, *Chemical Communications*, **51** (15), 3147–3150.
14. Yang, H., Yuan, B., Zhang, X., Scherman, O.A. (2014) Supramolecular chemistry at interfaces: host–guest interactions for fabricating multifunctional biointerfaces, *Accounts of Chemical Research*, **47** (7), 2106–2115.
15. Lagona, J., Mukhopadhyay, P., Chakrabarti, S., Isaacs, L. (2005) The Cucurbit[n]uril Family, *Angewandte Chemie International Edition*, **44** (31), 4844–4870.
16. Isaacs, L. (2014) Stimuli responsive systems constructed using cucurbit[n]uril-type molecular containers, *Accounts of Chemical Research*, **47** (7), 2052–2062.
17. Smeets, J.W.H., Sijbesma, R.P., Van Dalen, L., et al. (1989) Synthesis and binding properties of basket-shaped hosts, *Journal of Organic Chemistry*, **54** (15), 3710–3717.
18. Ma, D., Zavalij, P.Y., Isaacs, L. (2010) Acyclic cucurbit[n]uril congeners are high affinity hosts, *Journal of Organic Chemistry*, **75** (14), 4786–4795.
19. Burnett, C.A., Witt, D., Fettinger, J.C., Isaacs, L. (2003) Acyclic congener of cucurbituril: synthesis and recognition properties, *Journal of Organic Chemistry*, **68** (16), 6184–6191.
20. Zhou, B., Xu, B., Ma, L., et al. (2014) Synthesis, crystal structure of the receptor derived from diethoxycarbonyl glycoluril for aromatic amide guest, *Chinese Journal of Chemistry*, **32** (11), 1111–1115.
21. Rezaei-Seresht, E., Salemi, S., Ahmadi, M., Taghizadeh, S. (2012) Synthesis and computational study of two new glycoluril clips containing benzocrown ether side walls, *Journal of Inclusion Phenomena and Macrocyclic Chemistry*, **76** (1–2), 69–74.
22. Golari, N., Rahimizadeh, M., Bakavoli, M., Rezaei-Seresht, E. (2014) An easy purification of glycoluril clips by affinity chromatography, *Jounal of Inclusion Phenomena and Macrocyclic Chemistry*, **80** (3–4), 353–358.
23. Reek, J.N.H., Priem, A.H., Engelkamp, H., et al. (1997) Binding features of molecular clips. Separation of the effects of hydrogen bonding and $\pi-\pi$ interactions, *Journal of the American Chemical Society*, **119** (42), 9956–9964.

24. Sijbesma, R.P., Wijmenga, S.S., Nolte, R.J.M. (1992) A molecular clip that binds aromatic guests by an induced-fit mechanism, *Journal of the American Chemical Society*, **114** (25), 9807–9813.
25. Reek, J.N.H., Engelkamp, H., Rowan, A.E., et al. (1998) Conformational behavior and binding properties of naphthalene-walled clips, *Chemistry a European Journal*, **4** (4), 716–722.
26. Hardouin-Lerouge, M., Cotelle, Y., Legoupy, S., Hudhomme, P. (2014) Synthesis of glycoluril-tetrathiafulvalene molecular clips for electron-deficient neutral guests through a straightforward Diels–Alder strategy, *New Journal of Chemistry*, **38** (11), 5341–5348.
27. Cotelle, Y., Allain, M., Legoupy, S., Hudhomme, P. (2014) Fused glycoluril-tetrathiafulvalene molecular clips as receptors for neutral electron acceptor guests, *Organic Letters*, **16** (10), 2590–2593.
28. Norrehed, S., Polavarapu, P., Yang, W., et al. (2013) Conformational restriction of flexible molecules in solution by a semirigid bis-porphyrin molecular tweezer, *Tetrahedron*, **69** (34), 7131–7138.
29. Rezaei-Seresht, E., Maleki, B., Amiri-Moghaddam, Z., Taghizadeh, S. (2013) Selective methylation of phloroglucinol in the presence of a glycoluril clip, *Tetrahedron Letters*, **54** (29), 3855–3857.
30. Reek, J.N.H., Elemans, J.A.A.W., de Gelder, R., et al. (2003) Self-association and self-assembly of molecular clips in solution and in the solid state, *Tetrahedron*, **59** (2), 175–185.
31. Elemans, J.A.A.W., Slangen, R.R.J., Rowan, A.E., Nolte, R.J.M. (2003) Synthesis and hierarchical self-assembly of cavity-containing facial amphiphiles, *Journal of Organic Chemistry*, **68** (23), 9040–9049.
32. Wang, Z.-G., Zhou, B.-H., Chen, Y.-F., et al. (2006) Substituent effects control the self-association of molecular clips in the crystalline state, *Journal of Organic Chemistry*, **71** (12), 4502–4508.
33. Chen, Y., She, N., Meng, X., et al. (2007) Chiral molecular clips control orthogonal crystalline organization, *Organic Letters*, **9** (10), 1899–1902.
34. Wu, A., Mukhopadhyay, P., Chakraborty, A., et al. (2004) Molecular clips form isostructural dimeric aggregates from benzene to water, *Journal of the American Chemical Society*, **126** (32), 10035–10043.
35. Chiang, P.-T., Chen, N.-C., Lai, C.-C., Chiu, S.-H. (2008) Direct observation of mixed-valence and radical cation dimer states of tetrathiafulvalene in solution at room temperature: association and dissociation of molecular clip dimers under oxidative control, *Chemistry a European Journal*, **14** (21), 6546–6552.
36. Mukhopadhyay, P., Wu, A., Isaacs, L. (2004) Social Self-Sorting in Aqueous Solution, *Journal of Organic Chemistry*, **69** (19), 6157–6164.
37. Ghosh, S., Wu, A., Fettinger, J.C., et al. (2008) Self-sorting molecular clips, *Journal of Oranic Chemistry*, **73** (15), 5915–5925.
38. Wu, A., Isaacs, L. (2003) Self-Sorting: The exception or the rule? *Journal of the American Chemical Society*, **125** (16), 4831–4835.
39. Cao, L.-P., Meng, X.-G., Ding, J.-Y., et al. (2010) Nanotubular non-covalent macrocycle within non-covalent macrocycle assembly: (MeOH)12 encapsulated in a molecular clip cyclododecamer. *Chemical Communications*, **46** (25), 4508–4510.
40. Reek, J.N.H., Kros, A., Nolte, R,J.M, (1996) Novel water soluble molecular clips. Towards nanostructures with controlled shape, *Chemical Communications*, (**2**), 245–247.
41. Elemans, J.A.A.W., Rowan, A.E., Nolte, R.J.M., et al. (1998) Bipyridine functionalized molecular clips. Self-assembly of their ruthenium complexes in water, *Chemical Communications*, (**15**), 1553–1554.
42. Smeets, J.W.H., Van Dalen, L., Kaats-Richter, V.E.M., Nolte, R.J.M. (1990) Functionalized basket-shaped hosts. Synthesis and complexation studies with (alkali) metal and ammonium and diammonium ions, *Journal of Organic Chemistry*, **55** (2), 454–461.
43. Schenning, A.P.H.J., de Bruin, B., Rowan, A.E., et al. (1995) Strong binding of paraquat and polymeric paraquat derivatives by basket-shaped hosts, *Angewandte Chemie International Edition English*, **34** (19), 2132–2134.
44. Schenning, A.P.H.J., de Bruin, B., Feiters, M.C., Nolte, R.J.M. (1994) Molecular golf balls: vesicles from bowl-shaped host molecules, *Angewandte Chemie International EditionEnglish*, **33** (15–16), 1662–1663.
45. Van Nunen, J.L.M., Stevens, R.S.A., Picken, S.J., Nolte, R.J.M. (1994) Tunable supramolecular structures from clips and baskets derived from glycoluril, *Journal of the American Chemical Society*, **116** (19), 8825–8826.
46. Hu, S.-L., She, N.-F., Yin, G.-D., et al. (2007) Synthesis, structural characterization, and fluorescent chemosensory properties of novel molecular clips based on diethoxycarbonyl glycoluril, *Tetrahedron Letters*, **48** (9), 1591–1594.

47. Li, L., She, N., Fei, Z., et al. (2010) Novel fluorescent molecular clips: selective recognition towards Fe^{3+} in aqueous solution, *Journal of Fluorescence*, **21** (3), 1103–1110.
48. She, N., Gao, M., Cao, L., et al. (2009) Sensor for nitrophenol based on a fluorescent molecular clip, *Organic Letters*, **11** (12), 2603–2606.
49. Cao, L., Ding, J., Wang, J., et al. (2010) Colorimetric fluoride sensor based on a bisthiourea functionalized molecular clip, *Synlett*, **2010** (17), 2553–2556.
50. Cheng, P.-N., Chiang, P.-T., Chiu, S.-H. (2005) A switchable macrocycle–clip complex that functions as a NOR logic gate, *Chemical Communications*, (**10**), 1285–1287.
51. Bhat, S.S., Kumbhar, A.S., Kumbhar, A.A., Khan, A. (2012) Efficient dna condensation induced by ruthenium(II) complexes of a bipyridine-functionalized molecular clip ligand, *Chemistry a European Journal*, **18** (51), 16383–16392.
52. Deshpande, M.S., Kumbhar, A.A., Kumbhar, A.S., et al. (2009) Ruthenium(II) complexes of bipyridine – glycoluril and their interactions with DNA, *Bioconjugate Chemistry*, **20** (3), 447–459.
53. Van Nunen, J.L.M., Folmer, B.F.B., Nolte, R.J.M. (1997) Induction of liquid crystallinity by host – guest interactions, *Journal of the American Chemical Society*, **119** (2), 283–291.
54. Martens, C.F., Feiters, M.C., Blonk, H.L., et al. (1991) Biomimetic electrochemical behaviour of a semi-encapsulated $[4Fe\text{-}4S]^{2+}$-cluster. *Journal of Inorganic Biochemistry*, **43** (2–3), 244.
55. Martens, C.F., Bongers, M.M.G., Kenis, P.J.A,, et al. (1997) Characterization of a [4fe-4 s]-ferredoxin model based on a concave tetradentate thiol ligand system, *Chemische Berichte*, **130** (1), 23–34.
56. Feiters, M.C., Klein Gebbink, R.J.M.., Schenning, A.P.H.J., et al. (1996) Development of supramolecular metalloprotein mimics, *Pure & Applied Chemistry* **68** (11), 2163–2170.
57. Reek, J.N.H., Rowan, A.E., de Gelder, R., et al. (1997) Novel cleft-containing porphyrins as models for studying electron transfer processes, *Angewandte Chemie International Edition English*, **36** (4), 361–363.
58. Klein Gebbink, R.J.M., Martens, C.F., Feiters, M.C., et al. (1997) Novel molecular receptors capable of forming Cu2–O2 complexes, Effect of preorganization on O_2 binding, *Chemical Communications*, (**4**), 389–390.
59. Sprakel, V.S.I, Feiters, M.C.F., Nolte, R.J.M. (2002) A cell for combined UV-visible and X-ray absorption spectroscopy studies under low-temperature and air exclusion conditions, *Review of Scientific Instruments*, **73** (8), 2994–2998.
60. Sprakel, V.S.I., Feiters, M.C., Meyer-Klaucke, W., et al. (2005) Oxygen binding and activation by the complexes of PY2- and TPA-appended diphenylglycoluril receptors with copper and other metals, *Dalton Tranactions*, **2005** (21), 3522–3534.
61. Martens, C.F., Klein Gebbink, R.J.M., Feiters, M.C., Nolte, R.J.M. (1994) Shape-selective oxidation of benzylic alcohols by a receptor functionalized with a dicopper(II) pyrazole complex, *Journal of the American Chemical Society*, **116** (13), 5667–5670.
62. Coolen, H.K.A.C., van Leeuwen, P.W.N.M., Nolte, R.J.M. (1992) A RhI-centered cage compound with selective catalytic properties, *Angewandte Chemie International Edition English*, **31** (7), 905–907.
63. Coolen, H.K.A.C., Meeuwis, J.A.M., van Leeuwen, P.W.M.N., Nolte, R.J.M. (1995) Substrate selective catalysis by rhodium metallohosts, *Journal of the American Chemical Society*, **117** (48), 11906–11913.
64. Coolen, H.K.A.C., van Leeuwen, P.W.N.M., Nolte, R.J.M. (1996) Synthesis and conformational behavior of rhodium(I) metallohosts derived from diphenylglycoluril, *Journal of Organic Chemistry*, **61** (14), 4739–4747.
65. Lindbäck, E., Dawaigher, S., Wärnmark, K. (2014) Substrate-selective catalysis, *Chemistry a European Journal*, **20** (42), 13432–13481.
66. Van Dongen, S.F.M., Elemans, J.A.A.W., Rowan, A.E., Nolte, R.J.M. (2014) Processive Catalysis, *Angewandte Chemie International Edition*, **53** (43), 11420–11428.
67. Benkovic, S.J., Valentine, A.M., Salinas, F. (2001) Replisome-mediated DNA replication, *Annual Review of Biochemistry*, **70**, 181–208.
68. Wang, J., Sattar, A.K., Wang, C.C., et al. (1997) Crystal structure of a pol alpha family replication DNA polymerase from bacteriophage RB69, *Cell*, **89** (7), 1087–1099.
69. Rowan, A.E. (1998) Novel porphyrin–viologen rotaxanes, *Chemical Communications*, (**5**), 611–612.
70. Elemans, J.A.A.W., Claase, M.B., Aarts, P.P.M., et al. (1999) Porphyrin clips derived from diphenylglycoluril. synthesis, conformational analysis, and binding properties, *Journal of Organic Chemistry*, **64** (19), 7009–7016.

71. Deutman, A.B.C., Smits, J.M.M., de Gelder, R., et al. (2014) Strong induced-fit binding of viologen and pyridine derivatives in adjustable porphyrin cavities, *Chemistry a European Journal*, **20** (36), 11574–11583.
72. Deutman, A.B.C., Monnereau, C., Moalin, M., et al. (2009) Molecular recognition and self-assembly special feature: Squaring cooperative binding circles, *Proceedings of the National Academy of Science USA*, **106** (26), 10471–10476.
73. Veling, N., Thomassen, P.J., Thordarson, P., et al (2008) Construction of supramolecular multi-component assemblies by using allosteric interactions, *Tetrahedron*, **64** (36), 8535–8542.
74. Thordarson, P., Coumans, R.G.E., Elemans, J.A.A.W., et al. (2004) Allosterically driven multicomponent assembly, *Angewandte Chemie International Edition English*, **43** (36), 4755–4759.
75. Elemans, J.A.A.W., Bijsterveld, E.J.A., Rowan, A.E., Nolte, R.J.M. (2000) A host–guest epoxidation catalyst with enhanced activity and stability, *Chemical Communications*, (**24**), 2443–2444.
76. Elemans, J.A.A.W., Bijsterveld, E.J.A., Rowan, A.E., Nolte, R.J.M. (2007) Manganese porphyrin hosts as epoxidation catalysts – activity and stability control by axial ligand effects, *European Journal of Organic Chemistry*, **2007** (5), 751–757.
77. Thordarson, P., Bijsterveld, E.J.A., Rowan, A.E., Nolte, R.J.M. (2003) Epoxidation of polybutadiene by a topologically linked catalyst, *Nature*, **424** (6951), 915–918.
78. Monnereau, C., Ramos, P.H., Deutman, A.B.C., et al. (2010) Porphyrin macrocyclic catalysts for the processive oxidation of polymer substrates, *Journal of the American Chemical Society*, **132** (5), 1529–1531.
79. Coumans, R.G.E., Elemans, J.A.A.W., Nolte, R.J.M., Rowan, A.E. (2006) Processive enzyme mimic: Kinetics and thermodynamics of the threading and sliding process, *Proceedings of the National Academy of Science*, **103** (52), 19647–19651.
80. Deutman, A.B.C., Monnereau, C., Elemans, J.A.A.W., et al. (2008) Mechanism of threading a polymer through a macrocyclic ring, *Science*, **322** (5908), 1668–1671.
81. Deutman, A.B.C., Varghese, S., Moalin, M., (2015) Slippage of a porphyrin macrocycle over threads of varying bulkiness: implications for the mechanism of threading polymers through a macrocyclic ring, *Chemistry a European Journal*, **21** (1), 360–370.
82. Hidalgo Ramos, P., Coumans, R.G.E., Deutman, A.B.C., et al. (2007) Processive rotaxane systems. studies on the mechanism and control of the threading process, *Journal of the American Chemical Society*, **129** (17), 5699–5702.
83. Deutman, A.B.C., Cantekin, S., Elemans, J.A.A.W., (2014) Designing processive catalytic systems. threading polymers through a flexible macrocycle ring, *Journal of the American Chemical Society*, **136** (25), 9165–9172.
84. Coumans, R.G.E., Elemans, J.A.A.W., Rowan, A.E., Nolte, R.J.M. (2013) Interlocked porphyrin switches, *Chemistry a European Journal*, **19** (24), 7758–7770.
85. Cantekin, S., Markvoort, A.J., Elemans, J.A.A.W., et al. (2015) Allosterically controlled threading of polymers through macrocyclic dimers, *Journal of the American Chemical Society*, **137** (11), 3915–3923.
86. Rebek, J. (1996) Assembly and encapsulation with self-complementary molecules, *Chemical Society Reviews*, **25** (4), 255–264.
87. Conn, M.M., Rebek, J. (1997) Self-assembling capsules, *Chemical Reviews*, **97** (5), 1647–1668.
88. Rebek, J. (1999) Reversible encapsulation and its consequences in solution, *Accounts of Chemical Research*, **32** (4), 278–286.
89. Branda, N., Wyler, R., Rebek, J. (1994) Encapsulation of methane and other small molecules in a self-assembling superstructure, *Science*, **263** (5151), 1267–1268.
90. Grotzfeld, R.M., Branda, N., Rebek, J. (1996) Reversible encapsulation of disc-shaped guests by a synthetic, self-assembled host, *Science*, **271** (5248), 487–489.
91. Kang, J.M., Rebek, J. (1996) Entropically driven binding in a self-assembling molecular capsule, *Nature*, **382** (6588), 239–241.
92. Rivera, J.M., Martin, T., Rebek, J. (1998) Chiral spaces: Dissymmetric capsules through self-assembly, *Science*, **279** (5353), 1021–1023.
93. Martin, T., Obst, U., Rebek, J. (1998) Molecular assembly and encapsulation directed by hydrogen-bonding preferences and the filling of space, *Science*, **281** (5384), 1842–1845.
94. O'Leary, B.M., Grotzfeld, R.M., Rebek, J. (1997) Ring inversion dynamics of encapsulated cyclohexane, *Journal of the American Chemical Society*, **119** (48), 11701–11702.

95. Kang, J.M., Rebek, J. (1997) Acceleration of a Diels-Alder reaction by a self-assembled molecular capsule, *Nature*, **385** (6611), 50–52.
96. Kang, J.M., Santamaria, J., Hilmersson, G., Rebek, J. (1998) Self-assembled molecular capsule catalyzes a Diels–Alder reaction, *Journal of the American Chemical Society*, **120** (29), 7389–7390.
97. Ajami, D., Rebek, J. (2006) Expanded capsules with reversibly added spacers, *Journal of the American Chemical Society*, **128** (16), 5314–5315.
98. Ajami, D., Rebek, J. (2007) Longer guests drive the reversible assembly of hyperextended capsules, *Angewandte Chemie International Edition.*, **46** (48), 9283–9286.
99. Ajami, D., Rebek, J. (2007) Adaptations of guest and host in expanded self-assembled capsules, *Proceedings of the National Academy of Science U SA*, **104** (41), 16000–16003.
100. Ajami, D., Rebek, J. (2009) Compressed alkanes in reversible encapsulation complexes, *Nature Chemistry*, **1** (1), 87–90.
101. Ajami, D., Rebek, J. (2006) Coiled molecules in spring loaded devices, *Journal of the American Chemical Society*, **128** (47), 15038–15039.
102. Ajami, D., Rebek, J. (2008) Gas behavior in self-assembled capsules, *Angewandte Chemie International Edition*, **47** (32), 6059–6061.
103. Taira, T., Ajami, D., Rebek, J. (2012) Encapsulation of ion pairs in extended, self-assembled structures, *Journal of the American Chemical Society*, **134** (29), 11971–11973.
104. Ajami, D., Tolstoy, P.M., Dube, H., et al. (2011) Encapsulated carboxylic acid dimers with compressed hydrogen bonds, *Angewandte Chemie International Edition*, **50** (2), 528–531.
105. Ajami, D., Dube, H., Rebek, J. (2011) Boronic acid hydrogen bonding in encapsulation complexes, *Journal of the American Chemical Society*, **133** (25), 9689–9691.
106. Taira, T., Ajami, D., Rebek, J. (2012) Hydration of isocyanates in an expandable, self-assembled capsule, *Chemical Communications*, **48** (68), 8508–8510.
107. Reynhout, I.C., Cornelissen, J.J.L.M., Nolte, R.J.M. (2009) Synthesis of polymer-biohybrids: from small to giant surfactants, *Accounts of Chemical Research*, **42** (6), 681–692.
108. Kostiainen, M.A., Kasyutich, O., Cornelissen, J.J.L.M., Nolte, R.J.M. (2010) Self-assembly and optically triggered disassembly of hierarchical dendron–virus complexes, *Nature Chemistry*, **2** (5), 394–399.
109. Keats, J. (1817) Letter to Benjamin Bailey, Nov. 22, http://www.notable-quotes.com/k/keats_john.html.

13

From Lanthanide Shift Reagents to Molecular Knots: The Importance of Molecular and Mental Flexibility

Jeremy K.M. Sanders

University Chemical Laboratory, University of Cambridge, Cambridge, UK

13.1 Introduction: 1969–76

My research career has lurched from one happy accident to another. Only rarely has my research group succeeded in achieving our specific goals or targets, but we have often been fortunate to make discoveries that were more interesting than expected. That has been possible only because throughout my time in Cambridge I have been privileged to work with graduate students and postdocs of exceptional talent who have taken on projects that are unrealistically ambitious: they have taken my ideas in new and better directions than I had thought of, and then made unexpected discoveries that open previously unimagined doors. This flexible approach to science, inherited from Dudley Williams, has come from me, but the success described in this story is to the credit of my co-workers, and I dedicate this chapter with thanks to them all.

In 1968–9, in the final year of my Batchelor's degree studies at Imperial College, London, I was allocated a dissertation on the determination of stereochemistry from NMR spin–spin coupling constants. I read and was inspired by Dudley Williams's mid-1960s papers in this area, so I moved to Cambridge to work for a PhD with him. In 1969 Dudley was more interested in mass spectrometry than NMR, but as I arrived, he saw a short paper by Hinckley [1] on lanthanide shift reagents, about which he knew virtually nothing. However, he could see their potential, made a connection with the interests of a new student, and within a few months we had magically produced spectacular science which rapidly became highly cited and influential: adding increasing amounts of a coordinatively unsaturated Eu β-diketonate complex to a

Macrocyclic and Supramolecular Chemistry: How Izatt–Christensen Award Winners Shaped the Field, First Edition.
Edited by Reed M. Izatt.

solution of *n*-hexanol in a 100 MHz spectrometer gradually pulled out every one of the predicted multiplets, leading to what we called "10,000 MHz" spectra [2].

Different lanthanides led to different degrees of shifts and broadening, and all in a way that contained structural information. This was all the more extraordinary since we had all been taught that the presence of paramagnetics killed NMR spectra of organic solutions. Dudley gave me complete scientific freedom, while also ensuring that everything I did was worth doing. He encouraged me to think laterally and imaginatively, to challenge orthodox thinking and to have the courage to work in new areas. He insisted that having provocative and testable ideas that might turn out to be wrong was more important than pursuing boring details. He preferred his group to sit in the library and read, or go and play sport, until we had a really good idea, rather than just do yet another experiment. He taught by example how to write well, and showed me how good illustrations are the best way of communicating difficult ideas. In retrospect, one might say that this simple coordination chemistry with significant applications was my introduction to supramolecular chemistry.

I was hooked, and had to become an academic. For my postdoctoral year I moved to Stanford to work on the NMR of proteins with Oleg Jardetzky. There I learned a lot of biology but achieved nothing in the lab. Fortunately, a junior academic post became available back in my home department, and I was appointed without even an interview.

Back in Cambridge in 1973, I found myself sharing an office with Ian Baxter. He had discovered an unusual reaction of quinones that I decided to investigate by NMR. We quickly discovered that radical anions were being formed and that the electron exchange between them led to NMR line-broadening that was related to the hyperfine coupling constant at that carbon. This was known in the physical chemical literature, but had never been exploited by organic chemists. Around the same time it became clear that the radical cations of chlorophylls are key intermediates in the first few picoseconds following light absorption in photosynthesis, and I set John Waterton, my first graduate student, the task of using NMR to measure hyperfine couplings in $Chl^{+\cdot}$ [3]: this was our first experience of macrocycles, and along the way we also demonstrated the use of spin-lattice relaxation and nuclear Overhauser effects to assign the NMR spectra of complex molecules [4]. Our early work up to 1976 is detailed in a published lecture [5]. My first postdoc, Iain Denniss, also modified chlorophyll to give it an imidazole tail that could coordinate intramolecularly [6].

13.2 Metalloporphyrins

13.2.1 Model photosynthetic systems

In 1977, the arrival of a new graduate student, K.N. Ganesh from India, keen to do synthetic work, coincided with the availability of large amounts of a synthetic porphyrin *bis*-acid from Alan Battersby's synthetic heme program. We decided therefore to try to make covalent synthetic photosynthetic systems containing both a porphyrin as electron donor and a quinone as electron acceptor, and then to look at the picosecond spectroscopy to explore the electron-transfer aspects. This was very successful and in the following few years our group made a variety of increasingly sophisticated model photosynthetic systems, as shown in Figure 13.1 [7–10].

In collaboration with Godfrey Beddard, then at Leeds, we made some of the earliest experimental observations of the counterintuitive but long-predicted Marcus inversion: as the driving force in favor of electron transfer increases, the rate reaches a maximum and then decreases as solvent reorganization around the molecule becomes the rate-determining step [11, 12].

The synthetic effort required for some of these systems took years, while the photophysics and electrochemistry could be done in days. This mismatch led Chris Hunter to develop simpler systems built up through coordination chemistry (Figure 13.2); they were easy to prepare and just as effective as electron-transfer systems [13, 14].

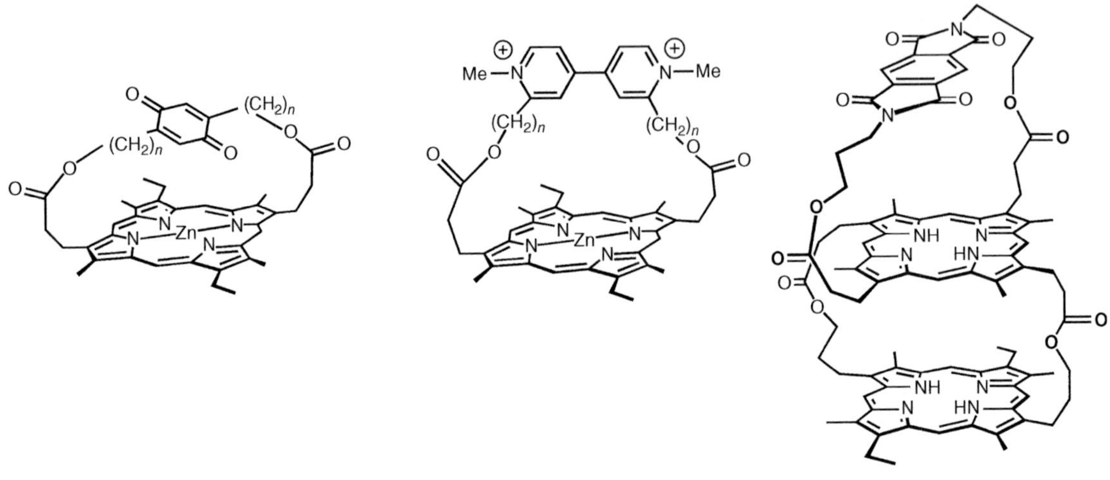

Figure 13.1 *Three model photosynthetic systems of the 1970s and 1980s*

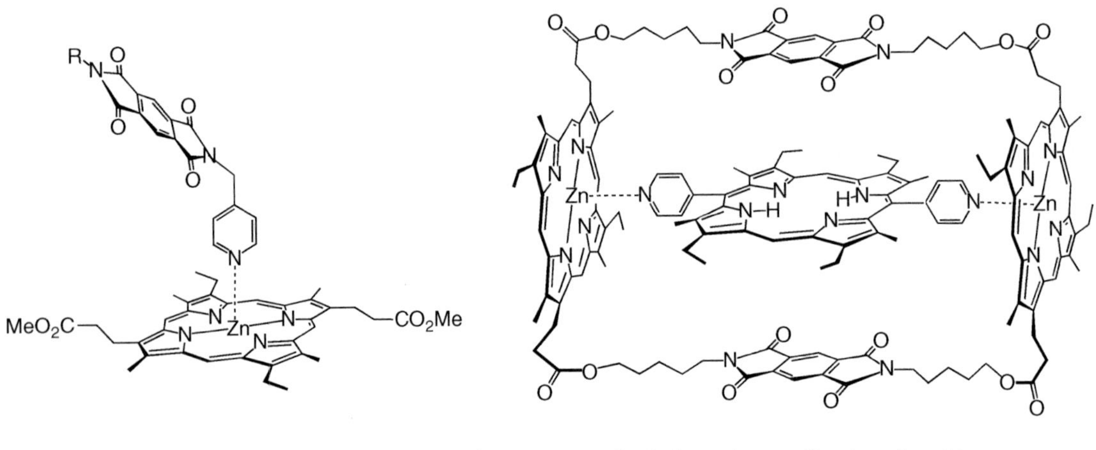

Figure 13.2 *Two model photosynthetic systems built through coordination chemistry*

By the late-1980s there were many electron-transfer systems published, each research group preferring a particular different architecture. Unfortunately, little by way of quantitative understanding or great overarching principle emerged from all these systems, beyond the recognition that one would need to follow nature's lead and build ever more complex molecules or assemblies containing an electrochemical gradient that could tease the photoexcited electron away from the hole it had left behind. It was also clear that one needed very close relationships with photophysicists. It was time to move on.

13.2.2 Flexible porphyrin dimers: π–π interactions and model enzymes

The metalloporphyrin moiety is spectroscopically eloquent, enabling binding to be measured through changes in electronic and NMR spectra, and geometries to be deduced from ring current shifts in the latter (Figure 13.3). The range of coordination selectivities displayed by different central metal ions added extra versatility and interest for supramolecular assembly. It rapidly became clear that porphyrin dimers connected by flexible linkers could accommodate a range of ligands within their cavities. That gave us the idea of creating model enzymes that could bind ligands within a cavity and catalyze reactions. However, in the absence of ligands, the cavities collapsed. Furthermore the porphyrins always took up the same offset geometrical relationship [15]: they were not stacked vertically on one another, even though this would optimize Van der Waals interactions (Figure 13.4).

Chris Hunter undertook to investigate this conundrum, and so was born his simple pictorial model of π–π interactions. By considering the electrostatic interactions between negatively charged lobes of p-orbitals and positively charged atomic centers he was able to predict not only the geometry of porphyrin–porphyrin interactions, but also the geometry of many other aromatic interactions in host–guest chemistry (Figure 13.5). Actually, he had rediscovered the distributed multipole approach previously published in a more mathematically rigorous form by Sally Price and Anthony Stone in our own department.

The resulting paper was published in JACS [16] because a referee stated that although they didn't believe any of the quantitative calculations (carried out on a primitive 1980s Mac), the paper should be published because its ideas would stimulate much experiment and theory, and a far-sighted editor agreed. Indeed, many of its counterintuitive predictions have been vindicated by experiment, and by late 2015, it had received over 3500 citations. Although there are now more rigorous theoretical treatments available it is still being cited at least twice every week because its simple pictorial message is so powerful.

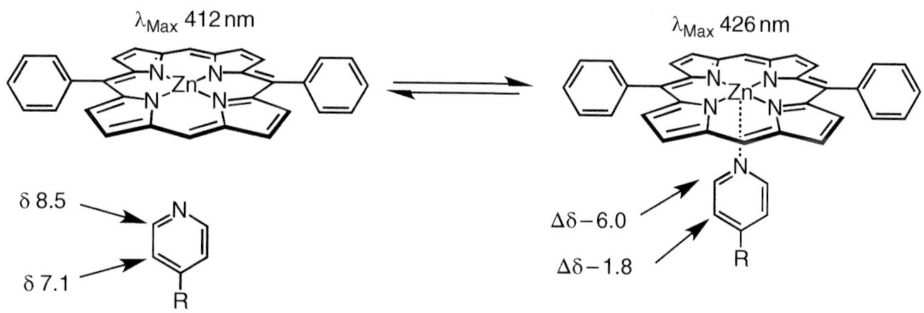

Figure 13.3 *The spectroscopic eloquence of porphyrin coordination chemistry illustrated by the example of Zn: δ values are ¹H chemical shifts in free pyridines; Δδ values are the shifts experienced on binding*

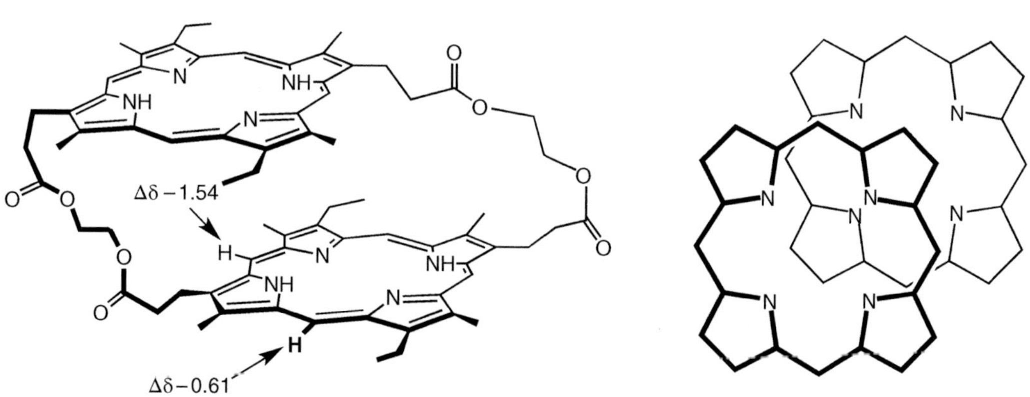

Figure 13.4 *A collapsed flexible porphyrin dimer and the geometry of porphyrin–porphyrin interactions. Δδ values are ¹H chemical shifts relative to a monomer*

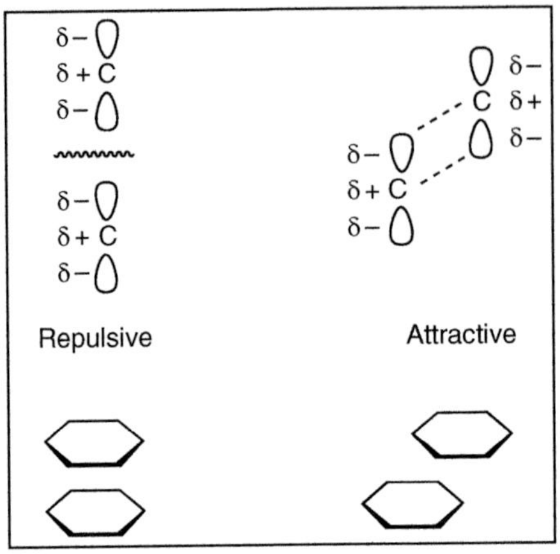

Figure 13.5 *Multipole model of π–π interactions*

Floppy dimers such as the one shown in Figure 13.4 had fascinating properties: the geometrical relationships of two chromophores could be explored via the exciton coupling that leads to shifts in their visible spectra [17], and when equipped with a different metal ion in each face [18] they could be used to investigate fundamental aspects of fast atom bombardment mass spectrometry [19]. Ligand binding that pulled the two porphyrins apart could be used to measure the strength of the π–π interaction between them [20, 21], but the cavity collapse did not make them promising as model enzyme systems. Fortunately at that moment Harry Anderson arrived to start his PhD studies.

13.2.3 The Anderson butadiyne system

When Harry saw our floppy structures and my suggested project, he thought for a few days and invented an entirely new system of porphyrin oligomers held together by butadiyne linkers that we supposed were sufficiently rigid to prevented collapse [22]. They also looked easy to prepare by oxidative coupling of a monomer equipped with terminal acetylenes (Figure 13.6). This trimer we named 2,2,2 for the number of alkyne groups in each linker. This system had the potential to bind multiple ligands within the cavity and so lead to catalysis, or at least accelerated bimolecular reactions due to proximity. In the event, the oxidation led efficiently to a mixture of cyclic dimer, trimer and tetramer that could be separated by extensive chromatography so that their binding properties could be studied [23].

These properties led us to imagine that we could use 4,4′-bipyridyl (Bipy), *s*-tripyridyltriazine (TPT), and tetrapyridylporphyrin as templates for the selective synthesis of the dimer, trimer and tetramer respectively (Figure 13.7). This approach worked spectacularly well, giving greatly improved access to these molecules [24]. Detailed studies by Sally Anderson revealed a far more subtle set of pathways and template roles than originally expected: Bipy acts positively to template the formation of cyclic dimer by bringing the two reactive ends of the linear dimer intermediate close together, thereby privileging the intramolecular reaction over intermolecular chain extension (Figure 13.8a). By contrast, TPT acts as a negative template, binding the linear dimer in such a way that the reactive ends cannot react intramolecularly and allowing intermolecular chain extension to compete more effectively (Figure 13.8b). When tetrapyridylporphyrin is used to template the synthesis of cyclic tetramer from preformed linear dimer, it simultaneously performs a variety of positive and negative roles [25]. The ideas around positive, negative (and scavenger) templates should have wider applicability than in these systems, but they have yet to be much explored by other groups [26]. In the following few years, these studies were extended to include a variety of shorter and longer alkyne linkers

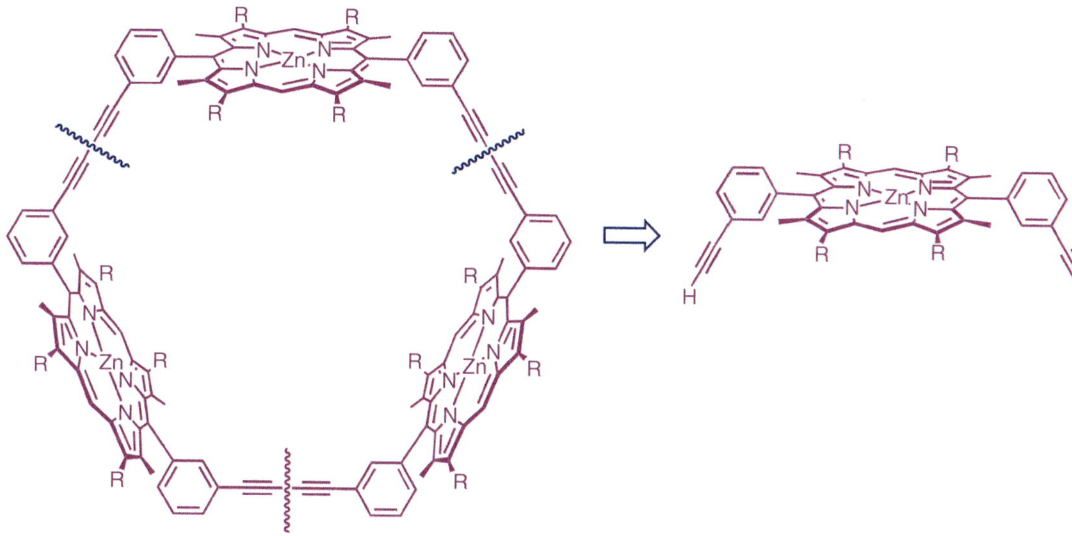

Figure 13.6 *Disconnection of 2,2,2 trimer suggesting a simple synthetic route*

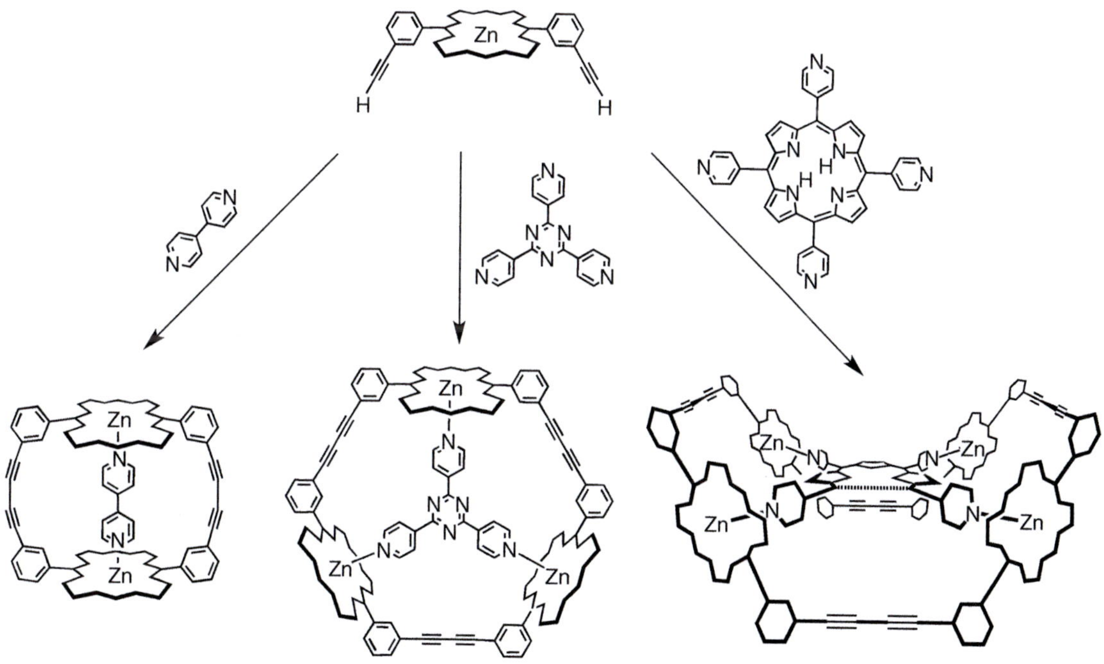

Figure 13.7 *Templated synthesis of cyclic 2,2-dimer, 2,2,2-trimer and 2,2,2,2-tetramer*

(from 1,1,1 to 4,4,4) and different types of porphyrins, generating a wide array of receptors whose potential for binding and catalysis could be explored [27, 28].

It was particularly productive to explore the fact that different central metal ions have complementary coordination properties [29], e.g. Sn(IV) porphyrins are O-selective [30]; Al(III) porphyrins bind one

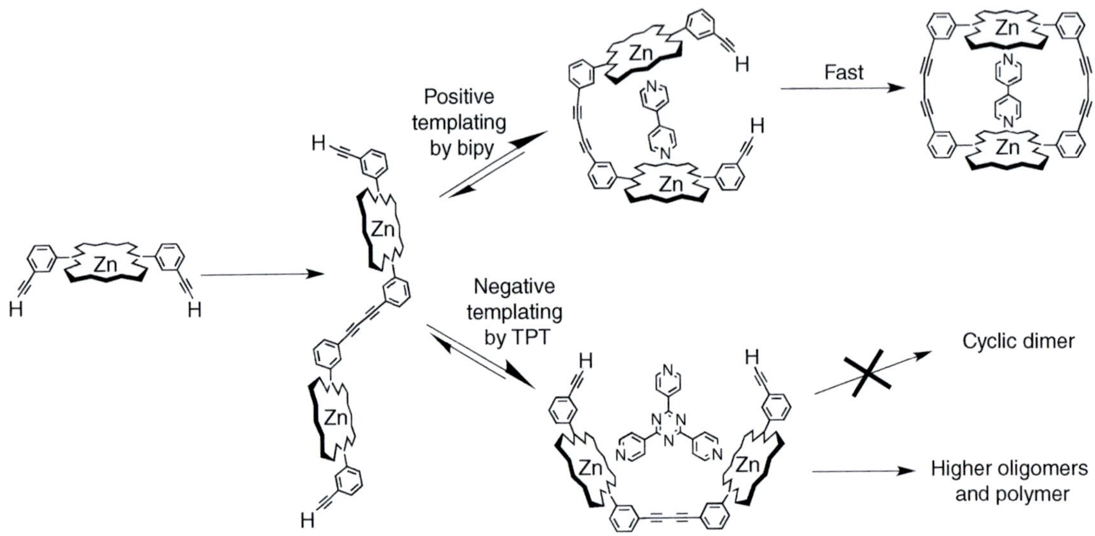

Figure 13.8 *Positive and negative templating in the synthesis of 2,2-dimer and 2,2,2-trimer*

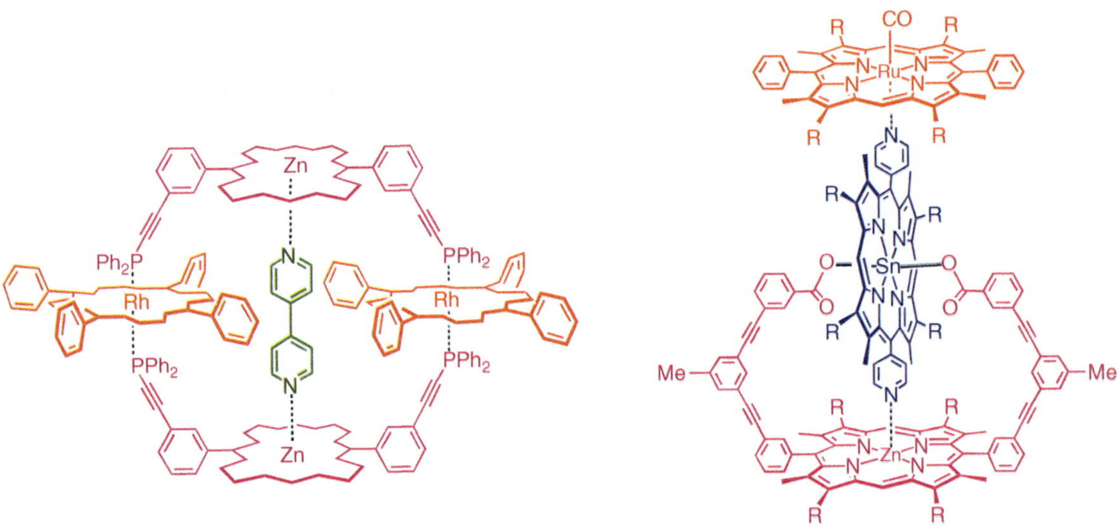

Figure 13.9 *Two multiporphyrin systems prepared by synergistic use of complementary coordination chemistries*

carboxylate and one pyridine [31]; and while Rh(III) porphyrins do bind pyridines strongly, they prefer phosphines even more strongly [32]. Combining these properties led to the efficient preparation of large and extraordinary systems from small(ish) porphyrin components, especially when geometrical complementarity was also programmed in [33–35]. Figure 13.9 shows just two examples of many. A dynamic combinatorial version of the template synthesis of porphyrin oligomers using disulfide chemistry was developed by Amy Kieran and also proved successful [36].

13.2.4 Chemistry inside cavities

The initial driving force for creating these super-macrocycles with cavities and binding sites was the prospect of enhanced chemistry between bound ligands within cavities. The Diels–Alder reaction was an obvious first target, given that the main barrier to reaction is entropic, i.e. bringing together the diene and dienophile in the correct geometry. Chris Walter chose the system shown in Figure 13.10 because when furan acts as dienophile the reaction is reversible and two different stereochemical outcomes are possible: the *endo*-adduct is the result of kinetic control and is familiar as the dominant product in irreversible Diels–Alder reactions, while the *exo*-adduct is the thermodynamic product which is accessible only because the reaction is reversible. Amazingly, the original Anderson large 2,2,2-trimer accelerated the *exo*-reaction 1000-fold and suppressed the *endo*-adduct [37], while the smaller and more rigid 1,1,2-trimer developed by Anton Vidal-Ferran selectively formed the *endo*-adduct [38]. We were able to demonstrate that the larger trimer was not a particularly good fit for the *exo*-adduct, but was able to flex sufficiently to bind the *exo*-transition state; it was not flexible enough to accommodate and doubly bind the *endo*-adduct. The smaller trimer was much more rigid, and could accommodate only the *endo*-transition state and product. Much later we discovered that the butadiyne-linked linear dimer shown as an intermediate in Figure 13.8 selectively accelerated the *exo*-reaction around 90-fold, a result that would have been really exciting if we had uncovered it earlier [38].

Inevitably the strong binding of the product within the cavity inhibited catalytic turnover. Moshe Nakash did explore other systems where the Diels–Alder reaction was accelerated, with the hope that the product might undergo further transformation and be released, allowing catalytic turnover; we had no success in the second step, but did demonstrate the role of host flexibility and distortion in rate acceleration [39]. Lindsey

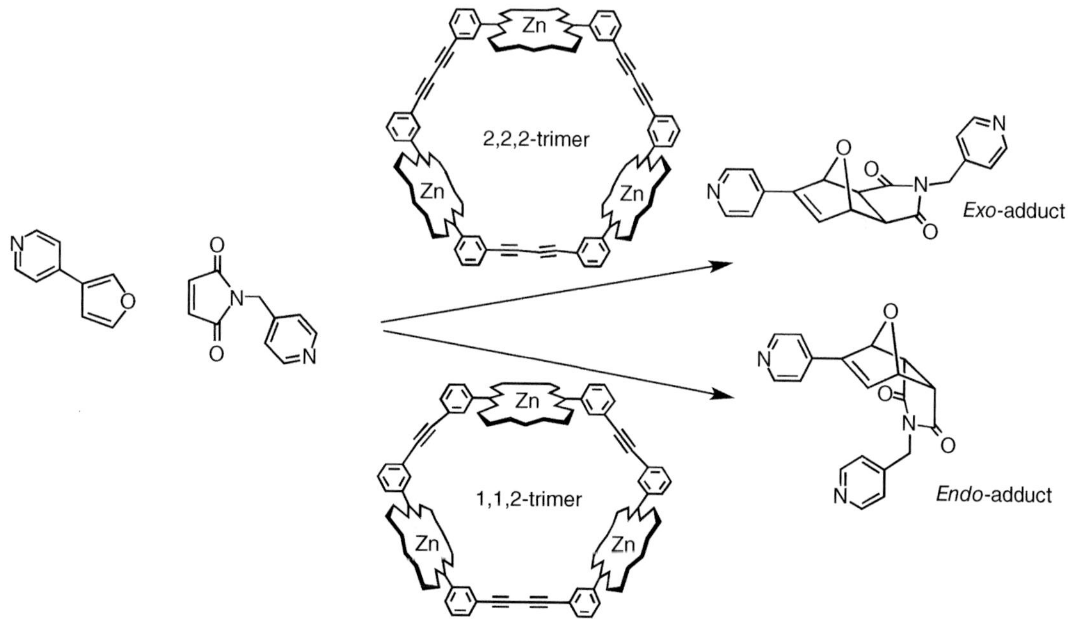

Figure 13.10 *Redirecting the stereochemical course of a Diels–Alder reaction*

Mackay discovered an acyl transfer system that showed turnover without inhibition [40] but the details of how this system worked proved too difficult to disentangle and exploit [41].

13.3 Macrocycles based on cholic acid

The arrival of Richard Bonar-Law, following on from his PhD work with Tony Davis, brought cholic acid expertise into the group. Tony had recognized in the 1980s that the cheap and readily available cholic acid possesses a very attractive combination of rigid and flexible regions, with potentially useful functionalization, and Richard's PhD work had developed a valuable set of synthetic tools for exploiting these. He quickly developed macrocyclic oligocholates [42], a self-assembling dimer [43], a bowl consisting of four cholates and a porphyrin [44], and wonderful cholate-capped porphyrins [45]. Two examples are shown in Figure 13.11. The bowl turned out to be capable of chiral recognition [46], and the capped porphyrins proved extremely versatile vehicles for understanding binding processes [47]. In both cases, the presence of hydroxyl groups

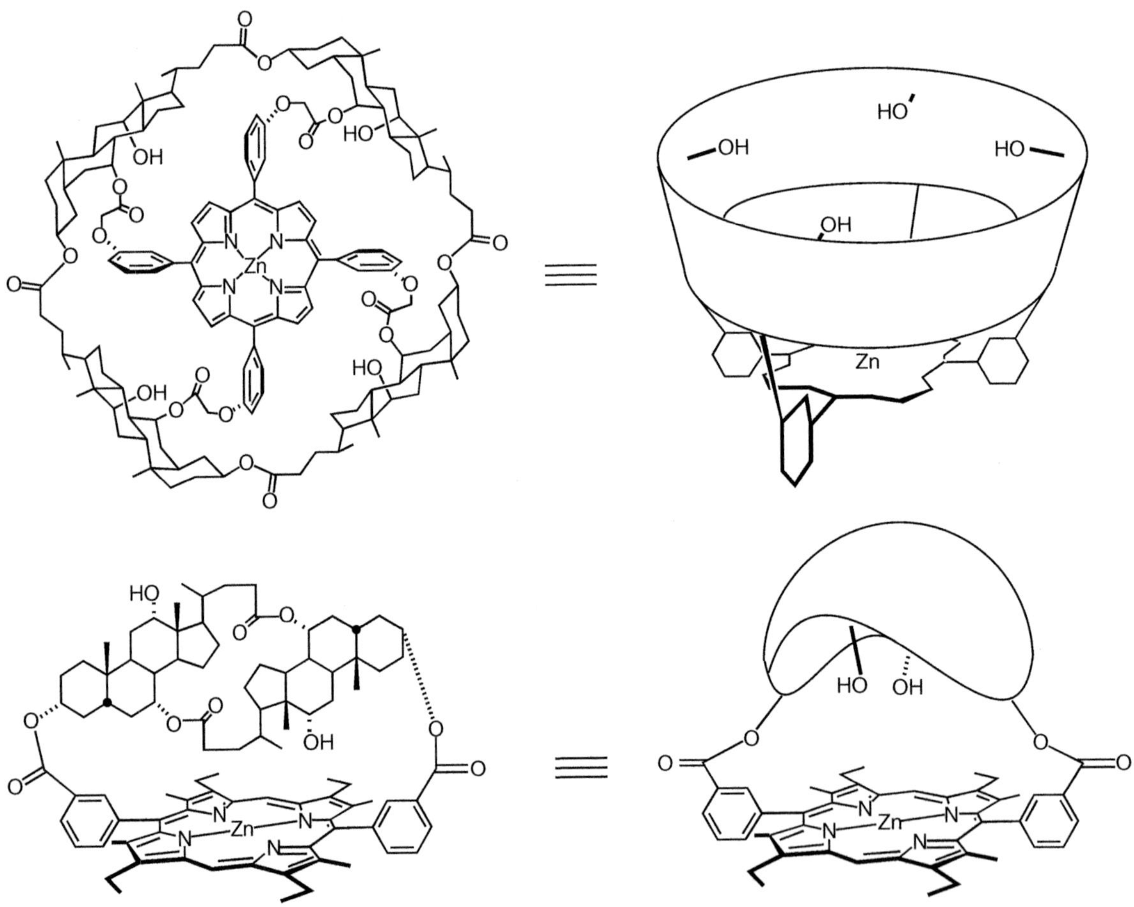

Figure 13.11 Two oligocholate–porphyrin systems with recognition properties

within a non-polar cavity endowed powerful binding properties. Most notably, Richard proved that the apparently selective recognition of sugars by macrocyclic hosts owes virtually nothing to the shape or detailed functionality of the host, but much more to the inherent "stickiness" of the sugar, and that is determined by the disposition of intra- and intermolecular hydrogen bonds in the sugar. Thus, most published receptors show the same order of selectivity, independent of structure. Richard also discovered, counterintuitively, that the binding of a sugar to the receptor in organic solvents could be strengthened by addition of small amounts of water or methanol. This is because there is a misfit between host and guest, and the hydrogen-bonding efficiency can be improved by insertion of a bridging water. This phenomenon is well known in sugar-recognizing antibodies, but we believe that our receptor is the first synthetic case to be recognized and characterized.

13.4 Designed donor–acceptor catenanes

In 1996 Darren Hamilton arrived with less than a year's funding left over from a fellowship, and the structures of two co-crystals containing stacked alternating donors and acceptors. We decided to combine this motif with our butadiyne linker to see if we could make Stoddart-like catenanes that would have the attraction of being neutral (Figure 13.12). The naphthalene substitution pattern needed to be changed to avoid chain entanglement, but this worked very nicely first time: mixing a donor and acceptor in solution led to a bright orange color that disappeared on heating as the complex dissociated, whereas the crude catenane solution stayed orange on heating because the D–A components could not escape from each other (Figure 13.13) [48]. This success led to a wide array of related new catenanes [49], but perhaps its most important role was to prepare us mentally for unexpected catenane discoveries in dynamic combinatorial chemistry which we were developing at the same time.

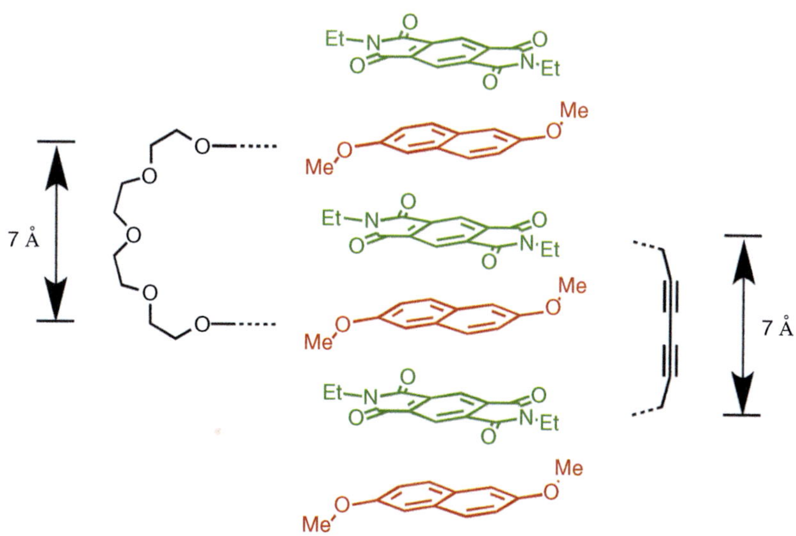

Figure 13.12 Donor–acceptor co-crystal showing supramolecular design thinking for catenane synthesis

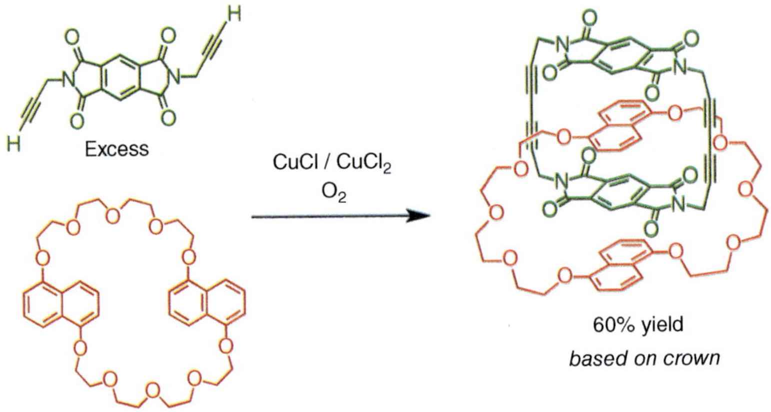

Figure 13.13 *Design and synthesis of a neutral donor–acceptor catenane*

13.5 Dynamic combinatorial chemistry

13.5.1 Conception and principles

The Cambridge version of dynamic combinatorial (or covalent) chemistry was conceived in Dublin, Ireland, on Thursday September 17, 1992. On that day, I gave a conference lecture describing the *exo*–selective acceleration of the bimolecular Diels–Alder reaction shown in Figure 13.10. Over lunch afterwards Chris Hunter and Fraser Stoddart pointed out the limitations of our approach: the complex multistep synthesis of the host followed by careful design of a highly specific diene–dienophile pair might have been successful in this case, but it was hardly a general solution to the problem of how to achieve effective molecular recognition and catalysis in supramolecular chemistry. We agreed that one had to build proof-reading into the receptor synthesis so that only successful molecules survived. This feedback effect would be reminiscent of the way that the mammalian immune system amplifies successful antibodies, although the mechanism would be fundamentally different. By the end of that day, most of the essential features of my early thinking in dynamic combinatorial chemistry – thermodynamically-controlled bond-breaking and -making, templating, selection and amplification – had been sketched out. Back in Cambridge, Richard Bonar-Law was already exploring the potential of base-catalyzed reversible transesterification for generating mixtures of macrocycles, but our first serious work aimed at achieving thermodynamically controlled templating was started by a new graduate student, Paul Brady, in October 1993. Progress was slow at first, but the first paper clearly describing the chemistry and articulating these ideas appeared in early-1996 [50]; the first templating results with this system appeared soon afterwards [51], and Stuart Rowan extended the chemistry to generate beautiful macrocycles based on cinchona alkaloids [52]. Around the same time Lehn independently conceived the dynamic combinatorial idea as a result of his work on metal helicates] [53].

The attraction of dynamic covalent chemistry (DCC) is that it combines the advantages of stable covalent bonds with the proof-reading advantages of supramolecular systems. It has found increasing application in many areas beyond supramolecular host–guest chemistry [54–57]. In dynamic combinatorial chemistry, simple molecular units (building blocks) are held together by reversible covalent bonds, generating a complex mixture of products which continuously interconvert: the composition of the mixture at equilibrium is thermodynamically controlled and is referred to as a dynamic combinatorial library. As a dynamic system, a library can respond to various stimuli that drive the constituents to reorganize in order to minimize the total free energy of the system. The stimuli that have been investigated include change of pH, temperature, light irradiation, and electric field, but the most exploited stimulus involves the introduction of a chemical template: libraries of macrocycles, for example, provide as many potential receptors as library members, and addition of a guest which binds strongly to one of the library components shifts the equilibrium towards its formation, resulting in the amplification of the successful receptor at the expense of the less successful. Allowing easy screening of potential receptors with different features, this approach has made DCC a powerful tool for the discovery of unpredictable new receptors. In essence, one designs the *experiment*, rather than the *molecule*. Figure 13.14 summarizes four conceptual uses of DCC that utilize molecular recognition. It has also provided us with a tool for gaining insights into how some chemical systems respond to external stimuli [58].

The reaction conditions required for successful base-catalyzed transesterification proved too rigorous and restrictive. Hydrazone chemistry in organic solvents was developed and explored by Sally-Ann Poulson, Graham Cousins and Ricardo Furlan [59, 60], while Sijbren Otto brought aqueous disulfide exchange expertise [61]. We and others have explored many other reversible chemistries [55–57] but acid-catalyzed hydrazone exchange and base-catalyzed disulfide exchange have proved the most successful in our hands. Another source of our success for DCC was the arrival in 2002 of Ana Belenguer after a 20-year career doing hplc in the pharmaceutical industry: her creation and management of a first class hplc laboratory has transformed our ability to analyze and understand complex libraries.

13.5.2 Hydrazones

Our most dramatic result from hydrazone chemistry was the discovery by Sarah Roberts and Ruby Lam of a peptide catenane that binds acetylcholine (Figure 13.15) [62]. In the absence of a template containing an Me_3N^+ group, a linear peptide with reactive ends leads to a dynamic library of cyclic dimer, trimer, and tetramer. Addition of template leads to the accumulation of over 70% yield of the previously undetectable catenane, which is effectively a hexamer composed of two interlocked trimers. The entropic cost of amplifying such a large species immediately demonstrated the strength and selectivity of recognition. Isolation of all the possible receptors indeed proved that the affinity of the catenane for acetylcholine was over 100 times greater than that of the simple macrocycles. The catenane itself, when stripped of template, has an NMR spectrum demonstrating the presence of multiple flexible interconverting conformations, but addition of the template leads to simplification and generation of a single dominant conformation which tightly grips the ammonium cation. Despite the entropic cost of this conformational restriction, the overall binding is entropically driven, apparently because of solvent release by the catenane.

More recently, Sophie Beeren developed an alternative set of building blocks and reaction conditions for hydrazone chemistry, leading to the discovery of exceptional linear receptors for phosphate anions [63].

13.5.3 Disulfides

Disulfide chemistry has proved to be even more successful and versatile. Early on, we were able to show the amplification of three different receptors from a single library in water by use of three different templates (Figure 13.16) [64, 65]. In the absence of template, the two library building blocks in water give a complex library (= mixture). Addition of the hydrophobic aromatic methylisoquinolinium cation efficiently amplifies a small receptor, while use of the larger cation derived from morphine amplifies the larger homotrimer. Surprisingly, the smallest template, Me_4N^+, led to stereospecific amplification of the largest macrocycle; it achieved this through an induced fit process in which only the correct *meso* stereoisomer could fold up into the appropriate conformation to give effective binding. These systems helped us to explore the fundamentals of how amplification correlated with binding strength and led to a new area of systems chemistry explored by Sijbren Otto when he embarked on his independent research career [58]. We had hoped to use transition-state analogues as templates to generate receptors that are effective catalysts, but along with other groups in this field, we achieved only modest success [66].

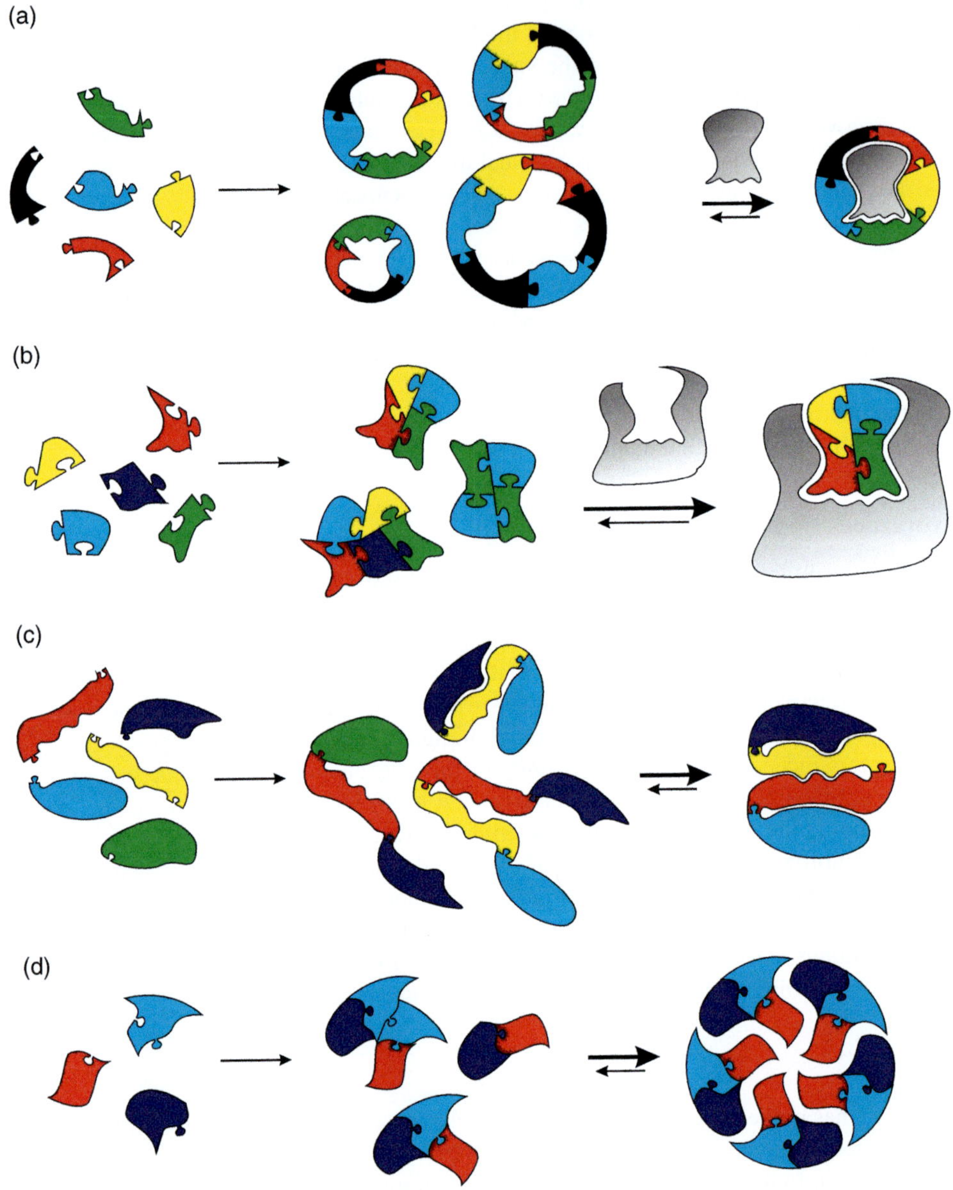

Figure 13.14 *Potential applications of Dynamic Combinatorial Chemistry. (a) Using a ligand template to create a complementary receptor; (b) Using a receptor template to create a complementary ligand, for example an enzyme inhibitor; (c) Optimizing a folding structure by self-templating; (d) Generating self-assembled aggregates. Adapted with permission from Corbett, P.T., Leclaire, J., Vial, L. et al. (2006) Dynamic combinatorial chemistry.* Chem. Rev., *106, 3652–3711. Copyright © 2006 American Chemical Society*

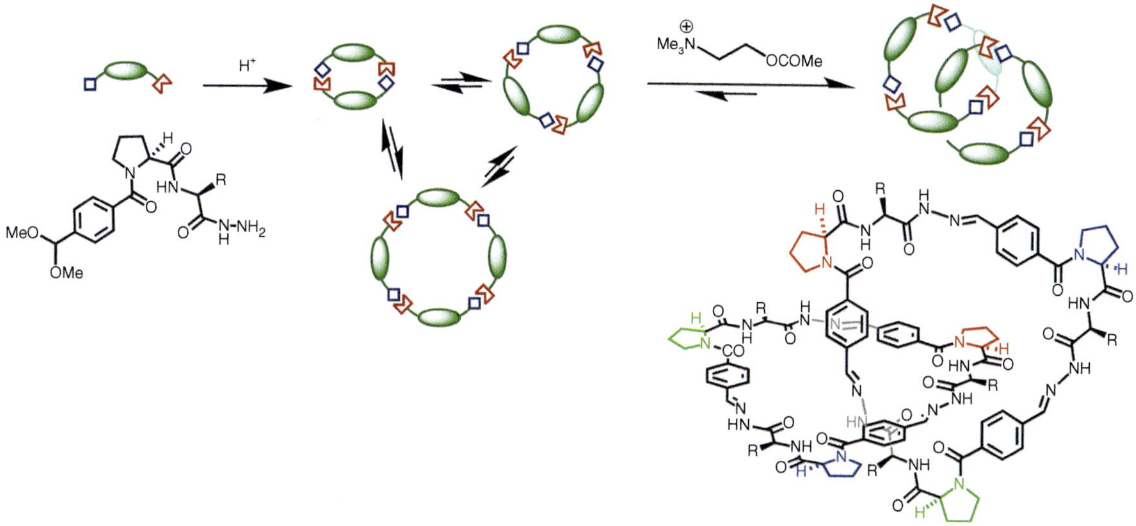

Figure 13.15 *A peptide hydrazone library and acetylcholine-templated formation of a catenane*

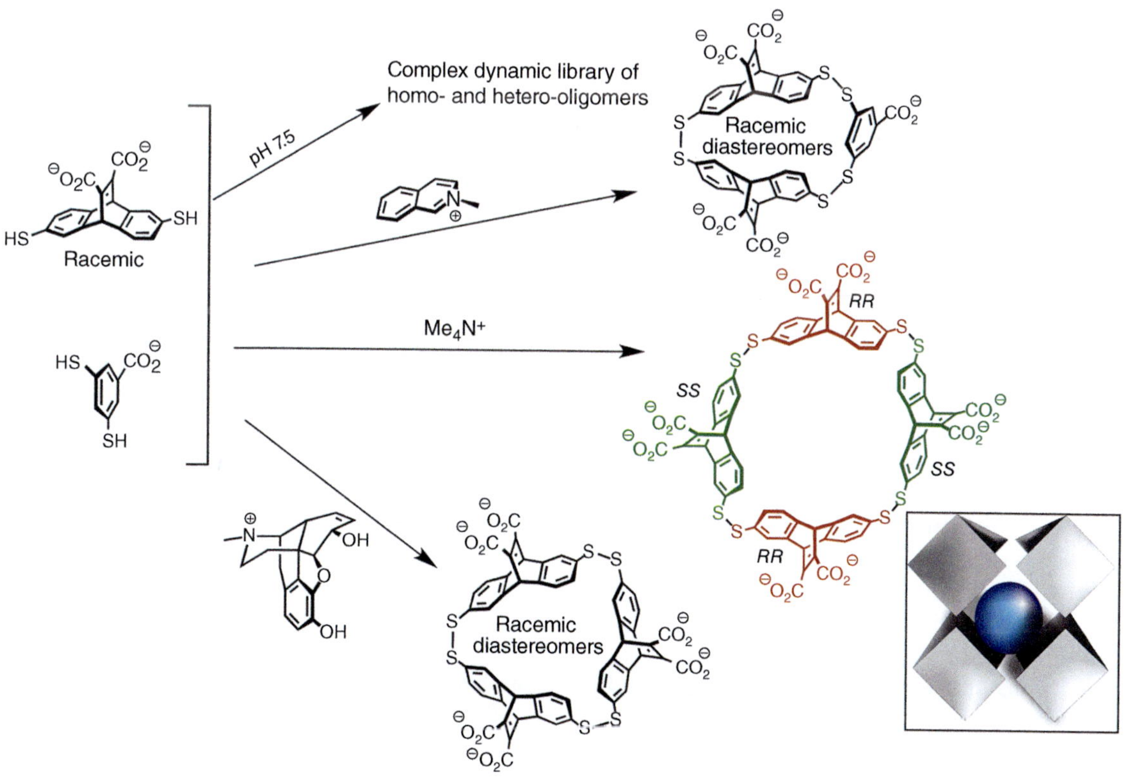

Figure 13.16 *Amplification of three different receptors from a single dynamic combinatorial library. (Adapted from ref. 55 with permission. Copyright © 2006 the American Chemical Society.)*

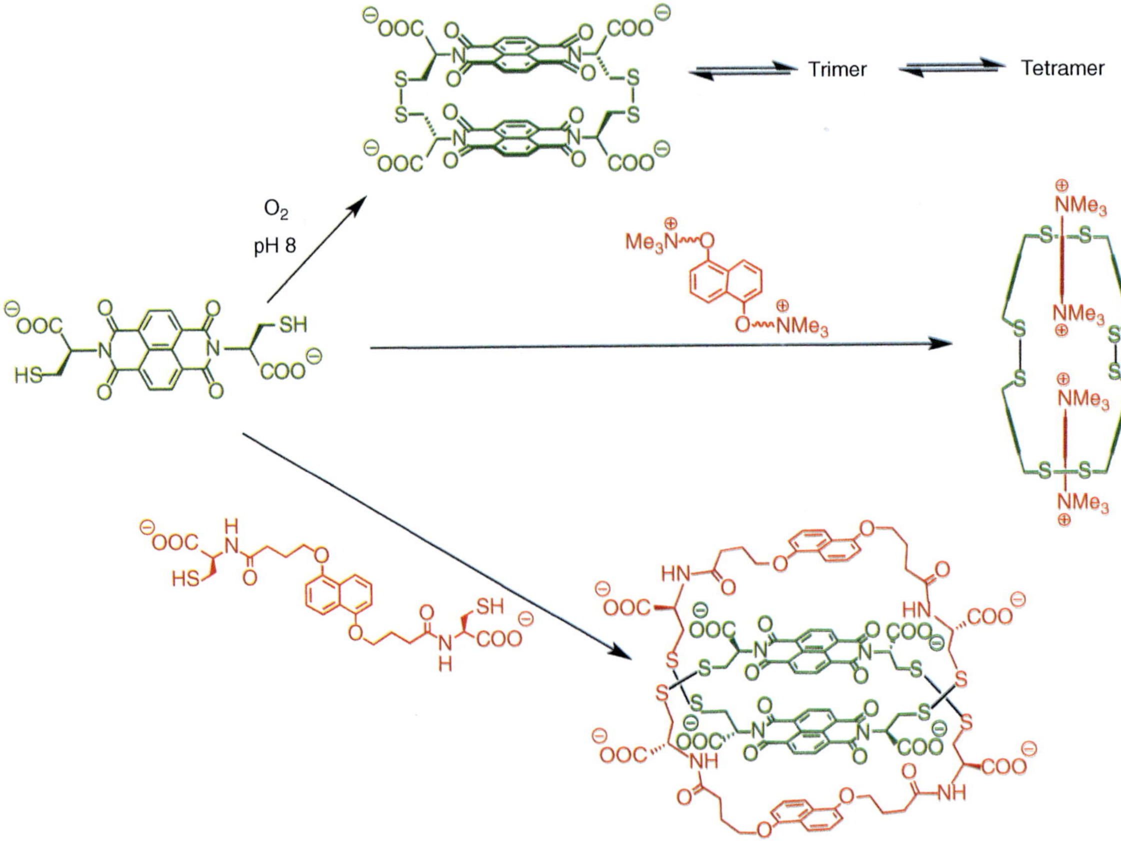

Figure 13.17 *Generation of receptors and a catenane from an NDI-cysteine building block in water*

Great success flowed from Paolo Pengo's development of a building block that combined the electron-deficient (acceptor) hydrophobic surface of naphthalenediimide (NDI) with the chirality, hydrophilicity and thiol functionality of cysteine (Figure 13.17) [67]. This exceptionally versatile building block, and its descendants, led to the discovery by Ho-Yu Au-Yeung, Fabien Cougnon and Nandhini Ponnuswamy of a remarkable range of water-soluble receptors and catenanes [68, 69]. When the unreactive electron-rich (donor) naphthalene derivative shown is used as a template, only the tetramer is amplified as it is sufficiently flexible to efficiently accommodate two template guest molecules [68], while if the donor contains reactive cysteines it is incorporated to give a catenane [69]. This catenane has the two acceptor NDI moieties stacked loosely on each other, in accordance with Chris Hunter's predictions [16], and its synthesis can be further improved by addition of the electron-rich template which can be accommodated in between them to give a D–A–D–A–D stack.

Variations on the structures of the donors and acceptors led to the discovery of a huge range of new catenanes [70], and culminated in the discovery of a series of molecular knots, some of which are shown in Figure 13.18 [71, 72]. The formation of these exquisite structures in water is driven by the need for the hydrophobic NDI surfaces to minimize their surface exposure. The knots are formed with great stereoselectivity, and only by building blocks with the correct chirality. Use of hplc to follow the kinetics and pathways of formation demonstrates that the knots are not only kinetically formed products but are also the

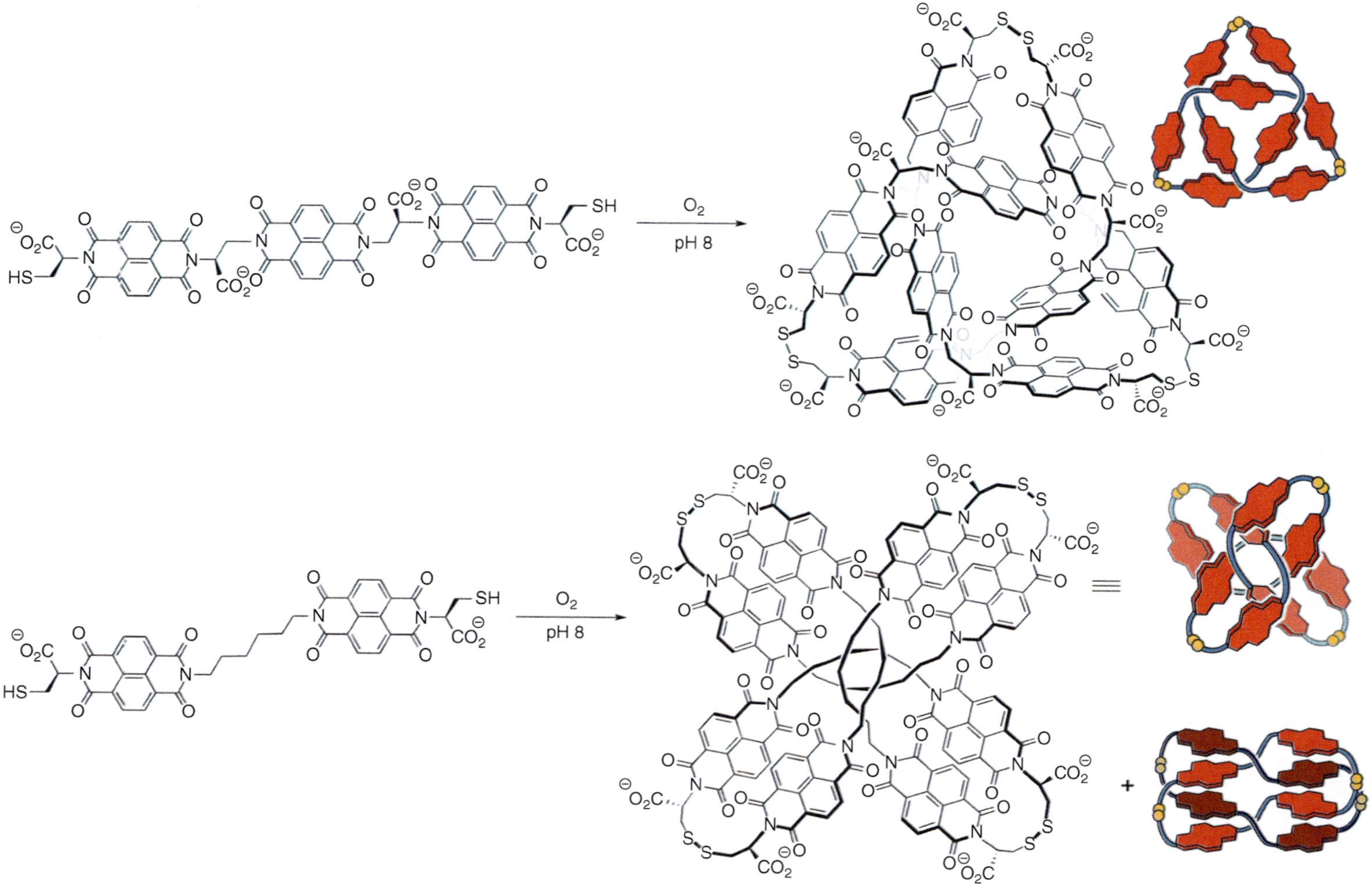

Figure 13.18 Synthesis of NDI-based molecular knots. In the three-fold symmetric trefoil knot shown at the top, some of the functional groups and linkers are hidden. In the lower reaction, only one of the four-fold symmetric structures is shown in detail

thermodynamically-stable products: even if a disulfide link is cleaved by an incoming thiolate nucleophile, it quickly snaps shut again because the cost of unpacking the closely-packed aromatic surfaces into water is so high. Intriguingly, the trimer NDI building block gives a trefoil knot with threefold symmetry, while the dimer NDI building block gives two different knots with fourfold symmetry. The future potential of this approach to knot synthesis appears to be huge.

An accidental by-product of the dynamic combinatorial project was the discovery by Dan Pantos of remarkable supramolecular NDI nanotubes, held together by hydrogen bonding in non-polar solvents. These are topologically linear non-covalent polymers rather than macrocycles, but they do have interesting host–guest properties [73].

13.6 Conclusions

At a molecular level, all our evidence suggests that a flexible molecule is more interesting and versatile than a rigid one. Donald Cram was, of course, correct that a perfectly-fitting rigid host–guest pair would generate the strongest possible binding and recognition, but as we are limited to molecules with integral numbers of bonds, we cannot in general have access to perfectly tuned structures; size increases have to be quantized on an Angstrom scale, and the rigid structures that are the wrong size are certainly not effective. I believe we have also learned that the supramolecular community's fear of entropy is misplaced, or at least exaggerated: we can use flexible systems that are responsive to the stimulus provided by a host or guest. Complete collapse needs to be avoided in general, so in our experience the optimum building blocks have turned out to have a combination of both rigid and flexible components. None of this should be a surprise, given what we know about the structure and function of proteins and nucleic acids. In relation to receptors discovered by dynamic combinatorial discovery, tetramers appear to be particularly successful in balancing flexibility against the entropic cost of creating even larger oligomers.

At a personal level, our supramolecular and macrocyclic science has ranged over organic, inorganic, and physical chemistry. Our NMR interests have also included biophysics, pharmacology, and metabolic studies, none of which are described here. Flexibility of approach – and the inability to concentrate on one project for more than a few years – has been essential, as has been a research group with an adventurous turn of mind. I have always been more interested in exploring the new, rather than filling in the details of the old, so right now my final postdoc, Ana Belenguer, is exploring dynamic combinatorial chemistry in the solid state, a completely new area and rich with potential [74]. My motto in the lab is always:

> *Expect the unexpected; welcome the unexpected; search for the unexpected. If every experiment gave exactly the answer we predicted, then we would never learn anything truly new.*

Or as Louis Pasteur put it:

> *In the fields of observation chance favours only the prepared mind [75].*

I would like to think that our attitude to science is perfectly captured by Matt Ridley [76]:

> *The fuel on which science runs is ignorance.*
> *Science... must be fed logs from the forests of ignorance that surround us.*
> *In the process, the clearing that we call knowledge expands, but the more it expands, the longer its perimeter and the more ignorance comes into view....*
> *A true scientist is bored by knowledge.... The forest is more interesting than the clearing.*

References

1. Hinckley, C.C. (1969) Paramagnetic shifts in solutions of cholesterol and dipyridine adduct of trisdipivalomethanatoeuropium(3). A shift reagent, *Journal of the American Chemical Society*, 91 (18), 5160–5161.
2. Sanders, J.K.M., Williams, D.H. (1970) A shift reagent for use in NMR spectroscopy. A first order spectrum of *n*-hexanol, *Journal of the Chemical Society, Chemical Communications*, 422–423.
3. Waterton, J.C., Sanders, J.K.M. (1978) Hyperfine coupling in chlorophyll radical cations. A nuclear magnetic resonance approach, *Journal of the American Chemical Society*, 100, 4044–4049.
4. Denniss, I.S., Waterton, J.C., Sanders, J.K.M. (1976) Assignment of metalloporphyrin and chlorophyll NMR spectra via spin-lattice relaxation times, *Journal of the Chemical Society, Chemical Communications*, 1049–1050.
5. Sanders, J.K.M. (1977) Meldola medal lecture. NMR spectral change as a probe of chlorophyll chemistry, *Chemical Society Reviews*, 6, 467–487.
6. Denniss, I.S., Sanders, J.K.M. (1978) Synthesis of a chlorophyll which does not aggregate, *Tetrahedron Letters*, 295–298.
7. Ganesh, K.N., Sanders, J.K.M. (1982) Quinone-capped porphyrins: synthesis and some chemical properties, *Journal of the Chemical Society, Perkin Transactions 1*, 1611–1615.
8. Leighton, P., Sanders, J.K.M. (1987) Synthesis of bipyridyl-, viologen-, and quinone-bridged porphyrins, *Journal of the Chemical Society, Perkin Transactions 1*, 2385–2393.
9. Cowan, J.A., Sanders, J.K.M. (1985) Pyromellitimide-bridged porphyrins as model photosynthetic systems. I. Synthesis and steady state fluorescence properties, *Journal of the Chemical Society, Perkin Transactions 1*, 2435–2437.
10. Cowan, J.A., Sanders, J.K.M., Beddard, G.S., Harrison, R.J. (1987) Modelling the photosynthetic reaction centre: photoinduced electron transfer in a pyromellitimide-bridged "special pair" porphyrin dimer, *Journal of the Chemical Society, Chemical Communications,* 55–58.
11. Irvine, M.P., Harrison, R.J., Beddard, G.S., *et al.* (1986) Detection of the inverted region in the photo-induced intramolecular electron transfer of capped porphyrins, *Chemical Physics*, 104, 315–324.
12. Harrison, R.J., Pearce, B., Beddard, G.S., *et al.* (1987) Photo-induced electron transfer in pyromellitimide-bridged porphyrins, *Chemical Physics*, 116, 429–448.
13. Anderson, H.L., Hunter, C.A., Sanders, J.K.M. (1989) Assembly of a photoactive supramolecule using porphyrin coordination chemistry, *Journal of the Chemical Society, Chemical Communications*, 226–227.
14. Hunter, C.A., Sanders, J.K.M., Beddard, G.S., Evans, S. (1989) A new approach to the assembly of electron donor-spacer-acceptor systems, *Journal of the Chemical Society, Chemical Communications*, 1765–1767.
15. Leighton, P., Cowan, J.A., Abraham, R.J., Sanders, J.K.M. (1988) The geometry of porphyrin–porphyrin interactions, *Journal of Organic Chemistry*, 53, 733–740.
16. Hunter, C.A., Sanders, J.K.M. (1990) The nature of π–π interactions, *Journal of the American Chemical Society*, 112, 5525–5534.
17. Hunter, C.A., Sanders, J.K.M., Stone, A.J. (1989) Exciton coupling in porphyrin dimers, *Chemical Physics*, 133, 395–404.
18. Cowan, J.A., Sanders, J.K.M. (1987) Synthesis and electrochemical properties of some metallated and peripherally substituted porphyrin cofacial dimers, *Journal of the Chemical Society, Perkin Transactions 1*, 2395–2402.
19. Naylor, S., Hunter, C.A., Cowan, J.A., *et al.* (1990) Metalloporphyrins in fast atom bombardment mass spectrometry: implications for processes occurring in the liquid matrix, *Journal of the American Chemical Society*, 112, 6507–6414.
20. Hunter, C.A., Meah, N.M., Sanders, J.K.M. (1990) DABCO-metalloporphyrin binding: ternary complexes, host-guest chemistry and the measurement of π–π interactions, *Journal of the American Chemical Society*, 112, 5773–5780.
21. Anderson, H.L., Hunter, C.A., Meah, N.M., Sanders, J.K.M. (1990) Thermodynamics of induced-fit binding inside polymacrocyclic porphyrin hosts, *Journal of the American Chemical Society*, 112, 5780–5789.
22. Anderson, H.L., Sanders, J.K.M. (1995) Enzyme mimics based on cyclic porphyrin oligomers: strategy, design and exploratory synthesis, *Journal of the Chemical Society, Perkin Transactions 1*, 2223–2229.

23. Anderson, H.L., Anderson, S., Sanders, J.K.M. (1995) Ligand binding by butadiyne-linked porphyrin dimers, trimers and tetramers, *Journal of the Chemical Society, Perkin Transactions 1*, 2231–2245.
24. Anderson, S., Anderson, H.L., Sanders, J.K.M. (1995) Template directed synthesis of linear and cyclic butadiyne-linked porphyrin oligomers up to a linear octamer, *Journal of the Chemical Society, Perkin Transactions 1*, 2247–2254.
25. Anderson, S., Anderson, H.L., Sanders, J.K.M. (1995) The roles of templates in the syntheses of porphyrin oligomers, *Journal of the Chemical Society, Perkin Transactions 1*, 2255–2267.
26. Anderson, S., Anderson, H.L., Sanders, J.K.M. (1993) Expanding roles for templates in synthesis, *Accounts of Chemical Research*, 26, 469–475.
27. Sanders, J.K.M. (1996) Templated chemistry of porphyrin oligomers, In *Comprehensive Supramolecular Chemistry*, Atwood, J.L., Davies, J.E.D., Macnicol, D.D., Vögtle, F. (Eds), Elsevier.
28. Sanders, J.K.M. (2000) The coordination chemistry of oligoporphyrins, In *The Porphyrin Handbook*, Kadish, K.M., Smith, K.M., Guilard, R. (Eds), Academic Press.
29. Sanders, J.K.M., Bampos, N., Clyde–Watson, Z., *et al.* (2000) The axial coordination chemistry of metalloporphyrins, In *The Porphyrin Handbook*, Kadish, K.M., Smith, K.M., Guilard, R. (Eds), Academic Press.
30. Hawley, J.C., Bampos, N., Sanders, J.K.M. (2003) Synthesis and characterization of carboxylate complexes of Sn(IV) porphyrin monomers and oligomers, *Chemistry – A European Journal*, 9, 5211–5222.
31. Davidson, G.J.E., Lane, L.A., Raithby, P.R., *et al.* (2008) Coordination polymers based on aluminum(III) porphyrins, *Inorganic Chemistry*, 47, 8721–8726.
32. Stulz, E., Scott, S.M., Ng, Y.-F., et al. (2003) Construction of multiporphyrin arrays using ruthenium and rhodium coordination to phosphines, *Inorganic Chemistry*, 42, 6564–6574.
33. Kim, H.-J., Bampos, N., Sanders, J.K.M. (1999) Assembly of dynamic heterometallic oligoporphyrins using cooperative zinc-nitrogen, ruthenium–nitrogen and tin-oxygen coordination, *Journal of the American Chemical Society*, 121, 8120–8121.
34. Redman, J.E., Feeder, N., Teat, S.J., Sanders, J.K.M. (2001) Rh(III)porphyrins as building blocks for coordination arrays: from dimers to heterometallic undecamers, *Inorganic Chemistry*, 40, 2486–2499.
35. Stulz, E., Scott, S.M., Bond, A.M., *et al.* (2003) Selection and amplification of mixed-metal porphyrin cages from dynamic combinatorial libraries, *Chemistry – A European Journal*, 9, 6039–6048.
36. Kieran, A.M., Bond, A.M., Belenguer, A.M., Sanders, J.K.M. (2003) Dynamic combinatorial libraries of metalloporphyrins: templated amplification of disulfide-linked oligomers, *Chemical Communications*, 2674–2675.
37. Walter, C.J., Anderson, H.L., Sanders, J.K.M. (1993) *Exo*-selective acceleration of an intermolecular Diels–Alder reaction by a porphyrin trimer host, *Journal of the Chemical Society, Chemical Communications*, 458–460.
38. Clyde–Watson, Z., Vidal–Ferran, A., Twyman, L.J., *et al.* (1998) Reversing the stereochemistry of a Diels–Alder reaction: use of metalloporphyrin oligomers to control transition state stability, *New Journal of Chemistry*, 22, 493–502.
39. Nakash, M., Clyde–Watson, Z., Feeder, N., *et al.* (2000) Product-induced distortion of a metalloporphyrin host: implications for acceleration of Diels-Alder reactions, *Journal of the American Chemical Society*, 122, 5286–5293.
40. Mackay, L.G., Wylie, R.S., Sanders, J.K.M. (1994) Catalytic acyl transfer by a cyclic porphyrin trimer: efficient turnover without product inhibition, *Journal of the American Chemical Society*, 116, 3141–3142.
41. Levy, E.G. and Sanders, J.K.M (1996) PhD thesis, University of Cambridge.
42. Bonar–Law, R.P., Sanders, J.K.M. (1992) Cyclocholates: synthesis and ion-binding, *Tetrahedron Letters*, 33, 2071–2074.
43. Bonar–Law, R.P., Sanders, J.K.M. (1993) Self-associating cyclocholates, *Tetrahedron Letters*, 34, 1677–1680.
44. Mackay, L.G., Bonar–Law, R.P., Sanders, J.K.M. (1993) Concise synthesis of a porphyrin–cyclocholate molecular bowl, *Journal of the Chemical Society, Perkin Transactions 1*, 1377–1378.
45. Bonar–Law, R.P., Sanders, J.K.M. (1995) Synthesis of cyclocholate-capped porphyrins, *Journal of the Chemical Society, Perkin Transactions 1*, 3085–3096.
46. Bonar–Law, R.P., Mackay, L.G., Sanders, J.K.M. (1993) Morphine recognition by a porphyrin–cyclocholate molecular bowl, *Journal of the Chemical Society, Chemical Communications*, 456–458.
47. Bonar–Law, R.P., Sanders, J.K.M. (1995) Polyol recognition by a steroid-capped porphyrin; enhancement and modulation of misfit binding by water or methanol, *Journal of the American Chemical Society*, 117, 259–271.

48. Hamilton, D.G., Sanders, J.K.M., Davies, J.E., et al. (1997) Neutral [2]catenanes from oxidative coupling of π-stacked components, *Chemical Communications*, 897–898.
49. Raehm, L., Hamilton, D.G., Sanders, J.K.M. (2002) From kinetic to thermodynamic assembly of catenanes: error checking, supramolecular protection and oligocatenanes, *Synlett*, 1743–1761.
50. Brady, P.A., Bonar–Law, R.P., Rowan, S.J., *et al.* (1996) "Living" macrolactonisation: thermodynamically-controlled cyclisation and interconversion of oligocholates, *Journal of the Chemical Society, Chemical Communications*, 319–320.
51. Brady, P.A., Sanders, J.K.M. (1998) Thermodynamically-controlled cyclisation and interconversion of oligocholates: metal-ion templated "living" macrolactonisation, *Journal of the Chemical Society, Perkin Transactions 1*, 3237–3254.
52. Rowan, S.J., Sanders, J.K.M. (1997) Building thermodynamic combinatorial libraries of quinine macrocycles, *Chemical Communications*, 1407–1408.
53. Lehn, J.-M. (1999) Dynamic combinatorial chemistry and virtual combinatorial libraries, *Chemistry – A European Journal*, 5, 2455–2463.
54. Rowan, S.J., Cantrill, S.J., Cousins, G.R.L., et al. (2002) Dynamic covalent chemistry, *Angewandte Chemie International Edition*, 41, 898–952.
55. Corbett, P.T., Leclaire, J., Vial, L., et al. (2006) Dynamic combinatorial chemistry, *Chemical Reviews*, 106, 3652–3711.
56. Reek, J.N.H., Otto, S. (Eds) (2010) *Dynamic Combinatorial Chemistry*, Wiley-VCH, Weinheim.
57. Cougnon, F.B.L., Sanders, J.K.M. (2012) Evolution of dynamic combinatorial chemistry, *Accounts of Chemical Research*, 45, 2211–2221.
58. Li, J.W., Nowak, P., Otto, S. (2013) Dynamic combinatorial libraries: from exploring molecular recognition to systems chemistry, *Journal of the American Chemical Society*, 135, 9222–9239.
59. Cousins, G.R.L., Poulsen, S.-A., Sanders, J.K.M. (1999) Dynamic combinatorial libraries of pseudo-peptide hydrazone macrocycles, *Chemical Communications*, 1575–1576.
60. Cousins, G.R.L., Furlan, R.L.E., Ng, Y.-F., et al. (2001) Identification and isolation of a receptor for *N*-methyl alkylammonium salts: molecular amplification in a pseudo-peptide dynamic combinatorial library, *Angewandte Chemie International Edition*, 40, 423–428.
61. Otto, S., Furlan, R.L.E., Sanders, J.K.M. (2000) Dynamic combinatorial libraries of macrocyclic disulfides in water, *Journal of the American Chemical Society*, 122, 12063–12064.
62. Lam, R.T.S., Belenguer, A., Roberts, S.L., et al. (2005) Amplification of acetylcholine-binding catenanes from dynamic combinatorial libraries, *Science*, 308, 667–669.
63. Beeren, S.R., Sanders, J.K.M. (2011) Discovery of linear receptors for multiple dihydrogen phosphate ions using dynamic combinatorial chemistry, *Journal of the American Chemical Society*, 133, 3804–3807.
64. Otto, S., Furlan, R.L.E., Sanders, J.K.M. (2002) Selection and amplification of hosts from dynamic combinatorial libraries of macrocyclic disulfides, *Science*, 297, 590–593.
65. Corbett, P.T., Tong, L.H., Sanders, J.K.M., Otto, S. (2005) Diastereoselective amplification of an induced-fit receptor from a dynamic combinatorial library, *Journal of the American Chemical Society*, 127, 8902–8903.
66. Brisig, B., Sanders, J.K.M., Otto, S. (2003) Selection and amplification of a catalyst from a dynamic combinatorial library, *Angewandte Chemie International Edition*, 42, 1270–1273.
67. Pengo, P., Pantoş, G.D., Otto, S., Sanders, J.K.M. (2006). Efficient and mild microwave-assisted stepwise functionalization of naphthalenediimide with alpha-amino acids, *Journal of Organic Chemistry*, 71 (18), 7063–7066.
68. Au-Yeung, H.Y., Cougnon, F.B.L., Otto, S., *et al.* (2010) Exploiting donor–acceptor interactions in aqueous dynamic combinatorial libraries: exploratory studies of simple systems, *Chemical Science*, 1, 567–574.
69. Au-Yeung, H.Y., Pantoş, G.D., Sanders, J.K.M. (2009) Dynamic combinatorial synthesis of a catenane based on donor–acceptor interactions in water, *Proceedings of the National Academy of Sciences of the United States of America*, 106, 10466–10470.
70. Cougnon, F.B.L., Ponnuswamy, N., Jenkins, N.A., *et al.* (2012) Structural parameters governing the dynamic combinatorial synthesis of catenanes in water, *Journal of the American Chemical Society*, 134, 19129–19135.
71. Ponnuswamy, N., Cougnon, F.B.L., Clough, J.M., et al. (2012) Discovery of an organic trefoil knot, *Science*, 338, 783–785.

72. Ponnuswamy, N., Cougnon, F.B.L., Pantoş, G.D., Sanders, J.K.M. (2014) Homochiral and meso figure-of-eight knots and a Solomon Link, *Journal of the American Chemical Society*, 136, 8243–8251.
73. Ponnuswamy, N., Pantoş, G.D., Smulders, M.M.J., Sanders, J.K.M. (2012) Thermodynamics of supramolecular naphthalenediimide nanotube formation: the influence of solvents, side chains, and guest templates, *Journal of the American Chemical Society*, 134, 566–573.
74. Belenguer, A.M., Lampronti, G.I., Wales, D.J., Sanders, J.K.M. (2014) Direct observation of intermediates in a thermodynamically controlled solid-state dynamic covalent reaction, *Journal of the American Chemical Society*, 136, 16156–16166.
75. Pasteur, L. (1854) Inaugural lecture, University of Lille.
76. Ridley, M. (1999) *Genome: The Autobiography of a Species in 23 Chapters*, Fourth Estate, London.

14

Texaphyrins: Life, Death, and Attempts at Resurrection

Jonathan L. Sessler

Department of Chemistry, The University of Texas at Austin, Austin, TX, USA

14.1 Introduction

This chapter seeks to detail the texaphyrin saga from a personal perspective. A more technical version of this report was published several years ago [1]. Like all history, the facts are subject to interpretation. Any factual errors or omissions are entirely my fault; I bear full responsibility. Nevertheless, in an effort to protect sensibilities, the names of key players will not generally be called out except in reference to specific historical events critical to the telling of the story. Moreover, all decisions, good or bad, are collectivized in an effort to avoid any ill-will. Of course, the contributions of coworkers, colleagues, and collaborators are duly referenced in the original publications. Their help, assistance, and dedication to the development of texaphyrin chemistry and the pre-ibrutinib days of Pharmacyclics, Inc. is greatly appreciated.

14.2 Early days

In August of 1984, I arrived in Austin, Texas to start work as an assistant professor at the University of Texas at Austin (UT Austin). In those days, as part of our orientation, we were required to take a class that I recall was entitled, "How to teach to Texans." Interspersed with pedagogical advice, was a crash course on Texas culture and history. Having set foot in the state previously only on the occasion of my job interview, this aspect of the four-day class was extremely useful to me. I learned, for instance, that the logo for Lone Star Beer was a five-pointed star, and that in Texas everything is bigger. Naturally, this inspired an effort to create

Macrocyclic and Supramolecular Chemistry: How Izatt–Christensen Award Winners Shaped the Field, First Edition.
Edited by Reed M. Izatt.

a macrocycle containing five nitrogen atoms and which was larger than the porphyrins on which I had worked as a graduate student and postdoc. The simplistic thinking was that creating a molecule that embodied key attributes of the state and which "chelated the flag" (cf. Figure 14.1) would be a good way to get tenure. This plan worked.

For reasons of synthetic ease, a decision was made to use imine linkages to connect an *ortho*-phenylenediamine with a diformyl tripyrrane. This condensation, shown in Scheme 14.1, worked well. Unfortunately, the resulting macrocycle, which resembled an expanded porphyrinogen, contained sp^3-hybridized methylene bridges. It did not benefit from the aromatic stability of porphyrins (18 π-electron systems) or display their high molar absorptivity or characteristic red-purple color. In the case of porphyrins, oxidative conversion from the porphyrinogen to porphyrin forms is facile. This did not prove to be the case for the tripyrrane-derived macrocycle. Fortunately, after a couple of nerve-racking years with the tenure clock running, the problem was solved by carrying out an oxidative metal insertion using the relatively large cation, Cd(II). The material obtained, a complex of the so-called sp^2 form, proved to be green, not red, with a lowest-energy, Q-like band above 700 nm. This was consistent with production of an aromatic, porphyrin-like species with a 22 (as opposed to 18) π-electron periphery. The complex was highly stable and the metal cation could not be readily removed. Structural evidence and good fortune allowed publication in the *Journal of the American Chemical Society* (*JACS*) [2], with a highlight in *C&E News* [3]. The latter write-up was where the term "texaphyrin" was first introduced into the literature, whereas the *JACS* paper was the first to contain the term "expanded porphyrin." Expanded porphyrins now represent a very active area of research with many practitioners world-wide making important contributions to the field. A review of this chemistry, including our own, non-texaphyrin-related efforts [4], is, unfortunately, beyond the scope of this chapter.

Early on, and thanks to the help of very talented coworkers, it proved possible to extend rapidly the coordination chemistry of the texaphyrins beyond Cd(II). Particular emphasis was placed on lanthanide cations since these trivalent species do not form stable 1:1 complexes with porphyrins. Moreover, we came to appreciate that the Gd(III) complex might have a role to play as a tumor-localizing MRI contrast agent. As a porphyrin chemist, I was aware that porphyrins typically localize well to tumors and other rapidly proliferating tissues. Indeed, at the time porphyrins were already recognized as highly promising photodynamic therapy (PDT) photosensitizers in part because of this localization ability, as well as their singlet oxygen sensitizing features [5]. Even today, PDT remains a vibrant area of investigation within the porphyrin community.

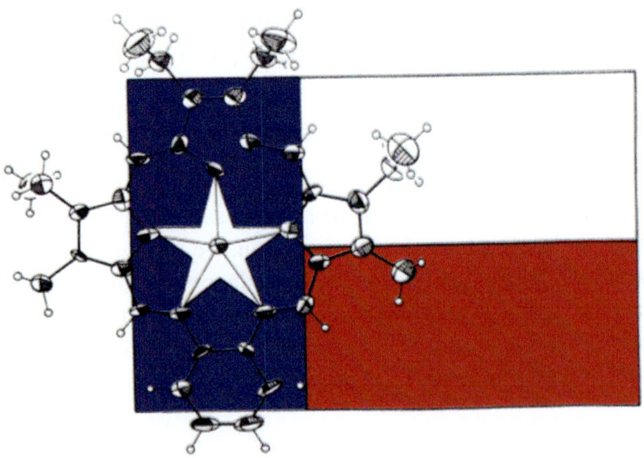

Figure 14.1 *Schematic representation of a texaphyrin molecule and its relationship to the Texas flag. (This figure was reproduced with permission from ref. [1]. Copyright © 2013 American Chemical Society.)*

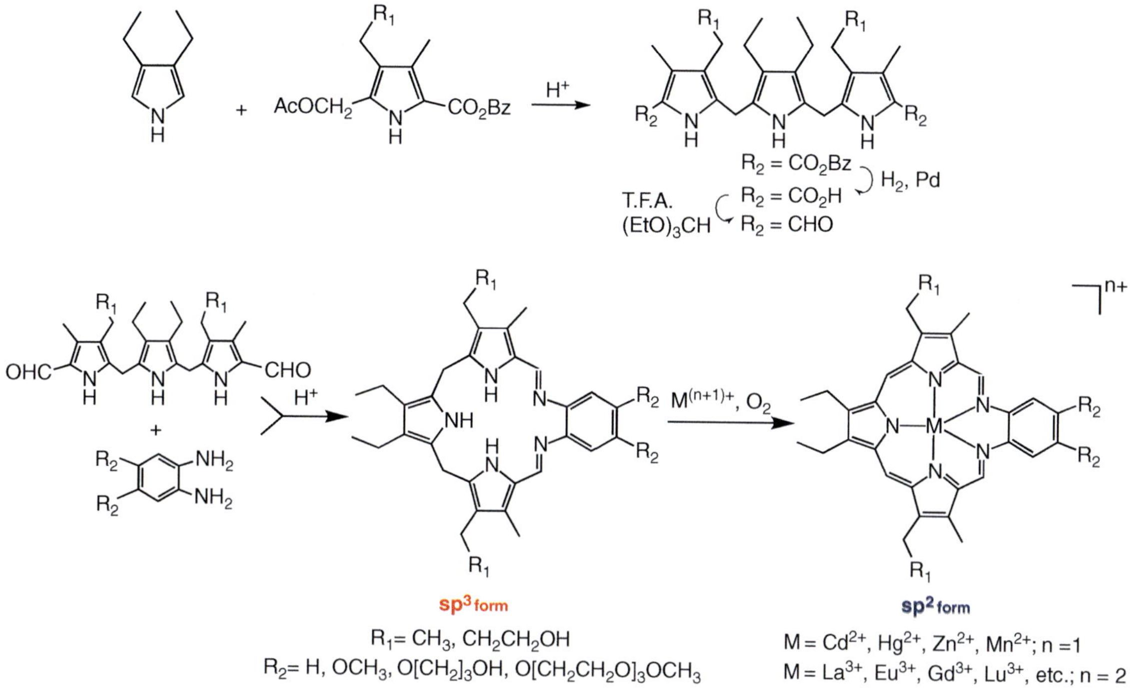

Scheme 14.1 *Generalized synthesis of texaphyrins*

In short order, complexes of all stable lanthanide(III) cations were obtained. A number were characterized by single-crystal X-ray diffraction analysis. This allowed the effects of the lanthanide contraction to be visualized in clear structural terms. Specifically, the extent of out-of-plane displacement and the overall coordination number were found to decrease as the lanthanide series was traversed (Figure 14.2) [6, 7]. About this time, the high relaxivity of the Gd(III) complex was confirmed through magnetic resonance measurements[8] and the ability of the diamagnetic texaphyrin complexes to produce singlet oxygen in high quantum yield was confirmed [9, 10].

14.3 Starting Pharmacyclics, Inc.

Recognizing that ostensibly similar compounds could be used for diagnostics (the Gd(III) complex via MRI) and therapy (the Lu(III) complex via PDT) led to the suggestion that the texaphyrins could serve as dual-function agents, useful for the diagnosis and subsequent treatment of cancerous lesions. This was before the term "theranostics" (also spelt theragnostics) had been coined. An effort to "sell" this concept to the NIH in the form of an RO1 proposal met with abject failure. The priority score received was abysmal (back then all NIH RO1 proposals received a score). In fact, the NIH priority score constituted the worst "grade" I had ever received, with the sole exception of the F given back in the 7th grade for not being able to shoot a basketball lay-up in gym class.

Despite the "feedback" from the NIH, I was convinced that being able to effect diagnostics and therapy in tandem would be useful in cancer drug development, and that texaphyrins would have a particularly useful role in advancing this concept. Fortunately, I had a follow-up appointment with my former oncologist,

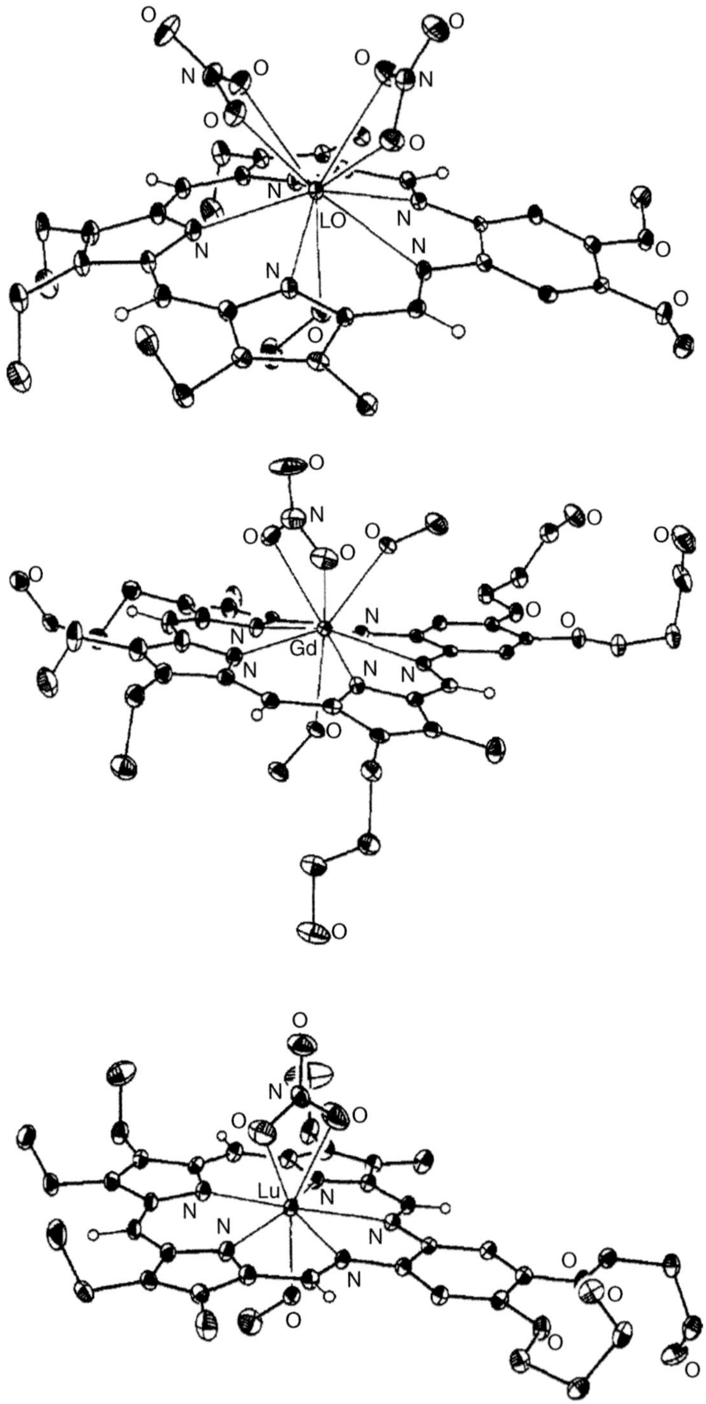

Figure 14.2 *From left to right: Views of the La(III), Gd(III), and Lu(III) complexes of texaphyrins with slightly different peripheral substituents. The bound anions are nitrate and the coordination spheres are completed with ligated methanol molecules. The metal centers are 0.914, 0.595, and 0.269 Å above the mean-square penta-aza plane in these three structures, respectively. (This figure was produced from material published in ref. [7], and used with permission. Copyright © 1994 American Chemical Society.)*

Dr. Richard A. Miller, for the Hodgkin disease that he helped manage back when I was in graduate school at Stanford. When Dr. Miller asked the routine, "How are you?" question I replied, "Physically fine, but I'm emotionally devastated." I explained, "I think I have a really good idea but the NIH pooh-poohed it." Luckily for me, Dr. Miller then asked if I had a copy of the proposal. I was hoping for this response and handed it over. That weekend I was visiting my mother, still alive and living in Berkeley in those days. Dr. Miller scared my mother witless by calling during dinner one evening and asking to speak to me. I still remember the opening lines of that conversation: "Jonathan, your tests are fine. I read your proposal. We are going to start a company."

It took tremendous effort and a couple of years to get the company going. I was given the honor of naming it, and chose "Pharmacyclics, Inc." The company opened for business in Silicon Valley in June of 1991, with Dr. Miller at the helm. As I recall, Dr. Gregory Hemmi, a recent doctoral graduate from my lab at UT Austin, was the first full-time employee. He remained full-time with Pharmacyclics, Inc. through late 2014. Two other doctoral students from my lab, Drs. Tarak D. Mody and Darren J. Magda, also joined Pharmacyclics in the early days. I had an active advisory role while maintaining my full-time appointment at UT Austin.

Early on, considerable effort was devoted to generating a water-soluble texaphyrin whose synthesis was scalable. This was achieved by appending two small PEG subunits to the phenyl moiety and using a tripyrrane bearing two hydroxypropyl substituents (see Figure 14.3). Tests of the basic attributes of the texaphyrin complexes derived from this particular core structure were ongoing actively at this time (early 1990s). These efforts were divided between the UT Austin labs and Pharmacyclics. They were also assisted through a number of collaborations.

Among the many accomplishments codified through this effort were demonstrations of the relatively high relaxivity of the first clinical candidate, a water solubilized gadolinium(III) complex, termed motexafin gadolinium (MGd), as well as the high singlet oxygen quantum yield of the second clinical candidate, a similarly functionalized lutetium texaphyrin complex, referred to as motexafin lutetium (MLu). The tumor localizing capability of these species was also established in animal models via a number of methods, including MRI and fluorescence emission (see below). Kilo-scale synthetic protocols under so-called GMP conditions were developed in conjunction with third parties. Ultimately, this led to MGd and MLu being prepared in 10 kg batches in good yield and in high purity.

On the fundamental research front, it was found that the texaphyrins were exceptionally easy to reduce compared to porphyrins and that the one-electron products could be reoxidized via treatment with oxygen [11]. This led to the consideration that MGd might function as a radiation sensitizer via production of so-called reactive oxygen species and, in due course, an appreciation that metallotexaphyrins could promote a cancer-specific increase in oxidative stress. The underlying chemistry is summarized in Scheme 14.2 and discussed in further detail in several original publications [12–15].

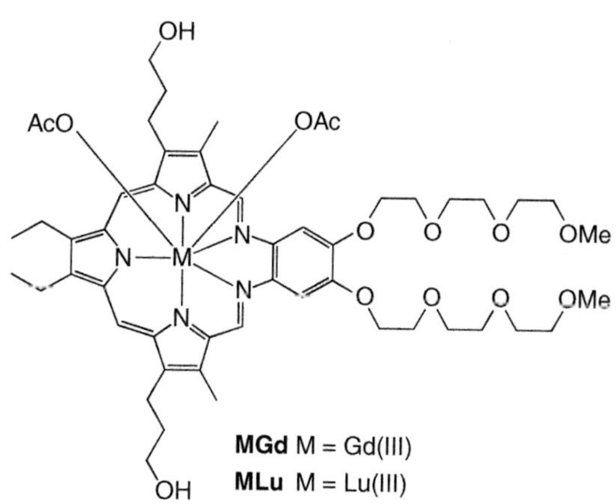

Figure 14.3 *Structures of MGd and MLu*

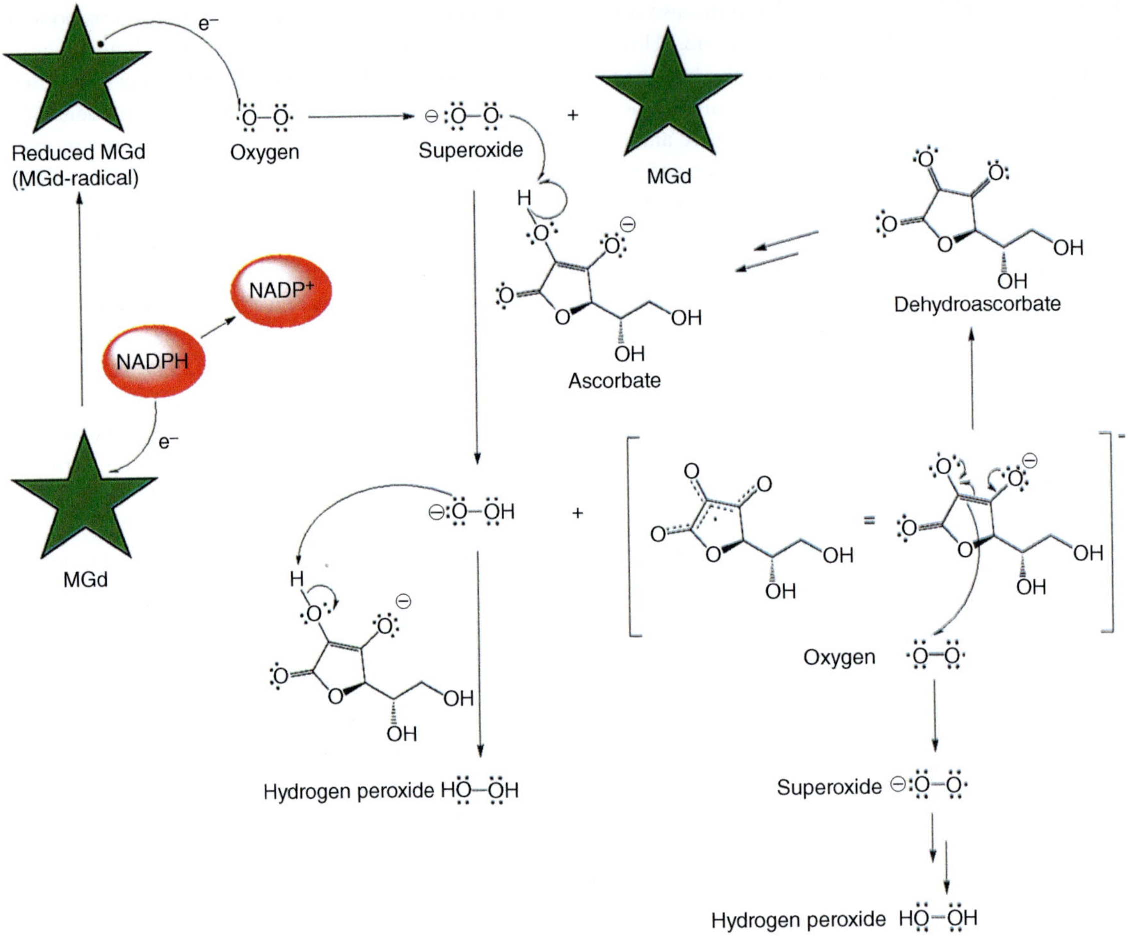

Scheme 14.2 Mechanistic representation of how MGd is thought to act as a redox mediator. (This figure is taken from ref. [1] and is reproduced with permission. Copyright © 2013 American Chemical Society.)

14.4 Early biological studies of texaphyrins

Some of the first biological tests with MGd involved monitoring its uptake with MRI. MGd contains a centrally bound lanthanide(III) cation with seven unpaired electrons. Its "top" and "bottom" faces are also relatively unhindered, allowing for the transient binding of exchangeable water molecules. Taken in aggregate, these attributes serve to enhance the effective spin lattice (T1) relaxation of the water near the MGd complex. In early work, Young et al. showed that gadolinium(III) texaphyrins could be used to enhance the contrast of MR images substantially [16]. This and subsequent studies served to confirm that MGd localized well in tumors, with no appreciable localization in adjacent normal tissues [17, 18]. The ratio of MGd in tumor cells to that in surrounding normal cells was calculated to be as high as 9:1 [19]. This ratio was found to increase to roughly 50:1 in the case of metastatic brain tumors, as inferred from MRI images of patients enrolled in Pharmacyclics' early clinical trials (Figure 14.4) [20, 21]. Presumably, this is the result of a breakdown in the blood–brain barrier.

In contrast to MGd, MLu is diamagnetic. The cancer-localization properties of this potential photodynamic agent were demonstrated early on *in vivo* through optical imaging (cf. Figure 14.5) [22–24]. Later they were

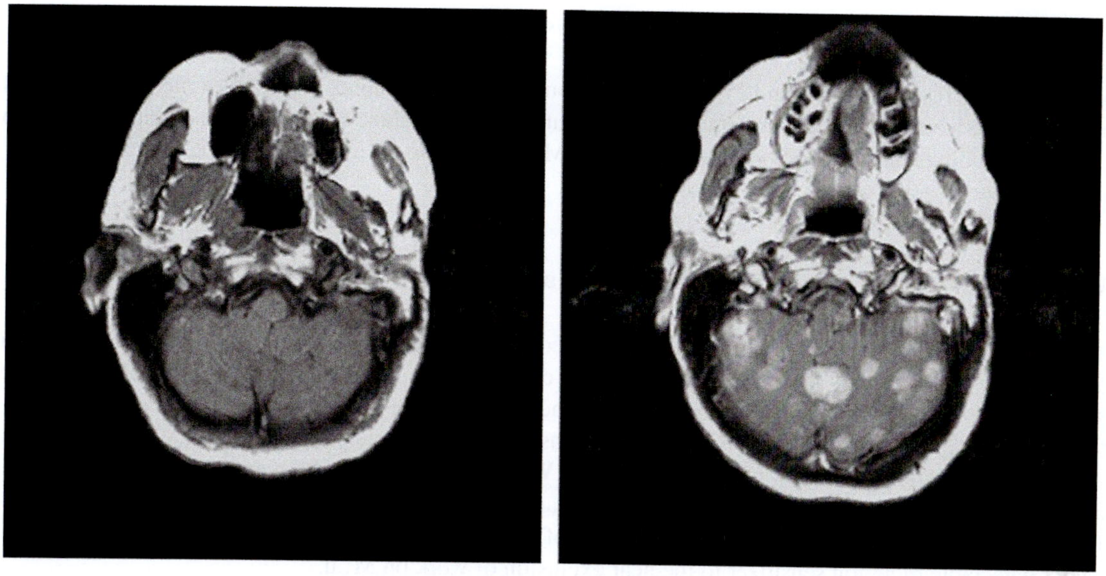

Figure 14.4 *T1-weighted brain MRI scans of a patient with non-small cell lung cancer. Left: non-contrast scan at baseline. Right: scan after the tenth administration of MGd at a dose level of 5 mg kg^{-1} per administration without use of other contrast agent. Note the many small lesions whose visualization indicates the presence of MGd. These data, but not this exact figure, were first published in ref. [21]*

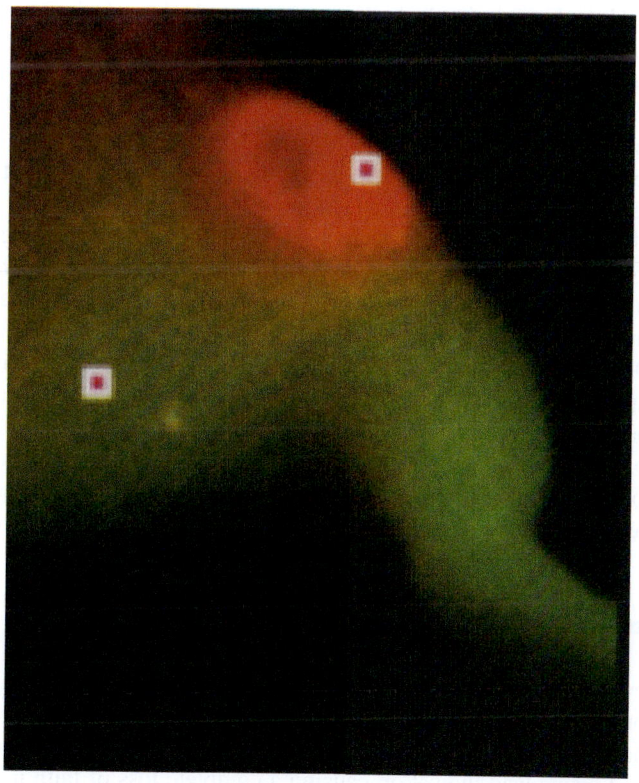

Figure 14.5 *Spectral bioimaging of Lu-Tex in an EMT6-bearing mouse. The mouse received a tail vein dose of 10 mmol Lu-Tex kg^{-1} and was imaged 15 h post-injection. Similar levels of uptake were observed between 1 and 27 hours. These data, but not this exact figure, were first published in ref. [22]*

confirmed using resonance Raman [25]. It was also found that this complex could be excited far to the red (i.e., >700 nm), where tissues are maximally transparent, and still produce singlet in high quantum yield (up to 70%, depending on conditions) [26]. This made MLu attractive as a potential photodynamic therapy photosensitizer [27–30].

14.5 Clinical studies of texaphyrins at Pharmacyclics, Inc.

By the mid-1990s the first human clinical trials were commenced. Pharmacyclics, Inc. then went public on Nasdaq under the call symbol PCYC in 1996. In the ensuing years a large number of clinical trials were instigated under Pharmacyclics' aegis. By the turn of the century, these included tests of MLu as a photosensitizer for photodynamic cancer therapy and catheter-based atherosclerotic plaque ablation, as well as studies of MGd as a radiation sensitizer for cancer therapy. While promising results were obtained using MLu, particularly in "photoangioplasty" (i.e., the photodynamic removal of plaque) [31], economic constraints arising in large part due to the so-called "tech wreck" of 2001, led Pharmacyclics, Inc. to emphasize MGd as an MRI-detectable radiation sensitizer to the near exclusion of work on MLu.

Based on a combination of results from earlier clinical trials and a recognized need for improved therapies, a decision was made at Pharmacyclics, Inc. to focus on testing MGd as a sensitizer for the whole-brain radiation therapy (WBRT)-based treatment of patients with primary non-small cell lung cancer (NSCLC) and metastatic brain tumors. With this clearly defined goal in Pharmacyclics' sights, a Phase III trial, code named the SMART Trial, was launched in 2003. Briefly, 550 patients were to be enrolled at 64 sites in North America and Europe. While more patients would have been desirable, both in prospect and, certainly, in retrospect, in fact, a number of "real world" factors constrained the number of patients that could be included in the SMART Trial. The trial as run was prospective and randomized across treatment (WBRT XRT + MGd) and control (WBRT XRT) arms. The primary endpoint was extending the time to neurological progression, basically the point when lifestyle changes, such as loss of mobility or an inability to eat unassisted, were triggered by the disease.

In late 2005, when the results of the trial became available, it was clear that the statistical requirements, namely a p value <0.05, set by the FDA were not met. A more detailed analysis of the data revealed that 2 of the 64 centers in the study delayed treatment by up two months. On the other hand, good efficacy was seen across the board, with a roughly 3× lengthening in the time to neurological progression being observed in the patients treated at North American centers (where the delays from diagnosis to treatment were minimal). Moreover, no dose-limiting toxicity events were noted. In spite of this detailed understanding of the data and efforts to educate the FDA, the new drug application (NDA) for MGd for use in this indication was rejected late in 2007 [32]. Subsequent independent (but unpublished) statistical analyses led to the suggestion that if as few as 50 additional patients had been included in the SMART Trial, MGd would have been approved. Of course, this is pure *ex post facto* speculation. Still, one cannot help wondering what might have been. There remains a cogent need for improved treatments for NSCLC patients, particularly those suffering from metastatic brain tumors. Currently, there is no FDA-approved radiation sensitizers for this and related indications.

14.6 Changes in direction at Pharmacyclics, Inc.

As a consequence of the SMART Trial results, and in an effort to ensure a backup pipeline, Pharmacyclics, Inc. in-licensed several programs from Celera, Inc. in 2006. The researchers working on these programs were hired as employees at Pharmacyclics, Inc. This helped establish a smooth technology transfer. By 2008, with the FDA rejection of MGd for its lead indication causing a complete loss of faith in texaphyrins across the investment community, the emphasis at Pharmacyclics, Inc. shifted to the development of the in-licensed programs. One of the compounds developed in this context was a covalent Bruton's tyrosine kinase (BTK) inhibitor that evolved to become the FDA-approved drug Imbruvica®. This agent shows particular promise in treating chronic lymphocytic leukemia (CLL), small lymphocytic lymphoma (SLL), mantle cell lymphoma (MCL), diffuse large B-cell lymphoma (DLBCL), and follicular lymphoma (FL). It may have a role in treating multiple myeloma (MM) and is being explored for a number of other indications.

With the demise of the texaphyrin program at Pharmacyclics, Inc. my association with the company ended. Most of the employees associated with the texaphyrin project were let go. This included my former students, Drs. Magda and Mody, both of whom contributed to the development of the Celera-derived technologies before moving on. At about this time, the CEO and President of Pharmacyclics, Inc., Dr. Richard Miller, was replaced by Mr. Robert Duggan under whose direction Imbruvica® was successfully brought to market. That success led to the acquisition of Pharmacyclics, Inc. by AbbVie for $21B in the spring of 2015 at a valuation of over $260/share. I still had a small amount of founders' stock at the time the deal closed (coincidentally, on my 59th birthday).

14.7 Current research efforts involving texaphyrin

For almost a decade now, no research or development work involving texaphyrins has been carried out at Pharmacyclics, Inc. All work with these compounds has been carried out at UT Austin or in collaboration with other academic labs or US National Labs.

Fortunately, well before Pharmacyclics, Inc. transitioned away from the development of MGd, efforts were ongoing to exploit the unique properties of the texaphyrins for drug discovery. Early on, it was found that certain lanthanide complexes of texaphyrin were effective at catalyzing phosphoester hydrolysis. This led to an effort in the 1990s, spearheaded by Magda and coworkers at Pharmacyclics, Inc., to exploit texaphyrins as potential antisense agents. By attaching texaphyrins to DNA sequences, it proved possible to effect cleavage of complementary RNA sequences with good site selectivity [33–36]. New antisense programs, designed to build on this early work, are now being instigated in the UT Austin laboratories.

The re-initiated antisense efforts are exciting. However, most research effort over the last decade has focused on exploiting the tumor-localizing properties of texaphyrins [1]. The ability of texaphyrins to localize to tumors is no longer a subject of conjecture. It has been clinically validated in roughly a thousand patients, as well as confirmed in a large number of published preclinical studies dating to the early days of Pharmacyclics, Inc. While the mechanism of this localization is yet to be elucidated fully, it is thought to reflect receptor-mediated endocytosis. On a purely operational level, and quite independent of mechanistic rationales, texaphyrins hold promise as small molecule, proprietary alternatives to classic localizing entities, such as antibodies and short peptide sequences (e.g. R-G-D), which may be used to effect tumor localization.

One avenue being explored in an effort to exploit the tumor-localizing capabilities of the texaphyrins is to use them to coordinate radionuclides. This could be beneficial in both radiodiagnostics and radiotherapy. With such an objective in mind, methods were worked out that allow for the rapid insertion of both lead and bismuth. Radioactive isotopes of these elements, particularly ^{212}Bi and ^{213}Bi, are attractive as α-emitters [37, 38]. These isotopes demonstrate a strong anticancer cell effect under hypoxic conditions, an effect ascribed to the high linear energy transfer radiation produced ($100\,keV\,\mu m^{-1}$) upon decay [39]. Also attractive as a radioisotope is ^{212}Pb [40]. It has a half-life of 10.64 hours and produces ^{212}Bi as its primary decay product along with a β-particle. Thus, ^{212}Pb could be conceivably used as an in-situ precursor of ^{212}Bi.

Unfortunately, the short half-lives of these lead and bismuth isotopes compels the use of ligands that form stable, non-labile complexes, and which do so rapidly. Texaphyrins are attractive in this regard. In initial work, it was found the use of polar organic solvents allowed for the rapid complexation of both Pb(II) and Bi(III) in their non-radioactive forms [41]. A single-crystal X-ray diffraction analysis of a PEG-free texaphyrin Bi(III) complex revealed a dimeric, μ-oxo structure (Figure 14.6). More recently, in collaboration with researchers at the Los Alamos National Lab, preliminary evidence has been obtained that insertion of radioisotopic forms of these cations is possible under aqueous conditions. Coupled with the tumor-selectivity properties of texaphyrins, these findings lead us to suggest that texaphyrins could emerge as useful candidates for radiotherapy.

14.8 Texaphyrin-platinum conjugates

At present it is considered that the most promising avenue for reviving the texaphyrins as possible therapeutic leads involves their use as tumor-localizing platinum drug delivery agents. Cisplatin and carboplatin (Figure 14.7) are among the most important antitumor agents in the clinic. They have demonstrated significant activity in several cancer types and are generally used in front line therapies by oncologists [42].

Although platinum drugs are curative against testicular cancer, their clinical utility against other cancers remains limited by intrinsic or acquired resistance, which in the case of non-small cell lung cancer (NSCLC), mesothelioma, and ovarian cancer is reflected in low 5-year survival rates of only 5–20% [43]. The biochemical and pharmacologic mechanisms of platinum resistance, particularly in the case of cisplatin (e.g., decreased drug uptake, increased glutathione production, and increased DNA adduct repair), are well

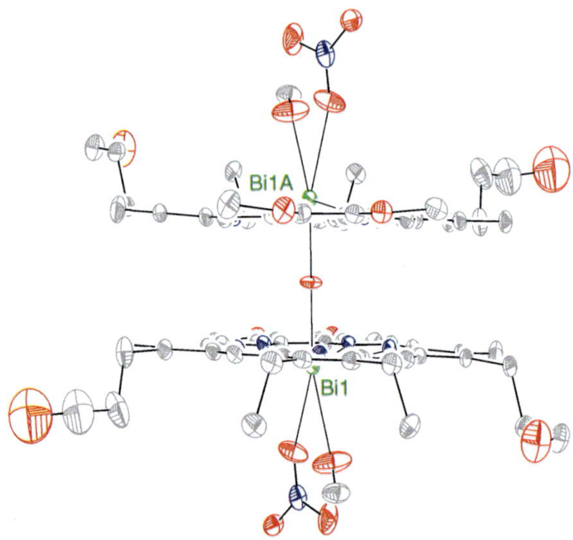

Figure 14.6 X-ray structure of a μ-oxo bismuth(III) texaphyrin complex. These data, but not this exact figure, were first published in ref. [41]

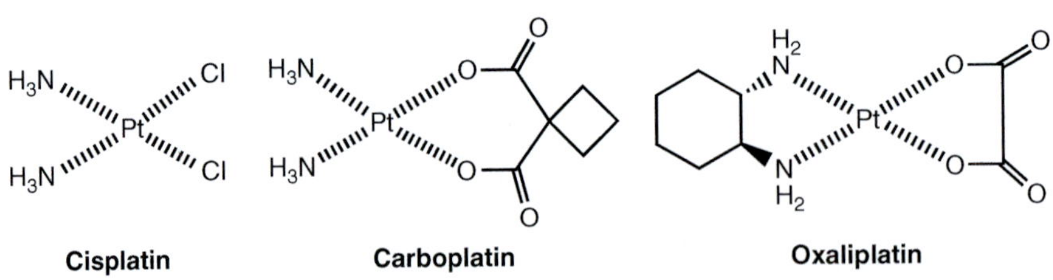

Figure 14.7 FDA-approved platinum drugs

established [42]. These mechanisms co-exist with molecular mechanisms of resistance and present treatment-limiting barriers to the effective application of platinum drugs in the clinic.

In conjunction with Dr. Zahid Siddik, a collaborator at the MD Anderson Cancer Center, we came to appreciate that if the inherent toxicity of a Pt agent could be reduced, more active Pt could be delivered to tumors and that this might translate into an improved clinical outcome. We proposed that these desirable objectives could be attained by developing suitable texaphyrin-Pt conjugates.

After considerable initial research effort and early studies of texaphyrin–platinum constructs that failed to display appropriate solubility characteristics [44], or which failed to overcome DNA-repair mechanisms of resistance [45, 46], an oxaliplatin-inspired conjugate was produced recently (Figure 14.8). This particular construct, termed oxaliTEX was found to be equally effective *in vitro* for both wild-type (A2780) and platinum-resistant (2780CP) ovarian cell lines (see Figure 14.9) [47]. No decrease in DNA–Pt lesions were seen as a function of time, which was taken as evidence that DNA repair mechanisms were being overcome.

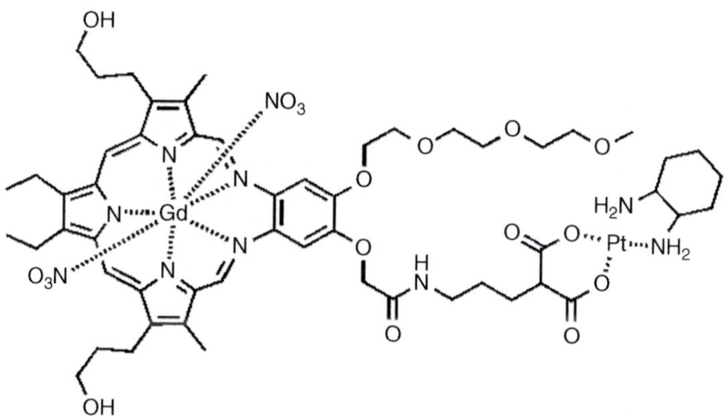

Figure 14.8 *Structure of oxaliTEX14*

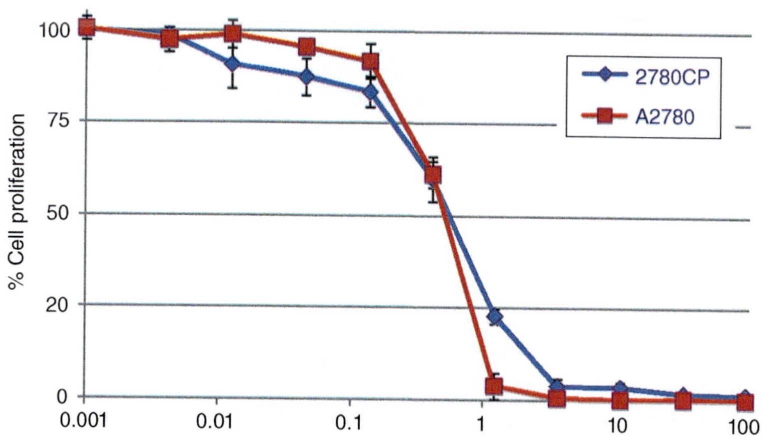

Figure 14.9 *Cytotoxicity profiles of oxaliTEX with the cisplatin-sensitive A2780 and cisplatin-resistant 2780CP cell lines. The complex was made up as a stock solution (for which the Pt concentration was confirmed by FAAS) and serially diluted before adding to cells, which were then incubated for 5 days at 37 °C in 5% CO_2. Error bars represent standard deviations. These data, but not this exact figure, were first published in ref. [47]*

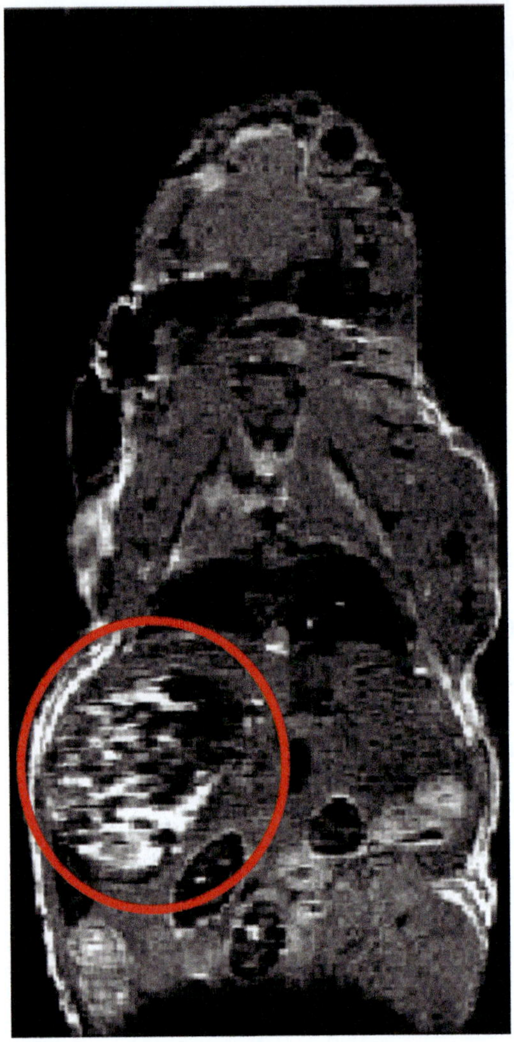

Figure 14.10 MRI of a tumor-bearing mouse recorded after administration of oxaliTEX and sacrificing the animal. The encircled region is the tumor

This latter finding is consistent with reports that diaminocyclohexyl-containing platinum complexes are able to overcome this particular aspect of platinum drug resistance [48, 49].

Currently, biological studies of oxaliTex are in progress. To date, support for the localization of oxaliTex in cancerous lesions has come from an MRI study of a tumor-bearing mouse post-sacrifice (Figure 14.10), while a maximum tolerated dose (MTD) in mice of >100 mg kg^{-1} has been determined. This later MTD value is >6× that of oxaliplatin on a per-platinum basis (>65 vs. 10 μmol kg^{-1}). Preliminary evidence of tumor growth inhibition has also been seen in xenograft mouse models.

New work with a Pt(IV) texaphyrin-conjugate (Figure 14.11) also looks quite promising [50], as do efforts to use texaphyrins to localize cytotoxins, such as methotrexate [51] and doxorubicin, to cancerous sites. Thus, with continued effort it is possible that the texaphyrins could see resurrection as potential clinical candidates. Not surprisingly, a major push is ongoing right now in the UT Austin laboratories with the hope of achieving

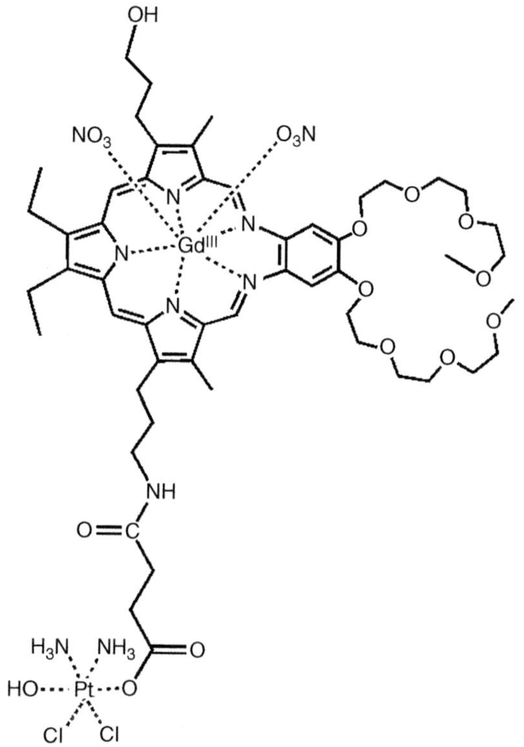

Figure 14.11 *Structure of a recently-prepared Pt(IV) texaphyrin conjugate. See ref. [49] for further details*

this objective. With good luck, hard work, strong financing, and all the intangibles that separate success from failure, the texaphyrins may once again be positioned to see some of their promise realized for the benefit of cancer patients.

14.9 Acknowledgements

This chapter is dedicated to all who have worked with texaphyrins and to the researchers, clinicians, investors, and patients who have given a part of themselves in the unending fight against cancer. Special thanks go to my three former students, Gregory Hemmi, Darren Magda, and Tarak D. Mody who voted with their feet and took jobs at Pharmacyclics, Inc. in the early days, to Richard A. Miller, without whom there would be no Pharmacyclics, and to Robert Duggan and his team who made the company I was lucky enough to co-found a success. Support for my lab from the National Institutes of Health (NCI) and the Cancer Prevention and Research Institute of Texas (CPRIT) is also gratefully acknowledged.

References

1. Preihs, C., Arambula, J., Magda, D. et al. (2013) Recent developments in texaphyrin chemistry and drug discovery, *Inorganic Chemistry*, **52**, 12184–12192.
2. Sessler, J.L., Murai, T., Lynch, V., Cyr, M. (1988) An “Expanded Porphyrin”: The synthesis and structure of a new aromatic pentadentate ligand, *Journal of the American Chemical Society*, **110**, 5586–5588.

3. Unusual porphyrin analog promises many applications, *Chemical & Engineering News,* August 8, 1988, 26–27.
4. Sessler, J.L., Seidel, D. (2003) Synthetic expanded porphyrin chemistry, *Angewandte Chemie*, **115**, 5292–5333; Sessler, J.L., Seidel, D. (2003) *Angewandte Chemie, International Edition*, **42**, 5134–5175.
5. Dougherty, T.J., Gomer, C.J., Henderson, B.W. et al. (1998) Photodynamic therapy. *Journal of the National Cancer Institute*, **90**, 889–905.
6. Sessler, J.L., Mody, T.D., Hemmi, G.W., Lynch, V. (1993) Synthesis and structural characterization of lanthanide(III) texaphyrins, *Inorganic Chemistry*, **32**, 3175–3187.
7. Sessler, J.L., Hemmi, G., Mody, T.D. et al. (1994) Texaphyrins: Synthesis and applications, *Accounts of Chemical Research*, **27**, 43–50.
8. Geraldes, C.F.G.C., Sherry, A.D., Vallet, P. et al. (1995) Nuclear magnetic relaxation dispersion studies of a water-soluble gadolinium(III)–texaphyrin complex, *Journal of Magnetic Resonance Imaging*, **5**, 725–729.
9. Harriman, A., Maiya, B.G., Murai, T. et al. (1989) Metallotexaphyrins: A new family of photosensitizers for efficient generation of singlet oxygen, *Journal of the Chemical Society, Chemical Communications*, 314–316.
10. Sessler, J.L., Hemmi, G., Maiya, B.G. et al. (1991) Tripyrroledimethine-derived ("texaphyrin"-type) macrocycles: Potential photosensitizers which absorb in the far-red spectral region, *Proceedings of the SPIE; The International Society for Optical Engineering*, **1426**, 318–329.
11. Sessler, J.L., Tvermoes, N.A., Guldi, D.M. et al. (2001) Pulse radiolytic studies of metallotexaphyrins in the presence of oxygen: Relevance of the equilibrium with superoxide anion to the mechanism of action of the radiation sensitizer motexafin gadolinium (Gd–Tex^{2+}, Xcytrin®), *Journal of Physical Chemistry B*, **105**, 1452–1457.
12. Guldi, D.M., Mody, T.D., Gerasimchuk, N.N. et al. (2000) Influence of large metal cations on the photophysical properties of texaphyrin, a rigid aromatic chromophore, *Journal of the American Chemical Society*, **122**, 8289–8298.
13. Magda, D., Lepp, C., Gerasimchuk, N. et al. (2001) Redox cycling by motexafin gadolinium enhances cellular response to ionizing radiation by forming reactive oxygen species, *International Journal of Radiation Oncology *Biology *Physics*, **51**, 1025–1036.
14. Magda, D., Lepp, C., Gerasimchuk, N.; Lecane, P. et al. (2002) Motexafin gadolinium reacts with ascorbic acid to produce reactive oxygen species, *Chemical Communications,* 2730–2731.
15. Magda, D.J., Sessler, J.L., Gerasimchuk, N., Miller, R.A. Gadolium(III) texaphyrin (Xcytrin®): A new class of redox active drug leads, In *Medicinal Inorganic Chemistry*, J.L. Sessler, S. Doctrow, T. McMurry, S.J. Lippard (Eds), American Chemical Society Symposium Series 903, Oxford, Oxford University Press, 2005.
16. Young, S.W., Sidhu, M.K., Qing, F. (1994) Preclinical evaluation of gadolinium(III) texaphyrin complex: A new paramagnetic contrast agent for magnetic resonance imaging, *Investigative Radiology*, **29**, 330–338.
17. Viala, J., Vanel, D., Meingau, P. et al. (1999) Phases IB and II multidose trial of gadolinium texaphyrin, a radiation sensitizer detectable at MR Imaging: preliminary results in brain metastases, *Radiology*, **3**, 755–759.
18. Rosenthal, D.I., Nurenberg, P., Becerra, C.R. et al. (1999) A phase I single-dose trial of gadolinium texaphyrin (Gd–Tex), a tumor-selective radiation sensitizer detectable by magnetic resonance imaging, *Clinical Cancer Research*, **5**, 739–745.
19. Miller, R.A., Woodburn, K., Fan, Q. et al. (1999) *In vivo* animal studies with gadolinium(iii) texaphyrin as a radiation enhancer, *International Journal of Radiation Oncology *Biology *Physics*, **45**, 981–989.
20. Carde, P., Timmerman, R., Mehta, M.P. et al. (2001) Multicenter phase Ib/II trial of the radiation enhancer motexafin gadolinium in patients with brain metastases, *Journal of Clinical Oncology*, **19**, 2074–2083.
21. Mehta, M.P., Shapiro, W.R., Glantz, M.J. et al. (2002) Lead-in phase to randomized trial of motexafin gadolinium and whole-brain radiation for patients with brain metastases: Centralized assessment of magnetic resonance imaging, neurocognitive, and neurologic end points, *Journal of Clinical Oncology.* **20**, 3445–3453.
22. Woodburn, K.W., Fan, Q., Miles, DR. et al. (1997) Localization and efficacy analysis of the phototherapeutic lutetium texaphyrin (PCI-0123) in the murine EMT6 sarcoma model, *Photochemistry and Photobiology*, **65**, 410–415.
23. Rockson, S.G., Lorenz, D.P., Cheong, W.-F., Woodburn, K.W. (2000) Photoangioplasty an emerging clinical cardiovascular role for photodynamic therapy, *Circulation*, **102**, 591–596.
24. Mody, T.D., Sessler, J.L. (2001) Texaphyrins: A new approach to drug development, *Journal of Porphyrins and Phthalocyanines*, **5**, 134–142.
25. Synytsya, A., Král, V., Matejka, P. et al. (2004) Biodistribution assessment of a lutetium(iii) texaphyrin analogue in tumor-bearing mice using NIR Fourier-transform Raman spectroscopy, *Photochemistry and Photobiology*, **79**, 453–460.

26. Sessler, J.L., Miller, R.A. (2000) Texaphyrins. New drugs with diverse clinical applications in radiation and photodynamic therapy, *Biochemical Pharmacology*, **59**, 733–739.
27. Young, S.W., Woodburn, K.W., Wright, M. et al. (1996) Lutetium texaphyrin (PCI-0123): A near-infrared, water-soluble photosensitizer, *Photochemistry and Photobiology*, **63**, 892–897.
28. Woodburn, K.W., Fan, Q., Kessel, D. et al. (1996) Phototherapy of cancer and atheromatous plaque with texaphyrins, *Journal of Clinical Laser Medicine & Surgery*, **14**, 343–348.
29. Sessler, J.L., Dow, W.C., O'Connor, D. et al. (1997) Biomedical applications of lanthanide(III) texaphyrins. Lutetium(III) texaphyrins as potential photodynamic therapy photosensitizers, *Journal of Alloys and Compounds*, **249**, 146–152.
30. Woodburn, K.W., Fan, Q., Kessel, D. et al. (1998) Photodynamic therapy of B16F10 murine melanoma with lutetium texaphyrin, *Journal of Investigative Dermatology*, **110**, 746–751.
31. Rockson, S.G., Kramer, P., Razavi, M. et al. (2000) Photoangioplasty for human peripheral atherosclerosis: Results of a phase I trial of photodynamic therapy with motexafin lutetium (Antrin), *Circulation*, **102**, 2322–2324.
32. *FDA News* Drug Daily Bulletin, Cancer drug not approvable, US FDA Says, December 27, 2007; Pharmacyclics, Inc. (PCYC) Cancer drug not approvable says US FDA, December 26, 2007, < http://www.biospace.com/news_print.aspx?NewsEntityId=81304> Accessed February 25, 2016.
33. Magda, D., Miller, R.A., Sessler, J.L., Iverson, B.L. (1994) Site-specific hydrolysis of RNA by europium(III)–texaphyrin conjugated to a synthetic oligodeoxynucleotide, *Journal of the American Chemical Society*, **116**, 7439–7440.
34. Magda, D., Wright, M., Miller, R.A., Sessler, J.L., Sansom, P.I. (1995) Sequence-specific photocleavage of DNA by an expanded porphyrin with irradiation above 700 nm, *Journal of the American Chemical Society*, **117**, 3629–3630.
35. Magda, D., Crofts, S., Lin, A. et al. (1997) Synthesis and kinetic properties of ribozyme analogues prepared using phosphramidite derivatives of dysprosium(III) texaphyrin, *Journal of the American Chemical Society*, **119**, 2293–2294.
36. Magda, D.M., Wright, M., Crofts, S., Lin, A., Sessler, J.L. (1997) Metal complex conjugates of antisense DNA which display ribozyme-like activity, *Journal of the American Chemical Society*, **119**, 6947–6948.
37. Kozak, R.W., Atcher, R.W., Gansow O.A. et al. (1986), Bismuth-212-labeled anti-TAC monoclinal antibody: α-particle emitting radionucleotides as modalities for radioimmunotherapy, *Proceedings of the National Academy of Sciences USA*, **83**, 474–478.
38. Brechbiel, M.W., Pippin, C.G., McMurry, T.J. et al. (1991) An effective chelating agent for labeling of monoclonal antibody with bismuth-212 for α-particle mediated radioimmunotherapy, *Journal of the Chemical Society, Chemical Communications*, **1991**, 1169–1170.
39. Zalutsky, M.R., Pozzi, O.R. (2004) Radioimmunotherapy with α-particle emitting radionuclides, *Quarterly Journal of Nuclear Medicine and Molecular Imaging*, **48**, 289–296.
40. Kumar K., Magerstaedt M., Gansow, O.A. (1989) Lead(II) and bismuth(III) complexes of the polyazacycloalkane-N-acetic acids 1,4,7-triazacyclononane-N,N′,N″-triacetic acid, 1,4,7,10-tetraazacyclododecane-N,N′,N″,N‴-tetraacetic acid, and 1,4,8,11-tetraazacyclotetradecanetetraacetic acid, *Journal of the Chemical Society, Chemical Communications*, 145–146.
41. Preihs, C., Arambula, J.F., Lynch, V.M., Siddik, Z.H., Sessler, J.L. (2010) Bismuth- and lead-texaphyrin complexes: Towards potential α-core emitters for radiotherapy. *Chemical Communications*, **46**, 7900–7902.
42. Siddik, Z.H. (2003) Cisplatin: mode of cytotoxic action and molecular basis of resistance, *Oncogene*, **22**, 7265–7279.
43. American Cancer Society, *Cancer Facts and Figures 2005*; ACS: Atlanta, GA; 2011.
44. Magda, D.J., Wang, Z., Gerasimchuk, N. et al. (2004) Synthesis of texaphyrin conjugates, *Pure and Applied Chemistry*, **76**, 365–374.
45. Arambula, J.F., Sessler, J.L., Fountain, M. et al. (2009), Gadolinium texaphyrin (Gd–Tex)–malonate–platinum conjugates: Synthesis and comparison with carboplatin in normal and Pt-resistant cell lines, *Journal of the Chemical Society, Dalton Transactions*, **48**, 10834–10840.
46. Arambula, J.F., Siddik, Z., Sessler, J.L. (2011), Overcoming biochemical pharmacologic mechanisms of platinum resistance with a texaphyrin–platinum conjugate, *Bioorganic & Medicinal Chemistry Letters*, **21**, 1701–1705.

47. Arambula, J., Sessler, J.L., Siddik, Z.H, (2012) A texaphyrin–oxaliplatin conjugate that overcomes both pharmacologic and molecular mechanisms of cisplatin resistance in cancer cells. *Medicinal Chemistry Communications*. **3**, 1275–1281.
48. Kelland, L.R., Sharp, S.Y., O'Neill, C.F. et al. (1999), Mini-review: discovery and development of platinum complexes designed to circumvent cisplatin resistance, *Journal of Inorganic Biochemistry*, **77**, 111–115.
49. Heiger-Bernays, W.J., Essigmann, J.M., Lippard, S.J. (1990) Effect of the antitumor drug *cis*-diamminedichloroplatinum(II) and Related platinum complexes on eukaryotic DNA replication, *Biochemistry*, **29**, 8461–8466.
50. Thiabaud, G., Arambula, J.F., Siddik, Z.H., Sessler, J.L. (2014) Photo-induced reduction of Pt(IV) within an anti-proliferative Pt(IV)–texaphyrin conjugate allows for controlled DNA ligation, *Chemistry – A European Journal*, **20**, 8942–8947.
51. Wei, W.-H., Fountain, M., Magda, D. et al. (2005), Gadolinium texaphyrin–methotrexate conjugates. Towards improved anticancer chemotherapeutic agents, *Organic & Biomolecular Chemistry*. **3**, 3290–3296.

15

Macrocyclic Coordination Chemistry of Resorcin[4]arenes and Pyrogallol[4]arenes

Harshita Kumari,[1] Carol A. Deakyne,[2] and Jerry L. Atwood[2]

[1] *James L. Winkle College of Pharmacy, University of Cincinnati, Cincinnati, OH, USA*
[2] *Department of Chemistry, University of Missouri–Columbia, Columbia, MO, USA*

15.1 Introduction

Self-assembled supramolecular nanocapsules were first originated in the late 1990s [1–6]. One of the remarkable discoveries during this decade was extension of metal-coordination chemistry to supramolecular synthons, an extension that led to construction of large spherical assemblies with discrete cavities [1–15]. The architectures of the early nanoframeworks conformed to Platonic and Archimedean solids and suggested endless possibilities with respect to formation of non-covalently bonded and metal-seamed nanoarchitectures. Pyrogallol[4]arenes and resorcin[4]arenes are two such synthons that have shown the potential to self-assemble in a variety of architectures, including capsules, tubes, toroids, ellipses and bilayers [16–59]. The observed architectures are similar in some cases and unique in others: for example, spherical capsules are primarily symmetric; however, asymmetric, oblate and ellipsoidal structures have been obtained. The type of metal, metal salt, and metal coordination sites dictate the outcome of nanoassembly. Other factors, such as solvent and temperature, are also crucial in forming nanoassemblies. The current book chapter discusses the metal-coordination chemistry of pyrogallol[4]arenes and resorcin[4]arenes.

Macrocyclic and Supramolecular Chemistry: How Izatt–Christensen Award Winners Shaped the Field, First Edition.
Edited by Reed M. Izatt.

15.2 History of hydrogen-bonded pyrogallol[4]arene- and resorcin[4]arene-based nanocapsules

The bowl-shaped macrocycles of pyrogallol[4]arenes and resorcin[4]arene are sister derivatives of calixarenes. Pyrogallol[4]arenes are synthesized from the acid condensation reaction of aldehydes with pryogallol/1,2,3-trihydroxybenzene in 95% ethanol. Similarly, resorcin[4]arenes are synthesized by the acid condensation reaction of aldehydes with resorcin/1,3-dihydroxybenzene in 95% ethanol, refluxed over a period of hours to days (depending on the type of aldehyde used) [60]. The unique hydroxyl groups at the upper rim of the bowls provide interaction sites for hydrogen bonding as well as metal complexation (Figure 15.1).

Metal complexation of pyrogallol[4]arenes was preceded by the formation of hydrogen-bonded frameworks [42]. The bowl shape of pyrogallol[4]arene and resorcin[4]arene was first utilized to form hydrogen-bonded nanocapsules. The first hydrogen-bonded nanocapsule of reseocin[4]arene was discovered in 1997 by Atwood and MacGillivray [2]. This unique first-generation nanocapsule is made up of 14 subunits (six *C*-methylresorcin[4]arenes/RsC$_1$ and eight water molecules), which are stitched with 60 hydrogen bonds. The RsC$_1$nanocapsular assembly adopts a snub cuboctahedron architecture and has an internal cavity of 1375 Å^3 [2]. This geometric arrangement led to the use of Platonic and Archimedean solids as a basis for the construction of discrete supramolecular nanoassemblies.

The second generation of nanocapsules was synthesized from the complexation of the sodium salt of a *p*-sulfonatocalix[4]arene with pyridine N-oxide and a lanthanide (III) nitrate/cadmium (II) salt [61]. Varying stoichiometries of the macrocycle, base and metal salt yielded spheroidal and tubular nanoassemblies. Substitution of 18-crown-6 with pyridine-N-oxide caused a structural alteration from an icosahedron to a cuboctahedron geometry. Unlike the first-generation capsule, which primarily consists of hydrogen bonds, the second-generation capsules consist of covalent coordinate bonds to metal ions.

The third generation of nanocapsules evolved from *C*-alkylpyrogallol[4]arenes. Mattay and co-workers synthesized the first hexameric nanocapsule of *C*-isobutylpyrogallol[4]arene [60]. Later on, several hexamers of pyrogallol[4]arene with different alkyl tail lengths were reported [17]. Compared to the resorcin[4]arene-based nanocapsule, which exhibited 60 hydrogen bonds, the pyrogallol[4]arene-based nanocapsule exhibited 72 hydrogen bonds. The PgC-hexameric nanocapsules were seamed by 24 intramolecular and 48 intermolecular hydrogen bonds. The hexamer had an internal volume of 1500 Å^3 and could accommodate a variety of guests, such as pyrene, pyrene butyric acid, water, acetone, etc [28, 62].

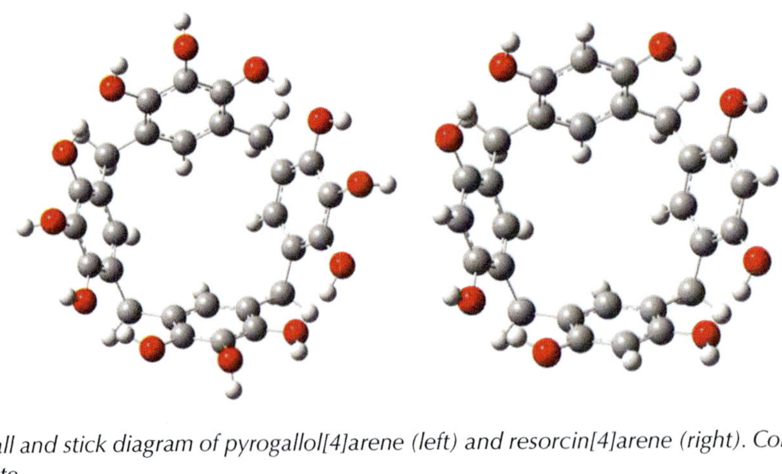

Figure 15.1 *Ball and stick diagram of pyrogallol[4]arene (left) and resorcin[4]arene (right). Color codes: O: Red, C: Grey, H: White*

The fourth generation of nanocapsules was built from mixed macrocycles of pyrogallol and resorcinol [22]. The synthesis involved an acid condensation reaction of isovaleraldehyde with an equimolar ratio of pyrogallol and resorcin in 95% ethanol at RT. The reaction yielded the mixed macrocycle pyrogallol[3]resorcin[1]arene, which was then crystallized from diethylether to form the hexameric hydrogen-bonded nanocapsule [22]. The resultant mixed macrocylic hydrogen-bonded hexamer has six diethyl ether molecules positioned inside and six positioned outside of the hexamer, and the nanoassembly has an internal cavity of 860 Å^3. Unlike pyrogallol[4]arene, which has 12 hydroxyls at the upper rim, the mixed macrocycle has 11 hydroxyl groups that allow for 66 hydrogen bonds within the hexamer. Of these, 48 hydrogen bonds are involved in intermolecular [24] and intramolecular [24] binding. The remaining 18 hydrogen bonds involve hydroxyl coordination with diethyl ether molecules (six oriented inside and six outside). The hexameric capsule conforms to the shape of a trigonal antiprism constructed from the centroids of the centers of the rings of the pyrogallol[3]resorcin[1]arenes [22].

Several researchers have studied the structures and properties of hydrogen-bonded pyrogallol[4]arene and resorcin[4]arene nanocapsules [63–67]. The properties that have been of the most interest include the orientation of guest molecules within the nanocapsule, stability of the host–guest complex in solution, mechanism of formation of the nanocapsule and transport of guest molecules across the nanocapsule. Solid-state geometries of hydrogen-bonded frameworks may not remain intact in solution upon dissolution. The solvent interactions could potentially compete with hydrogen-bonding interactions that hold the nanoassembly together [8]. For example, in the case of hydrogen-bonded resorin[4]arene nanocapsules, water or alcohol molecules are a part of the framework, as evidenced by diffusion NMR [63, 64, 68]. However, water molecules are not a part of the pyrogallol[4]arene framework in solution. Cohen and coworkers have done extensive work to elucidate the free and bound guest/solvent molecules and have shown the formation of hexameric nanocapsules both in the absence and presence of guests other than solvent molecules [63–67]. The escape of the guest from the resorcin[4]arene hexamer in solution causes the complete disappearance of the resorcin[4]arene moiety. In addition to guest escape, guest orientation and contortion have also been investigated by Rebek et al [9].

15.3 Metal-seamed pyrogallol[4]arene- and resorcin[4]arene-based complexes

Hydrogen-bonded nanocapsules assume the shape of a snub cube (resorcin[4]arene), a rhombicuboctahedron (pyrogallol[4]arene), or a trigonal antiprism (pyrogallol[3]resorcin[1]arene). Of these, the hydrogen-bonded hexameric nanocapsules of *C*-alkylpyrogallol[4]arenes (third generation) served as molecular blueprints for the copper-seamed pyrogallol[4]arene-based nanocapsules (fifth generation) and later on for several other types of metal-seamed hexamers [69].

15.3.1 Copper-seamed pyrogallol[4]arene hexamers

The hexameric copper-seamed nanocapsules were synthesized by mixing 1 equivalent of *C*-propan-3-ol-pyrogallol[4]arene with 4 equivalents of copper nitrate in an acetone/water mixture [69]. Slow evaporation of the solution yielded red-colored single crystals suitable for X-ray diffraction. The single-crystal structure of copper-seamed *C*-propan-3-ol-pyrogallol[4]arene revealed the formation of a hexamer composed of six macrocyclic units and 24 copper centers (Figure 15.2). Each copper center adopts an octahedral geometry and is coordinated with four of the hydroxyl groups of adjacently positioned macrocycles and two water molecules.

The hexameric nanocapsule is held together by eight Cu_3O_3 triads resulting from the deprotonation of 48 pyrogallol hydroxyl groups (Figure 15.3). The remaining 24 hydroxyl groups are hydrogen bonded to the pyrogallols of adjacent macrocycles, connecting the robust nanocapsular framework. The geometry of the hexamer is similar to that of a hypothetical truncated octahedron that consists of eight triangular and six octagonal facets. Each triangular facet of the octahedron is occupied by a Cu_3O_3 triad.

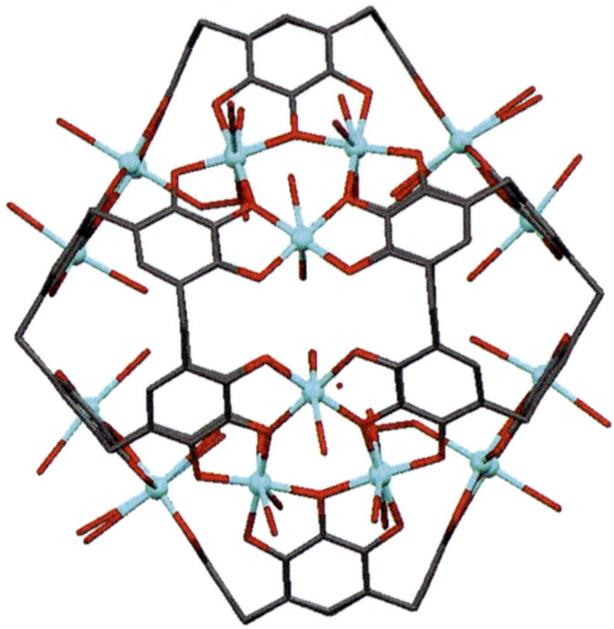

Figure 15.2 Copper-seamed pyrogallol[4]arene hexamer. For clarity, alkyl tails and hydrogen atoms have been deleted. Color codes: Cu: Turquoise; O: Red; C: Grey. Adapted from ref. [69]

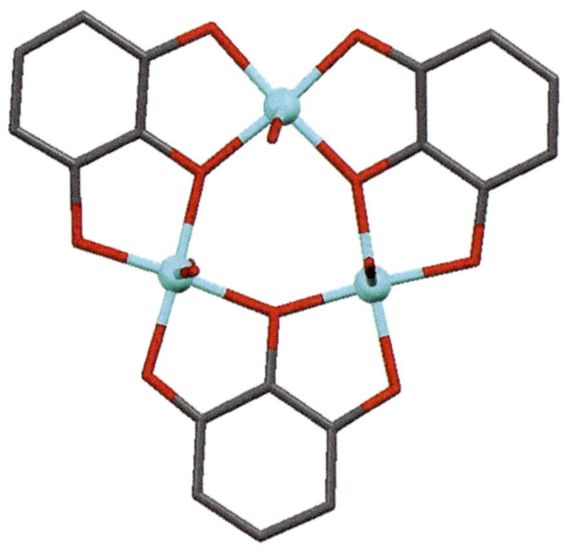

Figure 15.3 Cu_3O_3 triad of the copper-seamed pyrogallol[4]arene based hexamer. Color codes: Cu: Turquoise; O: Red; C: Grey. Adapted from ref. [69]

The copper-seamed pyrogallol[4]arene hexamer was the first metal-seamed pyrogallol[4]arene-based nanocapsule reported by Atwood et al. More recently, the solution stability of copper-seamed nanocapsules was analyzed using small-angle neutron scattering (SANS) [70].

SANS analyses revealed the structural parameters (shape and size) of the copper-seamed *C*-alkylpyrogallol[4]arene (PgC_nCu, n=alkyl tail length) nanocapsule in acetone [70]. For n=3, 6 and 9 in PgC_nCu, the solution geometries were spherical and the nanometric dimensions were close to those obtained in the solid state [70]. For n=11 or in $PgC_{11}Cu$, the solution geometry was ellipsoidal [71]. The geometric dimensions were large and indicated formation of a super-assembly of ~20 $PgC_{11}Cu$ nanocapsular units with an overall ellipsoidal arrangement facilitated by the interdigitation of the long alkyl tails [71].

15.3.2 Zinc-seamed pyrogallol[4]arene dimers

In 2007, a novel zinc-seamed dimeric nanocapsule of pyrogallol[4]arene was reported [40]. The dimeric nanocapsule was synthesized by treating a methanolic solution of pyrogallol[4]arene with a methanolic solution of a pre-prepared zinc pyridine complex in the molar ratio of 1:4. The resultant yellow precipitate was filtered and dissolved in DMSO, and the subsequent solution was slowly evaporated for several days to obtain single crystals of the PgC_nZn dimeric nanoassembly. The dimeric nanoassembly contains two pyrogallol[4]arene units, eight zinc centers and eight external pyridine ligands (Figure 15.4).

Unlike the octahedral coordination of the copper hexamer, the zinc centers in the dimer adopt a square pyramidal coordination through an external pyridine/DMSO ligand and four deprotonated hydroxyls, two from each of the oppositely positioned pyrogallol[4]arene subunits. The eight zinc centers form an equatorial belt that seams the two hemispheres of the nanocapsule. The internal volume of a dimeric nanocapsule (~250 Å^3) is much smaller than that of a hydrogen-bonded or metal-seamed hexamer; however, the dimer is shown to accommodate small guest molecules, such as pyridine. The zinc-seamed nanocapsule has also been tested for facile ligand exchange through dialysis and azeotropic vacuum distillation of DMSO with 3-methylpyridine [41]. The dimeric nanocapsule of $(PgC_n)_2Zn_8$(3-methylpyridine)$_8$ ⊂ (3-methylpyridine) was synthesized by the traditional method of treating the zinc-3-methylpyridine complex with the macrocycle in a 4:1 ratio and the resultant moiety was tested for ligand-exchange properties. ^{1}H NMR analysis showed a quick exchange of 3-methylpyridine ligands with DMSO and a higher affinity for DMSO over 3-methylpyridine upon titration. In addition, in most

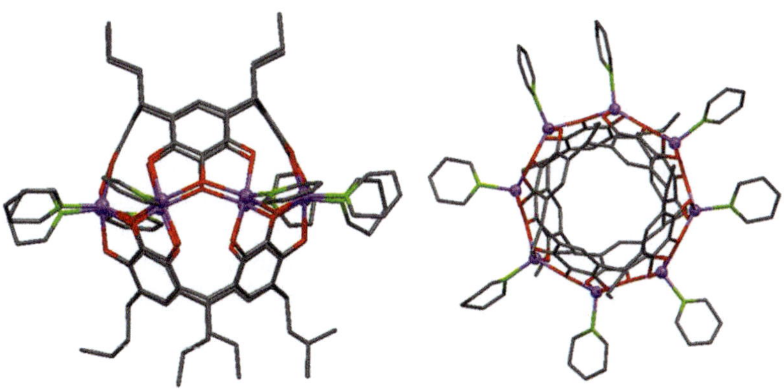

Figure 15.4 *Front view (left) and top view (right) of the zinc-seamed C-propylpyrogallol[4]arene dimer. For clarity, hydrogen atoms and solvent molecules have been deleted. Color codes: N: Green; Zn: Lavender; C: Grey; O: Red. Adapted from ref. [40]*

cases a pyridine guest molecule, in this case 3-methylpyridine, is shown entrapped within the dimeric nanocapsule. Ligand exchange does not affect the type of guest enclosed, indicating the robustness of the zinc-seamed dimers. Similarly to the copper-seamed PgC hexamer, the zinc-seamed dimers were found to be stable in solution (SANS). The PgC_3Cu hexamers were shown to have a radius of 10 Å, whereas a typical PgC_3M dimer has a radius of 7Å [52]. The zinc dimer has a slightly larger radius than a conventional dimer, yet is smaller than a hexamer, due to the presence of ligands around the equatorial belt and their fast exchange in solution.

Zinc–ligand complexation was also investigated through quantum chemical calculations. Given the large size of zinc-seamed nanocapsules (120 atoms, R=H, no axial ligands), electronic structure calculations were done on mononuclear and multinuclear zinc complexes, representing the metal environment in the capsules [72]. Initially, monomeric zinc hydroxide complexes of the form $Zn(OH)_2X_2Y$ and $Zn(OH)_2X_{1,2}Y_2$, where X and Y were combinations of H_2O, CH_3OH, NH_3 and C_5H_5N (with the exception of NH_3/C_5H_5N), were studied. A preference for nitrogen ligands in the inner shell of the complexes, and a $C_5H_5N > NH_3 > CH_3OH$ trend in inner-shell ligand-binding affinities, were observed from MP2/B2-pp//M05-2x/B2-pp calculations. The higher relative binding affinity of C_5H_5N versus that of CH_3OH is consistent with experiment, wherein a substitution of methanol with pyridine for zinc-seamed pyrogallol[4]arene is not observed during synthesis [72].

15.3.3 Zinc-seamed pyrogallol[4]arene tethering complexes

The zinc dimers were not only shown to exist as discrete nanocapsules but also were shown to act as building blocks that complexed with 4,4′-bipyridine to form 2D metal–organic frameworks [73]. The type of linker and solvent were guided by the electronic structure calculations on zinc model complexes. 4,4′-bipyridine and DMSO have similar Zn–L binding strengths; however, there is a slight reduction in binding strength when 4,4′-bipyridine is linked to a second Zn complex. The resultant MOF has several voids and channels as well as unusual coordination geometries along the periphery of the nanocapsule. The latter property leads to changes in the overall internal volume of the capsule and in the behavior of the enclosed guest (Figure 15.5).

15.3.4 Copper-seamed *C*-methylpyrogallol[4]arene dimer coordination polymers

In addition to the zinc-seamed dimers, copper-seamed *C*-alkylpyrogallol[4]arene-based dimers act as building blocks to form infinitely long coordination polymers, either by direct coordination or with additional bridging ligands (Figure 15.6) [49].

15.3.5 Gallium-seamed pyrogallol[4]arene complexes

The hexameric and dimeric capsules were not just limited to copper- and zinc-seamed pyrogallol[4]arenes. Soon after, a gallium-seamed pyrogallol[4]arene nanocapsule was discovered [47, 55]. Unlike the copper-seamed nanocapsule, which adopts a near spheriodal/truncated cuboctahedron geometry, the gallium-seamed-*C*-butylpyrogallol[4]arene (PgC_4Ga) hexamer adopts a rugby-ball shaped geometry in the solid state. The gallium-seamed pyrogallol[4]arene hexamer is made up of six pyrogallol[4]arene units and 12 gallium centers that seam four out of the six pyrogallol[4]arene subunits. The remaining two pyrogallol[4]arene units have water molecules bridging with hydroxyls and forming water gates of the hexamer. The gallium centers in the hexamer were shown to have octahedral geometry with two water coordination sites pointing in and out of the nanocapsule and four hydroxy-coordinated sites with pyrogallol[4]arenes [55].

The unique solid-state geometry of the gallium-seamed pyrogallol[4]arene nanocapsule led to the discovery of several other mixed-metal nanocapsules. Addition of a zinc nitrate solution to a solution of pre-formed gallium-seamed *C*-butylpyrogallol[4]arene hexamers led to the formation of a PgC_4GaZn hexamer, which adopted a spherical geometry similar to that of the PgC_nCu hexamers. SANS measurements on both PgC_4Ga

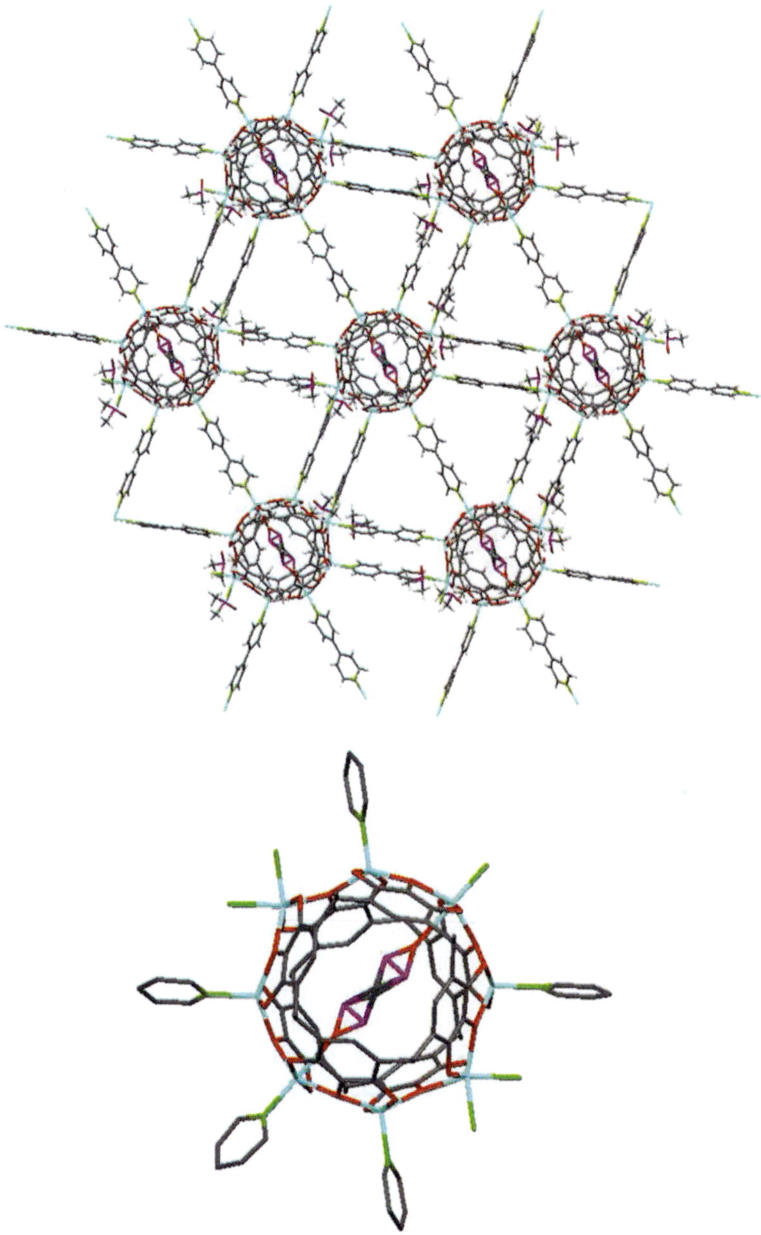

Figure 15.5 *Tethering of zinc dimers via 4,4′-bipyridine (Top); Coordination site of tethered zinc dimer showing enclosed guest and external ligands (Bottom). Adapted from ref. [73]*

and PgC_4GaZn, however, showed structural alterations in solution [74]. Both the PgC_4Ga (solid-state rugby ball) and PgC_4GaZn (solid-state sphere) hexamers changed to toroidal geometries of different dimensions in solution. The different arrangement of the bowls in PgC_4Ga and PgC_4GaZn in solution was confirmed via NMR spectroscopy [55]. In the case of PgC_4Ga, the bowls are arranged with the hydroxyl groups exposed to

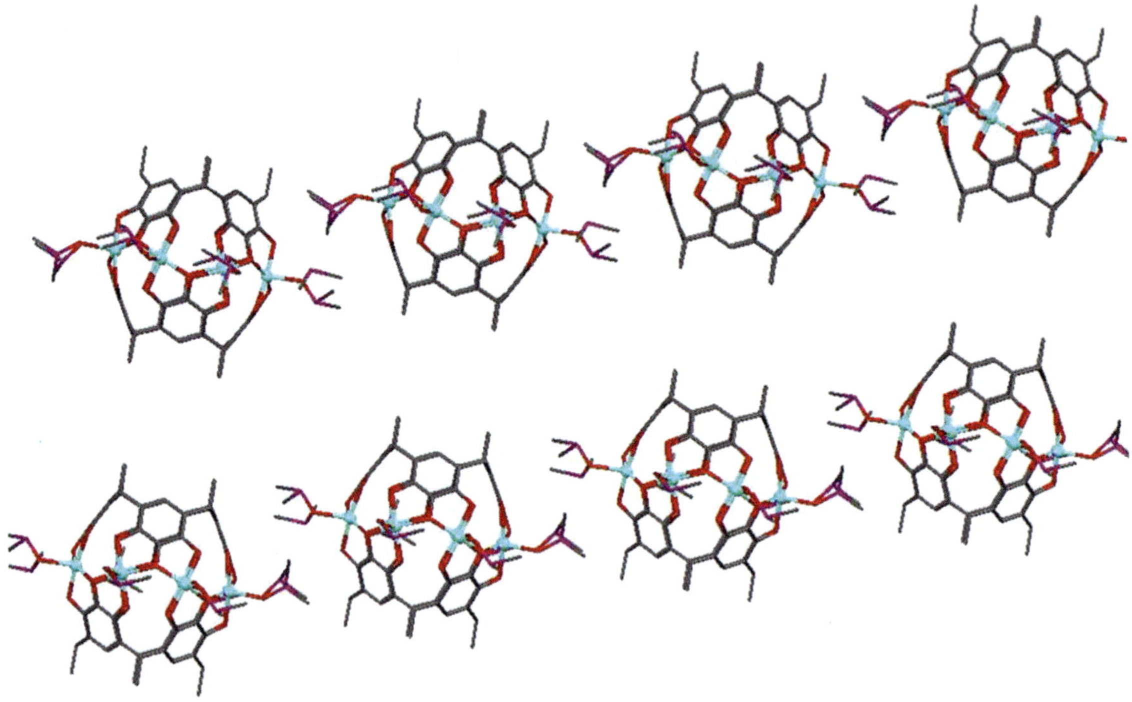

Figure 15.6 *Front view of copper-seamed C-methylpyrogallol[4]arene dimers arranged in the form of polymers. For clarity, hydrogen atoms and solvent molecules have been deleted. Color code: Cu: Turquoise; S: Pink; O: Red; C: Grey. Adapted from ref. [49]*

the solvent molecules, whereas in the case of PgC_4GaZn, the bowls are arranged with the hydroxyl groups forming the inner wall of the proposed toroidal nanoassembly. The bowls in PgC_4GaZn are connected via Ga ions and the nanoassembly has an enclosed acetone guest [55].

The transition in geometry of PgC_4Ga from a solid-state rugby ball to a solid-state sphere for the mixed-metal PgC_4GaZn hexamer was also observed for other transition metal centers (Cu, Co, Ni) [46, 47]. Unlike the case with transition metals, addition of cadmium nitrate to gallium-PgC_4 hexamers led to the formation of a unique canted spherical hexamer (Figure 15.7) [75]. The bigger size of the Cd centers causes one of the six pyrogallol[4]arene units to reside lower in the cavity than the other pyrogallol[4]arene units. Both the Cd and Ga centers have octahedral coordination sites. Ga coordinates with four pyrogallol hydroxyls and two water molecules, whereas Cd coordinates with four pyrogallol hydroxyls and two nitrates. The nitrate and water molecules are positioned outside and inside the nanocapsule (Figure 15.7).The unique cadmium(II)nitrate ion pairs contribute to the deviation from the usual spheroidal geometry that is found for the bimetallic Ga/Cd-PgC_4 hexamer [75].

These intriguing solid-state geometric transitions also led to the discovery of the heretofore elusive transport of cations (Ag^{+1}, Cs^{+1} and Tl^{+1}) into the PgC_4Ga nanocapsules. Complexes of $PgC_4Ga \subset AgNO_3$, $PgC_4Ga \subset Cs$ and $PgC_4Ga \subset Tl$ ions were obtained by treating an acetonic solution of pre-formed PgC_4Ga hexamers with silver nitrate, cesium nitrate or thallium nitrate ethanolic solutions (Figure 15.8) [44, 57, 58]. The rugby-ball-shaped, bimetallic, water-gated hexamers of $PgC_4Ga \subset Cs$ and $PgC_4Ga \subset Tl$ were then shown to be stitched by Zn/Rb/K ions to form tri-metallic nanocapsules that adopted a more spherical geometry [55].

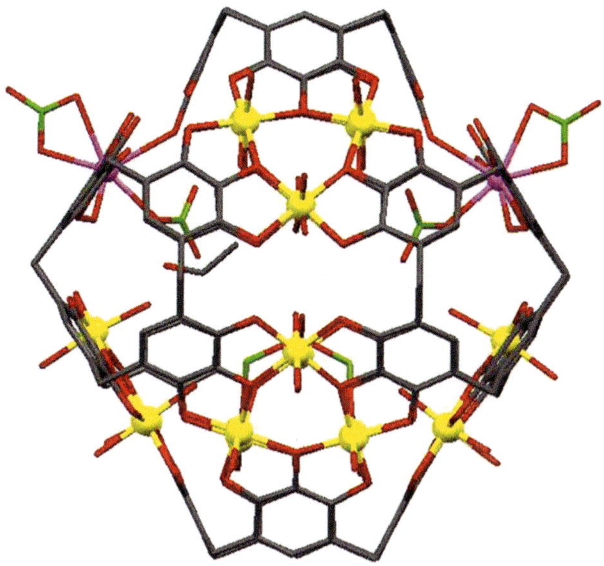

Figure 15.7 *Front view of the gallium/cadmium-seamed C-alkylpyrogallol[4]arene hexamer. For clarity, hydrogen atoms, alkyl tails and solvent molecules have been deleted. Color codes: Cd: Pink; Ga:Yellow; C: Grey; O: Red; N: Green. Adapted from ref. [75]*

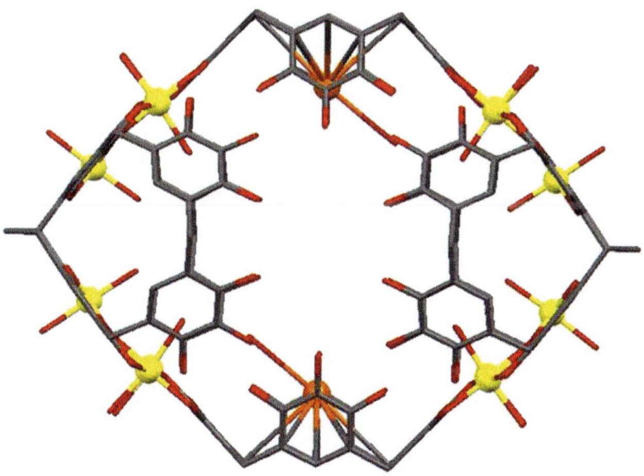

Figure 15.8 *Front view of the gallium-seamed C-alkylpyrogallol[4]arene hexamer with enclosed silver ion. For clarity, hydrogen atoms, alkyl tails and solvent molecules have been deleted. Color codes: Ag: Orange; Ga: Yellow; C: Grey; O: Red. Adapted from ref. [55]*

15.3.6 Nickel-seamed pyrogallol[4]arene dimers and hexamers

More recently, nickel-seamed hexamers and dimers were also formed from pryogallol[4]arene complexation. Treatment of a methanolic solution of nickel nitrate salts with a methanolic solution of pyrogallol[4]arenes and pyridine in the ratio of 4:1:14, respectively, led to the formation of dimers and hexamers. The nickel-seamed PgC_n

hexamer is structurally similar to the copper-seamed pyrogallol[4]arene hexamer, and consists of six pyrogallol[4]arene units and 24 nickel centers. The coordination environment of the nickel is also similar to that of the copper, with two water and four pyrogallol hydroxyl coordination sites per metal center. The copper- and nickel-seamed PgC dimers have the same geometry as that of the zinc dimer, with square pyramidal metal centers [52].

Notably, for both metals (Cu and Ni), a mixture of dimers and hexamers was obtained from the same synthesis protocol. This mixture led to the investigation of controlling the type of pyrogallol[4]arene-based nanocapsule formed in the synthesis [52]. A series of in-situ SANS experiments were performed at varying temperatures (RT, 50 °C, –40 °C), solvents (acetone and methanol) and metals (Cu and Ni) to deduce the dimer–hexamer equilibrium shifts. Detailed solution structural analyses using SANS revealed that the hexamers were preferentially obtained at low temperatures, whereas the dimers were obtained at higher temperature. Three-day extended stability studies on these systems showed a shift in volume fraction from hexamers to dimers for all conditions, suggesting that the hexamers are the kinetic product at low temperatures, whereas the dimers are the thermodynamic product at all temperatures (Figure 15.9). This study proved very crucial in hexamer-dimer separations as well as in synthesizing novel nanocapsular entities.

15.3.7 Ferrocene-containing pyrogallol[4]arene complexes

Apart from zinc-, copper-, and nickel-seamed dimers, hydrogen-bonded dimers and nanotubes containing ferrocene as the guest molecule were also reported [76, 77]. Ferrocene was used as a template and was encapsulated within pyrogallol[4]arene-based dimers and nanotubes. Albeit, the iron did not complex with pyrogallol[4]arene, the adjacent Fe metal centers for both nanotubes and dimers showed wave-function overlap and anti-ferromagnetic interactions at low temperatures (Figure 15.10) [78].

Similar magnetic behavior was observed for nickel-seamed hexamers and dimers [78]. The adjacent nickel centers in the dimers and hexamers were found to have weak anti-ferromagnetic interactions; however, the

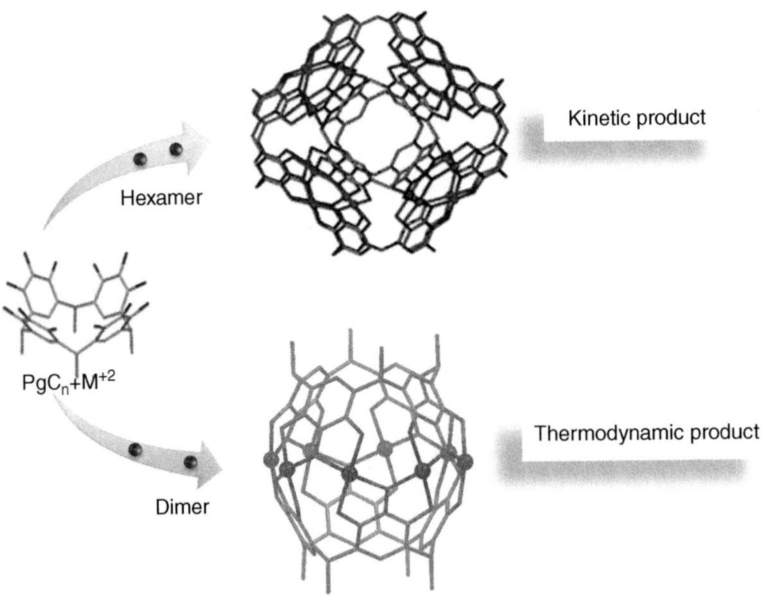

Figure 15.9 *Schematic for thermodynamic (dimer) versus kinetic (hexamer) product of metal-seamed pyrogallol[4]arene nanoassemblies*

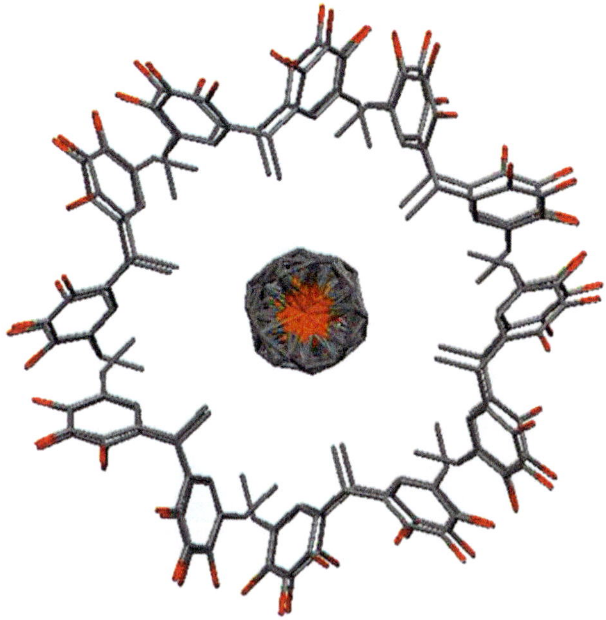

Figure 15.10 Top view of a ferrocene-enclosing C-methylpyrogallol[4]arene nanotube. For clarity, hydrogen atoms and solvent molecules have been deleted. Color codes: C: Grey, O: Red; Fe: Orange. Adapted from ref. [78]

interaction was stronger than that observed for the ferrocene-enclosing PgC_1 nanotubes and dimers. Interestingly, the ferrocene-enclosing PgC_1 dimers were found to be stable in solution but the ferrocene-enclosing hydrogen-bonded nanotubes rearranged to dimers in solution, indicative of weak intramolecular interactions in hydrogen-bonded nanotubes of pyrogallol[4]arenes.

15.3.8 Iron-seamed pyrogallol[4]arene nanotubes

The iron-containing nanotubes, on the other hand, were found to be stable in solution (Figure 15.11) [79]. However, in-situ SANS measurements on the iron complex of pyrogallol[4]arene showed that it undergoes a structural change from a nanotube to a dimer and then to a hexamer upon a change in pH. Also, centrifugation of iron nanotubes in solution leads to the shortening of the nanotubes.

15.3.9 Holmium-seamed pyrogallol[4]arene dodecamer

A holmium complex of pyrogallol[4]arene was synthesized by treating holmium nitrate with pyrogallol[4]arene and pyridine [59]. The resulting blue precipitate, although indicative of the formation of a nanoassembly, was difficult to crystallize. SANS studies revealed the solution structure of PgC_nHo. A sphere of 18.2 Å in radius was observed in solution. Based on the size of pyrogallol[4]arene, the coordination sites and the internal volumes of various Archimedean solids, a truncated cuboctahedron geometry was reported for the nanoassembly, which consisted of 12 pyrogallol[4]arene subunits. The dodecahedron of PgC_3Ho was shown to have three times the volume of a traditional hexameric nanocapsule. Each side of the hexagon in the proposed dodecahedron is about 8.5 Å, which corresponds to either the upper rim diameter of the

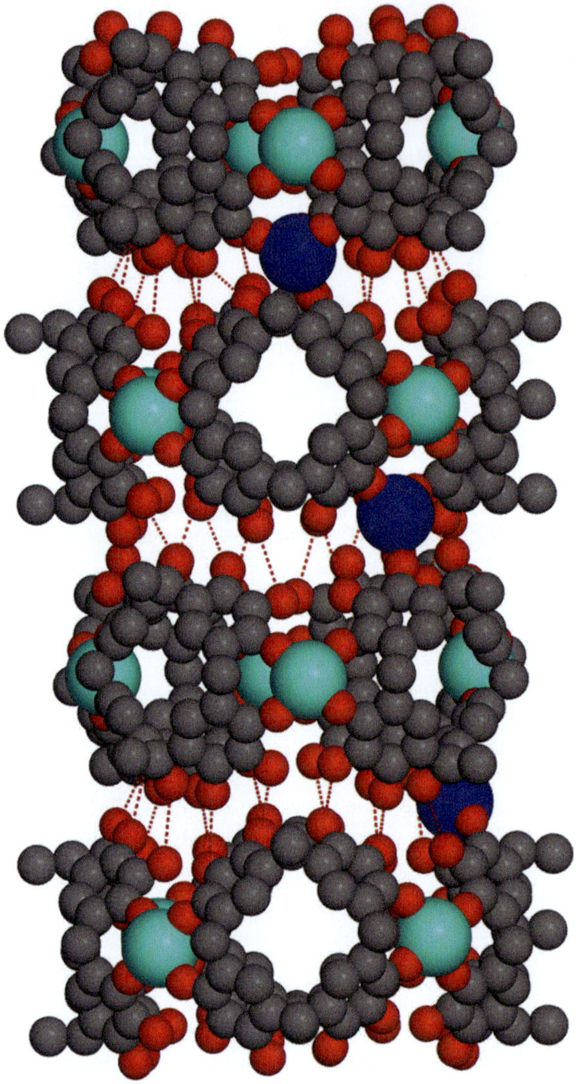

Figure 15.11 Front view of an iron-containing pyrogallol[4]arene nanotube. Blue filled circles: Fe. Adapted from ref. [79]

pyrogallol[4]arene or a pyrogallol-(H)O··Ho··NO_3^{-1}...O(H)-pyrogallol distance (Figure 15.12). Thus far,this is the largest observed metal-seamed nanocapsule of pyrogallol[4]arene in solution [59].

15.3.10 Insulin-enclosing hydrogen-bonded resorcin[4]arene nanocapsules

The largest nanocapsule of resorcin[4]arene, however, is a hydrogen-bonded insulin-encapsulating entity (with a radius of 16 Å) [80]. Insulin was used as a biotemplate to induce paneling of *C*-methylresorcin[4] arene around it. The spherical structure of the [RsC_1 ⊂ insulin monomer] nanocapsule was studied in solution by contrast variation experiments on SANS (Figure 15.13) [80].

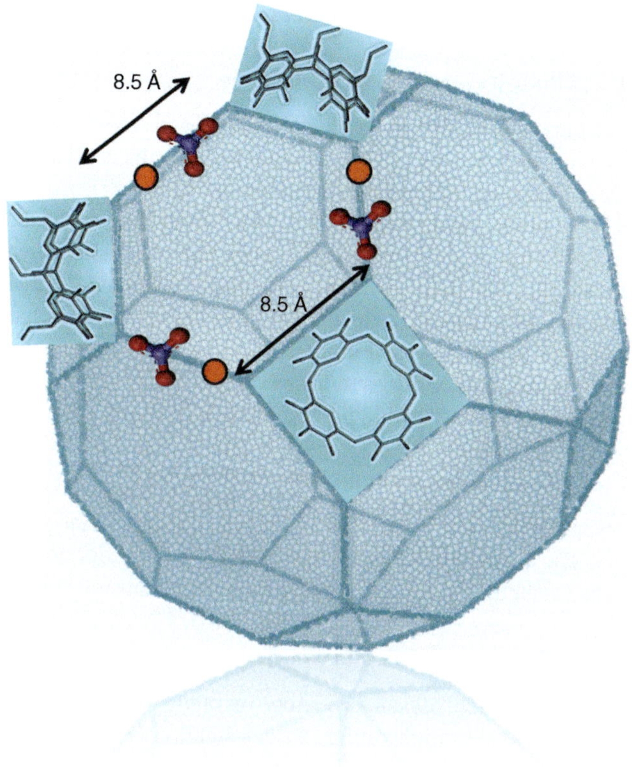

Figure 15.12 *Front view of a holmium-seamed pyrogallol[4]arene dodecamer. Adapted from ref. [59]*

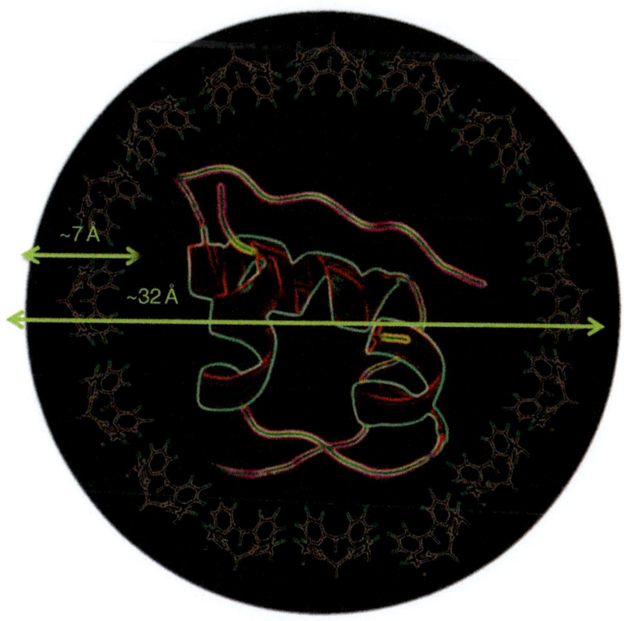

Figure 15.13 *Solution structure of the resorcin[4]arene based nanocapsule, showing an insulin monomer as an encapsulate. Adapted from ref. [80]*

15.3.11 Zinc-seamed pyrogallol[3]resorcin[1]arene dimer

Symmetric dimeric nanocapsules (Cu/Ni/Zn) of pyrogallol[4]arenes provide a robust framework; however, the magnetic properties of such complexes are not pronounced because the even number of metal atoms leads to cancellation of the dipolar moments of oppositely aligned metal centers. Efforts were driven towards introducing defects into dimeric/hexameric frameworks which could yield interesting magnetic properties. With this in mind, a novel zinc-seamed pyrogallol[3]resorcin[1]arene macrocycle was synthesized. The nanocapsule is composed of two units of a 3 pyrogallol:1 resorcinol mixed macrocycle and seven metal centers [81]. The septa-metallated zinc dimer has an opening close to the resorcin end that is charge-balanced by nitrates and could serve as a gate for exchange of ions (Figure 15.14). The external pyridine ligands on the seven metal centers give the metal centers a square pyramidal geometry; however, the two pyridines on the zinc centers close to the resorcin end of the macrocycle point outwards [81].

15.3.12 Ionic dimeric Cs-containing pyrogallol[4]arene nanocapsule

In 2007, Dalgarno and Power reported the formation of an ionic dimeric nanocapsule that hosts Cs atoms along the walls of the dimer. The alkyl tail of the pyrogallol[4]arene was functionalized with a chloride and then treated with cesium salts to obtain this unique ionic dimeric nanocapsule (Figure 15.15) [39].

15.3.13 Zirconium, silver, and iron-seamed resorcin[4]arene complexes

Compared to pyrogallol[4]arenes, which present a variety of metal-seamed complexes, resorcin[4]arenes have relatively few examples of metal complexation. Direct metal coordination with resorcin[4]arene has only been achieved with zirconium, iron, and silver; however, in all cases, discrete nanocapsular entities were not formed, due to the absence of the central hydroxyl group. The zirconium(IV) resorcin[4]arene complex was reported by Solari et al. in 1996 [82]. A square-shaped array of Zr(IV) and Fe(II) ions coordinated to a

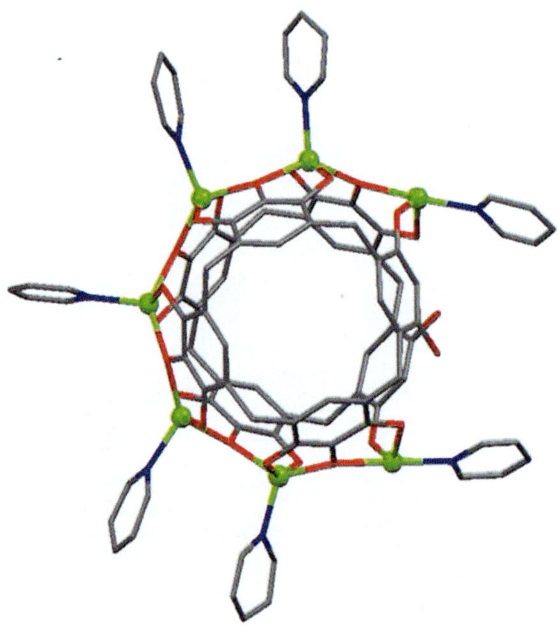

Figure 15.14 Top view of a zinc-seamed pyrogallol[3]resorcin[1]arene dimer with pyridine and nitrate as external ligands. For clarity, alkyl tails, hydrogen atoms and solvent molecules have been deleted. Adapted from ref. [81]

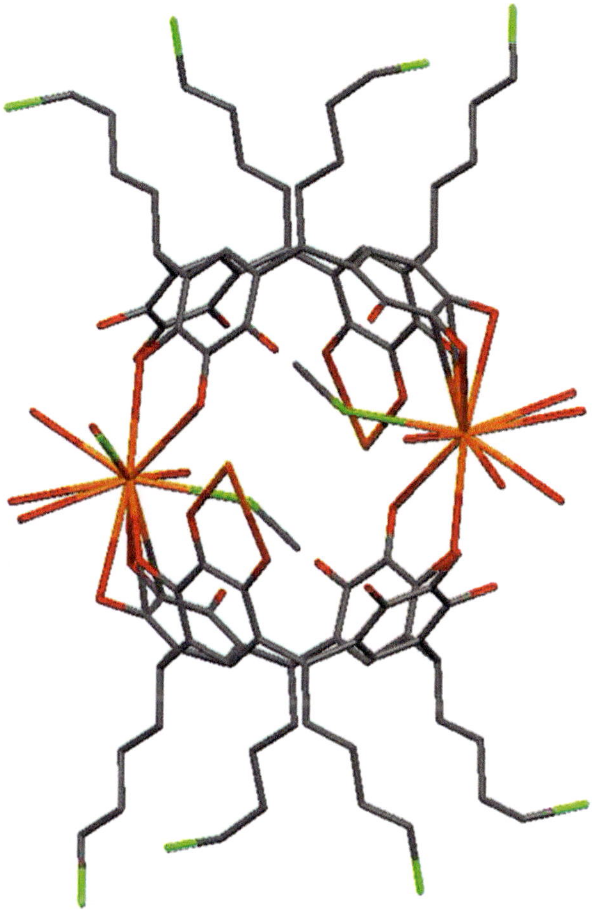

Figure 15.15 *Cs-seamed C-chlorobutylpyrogallol[4]arene dimeric nanocapsule. For clarity, hydrogen atoms and solvent molecules have been deleted. Color codes: Cs: Orange; Cl: Green; C: Grey; O: Red. Adapted from ref. [39]*

polyoxo matrix was obtained by the metal complexation ($[Zr(\eta\text{-}C_5H_5)_2Me_2]$ and $[Fe(phen)(mes)_2]$) with *C*-hexylresorcin[4]arenes. One macrocycle is coordinated to four zirconium metal centers, with each metal center coordinated to hydroxyl groups of adjacent resorcin units of the macrocycle (Figure 15.16) [82].

Munakata et al. reported a silver(I) complex of resorcin[4]arene with a metal:macrocycle ratio of 2:1. The metal centers coordinate with hydroxyl groups of oppositely facing resorcinol units of the same bowl [83]. Such a conformation is facilitated by the pinching of the boat stereoisomer of the macrocycle. The complex is formed via the dinuclearcation of $[Ag_2(L^1)(C_6H_6)_2][ClO_4]_2.5C_4H_8O$ and the hydroxyl groups of the resorcin[4]arenes. The silver metal center bridges between the resorcin[4]arene units, which leads to an infinite 2D framework, and the polyoxo anion connects between the silver ions and covers the cavity of the macrocycle (Figure 15.17) [83].

The other examples of direct metal coordination of resorcin[4]arene were assisted by the presence of additional ligands at the central carbon position of the resorcinol. The ligands include bromide, cyanide, sulfonate, carboxylic acid and pyridine [84]. One example is the complexation of *C*-hexylcyanoresorcin[4]arene with silver(I) ions, wherein the coordination of silver ions between layers of resorcin[4]arenes was facilitated by the cyano groups (Figure 15.18) [84].This coordination led to the formation of a unique 2D framework with channel spaces suitable for solvent and gas storage.

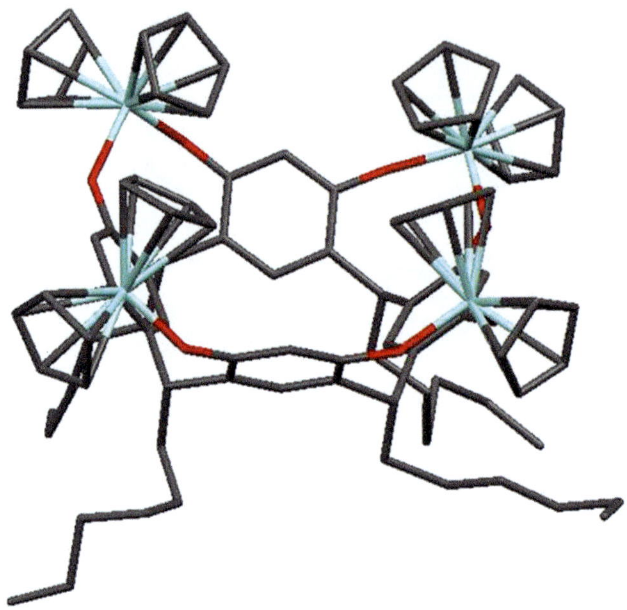

Figure 15.16 Front view of the zirconium-resorcin[4]arene complex showing metal coordination with the hydroxy groups of adjacent resorcin units. For clarity, hydrogen atoms and solvent molecules have been deleted. Color codes: Zr: Turquoise; O: Red; C: Grey. Adapted from ref. [82]

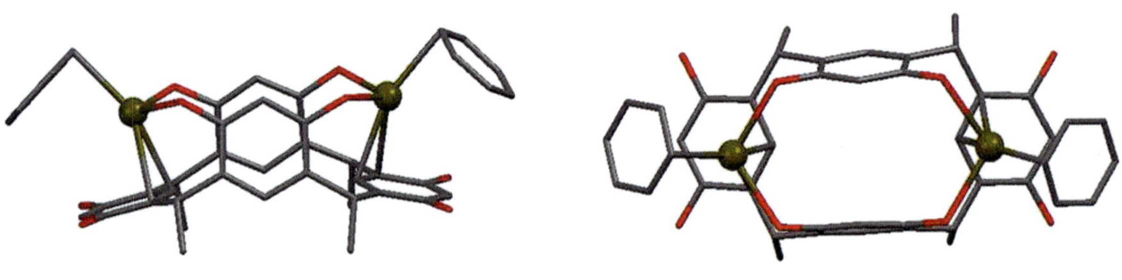

Figure 15.17 Front view (left) and top view (right) of silver complex of C-methylresorcin[4]arene. For clarity, solvent molecules and hydrogen atoms have been deleted. Color code: Ag: Olive green; O: Red; C: Grey Adapted from ref. [83]

15.3.14 Manganese and cobalt complexes enclosed within resorcin[4]arenes

Despite the absence of a central hydroxyl group, the bowl shape of resorcin[4]arene is suitable for enclosing guest molecules and forming solvent-bridged nanocapsules. For example, in a recent study, *C*-propylresorcin[4]arene was shown to host a manganese–DMSO complex, cobalt–DMSO complex and cobalt–DMSO–nitrobenzene complex [56]. The reaction of the macrocycle, RsC_3, with metal nitrates (manganese nitrate/cobalt nitrate) in a DMSO/nitrobenzene mixture yielded single crystals of $[(RsC_3)\subset(Mn(DMSO)_3)](DMSO)_2NO_3]$, $[(RsC_3)_2\subset(Co(DMSO)_3)(DMSO)_2NO_3]$, and $[(RsC_3)\subset(Co(DMSO)_3)](DMSO)_3NO_3(C_6H_5NO_2)]$. The metal centers, in all three cases, have an octahedral geometry and are coordinated with six DMSO molecules (Figure 15.19).

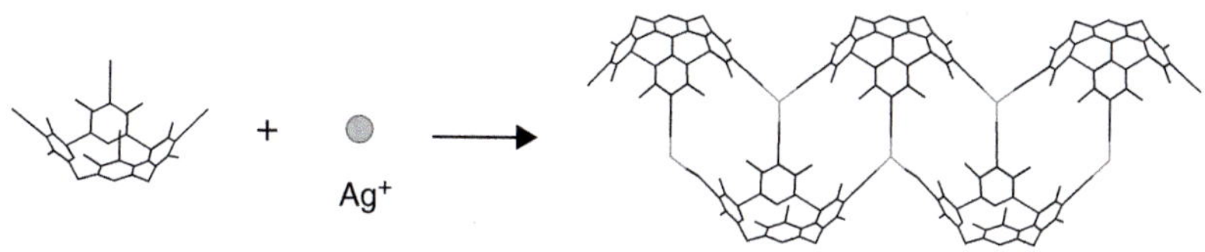

Figure 15.18 *Silver coordination complex of C-hexylcyanoresorcin[4]arene. For clarity, alkyl tails, solvent molecules, and hydrogen atoms have been deleted. Adapted from ref. [84]*

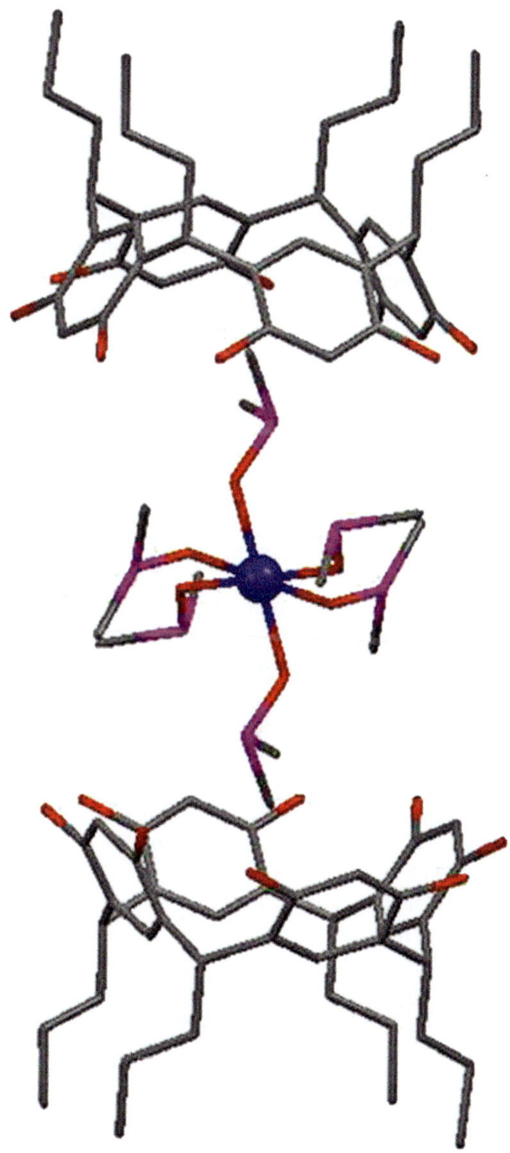

Figure 15.19 *Front view of the cobalt–DMSO-enclosing hydrogen-bonded C-propylresorcin[4]arene dimer. For clarity, hydrogen atoms and solvent molecules have been deleted. Color codes: Co: Blue; S: Pink; C: Grey; O: Red. Adapted from ref. [56]*

The manganese–DMSO-enclosing RsC_3 complex exists in only one form and is isostructural with the cobalt–DMSO complex of RsC_3. In both cases, a dimeric hydrogen-bonded nanocapsule of resorcin[4]arene was obtained, which is bridged by three solvent molecules along the dimer edges. Compared to these trisolvent-bridged complexes, the second form of the cobalt complex (with DMSO and nitrobenzene) exists as a disolvent-bridged hydrogen-bonded dimer. Although, the metal does not seam the frameworks of these nanocapsules, the metal complexes act as templates for the construction of the hydrogen-bonded dimeric nanocapsules.

15.4 Concluding remarks

The study of metal complexation of pyrogallol[4]arene- and resorcin[4]arene-based nanoassemblies is an important area of research. Efforts have been dedicated to constructing novel pyrogallol[4]arene- and resorcin[4]arene-based frameworks due to their intriguing magnetic and solution-phase properties. A variety of transition metal complexes of pyrogallol[4]arene of various shapes and sizes have been discovered over the last decade. These nanocapsules have a well-defined internal cavity that can be tuned to host different types of guest molecules. Although resorcin[4]arene has fewer metal-seamed complexes to date, functionalization of the central carbon with ligands such as bromide, cyanide, or sulfonate has enabled formation of metal-seamed RsC complexes. In addition, use of a template such as insulin has led to the formation of larger resorcin[4]arene-containing host molecules.

References

1. Johnson, C.P., Atwood, J.L., Steed, J.W., *et al.* (1996) Transition metal complexes of *p*-sulfonatocalix[5]arene, *Inorganic Chemistry*, **35** (9), 2602–2610.
2. MacGillivray, L.R., Atwood J.L. (1997) A chiral spherical molecular assembly held together by 60 hydrogen bonds, *Nature*, **389** (6650), 469–472.
3. Orr, G.W., Barbour, L.J., Atwood, J.L. (1999) Controlling molecular self-organization: formation of nanometer-scale spheres and tubules, *Science*, **285** (5430), 1049–1052.
4. Wyler, R., de Mendoza, J., Rebek, J. (1993) A synthetic cavity assembles through self-complementary hydrogen bonds, *Angewandte Chemie International Edition*, **32** (12), 1699–1701.
5. Shimizu, K.D., Rebek, J. Jr. (1995) Synthesis and assembly of self-complementary calix[4]arenes, *Proceedings of the National Academy of Sciences of the United States of America*, **92** (26), 12403–12407.
6. Meissner, R.S., Rebek, J. Jr., de Mendoza, J. (1995) Autoencapsulation through intermolecular forces: a synthetic self-assembling spherical complex, *Science*, **270** (5241), 1485–1488.
7. Shivanyuk, A., Rebek, J. Jr. (2003) Assembly of resorcinarene capsules in wet solvents, *Journal of the American Chemical Society*, **125** (12), 3432–3433.
8. Shivanyuk, A., Rebek, J. Jr. (2001) Reversible encapsulation by self-assembling resorcinarene subunits, *Proceedings of the National Academy of Sciences of the USA*, **98** (14), 7662–7665.
9. Rebek, J.J. (2007) Contortions of encapsulated alkyl groups, *Chemical Communications*, **27**, 2777.
10. Rebek, J. (2009) Molecular behavior in small spaces, *Accounts of Chemical Research*, **42** (10), 1660–1668.
11. Palmer, L.C., Rebek, J. (2005) Hydrocarbon binding inside a hexameric pyrogallol[4]arene capsule, *Organic Letters*, 7 (5), 787–789.
12. Conn, M.M., Rebek, J. (1997) Self-assembling capsules, *Chemical Reviews*, 97 (5), 1647–1668.
13. Biros, S.M., Rebek, J.J. (2007) Structure and binding properties of water-soluble cavitands and capsules, *Chemical Society Reviews*, 36 (1), 93.
14. Barrett, E.S., Dale, T.J., Rebek, J. Jr. (2007) Assembly and exchange of resorcinarene capsules monitored by fluorescence resonance energy transfer, *Journal of the American Chemical Society*, 129 (13), 3818–3819.

15. Ajami, D., Schramm, M.P., Volonterio, A., *et al.* (2007) Assembly of hybrid synthetic capsules, *Angewandte Chemie International Edition*, **46** (1 + 2), 242–244.
16. Atwood, J.L., Barbour, L.J., Hardie, M.J., *et al.* (2001) Metal sulfonatocalix[4,5]arene complexes: bi-layers, capsules, spheres, tubular arrays and beyond, *Coordination Chemistry Reviews*, 222 (1), 3–32.
17. Atwood, J.L., Barbour, L.J., Jerga, A. (2001) Hydrogen-bonded molecular capsules are stable in polar media, *Chemical Communications*, 22, 2376–2377.
18. Atwood, J.L., Barbour, L.J., Ness, T.J., *et al.* (2001) A well-resolved ice-like (H2O)8 cluster in an organic supramolecular complex, *Journal of the American Chemical Society*, 123 (29), 7192–7193.
19. Barbour, L.J., Atwood, J.L. (2001) Non-covalent interactions exert extraordinary influence over conformation and properties of a well-known supramolecular building block, *Chemical Communications*, 19, 2020–2021.
20. Atwood, J.L. (2002) Supramolecular chemistry and self-assembly special feature: organization of the interior of molecular capsules by hydrogen bonding, *Proceedings of the National Academy of Sciences of the USA*, 99 (8), 4837–4841.
21. Atwood, J.L., Barbour, L.J., Jerga, A. (2002) On the synthesis and structure of the very large spherical capsules derived from hexamers of pyrogallol[4]arenes, *Journal of Supramolecular Chemistry*, **1** (3), 131–134.
22. Atwood, J.L., Barbour, L.J., Jerga, A. (2002) Organization of the interior of molecular capsules by hydrogen bonding, Proceedings of the National Academy of Sciences of the USA, 99 (8), 4837–4841.
23. Atwood, J.L., Barbour, L.J., Dalgarno, S.J., *et al.* (2004) Toward mimicking viral geometry with metal-organic systems, *Journal of the American Chemical Society*, 126 (41), 13170–13171.
24. Cave, G.W., Antesberger, J., Barbour, L.J., *et al.* (2004) Inner core structure responds to communication between nanocapsule walls, *Angewandte Chemie International Edition*, 43 (39), 5263–5266.
25. Dalgarno, S.J., Hardie, M.J., Atwood, J.L., *et al.* (2004) Bilayers, corrugated bilayers, and coordination polymers of p-sulfonatocalix[6]arene, Inorganic Chemistry, 43 (20), 6351–6356.
26. Antesberger, J., Cave, G.W., Ferrarelli, M.C., *et al.* (2005) Solvent-free, direct synthesis of supramolecular nanocapsules, Chem*ical Communications*, **7**, 892–894.
27. Cave, G.W., Ferrarelli, M.C., Atwood, J.L. (2005) Nano-dimensions for the pyrogallol[4]arene cavity, *Chemical* Communications, 22, 2787–2789.
28. Dalgarno, S.J., Tucker, S.A., Bassil, D.B., *et al.* (2005) Fluorescent guest molecules report ordered inner phase of host capsules in solution, *Science*, 309 (5743), 2037–2039.
29. McKinlay, R.M., Cave, G.W., Atwood, J.L. (2005) Supramolecular blueprint approach to metal-coordinated capsules, *Proceedings of the National Academy of Sciences of the USA*, 102 (17), 5944–5948.
30. McKinlay, R.M., Thallapally, P.K., Cave, G.W., *et al.* (2005) Hydrogen-bonded supramolecular assemblies as robust templates in the synthesis of large metal-coordinated capsules, *Angewandte Chemie International Edition*, 44 (35), 5733–5736.
31. Dalgarno, S.J., Atwood, J.L., Raston, C.L. (2006) Sulfonatocalixarenes: molecular capsule and “Russian doll” arrays to structures mimicking viral geometry, *Chemical Communications*, 44, 4567–4574.
32. Guo, D.-S., Wang, K., Liu, Y. (2008) Selective binding behaviors of p-sulfonatocalixarenes in aqueous solution, *Journal of Inclusion Phenomena & Macrocyclic Chemistry*, 62(1–2), 1–21.
33. Dalgarno, S.J., Bassil, D.B., Tucker, S.A., *et al.* (2006) Cocrystallization and encapsulation of a fluorophore with hexameric pyrogallol[4]arene nanocapsules: structural and fluorescence studies, *Angewandte Chemie International Edition*, 45 (42), 7019–7022.
34. Dalgarno, S.J., Cave, G.W., Atwood, J.L. (2006) Toward the isolation of functional organic nanotubes, *Angewandte Chemie International Edition*, 45 (4), 570–574.
35. Dalgarno, S.J., Power, N.P., Antesberger, J., *et al.* (2006) Synthesis and structural characterisation of lower rim halogenated pyrogallol[4]arenes: bi-layers and hexameric nano-capsules, *Chemical Communications*, 36, 3803–3805.
36. Heaven, M.W., Cave, G.W., McKinlay, R.M., *et al.* (2006) Hydrogen-bonded hexamers self-assemble as spherical and tubular superstructures on the sub-micron scale, *Angewandte Chemie International Edition*, 45 (37), 6221–6224.
37. McKinlay, R.M., Thallapally, P.K., Atwood, J.L. (2006) Hexameric C-alkylpyrogallol[4]arene molecular capsules sustained by metal-ion coordination and hydrogen bonds, *Chemical Communications*, **28**, 2956–2958.
38. Bassil, D.B, Dalgarno, S.J., Cave, G.W., *et al.* (2007) Spectroscopic investigations of ADMA encapsulated in pyrogallol[4]arene nanocapsules, *Journal of Physical Chemistry B*, 111 (30), 9088–9092.

39. Dalgarno, S.J., Power, N.P., Atwood, J.L. (2007) Ionic dimeric pyrogallol[4]arene capsules, *Chemical Communications*, 33, 3447–3449.
40. Power, N.P., Dalgarno, S.J., Atwood, J.L. (2007) Robust and stable pyrogallol[4]arene molecular capsules facilitated via an octanuclear zinc coordination belt, *New Journal of Chemistry*, 31 (1), 17–20.
41. Power, N.P., Dalgarno, S.J., Atwood, J.L. (2007) Guest and ligand behavior in zinc-seamed pyrogallol[4]arene molecular capsules, *Angewandte Chemie International Edition*, 46 (45), 8601–8604.
42. Dalgarno, S.J., Power, N.P., Atwood, J.L. (2008) Metallo-supramolecular capsules, *Coordination Chemistry Reviews*, 252 (8+9), 825–841.
43. Dalgarno, S.J., Power, N.P., Warren, J.E., *et al.* (2008) Rapid formation of metal-organic nano-capsules gives new insight into the self-assembly process, *Chemical Communications*, 13, 1539–1541.
44. Jin, P., Dalgarno, S.J., Barnes, C., *et al.* (2008) Ion transport to the interior of metal-organic pyrogallol[4]arene nanocapsules, Journal of the American Chemical Society, 130 (51), 17262–17263.
45. Dalgarno, S.J., Szabo, T., Siavosh-Haghighi, A., *et al.* (2009) Exploring the limits of encapsulation within hexameric pyrogallol[4]arene nano-capsules, *Chemical Communications*, 11, 1339–1341.
46. Jin, P., Dalgarno, S.J., Warren, J.E., *et al.* (2009) Enhanced control over metal composition in mixed Ga/Zn and Ga/Cu coordinated pyrogallol[4]arene nanocapsules, *Chemical Communications*, 23, 3348–3350.
47. Jin, P., Dalgarno, S.J., Atwood, J.L. (2010) Mixed metal-organic nanocapsules, *Coordination Chemistry Reviews*, 254 (15–16), 1760–1768.
48. Maerz, A.K., Thomas, H.M., Power, N.P. *et al.* (2010) Dimeric nanocapsule induces conformational change, *Chemical Communications*, 46 (8), 1235–1237.
49. Fowler, D.A., Mossine, A.V., Beavers, C.M., *et al.* (2011) Coordination polymer chains of dimeric pyrogallol[4]arene capsules, *Journal of the American Chemical Society*, 133 (29), 11069–11071.
50. Kline, K.K., Fowler, D.A., Tucker, S.A., *et al.* (2011) Encapsulation of acenaphthene within C-propan-3-ol-pyrogallol[4]arene dimeric nanocapsules, *Chemistry – A European Journal*, 17 (39), 10848–10851.
51. Kumari, H., Kline, S.R., Wycoff, W.G., *et al.* (2012) Investigating structural alterations in pyrogallol[4]arene-pyrene nanotubular frameworks, *Small*, 8 (21), 3321–3325.
52. Kumari, H., Mossine, A.V., Kline, S.R., *et al.* (2012) Controlling the self-assembly of metal-seamed organic nano-capsules, *Angewandte Chemie International Edition*, 51 (6), 1452–1454.
53. Mossine, A.V., Kumari, H., Fowler, D.A., *et al.* (forthcoming) Utilizing aliphatic diols to selectively template tubular C-methylpyrogallol[4]arene inclusion complexes, *New Journal of Chemistry*, manuscript in preparation.
54. Fowler, D.A., Rathnayake, A.S., Kennedy, S., *et al.* (2013) Introducing defects into metal-seamed nanocapsules using mixed macrocycles, *Journal of the American Chemical Society*, 135 (33), 12184–12187.
55. Kumari, H., Jin, P., Deakyne, C., *et al.* (2013) Metal ion transport across metal-organic pyrogallol[4]arene-based nanocapsules, *Current Organic Chemistry*, 17 (14), 1481–1488.
56. Drachnik, A.M., Kumari, H., Barnes, C.L., *et al.* (2014) Encapsulation of manganese and cobalt complexes within resorcin[4]arene dimers, *CrystEngComm*, 16 (31), 7172–7175.
57. Kumari, H., Jin, P., Teat, S.J., *et al.* (2014) Entrapment of elusive guests within metal-seamed nanocapsules, *Angewandte Chemie International Edition*, 53 (48), 13088–13092.
58. Kumari, H., Jin, P., Teat, S.J., *et al.* (2014) Strong cation···π interactions promote the capture of metal ions within metal-seamed nanocapsule, *Journal of the American Chemical Society*, 136 (49), 17002–17005.
59. Kumari, H., Kline, S.R., Fowler, D.A., *et al.* (2014) Solution superstructures: truncated cubeoctahedron structures of pyrogallol[4]arene nanoassemblies, *Chemical Communications*, 50 (1), 109–111.
60. Gerkensmeier, T., Iwanek, W., Agena, C. *et al.* (1999) Self-assembly of 2,8,14,20-tetraisobutyl-5,11,17,-23-tetrahydroxyresorc[4]arene, *European Journal of Organic Chemistry*, 9, 2257–2262.
61. Orr, G.W., Barbour, L.J., Atwood, J.L. (1999) Controlling molecular self-organization: formation of nanometer-scale spheres and tubules, *Science*, 285 (5430), 1049–1052.
62. Whetstine, J.L., Kline, K.K., Fowler, D.A., *et al.* (2010) Spectroscopic investigations of pyrene butanol encapsulated in C-hexylpyrogallol[4]arene nanocapsules, *New Journal of Chemistry*, 34 (11), 2587–2591.
63. Avram, L., Cohen, Y. (2002) Spontaneous formation of hexameric resorcinarene capsule in chloroform solution as detected by diffusion NMR, *Journal of the American Chemical Society*, 124 (51), 15148–15149.

64. Avram, L., Cohen, Y. (2003) Hexameric capsules of lipophilic pyrogallolarene and resorcinarene in solutions as probed by diffusion NMR: one hydroxyl makes the difference, *Organic Letters*, 5 (18), 3329–3332.
65. Cohen, Y., Evan-Salem, T., Avram, L. (2008) Hydrogen-bonded hexameric capsules of resorcin[4]arene, pyrogallol[4]arene and octahydroxypyridine[4]arene are abundant structures in organic solvents: a view from diffusion NMR, *Supramolecular Chemistry*, 20 (1 + 2), 71–79.
66. Evan-Salem, T., Cohen, Y. (2007) Octahydroxypyridine[4]arene self-assembles spontaneously to form hexameric capsules and dimeric aggregates, *Chemistry – A European Journal*, 13 (27), 7659–7663.
67. Slovak, S., Evan-Salem, T., Cohen, Y. (2010) Self-assembly of a hexameric aggregate of a lipophilic calix[4]pyrrole-resorcinarene hybrid in solution: a diffusion NMR study, *Organic Letters*, 12 (21), 4864–4867.
68. Yamanaka, M., Shivanyuk, A., Rebek, J. (2004) Kinetics and thermodynamics of hexameric capsule formation, *Journal of the American Chemical Society*, 126 (9), 2939–2943.
69. McKinlay, R.M., Cave, G.W.V., Atwood, J.L. (2005) Supramolecular blueprint approach to metal-coordinated capsules, *Proceedings of the National Academy of Sciences of the United States of America*, 102 (17), 5944–5948.
70. Kumari, H., Kline, S.R., Schuster, N.J., *et al.* (2011) Solution structure of copper-seamed C-alkylpyrogallol[4]arene nanocapsules with varying chain lengths, *Chemical Communications*, 47 (45), 12298.
71. Kumari, H., Kline, S.R., Schuster, N.J., *et al.* (2011) Exploring the ellipsoidal and core–shell geometries of copper-seamed C-alkylpyrogallol[4]arene nanocapsules in solution, *Journal of the American Chemical Society*, 133 (45), 18102–18105.
72. Mayhan, C.M., Szabo, T.J., Adams, J.E., *et al.* (2013) Mononuclear and polynuclear 5-coordinate zinc(II) model complexes: a quantum chemical calibration study of their structure and energy, *Structural Chemistry*, 24 (6), 2089–2099.
73. Mossine, A.V., Mayhan, C.M., Fowler, D.A., *et al.* (2014) Zinc-seamed pyrogallol[4]arene dimers as structural components in a two-dimensional MOF, *Chemical Science*, 5, 2297–2303.
74. Kumari, H., Kline, S.R., Wycoff, W.G., *et al.* (2012) Solution-phase structures of gallium-containing pyrogallol[4]arene scaffolds, *Angewandte Chemie International Edition*, 51 (21), 5086–5091.
75. Jin, P., Kumari, H., Kennedy, S., *et al.* (2014) Structural alteration of the metal-organic pyrogallol[4]arene nanocapsule motif by incorporation of large metal centres, *Chemical Communications*, 50 (34), 4508–4510.
76. Mossine, A.V., Kumari, H., Fowler, D.A., *et al.* (2011) Ferrocene as a hydrophobic templating agent with pyrogallol[4]arenes, *Israel Journal of Chemistry*, 51 (7), 840–842.
77. Mossine, A.V., Kumari, H., Fowler, D.A., *et al.* (2012) Ferrocene species included within a pyrogallol[4]arene tube, *Chemistry – A European Journal*, 18 (33), 10258–10260.
78. Kumari, H., Dennis, C.L., Mossine, A.V., *et al.* (2013) Magnetic differentiation of pyrogallol[4]arene tubular and capsular frameworks, *Journal of the American Chemical Society*, 135 (19), 7110–7113.
79. Kumari, H., Kline, S.R., Dennis, C.L., *et al.* (2012) Solution-phase and magnetic approach towards understanding iron gall ink-like nanoassemblies, *Angewandte Chemie International Edition*, 51 (37), 9263–9266.
80. Kumari, H., Kline, S.R., Atwood, J.L. (2012) Investigation of encapsulation of insulin biotemplate within C-methylresorcin[4]arenes, *Chemical Communications*, 48 (30), 3599–3601.
81. Fowler, D.A., Rathnayake, A.S., Kennedy, S., *et al.* (2013) Introducing defects into metal-seamed nanocapsules using mixed macrocycles, *Journal of the American Chemical Society*, 135 (33), 12184–12187.
82. Solari, E., Lesueur, W., Klose, A., *et al.* (1996) A square array of metal ions from the full metallation of calix[4]resorcinarenes, Chemical Communications, 7, 807.
83. Munakata, M., Wu, L.P., Kuroda-Sowa, T., *et al.* (1999) Open dimeric and capped polymeric container molecules: calixarene and resorcinarene complexes of AgI co-ordinated by participation of the upper-rim carbon atoms, *Journal of the Chemical Society, Dalton Transactions*, 3, 373–378.
84. Mossine, A.V., Thavornyutikarn, P., Atwood, J.L. (2013) Conservation of a unique noncovalent motif in a tetracyanoresorcin[4]arene silver complex, *CrystEngComm*, 15 (9), 1673.

16

Dynamic Control of Recognition Processes in Host–Guest Systems and Polymer–Polymer Interactions

Seiji Shinkai

Institute for Advanced Study, Kyushu University, Fukuoka;
Nanotechnology Laboratory, Institute of Systems, Information Technologies and Nanotechnologies (ISIT), Fukuoka;
Department of Nanoscience, Faculty of Engineering, Sojo University, Kumamoto, Japan

16.1 Introduction

Ions and molecules are abundant in nature, and the need to measure the concentration of selected ions and small organic molecules both *in vivo* and *in vitro* can be critical. This field, called "Host–Guest Chemistry" or more generally "Molecular Recognition," has developed in relation to comprehension of the principle found commonly in nature, where enzymes, antibodies, and a plethora of other biological macromolecules bind ions and small organic molecules in order to perform their various physiological tasks. The key to this process is molecular recognition.

When I was a PhD student, I came across a new, very attractive concept proposed by Leffler [1], the "enthalpy–entropy compensation relationship." This concept was very useful in obtaining insights into thermodynamic data on association and kinetic processes. However, the relationship implies that high selectivity and high activity cannot appear in one equilibrium or reaction series. The question therefore arises, how can we create such an exceptional system with high selectivity and high activity that deviates from the enthalpy–entropy compensation relationship? This was the starting point of my chemistry. Eventually, I developed one potential breakthrough idea: as long as one association process or one kinetic process is treated independently, it is still restricted by the compensation relationship, whereas if two, or more than two, systems are

Macrocyclic and Supramolecular Chemistry: How Izatt–Christensen Award Winners Shaped the Field, First Edition.
Edited by Reed M. Izatt.

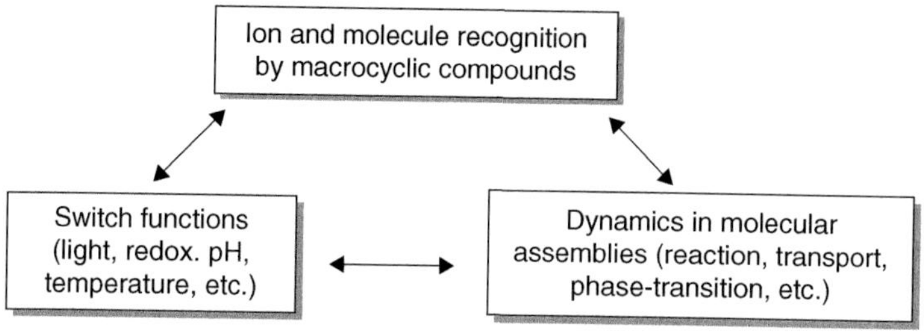

Figure 16.1 *Molecular design concept*

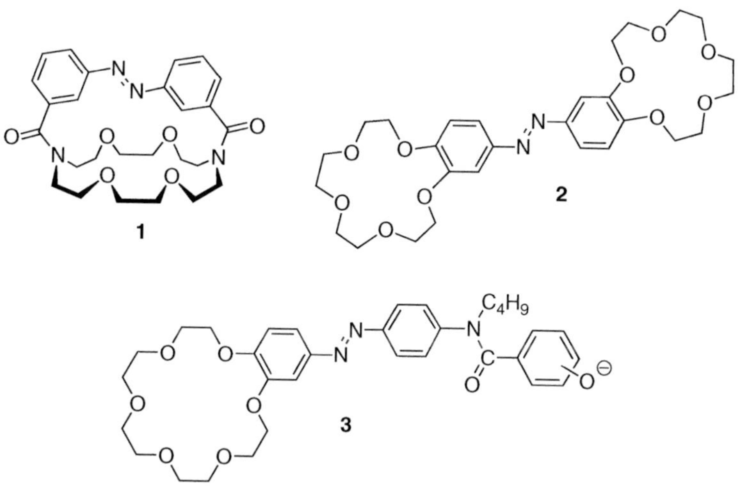

Scheme 16.1 *Structures of crown compounds **1–3***

linked reversibly, one may find an exceptional process diverted from the compensation relationship through "switching" to and from other conjugated systems. This original concept enabled me to create a number of new dynamic ion- and molecule-recognition systems that are combined with switch-functionalized systems and molecular-assembly systems (Figure 16.1).

We sought to extend this concept to bio-organic research fields. There, I learned that in biological systems dynamic and cooperative actions artfully combine and mesh to realize desirable biological functions. In this review article, I would like to introduce our research extension process from classic molecular machines via dynamic allosteric systems and aggregation-induced emission to polymer–polymer recognition systems. In the final section, I will describe a successful biochemical application in the design of non-toxic DNA drug carriers with helix-forming polysaccharides.

16.2 Dynamic control of crown ether functions by chemical and physical signals [2]

A crown ether has the ability to associate with charged and uncharged species, and its conformation is easily changeable because of the flexible nature of macrocyclic polyethers. Hence, it seemed to me that crown ethers were ideal compounds to realize the concept outlined in Figure 16.1, because one can easily change the structure that is the origin of their guest selectivity. The purpose of our investigation was to control their association phenomena by stimuli from the outside world.

Compound **1** is the first example of a "photoresponsive crown ether": in response to light-mediated *cis–trans* interconversion of the azobenzene cap, metal selectivity can be changed between Na^+ (*trans*-form) and K^+ (*cis*-form) (Scheme 16.1). Compound **2** is another photoresponsive crown ether, the UV light-generated *cis*-form of which shows K^+ selectivity because of the formation of an intramolecular 1:2 metal/crown sandwich complex. This compound was successfully applied to the photocontrol of metal extraction, liquid membrane transport, ion transport across polymer-liquid crystal composite membranes, etc. Compound **3**, which has an azobenzene-linked anionic cap, showed photocontrollable, monensin-mimetic functions in a liquid-membrane ion-transport system.

We moved on to design and synthesize a variety of photoresponsive crown ethers. We noticed, although much later, that some of them already included very important general chemical concepts proposed later. For example, compound **4** is one of a series of so-called "tail-biting" crown ethers (Figure 16.2). In a K^+-transport system across a membrane, active transport occurs only when both UV light irradiation and acidic aqueous out-phase are donated [3]. When either one of them is omitted, passive transport is still possible but active transport is no longer possible [3]. This result is exactly related to a "logic gate" concept proposed by A. P. deSilva [4]. Their solution properties are also interesting. When the spacer (n) is short, they tend to form pseudocyclic dimers, whereas the spacer (n) is long enough, they tend to form supramolecular polymers due to intermolecular "tail-biting." When exposed to UV light, the supramolecular polymers are converted to the intramolecular "tail-biting" monomers [5]. Supramolecular polymers are now widely investigated, but even now, this reversible monomer–supramolecular polymer interconversion is a very rare example. Compound **5** is a flexible dibenzo-24-crown-8 immobilized on rigid polystyrene resin. When the immobilization reaction is performed in the presence of Cs^+, the crown ring maintains a conformation suitable for Cs^+ binding [6]. When the azobenzene moieties are shortened by UV light-induced *trans*-to-*cis* isomerization, the crown ring

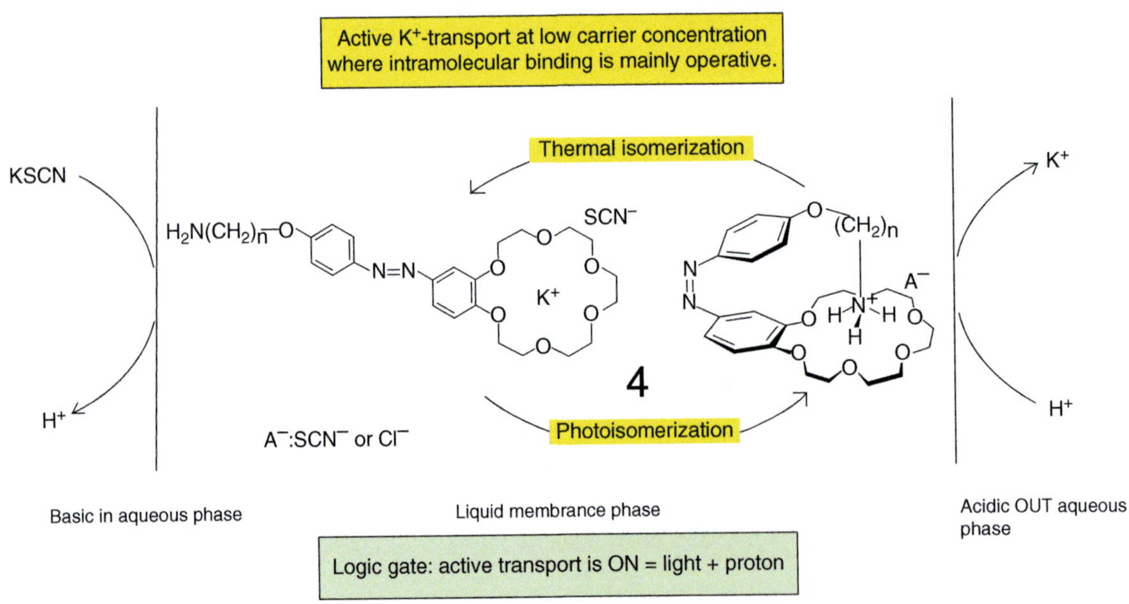

Figure 16.2 *Active transport of K^+ mediated by a tail-biting crown compound* ***4*** *supported by cooperative actions of light and proton*

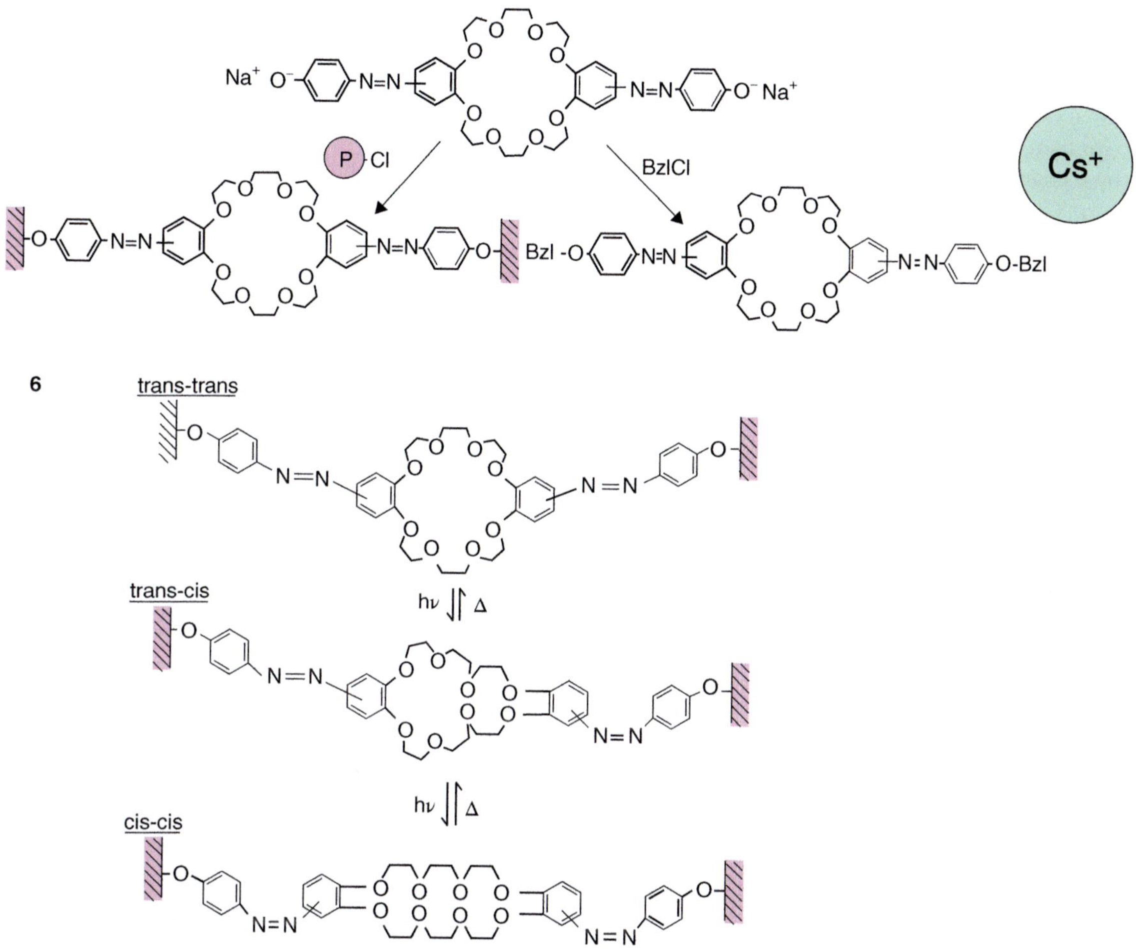

Figure 16.3 *Metal (Cs^+)-imprinting with a 24-crown-8 derivative* ***5*** *on a hard polystyrene resin surface and its reversible photocontrol*

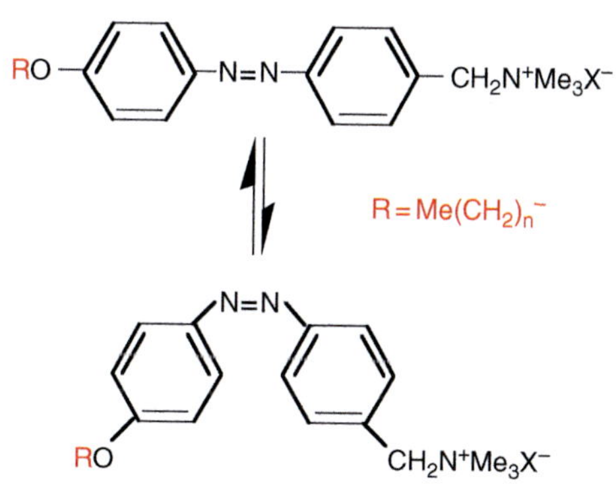

Figure 16.4 Cis–trans *transformation of an azobenzene-integrated surfactant compound* ***6***

is stretched and deformed and Cs^+ is kicked out of the ring (Figure 16.3). This result reminds us of a concept called "molecular imprinting." We also synthesized amphiphilic azobenzene compound **6** and demonstrated a photoinduced change in its aggregate morphology. It is well-known that later, many similar papers were written on the basis of the concept of "photocontrol of molecular assemblies" (Figure 16.4).

We then extended our research efforts to the new stimulus sources other than light. Disulfide-containing crown ether **7** was designed for the redox control of metal affinity (see Scheme 16.2). Amphiphilic crown ethers **8** and **9** were designed for selective ion-transport across polymer–liquid crystal composite membranes and the metal control of monolayer functions or organic gel functions, respectively. It was found that the K^+ transport ability of **8** changes in an all-or-nothing manner above and below the phase-transition temperature of the deposited liquid crystal and **9** creates chiral ion-binding sites in cholesterol-stacked monolayer and gel assemblies.

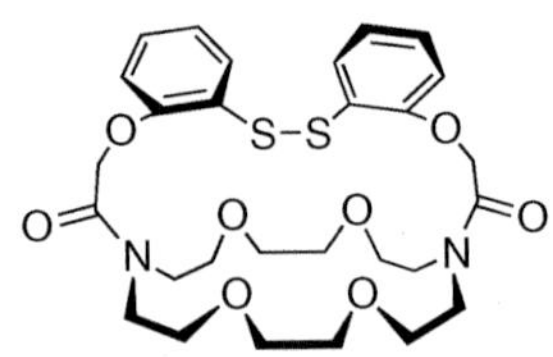

7

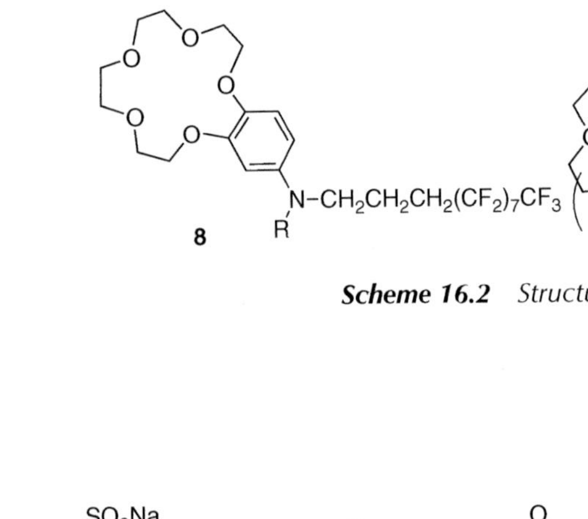

8

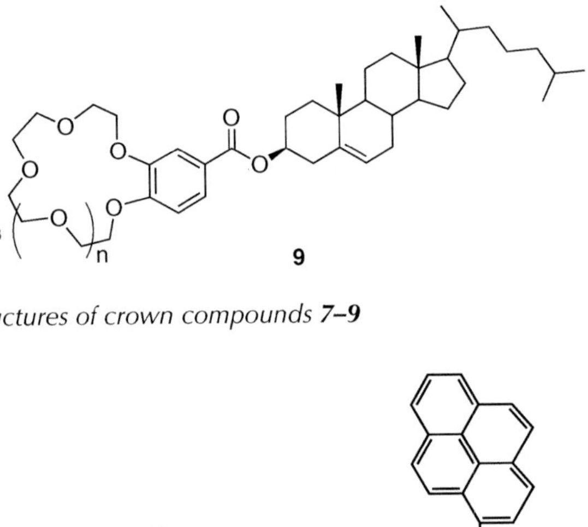

9

Scheme 16.2 *Structures of crown compounds **7–9***

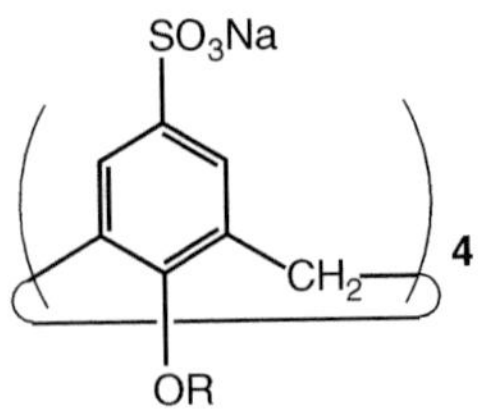

10

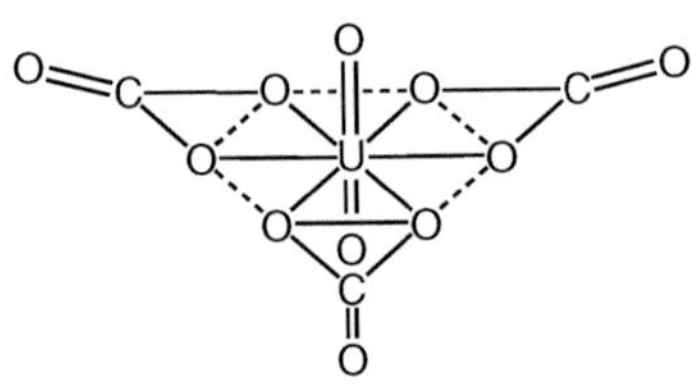

11

Uranyl tricarbonate with hexadentate coordination

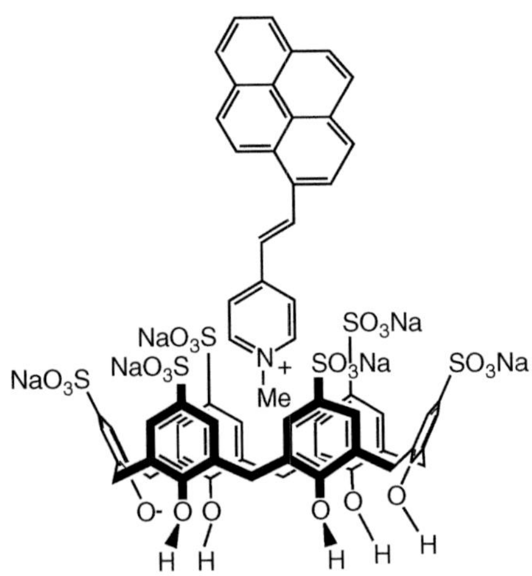

12

Scheme 16.3 *Structures of calix[n]arene compounds **10–12***

16.3 Stereochemical studies of calix[n]arene derivatives [7]

In the early 1980s, the structures and conformational motions of calix[n]arenes were totally unknown in solution, although there were a few reports on the solid-state structures studied by X-ray crystallography. We considered that without a basic knowledge of the solution structures, it is nearly impossible to design excellent host molecules for ion and molecule recognition from calix[n]arenes, or to understand their guest-binding properties. We thus devoted our primary research effort toward stereochemical studies of calix[n]arene derivatives using various spectroscopic methods. Major achievements can be summarized as follows.

1. It was shown by NMR spectroscopic and MM3 computational studies that unmodified calix[n]arenes adopt a "cone" conformation and ring inversion occurs with the activation of free energy = 14–17 kcal mol^{-1}. This computational simulation of the VT NMR spectra was reported for the first time by my group.
2. It was found that a propyl group is bulky enough as an *O*-substituent to suppress the oxygen-through-the-annulus rotation of calix[4]arene. The finding enabled us to classify all possible positional and conformational isomers derived from calix[4]arene (13 isomers). We developed all synthetic routes to these isomers.
3. When three different substituents are introduced into calix[4]arene and the oxygen-through-the-annulus rotation is inhibited, the resultant calix[4]arene products become "chiral." Based on this concept, we synthesized and optically resolved chiral calix[4]arenes for the first time. Subsequently, we classified all possible chiral calix[4]arenes derived from calix[4]arene. This is very useful as a guide for molecular design of chiral calix[4]arenes.
4. Ring inversion of calix[6]arene derivatives had long been a controversial problem: there was no clear evidence to determine whether their conformation is immobilized or still rotational, and this had long been the origin of chaos in calix[6]arene chemistry. We have settled this critical problem by one specific idea. Since ring inversion of chiral calix[6]arenes inevitably accompanies their racemization, the absence of racemization becomes firm evidence for the immobilization of the calix[6]arene ring. This idea really worked, and it became possible to distinguish between immobilized and rotational calix[6] arene derivatives.

16.4 Ion and molecule recognition by functionalized calix[n]arenes and their application to super Na^+-sensors and novel [60]fullerene isolation methods [8]

In the early 1980s, it was still unclear whether or not calix[n]arenes were really useful as host molecules for ion and molecule recognition in solution. In 1987, we synthesized sulfonated water-soluble calix[n]arenes **10** (see Scheme 16.3) and proved for the first time that they can include various organic guest molecules in water.

Furthermore, two exciting results were obtained from this class of water-soluble calix[n]arenes. Firstly, we found that calix[5]arene and calix[6]arene with unmodified –OH or modified with CH_2COOH not only have high, record-breaking stability constants (log K = 18.4–19.2) but also an usually high selectivity for UO_2^{2+} (selectivity factor = 10^{12}–10^{17}). These compounds are now called "super-uranophiles". These compounds can provide preorganized arrangements of ligand groups suitable to pseudoplanar penta- and hexacoordination of UO_2^{2+} complexes **11**. Secondly, it was found that π-basic cavities of cone-calix[n]arenes act as a receptor site for quaternary ammonium ions owing to cation–π and/or CH π interactions. This finding was successfully applied to molecular design of a light-signaling acetylcholine receptor **12**.

It was shown that calix[n]arenes are very useful as a platform to design highly ion-selective ionophores and their light-signaling devices. Ungaro et al. [9] and McKervey et al. [10] reported that calix[4]aryl esters and amides show high Na^+ selectivity. We noticed that upon Na^+-binding, the carbonyl groups on the lower

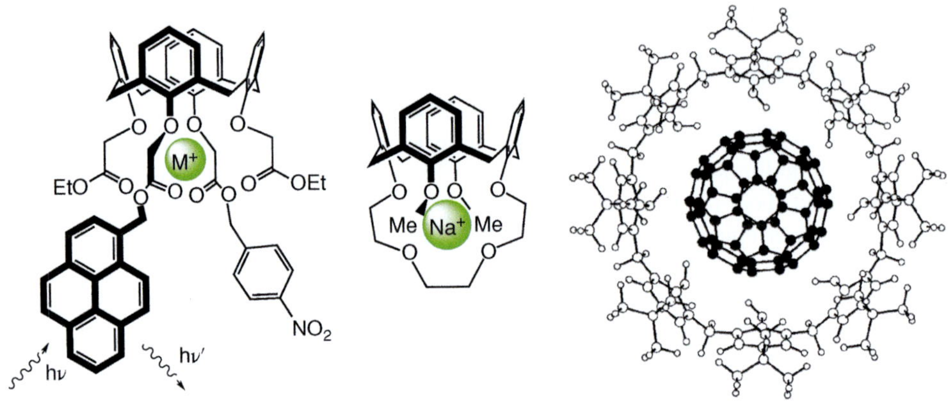

13 14 15

Scheme 16.4 *Inclusion properties of calix[n]arene compounds* ***13–15***

edge turn around from the outside to the inside. In compound **13**, this rotation caused a distance change between a fluorophore (pyrene) and a quencher (*p*-nitrophenyl) introduced into distal ester groups, resulting in a fluorescence intensity change upon Na^+-binding (Scheme 16.4). This is the first example of a Na^+-selective light-signaling calix[4]arene, featuring its dynamic action. A similar idea was reported by Jin et al. of Hokkaido University [11].

Reinhoudt et al. [12] have made calix[4]crown as well as highly preorganized calix[4]spherands. We have found that calix[4]crown-4 **14** shows a remarkably high Na^+/K^+ selectivity of $10^{5.0}$. This value is higher by two orders of magnitude than any other Na^+-directed ionophores developed so far. There is a long historical background for the development of **14**: valinomycin designed by nature for K^+/Na^+ recognition shows its highest affinity with Rb^+ rather than with K^+, suggesting that this shift is more effective in achieving high K^+/Na^+ selectivity. The ionophoric cavity in **14** is slightly smaller than Na^+, which is why the cavity can strongly reject K^+. This compound is in practical use as an Na^+ sensor.

Lastly, worthy of mention is our serendipitous discovery that calix[8]arenes selectively include [60]fullerene and form the precipitate **15** [8b]. This method is now widely used for isolation of [60]fullerene from [70]fullerene and other higher fullerenes in industrial processes. Surprisingly, the same method was concurrently discovered by Atwood's group on the opposite side of the earth [13].

16.5 Molecular design of novel sugar-sensing systems using boronic acid–diol macrocyclization [14]

To develop new receptor molecules that can precisely recognize saccharide molecules, we synthesized a number of diboronic acid derivatives. Their selectivity can be tuned by the relative spatial position of the two boronic acid groups, and the complexation event can be read out by circular dichroism spectroscopy: for example, **16** and **17** (see Scheme 16.5) show selectivity for glucose among monosaccharides and lactose among disaccharides, respectively. When a boronic acid group is combined intramolecularly with an aminomethyl-fluorophore, as in **18**, the complexation event can be conveniently read out by a turn-on-type fluorescence method. This is a novel application of a PET (photoinduced electron transfer) sensor. Fluorescent diboronic acid **19** shows remarkably high selectivity for glucose, which will be soon industrialized as a glucose sensor. It was also demonstrated using **20** with a chiral 1,1′-binaphthyl group as a fluorophore that even chiral

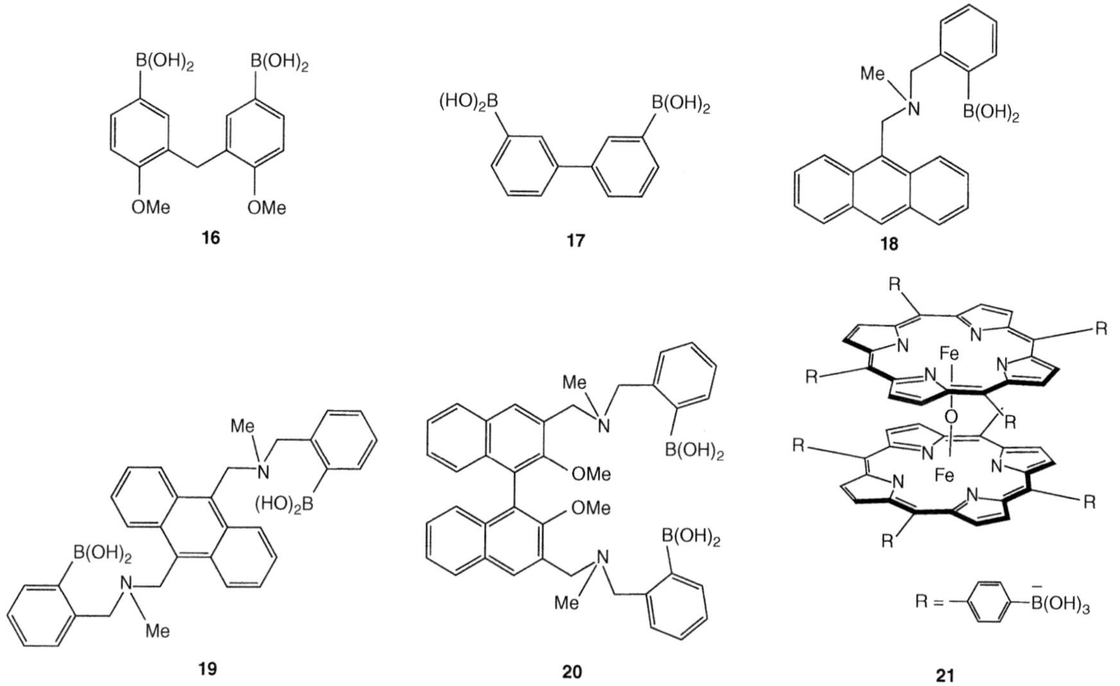

Scheme 16.5 *Structures of boronic acid derivatives* ***16–21***

recognition of saccharides becomes possible. The porphyrin skeleton is shown to be useful not only as a platform to control the spatial position of boronic acids but also as a chromophore for reading-out the absolute configuration of saccharides. The typical example is a μ-oxo-dimer, **21**, which shows remarkably high affinity only with glucose and galactose.

16.6 From molecular machines to allosteric effects [15]

Positive or negative allosterisms are ubiquitously seen in nature where biological events have to be efficiently regulated in response to chemical or physical signals from the outside world. Typical examples are observed for cooperative dioxygen binding to hemoglobin, hexamerization of arginine repressor, a cooperative effect with respect to the concentration of arachidonate-containing phospholipids in cytosolic phospholipase A_2, etc. Allosteric systems provide a means of obtaining chemical feedback, an essential factor in achieving total control over molecular-scale chemical processes. These systems are characterized by nonlinear binding which is quite different from conventional linear binding. Nonlinear binding, by definition, requires that the initial binding of a guest has a different effect to subsequent enzyme–substrate or host–guest interactions. It is certain, therefore, that such natural systems that feature allosteric contrivance must be very dynamic, like "molecular machines." The biomimetic design of such allosteric systems is of great significance in order to regulate the complexation ability or the catalytic activity of artificial receptors nonlinearly. The simplest mode of allosteric action takes the form of heterotropic allosterism, where the binding of one chemical species influences the binding of a second, different chemical species. Homotropic allosterism, however, is considerably more difficult to achieve because the initial binding of a guest species must have a different

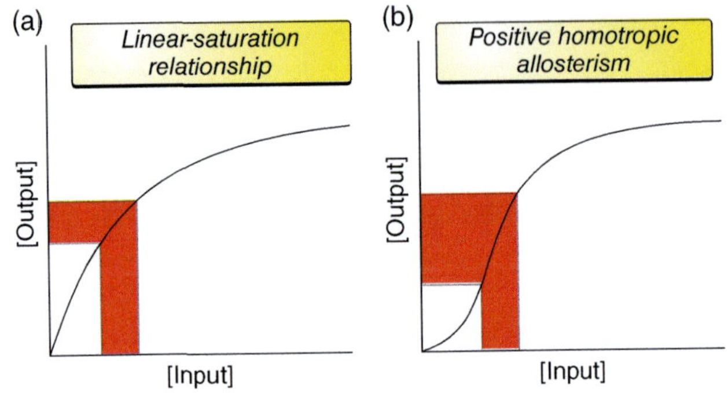

Figure 16.5 *Input–output correlation in linear and nonlinear bindings. (Reproduced from ref. [15a] with permission. Copyright © 2001 American Chemical Society)*

effect to that of the subsequent interactions with the same guest species. It is certain, however, that positive homotropic allosterism is very useful in amplifying and converting weak chemical or physical signals into other signals which are convenient for us to read out and record. In conventional binding, for example, the input signal is "linearly" transmitted to the output signal, followed by saturation in a high-input region (Figure 16.5a). This situation, useful as a calibration curve, is an important factor in sensing events. In the positive homotropic allosterism, on the other hand, a small change in the input signal is amplified into a large change in the output signal in the steep transition region (Figure 16.5b). In addition, one may surmise that the region lower than this transition is an "off" state whereas the region higher than this transition is an "on" state. Or, when the input signal is viewed from the output signal axis, one may take the view that this system possesses a buffer-like function.

We considered how one might design such intriguing positive allosteric systems in relation to molecular recognition for small molecules, metal ions, natural products, etc. Eventually, we found that double-decker porphyrins can provide an ideal skeleton to design an allosteric receptor system: therein, two porphyrins are freely rotating, but once frozen by the first guest-binding, the system has three geometrically equivalent binding sites. This situation cooperatively facilitates subsequent guest-binding processes, resulting in a homotropic binding cascade. It has been shown that the guest-binding events are mostly dynamic, which combine with the molecular recognition systems so that subsequent guest bindings can occur more favorably than the first. In addition, it has been suggested that positive homotropic allosterism can be utilized as a new strategy to attain high guest selectivity and guest affinity which cannot be attained by conventional 1:1 guest binding.

16.7 From allosteric effects to aggregation-induced emission (AIE)

The development of fluorescence sensing systems that recognize biologically important metals, molecules, and polymers has become an active research area because the fluorescence technique is highly simple and sensitive, facilitating its potential applications in bioprobes, imaging, and monitoring. As fluorophores, small organic dyes have been mainly used, in which recognition sites specific to concerned targets are integrated. The organic dyes' wavelength window, brightness, and photostability are optimized by appropriate chemical modifications. However, most organic dyes are intrinsically emissive and quenched at higher concentrations due to their aggregation (self-quenching). Various fluorescent probes have been designed on the basis of this principle,. However, this approach (i.e., turn-off switching) inevitably causes low response as well as low

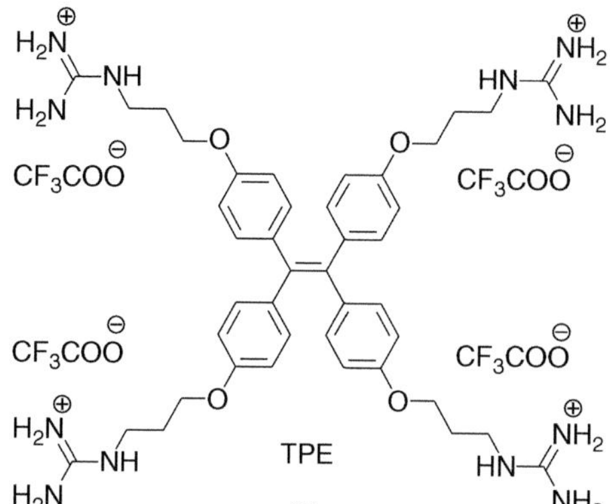

Scheme 16.6 *Structure of guanidinium-tethered tetraphenylethene (TPE: compound **22**) for ATP detection*

signal-to-background contrast in fluorescence detection because simultaneous satisfaction of the requirements for efficient emission and efficient quenching is rather difficult. As an alternate approach, a few examples have demonstrated target-specific molecular probes, in which the recognition-induced disassembling of the probes switches them "turn-on." In contrast, fluorophores exhibiting aggregation-induced emission (AIE) have been increasingly attracting attention because they behave in an opposite manner to intrinsic self-quenching dye molecules [16]. The design of AIE molecules can be realized by a modular-type structure composed of an emissive moiety, a spacer and a recognition site, and aggregation is triggered by the target-specific association. Most importantly, the AIE molecules are constructed to show "turn-on" switching in response to self-assembly with a target molecule. This cascade, "guest-binding–aggregation–fluorescence emission," is very similar to that of allosteric recognition, "guest-binding–suppression of molecular motion–improved binding ability."

In order to elicit the full potential of AIE properties, we have designed a novel fluorescent molecule **22** (Scheme 16.6). This is the marriage of AIE with an allosteric function attained by a self-assembly system [17]. This strategy was inspired from our previous work on the highly selective molecular recognition attainable via allosteric control. Since allosteric control realizes nonlinear, sigmoidal information transduction, a small change in the input signal can be amplified into a large change in the output signal. If such a system is set up in a fluorescence sensing arrangement, a steep turn-on fluorescence transition will be realized in a nonlinear fashion, leading to a significant enhancement of the signal-to-background contrast. In this context, the present system is intrinsically different from the linearly transmitted examples generally observed for conventional fluorescence sensing systems. We have thus focused on tetraphenylethene, known as an AIE molecule, whose emission mechanism is rationalized in terms of the restriction of C–C bond rotation. Our probe **22** was designed to target ATP, one of the most significant biomolecules – a universal energy source as well as an extracellular signaling mediator [17]. The designed probe, TPE, has the following three unique features: (1) multiple recognition sites to be required for a nonlinear response under the allosteric control; (2) guanidinium groups as a recognition site, known to bind with phosphate anions in a 1:1 molar ratio via a specific hydrogen-bond-assisted electrostatic interaction in water; and (3) short spacers consisting of a propyl group between TPE and guanidinium to avoid background fluorescence arising from micelle formation of TPE itself in water. We had confirmed that our TPE probe effects selective fluorescent detection of ATP through cooperative self-assembly that is characterized by a sensitive, nonlinear fluorescence response together with the high signal-to-background ratio. More recently, we found that AIE-type self-assembly of guanidinium-tethered oligophenylenevinylene is even effective in translating chiral information into fluorometric optical signals [18].

16.8 Extension of cooperative actions to polymeric and biological systems [19]

For centuries, some fungi have been recognized as potential medicines for the treatment of gynecological diseases. Modern chemistry has revealed that active ingredients of these fungi are polysaccharides belonging to β-1,3-glucan family. Among others, schizophyllan (SPG), curdlan (Figure 16.6a), and lentinan (Figure 16.6b) have been commercialized as medicines directed at uterine cancer in Japan. Immunology showed that these β-1,3-glucans can activate natural immunity by promoting secretion of interleukins.

The β-1,3-glucan polysaccharides have triple-stranded helical structures with their sense and pitch comparable to those of polynucleotides. We recently revealed that the β-1,3-glucans could interact with certain polynucleotides to form triple-stranded and helical macromolecular complexes consisting of two polysaccharide-strands and one polynucleotide-strand (Figure 16.7). This unique property of the β-1,3-glucans has made it possible to utilize these polysaccharides as potential carriers for various functional polynucleotides. In particular, cell-uptake efficiency of the resultant polysaccharide/polynucleotide complexes was remarkably enhanced when functional groups recognized in a biological system were introduced as pendent groups.

The β-1,3-glucans can also interact with various one-dimensional architectures, such as single-walled carbon nanotubes, to produce unique nanocomposites, in which the single-walled carbon nanotubes are entrapped within the helical superstructure of β-1,3-glucans. Various conductive polymers and gold nanoparticles are also entrapped within the helical superstructure in the similar manner. In addition, diacetylene monomers entrapped within the helical superstructure can be photopolymerized to afford the corresponding poly(diacetylene)-nanofibers with a uniform diameter. These findings indicate that the β-1,3-glucans are very attractive and useful materials not only in biotechnology but also in nanotechnology. These unique properties of the β-1,3-glucans undoubtedly originate from their inherent, very strong helix-forming character, which has never been observed for other polysaccharides.

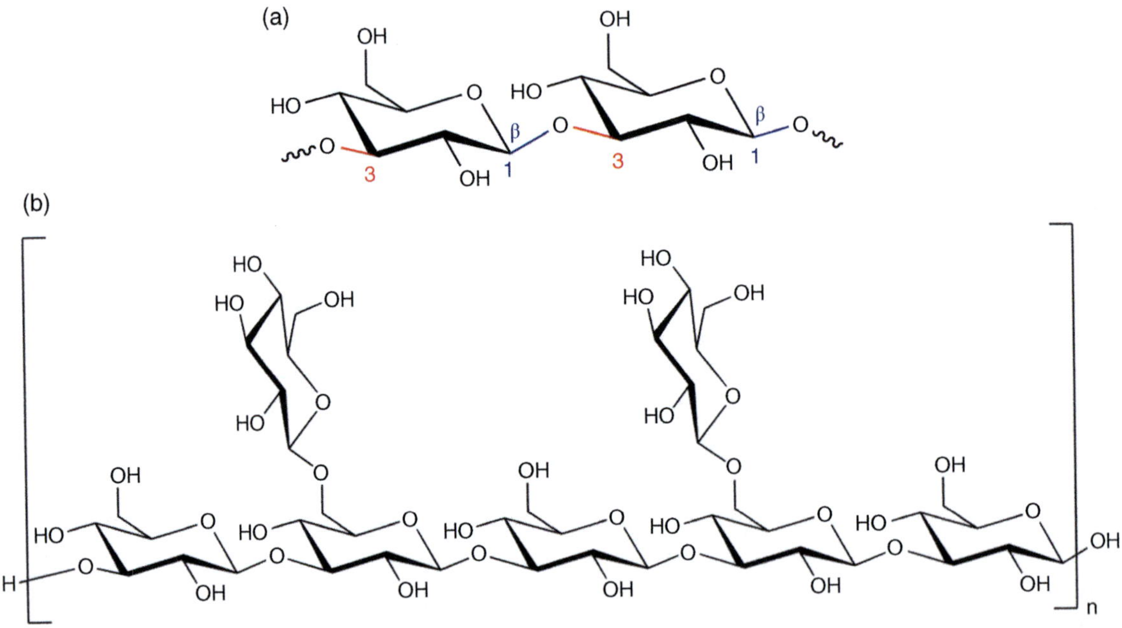

Figure 16.6 Chemical structures of curdlan and lentinan. (Reproduced from ref. [19a] with permission of The Royal Society of Chemistry)

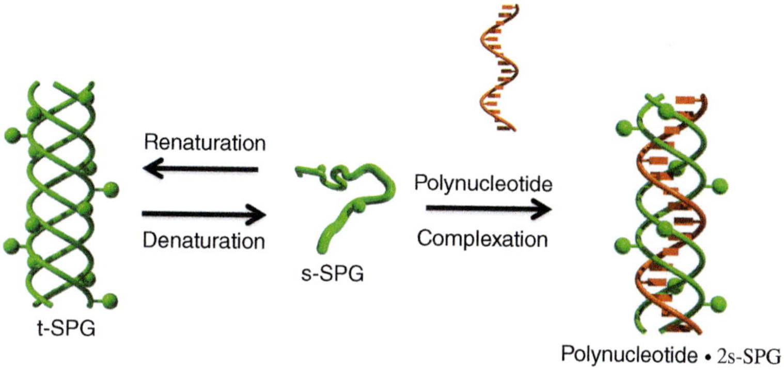

Figure 16.7 *Schematic illustration of the equilibria involved in the formation of the triple-helical complex between one polynucleotide chain and two schizophyllan (s-SPG) chains. (Reproduced from ref. [19c] with permission of John Wiley & Sons, Ltd)*

16.9 Summary

In summary, my research is not only to develop macrocyclic host molecules for "static" ion and molecule recognition but also to combine them with "dynamic" recognition processes in functionalized host molecules. This is a new basic concept leading to the design of "molecular machines" coupled with recognition processes. Furthermore, this dynamic concept can be extended to molecular assembly systems and helix-forming polymers. I believe that these dynamic systems bearing the molecular recognition ability or the stimulus-responsive ability developed in my research group may be fruitfully applied to many recognition systems, interfaces with biological systems, and industrial purposes.

16.10 Acknowledgements

I thank all the staff colleagues, postdoctoral fellows, and students at Nagasaki University, Kyushu University, Sojo University, and on JST-supported projects. Thanks to their ingenious inspirations and all-out efforts, the fruitful results summarized in this chapter have been achieved.

References

1. Leffler, L.E., Grunwald, E. (1963) *Rate and Equilibria of Organic Reactions*, John Wiley & Sons, New York.
2. References related to this chapter are summarized in the following articles: a) Shinkai, S., Manabe, O. (1984) Photocontrol of ion extraction and ion-transport by photofunctional crown ethers, *Topics in Current Chemistry*, **121**, 67–104. b) Shinkai, S. (1990) Functionalization of crown ethers and calixarenes: new applications as ligands, carriers, and host molecules, *Bioorganic Chemistry Frontier*, **1**, 161–195.
3. Shinkai, S., Ishihara, M., Ueda, K., Manabe, O. (1985) Photoresponsive crown ethers. Part 14. Photoregulated crown-metal complexation by competitive intramolecular tail(ammonium)-biting, *Journal of the Chemical Society, Perkin Transactions*, **2** (4), 511–518.
4. Desilva, A.P. (2013) Molecular logic-based computation, *Monographs in Supramolecular Chemistry*, **12**, 1–397.
5. Shinkai, S., Yoshida, T., Miyazaki, K., Manabe, O. (1987) Photoresponsive crown ethers. 19. Protocontrol of reversible association-dissociation phonomena in "tail(ammonium)-biting" crown ethers, *Bulletin of the Chemical Society of Japan*, **60** (5), 1819–1824.

6. Shinkai, S., Kinda, H., Manabe, O. (1982) Photoresponsive complexation of metal cations with an azobenzene-crown-azobenzene bridge immobilized in polymer supports, *Journal of the American Chemical Society*, **104** (10), 2933–2934.
7. References related to this chapter are summarized in the following articles: a) Shinkai, S. (1993) Calixarenes – the third generation of supramolecules, *Tetrahedron*, **49** (40), 8933–8968. b) Shinkai, S. (1993) Calixarenes as the third supramolecular host, *Advanced Supramolecular Chemistry*, **3**, 97–130. c) Linnane, P., Shinkai, S. (1994) Calixarenes: adaptable hosts par excellence, *Chemistry & Industry*, **20**, 811–814.
8. References related to this chapter are summarized in the following articles: a) Takeshita, M., Shinkai, S. (1995) Recent topics on functionalization and recognition ability of calixarenes: the "third host molecule", *Bulletin of the Chemical Society of Japan*, **68** (4), 1088–1097. b) Suzuki, T., Nakashima, K., Shinkai, S. (1994) Very convenient and efficient purification method for fullerene (C_{60}) with 5,11,17,23,29,35,41,47-octa-tert-butylcalix[8]arene-49,50,51,52,53,54,55,56-octol, *Chemistry Letters*, **4**, 699–702. c) Ikeda, A., Shinkai, S. (1997) Novel cavity design using calix[n]arene skeletons: toward molecular recogniton and metal binding, *Chemical Reviews*, **97** (5), 1713–1734.
9. Arduini A., Pochini, A., Reverberi, S., Ungaro, R. (1986) The preparation and properties of a new lipophilic sodium selective ether ester ligand derived from p-t-butylcalix[4]arene, *Tetrahedron*, **42** (7), 2089–2100.
10. McKervey, M.A., Seward, E.M., Ferguson, G., Ruhl, B., Harris, S.J. (1985) Synthesis, x-ray crystal structures, and cation transfer properties of alkyl calixaryl acetates, a new series of molecular receptors, *Journal of the Chemical Society, Chemical Communications*, **7**, 388–390.
11. Jin, T., Ichikawa, K., Koyama, T. (1992) A fluorescent calix[4]arene as an intramolecular excimer-forming Na^+ sensor in nonaqueous solution, *Journal of the Chemical Society, Chemical Communications*, **6**, 499–501.
12. Ghidini, E., Ugozzoli, F., Ungaro, R., Harkema, S., Abu El-Fadl, A., Reinhoudt, D.N. (1990) Complexation of alkali metal cations by conformationally rigid, stereoisomeric calix[4]arene crown ethers: a quantitative evaluation of preorganization, *Journal of the American Chemical Society*, **112** (19), 6979–6985.
13. Atwood, J.L., Koutsantonis, G.A., Raston, C.L. (1994) Purification of C^{60} and C^{70} by selective complexation with calixarenes, *Nature*, **368** (6468), 229–231.
14. References related to this chapter are summarized in the following articles: a) James, T.D., Sandanayake, S., Shinkai, S. (1995) Recognition of sugars and related compounds by "reading-out"-type interfaces, *Supramolecular Chemistry*, **6** (1–2), 141–157. b) James, T.D., Linnane, P., Shinkai, S. (1996) Fluorescent saccharide receptors: assembly and evaluation of boronic acid derived PET sensors, *Journal of the Chemical Society, Chemical Communications*, **3**, 281–288. c) James, T.D., Sandanayake, S., Shinkai, S. (1996) Saccharide sensing with molecular receptors based on boronic acid, *Angewandte Chemie International Edition*, **35** (17), 1910–1922. d) James, T.D., Shinkai, S. (2001) Artificial receptors as chemosensors for carbohydrates, *Topics in Current Chemistry*, **218**, 159–200.
15. References related to this chapter are summarized in the following articles: a) Shinkai, S., Ikeda, M., Sugasaki, A., Takeuchi, M. (2001) Positive allosteric systems designed on dynamic supramolecular scaffolds: toward switching and amplification of guest affinity and selectivity, *Accounts of Chemical Research*, **34** (6), 494–503. b) Takeuchi, M., Ikeda, M., Sugasaki, A., Shinkai, S. (2001) Molecular design of artificial molecular and ion recognition systems with allosteric guest responses, *Accounts of Chemical Research*, **34** (11), 865–873.
16. a) Hong, Y.N., Lam, J.W.Y., Tang, B.Z. (2011) Aggregation-induced emission, *Chemcal Society Reviews*, **40** (11), 5361–5388. b) Wu, J., Liu, W., Ge, J., Zhang, H., Wang, P. (2011) New sensing mechanisms for design of fluorescent chemosensors emerging in recent years, *Chemcal Society Reviews*, **40** (11), 3483–3495.
17. Noguchi, T., Shiraki, T., Dawn, A., Tsuchiya, Y., Lien, L.T.N., Yamamoto, T., Shinkai, S. (2012) Nonlinear fluorescence response driven by ATP-induced self-assembly of guanidinium-tethered tetraphenylethene, *Chemical Communications*, **48** (65), 8090–8092. b) Noguchi, T., Dawn, A., Yoshihara, D., Tsuchiya, Y., Yamamoto, T., Shinkai, S. (2013) Selective detection of NADPH among four pyridine-nucleotide cofactors by a fluorescent probe based on aggregation-induced emission, *Macromolecular Rapid Communications*, **34** (9), 779–784. c) Noguchi, T., Roy, B., Yoshihara, D., Tsuchiya, Y., Yamamoto, T., Shinkai, S. (2014) Cyclization-induced turn-on fluorescence system applicable to dicarboxylate sensing, *Chemistry – A European Journal*, **20** (2), 381–384.
18. Noguchi, T., Roy, B., Yoshihara, D., Tsuchiya Y., Yamamoto, T., Shinkai, S. (2014) Translation of dicarboxylate structural information to fluorometric optical signals through self-assembly of guanidinium-tethered oligophenylenevinylene, *Chemistry – A European Journal*, **20** (41), 13938–13944.

19. References related to this chapter are summarized in the following articles: a) Sakurai, K., Uezu, K., Numata, M., Hasegawa, T., Li, C., Kaneko, K., Shinkai, S. (2005) β-1,3-Glucan polysaccharides as novel one-dimensional hosts for DNA/RNA, conjugated polymers and nanoparticles, *Chemical Communications*, **35**, 4383–4398. b) Numata, M., Shinkai, S. (2011) Supramolecular wrapping chemistry by helix-forming polysaccharides: a powerful strategy for generating diverse polymeric nano-architectures, *Chemical Communications*, **47** (7), 1961–1975. c) Koumoto, K., Mizu, M., Sakurai, K., Kunitake, T., Shinkai, S. (2004) Complementary-strand-induced release of single-stranded DNA bound in the schizophyllan complex, *Chemistry & Biodiversity*, **1** (3), 520–529.

17

Cation Binders, Amphiphiles, and Membrane Active Transporters

George W. Gokel,[1] Saeedeh Negin,[1] Joseph W. Meisel,[1] Mohit B. Patel,[2] Michael R. Gokel,[1] and Ryan Cantwell[1]

[1] Center for Nanoscience, Departments of Chemistry & Biochemistry and Biology, University of Missouri–St. Louis, St. Louis, MO, USA
[2] Departments of Biology, University of Missouri–St. Louis, St. Louis, MO, USA

17.1 Introduction

This book is about the growth of the field of macrocyclic chemistry and some of the individuals who influenced its development. Although macrocycles had been prepared before the 1960s [1], the genesis of the field can be dated to Charles Pedersen's discovery of crown ethers [2]. This discovery was recognized at the time by a number of sagacious individuals as being a seminal development in the history of organic chemistry. Ultimately, Pederson shared the Nobel Prize with two other visionaries, Donald J. Cram [3] and Jean Marie Lehn [4].

A number of relatively young people were enticed into the field at an early stage in its development. These chemists had laboratories around the world and were certainly not concentrated in the United States. Many of us were unknown to each other except through the literature. An important contribution to the field was the BYU group's early organization of macrocycle meetings. The first few of these were organized by Reed Izatt and Jim Christensen and held in Provo, Utah. These meetings presented the opportunity for participants in this fledgling field to exchange information and to become acquainted with colleagues and competitors working in the area. The meetings grew in size and scope over time, and their impact was further enhanced by the Izatt–Christensen Award. Jerald Bradshaw and Steven Izatt initiated the Award: it was their idea and Steven was willing to commit to support it financially through IBC Advanced Technologies, Inc. The first award was given in 1991. Jim Christensen died in September 1987.

Macrocyclic and Supramolecular Chemistry: How Izatt–Christensen Award Winners Shaped the Field, First Edition.
Edited by Reed M. Izatt.

There can be little doubt that the pioneering work of people who participated in those early meetings, and those who received the Izatt–Christensen award, fostered the development of the field. Ultimately, the area of macrocyclic chemistry became supramolecular chemistry and is now an even broader area of research than was envisioned four decades ago.

One of the authors (GWG) was fortunate to have worked in the laboratory of Donald Cram. This was inspiring in two different ways. First, Cram was a consummate investigator and scientist, the most enthusiastic scientist that the author has known. His excitement about chemistry and chemical developments was infectious, and transferred to many of his students. An emblem of his greatness was that although many of the ideas and certainly the direction of his research came from him, he always referred to his coworkers as co-thinkers as well.

Second, Cram realized that the field of macrocyclic chemistry had enormous potential and he saw boundaries only in the far distance. He was proceeding in a direction that he called "pre-organization" [5]. His notion was that complexation and recognition could be enhanced by paying entropic costs in the synthesis of various receptor molecules. The success of this approach is obvious in his cavitand and container molecules [6].

The direction that we took in our early work was based on dynamics more than on enhancing the strength of weak-force interactions. The notion was that nature relies on weak-force interactions for recognition, but these forces cannot be so strong that they prevent dynamics. The hydrogen-bond network of DNA is a good example. Hoogsteen hydrogen bonding is generally considered more stable than the Watson–Crick mode. If DNA were very stable and did not readily separate into single-strand templates, where would life be?

Our efforts to develop dynamic complexing agents and transporters resulted in the lariat ethers [7], the hydraphile channel formers [8], and other synthetic amphiphiles [9]. Some of the work that we did is detailed below. Ironically, one of the most significant discoveries emerged not from our efforts to transport cations, but from looking for evidence for cation–π interactions between double and triple bonds and arenes and alkali metal cations [10].

17.2 Conceptual development of lariat ethers for transport

Immediately after the report of crown ethers by Pedersen, synthetic organic chemists began exploring the range of structures that was possible based on this new structural element [1]. Many aromatic rings and heterocycles were fused to the basic structure, and compounds were devised that had two rings or even three rings connected in various ways. In many instances the effort was made in order to define what sorts of compounds could be prepared and to discover the effects of structural variation on complexation strength.

Lariat ethers are crown ethers that have side-arms [7]. The synthesis of crown ethers having side-arms is really rather obvious in the context of structural variations. Indeed compounds with side-arms were already known when the lariat ether work began – such compounds had been prepared, for example, by Okahara and his coworkers at Osaka University [11]. Our interest was in developing compounds that not only had side-arms, but had donor elements within those side-arms. We recognized that crown ethers were excellent carrier molecules for various cations. Most of the compounds that had been devised to function as carriers, however, were essentially two dimensional. The remarkable cryptands reported by Lehn and Sauvage, were very powerful binders [12]. They could function as transport agents, but they had poor dynamics. Because the binding of cations by these compounds was so strong, release of the cation after transport was a problem.

Nature had solved the dynamics problem in carriers with structures such as valinomycin [13]. This cyclodepsipeptide is a remarkable structure in the sense that it is too large to bind a potassium cation and yet

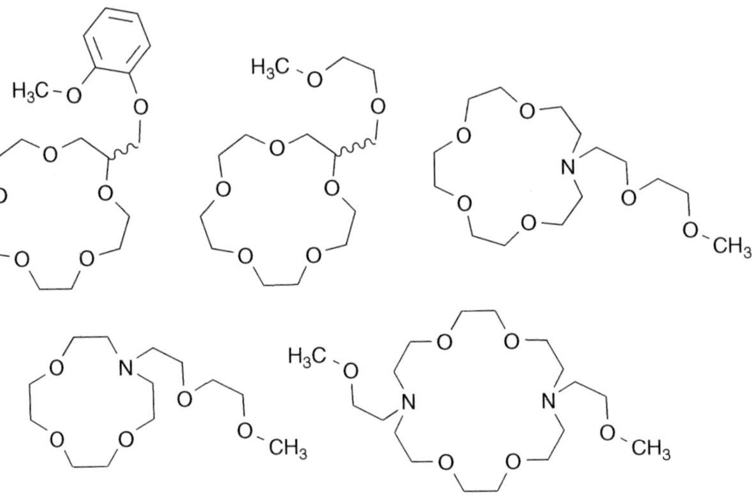

Figure 17.1 *Structures of one-armed carbon-pivot (upper left and center) and nitrogen-pivot lariat ethers. The structure in the lower right is a two-armed or bibracchial lariat ether*

it is very selective for it. Valinomycin is a 36-membered ring, but it has the ability to fold over into what has been called a "tennis ball seam" arrangement [14]. This folding imparts the three-dimensionality of cryptands to the bound complex while retaining high enough dynamics to release the cation after transport is completed. The development of lariat ethers was modeled on this principle [7]. The notion was that the macrocycle would bind a cation, and donor-containing side-arms would reach over and/or under the macrocycle to augment complexation in a dynamic but three-dimensional complex.

Initial studies were undertaken in which the side-arms were bound to carbon in an all-oxygen macrocycle. The attachment point was carbon, which presented a problem. Any side-arm attached to carbon in a typical 3n-crown-n structure would be inherently chiral. Moreover, the stereochemical issue is complicated by the lack of side-arm flexibility when it is bound to carbon. This meant that any complex that formed had a distinct sidedness to it. We quickly realized that it was necessary to attach the side-arm or arms to nitrogens within the azamacrocyclic rings. Nitrogens are typically poorer donors for alkali metal cations than are oxygens, but their presence was necessary in order to maintain dynamics. Of course, their donicity was enhanced some by alkyl substitution. We designated the lariat ethers as carbon- or nitrogen-pivot compounds depending on whether the side-arm or arms were attached at carbon or nitrogen (Figure 17.1).

One interesting result that emerged from a study of the *N*-pivot lariat ethers related to the so-called "hole size effect" [15]. A much discussed concept in the early stage of crown ether chemistry was that when a cation was bound by a macrocyclic ring of appropriate size its binding was at a maximum. In other words sodium would be bound more strongly by a 15-crown-5 macrocycle, whereas potassium would be bound more strongly by an 18-crown-6. The latter is generally true but the former is not: 18-crown-6 is a stronger binder for sodium cation (Na^+) than is the 15-membered ring for which it has a more appropriate size relationship. This is illustrated in the graph of Figure 17.2.

We designed molecules that had 12-, 15- and 18-membered rings and side-arms containing three donor oxygen atoms. We discovered that a lariat ether having a 12-membered ring and three donors in the side-arm exhibited a binding constant similar in magnitude to an 18-membered ring with a single donor oxygen in the side-arm [16].

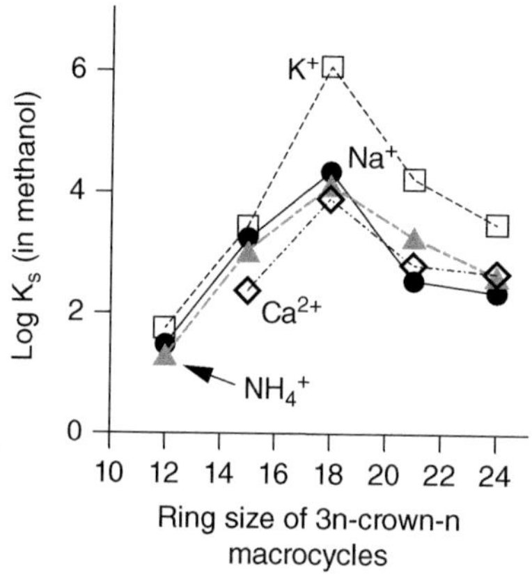

Figure 17.2 *A plot of binding constants measured in anhydrous methanol for Na^+, K^+, Ca^{2+}, and NH_4^+, with 12-crown-4, 15-crown-5, 18-crown-6, 21-crown-7, and 24-crown-8*

17.3 Recognition of the ability of lariat ethers to form membranes

As discussed above, the lariat ethers were designed so that the side-arms would have donor groups capable of augmenting cation binding by the macrocycle. When donor groups in the side-arms were lacking, however, the side-chains were nonpolar and the lariat ethers comprised amphiphiles. The crown ether macrocycle proved to be sufficiently polar that it could act as a head-group if non-polar side-arms were present [17]. It was well known that, for example, dibenzo-18-crown-6 could be made a better transport agent by *t*-butylating the benzene rings. The di-*t*-butyl compound was more hydrophobic than the parent and therefore more soluble in non-polar media through which alkali metal salts could be transported. Pedersen prepared a more hydrophobic version of dibenzo-18-crown-6 by reduction of the arenes to form dicyclohexano-18-crown-6 [18]. This isomer mixture was more soluble in non-polar solvents, and was a better alkali metal cation binder.

The lariat ethers were different in the sense that the hydrocarbon chains not only augmented transport; because of their hydrophobicity, they could form membranes. Compounds that have a polar head-group and a non-polar tail are amphiphiles. Amphiphiles form aggregates in aqueous solution when the concentration is appropriate. These aggregates may take various forms but typically, they will be either micelles or liposomes (also called vesicles) [19]. Typically, the relationship between head-group area and tail volume will dictate whether liposomes or micelles form. Twin-chained compounds such as natural phospholipids typically form bilayers and stable liposomes [20]. We experimented with a variety of lariat ethers to see if they would form either type of aggregate [21].

We recognized that cholesterol is a stabilizing and rigidifying element of many natural mammalian membranes. It therefore occurred to us to use the cholesteryl side-chain as the hydrophobic element in the lariat ether. Experience showed that cholestanyl was preferable for such studies as there was no possibility of double bond migration. Our initial efforts focused on making 2-carboxy-15-crown-5. This proved to be a cumbersome molecule to prepare. Instead we opted for the hydroxymethyl crown and allylated cholestanol. The allyl ether double bond was converted to the geminal diol. From this, we prepared the 15-crown-5 amphiphile. Although we were successful in obtaining this molecule, it was fraught with the difficulty of stereoisomers. We therefore turned our attention to the nitrogen-pivot lariat ethers.

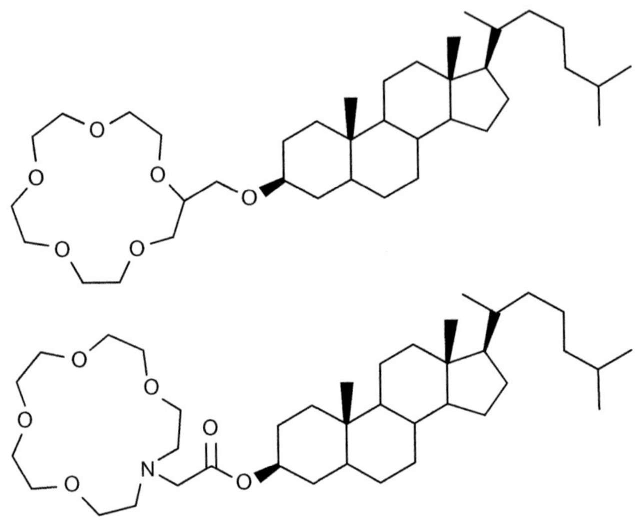

Figure 17.3 Carbon- and nitrogen-pivot steroidal lariat ethers that form stable aggregates in aqueous suspension. (Solid state steroidal lariat structure, CSD: FIXTEL.)

In order to prepare a steroidal lariat ether [22], cholestanol was allowed to react with chloroacetyl chloride. This formed the activated chloromethyl ester, which was then treated with aza-15-crown-5. The resulting compound is shown in Figure 17.3. This material was melted, dispersed, and sonicated in deionized water. The concentrations of amphiphile used in these experiments were in the range of 1–5 μM. A stable suspension was observed, was confirmed by the appearance of a bluish color, by aggregate sizing (dynamic light scattering, DLS), and by transmission electron microscopy, to be vesicles. The aggregates thus produced had an average size of 300 to 330 Å.

A number of other cholestanol-containing compounds were prepared and their aggregation behavior studied. In addition, a number of two- and three-armed di- and triaza lariats were studied. The results in these cases were more complex, and it was concluded that aggregate organization was strongly influenced by head-group interactions when the side-arms were linear alkyl chains [23].

The work was eventually extended to include twin macrocycles linked by a single hydrocarbon chain. Fuhrhop had suggested the name bolaamphiphiles for twin-headed amphiphiles [24]. The term has often been shortened to "bolyte." Of course, such structures occur naturally, especially in the membranes of organisms classified in the domain (kingdom) *Archaea*. Our notion was to see if two crown head-groups spanned by a single hydrocarbon chain could form stable aggregates. Our hope was that liposomal structures – bola-amphisomes – would result. This proved to be the case.

A dozen crown ether bolytes were prepared by treating an aza-15-crown-5 or aza-18-crown-6 compound with a linear diacid such as dodecanedioic acid as its dichloride [25]. The corresponding diamides were reduced with $LiAlH_4$ to give compounds of the type shown in Figure 17.4.

A surprising outcome of this study was that liposomes were formed only when the spacer chain ("n") consisted of 10 or 12 methylene groups. In all other cases, save one, micelles were formed. The single exception was an aza-18-crown-6 bolyte having a spacer structure of $CO(CH_2)_{11}S(CH_2)_{12}S(CH_2)_{11}CO$ [26]. These results were important in the development of our synthetic channel compounds, as described in Section 17.4. It should be noted that professors Luis Echegoyen, Angel Kaifer, and post-doctoral fellow, now professor, Servando Munoz were intimately involved in these studies.

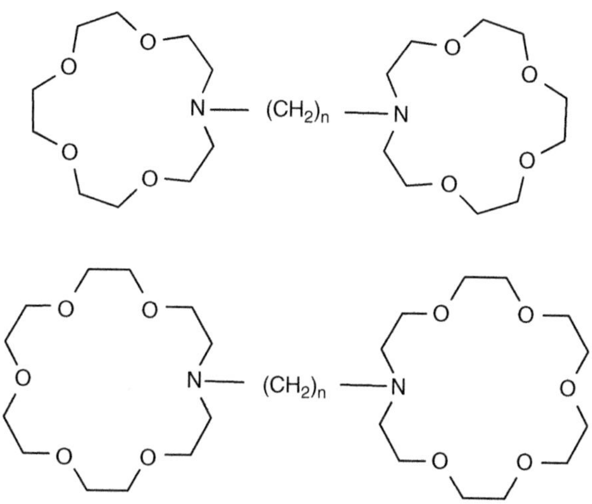

Figure 17.4 *Crown ether-based bola-amphiphiles, some of which formed micelles and others formed stable bilayer "bola-amphisomes"*

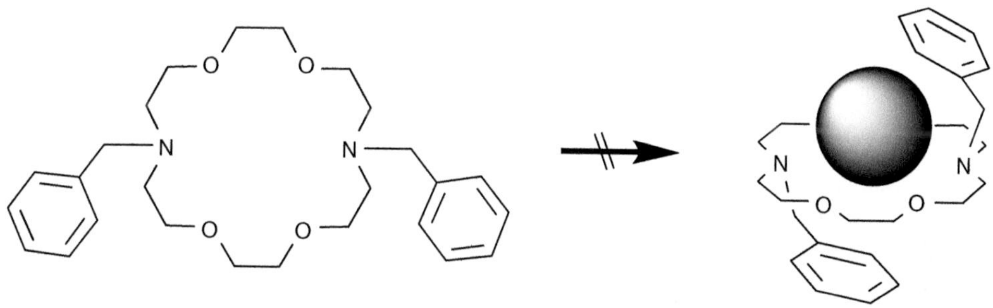

Figure 17.5 *Structure of N,N′-dibenzyl-4,13-diaza-18-crown-6 and a possible π-donor side-arm–alkali metal complex*

17.4 Use of lariat ethers to demonstrate cation–π interactions

From the earliest preparations of lariat ethers containing aromatic side-arms, the question of π–base participation in alkali metal complexation loomed. The notion was that an electron-rich aromatic side-arm could serve as an apical donor group for a ring-bound cation in much the same way that that oxygen donors present in lariat ether side-arms could augment binding by serving as apical donors. We crystallized *N,N′*-dibenzyl-4,13-diaza-18-crown-6 and various sodium and potassium salts from a range of solvents and solvent mixtures, but we were unable to document the formation of a complex such as the one shown in Figure 17.5 [27].

We were aware, of course, that Atwood and coworkers had crystallized a dibenzo-18-crown-6 complex of $K(BPh)_4$ in which that apical position was occupied by benzene solvent [28]. This was clearly a cation–π (K^+–arene) complex, but some uncertainty existed in our minds about whether packing forces were playing an important role. Fortuitously, a discussion about my disappointment in this area led two graduate students working in my lab, Steve DeWall and Eric Meadows, to suggest that the chains connecting the macro-ring to the arenes were simply not long enough. They offered an analogy with nature, in which the benzene ring of phenylalanine is two carbons away from the amine nitrogen.

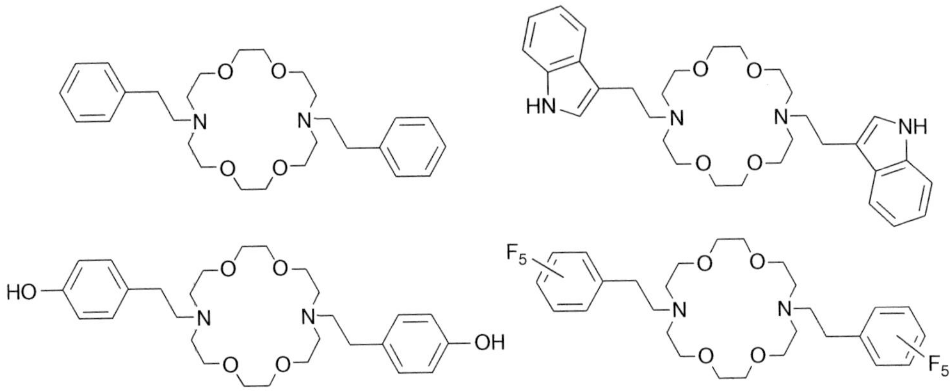

Figure 17.6 *Four bibracchial (two-armed) lariat ethers that were initially studied to determine if macro-ring-bound cations would exhibit apical cation–π coordination*

They prepared three new lariat ethers based on their concept of amino acids. The side-arm arenes were benzene, phenol, and indole, which represented the π-rich arenes present in the common amino acids phenylalanine, tyrosine, and tryptophan [29]. The three compounds are shown in Figure 17.6 (top row and lower left); each of them formed cation complexes in which both side-arm arenes clearly participated as π-donors. This work involved numerous solid state analyses and we were fortunate to collaborate with Len Barbour in this effort.

A careful examination of the solid state structures for the phenyl, hydroxyphenyl, and indolyl derivatives showed that packing was nearly identical in the solid state, even when the side-arm arene contained no donor group to coordinate to the external anion. We thus prepared the *bis*(pentafluorophenyl) shown at the lower right of Figure 17.6. Its arenes are clearly electron poor and, if anything, π-acids rather than π-bases. When complexed by KI, there was no side-arm arene involvement in the complex. Rather the complexes were in an infinite stack in which K–I–K–I... interactions organized the matrix. It should also be noted that solution (NMR) studies confirmed the formation of cation–π interactions in cases where solid state evidence was clear.

Our early failure to find cation–π interactions with allyl and propargyl side-chains was then tested by Jiaxin Hu, a post-doctoral fellow, who likewise extended the side-arms. He prepared numerous 15- and 18-membered ring compounds having double and triple bonds and showed that when these π-donors were appropriately placed they served as additional donors for the ring-bound cation [30].

There are four arenes among the 20 common amino acids. Benzene, hydroxybenzene, and indole are all electron-rich and all showed clear evidence for cation–π enhancement of complexation. See structures in Figure 17.7. Imidazole is an electron-poor arene and it also has nucleophilic nitrogen within its heterocyclic ring structure. When imidazole was attached as a side-arm to a macrocycle, the cation complex was stabilized by the side-arm, but the interaction was with the nitrogen donor rather than with the arene [31].

An interesting addendum to this study led us to question whether such interactions were common in solid state structures of proteins. Burley and Petsko had recognized this possibility as early as 1986 based on an analysis of existing structures [32]. We used our angle and distance data to compare data obtained by others and found that in a case in which a water molecule was thought to be in contact with a tryptophan in a protein active site, the angles and distances suggested that the contact was actually with Na^+. Of course, Na^+ and H_2O have the same number of electrons and scatter similarly, so either assignment could be correct [33].

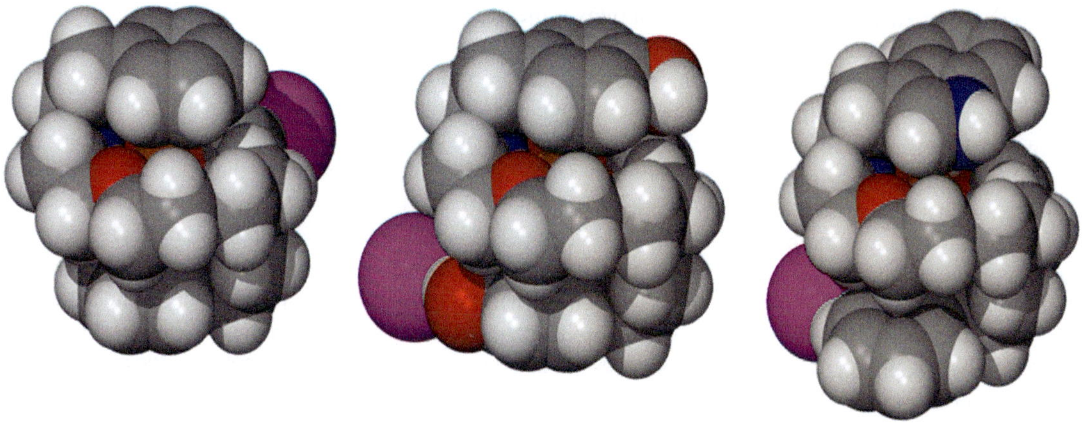

Figure 17.7 Solid state structures of K–I complexes between diazacrowns having (left) phenyl, (center) hydroxyphenyl, and (right) indolyl-terminated side-arms. The Cambridge Structural Database designations are OCABEZ, VEQTIU, GOXWIZ

17.5 Development of synthetic cation channels based on crown ethers

By the late 1980s, a vast number of crown ether structures were known. These had been studied as cation and molecule binders and many had proved to be transporters. The transport and detection of ions has a range of applications in analysis and commerce. Ion transport is also a natural phenomenon. In vital systems there are cation-complexing agents and carriers, to be sure, but most ion transport is mediated by protein channels. As noted above, the flexible and dynamic depsipeptide valinomycin was the inspiration for lariat ethers. The fact that most natural ion transport involves pores presented the synthetic ion channel challenge answered by several labs.

The time in which the first labs were working to design synthetic ion channels pre-dated any solid state structure. To be sure, there was considerable information available from a century of biochemical and biophysical study. Much was known about how channels functioned, but important structural information was lacking. For example, the open–close behavior characteristic of natural channels was analogized to a "swinging gate" for want of a more detailed structural model [34].

Our efforts were centered on crown ether compounds for obvious reasons. We knew that crowns bind ions in a selective fashion. We knew that they could be incorporated as head-groups in amphiphilic assemblies. We knew that various side-arms could be attached and that extended structures could be assembled from familiar modules. In fact, our initial design was, in a sense, a merger of lariat ether and bola–amphiphile studies.

It seemed clear to us that any synthetic channel would require openings – entry and exit portals – on either side of a bilayer membrane. These openings would, advantageously, be capable of serving as amphiphilic head-groups. The crown ether fit both of these requirements.

It seemed obvious that the channel or pore should be at least as long as the bilayer membrane is thick. What is somewhat less obvious is the actual thickness of a bilayer membrane. Numerous variables affect this question. Natural bilayers are extremely complex supramolecular assemblies that are often composed of phospholipids, ceramides, sphingolipids, proteins, steroids, and other compounds. Even considering only phosphatidylcholine derivatives, the two fatty acids may be the same or different, if different, they may have similar or very different chain lengths, and the fatty chains may be saturated or unsaturated. Indeed, the chains may exhibit multiple unsaturations and various bond geometries [35].

Even today, the structure and organization of bilayer membranes remains a challenge to define. Liposomes can be prepared from single phospholipids and are, therefore, a more manageable cell model. Even in this case, though, one can readily determine the size of the liposome – for example by DLS measurements – but that number does not reveal the bilayer thickness. One can double the length of the phospholipids but the extent of interpenetration (interdigitation) of the fatty acid chains is not precisely known. It is also not clear whether the insertion of a pore-forming amphiphile will alter the local thickness of the bilayer.

Another consideration is whether the synthetic amphiphile will insert into the bilayer from the bulk phase or if it must be pre-inserted into the bilayer in order to function. The latter method is often used in biophysics in the analysis of protein channels. The method chosen can alter the experimental results. An example is that the commercial detergent Triton X-100 shows channel-like activity when pre-inserted into a planar bilayer and studied by using a voltage clamp apparatus. It shows open–close behavior, as do natural protein channels [36]. In contrast, when added to a bulk phase, the detergent either does not insert into the bilayer or does not form the pores detected when it is pre-inserted [37]. Our strategy was to devise a compound that could self-insert into bilayer membranes.

At the time we were considering these issues and working on our design, both Jean-Marie Lehn [38] and Thomas Fyles [39] were doing likewise. It is interesting that their approaches both focused on a crown ether as a central scaffold from which the remainder of the molecule would emanate. Their central scaffold crowns were based on tartaric acid so that the emanating chains could be appropriately oriented. The polar headgroups were the termini of chains connected to the central crown.

Our design also incorporated a central macrocycle but it was included for a different reason. We estimated that the insulating regime of a bilayer membrane should be 30–40Å. This seemed rather a long distance for a poorly solvated cation within a pore to traverse unaided. We speculated that a central macrocycle would provide a shorter distance for the cation to "jump" from either the entry or the exit portal and thereby enhance transport. Incorporation of this third macrocycle was based largely on chemical intuition but proved to be fortuitous. A similar structural element was revealed in the KcsA voltage gated channel when MacKinnon's Nobel Prize-winning solid state structure appeared. The explanation offered for what we had called the "central relay" was as follows. "The narrow selectivity filter is only 12 angstroms long, whereas the remainder of the pore is wider and lined with hydrophobic amino acids. A large water-filled cavity and helix dipoles are positioned so as to overcome electrostatic destabilization of an ion in the pore at the center of the bilayer [40]."

The schematic for our channel is shown in Figure 17.8. We anticipated that the macrocyclic rings would all be 4,13-diaza-18-crown-6, the spacer chains would be dodecylene $[(CH_2)_{12}]$, and the distal side-arms would also be dodecyl groups. The decision to use dodecyl groups was based in some measure on the ready availability of precursor molecules.

Although we guessed correctly about the need for a central relay unit, the conformation initially envisioned was flawed [41]. The central relay was needed to create a polar element at the midplane of the bilayer. This would be accomplished if the central macrocycle was perpendicular to the distal macrocycles. This would have two beneficial effects. First, the polar crown in the conformation shown concentrated polarity at the midplane of the bilayer, the least polar part of the insulating regime of the membrane. Second, if perpendicular, the polarity would be dispersed over ~10Å and the transient ion would not be required to pass through the orifice. Passing "by" rather than passing "though" is surely a more dynamic process.

The first pore-former compound [42] (hydraphile) prepared in our lab was synthesized by professor Akio Nakano, who was on sabbatical leave from Toa University in Japan. Echegoyen and coworkers used an NMR exchange methodology developed originally by Riddell and coworkers [43] to show that Na^+ exchange through liposomal membranes occurred when the channel was added to the bulk phase. Numerous control experiments were conducted to determine that all of the structural elements designed into the channel were required for function. A summary of much of this work has appeared in a review in *Chemical Communications* [44], and will not be repeated here. We will note that as more elaborate structures were brought to hand, we coined the term "hydraphile" to replace the rather awkward "*tris*(macrocylic) pore-formers."

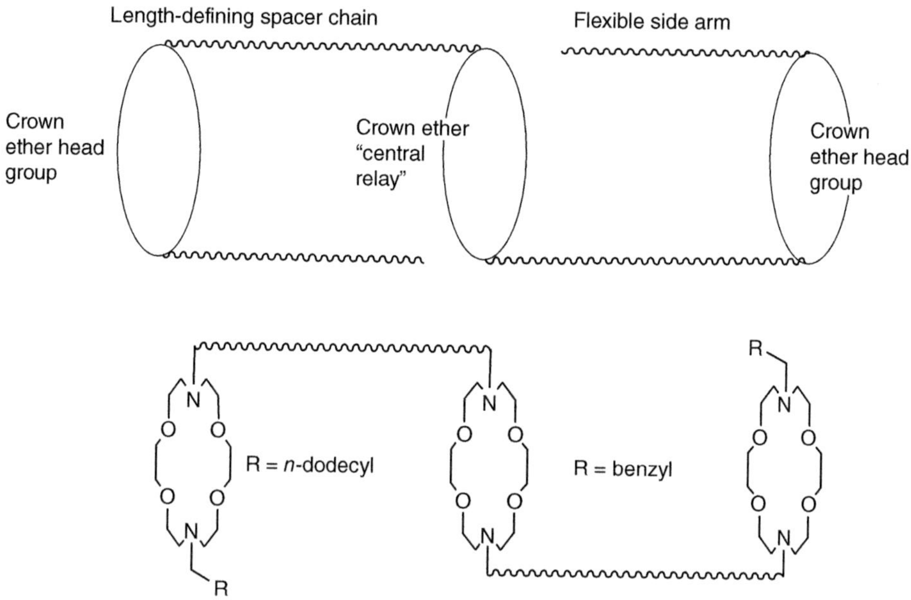

Figure 17.8 *Top. Schematic for the design of a tris(macrocycle) pore former or "hydraphile." Bottom. Partial schematic showing two different possible side-arms and the three 18-membered ring macrocycles*

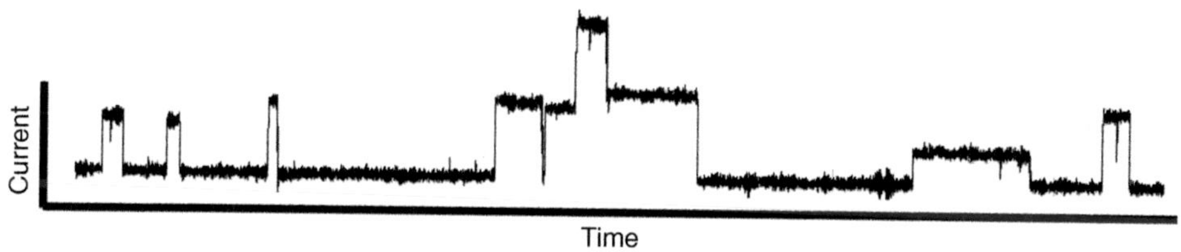

Figure 17.9 *Conductance traces obtained for a hydraphile synthetic pore-former by using the planar bilayer conductance voltage clamp method*

It is worth emphasizing that as useful as the Na^+ exchange method was in comparing the function of various channels and controls in liposomes, it was planar bilayer studies that showed the most convincing evidence of channel function. The initial data were obtained by (now professor) Ernesto Abel-Santos [45]. The observation of classic, open–close behavior for these compounds was gratifying. Two traces obtained for an early *tris*(macrocycle) pore-former are shown in Figure 17.9.

One of the interesting outcomes of our effort to obtain evidence for cation–π interactions led to a consideration of the cations present as head-groups in bilayer membranes. Natural membranes contain significant amounts of phosphatidylcholine, which has a positively charged quaternary ammonium head group. Investigation of a range of "hydraphiles" with differing side-chains – alkyl, aryl, aralkyl, and various substituted arenes –found that when the side-chains on the distal macrocycles were electron-rich, channels of the same length performed better than when the head-groups were electron-poor. It was inferred that favorable cation–π interactions stabilized the conductance pore [46].

17.6 Development of synthetic anion channels based on amphiphilic peptides

The successful development of a synthetic cation-conducting channel inspired us to consider the design of a synthetic anion-conducting channel. As was the case with the emerging area of crown ether chemistry, numerous new compounds were appearing that could complex such anions as Cl^- or Br^- and transport them across membranes of various sorts. A colleague, professor Paul Schlesinger, had experience with certain peptides that showed Cl^--transporter function and encouraged consideration of a new design.

More information was available concerning the remarkable ClC chloride channel [47] than had been available for the cation channel project. Even so, a number of design decisions were required. It seemed axiomatic that the first channels must have functioned in membranes. As there is no fossil record of early membranes, we chose to emulate the structure of a modern membrane. We devised a prototype structure that was based on the structure of modern phospholipids [48]. We used a number of modules within that would facilitate the overall design as well as synthetic access. For example, the two fatty acids present in a typical phospholipid were emulated by a dialkylamine. The glyceryl residue was modeled by diglycolic acid. This combination was advantageous because the reaction of a dialkylamines with diglycolic anhydride would give a major segment of the desired structure in a single step.

An additional advantage of this approach is that the chains and chain lengths in the dialkylamines can be varied, as can the diacid. Compounds having twin normal alkyl chains from methyl to octadecyl were prepared [49] as were examples having unequal chain lengths [50] and even arenes. Diacids included succinic, thiadiglycolic, glutaric, and others (see Figure 17.10) [51].

The most difficult decision concerned what residues would comprise the entry portal/selectivity filter in these structures. The putative selectivity filter of the ClC channel incorporated glycine and proline. We decided to use a heptapeptide sequence of $(Gly)_3Pro(Gly)_3$ as the amphiphilic head-group, fully expecting our amphiphile to self-assemble into a dimeric pore in order to function. The C-terminal glycine was esterified with a benzyl group to reduce polarity and to permit ready changes in that group, as it could be removed hydrogenolytically. The resulting initial structure is shown in Figure 17.11.

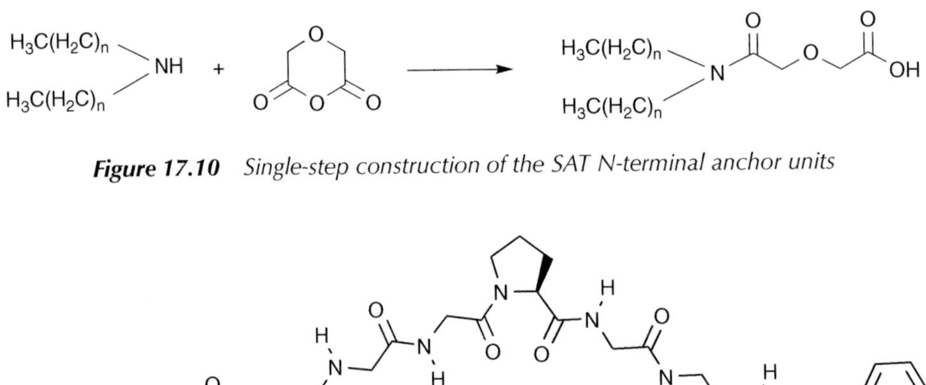

Figure 17.10 *Single-step construction of the SAT N-terminal anchor units*

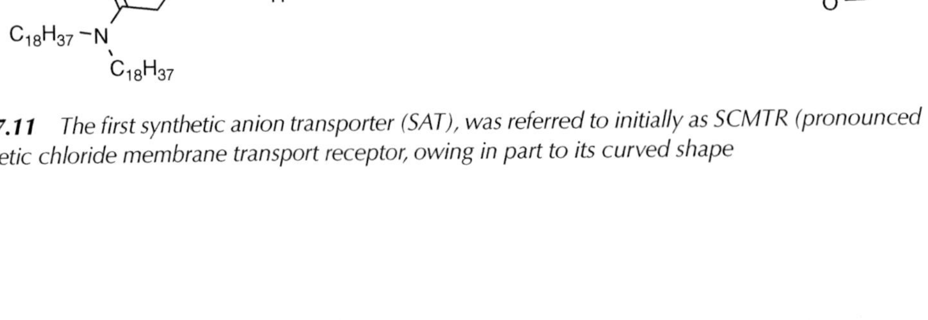

Figure 17.11 *The first synthetic anion transporter (SAT), was referred to initially as SCMTR (pronounced "scimitar") for synthetic chloride membrane transport receptor, owing in part to its curved shape*

We called these compounds synthetic anion transporters or SATs. They transported chloride anion in preference to potassium cation by a ratio of at least 10:1. Extensive study led us to conclude that SATs inserted into the bilayer from the bulk phase, although only about 35% of the available monomers entered the membrane. Once in the membrane, they self-assembled to form functional Cl^--transporting dimeric pores. Planar bilayer conductance studies revealed open–close behavior, and numerous structural variations within the framework altered transport rates, selectivity, or both.

One of the first control experiments conducted was to determine if the proline present in the ClC channel was required. The $Gly_3ProGly_3$ heptapeptide sequence was replaced by $Gly_3LeuGly_3$. The change was not expected to alter hydrophobicity but proline's ability to bend the peptide would be lost. In fact, Cl^- transport was dramatically diminished by the Pro → Leu mutation [52].

In a major synthetic effort, two monomeric SATs were linked head-to-head or tail-to-tail. The ability to transport Cl^- was assessed for each of these "*pseudo*-dimers." In both cases, function exceeded that of an equal concentration of monomers. This does not confirm dimer formation as the active conductor but it comports with it. Hill plots on the various monomer structures also showed cooperativity numbers of about 2.

17.7 Membrane active amphiphiles as biologically active and applicable compounds

The development of hydraphile ion channel formers was begun in part because of a desire to emulate nature and in part as an intellectual exercise. Once functioning synthetic ion channels were in hand, the obvious question seemed to be whether or not they were biologically active and/or compatible with vital organisms [53]. Matthew Leevy (now professor) reasoned that the non-rectifying hydraphiles would disrupt ion homeostasis in bacteria if they inserted into their barrier membranes and conducted ions.

We anticipated that the hydraphile molecules were inserting in the barrier membranes. This was tested by preparing a fluorescent hydraphile (Figure 17.12) and examining a suspension of *E. coli* to which it had been added [54]. Fluorescence micrographs showed clearly that the fluorescent dye was localized in the periphery of *E. coli*. A further experiment was undertaken with a non-fluorescent hydraphile, which was added to human embryonic kidney (HEK 293) cells. Addition to the bathing medium caused a rapid increase in membrane electrical activity as judged by a patch clamp study. Bathing with saline removed the hydraphile, diminished electrical activity, and left the HEK cells vital [55].

It occurred to us that if these various synthetic amphiphiles could penetrate membranes, they might alter the ability of other substances to pass through bilayers. We decided to examine the possible penetration of antibiotics into *E. coli* and *B. subtilis*. These organisms have very different membranes. We chose four FDA-approved antibiotics for our initial studies: erythromycin, kanamycin, rifampicin, and tetracycline. These antibiotics are widely prescribed and their structures (Figure 17.13) are very different. Moreover, their modes of action are well established and different in all four cases.

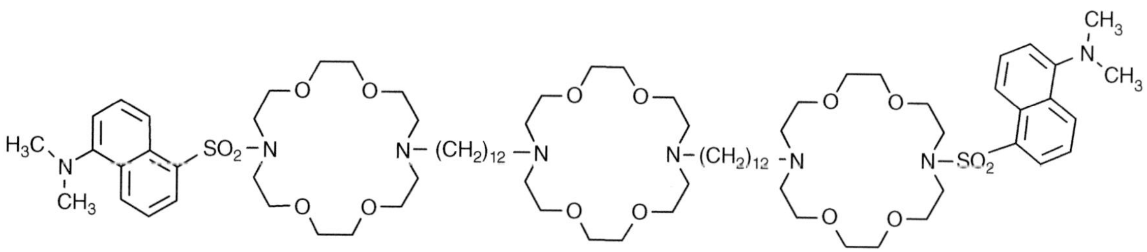

Figure 17.12 *The functional, fluorescent dansyl channel*

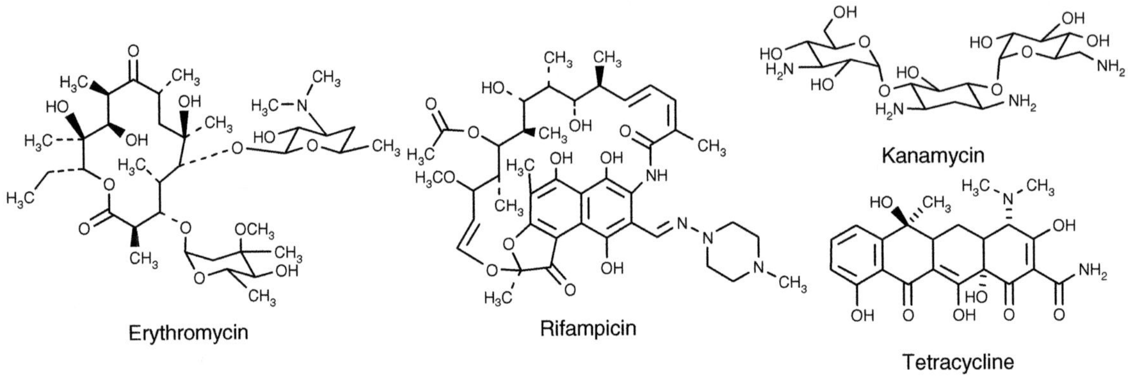

Figure 17.13 *Chemical structures of erythromycin, kanamycin, rifampicin, and tetracycline*

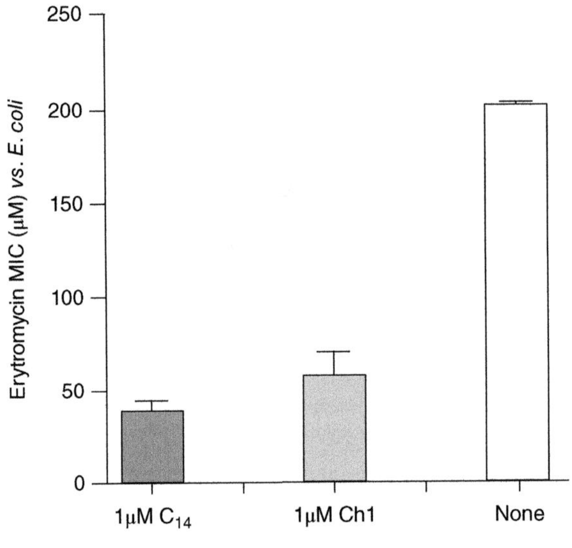

Figure 17.14 *Hydraphile-enhanced potency of erythromycin against* E. coli

Minimum inhibitory concentrations (MICs) were determined for hydraphile-treated Gram-negative *Escherichia coli*, Gram-positive *Bacillus subtilis*, and for the primary eukaryote (yeast) *Saccharomyces cerevisiae* (the barrier membranes of the three organisms are quite different) [56].

MIC values were determined for each organism under study in the presence of hydraphile and, separately, in the presence of antibiotic. Then the organism was exposed to a non-toxic concentration of hydraphile – typically half MIC – and the MIC for the antibiotic determined again. If the presence of the hydraphile increased antibiotic potency, the MIC for the antibiotic should be lower than in the absence of the amphiphile. Figure 17.14 shows data for two channels, each of which was added at 1 μM or half MIC or less. The right-hand column shows the MIC for erythromycin against a strain of *E. coli*. It is ineffective. However, adding benzyl C_{14} hydraphile or *n*-dodecyl C_{12} hydraphile enhances erythromycin potency by 8- or 4-fold, respectively.

The MIC values varied with the structure of the hydraphile. It was found that hydraphile molecules that were too short to effectively conduct ions showed very little toxicity. The very longest of the hydraphiles also showed reduced activity, perhaps because their conductance pores were either hard to form or too long to assume the ideal conformation within the membrane. The effective hydraphiles were potent at low micromolar to high nanomolar concentrations.

These results have been expanded to a range of organisms, antibiotics, and amphiphiles. The hydraphile amphiphiles have an astonishing ability to enhance the potency of antibiotics and a major effort is currently underway to understand the mechanism or mechanisms of action and to derive utility from these observations. During the course of these studies, it became apparent that the solvent DMSO could also play a role in altering antibiotic potency. We have striven to keep the amount of DMSO present in our experimental studies both low and constant. We have completed extensive controls to enhance our confidence in our results. In addition, though, we have queried whether the effect of DMSO is constant. We have recently reported that the influence of DMSO is variable and simple "rule of thumb" evaluations may be problematic [57].

In other recent work, we have found that the widely studied plant *Arabidopsis thaliana* shows significant changes in root morphology in the presence of certain hydraphiles [58]. Other synthetic amphiphiles are under study in this context and have already shown remarkable biological activity.

17.8 Conclusion

Part of the work summarized here resulted in the receipt of the Izatt–Christensen International Award in Macrocyclic Chemistry in 1996 by one of the authors (GWG). It was gratefully received. The work described herein is only a fraction of the results that have been produced by my talented coworkers, past and present. Some of the basic ideas described here were those of the awardee, but many more came from the creative minds of those in my laboratory that Cram would have called "co-thinkers." I am infinitely indebted to them.

References

1. Gokel, G.W., Korzeniowski, S.H. (1982) *Macrocyclic Polyether Syntheses*, Springer-Verlag, Berlin.
2. Pedersen, C.J. (1967) Cyclic polyethers and their complexes with metal salts, *Journal of the American Chemical Society*, **89**, 7017–7036.
3. Cram, D.J. (1988) The design of molecular hosts, guests, and their complexes, *Science*, **240**, 760–767.
4. (a) Lehn, J.-M. (1988) Supramolecular chemistry – scope and perspectives molecules, supermolecules, and molecular devices (Nobel lecture), *Angewandte Chemie International Edition*, **27**, 89–112. (b) Lehn, J.-M. (1995) *Supramolecular Chemistry*. Wiley-VCH, Weinheim.
5. Cram, D.J., Kaneda, T., Helgeson, R.C., Brown, S.B., Knobler, C.B., Maverick, E., Trueblood, K.N. (1985) Host–guest complexation. 35. Spherands, the first completely preorganized ligand systems, *Journal of the American Chemical Society*, **107**, 3645–3657.
6. Cram, D.J., Cram, J.M. (1994) *Container Molecules and Their Guests*, Royal Society of Chemistry, Cambridge.
7. Gokel, G.W., Dishong, D.M., Diamond, C.J. (1980) Lariat ethers. Synthesis and cation binding properties of macrocyclic polyethers possessing axially disposed secondary donor groups, *Journal of the Chemical Society, Chemical Communications*, 1053–1054.
8. Nakano, A., Xie, Q., Mallen, J.V., Echegoyen, L., Gokel, G.W. (1990) Synthesis of a membrane-insertable, sodium cation conducting channel: kinetic analysis by dynamic sodium-23 NMR, *Journal of the American Chemical Society*, **112**, 1287–1289.
9. Wang, W., Li, R., Gokel, G.W. (2009) "Aplosspan:" a bilayer-length, ion-selective ionophore that functions in phospholipid bilayers, *Chemical Communications*, 911–913.
10. Gokel, G.W., Barbour, L.J., Ferdani, R., Hu, J. (2002) Lariat ether receptor systems show experimental evidence for alkali metal cation-pi interactions, *Accounts of Chemical Research*, **35**, 878–886.

11. (a) Kuo, P.-L., Miki, M., Ikeda, I., Okahara, M. (1978) (a) New synthesis of N-alkyl monoaza crown ethers, *Tetrahedron Letters*, 4273–4276. (b) Kuo, P.-L., Miki, M., Ikeda, I., Okahara, M. (1978) New synthesis of N-alkyl monoaza crown ethers, *Tetrahedron Letters*, 4273–4276. (c) Ping-Lin, K., Miki, M., Okahara, M. (1978) General synthesis of substituted and unsubstituted crown ethers, *Chemical Communications*, 504–505.
12. (a) Dietrich, B., Lehn, J.M., Sauvage, J.P. (1969) Diaza-polyoxa-macrocyclic and macrobicyclic compounds, *Tetrahedron Letters*, 2885–2888. (b) Lehn, J.M. (1985) Supramolecular chemistry: receptors, catalysts, and carriers, *Science*, **227**, 849–856.
13. (a) Brockmann, H., Schmidt-Kastner, G. (1955) Antibiotics from actinomycetes. XXVII. Valinomycin I, *Chemische Berichte*, **88**, 57–61. (b) Moore, C., Pressman, B.C. (1964) Mechanism of action of valinomycin on mitochondria, *Biochemical and Biophysical Research Communications*, **15**, 562–567. (c) Pinkerton, M., Steinrauf, L.K., Dawkins, P. (1969) Molecular structure and some transport properties of valinomycin, *Biochemical and Biophysical Research Communications*, **35**, 512–518.
14. Truter, M.R., Bush, M.A. (1970) Crystal structures of three alkali-metal complexes with cyclic polyethers, *Journal of the Chemical Society D*, 1439–1440.
15. Gokel, G.W., Goli, D.M., Minganti, C., Echegoyen, L. (1983) Clarification of the hole-size cation diameter relationship in crown ethers and a new method for determining calcium cation homogeneous equilibrium binding constants, *Journal of the American Chemical Society*, **105**, 6786–6788.
16. Arnold, K.A., Hernandez, J.C., Li, C., Mallen, J.V., Nakano, A., Schall, O.F., Trafton, J.E., Tsesarskaja, M., White, B.D., Gokel, G.W. (1995) Analysis of sodium, potassium, calcium, and ammonium cation binding selectivity in one- and two-armed nitrogen-pivot lariat ethers, *Supramolecular Chemistry*, **5**, 45–60.
17. Echegoyen, L.E., Hernandez, J.C., Kaifer, A., Gokel, G.W., Echegoyen, L. (1988) Aggregation of steroidal lariat ethers: the first example of nonionic liposomes (niosomes) formed from neutral crown ether compounds, *Chemical Communications*, 836–837.
18. Pedersen, C.J. (1972) Macrocyclic polyethers. Dibenzo-18-crown-6 polyether and dicyclohexyl-18-crown-6 polyether, *Organic Syntheses*, **52**, 66–74.
19. Israelachvili, J.N., Mitchell, D.J., Ninham, B.W. (1976) Theory of self-assembly of hydrocarbon amphiphiles into micelles and bilayers, *Journal of the Chemical Society, Faraday Transactions* **2**, 72, 1525–1568.
20. Israelachvili, J. (1992) *Intermolecular and Surface Forces*, Academic Press, London.
21. Gokel, G.W., Echegoyen, L. (1990) In *Advances in Bio-organic Frontiers*, Dugas, H. (Ed), Springer-Verlag, Berlin.
22. Gokel, G.W., Hernandez, J.C., Viscariella, A.M., Arnold, K.A., Campana, C.F., Echegoyen, L., Fronczek, F.R., Gandour, R.D., Morgan, C.R., Trafton, J.E., Miller, S.R., Minganti, C., Elband, D., Schultz, R.A., Tamminen, M. (1987) Steroidal lariat ethers: a new class of macrocycles and the crystal structure of N-(cholesteryloxycarbonyl) aza-15-crown-5, *Journal of Organic Chemistry*, **52** (14), 2963–2968.
23. De Wall, S.L., Wang, K., Berger, D.L., Watanabe, S., Hernandez, J.C., Gokel, G.W. (1997) Azacrown ethers as amphiphile headgroups: formation of stable aggregates from two- and three-armed lariat ethers, *Journal of Organic Chemistry*, **62**, 6784–6791.
24. Fuhrhop J-H, David H-H, Mathieu J, Liman U, Winter H-J, Boekma E. (1986) Bolaamphiphiles and monolayer membranes made from 1,6,19,24-tetraoxa-3,21-cyclohexatriacontadiene-2,5,20,23-tetrone. *Journal of the American Chemical Society*, **108**:1785–1791.
25. Muñoz S, Mallén JV, Nakano A, Chen Z, Echegoyen L, Gay I., Gokel G. W. (1992) Lariat ether bola-amphiphiles: formation of crown ether based bola-amphisomes. *Chemical Communications*, 520–522.
26. Muñoz S, Mallén J, Nakano A, Chen Z, Gay I, Echegoyen L, Gokel GW. (1993) Ultrathin monolayer lipid membranes from a new family of crown ether-based bola-amphiphiles *Journal of the American Chemical Society*, **115**:1705–1711.
27. Arnold, K.A., Viscariello, A.M., Kim, M., Gandour, R.D., Fronczek, F.R., Gokel, G.W. (1988) N,N'-bis(Substituted)-4,13-diaza-18-crown-6 derivatives having pi-donor-group-side arms: correlation of thermodynamics and solid state structures, *Tetrahedron Letters*, 3025–3028.
28. Atwood, J.L., Hrncir, D.C., Rogers, R.D. (1983) The use of crown ethers to access new $M[Al_2R_6X]$ species. Synthesis and crystal structure of $[K{\cdot}dibenzo\text{-}18\text{-}crown\text{-}6][Al_2Me_6Cl]{\cdot}2C_6H_6$, *The Journal of Inclusion Phenomena and Macrocyclic Chemistry*, **1**, 199–205.
29. (a) De Wall, S.L., Meadows, E.S., Barbour, L.J., Gokel, G.W. (1999) Solution and solid state evidence for alkali metal, cation-pi interactions with indole, the sidechain of tryptophan, *Journal of the American Chemical Society*,

121, 5613–5614. (b) De Wall, S.L., Barbour, L.J., Gokel, G.W. (1999) Cation-pi complexation of potassium cation with the phenolic sidechain of tyrosine, *Journal of the American Chemical Society*, **121**, 8405–8406.

30. Gokel, G.W., Barbour, L.J., Ferdani, R., Hu, J. (2002) Lariat ether receptor systems show experimental evidence for alkali metal cation-pi interactions, *Accounts of Chemical Research*, **35**, 878–886.
31. Hu, J., Barbour, L.J., Ferdani, R., Gokel, G.W. (2002) Sodium cation complexation behavior of the heteroaromatic side chains of histidine and tryptophan, *Chemical Communications*, 1810–1811.
32. Burley, S.K., Petsko, G.A. (1986) Amino-aromatic interactions in proteins, *FEBS Letters*, **203**, 139–143.
33. De Wall, S.L., Meadows, E.S., Barbour, L.J., Gokel, G.W. (2000) Synthetic receptors as models for alkali metal cation-pi binding sites in proteins, *Proceedings of the National Academy of Sciences of the United States of America*, **97**, 6271–6276.
34. Hille, B. (1992) *Ionic Channels of Excitable Membranes*, 2nd edition. Sinauer Press, Sunderland, MA.
35. (a) Yeagle, P. (Ed) (1992) *The Structure of Biological Membranes*, CRC Press, Boca Raton, FL. (b) Yeagle, P. (1993) *The Membranes of Cells* (Second Edition), Academic Press, New York. (c) Yeagle, P.L. (Ed) (2012) *The Structure of Biological Membranes*, CRC Press, Boca Raton, FL.
36. Schlieper, P., De Robertis, E. (1977) Triton X-100 as a channel-forming substance in artificial lipid bilayer membranes, *Archives of Biochemistry and Biophysics*, **184**, 204–208.
37. Rostovtseva, T.K., Bashford, C.L., Lev, A.A., Pasternak, C.A. (1994) Triton channels are sensitive to divalent cations and protons, *Journal of Membrane Biology*, **141**, 83–90.
38. Jullien, L., Lehn, J.-M. (1988) The "chundle" approach to molecular channels. Synthesis of a macrocycle-based molecular bundle, *Tetrahedron Letters*, **29**, 3803–3806.
39. Carmichael, V.E., Dutton, P.J., Fyles, T.M., James, T.D., Swan, J.A., Zojaji, M. (1989) Biomimetic ion transport: a functional model of a unimolecular ion channel, *Journal of the American Chemical Society*, **111**, 767–769.
40. Doyle, D.A., Cabral, J.M., Pfuetzner, R.A., Kuo, A., Gulbis, J.M., Cohen, S.L., Chait, B.T., MacKinnon, R. (1998) The structure of the potassium channel: molecular basis of K^+ conduction and selectivity, *Science*, **280**, 69–77.
41. Murray, C.L., Shabany, H., Gokel, G.W. (2000) The central "relay" unit in hydraphile channels as a model for the water-and-ion "capsule" of channel proteins, *Chemical Communications*, 2371–2372.
42. Nakano, A., Xie, Q., Mallen, J.V., Echegoyen, L., Gokel, G.W. (1990) Synthesis of a membrane-insertable, sodium cation conducting channel: kinetic analysis by dynamic sodium-23 NMR, *Journal of the American Chemical Society*, **112**, 1287–1289.
43. Riddell, F.G., Hayer, M.K. (1985) The monensin-mediated transport of sodium ions through phospholipid bilayers studied by 23Na-NMR spectroscopy, *Biochimica et Biophysica Acta*, **817**, 313–317.
44. Gokel, G.W. (2000) Hydraphiles: design, synthesis, and analysis of a family of synthetic, cation-conducting channels, *Chemical Communications*, 1–9.
45. Abel, E., Meadows, E.S., Suzuki, I., Jin, T., Gokel, G.W. (1997) Unusually long open times, determined by planar bilayer conductance studies, for a synthetic tris(macrocycle) that functions as a transmembrane channel in a phospholipid bilayer, *Chemical Communications*, 1145–1146.
46. Weber, M.E., Elliott, E.K., Gokel, G.W. (2006) Activity of synthetic ion channels is influenced by cation-pi interactions with phospholipid headgroups, *Organic & Biomolecular Chemistry*, **4**, 83–89.
47. Estevez, R., Jentsch, T.J. (2002) CLC chloride channels: correlating structure with function, *Current Opinion in Structural Biology*, **12**, 531–539.
48. Schlesinger, P.H., Ferdani, R., Liu, J., Pajewska, J., Pajewski, R., Saito, M., Shabany, H., Gokel, G.W. (2002) SCMTR: a chloride-selective, membrane-anchored peptide channel that exhibits voltage gating, *Journal of the American Chemical Society*, **124**, 1848–1849.
49. Schlesinger, P.H., Djedovic, N.K., Ferdani, R., Pajewska, J., Pajewski, R., Gokel, G.W. (2003) Anchor chain length alters the apparent mechanism of chloride channel function in SCMTR derivatives, *Chemical Communications*, 308–309.
50. Djedovic, N., Ferdani, R., Harder, E., Pajewska, J., Pajewski, R., Weber, M.E., Schlesinger, P.H., Gokel, G.W. (2005) The C- and N-terminal residues of synthetic heptapeptide ion channels influence transport efficacy through phospholipid bilayers, *New Journal of Chemistry*, **29**, 291–305.
51. Pajewski, R., Pajewska, J., Li, R., Fowler, E.A., Gokel, G.W. (2007) The effect of midpolar regime mimics on anion transport mediated by amphiphilic heptapeptides, *New Journal of Chemistry*, **31**, 1960–1972.

52. Schlesinger, P.H., Ferdani, R., Pajewska, J., Pajewski, R., Gokel, G.W. (2003) Replacing proline at the apex of heptapeptide-based chloride ion transporters alters their properties and their ionophoretic efficacy, *New Journal of Chemistry*, **27**, 60–67.
53. Gokel, G.W., Negin, S. (2013) Synthetic ion channels: from pores to biological applications, *Accounts of Chemical Research*, **46**, 2824–2833.
54. Leevy, W.M., Donato, G.M., Ferdani, R., Goldman, W.E., Schlesinger, P.H., Gokel, G.W. (2002) Synthetic hydraphile channels of appropriate length kill Escherichia coli, *Journal of the American Chemical Society*, **124**, 9022–9023.
55. Leevy, W.M., Huettner, J.E., Pajewski, R., Schlesinger, P.H., Gokel, G.W. (2004) Synthetic ion channel activity documented by electrophysiological methods in living cells, *Journal of the American Chemical Society*, **126**, 15747–15753.
56. Leevy, W.M., Weber, M.E., Gokel, M.R., Hughes-Strange, G.B., Daranciang, D.D., Ferdani, R., Gokel, G.W. (2005) Correlation of bilayer membrane cation transport and biological activity in alkyl-substituted lariat ethers, *Organic & Biomolecular Chemistry*, **3**, 1647–1652.
57. Negin, S., Gokel, M.R., Patel, M.B., Sedinkin, S.L., Osborn, D.C., Gokel, G.W. (2015) The aqueous medium-dimethylsulfoxide conundrum in biological studies, *RSC Advances*, **5**, 8088–8093.
58. Patel, M.B., Stavri, A., Curvey, N.S., Gokel, G.W. (2014) Hydraphile synthetic ion channels alter root architecture in arabidopsis thaliana, *Chemical Communications*, **50**, 11562–11564.

18
Supramolecular Technology

David N. Reinhoudt

MESA+ Institute for Nanotechnology, University of Twente, Enschede, The Netherlands

18.1 Introduction

I still remember the first time I heard about crown ethers, when Howard Simmons, head of research at the DuPont Research Laboratories at Wilmington, visited the Organic Chemistry department at the Technical University, Delft, in 1968. At that time I was a graduate student in the group of Hugo Beyerman, who was a good friend of Simmons. In his lecture he discussed his own work on macrobicyclic amines together with the work on the macrocyclic polyethers (crown ethers) that had been published in 1967 by Charles Petersen [1]. I was fascinated by the ability of those compounds to complex selectively alkali cations because for me it was the first example of a molecule that could be used as a tool to achieve a function.

A year later, after having completed my PhD work at Delft, I started working as a research chemist at the Koninklijke/Shell Laboratories in Amsterdam. After two years I was given the opportunity to set up a program on host–guest chemistry. We developed a technology to solubilize barium sulfate-containing scale, at that time a serious problem in the oil industry. A combination of anionic aza-crown ethers and cryptands was able to dissolve up to 55 grams of BaSO4 in water in less than an hour [2]. Because the large scale synthesis of crown ethers from oligoethylene glycols required distillation, which had been reported to lead to explosions, we sought to develop a safe way to purify 18-crown-6. As was reported in a paper by Pedersen [3], crown ethers increased the solubility of certain polar organic molecules in nonpolar solvents. We systematically investigated the interaction of crown ethers with small organic compounds and found that there was a great preference for 18-crown-6 [4]. After my move to the University of Twente I continued this work. We published the X-ray structure of the 18-crown-6.(CH_3NO_2) 1:2 complex [5] and measured the thermodynamic stabilities of this and similar complexes in solution. It should be emphasized that in the early days of host–guest chemistry, the focus was entirely on cation and anion recognition. The interest in molecule-to-molecule interactions (supramolecular chemistry) followed later. When we published the first monograph

Macrocyclic and Supramolecular Chemistry: How Izatt–Christensen Award Winners Shaped the Field, First Edition.
Edited by Reed M. Izatt.

on crown ethers in 1980 only a few examples of complexes between crown ethers and neutral molecules had been reported in the literature [6], most of them without structural evidence. In the following years interest in molecular complexes began to dominate the field, and molecules of biological interest, such as urea were shown to complex with crown ethers [7].

18.2 Chemical sensing

From the beginning, host–guest (or, later, supramolecular) chemistry has been associated with potential applications – for instance, molecular catalysts (enzyme models), sensing, membrane separations, molecular materials etc. Our work in this direction started with chemical sensors based on Field Effect Transistors (FETs) as depicted in Figure 18.1.These had been used to measure pH values, as the charge density of the gate oxide surface will be reflected in the conductivity in the semiconducting channel between source and drain. We investigated the immobilization of ion receptors at the gate surface and measured the effect of different ions in the contacting solution on the conductivity of the semiconducting channel.

As early as1975 Moss et al. [8] described an ISFET with a polymeric PVC membrane containing valinomycin physically attached to the gate oxide, as a potassium-selective device. There were several shortcomings that prevented the development of such devices to commercial success. First, there were undesired variations with pH and CO_2 concentration, but the fundamental problem was that the interface between gate oxide and membrane was not defined in terms of a common chemical species. We solved this fundamental problem by introducing a chemically anchored polyhydroxyethyl methacrylate (polyHEMA) layer between the gate oxide and the polymer membrane containing the ion-selective receptor and lipophilic counter-ions The aqueous polyHEMA gel can not only be buffered to eliminate pH and CO_2 interference, but can also be filled with a defined concentration of the primary anion that is to be measured. As a consequence, the system is stable and thermodynamically defined [9]. The PVC membrane used in the literature did not chemically

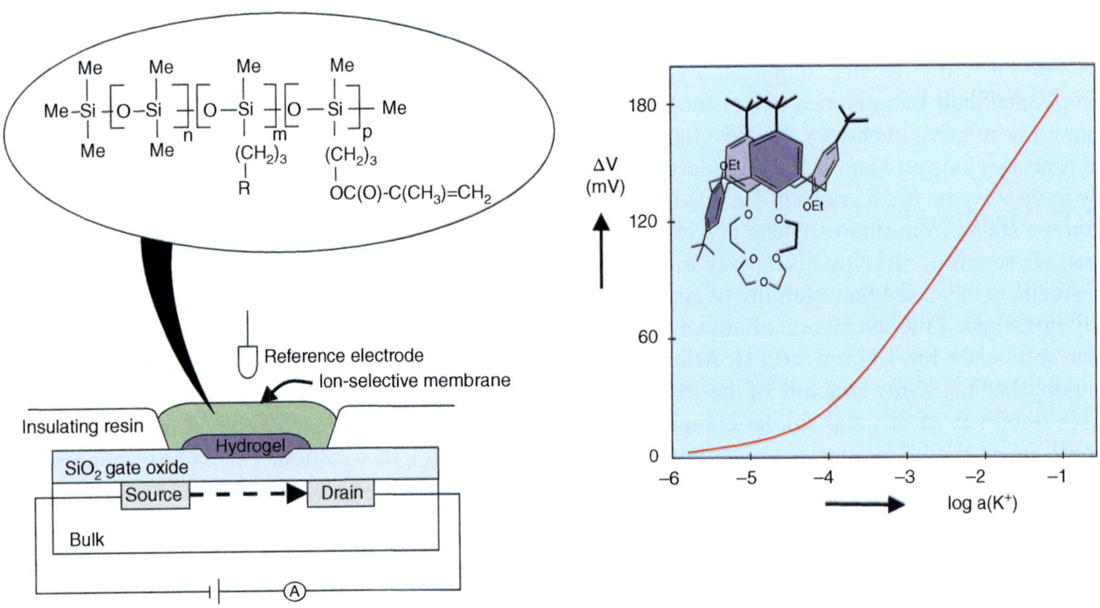

Figure 18.1 Chemfet design and K^+ response

Result CHEMFET measurements with different K^+- selective ionophores

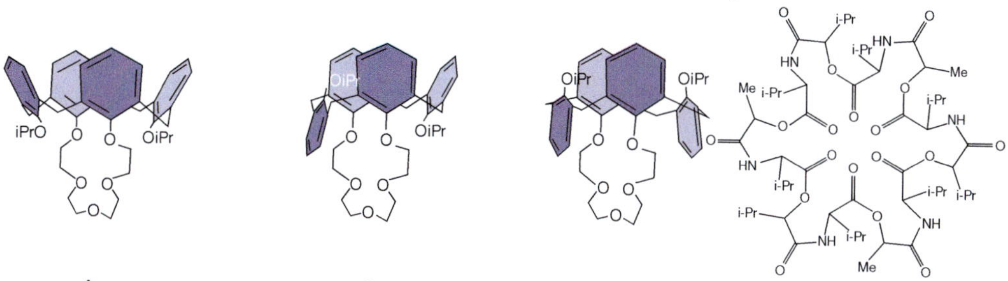

1 2 3 4

Ionophore	log $K_{K,J}^{pot}$ and slope [mV/dec][1,2]		
	0.1 M Li^+	0.1 M Ca^{2+}	0.1 M Na^+
1	−2.8 (56)	−3.9 (58)	−2.5 (54)
2	−3.6 (56)	−3.6 (55)	−3.2 (57)
3	−3.7 (56)	−4.0 (58)	−3.2 (55)
4	−3.5 (57)	−3.7 (59)	−3.1 (57)

Figure 18.2 *Chemfet measurements*

attach to the gate oxide and so we developed a series of new polysiloxane materials that fulfill all requirements of polarity, low glass transition temperature, chemical stability and the anchoring groups for ionophores and counter-ions [10].

In collaboration with Rocco Ungaro's group in Parma, we have developed in parallel different ionophores that can be anchored to the polymeric matrix and that have high ion selectivities for sodium, potassium, cesium, calcium, and heavy metals (Figure 18.2). Anions can be measured upon the introduction of anion selective receptors and immobilized cationic sites in these membranes [11].

18.3 Membrane transport

Transport of salts through liquid membranes is a technology with many possible applications. In biology, membranes provide a barrier between the intracellular and the extracellular aqueous environment. Synthetic ionophores have been tested in liquid membranes as a tool to study selectivities.

In our work we have used supported liquid membranes (Figure 18.3) in which a solution of ionophores is immobilized in the pores of a polymer [12]. Such membrane systems are much more defined than U-tube experimental set-ups (Bulk Liquid Membranes), allowing quantitative transport studies that provide the fundamental measures of the transport mechanism, the extraction and diffusion constants. Moreover, the amount of ionophores can be minimized, which is important for practical applications. The types of ionophores can be the same as we used in our CHEMFET work, but without the need to immobilize them to the matrix. One of our most important results was the uphill transport of cesium from a concentrated solution of sodium nitrate, a situation that occurs in the separation of nuclear waste. By taking advantage of the large difference in nitrate concentration we could remove more than 99.9% of the cesium in 24 hours [13].

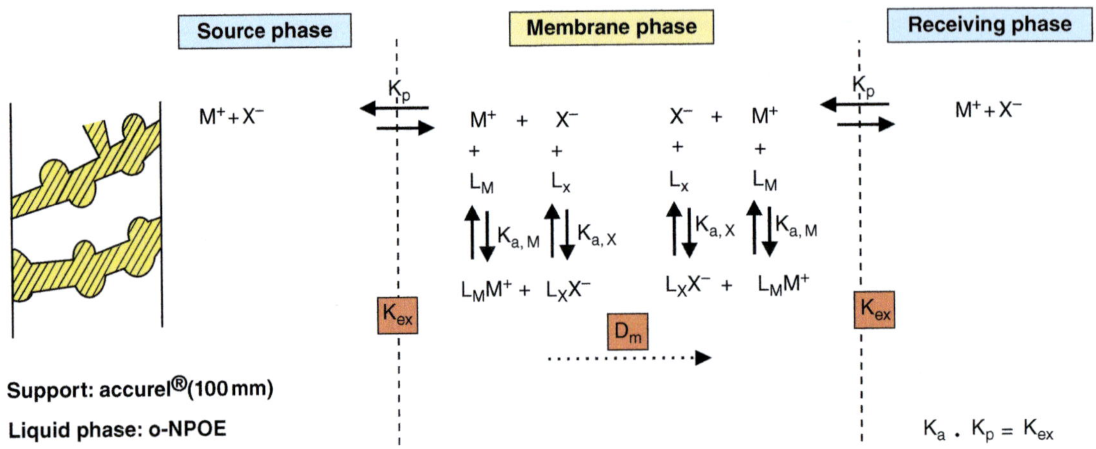

Figure 18.3 Schematic representation of a supported liquid membrane

We were the first to report the transport of a *neutral* molecule (urea) by a macrocycle through a supported membrane. A macrocyclic polyether receptor for urea also containing a Lewis acidic uranyl salenophene binding site, selectively complexes urea and takes it from an aqueous phase across the membrane phase [14].

18.4 Nonlinear optical materials

Over the years many materials have been investigated for application in nonlinear optics. The first materials were to be studied were inorganic crystals, but after the introduction of fiber optics, organic materials became of interest. NLO–active materials contain electron-donating and electron-accepting groups attached to conjugated backbones.The challenge in developing such materials is to optimize their molecular properties (β) and to organize them in a non-centrosymmetric supramolecular network.

We have reported a new class of multichromophoric, highly transparent NLO molecules based on the calix[4] arene skeleton [15]. Since the four NLO-phores that have a nitro-acceptor and an alkoxy-acceptor group can be fixed in a cone conformation, the molecules have a high β-value (Figure 18.4).We could organize them in a polar network by applying a strong electrical field on a film of the nitrocalixarenes at high temperature. Upon cooling, the polar order is maintained for a certain period of time, depending on the nature of the alkoxy substituent. The packing of the tetranitrotetrapropoxy calix[4]arenes in these films was elucidated by combining the results of X-ray diffraction, angle-dependent second harmonic generation, and atomic force microscopy (Figure 18.5). The structure was confirmed by solid state molecular dynamics calculations. The high stability of the non-centrosymmetric packing in this supramolecular material is caused by the electrostatic interactions between the electron-accepting nitro groups and the electron-donating alkoxy residues in consecutive layers.

18.5 Supramolecular technology for nanofabrication

Fabrication of nanostructures at surfaces started by micro contact printing or DPN (Dip Pen Nanolithography) on gold or other noble metals with molecules with a functional group that has an intrinsic affinity to those metals. Although these methods are attractive because of their simplicity, they also have some disadvantages, such as the lack of control of stability of these self-assembled monolayers (SAMs).

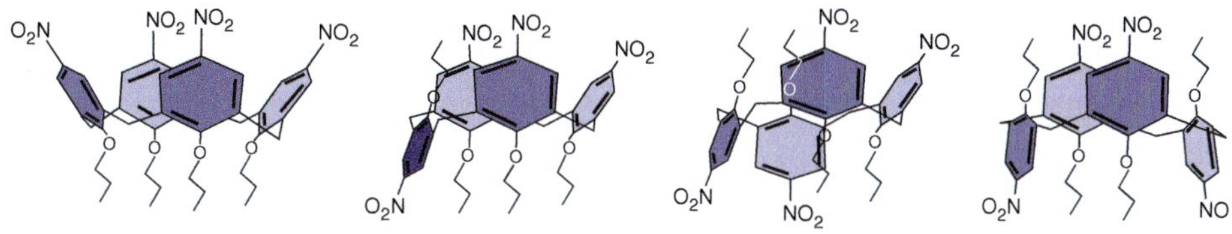

	Cone	Partial cone	1,2-Alternate	1,3-Alternate
m (D)	13.8	6.7	0	0
b_z (EFISH)[a]	30	27	–	–
b (NHRS)[a]	21[b]	22[b]	30[c]	8[c]
b (FHRS)[a]	36[b]	19[b]	20[c]	9[c]

Figure 18.4 *The four different calix conformations and NLO parameters*

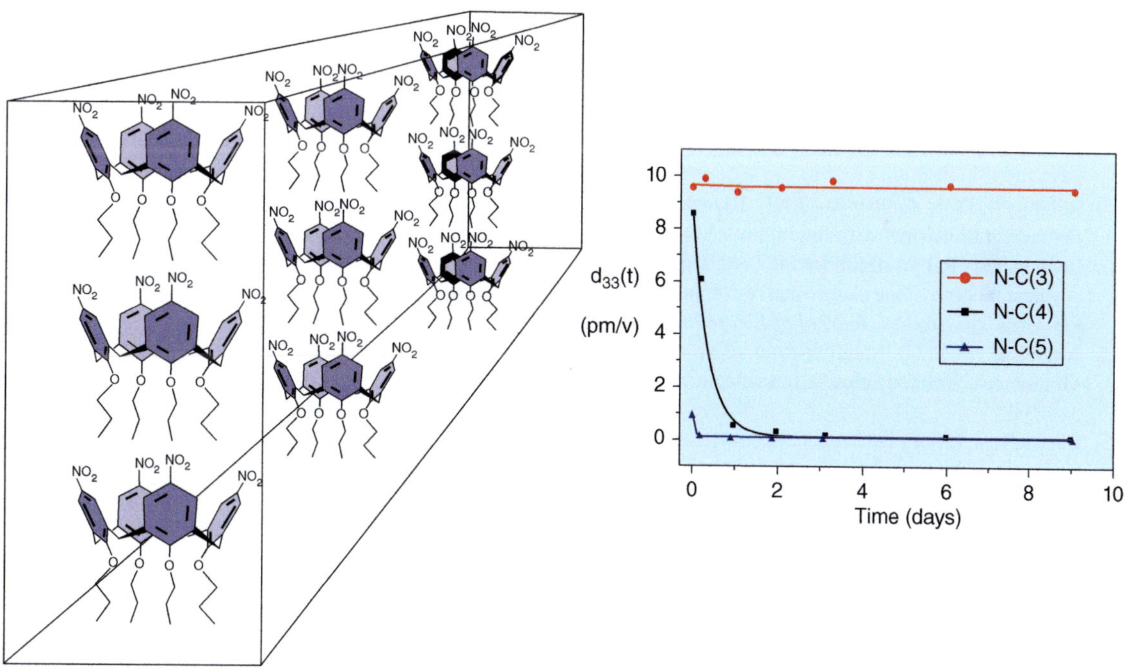

Figure 18.5 *Polar order in the tetranitrotetrapropoxy calix[4]arene*

Our work in this area has been inspired by the advantages of supramolecular chemistry, such as specificity of interaction, reversibility, and the possibility of using multivalency as a tool to obtain thermodynamically stable structures. We have prepared SAMs of molecular receptors like recorcenarenes, calixarenes, and cyclodextrins on gold and silica surfaces [16]. The latter have the advantage that the subsequent chemistry is compatible with aqueous conditions.On these monolayers, which we have named molecular print boards, we can attach molecules with organic residues that can complex into the cavity of the cyclodextrin.

In this way we have patterned molecular print boards with molecular guests and nanoparticles, showing that supramolecular chemistry is a very important tool for nanofabrication.

References

1. Pedersen, C.J. (1967) Cyclic polyethers and their complexes with metal salts, *Journal of the American Chemical Society*, **89** (26), 7017–7036.
2. De Jong, F.,Van Zon, A., Reinhoudt,D.N., Torny, G.J., Tomassen H.P. (1983) Chemistry of crown ethers (XIX). Functionalized crown ethers for the solubilization of barium sulfate, *Recueil des Traveaux Chimiques des Pays-Bas*, **102** (3), 164–173.
3. Pedersen, C.J. (1971) Crystalline complexes of macrocyclic polyethers with thiourea and related compounds, *Journal of Organic Chemistry* **36** (12), 1690–1693.
4. Van Zon, A., De Jong, F., Reinhoudt, D.N., Torny, G.J., Onwezen, Y. (1981) Chemistry of crown ethers (XVIII). Isolation and purification of 18-crown-6 by selective complexation with non-ionic compounds, *Recueil des Traveaux Chimiques des Pays-Bas* **100** (12), 453–459.
5 De Boer, J.A.A., Reinhoudt, D.N., Harkema,S., Van Hummel, G.J., de Jong, F. (1982) Thermodynamic constants of complexes of crown ethers and uncharged molecules and X-ray structure of the 18-crown-6.(CH3NO_2)$_2$ complex, *Journal of the American Chemical Society* **104**(15), 4073–4076.
6. De Jong, F., Reinhoudt, D.N. (1981) *Stability and reactivity of crown ether complexes*. Academic Press, New York.
7. Harkema, S., Van Hummel, G.J., Daasvatn, K., Reinhoudt, D.N. (1980) Complexes of crown ethers and neutral molecules and crystal structure of a urea 18-crown-6 (5:1) complex,. *Journal of the Chemical Society, Chemical Communications*, **1981**, 368–369.
8. Janata, J., Moss, S.D. (1976) Chemically sensitive field effect transistors, *Bio-Medical Engineering*, **11** (7), 241–245.
9. Reinhoudt, D.N., Engbersen, J.F.J., Brzozka, Z (1994) Development of durable K^+-selective chemically modified field effect transistors with functionalized polysiloxane membranes, *Analytical Chemistry*, **66** (21), 3618–3623.
10. Luchtenberg, R.J.W., Egberink, R.J.M., Van den Berg, A., Enbersen, J.F.W., Reinhoudt, D.N. (1998) The effects of covalent binding of the electro active components in durable CHEMFET membranes – impedance spectroscopy and ion-sensitivity studies, *Journal of Electroanalytical Chemistry*, **452**, 69–86.
11. Antonisse, M.M.G., Snellink-Ruel, B.H.M., Luchtenburg, R.J.W., Engbersen, J.F.J., Van den Berg, A. (2000) Membrane characterization of ion-selective CHEMFETs by impedance spectroscopy, *Analytical Chemistry*, **72** (2), 343–348.
12. Christoffels, L.A.J., De Jong, F.,Reinhoudt, D.N., Sivelli, S., Gazzola, L., Casnati, A., Ungaro, R. (1991) Facilitated transport of hydrophilic salts by mixtures of anion and cation carriers and by ditopic carriers, *Journal of the American Chemical Society*, **121** (43), 10142–10151.
13. Casnati, A., Pochini, A., Ungaro, R., Ugozzoli, F., Arnaud, F., Fanni, S., Schwing, M.-J., Egberink, R.J.M., de Jong, F., Reinhoudt, D.N. (1995) Synthesis, complexation and membrane transport studies of 1,3-alternate calix[4]arene-crown-6 conformers: A new class of cesium-selective ionophores. *Journal of the American Chemical Society*, **117** (10), 2767–2777.
14. Nijenhuis,W.F., Van Doorn, A.R., Reichwein, A.M., De Jong, F., Reinhoudt, D.N. (1991) Urea transport by macrocyclic carriers through a supported liquid membrane, *Journal of the American Chemical Society*, **113** (12), 3607–3608.
15. Kenis, P.J.A., Noordman, O.F.J., Houbrechts, S., Van Hummel, G.J., Harkema, S., Van Veggel, F.C.J.M., Clays, K., Engbertsen, J.F.J., Persoons, A., Van Hulst, N.F., Reinhoudt D.N. (1998) Second-order nonlinear optical properties of the four tetranitropropxycalix[4]arene conformers, *Journal of the American Chemical Society*, **120** (31), 7875–7883.
16. Schierbaum, K.D., Weiss, T., Thoden van Velzen, E.U., Engbersen, J.F.J., Reinhoudt, D.N., Gopel, W. (1994) Molecular recognition by self-assembled monolayers of cavitand receptors, *Science*, **265**, 1413–1415.

19

Synthesis of Macrocyclic Complexes Using Metal Ion Templates

Daryle H. Busch

Department of Chemistry, University of Kansas, Lawrence, KS, USA

19.1 Introduction

Editor's note: This section provides a first-hand account by Dr. Busch of events leading up to the synthesis of his first planned macrocycle.

I received my BA degree in chemistry from Southern Illinois University (SIU) in 1951. I enrolled at the University of Illinois-Champaigne for my PhD training (1951–1954) under Professor John Christian Bailar. Dr. Bailar was an accomplished organic chemist who had seen the vast unknown realms of inorganic chemistry, especially those in which both inorganic and organic chemistry principles are involved. I soon shared his enthusiasm and, like him, I was thinking about some of those questions that hadn't even been asked yet.

In a classroom lecture during my third year at SIU, Dr. Bailar came to our campus and presented a seminar in which he told us about the many coordination numbers metal ions and ligands can share. One example he gave was how four molecules of ammonia would bind to Cu^{2+}, but only two molecules of ethylenediamine would chelate with the same Cu^{2+}. Also, Cu^{2+} would bind only one molecule of the ligand triethylenetetramine. We understood this, but I was really curious about something that wasn't mentioned. What if the molecule with the four amine groups was a ring? It only needed a couple of methylene groups to become a complete ring. So I was surprised that the cyclic molecule hadn't been made. Today, chemists all know about natural product ligands that swallow up metal ions. Wouldn't these chemists like to make some useful synthetic ring-shaped molecules?

Think about this. Nature had probably developed many special cyclic ligands and probably we know only a fraction of these—some of them are precious to living creatures or plants, including us: i.e., dioxygen transport

Macrocyclic and Supramolecular Chemistry: How Izatt–Christensen Award Winners Shaped the Field, First Edition.
Edited by Reed M. Izatt.

by the iron-containing porphyrins, hemoglobin and myoglobin; the cobalt complex called corin that is part of Vitamin B_{12}; chlorophylls that play essential roles in the conversion of light to chemical energy; and others.

Given such well known examples of nature's macrocyclic ligands, it is easy to believe many other types of macrocycles are possible and some of them may already exist. Nature may have focused on the well-being of wild creatures and plants. Also, such natural macrocycles may exist and be known because of their use to people and their needs. It is certainly true that people should be able to find additional uses for macrocycles and then create new macrocycles for those uses. New chemicals can be useful for building materials for our homes, offices, vehicles, travel, tools and medicinal substances of many kinds and even military needs.

I had an interest in the concept of a macrocycle, but little time and no special funding to study macrocycles while at the University of Illinois. While I had not yet produced a new macrocycle, I could not forget the subject. In addition to my thesis work, I was editing chapters written by Dr. Bailar's former students and writing the 99 page introductory chapter for Dr. Bailar's 834 page book, *Chemistry of Coordination Compounds*. What a gift! While a student, I was paid and obliged to study a huge fraction of the current literature on the newly evolving inorganic chemistry of metal complexes with all manner of ligands, especially complicated organic molecules. Also, my family and I were supported during two summer quarters for editing/studying the current publications on coordination chemistry.

What a foundation for initiating a career researching the coordination chemistry of all kinds of ligands, especially organic molecules. And equally exciting, like much of the potential science over much of the world, this was a field long overdue for exploration, extensive revelations and exploitation. World War II had caused this extreme neglect. During my time at the University of Illinois, Dr. Bailar and I produced four publications together: (1) Stereochemistry of hexadentate EDTA; (2) Mirror image isomer of a beryllium complex; (3) Tetradentate and bidentate complexes of EDTA with Pd^{2+} and Pt^{2+}, and (4) Iron(II)–methine chromophore – this last a lovely colored iron compound.

My experiences at Illinois prepared me well in the fundamentals of coordination chemistry and led me to the study of macrocyclic compounds in my first permanent position as an Assistant Professor in the Department of Chemistry at Ohio State University. My research work at Ohio State focused on the synthesis, characterization, properties, and reactions of metallic complexes. I developed the habit of promptly publishing the work of my students with them as co-authors. My interest had come to focus more on the chemistry of metal coordination compounds and I began to get invitations to write reviews and similar documents, the first was for *Journal of Chemical Education* in 1956. More important, those talented early researchers in our Busch group represented the growing strength of our research team. Those first years at Ohio State enabled me to build the foundation and competitive position for us to grow into a lab group capable of overcoming the challenges Dr. Bailar had put before me years earlier in Illinois. Specifically, we developed the collective capability to conquer the design, synthesis, and applications of the general chemical compound category commonly labeled "macrocycle."

After the first few years at Ohio State, our expanded research team included a dozen or more graduate students and postdoctoral researchers. Macrocyclic chemistry became one of the top research subjects of our growing group. An important achievement was the first deliberate synthesis of a compound containing a new macrocyclic ligand. Our scientific progress made possible the focus of the group members on the design and synthesis of a wide range of macrocycles, as had indeed been our goal.

19.2 Macrocycle synthesis

19.2.1 The first planned macrocycle

Editor's Note: This section, prepared by Dr Kristin Bowman-James, describes three important milestones from the research enterprise of Daryle H. Busch during the first 25 years of his study of synthetic aza macrocycles. The interested reader is referred to an interview in *Coordination Chemistry Reviews* published in 2010, which provides a more extensive overview of the breadth of Busch's multifaceted career [1].

The first "planned" macrocycle synthesis by the Busch group was reported at the 142nd National American Chemical Society meeting in 1962, the highlight of which appeared in *Chemical & Engineering News* shortly thereafter [2], followed by the seminal paper in 1964 [3]. The cyclization of the Busch macrocycle, a 14-membered N_2S_2 ring (Figure 19.1), was accomplished by taking advantage of the metal ion, Ni(II), to hold the two ends of one of the reactants in place for the ring-closing condensation. The criteria for the template strategy to work were: (1) a tetradentate ligand with planar coordination; (2) terminal groups on the chelate that could undergo an addition reaction to close the ring; and (3) a suitable ring-closing reagent [3]. This synthetic strategy became known as the "coordination template hypothesis [3, 4]." The beginnings of the synthetic aza macrocyclic era, and early studies of a wide variety of synthetic macrocycles, are reviewed in an *Accounts of Chemical Research* article from 1978 [5].

19.2.2 Theoretical considerations

Early studies indicated that macrocyclic complexes were even more stable to dissociation than complexes with chelating ligands. Busch attributed this enhanced stability to "multiple juxtapositional fixedness" (MJF; Figure 19.2) [6]. Macrocycles contain multiple donor atoms within an enclosed ring. Thus, even when one of the coordinated donors detaches it is less likely to leave the vicinity due to the strong hold of the other donors still bound to the metal ion. Because of the cyclic structure and totally closed nature of the encircled macrocyclic ring, this enhanced stability was even greater than that seen with chelating ligands.

In parallel with Busch's MJF concept, Cabbiness and Margerum expanded on the thermodynamic and kinetic considerations of the enhanced stabilities, and introduced the term "macrocyclic effect" as a corollary to the well-known "chelate effect [7]."

19.2.3 Oxygen carriers and enzyme mimics

The Busch group is credited with one of the first reports of Fe(II) complexes with tetradentate synthetic macrocycles which paved the way to the bioinorganic chemists' dreams, at that time, of reversible dioxygen carriers [8, 9]. These first iron complexes were made from monocyclic ligands, but Busch realized that to stabilize Fe(II) complexes a more protective ligand system would be necessary.

In order to achieve a ligand system that allowed for reversible oxygen binding, Busch designed a macrocycle that contained two vertical rings to provide a scaffold for attaching a bridge to shield one of the axial sites of a bound metal ion. He referred to the resulting protected space above the metal ion as a "dry cave" (Figure 19.3a). These macrocycles fall into a class of molecules known as "lacunar" ligands, i.e., those that provide a superstructure for a cavity [10]. The new family of superstructure ligands, named cyclidenes, was reported in 1979 [11], and the synthesis of a totally synthetic functional coboglobin model was reported a year later [12, 13]. In 1981, Herron and Busch reported the first reversible dioxygen binding in non-porphyrin synthetic macrocycles where the other axial ligand can be either pyridine or N-methyl imidazole (Figure 19.3b) [14].

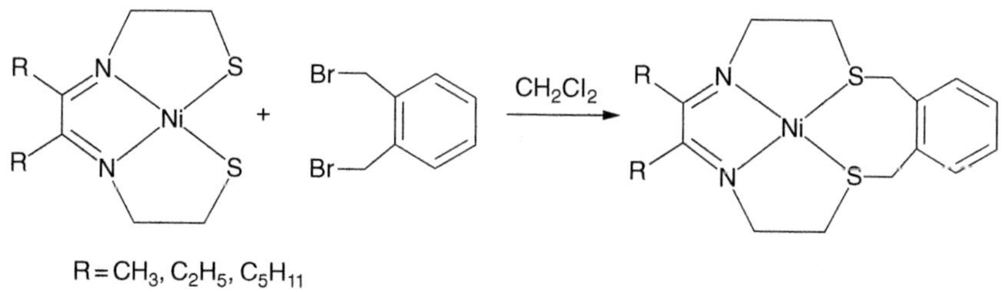

Figure 19.1 *Synthesis of the first N2S2 macrocycle using the coordination template effect*

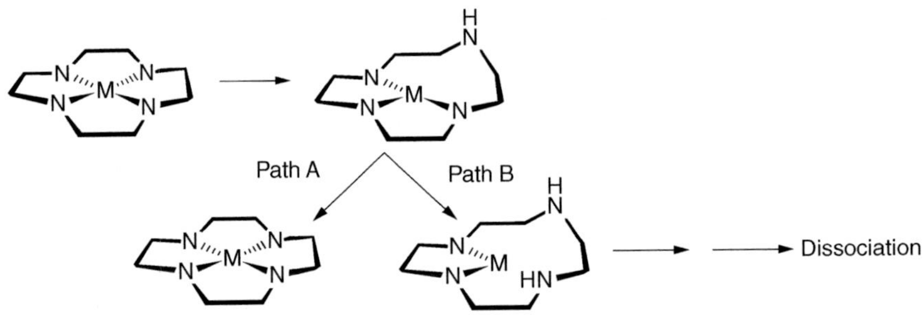

Figure 19.2 Dissociation possibilities for macrocyclic complexes illustrating the MJF effect

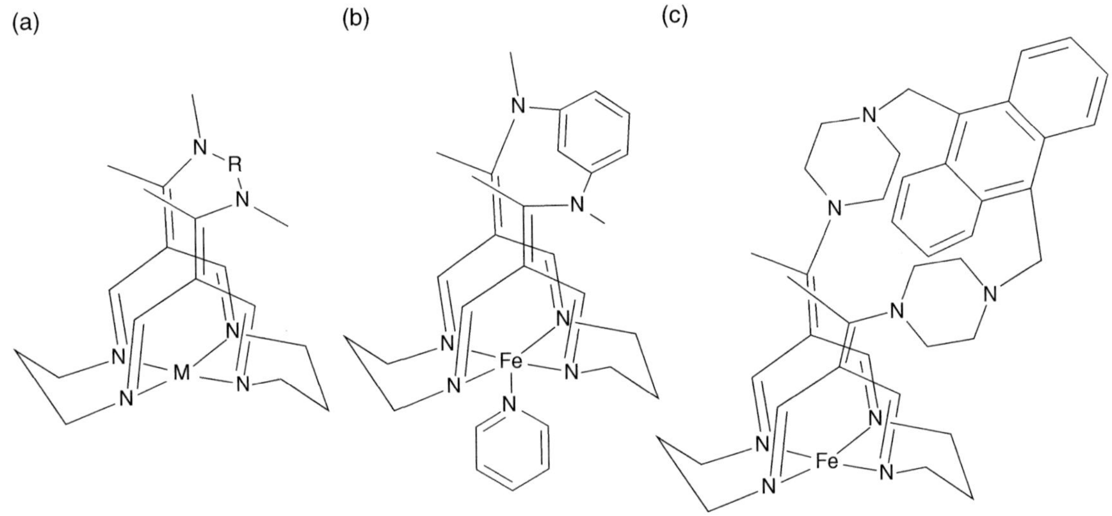

Figure 19.3 ChemDraw diagrams of three cyclidene ligands: (a) general structure; (b) iron complex capable of reversible binding of dioxygen; and (c) iron complex exhibiting cytochrome P450 behavior

By variation of the bridging ligand to produce larger "persistent voids," Takeuchi and Busch showed that the cyclidene can be tweaked to bind organic ligands [15–17]. Perhaps the culmination of these studies was the demonstration that the cyclidenes could provide the foundation for a ternary catalyst:oxidant:substrate complex by not only containing a molecule of dioxygen, but also an organic guest similar to cytochrome P450 models (Figure 19.3c) [18].

References

1. Heppert, J.A. (2010) Celebration of inorganic lives: Interview of Daryle H. Busch by Joseph A. Heppert, *Coordination Chemistry Reviews*, **254**, 1593–1606.
2. Thompson, M.C., Busch, D.H. (1962) Ligand reactions close chelate rings. Synthesis of completely encycled complexes uses metal ion coordination sphere as template, *Chemical &. Engineering News*, **40**, 57–61.
3. Thompson, M.C., Busch, D.H. (1964) Reactions of coordinated ligands. XI. Utilization of the template hypothesis to synthesize macrocyclic ligands *in situ*, *Journal of the American Chemical Society*, **86**, 3651–3656.
4. Blinn, E., Busch, D.H. (1968) Reactions of coordinated ligands. XV. Demonstration of the kinetic coordination template effect, *Inorganic Chemistry*, **7**, 820–824.

5. Busch, D.H. (1978) Distinctive coordination chemistry and biological significance of complexes with macrocyclic ligands, *Accounts of Chemical Research*, **11**, 392–400.
6. Busch, D.H. (1970) Metals and enzymes – Multiple juxtapositional fixedness, *Chemical & Engineering. News*, **48**, 9.
7. Cabbiness, D.K., Margerum, D.W. (1969) Macrocyclic effect on the stability of copper(II) tetramine complexes, *Journal of the American Chemical Society*, **91**, 6540–6541.
8. Goedken, V.L., Merrell, P.H., Busch, D.H. (1971) Complexes of iron(II) and iron(III) with the tetradentate macrocycle 5,7,7,12,14,14-hexamethyl-1,4,8,11-tetraazacyclotetradeca-4,11-diene, *Journal of the American Chemical Society*, **94**, 3397–3405.
9. Dabrowiak, J.C., Merrell, P.H., Busch, D.H. (1972) High- and low-spin six-coordinate complexes of iron(II) with a saturated tetradentate macrocyclic ligand, *Inorganic Chemistry*, **11**, 1979–1988.
10. Busch, D.H., Alcock, N.W. (1994) Iron and cobalt "lacunar" complexes as dioxygen carriers, *Chemical Reviews*, **94**, 585–623.
11. Schammel, W.P., Mertes, K.S.B., Christoph, G.G., Busch, D.H (1979) Totally synthetic heme protein models based on complexes with superstructure ligands, *Journal of the American Chemical Society*, **101**, 1622–1623.
12. Stevens, J.C., Jackson, P.J., Schammel, W.P., Christoph, G.G., Busch, D.H. (1980) Synthesis and structure of totally synthetic coboglobin models, *Journal of the American Chemical Society*, **102**, 3283–3285.
13. Stevens, J.C., Busch, D.H. (1980) Totally synthetic coboglobin models with oxygen binding capabilities equivalent to those of coboglobin, *Journal of the American Chemical Society*, **102**, 3285–3287.
14. Herron, N.; Busch, D.H. (1981) Reversible dioxygen binding by a totally synthetic non-porphyrin macrobicyclic iron(II) complex containing a persistent void, *Journal of the American Chemical Society*, **103**, 1236–1237.
15. Takeuchi, K.J., Busch, D.H. (1981) Guest–host association by transition-metal complexes containing permanent voids – progress toward models for the ternary complex of cytochrome P450, *Journal of the American Chemical Society*, **103**, 2421–2422.
16. Takeuchi, K.J., Busch, D.H., Alcock, N. (1983) Transition-metal host complexes designed for critical-locus accommodation of organic guest molecules: An approach to a new kind of coordination catalysts patterned after enzymes, *Journal of the American Chemical Society*, **105**, 4261–4270.
17. Takeuchi, K.J., Busch, D. H. (1983) Inclusion complex formation by transition-metal complexes containing a void in proximity to a site that can be made catalytically active, *Journal of the American Chemical Society*, **105**, 6812–6816.
18. Meade, T.J., Takeuchi, K.J., Busch, D.H. (1987) An inclusion complex containing the dioxygen molecule and an organic guest molecule cohabiting within a vaulted cobalt(II) cyclidene host: A rare kind of ternary complex, *Journal of the American Chemical Society*, **109**, 725–728.

20

Serendipity

Paul R. McGonigal[1] and J. Fraser Stoddart[2]

[1] *Department of Chemistry, Durham University, Durham, UK*
[2] *Department of Chemistry, Northwestern University, Evanston, IL, USA*

20.1 Serendipity in scientific discovery

Serendipity, as defined [1] by the Oxford English Dictionary, is the occurrence of happy and unexpected discoveries by accident. The term was first coined in 1754 by British politician Horace Walpole (the 4th Earl of Oxford) in a letter to his contemporary Sir Horace Mann. Walpole's usage of the word derived [2] from the fairytale "The Three Princes of Serendip." In the folklore of Serendip, known in English as Sri Lanka in modern times, the tale goes that the kingdom was threatened by fearsome dragons. The king said to his three sons that they should head out around the world in order to find potions that would kill the dragons. He made the point before they departed that they should keep their eyes open for adventures along the way and be prepared to pursue them, even if they did not relate directly to their original mission. True to form, they did not find anything to rid the island of the dragons, but they did solve mysteries that they had not imagined at the outset of their travels. They rescued a princess, saved a nation from starvation, slew a magic serpent, restored property to its rightful owners, and so on. Upon their return to Serendip, they were showered with gifts and by the king's own admission they had become wise enough and well qualified to succeed him in ruling the kingdom.

Despite the relatively long history of "serendipity", usage of the word was somewhat rare until the 20th century [1]. It has become far more widespread in recent decades. Be it through coincidence or causation, the rise in its use has proceeded in parallel to the emergence of modern science and all of the chance breakthroughs that have come with it. The manner of discovery demonstrated by the three princes is perhaps the opposite of scientific research that is highly directed toward a preconceived goal – using what is sometimes referred to as an "Edisonian" approach. Research of this kind is often designed in such a way that it produces

Macrocyclic and Supramolecular Chemistry: How Izatt–Christensen Award Winners Shaped the Field, First Edition.
Edited by Reed M. Izatt.

a desired output. Funding agencies, understandably – but perhaps unfortunately – often favor Edisonian-style research that is pursuant to their priorities in the short term. But of course, scientific advances come in many forms. Just as often, it is an unpredictable chance occurrence that serves as a starting point on the road to a previously unimagined area of research.

The late Ernest Eliel, stereochemist extraordinaire and a former President of the American Chemical Society, was an advocate of the important roles that fundamental research and serendipity have played in some of the most notable developments in the chemical sciences. His 1992 essay [3] on the topic, entitled *Science and Serendipity: The Importance of Basic Research*, recounts some of the serendipitous discoveries of the 20th century made through undirected basic research that have gone on to shape present day chemistry and society. Indeed, one can look back as far as 1856 to Sir William Perkin's synthesis [4] of the dye Mauveine (Figure 20.1a) and find that serendipity has been lending a helping hand since the very beginnings of chemical synthesis. At the time of its discovery, the 18-year old Perkin was attempting to synthesize quinine under the guidance of August von Hofmann. During one of his many failed attempts to do so, he noticed that a product of his reactions was bright purple in color. Rather than ignoring its presence and focusing on his originally intended goal, he isolated what transpired to be Mauveine and further refined its preparation. Perkin was astute enough to recognize the value in this mistakenly synthesized compound, which he went on to patent and commercialize as the first synthetic organic dye. Looking back on the discovery from the 21st century, it has the added significance of foreshadowing the industrialization of chemical synthesis, as well as the commercialization of scientific invention in general. As is so frequently the case, chance occurrence provided the initial spark, but it also took a sharp scientific mind to fan the flames of invention and realize the full potential of a discovery. Serendipity goes hand-in-hand with curious observation and diligent investigation – it is a pattern that emerges time and time again in research. This phenomenon is so familiar that David MacMillan [5] at Princeton University has even gone so far as to develop high-throughput experimentation techniques with the express purpose of minimizing the time between the serendipitous discovery of new chemical reactions.

A well-known example of a chance find from the early 20th century is that of penicillin [6], which was isolated by Sir Alexander Fleming (Figure 20.1c) while he was working in London at St Mary's Hospital next

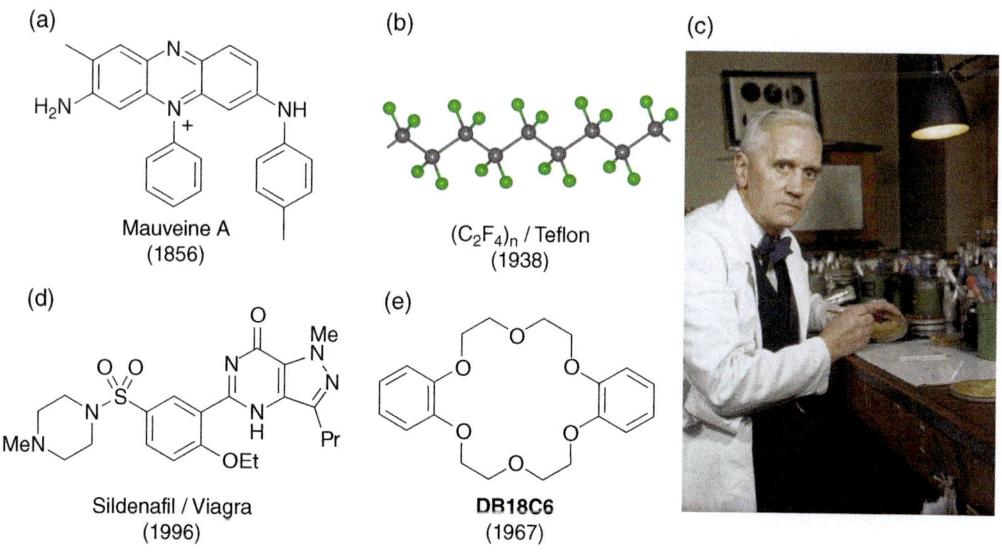

Figure 20.1 The structural formulas of chemical compounds whose synthesis or eventual application came as a surprise, and a photograph of Sir Alexander Fleming, who discovered penicillin in 1928

to Paddington Station. Many similar stories of serendipitous invention and discovery have followed, including [3] those of artificial sweeteners, cisplatin, Nylon, and Teflon (Figure 20.1b). In more recent times, Sildenafil (Figure 20.1d), commonly known by the brand name Viagra, was developed as a 5-phosphodiesterase inhibitor by Pfizer with the initial purpose of treating angina [7]. Phase I clinical trials revealed, quite unexpectedly, that it could be applied far more effectively as an on-demand oral treatment for erectile dysfunction. As a result, it went on to become a multi-million dollar industry, with annual sales peaking at almost 2 billion US dollars in 2008.

Of particular relevance to the motivation of Reed Izatt, macrocyclic thermodynamicist par excellence and one of the instigators of the Izatt–Christensen Award in Macrocyclic Chemistry, to commission and edit this book is a serendipitous observation made by Charles Pedersen in 1967 while working in the Elastomers Division at DuPont in Delaware. When attempting to synthesize a linear oligo(ethylene glycol) derivative capped with catecholates at either end, intended to complex the vanadyl ion and so afford a polymerization catalyst, he noticed that a template effect induced by K^+ ions favored the formation of a macrocyclic product [8, 9]. His discovery of synthetic routes to and the binding properties of crown ethers (Figure 20.1e) such as dibenzo[18]crown-6 (**DB18C6**) were not just seminal contributions to macrocyclic chemistry, but they also opened the door to the complexation of alkali metal cations, template-directed synthesis, and supramolecular chemistry in general, for which he shared the 1987 Nobel Prize in Chemistry.

This chapter describes some of the numerous occasions when serendipity has visited the Stoddart Laboratory in one form or another over the years. No doubt, this chapter is not the only one in this book that owes much to serendipity. In the first instance, we will consider the development of aromatic donor–acceptor (D–A) complexes and mechanically interlocked molecules (MIMs), lines of research which have led us recently to uncover the emergent ferroelectric properties of organic crystals formed as a result of Lock-Arm Supramolecular Ordering (LASO). Serendipity has also expressed itself twice over at the boundaries between coordination chemistry and materials science, a situation which will become evident in our description of a family of porous metal-organic frameworks (CD-MOFs) and a system for gold sequestration and recovery, both based on cyclodextrins (CDs).

20.2 Donor–acceptor charge transfer interactions

20.2.1 Introduction

It has been recognized [10] for many decades that attractive interactions develop between complementary electron-donating and electron-accepting molecules, be it in the gas, solution, or solid state. So broad is the definition of a donor–acceptor (D–A) complex, that, depending on the context, it has been applied [11] to species running the gamut from those that are interacting extremely weakly, such as argon–argon dimers, to those that are so strongly bound as to sometimes be classed as discrete molecules, such as NMe_3–BF_3. In 1949, interest in the topic of aromatic D–A complexes took on a new life after the discovery by Benesi and Hildebrand [12] of a new absorption band in the ultraviolet spectrum of I_2 dissolved in benzene. From early on, researchers in the area realized that the wavelength of this absorbance, labeled a charge transfer (CT) band, correlated with the strength of the non-covalent bonding interaction between the two species. The CT interactions are brought about by an electronic transition within the D–A pair ($D^{+\rho}$–$A^{-\rho}$, $0 \leq \rho \leq 1$), where ρ is the degree of ionicity. This observation paved the way for the investigation of a broad range of π–π D–A pairs, many of which remain commonplace in today's literature [13], e.g., perylene and tetracyano-*p*-quinodimethane (TCNQ).

In the knowledge (i) that crown ethers bind cations, including alkylammonium and transition metal ions; and (ii) that CT interactions can bring together electronically complementary π-systems, the question arose as to whether dibenzo crown ethers might form stable inclusion complexes with electron deficient aromatic cations. Our eventual involvement in answering this question came by way of studying the host–guest interactions between crown ethers and a series of transition metal complexes, including those with 2,2′-bipyridyl (2,2′-bpy) ligands [14]. Initial investigations, carried out in collaboration with Howard Colquhoun, now at the University of Reading, but then at the ICI Corporate Laboratory at Runcorn, and David Williams, erstwhile Professor of Structural Chemistry at Imperial College London, involved 1:1 adducts such as $[(Me_3P)PtCl_2(NH_3)]\subset$18-crown-6. X-Ray crystallography revealed that crown ethers participate in hydrogen bonding interactions with Pt-coordinated ammine ligands, exhibiting a form of second-sphere coordination [15, 16]. We then went on to discover that an analogous 1:1 inclusion complex formed from dibenzo[30]crown-10 (**DB30C10**) and $[Pt(2,2'\text{-bpy})(NH_3)_2]^{2+}$, with a complementary match of π-donors and a π-acceptor, in addition to the hydrogen bonding motif, underwent enormous changes in the ^1H NMR chemical shifts relative to those of the individual components. It transpired that a perfect triple π-stack emerges (Figure 20.2a) in $[Pt(2,2'\text{-bpy})(NH_3)_2]^{2+}\subset$**DB30C10**. The two electron-rich aromatic rings of the crown ether sandwich the electron-deficient 2,2′-bpy ligand between them.

The superstructural characteristics of $[Pt(2,2'\text{-bpy})(NH_3)_2]^{2+}\subset$**DB30C10** in the solid state led us to consider whether the bipyridinium herbicides Diquat ($\mathbf{DQT}^{2+}$) and Paraquat ($\mathbf{PQT}^{2+}$) might interact with **DB30C10** in an analogous fashion. At the time, we anticipated that CT interactions would still be present between the benzo rings of the crown ether and the electron-deficient bipyridium dication, while the second-sphere [N–H•••O] hydrogen bonding we had observed previously might be replaced by [C–H•••O] interactions involving the acidic methyl or methylene protons of $\mathbf{DQT}^{2+}$ or $\mathbf{PQT}^{2+}$. This hunch certainly turned out [17] to be true in the case of $\mathbf{DQT}^{2+}$, which forms (Figure 20.2b) the 1:1 complex $\mathbf{DQT}^{2+}\subset$**DB30C10**. The complexation was evident as soon as the two components were mixed, owing to the highly distinctive color change associated with the

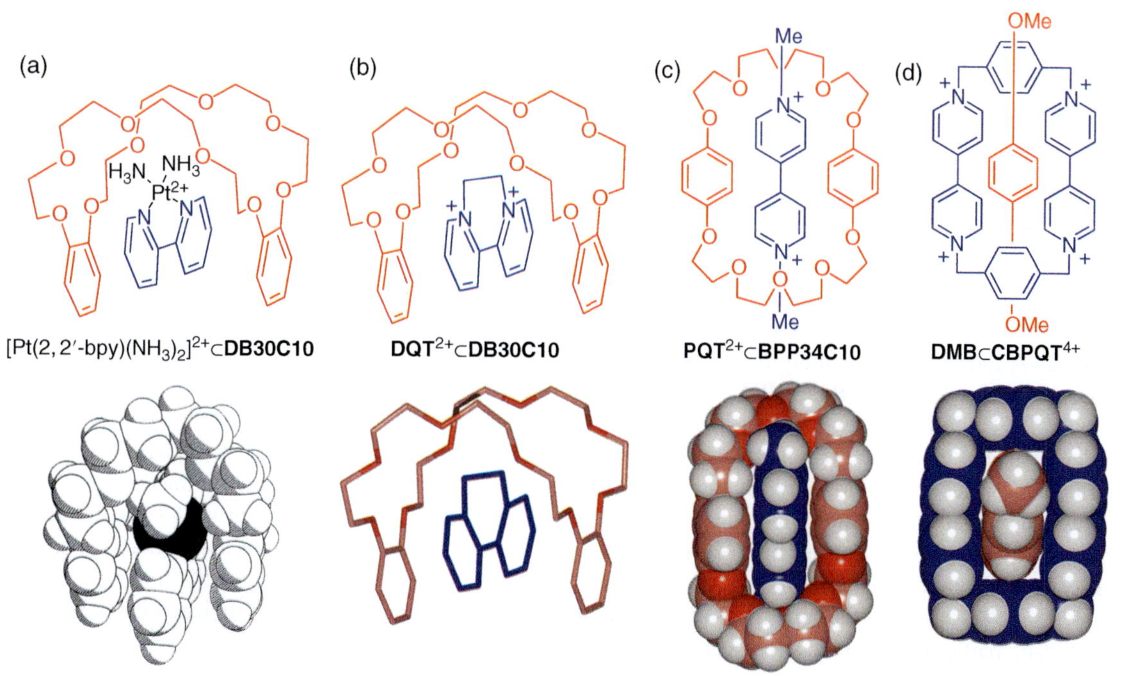

Figure 20.2 *Structural formulas and solid-state superstructures of D–A inclusion complexes. The electron-rich and electron-deficient aromatic systems interact through π–π charge transfer, [C–H•••O] hydrogen bonding and [C–H•••π] interactions. (Reprinted with permission from ref. [14]. Copyright © 1981 Wiley VCH.)*

formation of a CT complex. An association constant K_a of 17,500 M^{-1} was measured in Me_2CO for the hexafluorophosphate salt. Analysis of the solid-state superstructure revealed (Figure 20.2b) hydrogen-bonding interactions between the ether oxygen atoms and the methylene protons, in addition to the same sandwich-like arrangement of the aromatic rings that we had observed in the case of the 1:1 adduct with the transition metal ammine. Some subtle changes in the constitution of the crown ether were necessary before we arrived at an appropriate receptor for **PQT**$^{2+}$. Bisparaphenylene[34]crown-10 (**BPP34C10**) turned out to be the most effective member of the family of macrocycles that we screened, forming (Figure 20.2c) the complex **PQT**$^{2+}$⊂**BPP34C10** with a K_a value of 730 M^{-1} in Me_2CO for the hexafluorophosphate salt [18]. What really caught our attention, however, was the manner in which the **PQT**$^{2+}$ was threaded through the crown ether ring, hinting that we were on the cusp of making a mechanical bond. This possibility became even more apparent when we reversed the recognition motif. We synthesized the tetracationic cyclophane [19] cyclobis(paraquat-*p*-phenylene) (**CBPQT**$^{4+}$), whose rigid and rectangular shape creates a 7 Å-wide cavity between the two electron-deficient bipyridinium units that is ideally suited to hosting an electron-rich aromatic compound. We found that, although the binding interaction with 1,4-dimethoxybenzene (**DMB**) was rather weak (K_a = 17 M^{-1}) in MeCN solution, crystals of the **DMB**⊂**CBPQT**$^{4+}$ inclusion complex (Figure 20.2d) could be obtained readily [20]. The observation that the **DMB** was threaded through the aperture of the cyclophane, in what was later termed [21] a pseudorotaxane, served as our launching pad for exploring the nature of the mechanical bond.

20.2.2 Mechanically interlocked molecules

In essence, mechanically interlocked molecules (MIMs) are composed of two or more components with no covalent bond connecting them, but which cannot be separated from one another without first of all cleaving a covalent bond [22–27]. They possess a "mechanical bond" between their component parts. The term mechanical bond goes back to Harry Frisch [28], who coined it in 1953 before any examples had been made (in an abiotic setting, at least). A few years prior to our discovery of the recognition properties of **CBPQT**$^{4+}$, Jean-Pierre Sauvage [29, 30] had advanced the concept of template-directed synthesis of MIMs. [2]Catenanes and [2]rotaxanes are the archetypal examples of MIMs, the former being composed of two rings that are interlocked like the links of a chain. There is no doubt that when you have two interlocked rings, unless you break a covalent bond, you're not going to cleave them apart from one another, i.e., a catenane has the same properties of any conventional molecule in that respect. [2]Rotaxanes can also be considered as single molecules, consisting of a ring encircling a dumbbell component, from which it cannot escape on account of bulky groups at either end. Note that the "[2]" placed in front of catenane and rotaxane indicates the presence of two components within the molecules.

The D–A catenane [31], namely **1**$^{4+}$ (Figure 20.3), first saw the light of day in 1989, coming hot on the heels of the discovery of the **DMB**⊂**CBPQT**$^{4+}$ inclusion complex. Dication **2**$^{2+}$ reacts with xylylene bromide **3** to form **CBPQT**$^{4+}$ following two sequential nucleophilic substitutions, forming one fully fledged bipyridinium unit at each step. When the reaction is performed in the presence of a three-fold excess of **BPP34C10**, the first π-electron accepting bipyridinium unit that is formed worms its way into the cavity of the crown ether in order to maximize stabilizing non-covalent interactions. The second covalent-bond-forming step, therefore, occurs under the influence of the **BPP34C10**. More non-covalent-bonding interactions are gathered up as the CBPQT^{4+} ring clips itself around the hydroquinone (HQ) unit, trapping the topologically complex [2]catenane **1**$^{4+}$ in 70% yield. Although this mechanistic outcome was anticipated, its high efficiency came as quite a surprise, particularly when considering the low yields normally associated with macrocyclizations. Ultimately, it is the exquisite interplay between covalent, non-covalent, and mechanical bonds that is responsible for the catenation. The dication **2**$^{2+}$ does not interact strongly with the crown ether component in the beginning, yet once the first nucleophilic substitution leads to the formation of a bipyridinium unit, it is very happy to find a home inside the ring, since molecular recognition is being switched on all along the reaction pathway. Threading is driven by the [C–H•••O] and π–π stacking interactions. Another covalent bond is waiting to be made and more recognition, including a couple of weak [C–H•••π] interactions, is brought into play as the [2]catenane forms.

The weak interactions that template the formation of the [2]catenane "live on" within the molecule. It is clear from the solid-state structure (Figure 20.3), that the relative positions of the two rings, termed [32] their "co-conformation," is dictated by the same interactions which brought them together in the first place. In other words, once formed, the two rings of the [2]catenane are more than simply trapped together—they talk to one another. That's not to say that the two rings are static; far from it, in fact. Circumrotation of the rings allows for the exchange of degenerate forms as the rings undergo relatively unrestricted co-conformational dynamics under the influence of the non-covalent bonding interactions. Similarly large amplitude motion occurs in analogous degenerate [2]rotaxanes, which have been referred to as molecular shuttles [33]. Formation of a mechanical bond lifts the study of non-covalent interactions, dynamics, self-assembly, and template-directed synthesis from the supramolecular world (complexes) and brings them back into the realm of isolable and robust molecules. The dynamic behavior of the first D–A [2]catenane and [2]rotaxane opened the door to molecular switches (Section 20.2.2.1) once desymmetrized systems had been designed and produced.

From the point of view of synthesis and chemical topology [34], the strategies laid bare in the preparation of $\mathbf{1}^{4+}$ could be extended to ever more challenging targets – ones whose synthesis would have been almost inconceivable a few years earlier. By enlarging both the macrocyclic polyether and the tetracationic cyclophane, we were able to achieve the synthesis of [*n*]catenanes, where *n* indicates the number of rings mechanically interlocked together. As an exercise in pushing the template-directed synthesis protocols for the preparation of D–A catenanes to their limits, we synthesized (Figure 20.4a) a linear [5]catenane [35, 36] and also a branched [7]catenane [37]. We chose [38] the trivial name "olympiadane" for the [5]catenane on account of it having the same topology as the five-ringed symbol of the Olympic Games. The task of synthesizing olympiadane was completed in two key steps. Firstly, a [3]catenane intermediate was prepared by clipping a square-shaped tetracationic cyclophane (black, Figure 20.4a), containing bridging bitolyl linkers between the bipyridinium units, around two large macrocyclic polyethers (yellow and green, Figure 20.4a) each of which possess three 1,5-dioxynaphthalene (DNP) recognition sites. With two DNP units encircled by the central tetracationic cyclophane, the second step was to form a $CBPQT^{4+}$ ring at either end of the [3]catenane around one of the remaining DNP recognition sites in each of the two macrocycles in a manner analogous to the synthesis of $\mathbf{1}^{4+}$. Once again, David Williams' (Figure 20.4b) expertise was crucial in revealing the finer structural details of the catenane in the solid state. X-Ray crystallographic characterization (Figure 20.4a), coupled with

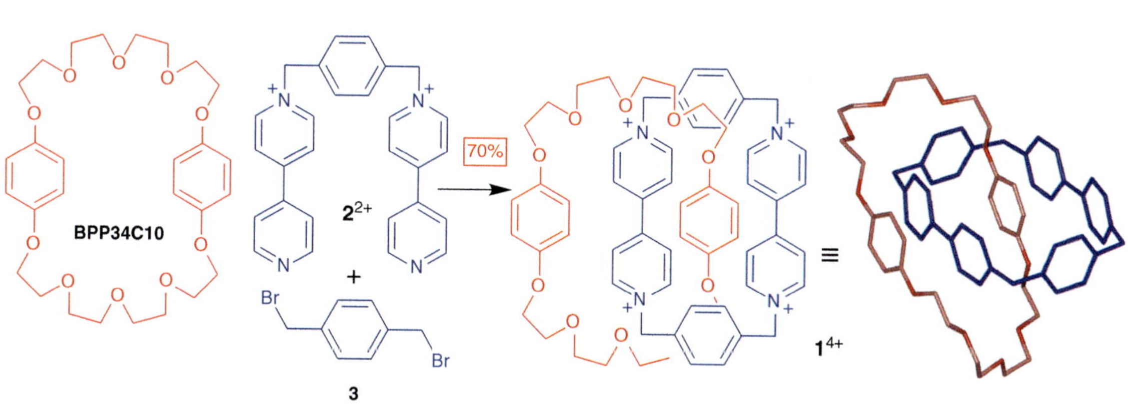

Figure 20.3 *The template-directed synthesis of an all-organic D–A [2]catenane. A stick representation of the solid-state structure of* $\mathbf{1}^{4+}$ *is shown with hydrogen atoms, counter-ions, and solvent omitted for the sake of clarity*

solution-phase spectroscopy gave insight (Figure 20.4c) into the non-covalent bonding interactions present in the molecule that aided and abetted its template-directed synthesis. Although aromatic D–A charge transfer and [C–H•••π] interactions both play their part in holding the components together and directing their relative positioning, it is [C–H•••O] hydrogen-bonding interactions between the ether oxygen atoms and acidic bipyridinium protons that are the major contributors to overall stabilization in an energetic sense.

20.2.2.1 *Molecular switches*

In conjunction with exploring how to improve upon the synthetic techniques to construct D–A MIMs, we also began to search for non-degenerate [2]catenanes and [2]rotaxanes – ones that act as rudimentary molecular switches. A molecular switch is a molecule that can be toggled back and forth between two or more distinct states by the application of a stimulus. The first such system that we developed was a bistable [2]rotaxane [39] with 4,4′-biphenol and benzidine recognition sites, or "stations," on which the $CBPQT^{4+}$ ring can reside. The electron-deficient $CBPQT^{4+}$ ring prefers to spend most of its time encircling the benzidine station. We discovered, however, that the ring is obliged to move to the 4,4′-biphenol station upon either protonation or electrochemical oxidation of the benzidine.

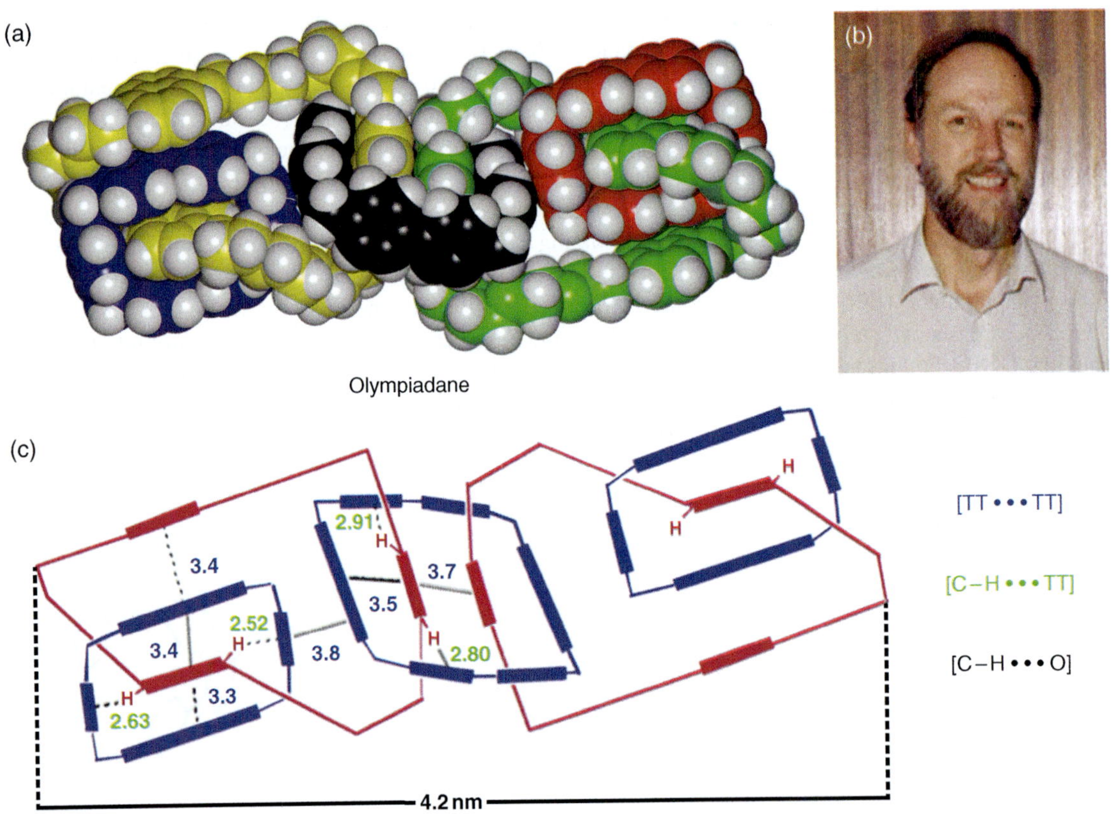

Figure 20.4 (a) The solid-state structure of the linear [5]catenane "olympiadane," which was revealed by (b) David Williams of Imperial College, London. (c) A schematic representation of the non-covalent bonding interactions present in the molecule. Distances in green and blue are given in Ångstroms. Solvent and counter-ions are omitted from the X-ray crystal structure for the sake of clarity

Later, we discovered that the combination (Figure 20.5) of a DNP station and an electrochemically switchable tetrathiafulvalene (TTF) donor is particularly appealing based on: (i) the excellent discrimination between ground-state geometries; (ii) the fidelity of switching; (iii) its particular ground-state kinetics [40]; and (iv) the reversibility of the redox-controlled processes. These features would become increasingly important as we pushed these switches forward toward applications in solid-state molecular electronic devices (MEDs) and on the surface of nanoparticles [41]. The [2]catenane $\mathbf{4}^{4+}$, incorporating both of these donors, was prepared [42]. In the solid state (Figure 20.5a) and in solution, the TTF-recognition site resides within the cavity of the $CBPQT^{4+}$ ring and the DNP interacts with one of the bipyridinium units in a side-on fashion. This state is termed the ground-state co-conformation (GSCC). Upon oxidation of the TTF to $TTF^{\bullet+}$, either chemically or electrochemically (Figure 20.5b), a new minimum-energy state is established in which the $CBPQT^{4+}$ ring encircles the DNP unit. The charged $TTF^{\bullet+}$ radical cation no longer has an affinity for the electron-deficient cavity of the $CBPQT^{4+}$ ring. A subsequent reduction step resets the switch to its original state, after first passing through (Figure 20.5b) a meta-stable co-conformation (MSCC) with a kinetic energy barrier of 16.7 kcal mol^{-1} to overcome before the GSCC is reached [43]. The molecular switches based on TTF and DNP are now, relatively speaking, well understood and have been studied extensively in device settings. In recent years, alternative sets of recognition motifs involving the radical–radical spin pairing interactions of

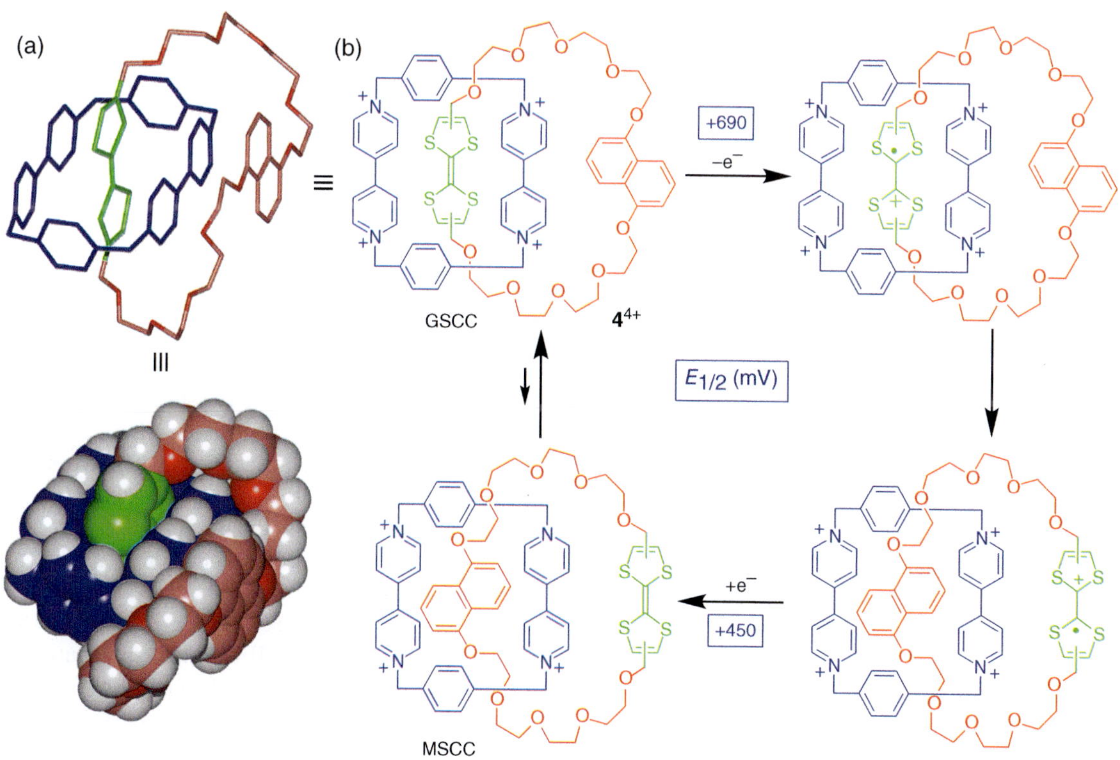

Figure 20.5 *(a) The solid-state structure of the molecular switch* $\mathbf{4}^{4+}$*. The structure is shown in stick representation with hydrogens atoms omitted (above) and space-filling with hydrogens included. Counterions and solvent are omitted for the sake of clarity. (b) Co-conformational switching of the [2]catenane between the TTF (green) and DNP (red) recognition sites in response to a redox stimulus. Although the crown ether component is a* cis–trans *isomeric mixture with respect to the TTF, only the* trans *isomer is observed in the solid state. Half-wave potentials are shown in blue*

reduced viologens [44] or oxidized TTF units [45] have emerged as powerful design elements for molecular switches. We have also begun to exert a higher level of control over the kinetics of co-conformational rearrangements [46], which has opened up the possibility of exploring artificial molecular pumps [47, 48] and non-equilibrium systems [49].

20.2.2.2 Molecular electronic devices

The bistability of molecular switches such as the [2]catenane **4**$^{4+}$, as well as their responsiveness to electrochemical stimulus, led us to consider whether they might be introduced [50–52] into MEDs – where one state corresponds to "ON" and the other to "OFF". In going from working with MIMs in solution to studying their properties in the solid state, one of the questions we sought to answer was: "How will the dynamics of switching be affected by moving from the relatively unhindered co-conformational motion in solution to that of increasingly ordered and congested environments?" Now, having probed [43] the dynamics of bistable MIMs in settings ranging from gels, self-assembled monolayers, and Langmuir–Blodgett films to molecular switch tunnel junctions [50] (MSTJs), it has become clear [52] that the thermodynamics that govern their switching remain more or less unchanged while the activation energy barriers associated with the kinetics of switching can vary by several kcal mol^{-1}. The rate of relaxation from the MSCC (ON) to the GSCC (OFF) is fast (seconds) in solution, where the molecules are relatively free, and slow in high-viscosity polymer matrices or closely packed monolayers, where the molecules are constrained. Experimental techniques, including time- and temperature-dependent electrochemistry and spectroscopy, have made it possible to monitor switching cycles and quantify the rates of co-conformational isomerism. Taking the amphiphilic bistable [2]rotaxane **5**$^{4+}$ (Figure 20.6a) as an example, the activation energy barrier to relaxing from its MSCC back to the GSCC rises from 16 kcal mol^{-1} in solution to 18 kcal mol^{-1} in a polymer matrix, and finally to around 22 kcal mol^{-1} in full two-terminal devices.

The fabrication, operation, performance, and analysis of devices with molecular switches as active elements is influenced profoundly by the choice of electrode material (Si wafer), as well as by the type of architecture (crossbar) used for the device. Working closely with James (Jim) Heath, firstly at UCLA and then at Caltech, we were able to develop a procedure to incorporate Langmuir–Blodgett monolayers of **5**$^{4+}$ at the MSTJs of crossbar memory devices with Si and Ti nanowire electrodes, which proved to be suitable for the fabrication (Figure 20.6b) of a molecular memory device [53]. The MSTJs of crossbar devices are two-terminal molecular junctions in which the active element – a monolayer of bistable molecules sandwiched between two electrodes – performs the electrochemically driven function of a resettable switch. Each of the MSTJs in the device can be addressed individually, and the switchable rotaxane components serve as the data-storage elements. We fabricated 160-kilobit memory circuits based on this architecture, with an impressive 10^{11} bits per square centimeter. To put those characteristics into context, the ultrahigh density means that the entire area taken up by the crossbar assembly is smaller than that of the cross-section of a white blood cell and, at the time of publication in 2007, it represented the dimensions analogous to those projected for a DRAM circuit available by 2020! Taken together, our forays into the development of MEDs have substantiated the idea that bistable catenanes and rotaxanes operate effectively as mechanical switches in soft-matter environments and, with appropriate handling and device design, can even withstand relatively harsh circuit-fabrication steps.

20.2.2.3 Drug delivery systems

In 2004, we began to put bistable rotaxanes through their paces at the surface of mesoporous silica nanoparticles (MSNPs), working in close collaboration with Jeffrey (Jeff) Zink at UCLA. Our goal was to modulate the uptake and release of cargo molecules, such as pharmaceuticals, from the pores of the MSNPs, allowing

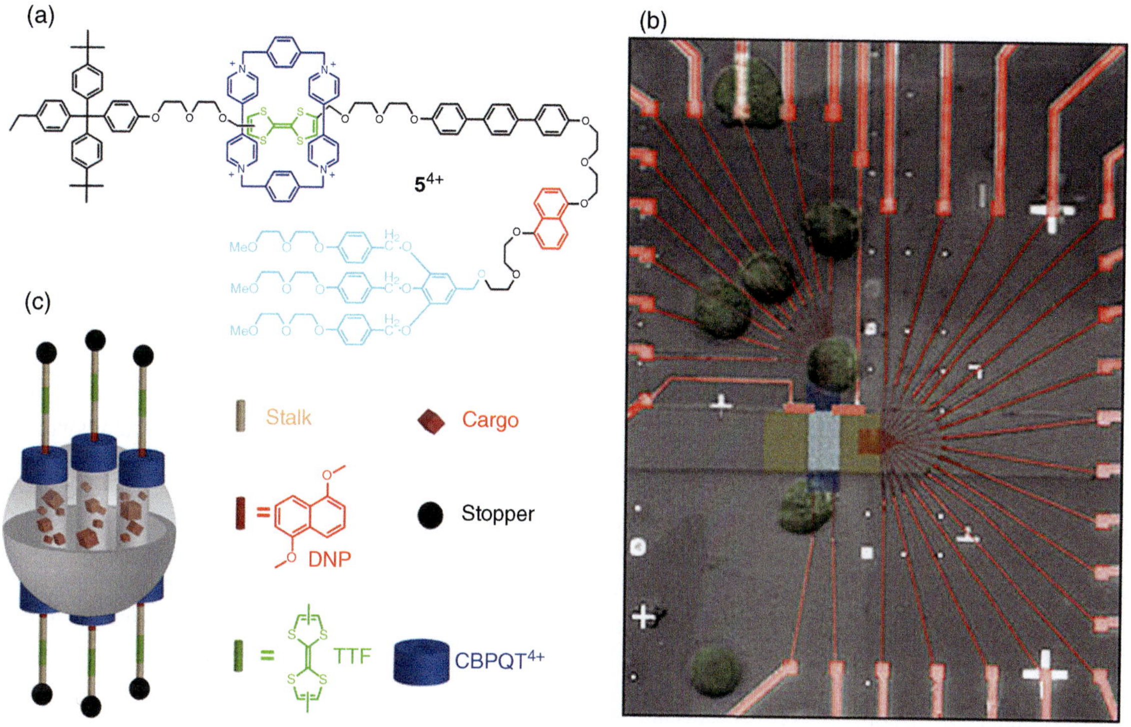

Figure 20.6 *(a) The structural formula of the amphiphilic bistable [2]rotaxane **5**$^{4+}$ employed as the active element in the fabrication of the memory device. (b) A 160-kilobit crossbar molecular memory, shown in orange at the center of the image, has a smaller cross-sectional area than that of a white blood cell (purple). (c) A graphical representation of a mechanized nanoparticle composed of a mesoporous silica bead (100–500 nm in diameter) covered with nanovalves based on a molecular switch similar to 54+. This mechanized nanoparticle can be used to deliver small molecules (drugs, dyes, perfumes, etc.) on command in prescribed environments. A key on the right defines the main structural elements of the device*

them to function as carrier vehicles in drug delivery [54]. The controlled release of drug molecules from within an inert carrier vehicle promises to enhance chemotherapeutics in a number of different ways: by (i) improving their pharmacokinetic properties; and (ii) achieving targeted release of the cargo; thereby (iii) concentrating active drugs within diseased tissues; and (iv) reducing their exposure to healthy cells. Employing bistable rotaxanes similar to those that supported our molecular electronics research program, we constructed [55] "mechanized nanovalves" that coat (Figure 20.6c) the exterior of MSNPs. Immobilization of the nanovalves on silica substrates has generally been achieved by first of all functionalizing the MSNPs with suitably reactive groups [56] – e.g., an isocyanate [55, 57, 58] or azide [59] – which can then be used as a handle to attach bistable rotaxanes under mild reaction conditions, such as carbamate formation or the copper-catalyzed azide–alkyne cycloaddition. The design of the reversible nanovalve system [57] shown in Figure 20.6c is such that the "stalks" attached to the MSNPs each have two stations for the CBPQT^{4+} rings – DNP stations close to the surface and, hence, close to the entrances to the nanopores, as well as TTF stations positioned further away. We demonstrated that when the fully assembled nanovalve is in its ground state, in which the rings encircle the TTF stations, the MSNPs can be loaded with a luminescent compound – tris(2,2′-phenylpyridyl)iridium(III) or rhodamine B. The luminescent cargo is then trapped inside the nanopores by adding $Fe(ClO_4)_3$ as an oxidant, which oxidizes the TTF units to TTF^{2+} dications and, consequently, causes the rings to shuttle to the DNP stations where they effectively cap the nanopores. Release of the cargo is subsequently initiated by reversal of the switching in response to an external stimulus. A mild reductant (sodium ascorbate) is added to restore the TTF stations to their initial oxidation states, which in turn causes the rings to return to their starting positions. Diffusion of the cargo from the nanopores can be monitored by the increase in their emission intensities. The strategy of blocking and unblocking the nanopores of MSNPs using reversible chemistry has proven to be a general one. Indeed, it is not restricted to D–A rotaxanes or even to MIMs. We have also shown that pores capped with inclusion complexes [58, 59] or polymers [60] are also effective as reversible blocking groups. Progress in the area continues apace, driven by the large number of research teams now working in the area [61, 62].

20.2.3 Lock-arm supramolecular ordering (LASO)

While attempting to synthesize a neutral cyclophane, starting from pyromellitimide and **3** in the presence of a template in the form of a naphthalene-1,5-diol substituted on its hydroxyl functions with diethylene glycol chains, we stumbled across an emergent crystal-growth phenomenon. At the outset, we had not predicted that simple donors and acceptors bearing diethylene glycol chains would give rise to a new class of robust co-crystals, but that turned out to be the case [63]. Serendipity was certainly at play in the initial discovery of what we have christened "Lock-Arm Supramolecular Ordering" (LASO). Even more surprisingly, and once again by chance rather than by design, the high aspect ratio crystals that are produced have proven to be more than just structurally interesting curiosities – they exhibit ferroelectric properties at room temperature, making them enticing candidates for applications in electronics [64].

The design elements of LASO co-crystals are: (i) a pair of D–A π-systems; which possess (ii) complementary hydrogen-bonding recognition units – in the form of flexible diethylene glycol arms on one substrate and polar functional groups (e.g., amino, hydroxyl, carbonyl) on the other – that participate in intermolecular hydrogen bonding. An example of such a D–A pair, **6** and **7**, is shown in Figure 20.7a. Crystallization is also guided by (iii) solvophobic forces, which contribute to the driving force for self-assembly. The cooperative network of CT interactions and intermolecular hydrogen bonding that develops between the two components leads to solvent-free co-crystals in which the donors and acceptors alternate in a highly ordered manner, forming (Figure 20.7b) a mixed aromatic stack. Although a predisposition for hydrogen bonding is engineered into the substrates, the exact pattern of intermolecular interactions that arises during crystallization cannot be predicted from first principles or chosen by design. Rather, the flexibility of the diethylene glycol arms allows the co-crystals to achieve energetically favorable intermolecular binding interactions under co-conformational control in the lattice, resulting in the emergence of robust and closely packed co-crystals.

As a strategy for forming robust mixed stacks, LASO has already proven to be highly modular. A series of four acceptors and eight donors has been studied with diverse aromatic cores and glycol chains located on either the donor or the acceptor. Moreover, the high aspect ratio crystals produced (Figure 20.7c) in the LASO protocol (i) grow quickly, typically over a period of just a few days; (ii) are large, growing up to several centimeters in length; and (iii) are extremely robust, remaining stable both in solution and in air, even after complete desolvation. These properties are highly desirable and bode well for finding real-world applications.

20.2.3.1 Emergent ferroelectric properties

Ferroelectrics are characterized by non-linear polarization in the presence of an electric field, some of which remains after the field is removed. The vector of this remnant polarization can be flipped by applying a sufficiently strong electric field in the opposite direction, making ferroelectrics useful for applications in random access memory. There is a drive toward developing practical all-organic ferroelectrics to replace or

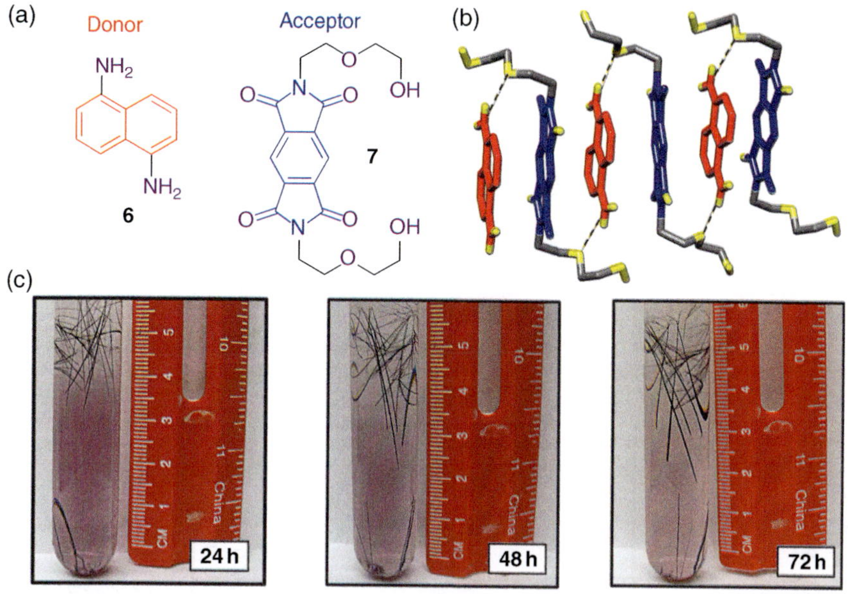

Figure 20.7 *(a) An example of a D–A pair that undergoes LASO to form (b) binary co-crystals that have an alternating arrangement of donors and acceptors extending along one axis sustained by a network of hydrogen bonds. (c) The high aspect ratio LASO crystals that are formed during liquid–liquid diffusion grow up to several centimeters in length*

complement traditional inorganic ferroelectrics because they have the potential to be lightweight, inexpensive, and easy to process in the fabrication of devices. Ordered networks of organic donors and acceptors have been known, in a small number of cases, to exhibit ferroelectric behavior. At their ferroelectric Curie temperature (T_C), a characteristic jump in ionicity is observed on account of the donors and acceptors dimerizing into pairs ($D^0\,A^0\,D^0\,A^0 \rightarrow D^{\delta+}\,A^{\delta-}\,D^{\delta+}\,A^{\delta-}$) below T_C. Before the development of LASO crystals, however, the practicality of known systems was hampered significantly by their low Curie transition temperatures of 81 K or below, restricting their use to cryogenic conditions.

Although complexes **6•7**, **8•7**, and **9•7** (Figure 20.8) are formed from symmetrical molecules, the LASO networks fulfill the requirements for a ferroelectric system on account of the symmetry-breaking dimerization of donors and acceptors, which occurs even at high temperature. Attempts were made to identify the T_C for each of the networks by measuring the dielectric constant of the crystals along their ferroelectric axis (the vector of the mixed aromatic stacks) as a function of temperature. Remarkably, no characteristic discontinuities were observed (Figure 20.8, top) within the window of 5 to 400 K, indicating that their T_Cs are only reached at higher temperature and that the ferroelectric phase persists at room temperature! The complexes maintain polar, non-centrosymmetric lattices at room temperature with ionicities, ρ, measured to be 0.68, 0.89, and 0.43, respectively. This high-temperature ferroelectricity was confirmed by measuring (Figure 20.8, bottom) hysteresis curves of polarization versus electric field along the ferroelectric axis. Remnant polarizations exceeding 1 μC cm^{-2} at room temperature were observed for all three of the LASO networks tested. Thus, materials made using the LASO strategy have consigned to the past the prevailing notion that room-temperature ferroelectrics are the sole preserve of inorganic materials. Organic D–A mixed-stack materials made using LASO hold promise as a supramolecular platform for electrically addressable hydrogels, ferroelectric catalysts, and CT-based sensitizers for photovoltaics.

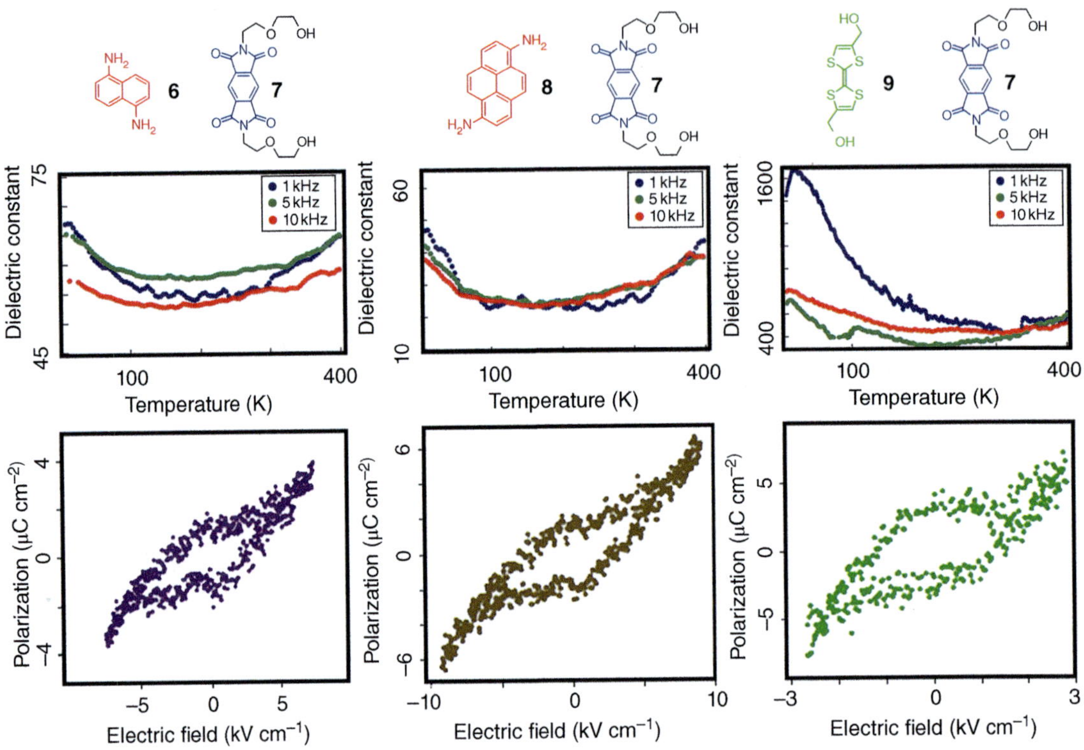

Figure 20.8 (top) Three D–A pairs that undergo LASO to form room-temperature ferroelectric materials as evidenced by (middle) measurement of their temperature-dependent dielectric constants, which do not show discontinuities between 5 and 400 K, and (bottom) polarization versus electric field curves, which exhibit hysteresis at 300 K

20.3 Cyclodextrins (CDs)

20.3.1 Introduction

The cyclodextrins (CDs) are a family of naturally occurring toroidal oligosaccharides composed (Figure 20.9a) of several α-D-glucopyranosyl units linked one after the other in a cyclic manner. Reports on CDs go back as far as 1891, although it was not until the 1960s that they began to be investigated actively as receptors. In terms of their structures, one face of the CD toroids is decorated (Figure 20.9b) with secondary hydroxyl groups and the other displays primary hydroxyl groups, while the cavity is lined by methine protons for the most part. This arrangement of functional groups leads to a relatively non-polar binding pocket at the interior of the CDs, while their exterior is highly hydrophilic, imparting good water solubility. Nowadays, the three most common members of the CD family, α-CD, β-CD, and γ-CD, comprising 6, 7, and 8 glucopyranosyl units respectively, are produced industrially by the microbial conversion of starch. All three have been mainstays of academic research for several decades, particularly in supramolecular chemistry where their properties as receptors for small hydrophobic molecules and as catalysts [65, 66] have been studied extensively. Their versatility, easy availability, and low cost has also led to them finding commercial applications, *inter alia*, as drug-solubilizing agents – encapsulating pharmaceuticals within their hydrophobic cavities and bringing them into aqueous solution. In some cases, complexation by a CD can also stabilize a drug against degradation *in vivo*.

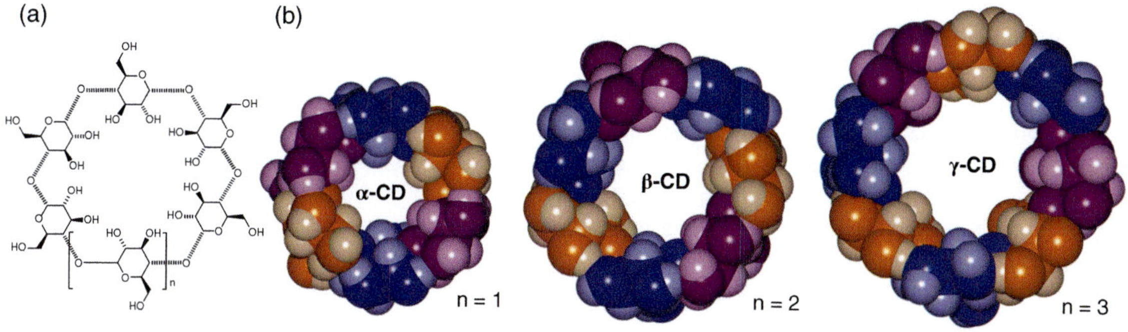

Figure 20.9 *(a) The structural formula and (b) solid-state single-crystal structures of α-CD [67], β-CD [68], and γ-CD [69]. The structures are shown in space-filling representations and viewed from the secondary face of the CD, looking through the center of the torus. Solvent molecules are omitted for the sake of clarity and the individual glucopyranosyl units are colored in blue, orange, and purple in order to allow them to be distinguished from one another*

20.3.2 Cyclodextrin Metal-Organic Frameworks (CD-MOFs)

The cavity of γ-CD, which is ~1 nm in diameter and ~0.8 nm in depth, is capable of housing two aromatic guests simultaneously, assisted by mutual π–π stacking and hydrophobic interactions between the aromatic rings of the molecules. We intended to exploit this molecular recognition phenomenon in the synthesis of heterocircuit Borromean rings, [70] by way, first of all, of forming a 2:1 complex in which two azobenzene derivatives would be held within the γ-CD torus. Upon mixing an excess of the K^+ salt of azobenzene-4,4′-dicarboxylate with γ-CD in aqueous solution and growing crystals by diffusion of MeOH into the solution during a few days, large cubic crystals were obtained. What was observed by X-ray crystallography, however, was completely unexpected. Rather than being the anticipated 2:1 inclusion complex, all that is visible (Figure 20.10) in the crystal is the γ-CD torus and the K^+ ions – the azobenzene-4,4′-dicarboxylate counterions are completely disordered. It transpired that the C_8 symmetrical γ-CD tori are linked together by coordination to the K^+ ions as part of an extended structure [71]. We had chanced upon an infinite nanoporous framework!

Having identified that the K^+ ions are the glue that holds the γ-CD tori together while the azobenzene-4,4′-dicarboxylate counterions were, to all intents and purposes, bystanders in the formation of the framework, we repeated the crystal growth using KOH as a simpler source of K^+ ions with the goal of forming an isoreticular framework [72, 73]. Taking (Figure 20.10a) a spoonful of sugar (γ-CD), a pinch of salt (KOH), and a swig of alcohol (aqueous MeOH), a CD metal-organic framework (CD-MOF) is created under the simplest of experimental conditions – so simple, in fact, that the preparation can be carried out in undergraduate teaching and high school laboratories [74]. This particular extended framework, produced using KOH, was labeled as CD-MOF-1. The recipe for the CD-MOF family of isoreticular materials is not only straightforward, it is also remarkably versatile. Substitution of the anionic partner to the K^+ ions with almost any common counterion, such as nitrate or benzoate, does not perturb the crystal growth, while crystals with isostructural frameworks have also been grown on gram scales from γ-CD with RbOH and CsOH, leading to CD-MOF-2 and -3, respectively [75].

X-Ray diffraction reveals that CD-MOF-1 has a cubic unit cell of space group *I*432. Its extended structure can be visualized as arising (Figure 20.10b, c) from a repeating unit of six γ-CD tori that pack into a body-centered cubic array. These $(\gamma\text{-CD})_6$ repeating units are held together by virtue of eight-coordinate K^+ ions, such that primary and glucopyranosidic hydroxyl groups coordinate to the alkali metal center. A spherical cavity is formed within the $(\gamma\text{-CD})_6$ cubes, which can be accessed by passing through the windows on each face that are shaped by the apertures of the γ-CD tori. Following these apertures along the *a*, *b*, or *c* axis, reveals (Figure 20.10c) circular, transverse channels that extend through the crystal with diameters of just under a nanometer, which are perforated (Figure 20.10b) by smaller triangular pores (0.4 nm in diameter) that are directed along the (111) planes of the crystal. Overall, the body-centered cubic array is highly porous, possessing (Figure 20.10d) a unique internal space that is shaped by the chiral γ-CD rings, their cavities, and the space between them. The CD-MOFs are robust enough to be removed from the aqueous alcohol in which they form and then to be activated by soaking in CH_2Cl_2 and evacuating under vacuum. Comparisons of the theoretical and experimentally obtained powder X-ray diffraction patterns show that the MOFs retain crystallinity after activation, while N_2, H_2, CO_2, and CH_4 adsorption isotherms have confirmed that the crystals are permanently nanoporous.

From the point of view of commercial applications, CD-MOFs hold significant promise – two potential applications are discussed in Sections 20.3.2.1 and 20.3.2.2. These applications are brought all the closer by the simplicity of the preparative procedure as well as the ready availability, safety, and low cost of the precursors. The MOFs can even be made from edible materials! Food grade γ-CD from Wacker, potassium benzoate (E212), and Everclear (95% grain alcohol) were found [71] to be compatible with the method of preparation. The result is an edible MOF that is not only easily made, but is also green, safe, and cheap – all characteristics that make it immensely appealing for applications on an industrial scale. There may be some potential in the pharmaceutical market, but its low price would also make the material amenable to applications in the food,

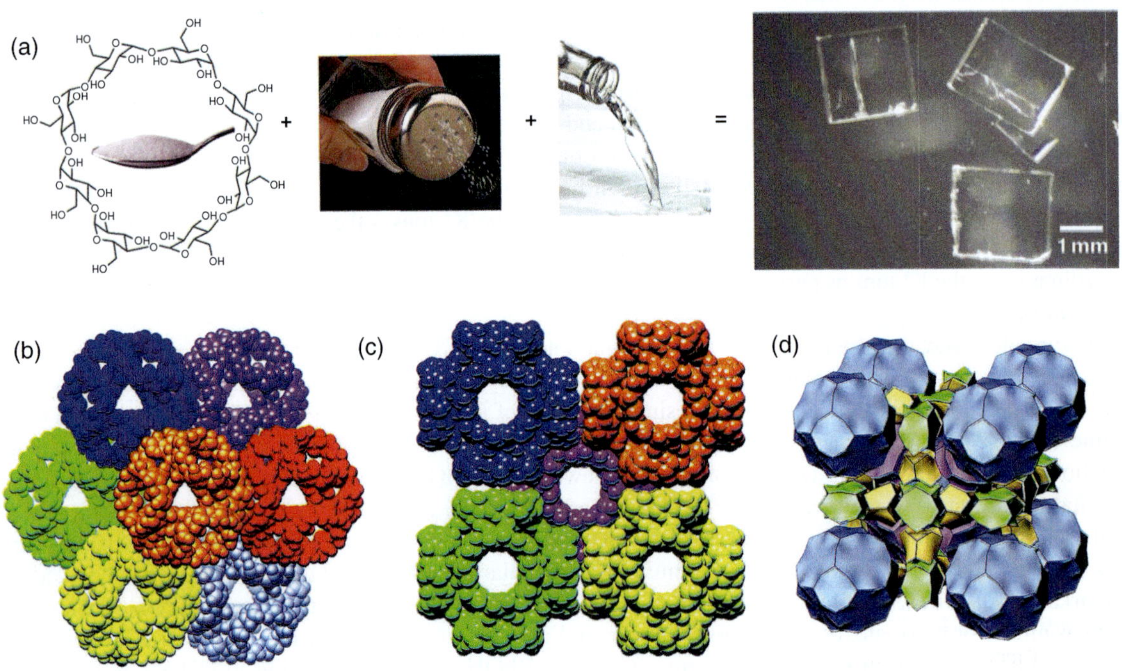

Figure 20.10 *(a) Cubic crystals of CD-MOFs are grown from the simple combination of a spoonful of sugar (γ-CD), a pinch of salt (an alkali metal salt), and a swig of alcohol (aqueous MeOH). Single-crystal X-ray diffraction reveals (b) triangular and (c) circular channels that run through the framework. Disordered solvent and anions occupy some of the void space in crystals. (d) A graphical representation that highlights the cavities and channels in CD-MOF-1 of different size and shape*

homecare, and personal care markets. Examples might include spoilage alerts, time released ripening agents, preservatives, toothpastes, deodorants, extended release air fresheners, or filters in air-conditioning systems. With a view to impacting upon these markets, we have formed a start-up company, PanaceaNano, to further the development of CD-MOFs and exploit them in a real-world setting.

20.3.2.1 CO_2 Sequestration and sensing

Reducing the levels of atmospheric CO_2 is of paramount importance for the planet and society at the start of the 21st century. As part of the diverse portfolio of tactics that are being devised in order to tackle this challenge, materials are required that detect CO_2 in the presence of other gases and sequester it selectively. In assessing the affinity of the CD-MOFs for different gases, we observed [76] that CD-MOF-2, which has an estimated BET surface area 1030 m^2 g^{-1}, shows a steep uptake in the low-pressure region of its CO_2 isotherm recorded at 298 K, indicative of a strong binding interaction between the framework and CO_2. This observation raised the question – why does the MOF take up CO_2 so rapidly? We speculated that the most likely culprit was chemisorption, in which CO_2 reacts reversibly with free hydroxyl groups of the γ-CD tori to form carbonic acid functions. In order to test this hypothesis, isotherms were measured (Figure 20.11a) for CH_4 and CO_2 uptake at different temperatures. Selectivity for CO_2 over CH_4 is high, particularly at low pressures where there is almost a 3000-fold preference for CO_2 adsorption. The uptake of CO_2 in the <1 torr region was found to be essentially independent of temperature, as was the position of a knee-point in the isotherm which

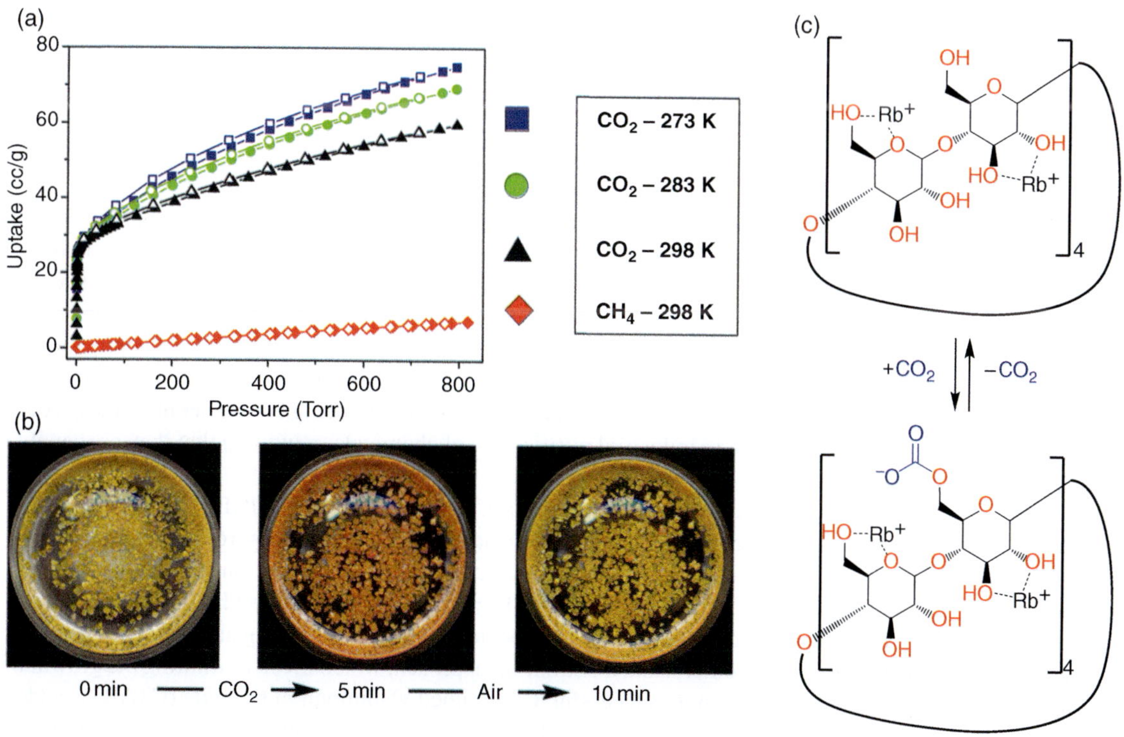

Figure 20.11 *(a) Gas adsorption isotherms for activated CD-MOF-2. Solid symbols indicate gas sorption and open symbols gas desorption. (b) Crystals of CD-MOF-2 with methyl red incorporated that display a change in color when exposed to* CO_2*, whereupon (c) the non-coordinated primary hydroxyl functions react with the* CO_2 *to form carbonic acid functions*

occurs at ~23 cm^3 g^{-1}. An enthalpy of –27.1 kcal mol^{-1} was measured for the low-pressure adsorption, suggestive of chemisorption [77]. At pressures >1 torr, on the other hand, adsorption of CO_2 is quite clearly (Figure 20.11a) temperature-dependent and is accompanied by a significantly lower enthalpy change of –15.6 kcal mol^{-1}, which is consistent with physisorption occurring in this region. Further confirmation that chemisorption is behind the high selectivity and affinity for CO_2 at low pressures came from cross-polarized magic angle spinning (CP-MAS) ^{13}C NMR spectra. Absorption of CO_2 resulted in (i) a change in the chemical shifts of the γ-CD carbon nuclei; and (ii) the appearance of a peak centered around 158 ppm that can be assigned to the carbonyl of the carbonic acid that is formed.

It is worth noting that we have not found any evidence for the formation of carbonic acid when samples of the parent γ-CD interact with CO_2. Rather, it seems that the unique environment of the MOF is responsible for rendering the hydroxyl groups of the γ-CD reactive towards CO_2, perhaps as a result of the high effective molarity that is brought about by nano-confinement of CO_2 in their vicinity when it enters the channels and cavities of the MOF. Visualizing the reversible CO_2 adsorption and desorption is made possible when the MOF has been laced with the pH indicator methyl red. The crystals are yellow (Figure 20.11b) in the absence of a high pressure of CO_2, but as soon as CO_2 is added, chemisorption occurs, which in turn generates (Figure 20.11c) carbonic acid functions and lowers the pH, causing the crystals to become orange. Once the CO_2 atmosphere is removed, the color of the framework returns to yellow again within 5 minutes. In addition to the measurable change in pH that occurs, a ~500-fold decrease in the ionic conductivity of CD-MOF-2 is observed [78] after CO_2 infusion in MeOH medium. This reduced conductivity can be ascribed to (i) the loss of coordinated hydroxide ions that can catalyze the deprotonation of the solvent; and (ii) blocking of the triangular pores, significantly reducing motility of the intermedium molecules. Based on these principles, we have been able to develop a device for the detection of CO_2 that works based on measuring the change in electrical conductivity of CD-MOF-2 crystals.

20.3.2.2 Chromatographic separations

The permanent porosity and robustness of MOFs has led researchers to investigate their application as stationary-phase materials in chromatographic separations. There is some evidence that otherwise challenging separations, e.g., of isomeric hydrocarbons, are facilitated by shape selectivity [79] and/or specific non-covalent bonding interactions [80] within the well-defined pores of MOFs. We speculated that the variety (Figure 20.10d) of pore shapes and sizes present in CD-MOFs might lend these materials to application as stationary phases for chromatography. Taking into account the many benefits associated with the preparation of CD-MOFs – their renewable feedstocks, low price, and environmentally benign components – they would be an attractive prospect for application in industrial purification of chemicals, providing that they are capable of separating appropriate analytes.

Before any chromatography could be performed, the first task was to produce CD-MOF crystals of a suitable particle size, and to do so in a manner that is reliable and scalable [81]. Attempts to process crystals of CD-MOF-2 in a top-down manner by grinding and sieving the crystals were met with only moderate success. Crystals that fell within a range of 10–37 μm in diameter could be isolated successfully and packed into columns, but the relatively broad distribution of particle size led to poor stationary-phase packing and suboptimal performance as a separation medium. On the other hand, a bottom-up approach, which involves limiting the growth of CD-MOF-1 crystals, proved to be much more successful. Cetyltrimethylammonium bromide (CTAB) is added as stabilizing agent to slow down nucleation, allowing the particle size to be controlled as a function of the crystal growth time. Using this protocol, a narrower particle size distribution in the range of 10–15 μm can be attained, which has proved to be far more effective when applied to chromatographic purification.

A column packed with CD-MOF-1 exhibited high shape selectivity in the separation of an industrially relevant mixture of benzenoid compounds, BTEX, which contains benzene, toluene, ethylbenzene, and the three xylene isomers. Separations of BTEX are commonplace during the refinement of crude oil [82], but the processes used in the chemical industry at present are energy intensive. By running high-performance liquid chromatography experiments in hexane, we have demonstrated [81] that CD-MOF-1 is capable of separating this petrochemical feedstock into its individual components with high resolution and good separation factors. Interestingly, the performance of the column was found to improve after several hours of operation, apparently as a result of tightly bound MeOH being gradually flushed out of the framework, which in turn frees up all of the channels and cavities to interact fully with the substrates. Grand canonical Monte Carlo simulations (Figure 20.12) and single component static vapor adsorption isotherms gave us some insight into the origin of the shape-selectivities at play in the separation of the xylene isomers. Initial absorption appears to take place preferentially in the transverse pores connecting the $(\gamma\text{-CD})_6$ cuboids before the large spherical cavities are filled. The small dimensions of the triangular channels renders them inaccessible to the xylene isomers. Simulations have revealed (Figure 20.12) a hierarchy of preferential adsorption to the stationary phase in the order *ortho* >> *meta* > *para*, in good agreement with the order of elution observed experimentally. The selectivity for the adsorption of *o*-xylene comes (Figure 20.12c) as a consequence of its highly efficient packing within the framework and favorable adsorbent–adsorbate interactions, in addition to relatively strong adsorbate–adsorbate interactions.

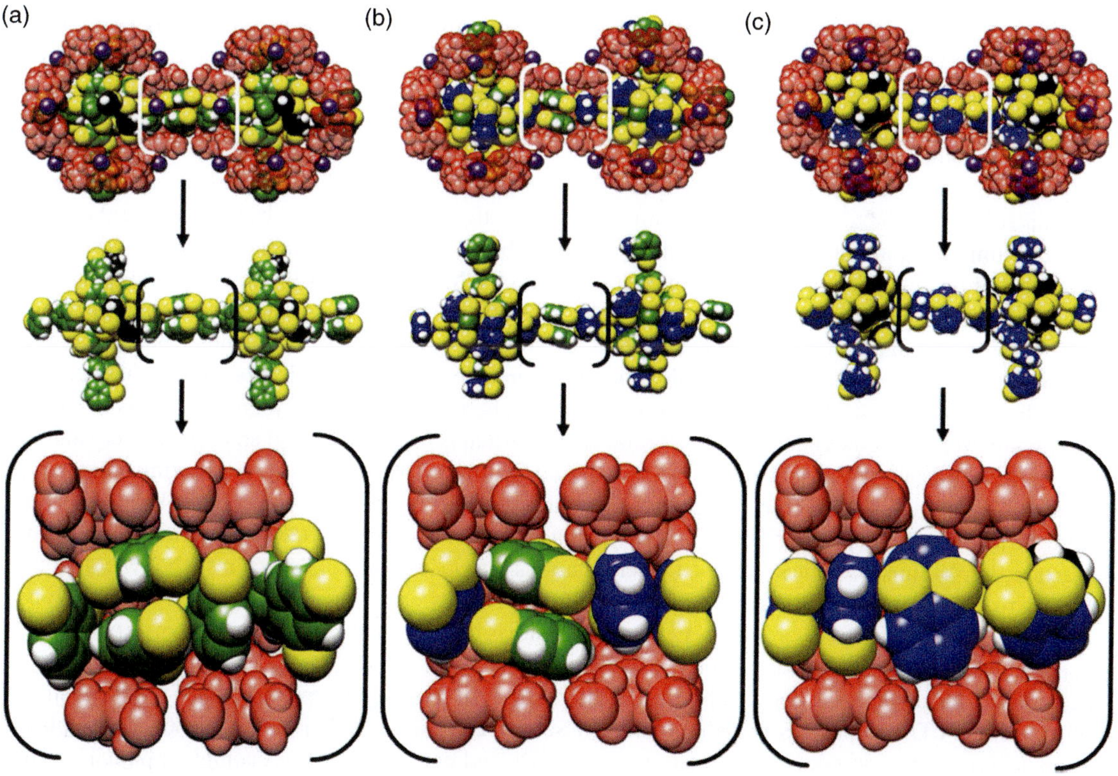

Figure 20.12 *Snapshots of simulations of binary mixtures of* o-xylene *(blue),* m-xylene *(green), and* p-xylene *(black) within the CD-MOF-2 framework viewed along the (100) axis. Equimolar mixtures of (a)* m-/p-, *(b)* o-/m-, *and (c)* o-/p-xylene. *Methyl groups are shown in yellow for the sake of clarity. (Reprinted with permission from ref. [81]. Copyright © 2015 American Chemical Society.)*

We extended the investigation to the vapor phase separation properties of CD-MOF-1 by studying the breakthrough of a mixture of *p*-, *m*-, and *o*-xylenes in N_2 at 343 K. The tendency of the framework to interact more strongly with *o*-xylene in the solution phase carries over to the vapor phase, where we observed an effluent of *para* and *meta* isomers for a short period before the *ortho* isomer broke through. Although *p*- and *m*-xylene breakthrough occurred simultaneously, the flow of gas was enriched in the *para* isomer at first, prior to the *m*-xylene concentration reaching a peak shortly afterward. Bearing in mind the green and economically viable starting materials used to make the CD-MOF-2 stationary phase, it is not a stretch of the imagination to think of it being applied as an *ortho*-selective adsorbent for small benzenoid hydrocarbons on an industrial scale. Moreover, when one considers the chirality that is inherent to CD-MOFs as well as the variety of pore shapes and sizes associated with the framework, the possibility comes to mind that these frameworks might also be exploited for the separation of enantiomers.

20.3.3 Selective interactions with gold salts

The choice of counterion in the preparation of CD-MOFs seemed, at first, to be of little consequence. Potassium salts with benzoate, hydroxide, chloride, nitrate, or other common counterions all lead to crystallization of the isoreticular frameworks without a hitch. Out of a curiosity to push the growth of CD-MOF to its limits, we employed ever more bulky anions and eventually came upon one, $[AuBr_4]^-$, which broke the mold. We could not have predicted, however, the path that this observation would lead us down – eventually allowing us to address a long-standing problem in the isolation of gold metal using a new environmentally benign procedure [83].

Although our curiosity was first piqued when we noticed that a combination of γ-CD and $KAuBr_4$ did not result in the large cubic CD-MOF-type framework, it was not until we attempted to carry out crystallization of $KAuBr_4$ with other members of the CD family, that our attention was well and truly captured. Upon mixing (Figure 20.13a) aqueous solutions of $KAuBr_4$ (20 mM, 1 ml) and α-CD (26.7 mM, 1.5 ml), a glossy suspension forms rapidly on the timescale of a few minutes – a situation which was not observed with either β-CD or γ-CD. Within just 5 minutes, an 80% yield of product, labeled **α•Br**, can be isolated as an off-white crystalline material by centrifugation and drying under vacuum. SEM (Figure 20.13b) and TEM reveal that the precipitate is made up of long needle-like crystals whose diameters are on the order of a few hundred nanometers, while their lengths exceed 10 μm! The cross sections of crystals grown from dilute solution on mica were measured (Figure 20.13c, d) by AFM to be approximately 1.3 nm, which closely matches with the external diameter of an α-CD ring. X-Ray crystallographic analysis of a single crystal allowed us to characterize (Figure 20.13e) the extended supramolecular structure that is behind the rapid and selective crystallization. A pseudorotaxane-like superstructure is formed from $[K(OH_2)_6][AuBr_4]\subset(\alpha\text{-CD})_2$ repeating units, which pack densely into an "infinite cable," extending down the longest dimension of the crystal. There are about 4000 of these 1D supramolecular chains lined up alongside one another in the cross-section of a 100 nm needle of **α•Br**. Within the cables, the α-CDs' orientations alternate (Figure 20.13f) in such a way that two secondary rims face one another at a distance of 7.15 Å and two primary rings face one another at a distance of 9.17 Å. The alignment of the toroids creates a pore, which is occupied entirely by an "ion wire" of hydrated K^+ cations with their $[AuBr_4]^-$ counter-ions.

The rapid co-precipitation witnessed between $KAuBr_4$ and α-CD is highly specific. Aside from the strong preference for α-CD, the process is also remarkably selective for the $[AuBr_4]^-$ anion. Other square planar noble metal complexes do not co-precipitate with the CDs and even another gold complex, $[AuCl_4]^-$, only co-crystallizes with the CDs somewhat reluctantly. Upon examination of the solid-state superstructures of crystals grown by combining α-, β-, or γ-CD with $KAuBr_4$ or $KAuCl_4$, labeled as **α•Br**, **β•Br**, **γ•Br**, **α•Cl**, **β•Cl**, and **γ•Cl**, respectively (Figure 20.14), the reason for this remarkable specificity becomes clear. The solid-state superstructure of **α•Br** is the only one in which: (i) the K^+ ions are surrounded in a octrahedral fashion by six water molecules; and (ii) the resulting ion pair fits perfectly within the CD channel entirely, packing in very densely. All of the other combinations result in comparably large void spaces. It appears that altering the size of the CD ring or the dimensions of the anion by even a small amount is enough to disturb the remarkable hand-in-glove fit of **α•Br** and change quite dramatically the propensity for a 2:1 complex to crystallize.

20.3.3.1 Second-sphere coordination

What forces are responsible for bringing about the $\{[K(OH_2)_6][AuBr_4]\subset(\alpha\text{-CD})_2\}_n$ pseudorotaxane present in **α•Br**? There is a subtle interplay of weak non-covalent bonding interactions that, acting cooperatively, add up to a powerful driving force for crystallization. In addition to (i) electrostatic interactions holding together the ion pair; and (ii) hydrophobic interactions between the interior of the α-CD tori and the relatively lipophilic $[AuBr_4]^-$ ions; there is also (iii) an extensive array of hydrogen bonds helping the assembly. The $[K(OH_2)_6]^+$

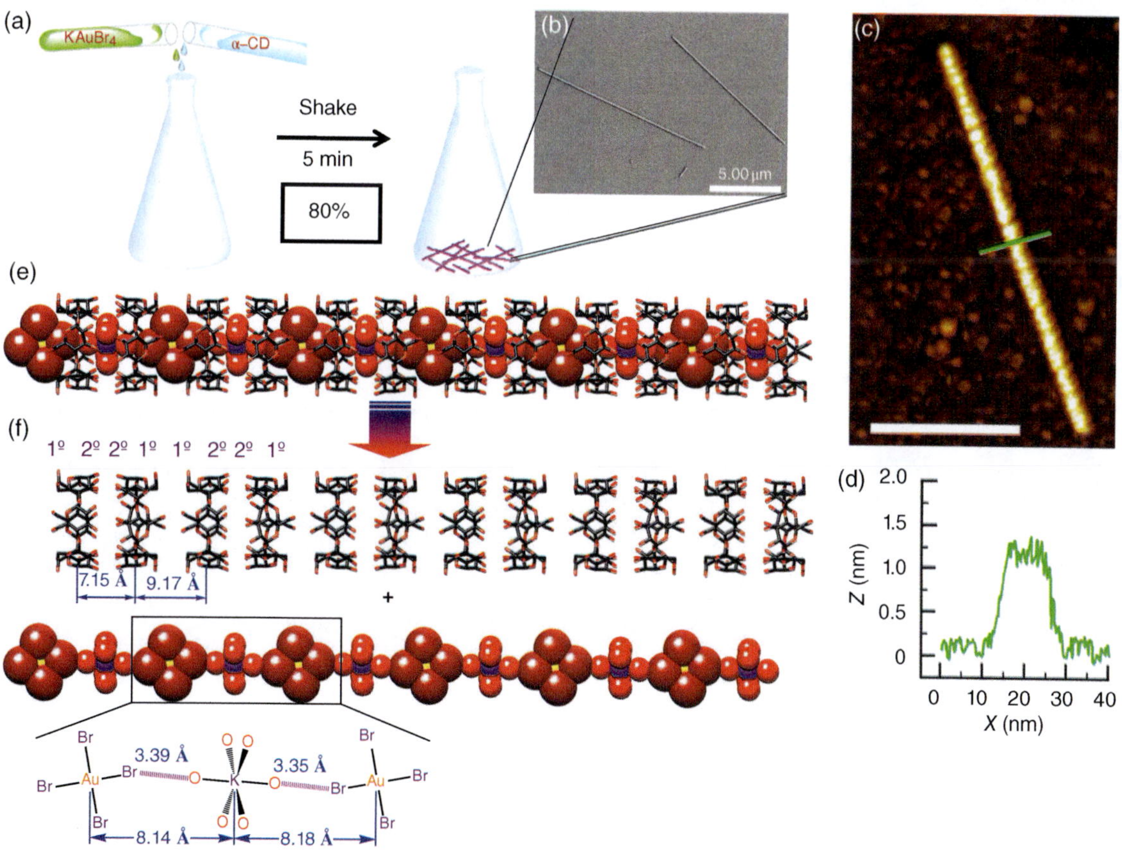

Figure 20.13 *(a) Selective and rapid formation of high aspect ratio crystals* ***α•Br*** *after mixing aqueous solutions of α-CD and* $KAuBr_4$*, which have been (b) observed by SEM to grow beyond 10 μm in length. (c) An AFM image of a crystal of* ***α•Br*** *whose (d) height and width of 1.3 nm correlate with the external diameter of α-CD. AFM Scale bar, 100 nm. (e) The single-crystal X-ray structure of the* $\{[K(OH_2)_6][AuBr_4]\subset(\alpha\text{-CD})_2\}_n$ *crystals reveals that (f) α-CD tori are aligned to form channels, which are filled by* $[K(OH_2)_6][AuBr_4]$ *cables*

and $[AuBr_4]^-$ ions are connected by [O–H•••Br] hydrogen bonding, as evidenced by the longitudinal Br–O distances of less than 3.4 Å, while the methylene protons at the primary face are involved in [C–H•••Br] interactions and, at the other rim, there is hydrogen bonding of the α-CD hydroxyl groups with water coordinated to the K^+ ion. Thus, the presence of the potassium-bound water molecules not only helps to fill the cavities of the α-CDs, it also brings second-sphere coordination to the fore [15, 16, 73, 84]. In some respects, the second-sphere coordination has brought us full-circle back to the types of interactions that are at play in the crown ether–metal complexes which we investigated (Section 20.2.1) in the early 1980s.

20.3.3.2 A greener solution in the isolation of gold

What is the use of this seemingly miraculous hand-in-glove fit? It would be of huge benefit to the gold mining industry and the communities that it affects if an environmentally benign method for gold recovery could be developed [85]. We are currently looking at the possibility of using the precipitation of **α•Br**, which would really take this technology into the realm of something that could be applied on a grand scale as a cheap and environmentally benign process in gold production. Gold is a commodity that the human race seeks with an appetite verging on addiction. It has been this way for millennia. China, Australia, Russia, the US, South Africa, and Peru constitute the countries around the globe where most (51.5% between the countries listed) of the gold is produced. Otherwise, gold is to be found in almost every other country, including our native Scotland! Its omnipresence even spills over into seawater. Unfortunately, cyanide leaching is the standard method used for recovering, accounting for 83% of the gold that is isolated throughout the world today. The leakage of toxic cyanide waste from tailing ponds into the local and then wider environment brings with it a real hazard for livestock and local residents. Improper disposal of tailings and cyanide produces heavy metals and acid mine waste that contaminate streams and groundwater and there have been countless spills that have affected communities as a result of the use of cyanide. A method based on selective precipitation with α-CD

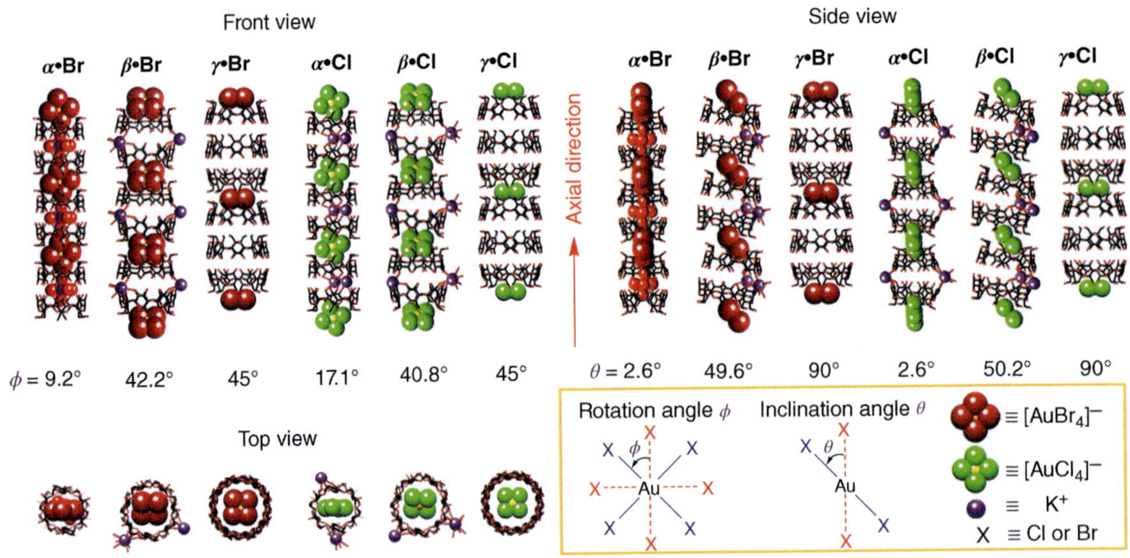

Figure 20.14 *Solid-state superstructures of crystals grown from combinations of α-, β-, or γ-CD with $KAuBr_4$ or $KAuCl_4$. φ is the rotation angle of the $[AuX_4]$ ion viewed from the front, ϑ is the inclination angle of the $[AuX_4]$ ion viewed perpendicular to the central axis of the CD tori. Image reproduced from ref. [73]*

offers an attractive alternative for the isolation of gold. In the first article which we have published on the topic [83], we have gone so far as to exemplify the process of enhancing the purity of a sample of gold alloy (58% gold, 14 K, containing other coinage metal impurities). A solution of $KAuBr_4$ was produced from the raw material by dissolving in an etchant solution of nitro-hydrobromic acid, producing $HAuBr_4$, and then adjusting the pH to around 4 or 5 with KOH. The gold complex was then precipitated in a highly selective manner from the milieu by the addition of α-CD and the result was that 91% of the gold that was introduced was captured in the supramolecular precipitate with 97% purity (almost 24 K).

If this new gold-recovery process is going to be commercially viable, however, we have to avoid the use of toxic and highly corrosive chemicals during the entire process. Nitro-hydrobromic acid, which we have employed so far to generate the $KAuBr_4$, is far from environmentally benign. A less toxic option is to use (Figure 20.15) either ozone or hydrogen peroxide in combination with KBr as the etchant solution. We are currently investigating this chemistry along with several other approaches. By comparison with traditional cyanide leaching, the α-CD method is fast, it is green, it is flexible, and importantly to make it an attractive prospect for adoption by industry, it promises to be profitable. *Fast* – the leaching time can be as little as two hours, compared to two days for cyanide. *Green* – it is cyanide-free and cyclodextrin, which is derived from corn starch, is environmentally benign and recyclable. *Flexible* – the binding interactions are so selective that the process can handle gold sources containing various base metals as well as noble metals, e.g., Pt and Pd. Aside from mining, this selectivity may make it useful for the recovery of gold from electronic waste that is steadily becoming a more and more valuable source of precious metals. *Profitable* – a process costing around a dollar per ounce of gold appears achievable with the α-CD method, which will be competitive with the cost

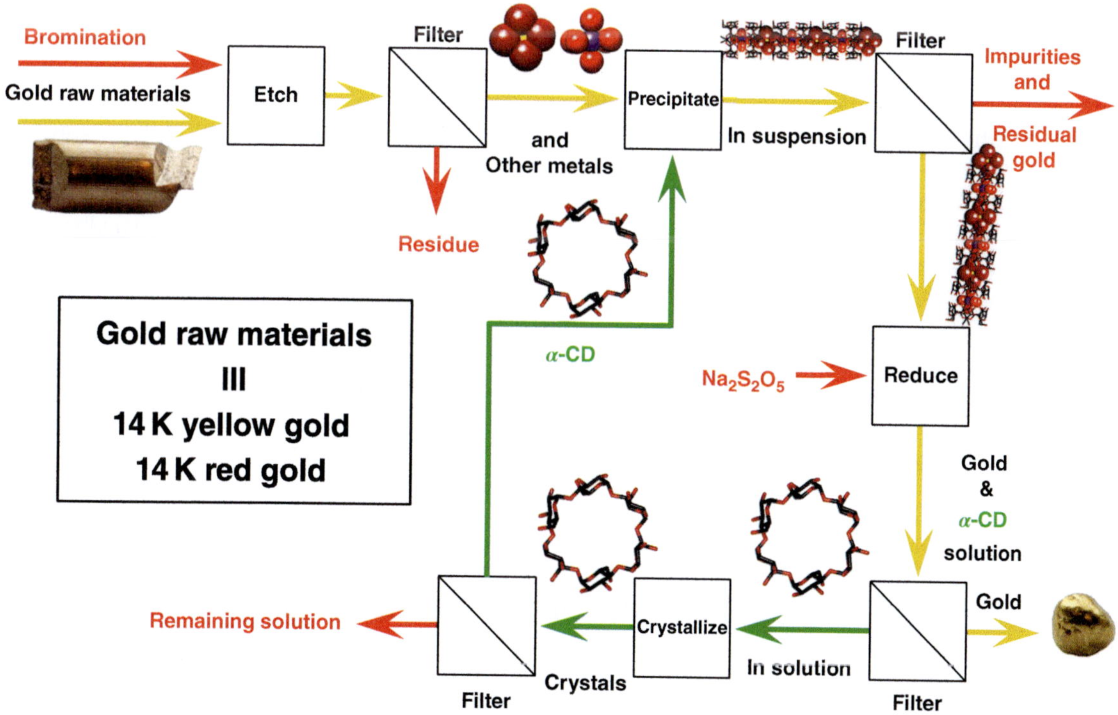

Figure 20.15 *Flow diagram for a cyanide-free gold recovery process employing the precipitation of* ***α•Br****. The effluent streams are marked in red along with the reagent inputs*

of cyanide leaching. A contributing factor to this low cost is the minimization of waste treatment costs. A start-up company called Cycladex has emerged from this particular serendipitous discovery with the aim of turning these potential benefits into a reality.

20.4 Conclusions and outlook

This chapter has thrown into sharp relief the stark contrast that exists between following an "Edisonian" approach to research and taking advantage of serendipitous discoveries. Research in the former category, e.g., in the pursuit of (i) artificial molecular machines and motors [46–49] which operate far-from-equilibrium [86–88]; (ii) molecular electronic devices [50–53] based on the electrochemically induced switching of bistable catenanes and rotaxanes; and (iii) drug delivery systems [60, 61] fabricated from mesoporous silica nanoparticles [54–59] adorned with molecular and supramolecular nanovalves [57, 58], while being intellectually rewarding, has been far more labor-intensive and is on a relatively protracted journey toward potential applications. It is a long haul! Our serendipitous discoveries involving hand-in-glove molecular recognition commensurate with one-step self-assembly processes that produce materials with emergent properties [89–94], on the other hand, have not altogether unreasonable prospects of finding worthwhile applications on a relatively short time-scale, e.g., (i) organic co-crystals incorporating donor-acceptor π–π stacks with complementary hydrogen-bonding recognition sites that display [63, 64] room temperature ferroelectric behavior; (ii) porous metal–organic frameworks [71–78] derived from γ-cyclodextrin and alkali metal cations; and (iii) the highly selective ability of α-cyclodextrin to form insoluble 2:1 complexes with potassium tetrabromoaurate [83, 84]. While this kind of "Bob's your Uncle" [95] chemistry may appear at first sight to be less satisfying intellectually in its pursuit, it is more often than not followed by formidable challenges when it comes to adapting it to work in the real worlds of industry and commerce where considerations, such as economy of scale and the disruption of current technologies, come into the picture with a vengeance.

A noteworthy feature that relates to two of the serendipitous discoveries highlighted in the chapter is that they are associated with macrocycles, in the form of cyclodextrins, that have been around longer than most – in fact, far more than a century [96]. This situation is perhaps hardly surprising, since carbohydrates, as a class or organic compounds, have a long and proud history when it comes to seminal advances involving fundamental phenomena, e.g., (i) the ingenious way in which Emil Fischer deduced the relative configurations of the aldoses and ketoses; (ii) the futuristic assignment of the chair conformation of the six-membered glucopyranose ring by Sir Norman Haworth in the mid-1920s; (iii) the prescient realization by Horace Isbell in the late 1930s that the orientations – now called axial and equatorial – of substituents on six-membered rings have a profound influence on their relative reactivities; and (iv) the observation by Raymond (Ray) Lemieux in the early 1950s of an electronic effect – which he dubbed the anomeric effect – that over-rides, at the anomeric centers of sugars, and at related constitutional positions in many other organic molecules, the normal expectations from conformational analyses based only on a consideration of steric effects. When it comes to looking at the cyclodextrins in a similar light, it is salutary to remember that, in the context of the discipline we now recognize under the banner of supramolecular chemistry [97, 98], these carbohydrate-derived macrocycles with their propensity to form inclusion complexes – and all that follows in the wake of this property – preceded the crown ethers [8, 9] and their numerous wholly synthetic followers by decades. Not only, however, have the cyclodextrins been employed in the period from 1950 up to the present day as vehicles to demonstrate the fundamental tenets of supramolecular chemistry, but it would also appear, from the research described in this chapter, that they have much still to offer in the realm of real-world applications.

The first International Izatt–Christensen Award was presented to Jean-Pierre Sauvage at the Symposium on Macrocyclic Chemistry held in Sheffield/UK in 1991. In characteristic style, Reed Izatt has reached out to

Jean-Pierre and all subsequent winners of this international prize and invited them to reflect on the past, comment about the present, and speculate about the future on the occasion of its silver jubilee. While the future of macrocyclic chemistry – broadly defined it is hoped – lies in the hands of the existing and the emerging generations of chemists, there is no lack of commentary [99] by those chemists with half a century of practice and experience behind them. While recently published essays [100, 101] on a "Platform for Change" and "Reinventing Chemistry" provide food for thought and suggestions for reflection, they cannot anticipate what is going to happen of a serendipitous flavor during the next half-century. There will no doubt be chance discoveries like Pedersen's of the crown ethers that will have a profound [8, 9] influence on the course of chemistry as it continues its youthful traits and contemporary qualities characteristic of the very nature of the science.

References

1. Definition: "Serendipity," http://www.oxforddictionaries.com/definition/english/serendipity (accessed February 24, 2016).
2. Merton, R.K., Barber, E.G. (2004) *The Travels and Adventures of Serendipity: A Study in Sociological Semantics and the Sociology of Science*, Princeton University Press, Princeton, NJ.
3. Eliel, E.L., Hodge, C., (1992) *Science and Serendipity: The Importance of Basic Research*, American Chemical Society, Washington, DC.
4. Travis, A.S. (1990) Perkin's mauve: Ancestor of the organic chemical industry, *Technology and Culture*, **31** (1), 51–82.
5. McNally, A., Prier, C.K., MacMillan, D.W.C. (2011) Discovery of an α-amino C–H arylation reaction using the strategy of accelerated serendipity, *Science*, **334** (6059), 1114–1147.
6. Bud, R. (2007) *Penicillin: Triumph and Tragedy*, Oxford University Press, Oxford.
7. Ghofrani, H.A., Osterloh, I.H., Grimminger, F. (2006) Sildenafil: From angina to erectile dysfunction to pulmonary hypertension and beyond, *Nature Reviews Drug Discovery*, **5** (8), 689–702.
8. Pedersen, C.J. (1967) Cyclic polyethers and their complexes with metal salts, *Journal of the American Chemical Society*, **89** (10), 2495–2496.
9. Pedersen, C.J. (1967) Cyclic polyethers and their complexes with metal salts, *Journal of the American Chemical Society*, **89** (26), 7017–7036.
10. Foster, R. (1969) *Organic charge-transfer complexes*. Academic Press, New York.
11. Foster, R. (1980) Electron donor-acceptor complexes, *Journal of Physical Chemistry*, **84** (17), 2135–2141.
12. Benesi, H.A., Hildebrand, J.H. (1949) A spectrophotometric investigation of the interaction of iodine with aromatic hydrocarbons, *Journal of the American Chemical Society*, **71** (8), 2703–2707.
13. Zhu, W.G., Yi, Y.P., Zhen, Y.G., Hu, W.P. (2015) Precisely tailoring the stoichiometric stacking of perylene-TCNQ co-crystals towards different nano and microstructures with varied optoelectronic performances, *Small*, **11** (18), 2150–2156.
14. Colquhoun, H.M., Stoddart, J.F., Williams, D.J., Wolstenholme, J.B., Zarzycki, R. (1981) Second sphere coordination of cationic platinum complexes by crown ethers – The X-ray crystal-structure of $[Pt(bpy)(NH_3)_2\cdot$dibenzo[30]crown-10$]^{2+}[PF_6]^{2-}\cdot xH_2O$ ($x \approx 0.6$), *Angewandte Chemie International Edition in English*, **20** (12), 1051–1053.
15. Colquhoun, H.M., Stoddart, J.F., Williams, D.J. (1986) Second-sphere coordination – A novel role for molecular receptors, *Angewandte Chemie International Edition in English*, **25** (6), 487–507.
16. Raymo, F.M., Stoddart, J.F. (1996) Second-sphere coordination, *Chemische Berichte*, **129** (9), 981–990.
17. Colquhoun, H.M., Goodings, E.P., Maud, J.M., et al. (1983) Complex-formation between dibenzo-3*n*-crown-*n* ethers and the diquat dication, *Journal of the Chemical Society, Chemical Communications*, (**20**), 1140–1142.
18. Allwood, B.L., Spencer, N., Shahriari-Zavareh, H., Stoddart, J.F., Williams, D.J. (1987) Complexation of paraquat by a bisparaphenylene-34-crown-10 derivative, *Journal of the Chemical Society, Chemical Communications*, (**14**), 1064–1066.
19. Odell, B., Reddington, M.V., Slawin, A.M.Z., et al. (1988) Cyclobis(paraquat-*p*-phenylene) – A tetracationic multipurpose receptor, *Angewandte Chemie International Edition in English*, **27** (11), 1547–1550.

20. Ashton P.R., Odell B., Reddington M.V., et al. (1988) Isostructural, alternately-charged receptor stacks - The inclusion complexes of hydroquinone and catechol dimethyl ethers with cyclobis(paraquat-*p*-phenylene), *Angewandte Chemie International Edition in English*, **27** (11), 1550–1553.
21. Ashton, P.R., Philp, D., Spencer, N., Stoddart, J.F. (1991) The self-assembly of [*n*]pseudorotaxanes, *Journal of the Chemical Society, Chemical Communications*, (**23**), 1677–1679.
22. Stoddart, J.F., Colquhoun, H.M. (2008) Big and little meccano, *Tetrahedron*, **64** (36), 8231–8263.
23. Sauvage, J.-P., Dietrich-Buchecker, C. (eds) (2008) *Molecular Catenanes, Rotaxanes and Knots: A Journey Through the World of Molecular Topology*, Wiley-VCH, Weinheim.
24. Stoddart, J.F. (2009) The chemistry of the mechanical bond, *Chemical Society Reviews*, **38** (6), 1802–1820.
25. Beves, J.E., Blight, B.A., Campbell, C.J., Leigh, D.A., McBurney, R.T. (2011) Strategies and tactics for the metal-directed synthesis of rotaxanes, knots, catenanes, and higher order links, *Angewandte Chemie International Edition*, **50** (40), 9260–9327.
26. Neal, E.A., Goldup, S.M. (2014) Chemical consequences of mechanical bonding in catenanes and rotaxanes: Isomerism, modification, catalysis and molecular machines for synthesis, *Chemical Communications*, **50** (40), 5128–5142.
27. Xue, M., Yang, Y., Chi, X., Yan, X., Huang, F. (2015) Development of pseudorotaxanes and rotaxanes: From synthesis to stimuli-responsive motions to applications, *Chemical Reviews*, **115** (15), 7398–7501.
28. Frisch, H., Martin, I., Mark, H. (1953) Zur struktur der polysiloxene. 1, *Monatshefte für Chemie*, **84** (2), 250–256.
29. Dietrich-Buchecker, C.O., Sauvage, J.-P., Kintzinger, J.-P. (1983) A new family of molecules – Metallo-catenanes, *Tetrahedron Letters*, **24** (46), 5095–5098.
30. Stoddart, J.F. (2009) The master of chemical topology Jean-Pierre Sauvage, *Chemical Society Reviews*, **38** (6), 1521–1529.
31. Ashton, P.R., Goodnow, T.T., Kaifer, A.E., et al. (1989) A [2]catenane made to order, *Angewandte Chemie International Edition in English*, **28** (10), 1396–1399.
32. Collier, C.P., Mattersteig, G., Wong, E.W., et al. (2000) A [2]catenane-based solid state electronically reconfigurable switch, *Science*, **289** (5482), 1172–1175.
33. Anelli, P.-L., Spencer, N., Stoddart, J.F. (1991) A molecular shuttle, *Journal of the American Chemical Society*, **113** (13), 5131–5133.
34. Forgan, R.S., Sauvage, J.-P., Stoddart, J.F. (2011) Chemical topology: Complex molecular knots, links, and entanglements, *Chemical Reviews*, **111** (9), 5434–5464.
35. Amabilino, D.B., Ashton, P.R., Reder, A.S., Spencer, N., Stoddart, J.F. (1994) Olympiadane, *Angewandte Chemie International Edition in English*, **33** (12), 1286–1290.
36. Amabilino, D.B., Ashton, P.R., Balzani, V., et al. (1998) Oligocatenanes made to order, *Journal of the American Chemical Society*, **120** (18), 4295–4307.
37. Amabilino, D.B., Ashton, P.R., Boyd, S.E., et al. (1997) The five-stage self-assembly of a branched heptacatenane, *Angewandte Chemie International Edition in English*, **36** (19), 2070–2072.
38. van Gülick, N. (1993) Theoretical aspects of the linked ring problem, *New Journal of Chemistry*, **17** (Special Issue) 619–625.
39. Bissell, R.A., Córdova, E., Kaifer, A.E., Stoddart, J.F. (1994) A chemically and electrochemically switchable molecular shuttle, *Nature*, **369** (6476), 133–137.
40. Fahrenbach, A.C., Bruns, C.J., Li, H., et al. (2014) Ground-state kinetics of bistable redox-active donor–acceptor mechanically interlocked molecules, *Accounts of Chemical Research*, **47** (2), 482–493.
41. Klajn, R., Stoddart, J.F., Grzybowski, B.A. (2010) Nanoparticles functionalised with reversible molecular and supramolecular switches, *Chemical Society Reviews*, **39** (6), 2203–2237.
42. Asakawa, M., Ashton, P.R., Balzani, V., et al. (1998) A chemically and electrochemically switchable [2]catenane incorporating a tetrathiafulvalene unit, *Angewandte Chemie International Edition*, **37** (3), 333–337.
43. Choi, J.W., Flood, A.H., Steuerman, D.W., et al. (2006) Ground-state equilibrium thermodynamics and switching kinetics of bistable [2]rotaxanes switched in solution, polymer gels, and molecular electronic devices, *Chemistry – A European Journal*, **12** (1), 261–279.
44. Fahrenbach, A.C., Zhu, Z.X., Cao, D., et al. (2012) Radically enhanced molecular switches, *Journal of the American Chemical Society*, **134** (39), 16275–16288.
45. Spruell, J.M., Coskun, A., Friedman, D.C., et al. (2010) Highly stable tetrathiafulvalene radical dimers in [3]catenanes, *Nature Chemistry*, **2** (10), 870–879.

46. McGonigal, P.R., Li, H., Cheng, C., et al. (2015) Controlling association kinetics in the formation of donor–acceptor pseudorotaxanes, *Tetrahedron Letters*, **56** (23), 3591–3594.
47. Cheng, C., McGonigal, P.R., Liu, W.G., et al. (2014) Energetically demanding transport in a supramolecular assembly, *Journal of the American Chemical Society*, **136** (42), 14702–14705.
48. Cheng, C., McGonigal, P.R., Schneebeli, S.T., et al. (2015) An artificial molecular pump, *Nature Nanotechnology*, **10** (6), 547–553.
49. Li, H., Cheng, C., McGonigal, P.R., et al. (2013) Relative unidirectional translation in an artificial molecular assembly fueled by light, *Journal of the American Chemical Society*, **135** (49), 18609–18620.
50. Heath, J.R. (2009) Molecular electronics, *Annual Reviews of Materials Research*, **39**, 1–23.
51. Coskun, A., Spruell, J.M., Barin, G., et al. (2012) High hopes: Can molecular electronics realise its potential? *Chemical Society Reviews*, **41** (14), 4827–4859.
52. Stoddart, J.F. (2005) From a meccano set to nano meccano, *Pure and Applied Chemistry*, **77** (7), 1089–1106.
53. Green, J.E., Choi, J.W., Boukai, A. et al. (2007) A 160-kilobit molecular electronic memory patterned at 10^{11} bits per square centimetre, *Nature*, **445** (7126), 414–417.
54. Li, Z., Barnes, J.C., Bosoy, A., Stoddart, J.F., Zink, J.I. (2012) Mesoporous silica nanoparticles in biomedical applications, *Chemical Society Reviews*, **41** (7), 2590–2605.
55. Chia, S.Y., Cao, J.G., Stoddart, J.F., Zink, J.I. (2001) Working supramolecular machines trapped in glass and mounted on a film surface, *Angewandte Chemie International Edition*, **40** (13), 2447–2448.
56. Ferris, D.P., McGonigal, P.R., Witus, L.S., et al. (2015) Oxime ligation on the surface of mesoporous silica nanoparticles, *Organic Letters*, **17** (9), 2146–2149.
57. Nguyen, T.D., Tseng, H.R., Celestre, P.C., et al. (2005) A reversible molecular valve, *Proceedings of the National Academy of Sciences of the United States of America*, **102** (29), 10029–10034.
58. Zhao, Y.L., Li, Z.X., Kabehie, S., et al. (2010) pH-Operated nanopistons on the surfaces of mesoporous silica nanoparticles, *Journal of the American Chemical Society*, **132** (37), 13016–13025.
59. Patel K., Angelos S., Dichtel W.R., Coskun A., Yang Y.W., Zink J.I., et al. (2008) Enzyme-responsive snap-top covered silica nanocontainers, *Journal of the American Chemical Society*, **130** (8), 2382–2383.
60. Fernando, I.R., Ferris, D.P., Frasconi, M., et al. (2015) Esterase- and pH-responsive poly(β-amino ester)-capped mesoporous silica nanoparticles for drug delivery, *Nanoscale*, **7** (16), 7178–7183.
61. Ang, C.Y., Tan, S.Y., Zhao, Y.L. (2014) Recent advances in biocompatible nanocarriers for delivery of chemotherapeutic cargoes towards cancer therapy, *Organic and Biomolecular Chemistry*, **12** (27), 4776–4806.
62. Yang, Y.W., Sun, Y.L., Song, N. (2014) Switchable host-guest systems on surfaces, *Accounts of Chemical Research*, **47** (7), 1950–1960.
63. Blackburn, A.K., Sue, A.C.-H., Shveyd, A.K., et al. (2014) Lock-arm supramolecular ordering: A molecular construction set for cocrystallizing organic charge transfer complexes, *Journal of the American Chemical Society*, **136** (49), 17224–17235.
64. Tayi, A.S., Shveyd, A.K., Sue, A.C.-H., et al. (2012) Room-temperature ferroelectricity in supramolecular networks of charge-transfer complexes, *Nature*, **488** (7412), 485–489.
65. Breslow, R. (1995) Biomimetic chemistry and artificial enzymes – Catalysis by design, *Accounts of Chemical Research*, **28** (3), 146–153.
66. Hou, X.S., Ke, C.F., Bruns, C.J., et al. (2015) Tunable solid-state fluorescent materials for supramolecular encryption, *Nature Communications*, **6**, Article 6884.
67. Chacko, K.K., Saenger, W. (1981) Topography of cyclodextrin inclusion complexes. 15. Crystal and molecular-structure of the cyclohexaamylose–7.57 water complex, form-III. Four and six-membered circular hydrogen-bonds, *Journal of the American Chemical Society*, **103** (7), 1708–1715.
68. Aree, T., Schulz, B., Reck, G. (2003) Crystal structures of β-cyclodextrin complexes with formic acid and acetic acid, *Journal of Inclusion Phenomena and Macrocyclic Chemistry*, **47** (1-2), 39–45.
69. Harata, K. (1987) The structure of the cyclodextrin complex. X. Crystal-structure of uncomplexed hydrated γ-cyclodextrin, *Bulletin of the Chemical Society of Japan*, **60** (8), 2763–2767.
70. Chichak, K.S., Cantrill, S.J., Pease, A.R., et al. (2004) Molecular Borromean rings, *Science*, **304** (5675), 1308–1312.
71. Smaldone, R.A., Forgan, R.S., Furukawa, H., et al. (2010) Metal-organic frameworks from edible natural products, *Angewandte Chemie International Edition*, **49** (46), 8630–8634.

72. Gassensmith, J.J., Smaldone, R.A., Forgan, R.S., et al. (2012) Polyporous metal-coordination frameworks, *Organic Letters*, **14** (6), 1460–1463.
73. Liu, Z.C., Stoddart, J.F. (2014) Extended metal-carbohydrate frameworks, *Pure and Applied Chemistry*, **86** (9), 1323–1334.
74. Smith, M.K., Angle, S.R., Northrop, B.H. (2015) Preparation and analysis of cyclodextrin-based metal-organic frameworks: Laboratory experiments adaptable for high school through advanced undergraduate students, *Journal of Chemical Education*, **92** (2), 368–372.
75. Forgan, R.S., Smaldone, R.A., Gassensmith, J.J., et al. (2012) Nanoporous carbohydrate metal-organic frameworks, *Journal of the American Chemical Society*, **134** (1), 406–417.
76. Gassensmith, J.J., Furukawa, H., Smaldone, R.A., et al. (2011) Strong and reversible binding of carbon dioxide in a green metal-organic framework, *Journal of the American Chemical Society*, **133** (39), 15312–15315.
77. Wu, D., Gassensmith, J.J., Gouvea, D., et al. (2013) Direct calorimetric measurement of enthalpy of adsorption of carbon dioxide on CD-MOF-2, a green metal-organic framework, *Journal of the American Chemical Society*, **135** (18), 6790–6793.
78. Gassensmith, J.J., Kim, J.Y., Holcroft, J.M., et al. (2014) A metal–organic framework-based material for electrochemical sensing of carbon dioxide, *Journal of the American Chemical Society*, **136** (23), 8277–8282.
79. Herm, Z.R., Wiers, B.M., Mason, J.A., et al. (2013) Separation of hexane isomers in a metal-organic framework with triangular channels, *Science*, **340** (6135), 960–964.
80. Lee, C.Y., Bae, Y.S., Jeong, N.C., et al. (2011) Kinetic separation of propene and propane in metal-organic frameworks: Controlling diffusion rates in plate-shaped crystals via tuning of pore apertures and crystallite aspect ratios, *Journal of the American Chemical Society*, **133** (14), 5228–5231.
81. Hartlieb, K.J., Holcroft, J.M., Stoddart, J.F. (2015) Carbohydrate–mediated purification of petrochemicals, *Journal of the American Chemical Society*, **137** (17), 5706–5719.
82. Othmer, K. (2008) *Separation Technology*, two volume set, second edition. Wiley-VCH, Weinheim.
83. Liu, Z.C., Frasconi, M., Lei, J.Y., et al. (2013) Selective isolation of gold facilitated by second-sphere coordination with α-cyclodextrin, *Nature Communications*, **4**, Article 1855.
84. Liu, Z.C., Schneebeli, S.T., Stoddart, J.F. (2014) Second-sphere coordination revisited, *Chimia*, **68** (5), 315–320.
85. Syed, S. (2012) Recovery of gold from secondary sources – A review, *Hydrometallurgy*, **115–116**, 30–51.
86. Astumian, R.D. (2012) Microscopic reversibility as the organizing principle of molecular machines, *Nature Nanotechnology*, **7** (11), 684–688.
87. Cheng, C., McGonigal, P.R., Stoddart, J.F., Astumian, R.D. (2015) Design and synthesis of non-equilibrium systems, *ACS Nano*, **9** (9), 8672–8688.
88. Kay, E.R., Leigh, D.A. (2015) Rise of the molecular machines, *Angewandte Chemie International Edition*, **54** (35), 10080–10088.
89. Gibb, B.C. (2009) Reaching out to complexity, *Nature Chemistry*, **1** (4), 252–253.
90. Balazs, A.C., Epstein, I.R. (2009) Emergent or just complex? *Science*, **325**, 1632–1634.
91. Gibb, B.C. (2011) The emergence of emergence, *Nature Chemistry*, **3** (1), 3–4.
92. Vicens, J., Vicens, Q. (2011) Emergences of supramolecular chemistry: From supramolecular chemistry to supramolecular science, *Journal of Inclusion Phenomena and Macrocyclic Chemistry*, **71** (3-4), 251–274.
93. Epstein, I.R. (2014) Coupled chemical oscillators and emergent system properties, *Chemical Communications*, **50** (74), 10758–10767.
94. Taylor, E. (2015) An explication of emergence, *Philosophical Studies*, **17**, (3), 653–669.
95. Wikipedia: "Bob's your Uncle," https://en.wikipedia.org/wiki/Bob%27s_your_uncle (accessed February 27, 2016)
96. Stoddart, J.F. (1989) A century of cyclodextrins, *Carbohydrate Research*, **192**, xii–xv.
97. Stoddart, J.F. (2009) Thither supramolecular chemistry? *Nature Chemistry*, **1** (1), 14–15.
98. Stoddart, J.F (2012) From supramolecular to systems chemistry: Complexity emerging out of simplicity, *Angewandte Chemie International Edition*, **51** (52), 12902–12903.
99. Stoddart, J.F. (2014) Putting mechanically interlocked molecules (MIMs) to work in tomorrow's world, *Angewandte Chemie International Edition*, **53** (42), 11102–11104.
100. Stoddart, J.F. (2015) A platform for change, *Supramolecular Chemistry*, **27**, 567–570.
101. Whitesides, G.M. (2015) Reinventing chemistry, *Angewandte Chemie International Edition*, **54** (11), 3196–3209.

21

Evolution of Zn^{II}–Macrocyclic Polyamines to Biological Probes and Supramolecular Assembly

Eiichi Kimura,[1] Tohru Koike,[2] and Shin Aoki[3]

[1] Professor Emeritus, Hiroshima University, Hiroshima, Japan
[2] Institute of Biomedical and Health Sciences, Hiroshima University, Hiroshima, Japan
[3] Faculty of Pharmaceutical Sciences, Tokyo University of Science, Noda, Japan

21.1 Introduction

From 1974 to the 1990s we synthesized dozens of new macrocyclic polyamines to find new chemical and biological functions for the new concept of host–guest chemistry. Scheme 21.1 shows the name and structure of a few relevant polyamines [1, 2]. Since the 1990s we have turned to new zinc enzyme models for the specific molecular recognition in aqueous media using Zn^{II}–macrocyclic polyamine complexes. This article reviews how the Zn–enzyme model studies have evolved to the discovery of new biological probes and supramolecular chemistry.

21.2 Zinc enzyme models from Zn^{II} macrocyclic polyamine complexes

Zinc enzymes are involved predominantly in hydrolysis of carboxylic esters, amides, peptides, and phosphates, or the hydration of CO_2. What are the chemical specialties of Zn^{II} ion that fits it these biological roles? Among the zinc enzymes, carbonic anhydrase (CA) appeared to be an accessible target for chemical model

Macrocyclic and Supramolecular Chemistry: How Izatt–Christensen Award Winners Shaped the Field, First Edition.
Edited by Reed M. Izatt.

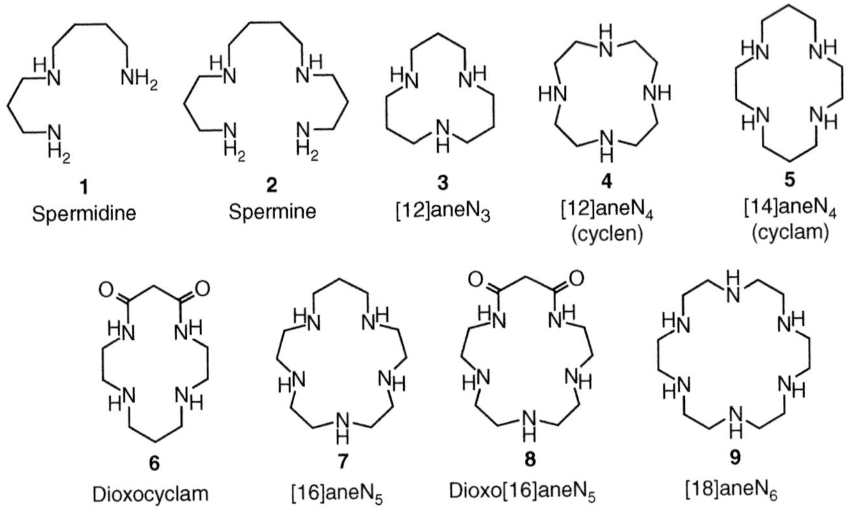

Scheme 21.1

study, because of the simple structure of the active center; Zn^{II} is tetrahedrally coordinated to three N(imidazole)s and an H_2O [2].

In 1975, Woolly reported a Zn^{II}–macrocyclic tetraamine complex, **10**, as a model of CA [3]. This work revealed the role of Zn^{II} as the generation of a nucleophilic reactant Zn^{II}–OH^- with a pK_a value of 8.7 for the catalytic hydration of acetaldehyde, a reaction which CA also catalyzes. This study left some CA properties unanswered, such as the mechanism of CA inhibition by anions and aromatic sulfonamides.

In 1990, we showed a Zn^{II} complex, **11**, of [12]aneN_3 (**3**) to possess a CA-mimicking tetrahedral structure, with three Ns and a H_2O, that has a pK_a value of 7.3 in aqueous solution at I=0.1 ($NaClO_4$) and 25 °C, close to the figure of ~7 for CA (Scheme 21.2). Like CA, **11** catalyzed methylacetate hydrolysis and acetaldehyde hydration, where the rate-determining steps are the attack of the Zn^{II}–OH^- species (**11b**) to carbonyl groups [4]. The maximum hydration rate constant (at 0 °C) of 200 M^{-1} s^{-1} was similar to 196 M^{-1} s^{-1} for **10**, which is one-seventh the value (1400 M^{-1} s^{-1}) for CA.

Later, the kinetics and mechanism of the hydration of CO_2 and the reverse dehydration of HCO_3^-, catalyzed by **11** were studied [5]. The pH-dependence of these reaction rates demonstrated that Zn^{II}–OH^- (a hydroxide-bound Zn^{II}) **11b** catalyzes the hydration of CO_2, whereas it is Zn^{II}–OH_2 (a water-bound Zn^{II})**11a** that catalyzes the dehydration, in analogy to the characteristic pH-dependence reported for CA. The second-order rate constants are 581 and 4.8 M^{-1} s^{-1} at 25 °C for hydration and dehydration, respectively. The four orders of magnitude greater rate for hydration by CA is rational if one assumes the pre-association of CO_2.with a binding constant of 10^4 M^{-1}. within the hydrophobic pocket in CA.

Our second zinc enzyme model was Zn^{II}–[12]aneN_4 (or cyclen, **4**) complex, **12**, with a pK_a of 8.0 in aqueous solution at I=0.1 ($NaClO_4$) and 25 °C for generation of Zn^{II}–OH^- (**12b**) [6]. The 5-coordinate, **12**, is more stable and easier to work with than **11** (the 1:1 complex stability constant is $10^{15.3}$ M^{-1} for **12** vs. $10^{8.4}$ M^{-1} for **11**), allowing us to study more in detail the unknown intrinsic properties of Zn^{II}.

Together with subsequent model studies of phosphatase [6], alcohol dehydrogenase [7], β-lactamase [8], and aldolase[9] using **11** and/or **12**, it may be concluded that the *common and essential role of* Zn^{II} *is to generate* Zn^{II}*–OH– species at physiological pH*, where H_2Os are activated to stronger nucleophiles to attack electrophilic substrates or, in general, bases, to deprotonate and activate neutral substrates such as alcohol and aldose. Among biological metal ions, the amphoteric Zn^{II} ion would best fit those functionalities at physiological pH.

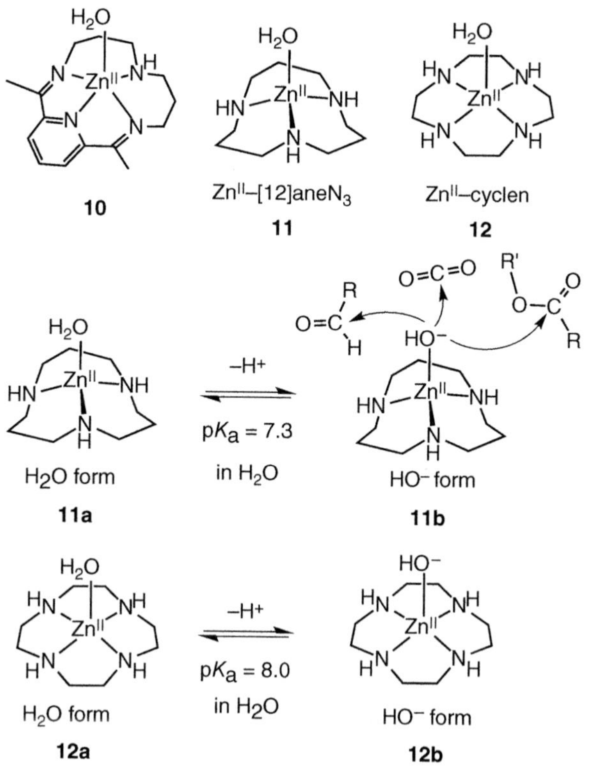

Scheme 21.2

Recently, Lightstone, Aines et al. studied Zn^{II} complexes of macrocyclic polyamines, [9]aneN$_3$, **3**, **4**, [13]aneN$_4$, and **5** theoretically and experimentally, looking for ligand effects on reversible CO_2 hydration kinetics [10]. It is of interest to note that Cd^{II}–cyclen complex, having a higher pK_a value of 10.8 for $Cd^{II}–OH_2$, showed a higher nucleophilicity to 4-nitrophenyl ester than a Zn^{II}–cyclen complex at pH 8.4 [11]. That is, in comparison to Zn^{II}, the less acidic Cd^{II} deprotonates the $Cd^{II}–OH_2$ at higher pH, but, once formed, the $Cd^{II}–OH^-$ is the stronger nucleophile [12].

21.2.1 The CA-mimicking anion affinity of model 11

Carbonic anhydrase and other zinc enzymes such as alkaline phosphatases are susceptible to inhibition by anions (e.g. Cl^-, I^-, SCN^-, CH_3COO^-) or organic molecules (e.g. aromatic sulfonamides). Moreover, the substrates and products are often anions (e.g. HCO_3^-, phosphate). Questions arose as to how CA manages to reversibly catalyze CO_2 hydration and HCO_3^- dehydration in the presence of those inhibitors.

We already knew the similarly strong OH^- affinities of CA (log K=6.5) and our model **11** (log K=6.4 at 25 °C) calculated from the similar pK_a values for $Zn^{II}–OH_2 \leftrightarrow Zn^{II}–OH^-$. The 1:1 anion affinity constants log $K(A^-)$ for **11** were determined at 25 °C in aqueous solution; HCO_3^- 4.0 > CH_3COO^- 2.6 > SCN^- 2.4 > I^- 1.6 > Br^- 1.5 > Cl^- 1.3 > F^- 0.8, to be compared with those reported for CA; SCN^- 3.2 > HCO_3^- 1.6 > I^- 1.2 > CH_3COO^- 1.1 > Br^- 1.1 > Cl^- 0.7 > F^- 0.1 [13]. It is instructive to find that among the biological anions, HCO_3^- has the highest 1:1 affinity (second to OH^-) with Zn^{II}. This chemical property of Zn^{II} is physiologically rational: HCO_3^- binding to Zn^{II} in blood (for dehydration) is favorable at acidic pH, unhindered by more abundant

Cl^- at $[HCO_3^-]=27$ mM and $[Cl^-]=105$ mM. However, Zn^{II}-bound HCO_3^-, a product of CO_2 hydration, is readily replaced by OH^- generated on Zn^{II} at alkaline pH. Thus, catalytic HCO_3^- dehydration or CO_2 hydration can proceed in either way depending on delicate pH changes.

21.2.2 The CA-mimicking sulfonamide inhibition of model 11

The strong CA inhibition by aromatic sulfonamides has been biochemically and medicinally (acetazolamide, **13**, is prescribed as a diuretic drug) well studied; direct binding of the sulfonamides to the active center Zn^{II} was suggested. Chemically, however, there was no CA model to support such a Zn^{II}–sulfonamide binding in neutral aqueous solution.

The binding of a series of aromatic sulfonamides – acetazolamide, 4-nitrobenzenesulfonamide, and *p*-toluenesulfonamide – to CA model **11** was investigated in 10% CH_3CN aqueous solution at $I=0.1$ ($NaClO_4$) and 25 °C [13]. The apparent 1:1 binding constants log K_i (3.6, 2.6, and 2.4, respectively, at pH 8.4) were proportional to the ease of the sulfonamide deprotonation as measured by pK_a values (7.5, 9.4, and 10.5), implying that the sulfonamide anions bind to the fourth coordination site of **11**. These log K_i values for the model were parallel to those for CA (7.9, 7.2, and 6.3), suggesting a similar inhibition mechanism for **11** and CA. Assuming the sulfonamide affinities log K_i ~3 for apo-CA (Zn^{II}-less CA), the intrinsic affinities to Zn^{II} in CA would be close to our model values. It is also of interest to find in this model study that the sulfonamide bindings (log $K_i=2.4\sim3.6$) are competitive to that of the HCO_3^- binding (log $K=2.8$) at pH 8.4.

We synthesized a tosylamide-appended **14**, which demonstrated that in the presence of Zn^{II}, the tosylamide (pK_a ~11.2) deprotonates at pH<7, yielding a stable tosylamide N^-–coordinating, tetrahedral complex, **15** (Scheme 21.3). Since the sulfonamide N^- coordination inhibits formation of the Zn^{II}–OH^- species, **15** showed no catalytic activity.

21.2.3 Development of new Zn^{II} fluorophores and apoptosis sensors

The Zn^{II}–sulfonamide binding complex, **15**, inspired us to synthesize a new fluorescent Zn^{II} sensor, **16** (Scheme 21.3) [14, 15]. Dansylamide is known to strongly fluoresce with CA [14]. The dansylamide-attached cyclen, **16**, forms a far more stable 1:1 Zn^{II} complex, **17** ($K_d=5.5\times10^{-13}$ M in aqueous solution at pH 7.8 and 25 °C) [14a], than any previous Zn^{II} fluorophores such as Zinquin, **18**, which forms mixed 1:1 ($K_d=3.7\times10^{-7}$ M) and 1:2 complexes ($K_d=8.5\times10^{-8}$ M) at pH 7.4 and 25 °C [14a,16]. The pendant dansylamide-deprotonation by Zn^{II} in **17** remarkably increases fluorescence 4.9-fold at 540 nm and 10-fold at 490 nm on excitation at 330 nm at pH 7.8. Fluorescence with 5 μM **16** linearly responds to $[Zn^{II}]=0.01\sim0$ 5 μM, which is unaffected by the presence of mM concentrations of other biological metal ions such as Na^+, K^+, Ca^{2+}, Mg^{2+}, Fe^{II}, or Fe^{III}. In our continuing study, we have new cyclen fluorophores, **19** [17], **20** [18], and **21** [19–21].

The macrocyclic sensor **16** is cell membrane-permeable, emitting weak fluorescence (at 550 nm) by itself in cells [22]. When Zn^{II} is incorporated into the cells using a Zn^{II} carrier, pyrithione, green fluorescence (at 530 nm) strengthens due to the formation of **17**. Compared with the Zn^{II}–Zinquin complexes, the more stable **17** stays intact in cells to keep emitting fluorescence for longer hours. Currently, fluorescent Zn^{II} sensing is attracting wide interest in chemical biology studies [23].

We also found **16**, like Zinquin [24], serving as a useful fluorescent sensor of apoptosis [22]. In apoptosis, free Zn^{II} ions are released from Zn^{II} storage vesicles. When apoptosis of cancer cells HeLa or HL60 is artificially induced (by an anticancer drug, etoposide or H_2O_2), **16** selectively stains cells only at early apoptotic stages with the green fluorescence. Reagent **16** is commercially available and its use is recorded in the literature [25]. The recently synthesized Zn^{II} sensor **21** was also found to be useful as an apoptosis sensor [19–21].

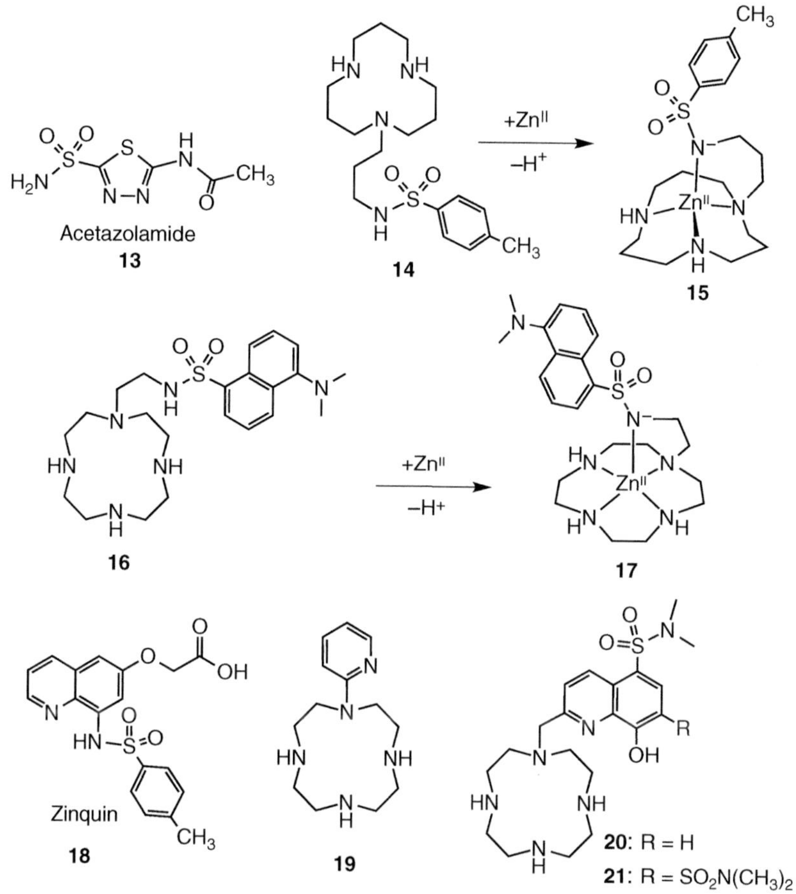

Scheme 21.3

21.2.4 From the Zn^{II}-aldolase model to aldol synthesis catalyst

Aldolase type-II, which catalyzes reversible aldol reactions, is a Zn^{II} enzyme, where Zn^{II} catalyzes enolization of the substrate carbonyl group for subsequent aldol C–C bond formation. However, there had been little elucidation of enolization by Zn^{II} in aqueous media; e.g. how was the enol deprotonation facilitated?

A carbonyl-bound Zn^{II} complex **23** with 1-(4-bromophenacyl)cyclen **22** was synthesized to study the role of Zn^{II} in carbonyl enolization in aqueous solution [9]. We determined the dissociation of a proton from **23** with an extremely low pK_a value of 8.4 in aqueous solution at I=0.1 ($NaClO_4$) and 25 °C, resulting in an enolate-coordinating complex **24a**, equilibrating with a Zn^{II}–OH^- species **24b** in a ratio of 23:77, as determined by NMR study (Scheme 21.4). This model implies that aldolase would similarly promote enolate formation of the substrates by lowering the high pK_a of *ca.* 19 of the carbonyl α proton to a physiologically feasible pK_a value.

Recently this knowledge has been applied to organic synthesis [26–30]. Aldol reactions between acetone and benzaldehydes **25a–c** in the presence of catalytic amounts of Zn^{II} complexes of L-amino acid-attached cyclen **26a–d** in acetone/H_2O gave the aldol products **27a–c** in good chemical yields with high enantioselectivity up

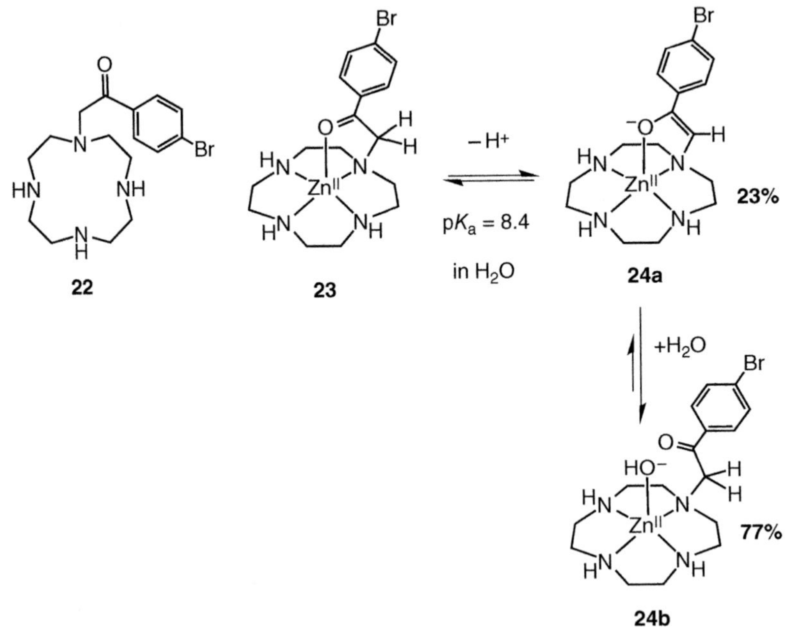

Scheme 21.4

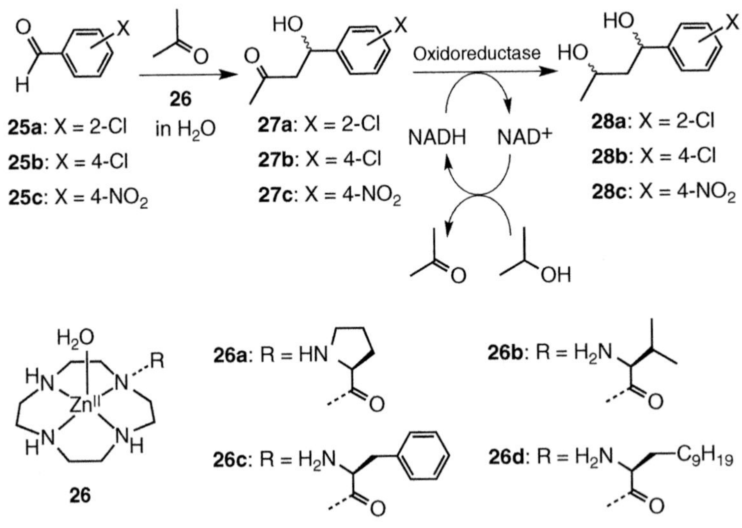

Scheme 21.5

to 96% *ee* (*R*) (Scheme 21.5). Mechanistic studies concluded that the chiral amino acid part and the Zn^{II}–cyclen part catalytically function in a cooperative manner to permit the Zn^{II}–enolate of acetone to react with the benzaldehydes in enantioselective C–C bond formation [27]. One-pot chemoenzymatic synthesis in combination with successive reduction of (**27a–c**) by a recombinant oxidoreductase gave optically active 1,3-diols (**28a–c**) in high chemical yields [28].

21.2.5 Reaction of Zn^{II}-macrocyclic polyamines with phosphomono-, di-, and triesters

Our CA models **11** and **12**, which are coordinately rigid and yet have open sites, were also useful in elucidating the role of Zn^{II} in zinc phosphatases or dianionic phosphate inhibition of zinc enzymes [6a]. A neutral phophotriester, *tris*(4-nitrophenyl) phosphate (TNP) and a monoanionic phosphodiester, *bis*(4-nitrophenyl) phosphate (BNP^-) were hydrolyzed by **11**and **12**. Kinetic studies in 33% EtOH aqueous solution showed the reactive species to be, commonly, Zn^{II}–OH^-. The reactivities of OH^- (pK_a 15.7), **11b** (pK_a 7.3), and **12b** (pK_a 8.0) for neutral TNP hydrolysis are indicated by second-order rate constants – 10.7, 7.0, and 3.7 M^{-1} s^{-1} at $I = 0.2$ ($NaClO_4$) and 25 °C, respectively, signifying that OH^- is the most reactive. On the other hand, for anionic BNP^- hydrolysis, the rate constants are 2.4, 8.5 and 2.1 ($\times 10^{-5}$ M^{-1} s^{-1} at 35 °C), respectively. A bifunctional role of Zn^{II} in **11** toward the anionic diphosphate, in which Zn^{II}–OH^- **11b** acts as a nucleophile and the open Zn^{II} coordination site simultaneously anchors the substrate P–O^- group, was made further apparent by comparing these results with the second-order rate constants of 4-nitrophenylacetate hydrolysis – 9.5, 4.1×10^{-2}, and 1.1×10^{-1} M^{-1} s^{-1} at 25 °C, respectively, where all the OH^- species act monofunctionally as nucleophiles. In comparison to TNP and BNP^-, dianionic phosphomonoesters, phenyl phosphate PP^{2-} and 4-nitrophenyl phosphate NPP^{2-} were not hydrolyzed by **11** or **12**, but bind to them to make 1:1 complexes. Binding constants log $K = 3.5$ and 3.1 for **11a**–PP^{2-} and **11a**–NPP^{2-}, respectively, are greater than 2.6 for the **11a**–CH_3COO^- complex, due to their –2 charge and the bidentate nature of the phosphomonoesters. Monoanionic phosphodiesters, having only weak associations with **11** (log $K < 0.5$ for BNP^-), rather act as substrates. This acquired knowledge set the basis for the following studies of various types of phosphate recognitions by Zn^{II}.

21.2.6 A dinuclear Zn^{II} phosphomonoesterase model 29

While the mononuclear Zn^{II} complexes **11** and **12** failed to hydrolyze dianionic phosphomonoesters, a new dinuclear Zn^{II} cryptate **29** has offered a rare phosphomonoesterase (e.g. alkaline phosphatase) model [31]. Initially, **29** was thought to be non-reactive to any phosphates, because both Zn^{II} ions appeared coordinately saturated in a five-coordinate, trigonal–bipyramidal environment. To our surprise, **29** exclusively hydrolyzed NPP^{2-}, but not BNP^- or TNP, nor a more reactive 4-nitrophenyl acetate. Thus, **29** is a special Zn^{II} complex that selectively reacts with phosphomonoesters (and ATP^{4-}, too), serving as a useful model to understand the mechanism of dinuclear Zn^{II} phosphomonoesterases. An X-ray structure analysis of **29** (and the product **30**) confirmed two identical Zn^{II} centers in trigonal bipyramidal coordination spheres. The reaction of **29** with NPP^{2-} in pH = 5–7 aqueous solutions at 55 °C yielded a phosphoryl amide complex **30** as the sole product (Scheme 21.6). The reaction of **29** with ATP^{4-} also gave **30** (plus ADP^{3-}) with much slower rate. Second-order dependence of the rate (first order in [**29**] and in [NPP^{2-}]) was observed. We also found that this reaction was competitively inhibited by inorganic phosphate dianion $HOPO_3^{2-}$, but not by the strongly mononuclear Zn^{II} binding SCN^-. The reaction mechanism of the P–O ester bond-cleavage is proposed in Scheme 21.6. Two Zn^{II} ions appropriately separated (by an alkoxide-bridge) allowed the two O^- anions of NPP^{2-} to bind to the sixth coordination sites of each Zn^{II}, activating NPP^{2-} to become susceptible to subsequent nucleophilic attack. Concomitantly, the two axial Zn^{II}–N bonds loosened and one of the leaving N's attacked the activated NPP^{2-} to displace 4-nitrophenolate to yield **30**.

This phosphoryl-transfer mechanism would be relevant to the mechanism of dimetallic phosphomonoesterases in general [32]: first, the strong bidentate association of the phosphomonoesters will trigger the dissociation of the OH^- species from the M^{n+}–OH^- or the bridging μ-OH^-; and secondly, this leaving OH^- species will attack at the electrophilic site of the metal-interacting (and thus activated) phosphoesters.

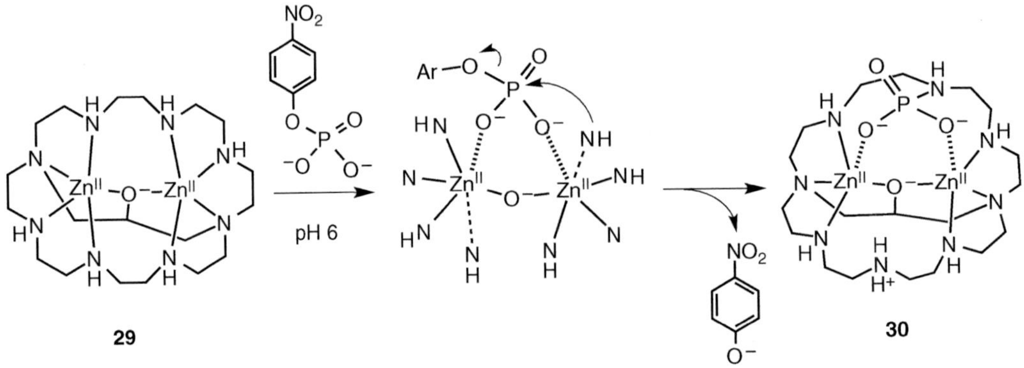

Scheme 21.6

21.2.7 Development of "Phos-tag" for protein phosphorylation analysis

The above phosphatase models have led us to a new alkoxide-bridging dinuclear Zn^{II} (or Mn^{II}) complex **31** – a practical reagent for protein phosphorylation analysis (called "Phos-tag") [33] (Scheme 21.7). The Zn^{II}–Phos-tag complex strongly binds to phenyl phosphate dianion PP^{2-} ($K_d = 2.5 \times 10^{-8}$ M) at neutral pH and 25 °C in aqueous solution. The selectivity indexes against other anions such as SO_4^{2-}, CH_3COO^-, Cl^-, and bisphenyl phosphodiester BPP^- are very high 5.2×10^3, 1.6×10^4, 8.0×10^5, and $>2 \times 10^6$, respectively. Phos-tag is useful in assaying protein phosphorylation (at serine, threonine, or tyrosine residues) by kinases, and dephosphorylation by phosphatases, which are the most important mechanisms to regulate protein functions or transmit signals throughout cells. Phosphate-affinity column chromatography using Phos-tag-attached agarose **31a** can separate phosphorylated proteins from a complex mixture of unphosphorylated proteins [34]. Another analytical method is based on the mobility shift of phosphorylated proteins in SDS-PAGE with Phos-tag-attached polyacrylamide gel **31b** [35], where the phosphorylated proteins on the gel are separated as retarded migration bands from the unphosphorylated proteins. This method allowed easy, yet quantitative, assay of protein kinase activities. Conventional methods use expensive apparatus and enzymes, and cumbersome labeling with a dye or radioactive isotopes [36]. The Phos-tag-attached biotin **31c** was also exploited for specific detection of phosphorylated proteins blotted on a PVDF membrane [37]. These Phos-tag tools are now all commercially available for routine assays, contributing to wide range of phosphoproteome studies [38].

21.2.8 Cooperative phosphate dianion recognition by multiple Zn^{II}–cyclens

A new centrosymmetirc *tris*(Zn^{II}–cyclen) **32** is another good receptor of phosphomonoester dianions in aqueous solution (Scheme 21.8) [39]. The design of **32** was based on the X-ray crystal structure of a C_3-symmetric1:3 complex **34** of NPP^{2-} with **12.** The ^{31}P NMR titration of **32** with PP^{2-} in D_2O confirmed the formation of a 1:1 **32**–PP^{2-} complex **33** (log K_{aff} = 5.8) (K_{aff}: affinity constant). Thus, **32** is a much better selective host for phosphate dianions than the parent **12** (log K_{aff} = 3.3 with PP^{2-}) or *bis*(Zn^{II}-cyclen) linked with a *m*-xylene spacer **35** (log K_{aff} = 4.0) [40].

To cite a few relevant references, Lamb et al. synthesized a resorcinarene bowl bearing four cyclen arms ("cyclen bowl") [41a] and its Zn^{II}_4 complex [41b] for anion-exchange chromatography. König et al. synthesized a **35**-homologue with a rigid triazine bearing polyaromatic functions **36** for new luminescent receptor of phosphates [42]. The **35** moiety was used for another mechanism of apoptosis sensor **37a** that senses dianionic phophatidylserine coming off from degraded apoptotic cell membrane [43]. The same **35** moiety labeled

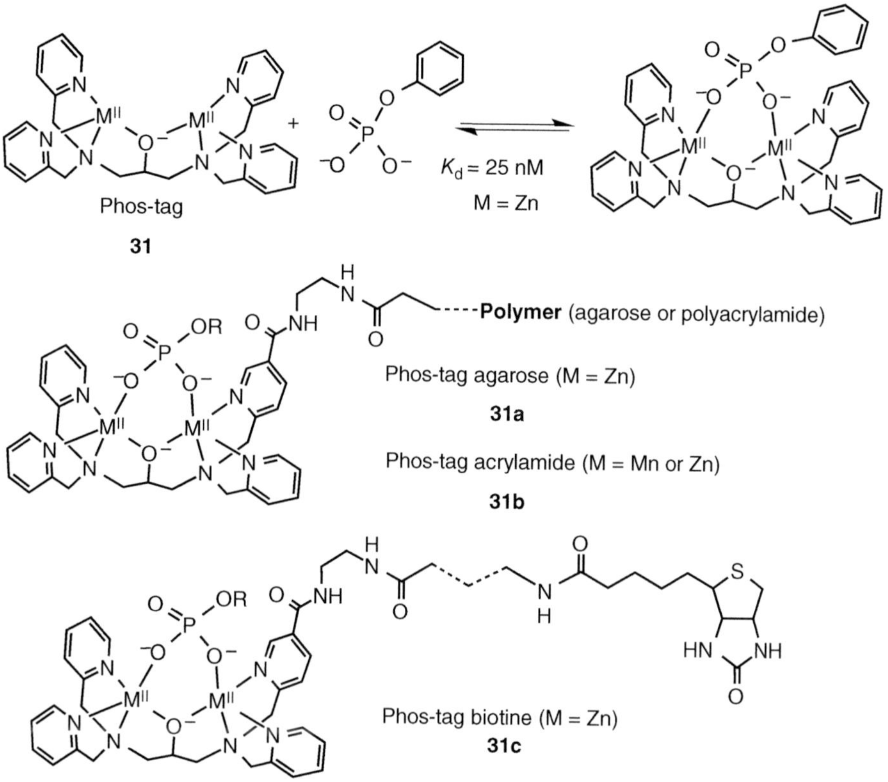

Scheme 21.7

with ^{18}F linked to the benzene ring in **37b** is used for PET imaging after radiation therapy or cytostatic drug administration for cancerous Hela cells [43].

A multiple Zn^{II}–cyclens supermolecule **39** composed of three **38**s assembled by Ru^{II} is a new luminescent sensor for inositol 1,4,5-trisphosphate (IP_3, an important second messenger in intracellular signal transduction) and its model compound, *cis,cis*-1,3,5-cyclohexanetriol trisphosphate (CTP_3) (Scheme 21.9) [44]. X-ray crystal structure of **39** (a racemic mixture) showed each terminal three Zn^{II}–cyclen units to orient toward the incoming guest molecule. The ^1H NMR and UV titrations of **39** indicated the formation of a 1:2 complex with CTP_3 in D_2O solution. Maximum emission at 610 nm (excitation at 300 nm at pH 7.4) of **39** increased 4.2-fold at 584 nm in the 1:2 complex. This supramolecular sensor selectively responds only to CTP_3 and IP_3, discriminating against monophosphates (inorganic phosphate, PP^{2-} and D-glucose-6-phosphate) and diphosphates (*cis,cis*-cyclohexanediol diphosphate and D-glucose-1,6-diphosphate).

21.2.9 Discovery that guanidine is a Zn^{II}-binding ligand at neutral pH

Arginine residues, having a pK_a of 12.5 in Zn enzymes, had been perceived as functioning merely as protonated guanidinium cations in neutral solution, which sometimes interacted with anions (e.g. phosphates) [45]. Guanidine coordination to Zn^{II} had not been conceived of. We synthesized a guanidine-pendant cyclen, **40**, which yielded a Zn^{II} complex, **42,** in neutral aqueous solution [46]. The X-ray crystal structure of **42** revealed distinct coordination of the pendant guanidine through an imine nitrogen (Scheme 21.10). We then studied

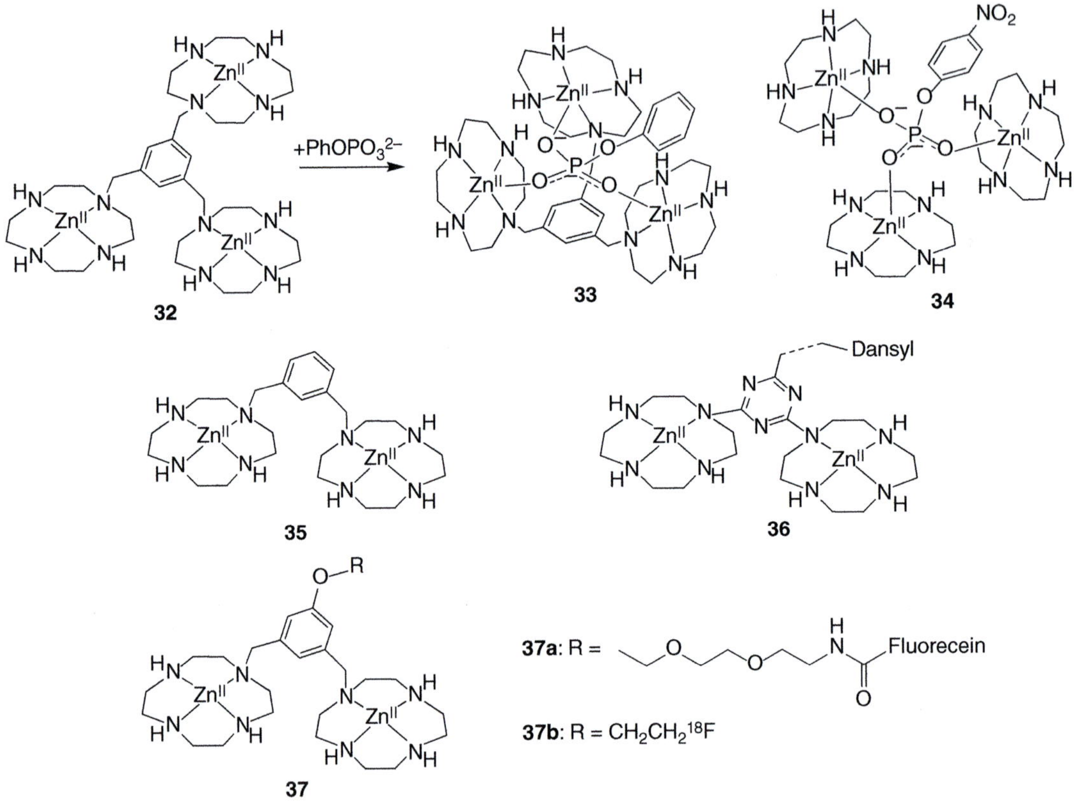

Scheme 21.8

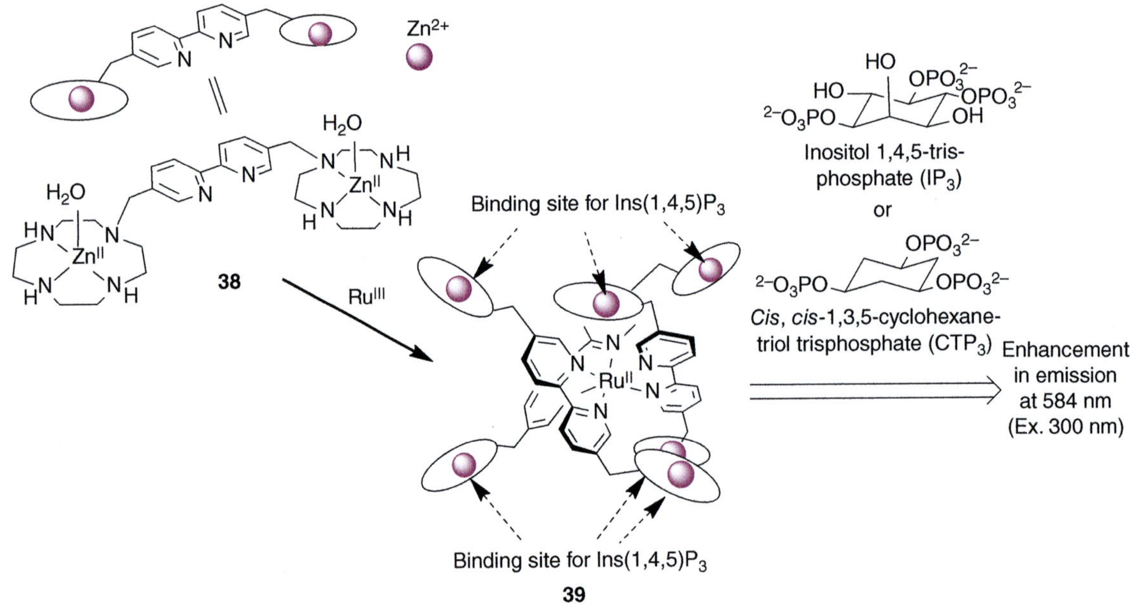

Scheme 21.9

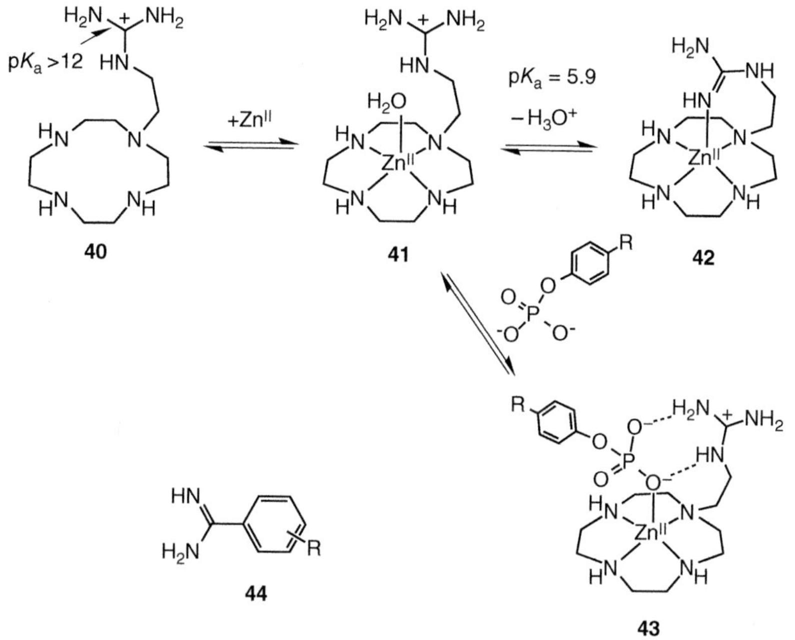

Scheme 21.10

how the pendant guanidine prefers Zn^{II} over protons in aqueous solution. Potentiometric pH titration analysis disclosed the equilibrium between coordination (**42**) and protonation (**41**) with an extremely low guanidinium deprotonation constant of pK_a = 5.9 at I = 0.1 (NaClO$_4$) and 25 °C in aqueous solution. The speciation diagram for a mixture of 1 mM **40** and 1 mM Zn^{II} as function of pH (at 25 °C) indicates that **41** is the more abundant (75%) at pH 5.5. Above pH 5.9, **42** becomes predominant.

Dianionic phosphomonoesters, being good ligands to Zn^{II} as well as to guanidinium cation [46], may displace the Zn^{II}-coordinating guanidine. Potentiometric and ^{31}P NMR titrations showed the NPP^{2-}-coordinating complex **43** to be predominant at pH < 7 and **42 to be** predominant at pH > 7. The PP^{2-}-bound complex **43** was later isolated.

Our model finding may imply a new role for arginine residues in metalloenzymes, which has recently inspired an argument as to whether or not arginine functions as a metal ligand in metalloenzymes [47]. In recent research, amidine (a guanidine homologue) compounds, **44**, were found to inhibit anthrax lethal factor (a Zn enzyme). A proposed mechanism is the strong binding of amidines to Zn^{II} [48], which was theoretically tested using the Zn^{II}–guanidine bonding of **42** as a model. The quantum chemical calculations supported a robust Zn^{II}–N(amidine) bond; the amidine pK_a was estimated to decrease from 12.5 to ~7 in the lethal factor center. In another recent implication, Tierney, Iverson et al. proposed the coordination of the substrate arginine to one of the two metal ions in human arginase I to explain enhanced arginine hydrolysis at neutral pH when displacing native Mn^{II} by the more acidic Co^{II} [49].

21.2.10 ^{11}B NMR sensing of Zn^{II} ion *in vitro* and *in vivo* using phenylboronic acid-pendant cyclen

Non-invasive magnetic resonance imaging (MRI) including “Chemical Shift Imaging” (CSI) technique is a powerful method for the imaging of intracellular molecules *in vivo* or for monitoring various biological

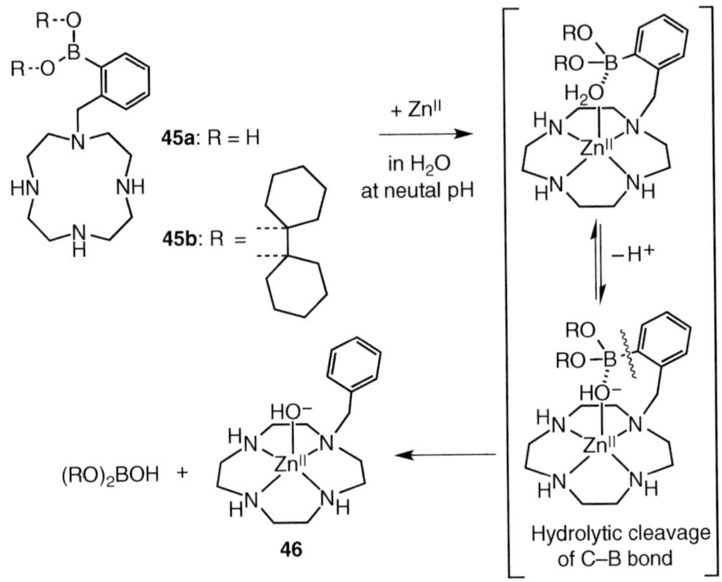

Scheme 21.11

events. ^{11}B NMR may be useful for the CSI technique, since boron is an ultratrace element in animal cells and tissues. The ^{11}B NMR signals of phenylboronic acid-pendant cyclens **45** were examined to sense cellular Zn^{II} (Scheme 21.11) [50]. Initially with **45a**, the carbon–boron bond was cleaved to produce **46** upon addition of Zn^{II}; the broad ^{11}B NMR signal at 31 ppm of **45a** shifted upfield to 19 ppm of $B(OH)_3$. ^{1}H NMR, X-ray single-crystal structure analysis, and UV absorption spectra also supported the formation of product **46**. Because the cellular uptake of **45a** was very limited, a more cell-membrane permeable ligand **45b** was synthesized for successful Zn^{II} sensing in Jurkat T cells.

21.2.11 From the β-lactamase model to development of new magnetic beads device for purification of thiol-containing biomolecules

The CA model **12** also catalyzed hydrolysis of *N*-benzylpenicillin **47**, mimicking mononuclear zinc(II) β-lactamase [8]. Interestingly, model **12** concomitantly catalyzed epimerization of the hydrolysis product, (5*R*)-benzylpenicilloate to (5*S*)-benzylpenicilloate (Scheme 21.12). During the mechanistic study of this epimerization, we found an outstanding affinity of **12** for thiolate anions *R*–S$^-$. For instance, a thiol-containing hypertension drug, captopril **48** (an inhibitor of a Zn enzyme, angiotensin II) showed a 1:1 binding constant *K* of 10^7 M^{-1} ($K = [R\text{–}S^-\text{–}\mathbf{12a}]/[R\text{–}S^-][\mathbf{12a}]$) at 25 °C in aqueous solution. The *R*–S$^-$ ligand coordinated to Zn^{II}–cyclen was found stable to air oxidation, in contrast to the non-coordinated ligand.

This finding recently has led to a new device using the magnetic-bead technology for simple and efficient analysis, separation or enrichment of thiol-containing biomolecules, such as L-glutathione, its oxidized SS-dimer, or cysteine-containing peptides [51]. The **12** moiety is linked to a hydrophilic cross-linked agarose coating on a particle that has a magnetic core (called "Thio-tag MG bead", **49**, Scheme 21.13). The entire separation protocol for thiol compounds, from sample addition to elution, requires less than one hour per sample. By using mild buffers at neutral to weakly acidic pH, there is little damage to the thiol samples, and thus purified samples are suitable enough for biological research. This new method is much simpler than the current thiol-selective covalent chromatography technique involving reversible disulfide bond formation.

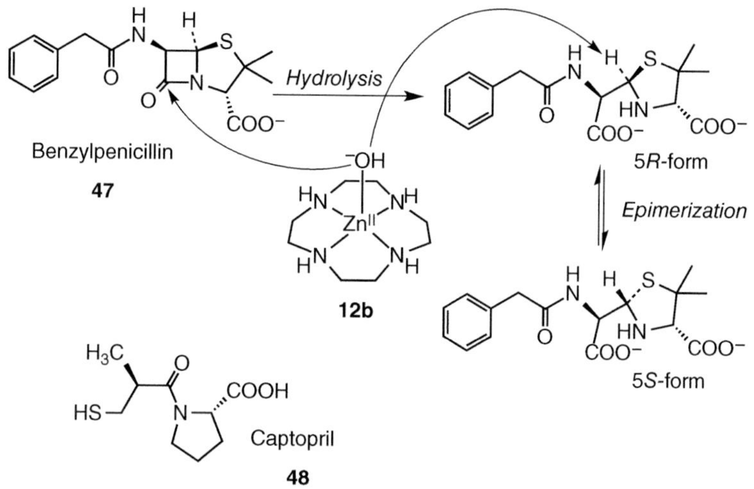

Scheme 21.12

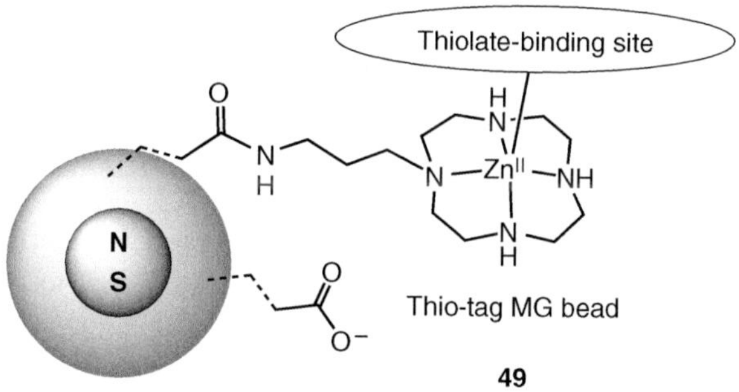

Scheme 21.13

21.3 Zn^{II}–cyclens for selective recognition of nucleobases (thymine and uracil) and manipulation of genes

Our basic knowledge derived from zinc enzyme models has advanced to new molecular recognition of nucleobases, thymine and uracil [52]. In 1993, we were lucky to find **12** to interact with azidothymidine (AZT, **50**; Scheme 21.14) in pH 8.5 aqueous solution to yield a 5-coordinate, square-pyramidal 1:1 complex **51**, where an interaction occurs between Zn^{II} and the N(3)-deprotonated anion of the pyrimidine ring, reinforced by two complementary hydrogen bonds between the two imide oxygens of thymine and two NH's of cyclen [53]. Complex **51** is fairly stable ($K_{app}=3.2\times10^{3}\,M^{-1}$ at pH 8 and 25 °C in aqueous solution), but kinetically labile, as seen in its NMR chemical shifts behaviors in D_2O solution. Potentiometric titrations with all the nucleosides and homologues disclosed selective interactions of **12** with thymidine (dT), AZT, uridine (U), or riboflavin, all having in common "imide" functionalities (pK_a values of 9.7, 9.7, 9.2, and 9.9, respectively).

With the cyclen ligand appended with an acridinylmethyl group **52** (Scheme 21.15), dT recognition is enhanced (1:1 dissociation constant $K_d=8\,\mu M$ vs. 0.79 mM for **12** at pH 7.4 and 25 °C in aqueous solution) and becomes kinetically inert in D_2O solution in NMR timescale [54]. X-ray crystal structure of the **52**–(1-methylthymine) complex, **53**, showed an additional π–π stacking interaction between the acridine and

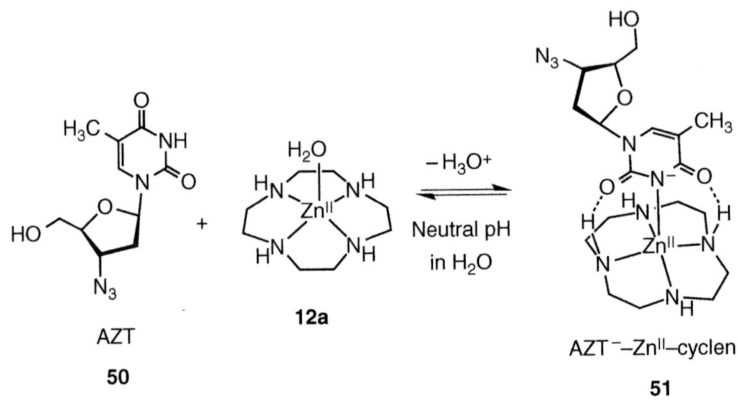

Scheme 21.14

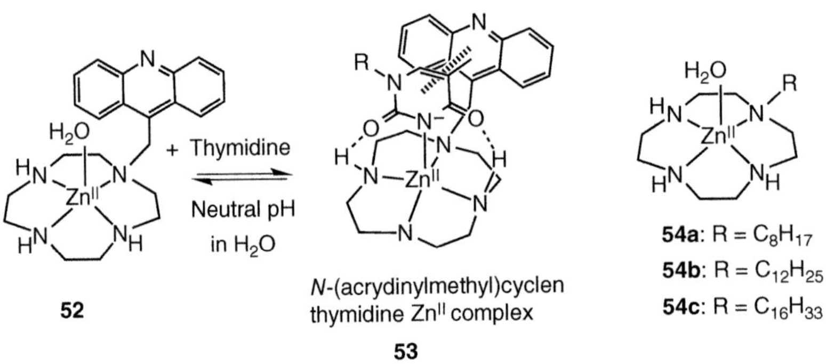

Scheme 21.15

thymine, an observation supported by ^{1}H NMR study. By replacing the 9-acridinylmethyl part of **52** by a redox-active 9,10-anthraquinonyl-2-methyl group, dT and U became electrochemically recognizable [55].

Lipophilic Zn^{II}–cyclen complexes **54a–c** are carriers of T and U nucleosides and nucleotides for selective transport from an aqueous phase to a $CHCl_3$ phase [56]. The transport mechanism involves the formation of lipophilic **54a–c**–T$^-$(or U$^-$) complexes. After $CHCl_3$-extraction, the lipophilic complexes dissociated to release nucleosides or nucleotides into acidic pH 6 aqueous solution.

21.3.1 Recognition of thymine and uracil in single- and double-stranded synthetic nucleic acids

The recognition by **12** and **52** of nucleobases thymine (T) and uracil (U) in poly(U), poly(U)–poly(A), poly(dT)–poly(dA) was studied using UV and CD spectral titrations or thermal denaturation (T_m) measurements [57, 58]. The apparent affinity constant of **12** with each U in poly(U) was $K = 10^{3.0}$ M^{-1} at pH 8 and 25 °C in aqueous solution, almost the same as the value for the interaction of **12** with U, a fact suggesting that **12**–U interaction is little perturbed by the phosphodiester backbone of poly(U). Moreover, **12** inhibited poly(U)–poly(A) hybridization [57]. Conversely the double-stranded poly(U)–poly(A) and poly(dT)–poly(dA) were disrupted, as demonstrated by the decreasing T_m of the duplexes at pH 7.6 with an increase in the concentration of **12** and **52** (the efficiency order is **52** > **12**, correlated with their affinity to U or T) [58].

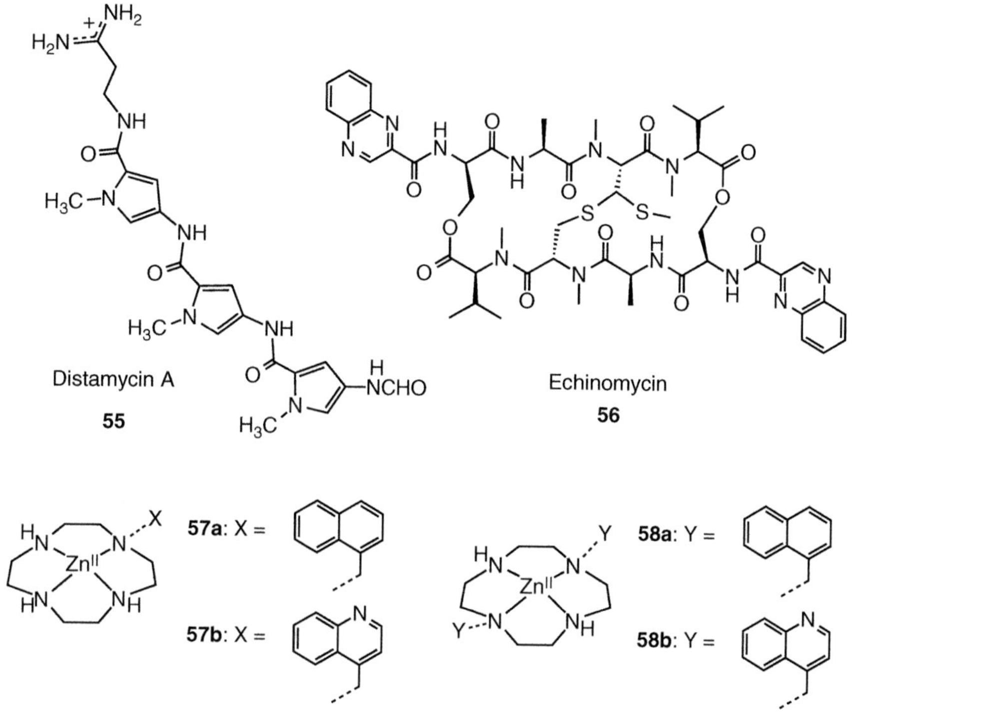

Scheme 21.16

The recognition of T by **52** was studied in detail using dinucleotides, TpT, GpT, CpT, ApT, a single-stranded octanucleotide d(GTGACGCC), and a double-stranded octanucleotide $d(CGCTAGCG)_2$ [59]. Complex **52** binds to T in the dinucleotides and the single-stranded octanucleotide with similar 1:1 binding constants, log K_{app} of ca. 5 (at pH 8 and 25 °C in aqueous solution) regardless of the adjacent bases (K_{app}: apparent complexation constant). The resulting complexes are all inert over the NMR timescale. When **52** (0.6 mM, 1.0 mM, and 1.2 mM) was added to the single-stranded octanucleotide (1 mM) at pD 8.4 and 5 °C, 50%, 80%, and 90%, respectively, of the d(G-(**52**–T^-)-GACGCC) were formed, as shown by NMR study. NMR imino proton exchange measurements showed the efficient disintegration of the duplex, $d(CGCTAGCG)_2$ due to **52** binding to T^-.

The genomic effect of **12** or **52**–U (or T) recognition was revealed. First, we studied polyphenylalanine synthesis *in vitro*, in cell-free extracts, where poly (U) was used as a messenger RNA (the sequence UUU is a codon for phenylalanine) [57]. When **12** (at $[\mathbf{12}]_{added}/[U]_{poly(U)} = 1$ and 10) was added, poly(Phe) synthesis was inhibited by 15% and 26%, respectively. Secondly, in *E. coli* RNA polymerase reaction *in vitro*, **52** inhibited $poly(dA–dT)_2$ or poly(dA)–poly(dT)-directed transcription to a greater degree than $poly(dG–dC)_2$-directed transcription [60]. These facts implied genomic applicability of the Zn^{II}–cyclen complexes.

21.3.2 Recognition of thymine in native DNA

The binding sites of the Zn^{II}–cyclen complexes in native DNA (arbitrary AT-rich 150-base pairs from plasmid pUC19) were identified by combination of the DNase I and micrococcal nuclease footprinting techniques [61]. A typical A–T binder, distamycin A **55** and a G–C binder, echinomycin **56** (Scheme 21.16) were references on the same gel plate. There were three remarkable discoveries. (i) The Zn^{II}–cyclens with aromatic pendants (**52, 57a, 57b, 58a**, and **58b**) will bind to homopolymeric A–T regions which are nearly overlapped by those bound by the A-T binder **55**. The IC_{50} (for 50% inhibition of the DNase I digestion at position 45–50

(TpTpTpTpTpTpT) in the upper strand were 8–30 μM *vs*. 0.5 μM with **55**. The most effective Zn^{II} inhibitors were the double aromatic-appended derivatives **58a** and **58b**. (ii) While **55** interacts simultaneously with A and T in A–T pairs to stabilize the double helix, the Zn^{II}–cyclens interact only with its T partners to disrupt the A–T pairs. (iii) The A-T binder **55** and the Zn^{II}–cyclens, despite their different binding mechanisms, reversibly compete for the same A–T regions.

Recently, **57b** has been applied as an additive suppressing non-specific, undesirable products for selective amplification of target DNA in DNA polymerase chain reaction (PCR) [62]. The selectivity of PCR is due to a greater decrease in T_m of the heteroduplex primer/template DNA than that of homoduplex primer/DNA.

To cite recent literature, Morrow et al. reported that **57b** binds with μM affinity to thymine (or uracil) bulges in DNA (or RNA) duplexes or in the stems of hairpin with good selectivity over fully complementary stems or bulges containing A, G, or C nucleotides [63]. Bond et al. showed the perturbation of A–T stacking in duplex DNA by a ferrocene-attached cyclen complex, whereupon the yield of DNA-mediated charge transport diminishes to sense single-base sequence mismatches on voltammetric redox waves [64].

21.3.3 Recognition of AT-rich TATA box DNA

The AT–rich DNA sequence ("TATA box") located 25 to 30 bases upstream from the transcriptional start site is an essential element of the promoter for eukaryotic RNA polymerase II. TATA binding protein TBP must bind to the TATA box for initiation of the transcription. The A–T binder **55** strongly binds to the TATA box to inhibit the binding of TBP. We found that **58b** [65] selectively and efficiently binds to the TATA box region of a 230-bp SV40 early promoter sequence containing an AT-rich TATA box (TATTAT) and a GC-rich GC box [66]. The inhibitory effect on the binding of human TBP to the TATA box was then studied by a gel mobility assay using a 25-bp TATA consensus DNA fragment, which showed IC_{50} values of 15 μM (for **52**) and 2.5 μM (for **58b**) *vs*. 0.4 μM (for **55**).

A genomic effect of the Zn^{II}–cyclen recognition of the AT-rich DNA was tested for the inhibition of calf thymus DNA-directed transcription by measuring the incorporation of [^{3}H]-UTP into the transcribed RNA in the presence of *E. coli* RNA polymerase [60]. The reactants contained all other nucleotide substrates needed for RNA synthesis. The Zn^{II}–cyclens having higher binding affinities to calf thymus DNA showed the higher inhibition of the [^{3}H]-UTP incorporation; i.e. the higher inhibition of RNA transcription. Among the Zn^{II}–cyclen series, the most potent inhibitor was **58a** (IC_{50} = 55 μM).

21.3.4 Simple detection of single nucleotide polymorphisms (SNP) by Zn^{II}–cyclen polyacrylamide gel electrophoresis (PAGE)

The discovery of the modification of DNA conformation by Zn^{II}–cyclens has led to development of a new method for detection of SNP by PAGE using **12** as an additive [67]. A wild type and nine mutant DNA's (99-bp) showed almost identical mobility in ordinary PAGE, whereas by adding **12** (1 mM) the mobilities became varied, due to the different local DNA conformation changes, resulting in varied SNP mobilities. Furthermore, in combination with a PCR-based heteroduplex method, the PAGE detection became more sensitive for otherwise undetectable SNP such as A–T to T–A or G–C to C–G mutation [68]. The heteroduplex DNA's are more subject to the effect of the **12**–T bindings to give the slower migration bands. The disruption of A–T base pairing by **12** binding to T affects a single mutation site (G–C to C–G) that is one to even ten bases away, causing PAGE-detectable local conformation changes in the appropriate DNA length (176–241 bp). This technique was applied to the SNP mapping and heterozygosity screening of a human cardiac Na^+ channel gene (SCN5A) that relates to inherited arrhythmia syndromes [69]. This new method also has been applied to genotyping and mapping assay of human cytochrome P450 (CYP3A5) [70], which is known to be subject to frequent SNP, resulting in the formation of truncated proteins and extremely low enzymatic activity in drug metabolism.

21.3.5 Recognition of TpT with *bis*(Zn^{II}–cyclen) 59 and TpTpT with *tris*(Zn^{II}–cyclen) 60

A *p-bis*(Zn^{II}–cyclen), **59**, and a *tris*(Zn^{II}–cyclen), **60**, bind with thymidylthymidine (TpT) and thymidylthymidylthymidine (TpTpT), yielding very stable 1:1 complexes **61** and **62**, respectively, in pH 7.4 aqueous solution (Scheme 21.17) [71].

The 1:1 dissociation constants, K_d, are 6.3×10^{-7} M for **61** and 8.0×10^{-10} M for **62** (at pH 7.4 and 25 °C), to be compared with mM order for **12**–T binding. Later, Bond et al. reported 1:1 complexation of TpT with a ferrocene bearing *bis*(Zn^{II}–cyclen) ($K = 0.89\times10^{6}$ M^{-1} at pH 7.4) for electrochemical sensing [72].

The *bis*(Zn^{II}–cyclen)s, **35** and **59** are also efficient ditopic receptors for T and U nucleotides including AZT monophosphate and AZT diphosphate [73]. The resulting 1:1 ternary complexes, **63a** and **63b**, are kinetically inert and thermodynamically more stable (the dissociation constants are μM order at pH 7.6) than the 1:1 T^{-}–(Zn^{II}–benzylcyclen) complex, due to simultaneous bindings to T moieties and terminal dianionic phosphate moieties. It is pertinent to note that we earlier found the *m*- and *p*-bis(Zn^{II}–cyclen)s, **35** and **59**, to possess potent anti-HIV activities [74]. It would be interesting to see if the combination of the two mechanistically different kinds of anti-HIV active agents, AZT nucleotides and **35** or **59**, may produce a new cocktail for AIDS treatment.

DNA-directed transcription *in vitro* in the presence of **59** and **60** was studied to examine inhibition of the incorporation of [^{3}H]-ATP and [^{3}H]-UTP into RNA [75]. The clearest effect of **60** was found when poly(dA)·poly(dT) was used as a template. The IC_{50} value for [^{3}H]-ATP incorporation is 0.8 μM *vs.* 40 μM for [^{3}H]-UTP, indicating that **60** more efficiently inhibits transcription from the poly(dT) strand than from the poly(dA) in the template. In comparison, **12** and **52** showed little activities.

21.3.6 Inhibition of photo[2+2] cycloaddition and promotion of photosplitting of the *cis*-syn-cyclobutane thymine dimer

Exposure of cellular nucleic acids to UV irradiation leads to a variety of lesions, which, unless repaired, are potentially carcinogenic, mutagenic, or cytotoxic. Among the photoproducts of DNA, the major ones are *cis,syn* (T[*c,s*]T) (**65**), and *trans,syn* (T[*t,s*]T) cyclobutane thymine dimers, which result from the photo[2+2] cycloaddition of two adjacent thymidilyl(3′-5′)thymidine site **64** (Scheme 21.18). In the aforementioned stable TpT–**59** complex, **61**, the two thymine rings would be separated far more remotely (*ca.* 10 Å, according to energy-minimization calculation) than ordinary 3~4 Å in intact DNA. Indeed, **59** (and **35**) inhibited the photo[2+2]cycloaddition and/or promoted the reverse photosplitting of the thymine dimers **65** to protect the TpT sites **64** against UV light [76]. An equimolar amount of **35** and **59** more effectively inhibited the photodimerization (85% inhibition of the control reaction) than two equimolar **12** during the UV exposure for 20 min (where possible secondary reverse photosplitting is insignificant). Even after the photodimerization and photosplitting reached an equilibrium at prolonged irradiation for 3 h, 90% of TpT remained. The presence of 1 mM $ZnSO_4$ without cyclen ligands did not affect the reactions. From the dissociation constant of *ca.* 1 μM at pH 7.6, 95% of the 1:1 complex **61** is calculated to be present at [TpT] = [**59**] = 0.2 mM at pH 7.6 and 25 °C. Under the same conditions, TpT (0.2 mM) and **12** (0.4 mM) form only 4% of the 1:2 complex.

Later, a different type of artificial DNA photolyase, **66**, was designed, using the Zn^{II}–cyclen platform appended with a tryptophan linking to a lumiflavin unit that facilitates the photoreversion of **65** to **64** via radical anion species [77]. Another photorepairing Zn^{II}-cyclen complex is *N*-dansylcyclen **67**, which undergoes S–N bond cleavage reaction upon photoirradiation at 365 nm to give a sulfonyl radical **68** to abstract an electron from **65** and reproduce **64** catalytically via radical cation species [78].

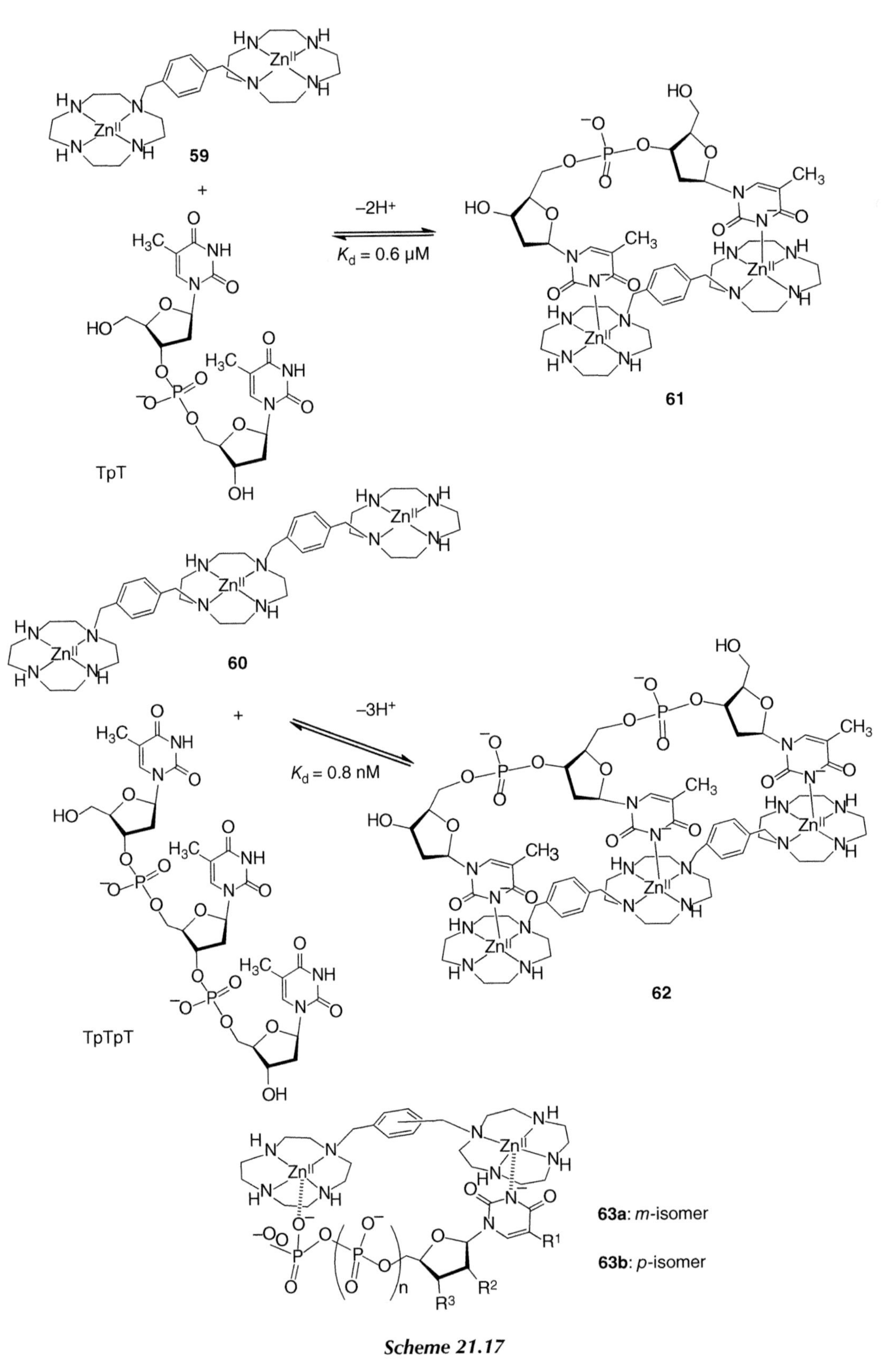
59
+
TpT
−2H+
Kd = 0.6 μM
61
60
+
TpTpT
−3H+
Kd = 0.8 nM
62
63a: m-isomer
63b: p-isomer

Scheme 21.17

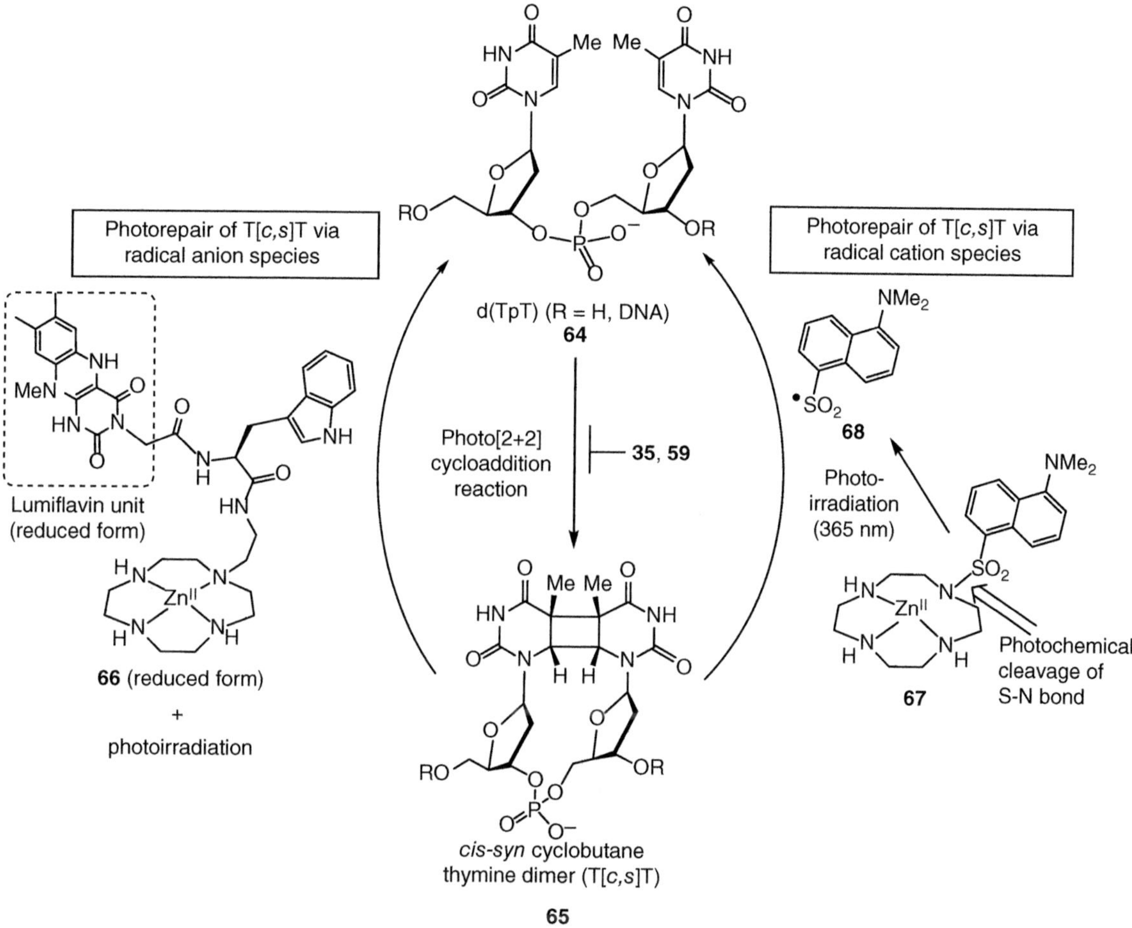

Scheme 21.18

21.3.7 Potent inhibition of HIV-1 TAR RNA –TAT peptide binding

The search for small molecules that interact with RNA is attractive for drug discovery. One example is AIDS therapeutics which target viral RNA sites to prevent the formation of key RNA–protein complexes. Transcription of HIV-1 genome is facilitated by a HIV-1 regulatory protein, TAT, which activates synthesis of full-length HIV-1 mRNA by its binding to TAR RNA. The TAR element adopts a hairpin structure with a U-rich bulge (UUU or UCU) located four base pairs below a six-nucleotide loop. The bulge is the TAT binding site.

We found the *tris*(Zn^{II}–cyclen) **60** to be the most potent inhibitor ever reported for the TAR–TAT interaction, due to its extremely strong binding to the UUU bulge (Scheme 21.19) [79]. Footprinting analysis using micrococcal nuclease, RNase A, and phosphodiesterase I revealed the UUU bulge to be strongly bound by **60** in the TAR model RNA sequence TAR_{33} **69**. We then studied inhibition of the TAR–TAT-consensus peptide TAT_{47-86} **70** binding using a gel mobility shift assay. Among all the Zn^{II}–cyclen complexes tested, **60** was the most potent inhibitor with an extremely low IC_{50} value of 20 nM. Under the same conditions, the currently

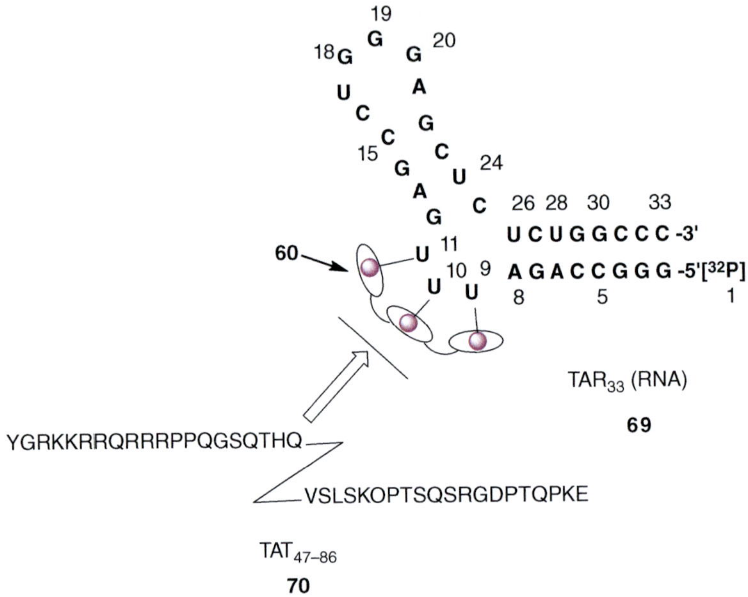

Scheme 21.19

most potent antibiotic inhibitor, neomycin, showed $IC_{50} = 50\,\mu$M. Our present findings imply that Zn–cyclen complexes are useful binders to U-rich elements in single-stranded RNA. It is of pertinence to add that **60** alone, like **35** and **59**, possesses a strong anti-HIV activity.

21.4 New supramolecular assemblies with Zn^{II}–cyclen

Following the discovery of **12** binding to the imide functions of T and U with mM affinity, we found the *bis*(Zn^{II}–cyclen) **59** (and **35**) binding to barbital containing di-imide moiety with μM affinity in 1:1 Bar^{2-} (dianionic barbital form) complex **71** at pH 8 (Scheme 21.20) [40, 80]. We then tested the centrosymmetric *tris*(Zn^{II}–cyclen) **32** binding with centrosymmetric tri-imide-bearing cyanuric acid (CyA), anticipating formation of a 1:1 complex **72** [81]. Instead, a startling product crystallized out as a 4:4 **32**–CyA^{3-} complex **74** in 77% yield from strong alkaline (pH > 11.5) aqueous solution. In neutral solution, however, a different 2:3 **32**–CyA^{2-} complex **73** crystallized in a quantitative yield.

21.4.1 Supramolecular cuboctahedron cages

The X-ray crystal analysis of **74** revealed a highly symmetric 4:4 cuboctahedron-like cage structure; an assembly of four triangular *tris*(Zn^{II}–cyclen) **32**s linked with four smaller triangular CyA^{3-} with 12 zinc atoms at 12 vertices, together with six open rectangular planes (Scheme 21.21a). The cage molecule is stabilized by the Zn^{II}–N^- bonds and the hydrogen-bond network between Zn^{II}–N–H···O=C of CyA^{3-} (totally 24 hydrogen bonds) (Scheme 21.21b). The cage **74** has a well-defined inner space, which is separated from outer space by four phenyl groups and CyA^{3-}. The inner cavity is visualized as a truncated tetrahedron with approximate volume of 140 Å^3, suggesting possible encapsulation space for guest molecules. The 4:4 assembly **74**,

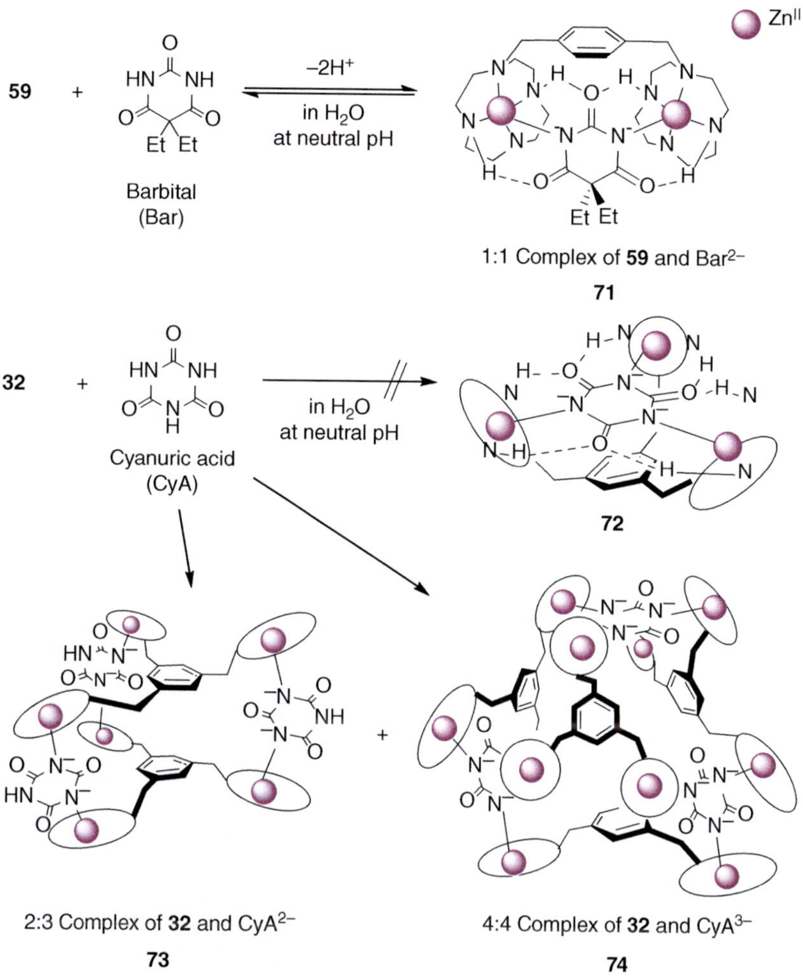

Scheme 21.20

stable only in solid or in DMSO solution, tends to be disproportionate to the 2:3 complex **73** in neutral D_2O, as demonstrated by the ^{1}H NMR study. The 2:3 complex **73** is stable at neutral pH and well characterized, including by X- ray crystal analysis.

To build a more robust and practical cage molecule at neutral pH, CyA was replaced by trithiocyanuric acid (TCyA) [82]. Two advantages of TCyA were expected: (1) three imide deprotonation constants (pK_a) of TCyA are *ca.* twop orders lower than those for CyA, to facilitate $TCyA^{3-}$ at neutral pH; and (2) $TCyA^{3-}$ tends to aromatize to a 1,3,5-triazine structure so that the anions should localize more on the exocyclic sulfur atoms to favor stronger Zn^{II}–S^- bonding. Indeed, mixing of **32**, TCyA, and $NaNO_3$ in pH 8 aqueous solution quantitatively yielded a new 4:4 assembly product **75** (Scheme 21.22). The X-ray structure showed that the exterior frame is a twisted (to right or left) cuboctahedron. Thus, the 4:4 complex **75** is chiral. Each triangular $TCyA^{3-}$ links to the three Zn^{II}–cyclen moieties mainly by Zn^{II}–S^- coordinate bonds. Since the three C–S–Zn bonds around all of the four TCA^{3-} units are bent by 100° clockwise or anticlockwise, **75** is chiral. As hoped, **75** is thermodynamically stable and kinetically (on the NMR timescale) inert.

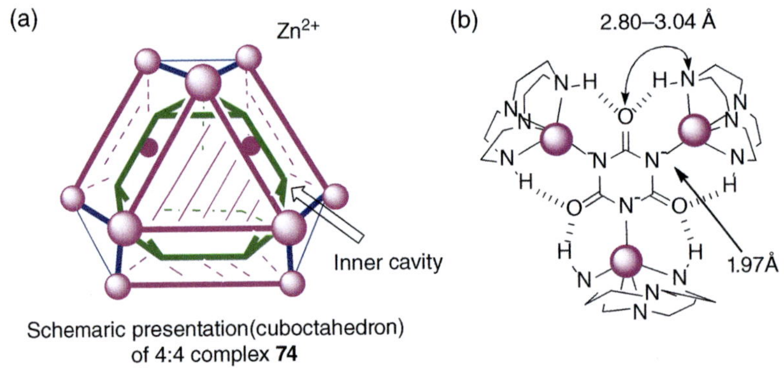

Scheme 21.21

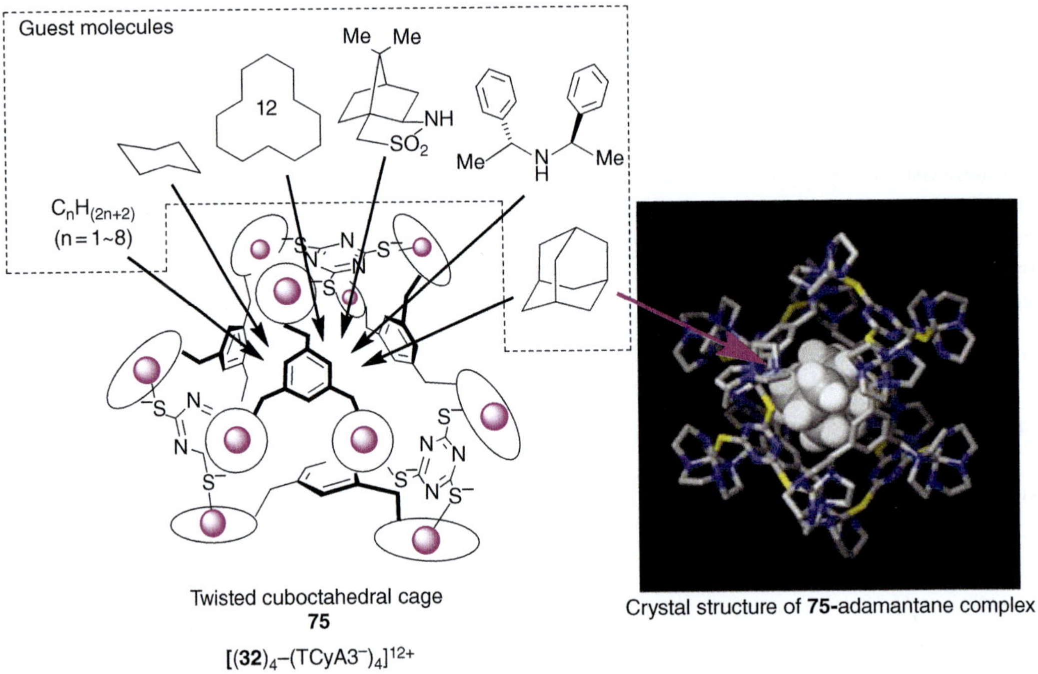

Scheme 21.22

The most interesting property of **75** is its inclusion of various organic compounds as guests. A **75**–adamantane (ADM) complex crystallized from a mixed solution at pH 8. Its X-ray structure analysis showed the encapsulation of ADM in the inner cavity, where the four composite cyclohexane rings of the guest ADM squarely face the tetrahedrally surrounding phenyl rings of the host **75**. The guest encapsulation in D_2O solution was confirmed by the ^1H upfield shifts of the guests due to the shielded magnetic field by the surrounding phenyl rings. The guests are size-matching hydrophobic molecules such as ADM, nitrophenols, ibuprofen, biadamantane, or tetraethylammonium cation. Although the capsule **75** becomes more stabilized by guest-inclusion, the guest molecule is slowly replaced by other guests over a long period at room temperature in aqueous solution.

Recently, guest inclusion by **75** has been studied in more detail [83]. The ^{1}H NMR study, GC measurements, or crystal structure analysis confirmed inclusion of other guests such as hydrocarbons $C_nH_{(2n+2)}$ ($n = 1 \sim 8$), *cis*- and *trans*-decaline, or cyclododecane. Computational simulations of **75**–guest complexes using Amber* of the MacroModel indicate **75** to have a flexible structure due to rotation about the C–S^- bonds of TCA^{3-}. The *ΔSASA* values (ΔSASA: difference in solvent-accessible surface area) were calculated as parameters to evaluate the surface complementarity between the guest molecules and the inner cavity of the host to explain the order of stability estimated by the guest-replacement experiments. In addition, the storage and release of volatile guest hydrocarbons in aqueous solution and in solid state have been studied.

21.4.2 Supramolecular trigonal prisms

The strong Zn^{II}–imide interaction was further applied to the synthesis of prismatic supermolecules (Scheme 21.23) [84]. By replacing the centrosymmetric *tris*(Zn^{II}–cyclen), **32**, for the linear *tris*(Zn^{II}–cyclen) **60**, the reaction with TCyA quantitatively yielded the 3:3 complex **76** in pH 8 aqueous solution. It has a trigonal prism structure by X-ray crystal analysis; its prism length is 2.9 Å, the distance between two adjacent $TCyA^{3-}$ is 1.2 Å. Of the three $TCyA^{3-}$ units, the two terminal $TCyA^{3-}$ units bind to Zn^{II} ions by Zn^{II}–S^- bonds. On the other hand, the central $TCyA^{3-}$ has less aromatic triazine structure with the imide N^- binding to Zn^{II}. The prism complex **76** is thermodynamically stable and kinetically inert (on NMR timescale) at μM concentrations in neutral aqueous solution.

A linear tetrakis(Zn^{II}–cyclen) **77** forms a similar trigonal prism 3:4 complex **78** with $TCyA^{3-}$, which is very stable at nM concentrations in neutral aqueous solution [85]. The extremely stable **78** did not dissociate in the presence of other Zn^{II}–binding anions such as phosphates or DNA. The interaction studies with calf-thymus DNA by thermal melting (T_m) experiments, gel mobility shift assays, or dynamic light scattering data indicated **78** acting as a polycationic template to induce DNA aggregation.

21.4.3 Supramolecular catalysts mimicking dimetallic phosphomonoesterases

The 4:4:4 mixing of **38,** CyA and Cu^{II} yielded the novel supramolecular catalyst **79** quantitatively (Scheme 21.24) [86, 87]. Most interestingly, **79** contains two sets of $Cu_2(\mu\text{-OH})_2$ groups at central cores and selectively catalyzes hydrolysis of phosphomonoester, NPP^{2-}, at neutral pH. Hence, **79** represents the first

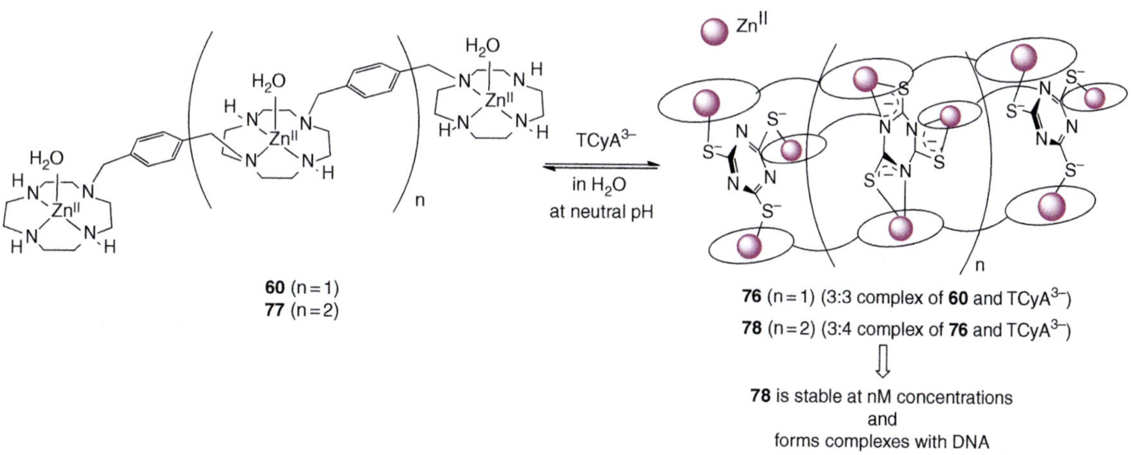

Scheme 21.23

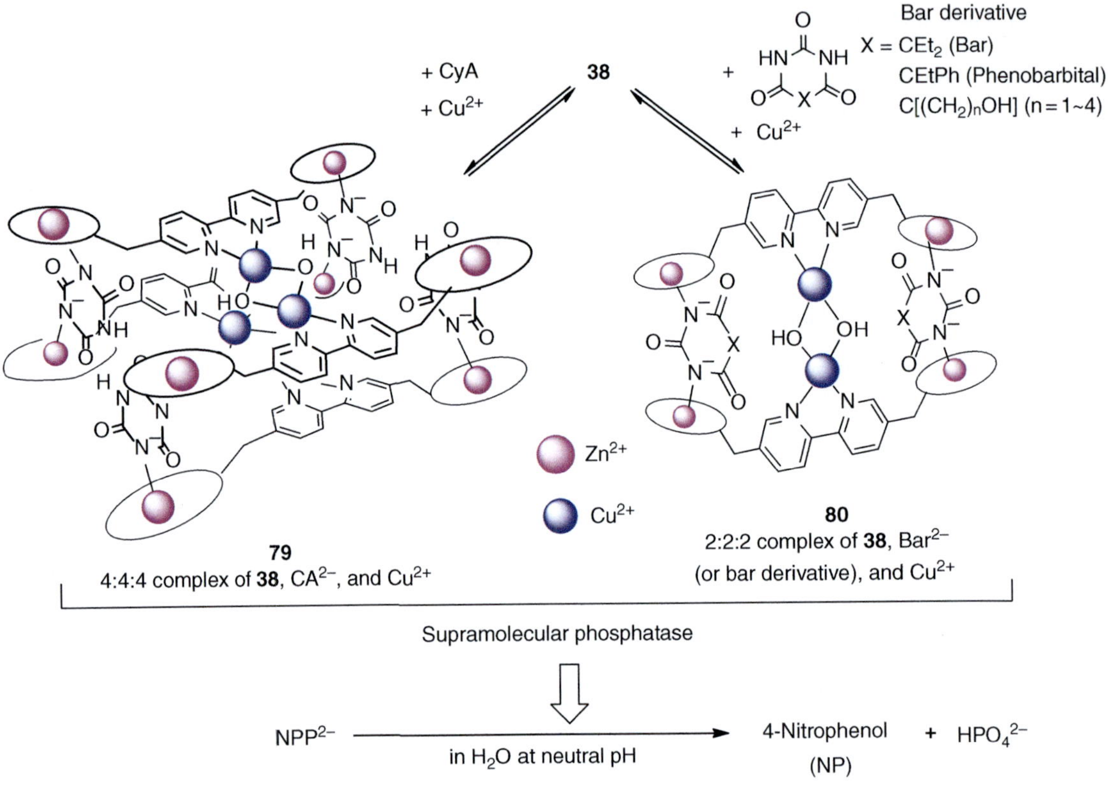

Scheme 21.24

supramolecular phosphomonoesterase model which mimics the active centers of dimetallic phosphatases, although the catalytic turnover of NPP^{2-} hydrolysis is yet *ca.* 3 per second.

As a further extension of this work, **38** was mixed with Cu^{II} and barbital (Bar) to obtain a 2:2:2 complex **80** also containing the $Cu_2(\mu\text{-}OH)_2$ function, which catalyzed the hydrolysis of NPP^{2-} [88]. On the basis of the X-ray crystal structure of **80**, various barbitals were employed (e.g., $X=C[(CH_2)_nOH]_2$) for derivatization *in situ* [88, 89]. The present self-assembling strategy would make a useful prototype for construction of supramolecular catalysts.

21.5 Acknowledgements

We wish to thank laboratory staff, coworkers, students, and collaborators for their great contributions to development of the macrocyclic polyamine chemistry. It has always been a joy, honor, and privilege to be a member of the international community of macrocyclic and supramolecular chemistry, which has kept us well-informed, inspired, and motivated ever since we started macrocyclic studies in 1970s. EK is grateful to Prof. Reed Izatt for giving him an opportunity to review his lifework for this permanent landmark book.

References

1. Reviews: a) Kimura, E. (1986) Distinctive coordination chemistry and biological relevance of complexes with macrocyclic oxo polyamines, *Journal of Coordination Chemistry*, **15**: 1–28. b) Kimura, E. (1992) Design of new macrocyclic ligands beyond cyclam to control redox properties of metal ions, In *Crown Compounds toward Future*

Applications, Cooper, S. R. (Ed), VCH Publisher, Weinheim, Chap. 6, 81–98. c) Kimura, E. (1992) New developments in macrocyclic polyamine chemistry, *Crown Ethers and Analogous Compounds*, Hiraoka, M. (Ed), Elsevier, Amsterdam, Chap. 8, 351–478. d) Kimura, E. (1992) Macrocyclic polyamines with intelligent functions, *Tetrahedron*, **48**, 6175–7217. e) Kimura, E. (2012) Evolution of macrocyclic polyamines from molecular science to supramolecular science, *Bulletin of Japanese Society of Coordination Chemistry*, **59**, 26–47.

2. Reviews: a) Kimura, E. (1993) Macrocyclic polyamine zinc(II) complexes as advanced models for Zn(II) enzymes, In *Progress in Inorganic Chemistry*, Karlin, K. D. (Ed), John Wiley and Sons, **41**, 443–492. b) Kimura, E., Koike, T. (1996) Intrinsic properties of zinc(II) ion pertinent to zinc enzymes, *Advances in Inorganic Chemistry*, Sykes, A. G. (Ed), Academic Press, New York, **44**, 229–261. c) Kimura, E., Koike, T., Shionoya, M. (1997) Advances in zinc enzyme models by small mononuclear zinc(II) complexes, *Structure and Bonding*, Sadler, H. P. (Ed), Springer, Berlin, **8**, 1–28. d) Kimura, E. (2001) Model studies for molecular recognition of carbonic anhydrase and carboxypeptidase, *Accounts of Chemical Research*, **34**, 171–179. e) Aoki, S., Kimura, E. (2004) Zinc hydrolases, In *Comprehensive Coordination Chemistry II*, Que. L. Jr, Tolman, W. B. (Eds), Elsevier, Amsterdam, **8**, 601–640.
3. Woolly, P. (1975) Models for metal ion function in carbonic anhydrase, *Nature*, **258**, 677–682.
4. Kimura, E., Shiota, T., Koike, T., et al. (1990) A Zinc(II) Complex of 1,5,9-triazacyclododecane ([12]aneN_3) as a model for carbonic anhydrase, *Journal of the American Chemical Society*, **112**, 5805–5811.
5. Zhang, X., Eldik, R., Koike, T., Kimura, E. (1993) Kinetics and mechanism of the hydration of CO_2 and dehydration of HCO_3^- catalyzed by a zinc(II) complex of 1,5,9-triazacyclododecane as a model for carbonic anhydrase, *Inorganic Chemistry*, **32**, 5749–5755.
6. a) Koike, T., Kimura, E. (1991) Roles of zinc(II) ion in phosphatases. A model study with zinc(II)-macrocyclic polyamine complexes, *Journal of the American Chemical Society*, **113**, 8935–8941. b) Kimura, E., Nakamura, I., Koike, T., et al. (1994) Carboxyester hydrolysis promoted by a new zinc(II) macrocyclic triamine complex with an alkoxide pendant: A model for the serine alkoxide nucleophile in zinc enzymes, *Journal of the American Chemical Society*, **116**, 4764–4771. c) Koike, T., Kajitani, S., Nakamura, I., et al. (1995) The catalytic carboxyester hydrolysis by a new zinc(II) complex with an alcohol-pendant cyclen: A novel model for indirect activation of the serine nucleophile by zinc(II) in zinc enzymes, *Journal of the American Chemical Society*, **117**, 1210–1219. d) Kimura, E., Kodama, Y., Koike, T., Shiro, M. (1995) Phosphodiester hydrolysis by a new zinc(II) macrocyclic tetraamine complex with an alcohol pendant: Elucidation of the roles of Ser-120 and Zn(II) in alkaline phosphatase, *Journal of the American Chemical Society*, **117**, 8304–8311.
7. Kimura, E., Shionoya, M., Hoshino, A., et al. (1992) A model for catalytically active zinc(II) ion in liver alcohol dehydrogenase: A novel "hydride transfer" reaction catalyzed by zinc(II)-macrocyclic polyamine complexes, *Journal of the American Chemical Society*, **114**, 10134–10137.
8. Koike, T., Takamura,M., Kimura, E. (1994) Role of Zn(II) in β-lactamase II: A model study with a Zn(II)–cyclen complex, *Journal of the American Chemical Society*, **116**, 8443–8449.
9. Kimura, E., Gotoh, T., Koike, T., Shiro, M. (1999) Dynamic enolate recognition in aqueous solution by zinc(II) in a phenacyl-pendant cyclen complex: Implications for the role of zinc(II) in Class II aldolases, *Journal of the American Chemical Society*, **121**, 1267–1274.
10. Koziol, C. A., Valdez, S. E., Baker, E. et al. (2012) Toward a Small molecule, biomimetic carbonic anhydrase model: Theoretical and experimental investigations of a panel of zinc(II) aza-macrocyclic catalysts, *Inorganic Chemistry*, **51**, 6803–6812.
11. Aoki, S., Sakurama, K., Ohshima, R., et al. (2008) Design and synthesis of a caged Zn^{2+} probe and its hydrolytic uncaging upon complexation with Zn^{2+}, *Inorganic Chemistry*, **47**, 2747–2754.
12. For the acidities of Zn^{II} *vs.* Cd^{II}: a) Kimura, E., Koike, T. (1991) Macrocyclic polyamines as a probe for equilibrium study of the acid function of zinc(II) ion in hydrolysis enzymes, *Comments in Inorganic Chemistry*, **11**, 285–301. b) Kimura, E., Koike, T., Shiota, T., Iitaka, Y. (1990) Acid properties of zinc(II) and cadmium(II) in complexation with macrocyclic oxopolyamine ligands, *Inorganic Chemistry*, **29**, 4621–4629.
13. Koike, T., Kimura, E., Nakamura, I., et al. (1992) The first anionic sulfonamide binding zinc(II) complexes with a macrocyclic triamine: Chemical verification of the sulfonamide inhibition of carbonic anhydrase, *Journal of the American Chemical Society*, **114**, 7338–7345.
14. Reviews: a) Kimura, E., Koike, T. (1998) Recent development of zinc-fluorophores, *Chemical Society Reviews*, **27**, 179–184. b) Kimura, E., Aoki, S. (2001) Chemistry of zinc(II) fluorophore sensors, *BioMetals*, **14**, 191–204.

15. Koike, T., Watanabe, T., Aoki, S., et al. (1996) A novel biomimetic Zinc(II)-fluorophore, dansylaminoethyl-pendant cyclen, *Journal of the American Chemical Society*, **118**, 12696–12703.
16. Zalewski, P. D., Forbes, I. J., Betts, W. H. (1993) Correlation of apoptosis with change in intracellular labile Zn(II) using zinquin, a new specific fluorescent probe for Zn(II), *Biochemical Journal*, **296**, 403–408.
17. Aoki, S., Kaido, S., Fujioka, H., Kimura, E. (2003) A new zinc(II) fluorophore, 2-(9-anthrylmecthylamino)ethyl-appended 1,4,7,10-tetraazacyclododecane, *Inorganic Chemistry*, **42**, 1023–1030.
18. Aoki, S., Kagata, D., Shiro, M., et al. (2004) Metal-chelation-controlled twisted intramolecular charge transfer and its application to fluorescence sensing of metal ions and anions, *Journal of the American Chemical Society*, **126**, 13377–13390.
19. Aoki, S., Sakurama, K., Matsuo, N., et al. (2006) A new fluorescent probe for Zinc(II); An 8-Hyroxy-5-*N,N*-dimethylaminosulfonylquinoline-Pendant 1,4,7,10-Tetraazacyclododecane, *Chemistry A European Journal*, **12**, 9066–9080.
20. Aoki, S., Sakurama, K., Ohshima, R., et al. (2008) Design and synthesis of a caged Zn^{2+} probe, 8-benzenesulfony-loxy-5-*N,N*-dimethylaminosulfonylquinolin-2-ylmethyl-pendant cyclen and its hydrolytic uncaging upon complexation with Zn(II), *Inorganic Chemistry*, **47**, 2747–2754.
21. Ohshima, R., Kitamura, M., Morita, A., et al. (2010) Design and synthesis of a fluorescent probe for Zn(II), 5,7-bis(*N,N*-dimethylaminosulfonyl)-8-hydroxyquinolin-pendant cyclen and Zn(II)-dependant hydrolytic and Zn(II)-independent photochemical reactivation of its benzensulfonyl-caged derivative, *Inorganic Chemistry*, **49**, 888–899.
22. (a) Kimura, E., Aoki, S., Kikuta, E., Koike, T. (2003) A macrocyclic zinc(II) fluorophore as a detector of apoptosis, *Proceedings of the National Academy of Sciences*, **100**, 3731–3736. (b) Kimura, E., Takasawa, R., Tanuma, S., Aoki, S. (2004) Monitoring apoptosis with fluorescent Zn^{2+} – indicators, *Science STKE*, **223**, 7.
23. Reviews: a) Jiang, P., Guo, Z. (2004) Fluorescent detection of Zn(II) in biological systems; recent development in the design of chemosensors and biosensors, *Coordination Chemistry Reviews*, **248**, 205–229. b) Nolan, E. M., Lippard, S. J. (2009) Small molecule fluorescent sensors for investigating zinc metalloneurochemstry, *Accounts of Chemical Research*, **42**, 193–203. c) Xu, Z., Yoon, J., Spring, D. R. (2010) Fluorescent chemosensors for Zn(II), *Chemical Society Reviews.*, **39**, 1996–2006.
24. a) Zalewski, P. D., Forbes, I. J., Seamark, R. F., et al. (1994) Flux of intracellular labile zinc during apoptosis (gene-directed cell death) revealed by a specific chemical probe, Zinquin, *Chemical Biology*, **1**, 153–161. b) Zalewski, P. D., Millard, S. H., Forbes, I. J., et al. (1994) Video image analysis of labile zinc in viable pancreatic islet cells using a specific fluorescent probe for zinc, *Journal of Histochemistry and Cytochemistry*, **42**, 877–884.
25. For example: Helmersson, A., von Arnold, S., Bozhkov, P. V. (2008) The level of free intracellular zinc mediates programmed cell death/cell survival decisions in plant embryos, *Plant Physiology*, **147**, 1158–1167.
26. Itoh, S., Kitamura, M, Yamada, S, Aoki, S. (2009) Chiral catalysts dually functionalized with amino acids and zinc(II) complexes for enantioselective direct aldol reactions inspired by natural aldolases: design, synthesis, complexation properties, catalytic activity, and mechanistic study, *Chemistry A European Journal*, **15**, 10570–10584.
27. Itoh, S., Tokunaga, T., Sonoike, S., et al. (2013) Asymmetric aldol reactions between acetone and benzaldehydes catalyzed by chiral Zn(II) complexes of aminoacyl 1,4,7,10-tetraazacyclododecane: fine-tuning of the amino-acid side chains and revised reaction mechanism, *Chemistry – An Asian Journal*, **8**, 2125–2135.
28. Sonoike, S., Itakura, T., Kitamura, M., Aoki, S. (2012) One-pot chemoenzymatic synthesis of chiral 1,3-diols using an enantioselective aldol reaction with chiral zinc(II) complex catalysts and enzymatic reduction using oxidoreductases with cofactor regeneration, *Chemistry – An Asian Journal*, **7**, 64–74.
29. Itoh, S., Tokunaga, T., Kurihara, M., Aoki, S. (2013) Asymmetric aldol reactions between cyclic ketones and benzaldehyde catalyzed by chiral Zn(II) complexes of aminoacyl-1,4,7,10-tetraazacyclododecane: Effects of solvent and additives on controlling the stereoselectivities of the aldol products, *Tetrahedron Asymmetry*, **24**, 1583–1590.
30. Itoh, S., Sonoike, S., Kitamura, M., Aoki, S. (2014) Design and synthesis of chiral Zn(II) complexes mimicking natural aldolases for catalytic C–C bond forming reactions in aqueous solution, *International Journal of Molecular Sciences*, **15**, 2087–2118.
31. Koike, T., Inoue,M., Kimura, E., Shiro, M. (1996) Novel properties of cooperative dinuclear zinc(II) ions. The selective recognition of phosphomonoesters and their P–O ester bond cleavage by a new dinuclear zinc(II) cryptate, *Journal of the American Chemical Society*, **118**, 3091–3099.

32. Kimura, E. (2000) Dimetallic hydrolases and their models, *Current Opinion in Chemical Biology*, **4**, 207–213.
33. Kinoshita, E., Takahashi, M., Takeda, H., et al. (2004) Recognition of phosphate monoester dianion by an alkoxide-bridged dinuclear zinc(II) complex *Dalton Transactions*, 1189–1193.
34. Kinoshita-Kikuta, E., Kinoshita, E., Yamada, A., et al. (2006) Enrichment of phosphorylated proteins from cell lysate using a novel phosphate-affinity chromatography at physiological pH, *Proteomics*, **6**, 5088–5095.
35. Kinoshita, E., Kinoshita-Kikuta, E., Takiyama, K., Koike, T. (2006) Phosphate-binding tag. A new tool to visualize phosphorylated proteins, *Molecular and Cellular Proteomics*, **5**, 749–757.
36. Kinoshita, E., Kinoshita-Kikuta, E., Koike, T. (2012) Phos-tag affinity electrophoresis for protein kinase profiling, in protein kinase technology, *Neuromethods*, **68**, 13–34.
37. Kinoshita, E., Kinoshita-Kikuta, E., Sugiyama, Y., et al. (2012) Highly sensitive detection of protein phosphorylation by using improved Phos-tag Biotin, *Proteomics*, **12**, 932–937.
38. Kinoshita, E., Kinoshita-Kikuta, E., Koike, T. (2015) Advances in Phos-tag-based methodologies for separation and detection of the phosphoproteome, *Biochimica et Biophysica Acta*, **1854**, 601–608.
39. Kimura, E., Aoki, S., Koike, T., Shiro, M. (1997) A Tris(Zn^{II}–1,4,7,10-tetraazacyclododecane) complex as a new receptor for phosphate dianions in aqueous solution, *Journal of the American Chemical Society*, **119**, 3068–3076.
40. Fujioka, H., Koike, T., Yamada, N., Kimura, E. (1996) A new bis(Zn^{II}–cyclen) complex as a novel chelator for barbiturates and phosphates, *Heterocycles*, **42**, 775–787.
41. a) Wang, J., Harrison, R. G., Lamb, J. D. (2009) Anion separation and preconcentration with cyclen and cyclen-resorcinarene derivatives, *Journal of Chromatographic Science*, **47**, 510–515. b) Wang J., Lamb J. D., Hansen L. D., Harrison, R. G. (2010) Multiple anion binding by a Zn-containing tetratopic cyclen-resorcinarene, *Journal of Inclusion Phenomena and Macrocyclic Chemistry*, **67**, 55–61.
42. Bhuyan, M., Katayev, E., Stadlbauer, S., et al. (2011) Rigid luminescent bis-Zn^{II}–bis-cyclen complexes for the detection of phosphate anions and non-covalent protein labeling in aqueous solution, *European Journal of Organic Chemistry*, **15**, 2807–2817.
43. Oltmanns, D., Zitxmann-Kolbe, S., Mueller, A., et al. (2011) Zn(II)–bis(cyclen) complexes and the imaging of apoptosis/necrosis, *Bioconjugate Chemistry*, **22**, 2611–2624.
44. a) Aoki, S., Zulkefeli, M., Shiro, M. et al. (2005) A luminescence sensor of inositol 1,4,5-triphosphate and its model compound by ruthenium templated assembly of bis(Zn^{2+}–cyclen) complex having a 2,2-bipyridyl linker, *Journal of the American Chemical Society*, **127**, 9129–9139. b) Kitamura, M.; Nishimoto, H.; Aoki, K., et al. (2010) Molecular recognition of inositol 1,4,5-trisphosphate and model compounds in aqueous solution by ditopic Zn^{2+} complexes containing chiral linkers, *Inorganic Chemistry*, **49**, 5316–5327.
45. Reviews: a) Aoki, S., Kimura, E. (2002) Recent progress in artificial receptors for phosphate anions in aqueous solution, *Reviews in Molecular Biotechnology*, **90**, 129–155. b) Tamaru, S., Hamachi, I. (2008) Recent progress of phosphate derivative recognition utilizing artificial small molecular receptors in aqueous Media, *Structure and Bonding*, **129**, 95–125. c) Hargrove, A. E., Nieto, S., Zhang, T., et al. (2011) Artificial receptors for the recognition of phosphorylated molecules, *Chemical Review*, **111**, 6603–6782.
46. Aoki, S., Iwaida, K., Hanamoto, N., et al (2002) Guanidine is a Zn^{2+}-binding ligand at neutral pH in aqueous solution, *Journal of the American Chemical Society*, **124**, 5256–5257.
47. Costanzo, L. D., Flores Jr, L. V., Christianson, D. W. (2006) Stereochemstry of guanidine-metal interactions; Implications for L-arginine-metal interactions in protein structure and function, *Proteins: Structure, Function, and Bioinformatics*, **65**, 637–642.
48. Nguyen, T. L., Panchal, R. G., Topol, I. A., et al. (2007) A theoretical study of anthrax lethal factor inhibition by a set of novel carbamimidolyl-aryl-vinyl-carboxamidines: A possible mechanism involving zinc-ligation by amidine, *Journal of Molecular Structure: THEOCHEM*, **821**, 139–144.
49. Stone, E. M., Glazer, E. S., Chantranupong, L., et al. (2010) Replacing Mn^{2+} with Co^{2+} in human arginase I enhances cytotoxicity toward L-arginine auxotrophic cancer cell lines, *ACS Chemical Biology*, **5**, 333–342.
50. Kitamura, M., Suzuki, T., Abe, R., et al. (2011) ^{11}B NMR sensing of d-block metal ions in vitro and in cells based on a carbon-boron bond cleavage of phenylboronic acid-pendant cyclen, *Inorganic Chemistry*, **50**, 11568–11580.
51. Fujioka, H., Tsunehiro, M., Kawaguchi, M., et al. (2014) Simple enrichment of thiol-containing biomolecues by using zinc(II)–cyclen functionalized magnetic beads, *Journal of Separation Science*, **37**, 1601–1609.

52. Reviews: a) Kimura, E., Kikuta, E. (2000) Why zinc in zinc enzymes? From biological roles to DNA base-selective recognition, *Progress in Reaction Kinetics and Mechanism*, **25**, 1–64. b) Aoki, S., Kimura, E. (2004) Zinc–nucleic acid interaction, *Chemical Review*, **104**, 769–787.
53. Shionoya, M., Kimura, E., Shiro, M. (1993) A new ternary zinc(II) complex with [12]aneN_4 and AZT. Highly selective recognition of thymidine and its related nucleosides by a zinc(II) macrocyclic tetraamine complex with novel complementary associations, *Journal of the American Chemical Society*, **115**, 6730–6737.
54. Shionoya, M., Ikeda, T., Kimura, E., Shiro, M. (1994) Novel "multipoint" molecular recognition of nucleobases by a new zinc(II) complex of acridine-pendant cyclen, *Journal of the American Chemical Society*, **116**, 3848–3859.
55. Tucker, J. H. R., Shionoya, M., Koike, T., Kimura, E. (1995) A zinc(II)–cyclen complex attached to an anthraquinone moiety that acts as a redox-active nucleobase receptor in aqueous solution, *Bulletin of the Chemical Society of Japan*, **68**, 2465–2469.
56. Aoki, S., Honda, Y., Kimura, E. (1998) The first selective and efficient transport of imide-containing nucelosides and nucleotides by lipohilic cyclen-zinc(II) complex, *Journal of the American Chemical Society*, **120**, 10018–10026.
57. Shionoya, M., Sugiyama, M., Kimura, E. (1994) Uracil-targeted inhibition of poly(A)–Poly(U) hybridization by a zinc(II)–cyclen complex, *Journal of the Chemical Society, Chemical Communications,* 1747–1748.
58. Kimura, E., Ikeda, T., Aoki, S., Shionoya, M. (1998) Macrocyclic zinc(II) complexes for selective recognition of nucleobases in single- and double-stranded polynucleotides, *Journal of Biological Inorganic Chemistry*, **3**, 259–267.
59. Kimura, E., Kitamura, H., Ohtani, K., Koike, T. (2000) Elaboration of selective and efficient recognition of thymine base in dinucleotides (TpT, ApT, CpT, and GpT), Single-stranded d(GTGACGCC), and double-stranded $d(CGCTAGCG)_2$ by Zn^{2+}–acridinylcyclen, *Journal of the American Chemical Society*, **122**, 4668–4677.
60. Kikuta, E., Katsube, N., Kimura, E. (1999) Natural and synthetic double-stranded DNA binding studies of macrocyclic tetraamine Zn(II) complexes with polyaromatic groups, *Journal of Biological Inorganic Chemistry*, **4**, 431–440.
61. Kikuta, E., Murata, M., Katsube, et al. (1999) Novel recognition of thymine base in double-stranded DNA by zinc(II)–macrocyclic tetraamine complexes appended with aromatic groups, *Journal of the American Chemical Society*, **121**, 5426–5436.
62. Kinoshita, E., Kinoshita-Kikuta, E., Koike, T. (2005) A heteroduplex-preferential T_m depressor for the specificity-enhanced DNA polymerase chain reactions, *Analytical Biochemistry*, **337**, 154–160.
63. del Mundo, I. M. A., Siters, K. E., Fountain, M. A., Morrow, J. R. (2012) Structural basis for bifunctional Zn(II) macrocyclic complex recognition of thymine bulges in DNA, *Inorganic Chemistry*, **51**, 5444–5457.
64. Shiddiky, M. J. A., Torriero, A. A. J., Zeng, Z., et al. (2010) Highly selective and sensitive DNA assay based on electrocatalytic oxidation of ferocene bearing Zn^{2+}–cyclen with diethylamine, *Journal of the American Chemical Society*, **132**, 10053–10063.
65. Kimura, E., Katsube, N., Koike, T., et al. (2002) Effects of bis(aromatic) pendants on recognition of nucleobase thymine by Zn^{2+}–cyclen, *Supramolecular Chemistry*, **14**, 95–1002.
66. Kikuta, E., Koike, T., Kimura, E. (2000) Controlling gene expression by zinc(II)–macrocyclic tetraamine complexes, *Journal of Inorganic Biochemistry*, **79**, 253–259.
67. Kinoshita-Kikuta, E., Kinoshita, E., Koike, T. (2002) A novel procedure for simple and effcient genotyping of single nucleotide polymorphisms by using the Zn(II)–cyclen complex, *Nucleic Acids Research*, **30**, e126/1–6.
68. Kinoshita, E., Kinoshita-Kikuta, E., Koike, T. (2005) Reliable and cost-effective screening of inherited heterozygosity by using Zn^{2+}–cyclen polyacrylamide gel electrophoresis, *Clinical Chemistry*, **51**, 2195–2198.
69. Kinoshita, E., Kinoshita-Kikuta, E., Koike, T. (2008) Detection of the Gua/Cyt- to-Cyt/Gua mutation in a Gua/Cyt-lined sequence using Zn^{2+}–cyclen polyacrylamide gel electrophoresis, *Analytical Biochemistry*, **380**, 122–127.
70. Kinoshita, E., Kinoshita-Kikuta. E, Nakashima, H., Koike, T. (2010) Genotyping and mapping assay of single-nucleotide polymorphisms in CYP3A5 using DNA-binding zinc(II) complexes, *Clinical Biochemistry*, **43**, 302–306.
71. Kimura, E., Kikuchi, M., Kitamura, H., Koike, T. (1999) Selective and efficient recognition of thymidylylthymidine (TpT) by bis(Zn^{2+}–cyclen) and thymidylylthymidylylthymidine (TpTpT) by Tris(Zn^{2+}–cyclen) at neutral pH in aqueous solution, *Chemistry A European Journal*, **5**, 3113–3123.

72. Zeng, Z., Torriero, A. A. J., Belousoff, M. J., et al. (2009) Synthesis, X-ray structure of ferrocene bearing bis(Zn^{2+}–cyclen) complexes and selective electrochemical sensing to TpT, *Chemistry A European Journal*, **15**, 10988–10996.
73. Aoki, S., Kimura, E. (2000) Highly selective recognition of thymidine mono- and diphosphate nucleotides in aqueous solution by ditopic receptors zinc(II)–bis(cyclen) complexes, *Journal of the American Chemical Society*, **122**, 4542–4548.
74. Review: Kimura, E., Koike, T., Inoue, Y. (1999) Macrocyclic polyamines and their metal complexes: A novel type of anti-HIV agents, In *Perspectives in Bioinorganic Chemistry*, Hay, R. W., Dilworth, J. R., Nolan, K. B. (Eds), JAI Press, Inc., **4**, 145–164.
75. Kikuta, E., Aoki, S., Kimura, E. (2002) New potent agents binding to poly(dT) sequence in double-stranded DNA: Bis(Zn^{2+}–cyclen) and Tris(Zn^{2+}–cyclen) complexes, *Journal of Biological Inorganic Chemistry*, **7**, 473–482.
76. Aoki, S., Sugimura, C., Kimura, E. (1998) Efficient inhibition of photo [2+2] cycloaddition of thymidilyl (3′–5′) thymidine and promotion of photosplitting of the *cis-syn*-cyclobutane thymine dimer by dimeric Zn(II)–cyclen complexes containing *m*- and *p*-xylyl spacers, *Journal of the American Chemical Society*, **120**, 10094–10102.
77. Yamada, Y., Aoki, S. (2006) Efficient cycloreversion of cis,syn-thymine photodimer by a Zn^{2+}–1,4,7,10-tetraazacyclododecane complex bearing a lumiflavin and tryptophan by chemical reduction and photoreduction of a lumiflavin unit, *Journal of Biological Inorganic Chemistry*, **11**, 1007–1023.
78. Aoki, S., Tomiyama, Y., Kageyama, Y., et al. (2009) Photolysis of the sulfonamide bond of metal complexes of N-Dansyl-1,4,7,10-tetraazacyclododecane in aqueous solution: A mechanistic study and application to the photorepair of cis,syn-cyclobutane thymine photodimer, *Chemistry – An Asian Journal*, **4**, 561-573.
79. Kikuta, E., Aoki, S., Kimura, E. (2001) A new type of potent inhibitors of HIV-1 TAR RNA-Tat peptide binding by zinc(II)–macrocyclic tetraamine complexes, *Journal of the American Chemical Society*, **123**, 7911–7912.
80. Koike, T., Takashige, M., Kimura, E., et al. (1996) Bis(Zn^{II}–cyclen) complex as a novel receptor of barbiturates in aqueous solution, *Chemistry A European Journal*, **2**, 617–623.
81. Aoki, S., Shiro, M., Koike, T., Kimura, E. (2000) Three-dimensional supermolecules assembled from a tris(Zn^{2+}–cyclen) complex and di- and trianionic cyanuric acid in aqueous solution, *Journal of the American Chemical Society*, **122**, 576–584.
82. Aoki, S., Shiro, M., Kimura, E. (2002) A cuboctahedral supramolecular capsule by 4:4 assembly of tris (Zn^{2+}–cyclen) and trianionic trithiocyanurate in neutral pH aqueous solution, *Chemistry A European Journal*, **8**, 929–939.
83. Aoki, S., Suzuki, S., Kitamura, M., et al. (2012) Molecular recognition of hydrocarbon guests by supramolecular cuboctahedral cage formed by self-assembly of tris(Zn^{2+}–1,4,7,10-tetraazacyclododecane) complexes with tricycnuric acid in aqueous solution, *Chemistry – An Asian Journal*, **7**, 944–956.
84. Aoki, S., Zulkefeli, M., Shiro, M., Kimura, E. (2002) New supramolecular trigonal prisms from zinc(II)–cyclen complex and trithiocyanate in aqueous solution, *Proceedings of the National Academy of Sciences*, **99**, 4894–4899.
85. Zulkefeli, M., Sogon, T., Takeda, K., et al (2009) Design and synthesis of a stable supramolecular trigonal prism formed by the self-assembly of a linear tetrakis(Zn^{2+}–cyclen) complex and trianionic trithiocyanuric acid in aqueous solution and its complexation with DNA, *Inorganic Chemistry*, **48**, 9567–9578.
86. Zulkefeli, M., Suzuki, A., Hisamatsu, Y., et al. (2011) Selective hydrolysis of phosphate monoester by a supramolecular phosphatase formed by self-assembly of a bis(Zn^{2+}–cyclen) complex, cyanuric acid and copper in aqueous solution, *Inorganic Chemistry*, **50**, 10113–10123.
87. Aoki, S., Hisamatsu, Y., Kitamura, M., Yamada, Y. (2014). Supramolecular host and catalysts formed by the synergistic molecular assembly of multinuclear zinc(II) complexes in aqueous solution, *Synergy in Supramolecular Chemistry* (ed. Nabeshima,T) CRC Press, 33–56.
88. Zulkefeli, M., Hisamatsu, Y., Suzuki, A., et al. (2014) Supramolecular phosphatases formed by the self-assembly of the bis(Zn^{2+}–cyclen) complex, copper(II) and barbital derivatives in water, *Chemistry – An Asian Journal*, **9**, 2831–2841.
89. Hisamatsu, Y., Miyazawa, Y., Yoneda, K., et al. (2016) Supramolecular complexes formed by the self-assembly of hydrophobic bis(Zn^{2+}–cyclen) complexes, copper and di- or triimide units for the hydrolysis of phosphate mono- and diesters in two-phase solvent systems (cyclen = 1,4,7,10-tetraazacyclododecane), *Chemical and pharmaceutical Bulletin*, **64**, 451–464.

22

Contractile and Extensile Molecular Systems: Towards Molecular Muscles

Jean-Pierre Sauvage, Vincent Duplan, and Frédéric Niess

Institut de Science et d'Ingénierie Supramoléculaires (ISIS), Université de Strasbourg et CNRS, Strasbourg, France

This chapter is dedicated to the memory of Dr. Christiane Dietrich-Buchecker, an extremely talented organic chemist who played a determining role in the development of catenanes and molecular knots from the beginning of the 1980s.

22.1 Preamble: the Izatt–Christensen award and Jean-Pierre Sauvage

In 1991, the senior author of this chapter, Jean-Pierre Sauvage, was awarded the Izatt–Christensen Award in Macrocyclic Chemistry. This event had very important consequences for his research and his career. Obviously, the work done in his group became more visible to the community. In addition, this recognition convinced him that he had been right 7 or 8 years earlier when he reoriented his research themes to follow contemporary research directions in relation to interlocking ring compounds.

Sauvage studied in Strasbourg, both at the university and the engineering school, Ecole Nationale Supérieure de Chimie de Strasbourg (ENSCS) before starting his PhD thesis in the young, dynamic, and very attractive group led by Professor Jean-Marie Lehn. He passed his thesis in 1971 on the synthesis and the study of cryptands and cryptates, work mostly done in collaboration with his best friend, Bernard Dietrich, who very sadly passed away in 2004. There can be no doubt that the enthusiasm, the curiosity, the extremely wide culture, and the friendly nature of Jean-Marie had a profound influence on Sauvage's scientific career and on the relationship he had later on in his own group with the many people who worked in it.

Macrocyclic and Supramolecular Chemistry: How Izatt–Christensen Award Winners Shaped the Field, First Edition.
Edited by Reed M. Izatt.

As a postdoctoral fellow in Oxford, in the research team led by Professor Malcolm L.H. Green, a very successful and still very young organometallic chemist at the time of Sauvage's visit, he acquired a good background in the field of transition metal chemistry and even inorganic photochemistry. He went back to Strasbourg in 1974 and, with Jean-Marie Lehn, he worked on an ambitious project aiming at splitting the water molecule to H_2 and O_2 under the action of visible light. The 1973 oil crisis was severe but, paradoxically, it had positive effects. The financial decision-makers of the time became relatively generous with scientists whose projects were linked to solar energy and, in particular, to conversion of light energy into chemical energy. One of the first multicomponent systems for photochemically generating molecular hydrogen from water was found in Strasbourg. It was published in 1977 [1]. Based on the famous $Ru(bipy)_3^{2+}$ complex, a rhodium complex acting as electron acceptor and electron relay, it needed a "sacrificial agent" as well as an hydrogen evolution catalyst (in particular, colloidal platinum, among other catalysts). It is interesting to note that this field of research experienced a big boom at the end of the 1970s and the beginning of the 1980s but, after a few years, it more or less vanished before coming back to center-stage during the 1990s. Today, many groups are still working on principles which are not so distant from those which were very popular 35 years ago or more.

Sauvage founded his independent research group in 1980, after being promoted to the rank of Research Director by the CNRS. In association with a few wonderful people such as Christiane Dietrich-Buchecker, an extremely talented organic chemist, Jean-Paul Collin, an expert in the field of electrochemistry, and two enthusiastic PhD students, Pascal Marnot and Romain Ruppert, he tackled several problems which were quite remote from one another. The group thus obtained significant results in various fields: (1) iridium-based molecular catalysts for the water gas shift reaction; (2) electrocatalytic reduction of CO_2 in water using nickel–cyclam as catalyst, with an amazing selectivity since water was not reduced although it was the solvent; (iii) and photochemistry of copper(I) complexes. This last project originated mostly from discussions with David R. McMillin, a real pioneer in the field of copper(I) complex photochemistry. Indeed, his first paper on this topic was published in 1978 [2]. David spent a sabbatical year in Strasbourg at the beginning of the 1980s, his host laboratory being that of Jean-Marie Lehn. He suggested that we prepare $Cu(dpp)_2^+$, a copper(I) complex containing two dpp ligands (dpp : 2,9-diphenyl-1,10-phenanthroline), since he anticipated novel and exciting photochemical properties for such a complex. He also offered to study it from a photochemical and photophysical point of view. From his previously published work, $Cu(dpp)_2^+$ was very likely to be strongly luminescent compared to the other copper(I) complexes of the same type already investigated. This is what we did, and a pleasant collaboration between Purdue University and our group started [3].

This event had dramatic consequences for our research. We decided to switch completely from rather inorganic chemistry projects, and we embarked on the synthesis of interlocking ring compounds (catenanes). The reason for this change of direction is quite obvious: playing with space-filling models, we realized that $Cu(dpp)_2^+$ contains two intertwined organic fragments and it is thus the ideal precursor to a [2]catenane. This, combined with the fact that Christiane Dietrich-Buchecker was such a skillful synthetic chemist, led to the synthesis of a [2]catenane following an extremely simple three-step strategy [4, 5]. The two strategies developed by the group are represented in a schematic fashion in Figure 22.1. This figure closely resembles the original drawing published in the 1983 paper [3].

Catenanes and rotaxanes had already attracted several laboratories, including the famous team led by Professor Schill [6]. These molecules had also been discussed and even obtained in minute quantities by Professor Wasserman at the very beginning of the 60s [7]. However, in spite of the beauty and cleverness of the approaches proposed by these teams, catenanes were generally considered to be very exotic and almost impossible-to-make molecular species.

To a large extent, the field of catenanes experienced a new birth thanks to David McMillin and to the fact that we collaborated with him. We had thus reoriented our main interests from inorganic chemistry and photochemistry to macrocyclic and topological chemistry. The first paper on our new field appeared in 1983. It was written in French and was published in *Tetrahedron Letters* [4]. Since the 1983 publication, many more complex catenanes, rotaxanes and molecular knots have been prepared by the team [8]. It is this particular

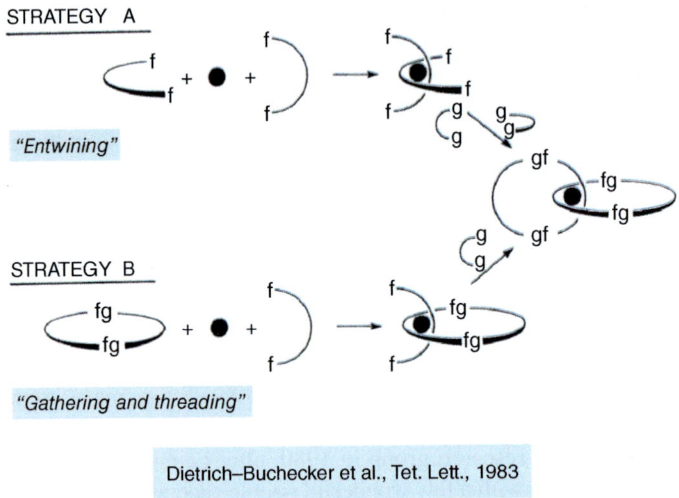

Figure 22.1 *General transition metal-templated strategy; the black dot represents a transition metal (copper(I)); the arcs of a circle and the rings contain coordinating fragments. The success of the approach relies on the formation of an "entanglement" before the cyclization reaction(s) is(are) carried out [4]. Note that the drawings were made using Indian ink and a lot of patience. The figure belongs to the pre-ChemDraw era*

piece of work which very probably attracted the attention of the Izatt–Christensen Award committee. In 1991, there was no precedent for this distinction so that its highly prestigious nature was not immediately obvious. Looking at the list of laureates nowadays, it is clear that this award means a major recognition for the people working in the field of macrocyclic and supramolecular chemistry, these two terms being taken in a very broad sense. Incidentally, it is interesting to note that the list of awardees for the 1991 to 2016 period contains at least ten scientists who are considered as experts in the field of catenanes and rotaxanes, in addition to their own research area. After the two early publications from the Strasbourg group [4, 5], several highly creative research teams reported novel and efficient strategies for making interlocking ring compounds. In particular, Stoddart (1989) [9], Hunter (1992) [10], Voegtle (1992) [11], and Fujita (1994) [12] proposed at a very early stage new templating principles for preparing catenanes.

The field of molecular machines, mostly initiated by Balzani, Stoddart and their co-workers [13], was to a large extent an extension of that of interlocking ring molecules, at least at the beginning. Our group reported the first catenane-based system able to undergo a complete electrochemically-driven rearrangement in 1994 [14]. We named the system a "swinging catenane." A ring could be forced to rotate inside the other ring by playing with the copper(I)/copper(II) redox couple. Since that time, molecular machinery has proliferated in a very impressive way thanks to the great contributions of various groups, either working with catenanes or rotaxanes, or using non-interlocking compounds.

22.2 Introduction

The central topic of the present chapter is a special type of molecular machine related to contraction and extension of switchable species. The first family of compounds (rotaxane dimers) discussed can be regarded as synthetic analogues of natural muscles, in the sense that, in analogy to striated muscle, the length of the object is modified by gliding filaments along one another rather than by introducing strain in a molecular system by modifying its length. The second family (molecular figures-of eight) is markedly different in the sense that the transformation of the molecule from one form to the other involves both stretching in a given direction and swelling in an orthogonal direction.

The area of research related to "molecular machines" has experienced a spectacular development in the course of the two last decades, as evidenced by the profusion of scientific papers, review articles, and books devoted to it [15]. The fast growth of the field is very probably related to the discovery of a large number of dynamic biological systems behaving as motors. The number of motor proteins which have been identified in the last two or three decades is impressive. The function of these natural systems is of utmost importance. Among the most classical examples ATP synthase [16], occupies a special place. It is a ubiquitous rotary motor responsible for the synthesis of ATP in all living organisms. The actin–myosin complex [17] of striated muscle is also well known and its mode of functioning is reasonably well understood. These fascinating biological systems represent an inexhaustible source of inspiration for synthetic chemists. Another important motivation for making and studying artificial molecular machines is related to future applications in various fields such as information storage and processing or nanoscale electro- or photo-chemically driven mechanical devices [18].

Catenanes and rotaxanes [19] are nowadays considered as normal and accessible molecules. They have triggered much interest in several research fields, also in relation to the fact that they are reasonably easy to prepare. The Strasbourg research team has mostly been working on transition metal-complexed interlocking ring systems [20]. These compounds are important in relation to molecular machine prototypes [14, 21]. It is quite obvious that molecular systems whose constitutive elements are held together by "mechanical" bonds instead of covalent bonds are prompted to undergo large-amplitude motions. One can easily imagine that a ring can rotate around the axis on which it is threaded (rotary motor) or can move along the same axis in a given direction or in the other, leading to a linear motor whose behavior is similar to that of a piston and a cylinder. In the course of the last 20 years, numerous interlocking compounds have led to the deliberate synthesis of dynamic systems of the molecular machine family. Purely organic [2]rotaxanes or transition metal-based rotaxanes behaving as molecular shuttles represent an important family of machine-like species [22]. Many copper-complexed catenanes and rotaxanes have been prepared and studied in Strasbourg since the report of a "swinging" catenane (1994), which was set in motion using electrochemical signals [14, 21b,d, 22g–i, 23].

More complex and promising systems can also be envisaged, such as molecular "muscles" [24]. The realm of molecular machines is not restricted to catenanes and rotaxanes. Fascinating work has been performed in this field using photochemically isomerizable chiral alkenes, leading to light-driven rotary motors [18h,i].

The purpose of this chapter is to discuss the synthesis and properties of dynamic systems reminiscent of biological muscles, systems which can be lengthened or contracted at will using a chemical or electrochemical signal. The first example of such molecular machine prototypes was reported 15 years ago [24a] by our group. More recently, spectacular systems of the same type have been described, based on different functioning principles, including surface-deposited species [24d] and macroscopic systems obtained by polymerizing extensible or contractile compounds [24c].

In biology, the functioning of skeletal muscles is one of the best understood linear motors [25]. The actin–myosin linear motor, which constitutes the most important part of muscle, has been extensively studied in the past three decades. There is obviously some analogy between the motion of actin filaments with respect to myosin-containing filaments (thick filament) and classical motors consisting of a piston moving in a cylinder (see Figure 22.2).

22.3 Interlocking ring compounds

22.3.1 Transition metal-based system

Since real muscles mostly consist of filaments able to glide along one another, a molecular assembly in which two string-like molecular fragments can glide along one another was designed. This is the process taking place in the sarcomere, in which the thick filament (containing myosin) moves along the thin filament (actin polymer) in one direction or the other so as to induce contraction or elongation.

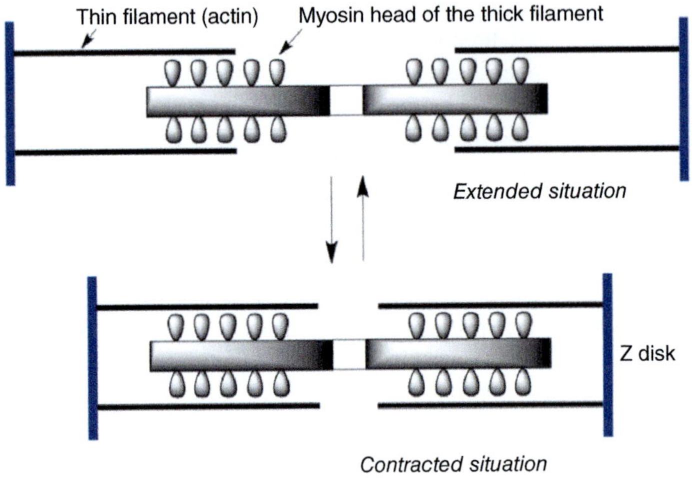

Figure 22.2 *Highly simplified scheme of a muscle in action. From the extended state, the contracted situation is obtained by a gliding process of the thick filament (myosin) along the thin filament (actin, a self-assembled dynamic polymer). Two Z disks delimit a sarcomere, i.e. the elemental unit of a muscle*

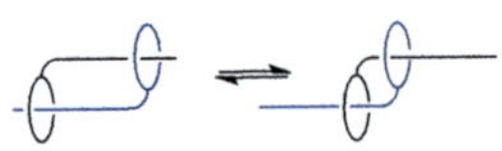

Figure 22.3 *The topology and the shape of a linearly arranged pseudo[2]rotaxane dimer are adapted to contraction/extension motion, mimicking the dynamic properties of a muscle*

Among the many possibilities offered by catenanes and rotaxanes in terms of dynamic objects, a [2]rotaxane dimer as represented in a schematic fashion in Figure 22.3 is particularly attractive. In this doubly threaded object, the motion is easy to visualize. Both strings (mimicking the muscle filaments) move along one another, but stay together thanks to the rotaxane nature of the system once bulky stoppers have been attached at the ends of the filament-like components.

As usual in our group, Cu(I) was used as gathering and templating metal. The movement was induced by a chemical reaction, namely a metal exchange [24a]. As shown in Figure 22.4, the doubly threaded compound discussed here can interact simultaneously with two metal centers, either in four- or five-coordinate geometry.

Coordination of Cu(I) to two 1,10-phenanthroline units results in a four-coordinate situation, corresponding to an extended geometry (upper part of Figure 22.4), whereas coordination of a divalent metal ion (Cu(II) or Zn^{2+}) affords a five-coordinate situation corresponding to interaction of each metal center with one 1,10-phenanthroline and one 2,2′,6′,2′′-terpyridine. It corresponds to a contracted geometry (bottom of Figure 22.4).

The synthesis of the muscle-like compounds will be briefly discussed in the following paragraph.

A 31-membered ring incorporating a dap fragment (dap: 2,9-diaryl-1,10-phenanthroline), prepared in several steps from commercially available compounds, was covalently linked to a linear fragment incorporating a 2,9-dimethyl-3,8-*bis*(p-hydroxyphenyl)-1,10-phenanthroline so as to afford the *bis*-chelating target consisting of a coordinating ring attached to a lateral arm containing another bidentate chelating group. The reaction of the ring-and-string conjugate with a stoichiometric amount of $[Cu(CH_3CN)_4]PF_6$ in CH_3CN/CH_2Cl_2 at room temperature led to the doubly threaded species represented on the bottom row of Figure 22.5. The success of this latter reaction was certainly far from certain, considering the number of Cu(I) complexes which could be formed by coordinating the two different chelating fragments (endocyclic and lateral coordination sites) of the ligand to the metal center. Nevertheless, we were happy to observe that formation of the

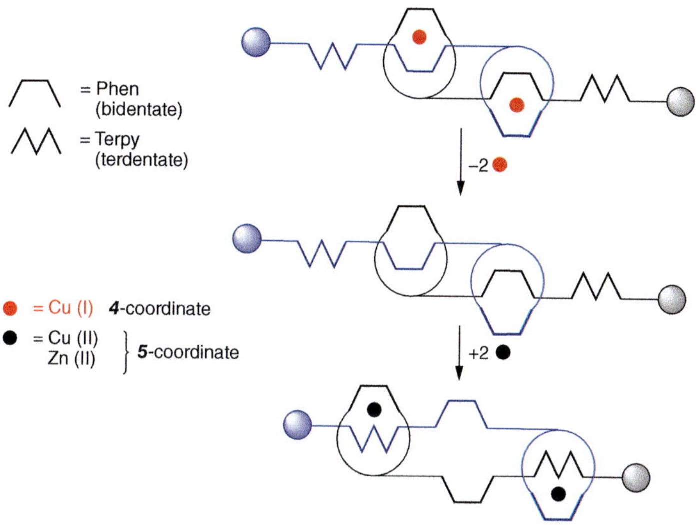

Figure 22.4 *Chemically induced motions (contraction/extension) of a [2]rotaxane dimer. The W-shaped fragment represents a terdentate chelating group such as 2,2′,6′,2′′-terpyridine (terpy), whereas the U-shaped species is a bidentate ligand of the 1,10-phenanthroline (phen) type*

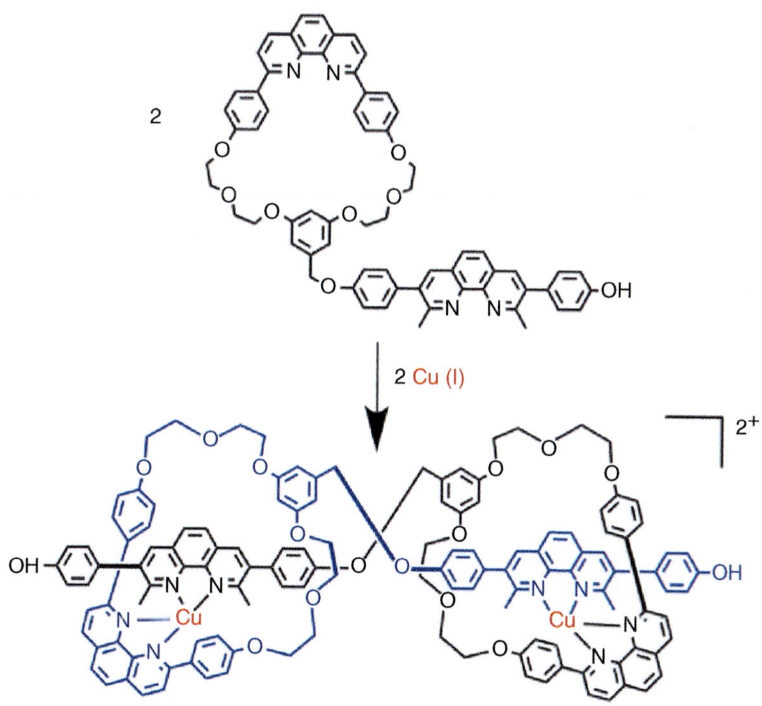

Figure 22.5 *Cu(I)-induced double-threading process leading to a dicopper(I)rotaxane dimer. The reaction is quantitative*

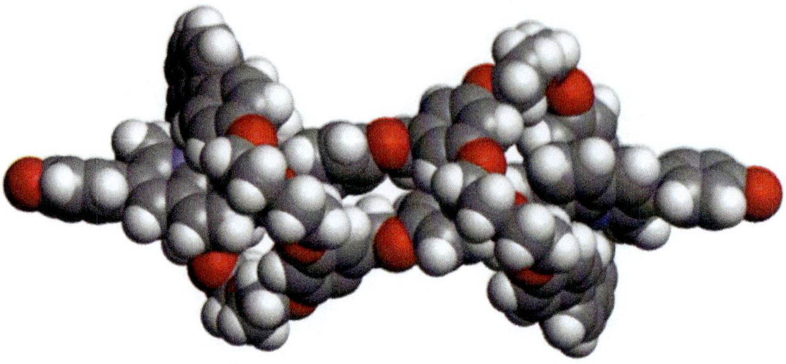

Figure 22.6 *X-ray structure of the dicopper(I) rotaxane dimer*

doubly threaded complex was quantitative. In fact, the complexity of the reaction leading to the doubly threaded species is attested by its slowness (several days at room temperature). Many coordination/decoordination steps have to occur before the system finds its thermodynamic well. The desired [2]pseudorotaxane was formed under thermodynamic equilibrium and thus it was obtained pure as a deep red crystalline solid without further purification. X-ray quality crystals were grown from acetone/diethyl ether by diffusion. The crystallographic structure of the dicopper(I) [2]rotaxane dimer is represented in Figure 22.6.

The functionalized terpy represented on the upper part of Figure 22.7 incorporates a tridentate chelating group and is attached to a bulky stopper. It was prepared separately and subsequently attached to both ends of the dicopper(I) precursor of Figure 22.5. This reaction afforded the muscle-like compound depicted in Figure 22.7 (bottom) as a dark red solid in 60% yield. This dicopper(I) "muscle" is in an extended form. The last coupling step not only allowed the introduction of terdentate sites on both sides of the molecular assembly but also afforded a real rotaxane in which the two bulky stoppers are able to completely inhibit potential unthreading reactions.

By treating the dicopper(I) pseudorotaxane dimer (top of Figure 22.8) with a large excess of KCN, the metal-free ligand was obtained in quantitative yield (not shown on the figure). After remetalation using $Zn(NO_3)_2$, the colorless dizinc complex depicted on Figure 22.8 (bottom) was obtained quantitatively. It is now in the contracted form. The extended dicopper(I) dimer could be regenerated by the reaction of the dizinc complex with an excess of $[Cu(CH_3CN)_4]PF_6$ at room temperature. Using a metal exchange reaction, the muscle-like compound is set in motion. The Cu(I) complex corresponds to the extended form; the Zn(II)-complexed species is the contracted state.

The two rotaxane dimers of Figure 22.8 represent one of the first examples of a unimolecular linear array able to elongate and contract at will under the action of a given stimulus. From Corey–Pauling–Koltun (CPK) model, one can estimate that the length of the compound changes from 83 to 65 Å between both forms (Figure 22.8). This is roughly the same relative contraction as that found in natural muscles (~27%).

It is noteworthy that several muscle-like compounds have been proposed in recent years. In most cases, they were set in motion using principles which were very different from ours. It is not the place in this chapter to review all the work done by other groups in this area, but we would like to mention the most significant contributions.

22.3.2 Cyclodextrin-type molecular muscles

Several groups prepared dynamic [2]rotaxane dimers based on cyclodextrins which, under the action of a given signal, could be contracted or elongated [26–29]. Reversible molecular motions mimicking natural muscles proceed through hydrophobic forces, size constraints, solvent polarity changes, or photochemical isomerization of azobenzene or stilbene derivatives.

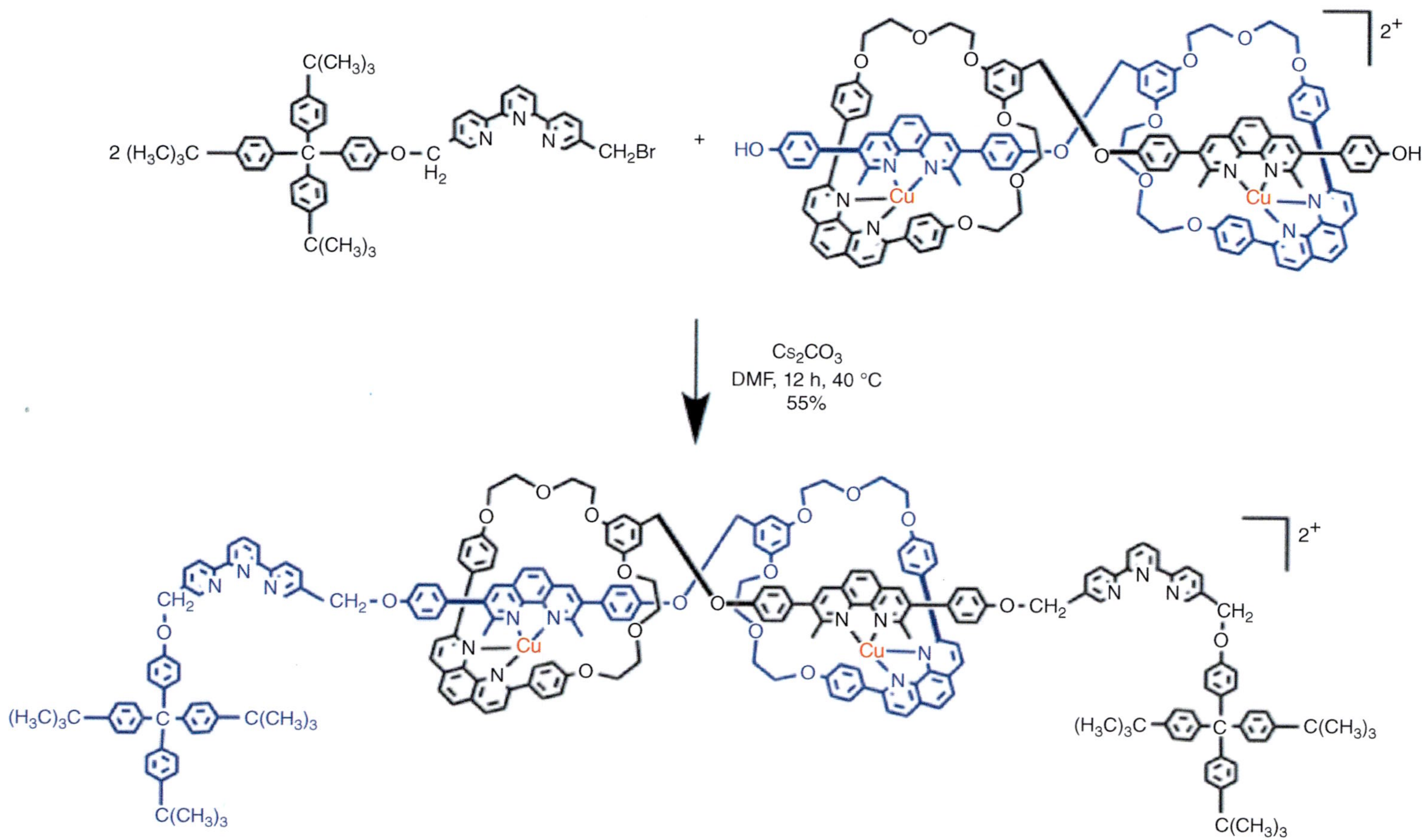

Figure 22.7 Functionalization of the [2]pseudorotaxane dimer, precursor to the "muscle", by attaching a terpy bulky stopper conjugate to both its ends. The "muscle" represented at the bottom of the figure is in its extended form

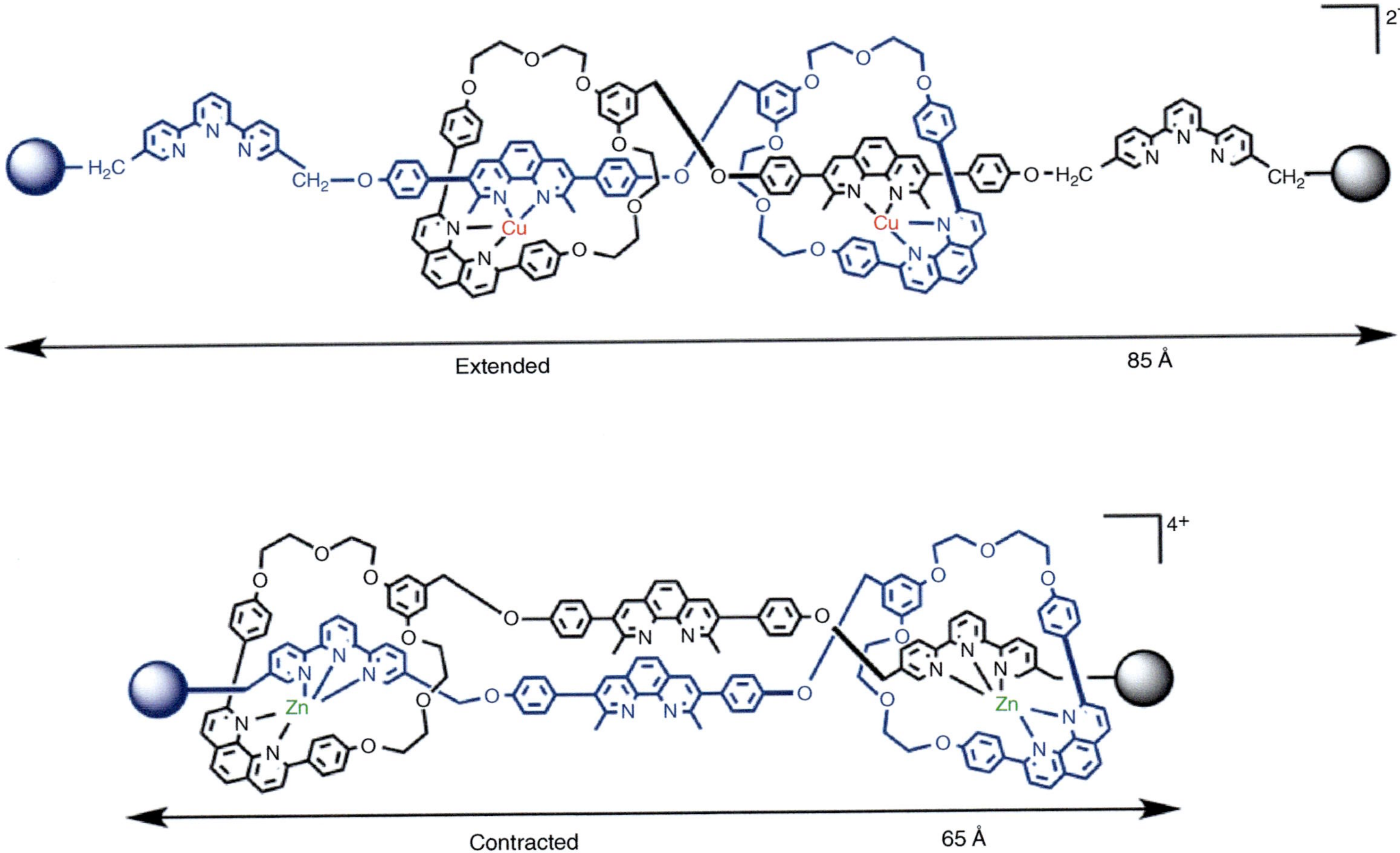

Figure 22.8 *The two forms of the muscle-like [2]rotaxane dimer. Whereas the interconversion between both forms is easily induced by metal exchange, oxidation of copper(I) to copper(II) (upper form of the figure) was unfortunately not followed by rearrangement of the complex to the extended situation, at least on a reasonable timescale. Due to the rigid structure of the molecule, the gliding motions of both threads along one another have to take place simultaneously and in a concerted manner. The kinetic barrier to global motion is thus very high, which explains why the motion is so slow when no demetalation step is involved*

22.3.3 Donor-acceptor based systems coupled to protonation/deprotonation processes

Another family of muscle-like compounds was developed based on interlocking architectures. These systems contain two types of recognition sites, one of them being based on hydrogen bonding (and thus pH sensitive) and the other operating via acceptor–donor interactions. In a recent publication, Stoddart and co-workers showed that a contraction and expansion motion was triggered by changing the pH of the solution containing the rotaxane [30]. They described a new bistable rotaxane dimer architecture in 2008, namely a two-component [c2]daisy chain topology [30]. This system consists of two self-complementary interlocking molecular fragments in which a secondary ammonium ion-containing lateral arm is attached to a crown ether (24-membered ring). The end of each arm is connected to a bulky stopper. Each dibenzo[24]-crown-8 ring (DB24C8) glides between the two specific binding sites incorporated in the thread, the motion being triggered by modifying pH. The gliding motions result in contracting or elongating the rotaxane dimer. The principle and its experimental realization are relatively simple, since a protonation/deprotonation reaction is responsible for the motion, as shown in Figure 22.9.

At the same time, Coutrot et al. published another example of pH-switchable molecular muscle[24b] based on a related principle. The system incorporates a cyclic rotaxane dimer consisting of two interlocking monomers. Each monomer incorporates a crown ether attached to a linear fragment. This fragment contains two stations: an ammonium function and a triazolium group. Two mannosidic units were used as bulky stoppers. The macrocycles interact with the N-benzylammonium stations of the threads by dipolar interaction and hydrogen bonds at acidic pH. The molecular muscle is thus in its extended mode (upper drawing of

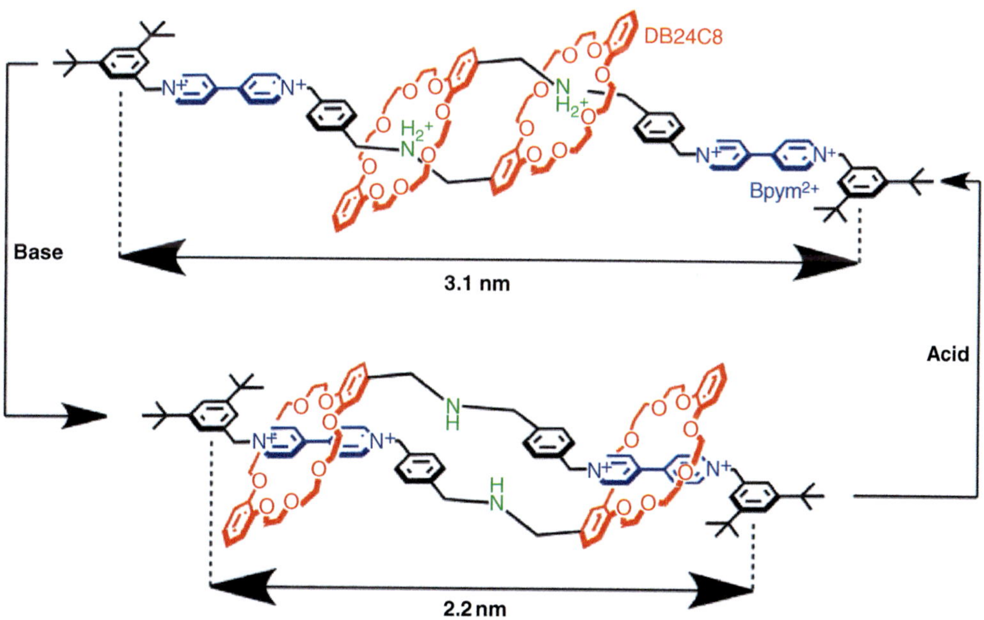

Figure 22.9 The two-component [c2]daisy chain rotaxane prepared and studied by Stoddart and co-workers [30]. Addition of base to the protonated form (upper form) destroys the hydrogen bonding interaction between the ammonium ion and the crown ether, allowing the crown ether (electron-rich ring, in red) to interact with a bipyridinium groups (strong electron acceptor, in blue). This process triggers contraction of the compound from ~3.1 nm to ~2.2 nm. The process is reversible since protonation restores the hydrogen bonding interaction between the crown ethers and the ionic $-NH_2^+$

Figure 22.10). At basic pH, deprotonation of the N-benzylammonium station takes place to generate benzylamine. This amine is of course unable to interact with the crown ether. By contrast, the electron accepting triazolium group is not affected by the pH increase and is thus available to interact with the ring of the other ring-and-string component in a donor–acceptor mode. The molecular muscle-like compound is thus in contracted mode (lower part of Figure 22.10).

A remarkable extension of Coutrot's work was published in 2012 [24c]. Giuseppone, Buhler and co-workers designed and synthetized a pH-driven muscle-like system in which bistable [c2]daisy chain rotaxanes are linked together linearly by taking advantage of a remarkably efficient coordination chemistry-based polymerization process [24c]. The Coutrot [c2]daisy chain monomer was end-functionalized with 2,2′,6′,2′′-terpyridine (terpy) groups and polymerized by coordinating the terpy units to either zinc(II) or iron(II) metal centers. A remarkably high degree of polymerization (DP) was obtained (~3,000 units) for both the contracted and extended forms.

As in Coutrot's work, at low pH, the N-benzylammonium stations are preferentially encircled by the crown ether. As a consequence, the doubly threaded dimer adopts an extended geometry (upper line of Figure 22.11). At basic pH, the N-benzylammonium groups are deprotonated and the rings move to interact with the triazolium stations acting as electron acceptors, thus adopting the contracted form shown on the lower part of Figure 22.10. These two states were very convincingly characterized by dynamic light-scattering (DLS) and small angle neutron scattering (SANS) experiments. For the iron(II)-assembled system, an increase in hydrodynamic radius from $R_H = (167 \pm 17)$ nm to $R_H = (189 \pm 19)$ nm, in contour length (that is, the total length of

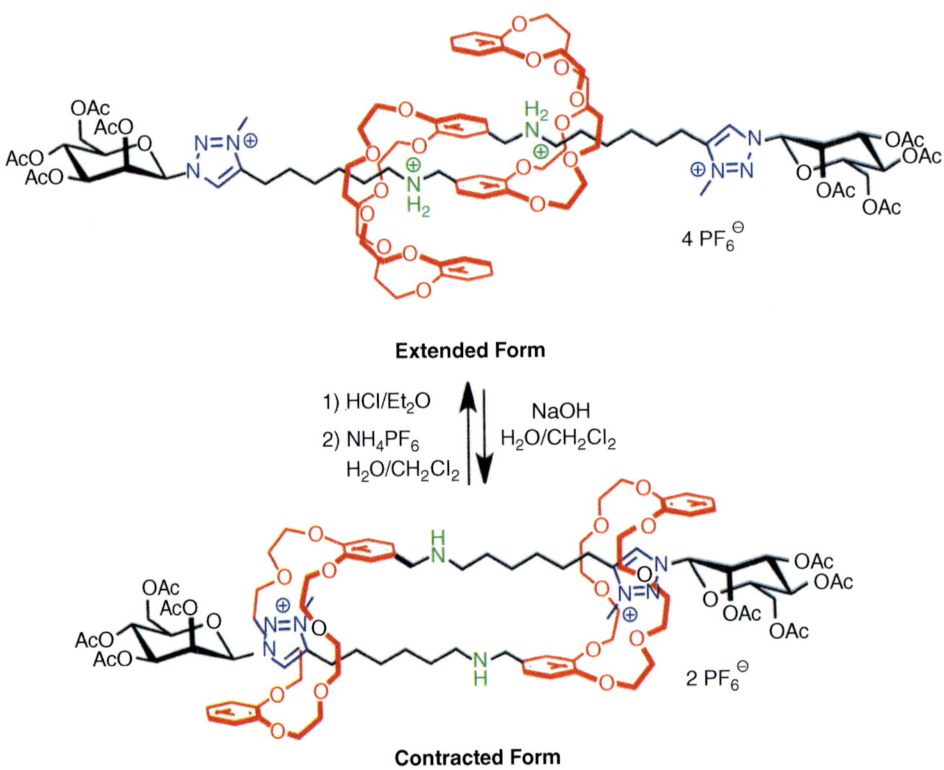

Figure 22.10 Coutrot and co-workers can mimic the contraction/extension motion of a muscle using a [2]rotaxane whose threads incorporate a triazolium station and a pH-sensitive benzylammonium group [24b]

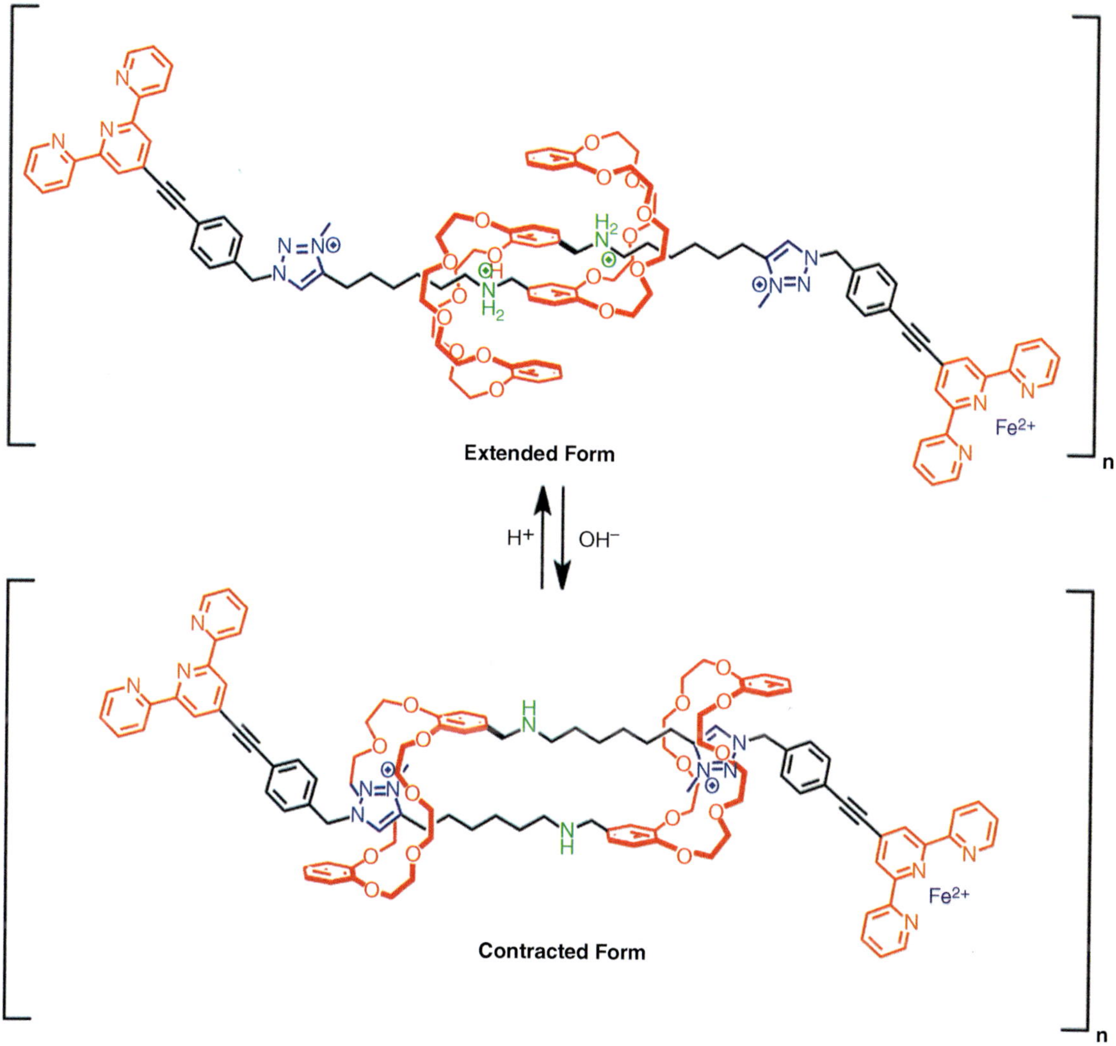

Figure 22.11 *The pH-switchable coordination rotaxane polymer prepared and studied by Giuseppone, Buhler and their co-workers [24c]*

the polymer at full extension) from $L_c = 9398$ nm to 15861 nm, and in molecular weight (the average mass of the polymer per unit of length) from M_{wfit} [$\times 10^6$ g mol^{-1}] = 7.030 to 7.890 was observed between the contracted and the extended form.

22.3.4 Other muscle-like systems based on the mechanical bond

Ten years ago, Stoddart, Ho and co-workers described a new artificial molecular muscle based on a switchable [3]rotaxanes [24d]. After attaching this switchable compound to a gold surface with disulfide tethers, they showed that the surface could be bent by chemically oxidizing or reducing the electroactive component of the [3]rotaxane. This system is really remarkable since it is one of the first molecular devices able to exert a force on a large object (micrometer size), sufficient to modify its shape thanks to the simultaneous action of numerous individual molecular muscle-like compounds.

Another interesting report describing a contraction–elongation motion was published in 2009 by the group of Chiu [31]. The compound is a molecular cage-based [2]rotaxane incorporating secondary ammonium ions. It is set in motion by modifying the nature of the counter-ions so as to modify the strength of hydrogen bonds involving the ammonium ions and crown ethers.

22.4 Non-interlocking compounds

The role of interlocking ring compounds in the development of molecular machines in general and muscle-like compounds in particular has been determining. However, non-interlocking compounds have also led to spectacular results, as already mentioned. As far as contraction/elongation processes are concerned, several families of compounds have been of particular interest.

Films of polymers have been used in the past to elaborate macroscopic devices whose behavior is reminiscent of muscles. A remarkable example of an electrochemically driven actuator is that of polypyrrole film (a redox-active and conducting polymer). By oxidizing or reducing the polymer matrix, counter-ions and solvent molecules are exchanged with the solvent. This process results in important volume changes which lead to large amplitude motions of the film once deposited onto another flexible but redox inactive organic film [32].

Non electroactive polymers have also shown to exhibit promising properties as muscle mimics. A family of "chemomechanical polymers" was introduced in 2003 by Schneider and his co-workers. In their work, the authors demonstrated that contraction or expansion processes can be induced by incorporation of given guests in polymers incorporating various binding sites [33, 34].

Another strategy, which should in principle also lead to electromechanical devices, relies on annulenes and their polymers. By oxidizing an [8]annulene, flattening of the ring should lead to significant increase of the overall length of the compound, opening the gate to new electromechanical actuators [35]. Contractile helical coordination compounds have also been investigated in relation to artificial muscles. In this case, the system is set in motion using a chemical signal. It is reminiscent of a spring rather than a muscle and is certainly very promising in terms of a mechanical device [36].

22.4.1 A contractile and extensible molecular figure-of-eight [38]

Non-interlocking cyclic systems undergoing controlled large-amplitude motions and thus behaving as molecular machine prototypes are very uncommon, although large rings are nowadays very accessible from the viewpoint of synthesis. They can occupy several markedly different shapes that can be interconverted using an appropriate signal [37]. It is thus to be expected that such switchable macrocycles will be used in the future as components of muscle-like compounds able to contract or elongate, in a similar manner to the few [2] rotaxane dimers discussed above [24].

Our group has recently been interested in switchable large rings of the figure-of-eight family. We recently reported the interconversion between a vertically oriented transition metal-complexed figure-of-eight and a horizontally disposed one. Such a complex is certainly a potential muscle-like system since the molecule can be contracted in a given direction (horizontal, for example) while it is elongated in a perpendicular direction (vertical). The switchable macrocycle is a 78-membered ring incorporating two different pairs of coordinating fragments. It can interact with transition metals in a bimodal way. The nature of the metal or its oxidation state determines its binding mode to the ring. In addition, each form of the complexed ring has the shape of a figure-of-eight, with one or two clear crossing-points between the two loops of the cycle.

The chemical structure of the ring as well as a schematic diagram of the compound is represented in Figure 22.12. The macrocycle incorporates bidentate and tridentate chelating groups. These coordinating fragments are arranged in an alternating situation: dpp (dpp: 2,9-diphenyl-1,10-phenanthroline) – terpy (terpy: 2,2′,6′,2″-terpyridine) – dpp –terpy. Both chelating groups, dpp and terpy, are classical coordinating species, well adapted to tetrahedral or octahedral coordination spheres respectively. It is thus expected that Cu(I) will form a *bis*-dpp complex by taking up the two dpp fragments whereas Cu(II) or Fe(II) will accommodate the two terpy groups, as represented in a schematic fashion in Figure 22.13.

The macrocyclic compound was prepared following a synthesis sequence, which is far from being trivial. As represented in Figure 22.14, its copper(I) complex was prepared quantitatively as a deep red solid by the reaction of the macrocycle with $Cu(CH_3CN)_4PF_6$. The iron(II) complex was obtained by stoichiometric addition of $Fe(OTf)_2$ to a solution of free ligand. The iron(II) complex was isolated as an intense purple solid. Both complexes were fully characterized in the usual way (^1H NMR, HR-MS(ES), electronic spectroscopy and electrochemistry). They could easily be demetalated to afford the metal-free compound (Figure 22.14). Treatment of the copper(I) complex by a KCN/water-acetonitrile solution at room temperature led to the free macrocycle in quantitative yield, whereas for the iron(II) complex, demetalation occurred by reaction of a Cs_2CO_3/DMF solution at 70 °C [39]. Remetalation was also carried out quantitatively. Metal exchange is nearly quantitative and can be carried out under mild chemical conditions. However, it must be noted that a metal-exchange process is certainly less promising in terms of molecular machinery than an electrochemically or photochemically driven process.

The copper complex was studied by cyclic voltammetry. The cyclic voltammogram of Figure 22.15 clearly indicates that after oxidation of the 4-coordinate Cu(I) complex at +0.65 V in the course of the −0.6 V to +1 V scan, the unstable copper(II) complex rearranges. Indeed, by inverting the scan potential at +1 V and by

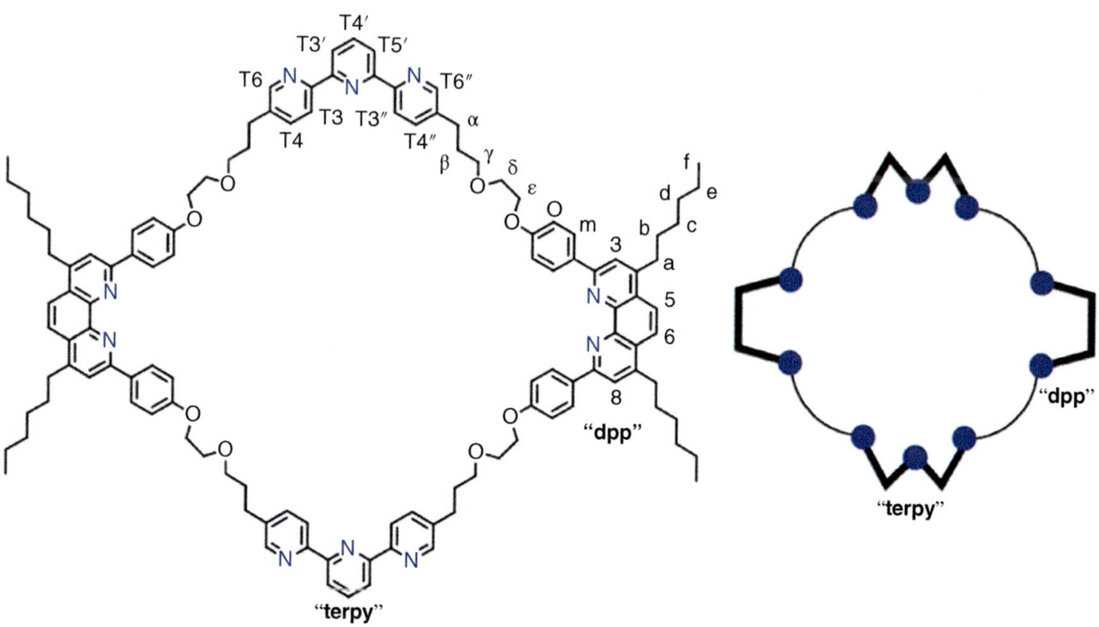

Figure 22.12 Chemical structure and schematic representation of the ring. Arbitrarily, the two terpy fragments are disposed on the vertical axis. The nitrogen atoms are symbolized by blue disks in the right-hand drawing. The target macrocyclic compound is a 78-membered ring

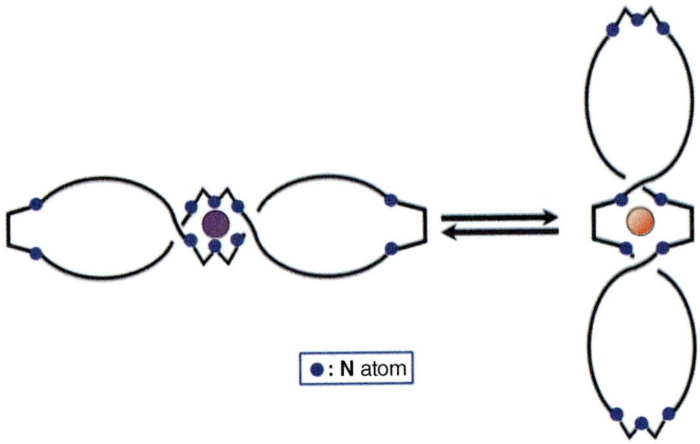

Figure 22.13 The two states of the figure-of-eight. On the left, the octahedral complex (the metal center, either Fe(II) or Cu(I), is a purple disk) implies that the long axis is horizontal whereas formation of the tetrahedral Cu(I) complex, indicated as a red disk, imposes a long vertical axis. The central entanglement is represented with two crossings, which is in agreement with the expected geometries of the respective complexes involved. The coordinating nitrogen atoms are represented as small blue disks. The conversion of the tetrahedral copper(I) complex to the iron(II) complex via metal exchange involves compression of the ring along the vertical axis and elongation of the horizontal dimension. The same holds true for the oxidation of copper(I) to copper(II). Arbitrarily, the vertical axis for both forms has been selected as the straight line between the central nitrogen atoms of the two terpy nuclei

scanning from +1 V to –0.6 V, the main wave is observed at a negative potential (–0.25 V). It corresponds to reduction of a 6-coordinate copper(II) complex to copper(I). A similar behavior has already been observed with copper-complexed catenanes [14, 21b]. although in these previous examples, rearrangement of the thermodynamically unstable copper(II) complexes obtained as transient species was much slower than in the present non-interlocking system. A similar behavior was observed by increasing the scan rate from 20 mV s^{-1} to 200 mV s^{-1}. These electrochemical experiments clearly establish that a fast rearrangement of the *bis*-dpp copper(II) complex to a *bis*-terpy complex takes place. In other words, contraction or extension of the copper-complexed figure-of-eight can be driven by an electrochemical signal. The general shape of the copper(II) complex is certainly very similar to that of the Fe(II) complex. Its height is thus expected to be small (~11 Å) whereas that of the copper(I) complex is probably close to 30 Å.

In the previous section, we discussed the properties of a new metal-complexed figure-of-eight as a potential elemental unit for generating muscle-like assemblies or materials. The compound can be contracted or elongated using a chemical signal (metal exchange) or an electrochemical process (Cu^{II}/Cu^{I}). The contraction and elongation processes are concerted in the sense that each process triggers an opposite process in an orthogonal direction. Interestingly, figure-of-eight complexes can also be used in contraction/extension of oligomers or polymers, thus leading to muscle-like species. The principle is depicted in Figure 22.16.

22.5 Conclusion

As shown in this chapter, a large variety of molecular structures can be used for making contractile and extensible compounds or assemblies. Interlocking and non-interlocking systems can be designed, prepared and studied in relation to artificial muscles.

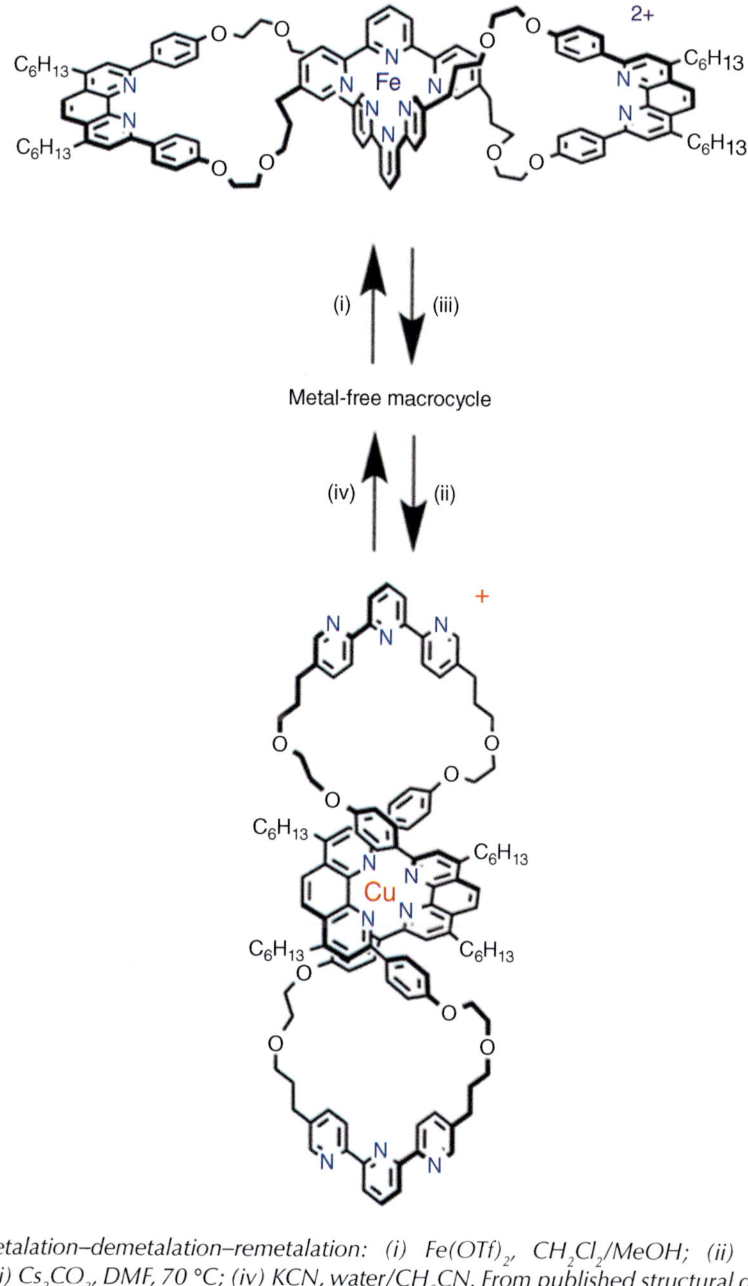

Figure 22.14 *Metalation–demetalation–remetalation: (i) $Fe(OTf)_2$, $CH_2Cl_2/MeOH$; (ii) $[Cu(CH_3CN)_4]PF_6$, CH_2Cl_2/CH_3CN; (iii) Cs_2CO_3, DMF, 70 °C; (iv) KCN, water/CH_3CN. From published structural data or space-filling models one can estimate the dimensions of the figure-of-eight for each situation. As far as the vertical axis is concerned, the "thickness" of the iron(II) complex can be estimated as the distance between the two 4′ hydrogen atoms of the terpy fragments: d(4′-4′) ~11 Å. The height of the copper(I) complex is much greater: d(4′-4′) ~30 Å according to CPK models*

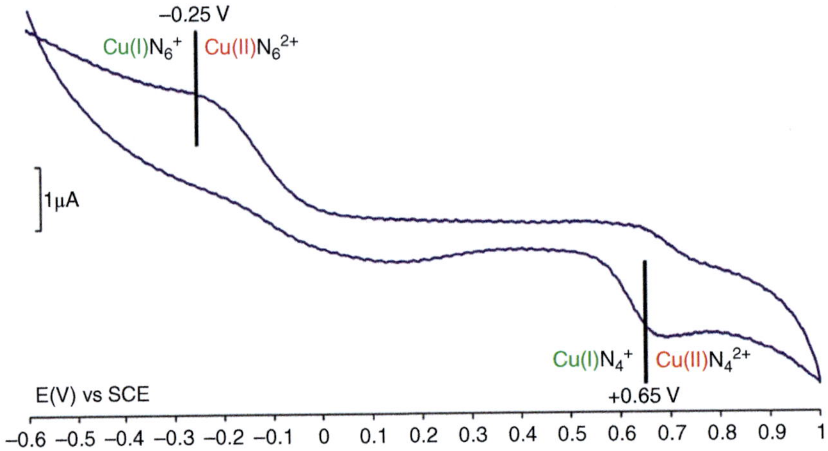

Figure 22.15 *Cyclic voltammogram of the macrocyclic copper(I) complex in CH_3CN (0.1 M $nBu_4N+.PF_6-$; Pt electrode; SCE, 20 mV s^{-1}; −0.6 V to +1 V). The anodic scan shows mostly oxidation of the 4-coordinate copper(I) complex to copper(II) with an oxidation wave around +0.65 V. The reductive scan (+1 V to −0.6 V) shows a relatively intense wave around −0.25 V in agreement with reduction of a 6-coordinate copper(II) complex*

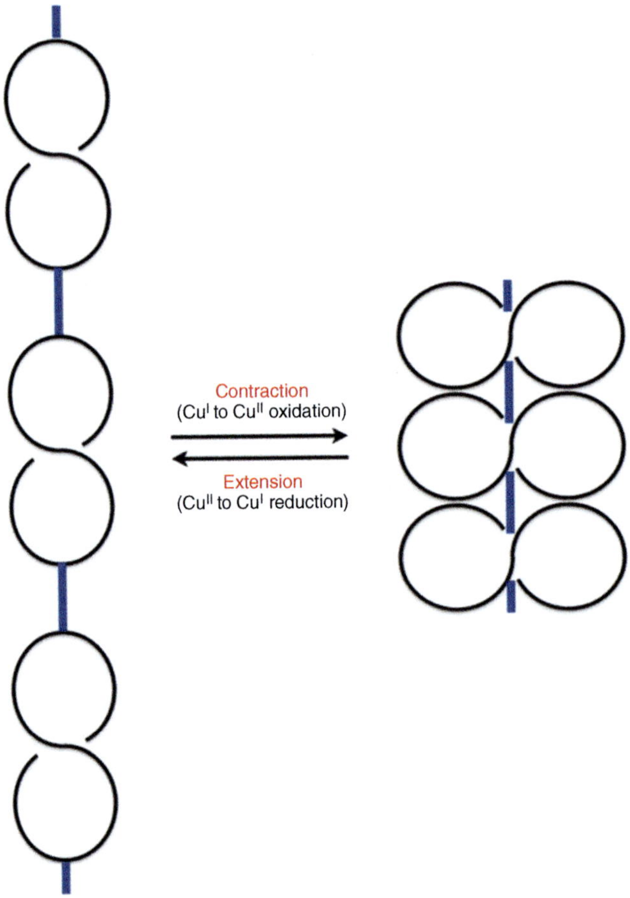

Figure 22.16 *Principle of a hypothetical polymer or oligomer with a figure-of-eight similar to the system discussed in this section as constitutive unit. The blue segments represent intermolecular spacers connecting the terpy fragments to one another via their 4′ carbon atoms. The thick and short structure represented on the right is converted to a long and thin species (left) electrochemically and vice versa*

Generally speaking, motivations for research in the field of controlled dynamic systems of the muscle-like family are manifold. The most important one is related to the scientific challenge that the synthesis and study of complex, functional molecular systems represent. Making contractile and extensible molecules of the rotaxane dimer type, or of the figure-of-eight family, was probably not even envisioned a few decades ago due to expected synthetic and analytical difficulties. Making and studying artificial muscles is related to the fascination that biology exerts on chemists and engineers. It is very challenging for a synthetic chemist to make a molecule or molecular system whose function and mode of action are reminiscent of those of biological systems. The few examples discussed in the present chapter are certainly very remote from natural muscles, and are undoubtedly very primitive compared to their natural analogs.

The field of molecular machines is now more than 20 years old; it is certainly time to address the important question of applications, either in a general fashion for machine-like compounds or more precisely for contractile and extensible molecular species. In which field of application will molecular muscles be most important in the future? Today it would be very risky to offer an answer to this question. Many types of nanodevices and nanomachines can be envisaged in relation to chemical applications (for example, sorting and transport of molecules in solution or through membranes) but also to purely mechanical applications. In the future, nano- and microrobots able to fulfill a large variety of functions (from medicine to everyday life) will be articulated. The use of molecular components to control their movements is certainly a promising possibility. In a relatively long-term perspective, artificial muscles of various length (from microns to millimeters or even centimeters) will be needed for various applications such as humanoid robots, actuators for microfluidic science and technology and prosthetic organs.

22.6 Acknowledgements

The work on artificial muscles performed in Strasbourg has been financially supported by the CNRS, the European Commission, the International Center for Frontier Research in Chemistry (ICFRC) and the Labex "Chimie des Systèmes Complexes" (CSC). We would like to thank these organizations. We also acknowledge the important work of Dr Christiane O. Dietrich-Buchecker (deceased in 2008) and Dr Maria Consuelo Jimenez (Valencia, Spain; mcjimene@qim.upv.es). Their contribution to the early work performed in Strasbourg on molecular muscle prototypes was determining.

We would also like to thank IBC Advanced Technologies, Inc. for being so supportive of the macrocyclic and supramolecular research field since the beginning of the 1990s.

References

1. Lehn, J.-M., Sauvage, J.-P. (1977) Chemical storage of light energy: Catalytic generation of hydrogen by visible light or sunlight irradiation of neutral aqueous solutions, *Nouveau Journal de Chimie*, **1**, 449.
2. Buckner, M.T., McMillin, D.R. (1978) Photoluminescence from copper(I) complexes with low-lying metal-to-ligand charge transfer excited states, *Journal of the Chemical Society, Chemical Communications*, **17**, 759.
3. Dietrich-Buchecker, C.O., Marnot, P.A., Sauvage, J.P. et al. (1983) Bis(2,9-diphenyl-1,10-phenanthroline)copper(I): a copper complex with a long-lived charge-transfer excited state, *Journal of the Chemical Society, Chemical Communications*, **9**, 513.
4. Dietrich-Buchecker, C.O., Sauvage, J.-P., Kintzinger, J.-P. (1983) Une nouvelle famille de molécules : les métallo caténanes, *Tetrahedron Letters*, **24**, 5095.
5. Dietrich-Buchecker, C.O., Sauvage, J.-P., Kern, J.-M. (1984) Templated synthesis of interlocked macrocyclic ligands: The catenands, *Journal of the American Chemical Society*, **106**, 3043.
6. Schill, G. (1971) *Catenanes, Rotaxanes and Knots*, Academic Press, New York.
7. Frisch, H.L., Wasserman, E. (1961) Chemical topology, *Journal of the American Chemical Society*, **83**, 3789.

8. Dietrich-Buchecker, C.O., Colasson, B. X., Sauvage, J.-P., (2005) Molecular knots, *Topic in Current Chemistry*, 249, 261 and references; Blanco, M.-J., Chambron, J.-C., Jiménez M.-C., Sauvage, J.-P (2003) Transition-metal-templated synthesis of rotaxanes, In *Topics in Stereochemistry*, Scott E. Denmark (Ed), Wiley Intersciences, New York, vol. **23**, pp 125; Chambron, J.-C., Dietrich-Buchecker, C.O., Sauvage, J.-P. (1996) Catenanes and knots, In *Large Ring Molecules*, J.A. Semlyen (Ed), p. 155.
9. Ashton, P. R., Goodnow, T. T., Kaifer, A. E. et al. (1989) A [2] catenane made to order, *Angewandte Chemie International Edition*, **101**, 1404.
10. Hunter, C. A. (1992) Synthesis and structure elucidation of a new [2]-catenane, *Journal of the American Chemical Society*, **114**, 5303.
11. Voegtle, F., Meier, S., Hoss, R. (1992) One-step synthesis of a fourfold functionalized catenane, *Angewandte Chemie International Edition*, **31**, 1619.
12. Fujita, M., Ibukuro, F., Hagihara, H., Ogura, K. (1994) Quantitative self-assembly of a [2]catenane from two preformed molecular rings, *Nature*, **367**, 720.
13. Ballardini, R., Balzani, V., Gandolfi, M. T. et al. (1993) A photochemically driven molecular machine, *Angewandte Chemie International Edition*, **32**, 1301.
14. Livoreil, A., Dietrich-Buchecker, C.O., Sauvage, J.-P. (1994) Electrochemically triggered swinging of a [2]-catenane, *Journal of the American Chemical Society*, **116**, 9399.
15. (a) J.-P. Sauvage (Ed) (2001) *Structure & Bonding: Molecular machines and motors*. Springer, Berlin; (b) Balzani, V., Venturi, M., Credi, A. (2008) *Molecular Devices and Machines – Concepts and perspectives for the Nanoworld*. Wiley-VCH, Weinheim; (c) Feringa, B. L. (2000) Nanotechnology: In control of molecular motion, *Nature*, **408**, 151; (d) Kay, E. R., Leigh, D. A., Zerbetto, F. (2007) Synthetic molecular motors and mechanical machines, *Angewandte Chemie International Edition*, **46**, 72; (e) J.-P. Sauvage, P. Gaspard (Eds) (2011) *From Non-Covalent Assemblies to Molecular Machines*, Solvay Institute, Wiley-VCH, Weinheim.
16. Walker, J. E. (1998) ATP synthesis by rotary catalysis (Nobel Lecture), *Angewandte Chemie International Edition*, **37**, 2308 and references therein.
17. Rayment, I., Holden, H. M., Whittaker, M. et al. (1993) Structure of the actin–myosin complex and its implications for muscle contraction, *Science*, **261**, 58.
18. The field of molecular machines is extremely active and it is thus impossible to cite all the important contributions reported by the many groups working in this research field. A few representative contributions are listed here: (a) Balzani, V., Clemente-Leon, M., Credi A., et al. (2006) Autonomous artificial nanomotor powered by sunlight, *Proceedings of the National Academy of Sciences USA*, **103**, 1178; (b) Berna, J., Leigh, D. A., Lubomska, M. et al. (2005) Macroscopic transport by synthetic molecular machines, *Nature Materials*, **4**, 704; (c) Collier, C. P., Mattersteig, G., Wong, E. W. et al. (2000) A [2]catenane-based solid state electronically reconfigurable switch, *Science*, **289**, 1172; (d) Eelkema, R., Pollard, M. M., Vicario, J. et al. (2006) Molecular machines: Nanomotor rotates microscale objects, *Nature*, **440**, 163; (e) Green, J. E., Wook Choi, J., Boukai, A. et al. (2007) A 160-kilobit molecular electronic memory patterned at 10^{11} bits per square centimetre, *Nature*, **445**, 414; (f) Harada, A. (2001) Cyclodextrin-based molecular machines, *Accounts of Chemical Research*, **34**, 456; (g) Kinbara, K., Aida, T. (2005) Toward intelligent molecular machines: Directed motions of biological and artificial molecules and assemblies, *Chemical Reviews*, **105**, 1377; (h) Koçer, A., Walko, M., Meijberg, W., Feringa, B. L. (2005) A light-actuated nanovalve derived from a channel protein, *Science*, **309**, 755; (i) Koumura, N., Zijlstra, R. W. J., van Delden, R. A. et al. (1999) Light-driven monodirectional molecular rotor, *Nature*, **401**, 152; (j) Leigh, D. A., Wong, J. K. Y., Dehez, F., Zerbetto, F. (2003) Unidirectional rotation in a mechanically interlocked molecular rotor, *Nature*, **424**, 174; (k) Nguyen, T. D., Liu, Y., Saha, S. et al. (2007) Design and optimization of molecular nanovalves based on redox-switchable bistable rotaxanes, *Journal of the American Chemical Society*, **129**, 626.
19. (a) Amabilino, D. B., Stoddart, J. F. (1995) Interlocked and intertwined structures and superstructures, *Chemical Reviews*, **95**, 2725; (b) Dietrich-Buchecker, C. O., Sauvage, J.-P. (1987) Interlocking of molecular threads: from the statistical approach to the templated synthesis of catenands, *Chemical Reviews*, **87**, 795; (c) Dietrich-Buchecker, C., Sauvage, J.-P (1999) *Catenanes, Rotaxanes and Knots. A Journey Through the World of Molecular Topology*. Wiley-VCH, Weinheim; (d) Fioravanti, G., Haraszkiewicz, N., Kay, E. R. et al. (2008) Three state redox-active molecular shuttle that switches in solution and on a surface, *Journal of the American Chemical Society*, **130**, 2593; (e) Schill G (1971) *Catenanes, Rotaxanes and Knots*. Academic Press, New York; (f) Vögtle, F., Dünnwald, T., Schmidt, T. (1996) Catenanes and rotaxanes of the amide type, *Accounts of Chemical Research*, **29**, 451.

20. (a) Chambron, J.-C., Collin, J.-P., Heitz, V. et al. (2004) Rotaxanes and catenanes built around octahedral transition metals, *European Journal of Organic Chemistry*, **8**, 1627; (b) Collin, J.-P., Durola, F., Mobian, P., Sauvage, J.-P. (2007) Pirouetting copper(I)-assembled pseudo-rotaxanes: Strong influence of the axle structure on the motion rate, *European Journal of Inorganic Chemistry*, **17**, 2420; (c) Dietrich-Buchecker, C.O., Khemiss, A., Sauvage, J.-P. (1986) High-yield synthesis of multiring copper(I) catenates by acetylenic oxidative coupling, *Journal of the Chemical Society, Chemical Communications*, **17**, 1376; (d) Dietrich-Buchecker, C. O., Sauvage, J.-P. (1983) Synthese de composes polyethers macrocycliques derives de la phenanthroline-1,10 diphenyl-2,9, *Tetrahedron Letters*, **24**, 5091; (e) Durola, F., Russo, L., Sauvage, J.-P. et al. (2007) Three-component entanglements consisting of three crescent-shaped bidentate ligands coordinated to an octahedral metal centre, *Chemistry – A European Journal*, **13**, 8749; (f) Faiz, J. A., Heitz, V., Sauvage, J.-P. (2009) Design and synthesis of porphyrin-containing catenanes and rotaxanes, *Chemical Society Reviews*, **38**, 422; (g) Mohr, B., Sauvage, J.-P., Grubbs, R. H., Weck, M. (1997) High-yield synthesis of [2] catenanes by intramolecular ring-closing metathesis, *Angewandte Chemie International Edition*, **36**, 1308; (h) Prikhod'ko, A. I., Durola, F., Sauvage, J.-P. (2008) Iron(II)-templated synthesis of [3]rotaxanes by passing two threads through the same ring, *Journal of the American Chemical Society*, **130**, 448.
21. (a) Balzani, V., Credi, A., Silvi S., Venturi, M. (2006) Artificial nanomachines based on interlocked molecular species: recent advances, *Chemical Society Reviews*, **35**, 1135; (b) Cárdenas, D. J., Livoreil, A., Sauvage, J.-P. (1996) Redox control of the ring-gliding motion in a Cu-complexed catenane: A process involving three distinct geometries, *Journal of the American Chemical Society*, **118**, 11980; (c) Cavallini, M., Biscarini, F., Leon, S. et al. (2003) Information storage using supramolecular surface patterns, *Science*, **299**, 531; (d) Champin, B., Mobian, P., Sauvage, J.-P. (2007) Transition metal complexes as molecular machine prototypes, *Chemical Society Reviews*, **36**, 358; (e) Chuang, C.-J., Li, W.-S., Lai, C.-C. (2009) A molecular cage-based [2]rotaxane that behaves as a molecular muscle, *Organic Letters*, **11**, 385; (f) Dawson, R. E., Lincoln, S. F., Easton, C. J. (2008) The foundation of a light-driven molecular muscle based on stilbene and α-cyclodextrin, *Chemical Communications*, **34**, 3980; (g) Wurpel, G. W. H., Brouwer, A. M., van Stokkum, I. H. M. et al. (2001) Enhanced hydrogen bonding induced by optical excitation: Unexpected subnanosecond photoinduced dynamics in a peptide-based [2]rotaxane, *Journal of the American Chemical Society*, **123**, 11327.
22. (a) Bissel, R. A., Cordova, E., Kaifer, A. E., Stoddart, J., F. (1994) A chemically and electrochemically switchable molecular shuttle, *Nature*, **369**, 133; (b) Keaveney, C. M., Leigh, D. A. (2004) Shuttling through anion recognition, *Angewandte Chemie International Edition*, **43**, 1222; (c) Murakami, H., Kawabuchi, A., Kotoo, K. et al. (1997) A light-driven molecular shuttle based on a rotaxane, *Journal of the American Chemical Society*, **119**, 7605; (d) Pérez, E. M., Dryden, D. T. F., Leigh, D. A. (2004) A generic basis for some simple light-operated mechanical molecular machines, *Journal of the American Chemical Society*, **126**, 12210; (e) Stanier, C. A., Alderman, S. J., Claridge, T. D. W. et al. (2002) Unidirectional photoinduced shuttling in a rotaxane with a symmetric stilbene dumbbell, *Angewandte Chemie International Edition*, **41**, 1769; (f) Wang, Q.-C., Qu, D.-H., Ren, J. et al. (2004) A lockable light-driven molecular shuttle with a fluorescent signal, *Angewandte Chemie International Edition*, **43**, 2661; (g) Durola, F., Sauvage, J.-P. (2007) Fast electrochemically induced translation of the ring in a copper-complexed [2]rotaxane: The biisoquinoline effect, *Angewandte Chemie International Edition*, **46**, 3537; (h) Durola, F., Lux, J., Sauvage, J.-P. (2009) A fast-moving copper-based molecular shuttle: Synthesis and dynamic properties, *Chemistry – A European Journal*, **15**, 4124; (i) Collin, J.-P., Durola, F., Lux, J., Sauvage, J.-P. (2009) A rapidly shuttling copper-complexed [2]rotaxane with three different chelating groups in its axis, *Angewandte Chemie International Edition*, **48**, 8532; (j) Collin, J.-P., Durola, F., Lux, J., Sauvage, J.-P. (2010) A copper-based shuttling [2]rotaxane with two bidentate chelates in the axis: steric control of the motion, *New Journal of Chemistry*, **34**, 34.
23. (a) Colasson, B. X., Dietrich-Buchecker, C., Jiménez-Molero, M. C. (2002) Towards molecular machines and motors based on transition metal complexes, *Journal of Physical Organic Chemistry*, **15**, 476; (b) Collin, J.-P., Gaviña, P., Sauvage, J.-P (1997) Electrochemically induced molecular motions in copper-complexed threaded systems: from the unstoppered compound to the semi-rotaxane and the fully blocked rotaxane, *New Journal of Chemistry*, **21**, 525.
24. (a) Jiménez-Molero, M. C., Dietrich-Buchecker, C., Sauvage, J.-P. (2000) Towards synthetic molecular muscles: Contraction and stretching of a linear rotaxane dimer, *Angewandte Chemie International Edition*, **39**, 3284; (b) Coutrot, F., Romuald, C., Busseron, E. (2008) A new pH-switchable dimannosyl[c2]daisy chain molecular machine, *Organic Letters*, **10**, 3741; (c) Du, G., Moulin, E., Jouault, N. et al. (2012) Muscle-like supramolecular

polymers: Integrated motion from thousands of molecular machines, *Angewandte Chemie International Edition*, **51**, 12504; (d) Liu, Y., Flood, A. H., Bonvallet, P. A. et al. (2005) Linear artificial molecular muscles, *Journal of the American Chemical Society*, **127**, 9745. (e) Niess, F., Duplan, V., Sauvage, J.-P. (2014) Molecular muscles: From species in solution to materials and devices, *Chemistry Letters*, **43**, 964.

25. Kitamura, K., Tokunaga, M., Iwane, A. H. (1999) A single myosin head moves along an actin filament with regular steps of 5.3 nanometres, *Nature*, **397**, 129.
26. Onagi, H., Easton, C. J., Lincoln, S. F. (2001) An hermaphrodite [2]rotaxane: Preparation and analysis of structure, *Organic Letters*, **3**, 1041.
27. Dawson, R. E., Lincoln, S. F., Easton, C. J. (2008) The foundation of a light-driven molecular muscle based on stilbene and α-cyclodextrin, *Chemical Communications*, **34**, 3980.
28. Miyauchi, M., Takashima, Y., Yamaguchi, H., Harada, A. (2005) Chiral supramolecular polymers formed by host–guest interactions, *Journal of the American Chemical Society*, **127**, 2984.
29. Tsukagoshi, S., Miyawaki, A., Takashima, Y. et al. (2007) Contraction of supramolecular double-threaded dimer formed by α-cyclodextrin with a long alkyl chain, *Organic Letters*, **9**, 1053.
30. Wu, J., Leung, K.-F., Benítez, D. et al. (2008) An acid–base-controllable [c2]daisy chain, *Angewandte Chemie International Edition*, **47**, 7470.
31. Chuang, C.-J., Li, W.-S., Lai, C.-C, et al. (2009) A molecular cage-based [2]rotaxane that behaves as a molecular muscle, *Organic Letters*, **11**, 385.
32. (a) Otero, T. F., Sansiñena, J. M. (1998) Soft and wet conducting polymers for artificial muscles, *Advanced Materials*, **10**, 491; (b) Otero, T.F., Martinez, J.G., Arias-Pardilla, J. (2012) Biomimetic electrochemistry from conducting polymers, a review: Artificial muscles, smart membranes, smart drug delivery and computer/neuron interfaces, *Electrochimica Acta*, **84**,112.
33. Schneider, H.-J.; Liu, T.; Lomadze, N. (2003) Molecular recognition in a supramolecular polymer system translated into mechanical motion, *Angewandte Chemie International Edition*, **42**, 3544.
34. Schneider, H.-J., Strongin, R. M. (2009) Supramolecular interactions in chemomechanical polymers, *Accounts of Chemical Research*, **42**, 1489.
35. (a) Marsella, M. J., Reid, R. J. (1999) Toward molecular muscles: Design and synthesis of an electrically conducting poly[cyclooctatetrathiophene], *Macromolecules*, **32**, 5982; M; (b) Marsella, J., Reid, R. J., Estassi, S., Wang, L.-S. (2002) Tetra[2,3-thienylene]: A building block for single-molecule electromechanical actuators, *Journal of the American Chemical Society*, **124**, 12507.
36. Stadler, A.-M.; Kyritsakas, N.; Graff, R.; Lehn, J.-M. (2006) Formation of rack- and grid-type metallosupramolecular architectures and generation of molecular motion by reversible uncoiling of helical ligand strands, *Chemistry – A European Journal*, **12**, 4503.
37. Two early examples of controlled ring expansion or contraction are: (a) Ueno, A.; Yoshimura, H.; Saka, R.; Osa, T. (1979) Photocontrol of binding ability of capped cyclodextrin, *Journal of the American Chemical Society*, **101**, 2779. (b) Shinkai, S.; Honda, Y.; Kusano, Y.; Manabe, O. A photoresponsive cylindrical ionophore, *Journal of the Chemical Society, Chemical Communications*, **15**, 848. (c) A remarkable example of a single ring able to fold up in two distinct ways under the action of given chemical signals was reported ten years ago. It is based on the pH-driven translocation of Cu^{2+} ions between different coordination sites. The same process can occur by coordinating imidazole to the dicopper(II) complex of the ring : Fabbrizzi, L., Foti, F., Patroni, S. et al. (2004) A sleeping host awoken by its guest: Recognition and sensing of imidazole-containing molecules based on double Cu^{2+} translocation inside a polyaza macrocycle, *Angewandte Chemie International Edition*, **43**, 5073.
38. Niess, F., Duplan, V., Sauvage, J.-P. (2014) Interconversion between a vertically oriented transition metal-complexed figure-of-eight and a horizontally disposed one, *Journal of the American Chemical Society*, **136**, 5876.
39. Prikhod'ko, A. I.; Sauvage, J.-P. (2009) Passing two strings through the same ring using an octahedral metal center as template: A new synthesis of [3]rotaxanes, *Journal of the American Chemical Society*, **131**, 6794.

Index

Page numbers in *italics* refer to figures; page numbers in **bold** refer to tables/boxes

Macrocyclic and Supramolecular Chemistry: How Izatt–Christensen Award Winners Shaped the Field, First Edition.
Edited by Reed M. Izatt.

Printed and bound by CPI Group (UK) Ltd, Croydon, CR0 4YY
15/09/2026
14962511-0001